Hazard and Risk Assessment in Civil Engineering Design

Hossein Bonakdari · Isa Ebtehaj · Joseph Ladouceur

Hazard and Risk Assessment in Civil Engineering Design

Hossein Bonakdari
Department of Civil Engineering
University of Ottawa
Ottawa, ON, Canada

Isa Ebtehaj
Soil and Environment Department, Faculty of Agriculture and Food Science
Laval University
Québec, QC, Canada

Joseph Ladouceur
Department of Civil Engineering
University of Ottawa
Ottawa, ON, Canada

ISBN 978-3-032-23599-2 ISBN 978-3-032-23600-5 (eBook)
https://doi.org/10.1007/978-3-032-23600-5

This Springer imprint is published by the registered company Springer Nature Switzerland AG
The registered company address is: Gewerbestrasse 11, 6330 Cham, Switzerland

Deepest thanks to our amazing families, who have shown endless support.

Preface I

Over the past several decades, civil and environmental engineering has evolved from a discipline primarily concerned with deterministic structural design into one fundamentally shaped by uncertainty, risk, and long-term system resilience. Infrastructure systems now operate within an environment characterized by climate change, rapid urbanization, aging assets, economic volatility, regulatory transformation, and technological interdependence. These forces have altered the assumptions that historically governed engineering design and have elevated risk-informed decision-making to a central position in modern practice.

Across the world, infrastructure networks, transportation corridors, water and wastewater systems, dams and levees, buildings, energy facilities, and environmental protection systems, face increasing exposure to natural and human-induced hazards. Rising temperatures, shifting precipitation patterns, intensifying extreme events, and expanding urban footprints have reduced traditional safety margins embedded in earlier design codes. Infrastructure systems are also increasingly interconnected, meaning that localized disruptions can propagate across sectors, amplifying economic and social consequences. In such an environment, deterministic approaches based solely on "best estimate" inputs are no longer sufficient. Instead, engineers must systematically identify hazards, quantify uncertainty, evaluate vulnerability, and make defensible decisions under incomplete information.

The discipline of risk analysis provides the conceptual framework for understanding how hazards interact with vulnerable systems to generate consequences. However, hazard identification alone cannot guarantee resilient outcomes. Engineers must also employ statistical and probabilistic tools to represent variability in materials, environmental loads, and system behavior. They must interpret data using rigorous descriptive statistics, model uncertainty through probability distributions, update beliefs using Bayesian reasoning, and assess diagnostic reliability through measures such as sensitivity and specificity. These probabilistic foundations are essential for moving beyond intuition toward structured and transparent decision-making.

This book is a systematic integration of these interconnected themes. It begins with the fundamentals of hazard and risk assessment in civil engineering, clarifying essential terminology and establishing the conceptual relationship between hazard, exposure, vulnerability, and consequence. It then explores how climate change and urban development intensify infrastructure risk, demonstrating that modern engineering challenges must be understood within broader environmental and societal contexts. Subsequent chapters develop the mathematical foundations of probability and Bayesian methods, descriptive statistics, uncertainty characterization, distribution selection, and goodness-of-fit testing, providing the analytical tools necessary for rigorous risk quantification.

Building upon this foundation, the book advances to structured decision methodologies. Decision trees and expected utility theory are introduced as systematic tools for modeling sequential engineering decisions under uncertainty. Sensitivity analysis techniques are presented to reveal parameter influence and identify tipping points that may alter preferred strategies. Scenario-based decision analysis extends this framework to address deep uncertainty, emphasizing robustness, regret minimization, and adaptive planning rather than reliance on single-point forecasts. Together, these approaches form a coherent methodology for selecting infrastructure strategies that remain acceptable across diverse plausible futures.

The contents of this book are the result of sustained research, teaching, and professional engagement in the areas of risk analysis, uncertainty modeling, resilience assessment, and infrastructure decision support. Many of the concepts and examples have been refined through graduate-level instruction and research supervision, as well as through collaboration with students and colleagues working on flood risk assessment, climate adaptation, water system planning, sustainable construction, and infrastructure resilience. Over time, it became clear that a unified framework linking hazard analysis, probability theory, uncertainty modeling, decision trees, sensitivity analysis, and scenario-based planning was needed to bridge gaps between theory and engineering practice.

The book is organized progressively to guide readers from conceptual foundations to advanced decision methodologies. Each chapter is interconnected yet sufficiently self-contained to allow focused study of particular topics. The examples are drawn from diverse civil engineering applications, including flood protection design, foundation selection under soil uncertainty, stormwater system sizing, wastewater treatment technology evaluation, seismic retrofitting, pavement material selection, and sustainable infrastructure planning, demonstrating how theoretical models translate into practical engineering decisions. Mathematical rigor is combined with implementation guidance to support both academic instruction and professional application.

Compared with conventional texts that treat risk, probability, or decision analysis in isolation, this book integrates these domains into a unified and application-oriented framework. Its distinctive contribution lies in connecting foundational hazard concepts with Bayesian updating, uncertainty quantification, decision tree analysis, sensitivity diagnostics, scenario-based robustness evaluation, and adaptive planning strategies. This integrated perspective reflects the evolving demands placed upon civil engineers, who must now design infrastructure systems that are not only efficient but also resilient, sustainable, and accountable to society.

It is my sincere hope that this book will contribute to advancing risk-based and resilience-oriented engineering practice. By equipping engineers, researchers, and graduate students with systematic tools for analyzing uncertainty and structuring complex decisions, it aims to support the development of infrastructure systems capable of performing reliably across uncertain and evolving futures.

On the occasion of the publication of this book, I would like to express my deep gratitude to colleagues, students, collaborators, and institutions that have supported and enriched the research and teaching activities that made this work possible. The intellectual exchange with graduate students and research partners has significantly shaped many of the ideas presented in these chapters. I am also grateful to the academic reviewers and editors who provided valuable feedback during the preparation of this manuscript. Finally, I extend my sincere appreciation to my family for their encouragement and patience throughout the development of this work.

We gratefully acknowledge Dr. Marek J. Druzdzel and BayesFusion for providing an academic license of the GeNIe software, which supported the development of Bayesian network examples and decision-analysis models presented in this book. Any comments and suggestions from readers are warmly welcomed. It is my hope that this book will serve not only as a reference but also as a foundation for continued innovation in risk-informed and resilience-oriented civil engineering.

Ottawa, Canada — Hossein Bonakdari
Québec, Canada — Isa Ebtehaj
Ottawa, Canada — Joseph Ladouceur

Preface II

In recent decades, civil and environmental engineering has undergone a profound transformation. Infrastructure systems are no longer evaluated solely on structural strength or deterministic safety factors. Instead, they are assessed within a broader context of uncertainty, climate variability, urban expansion, technological interdependence, and societal expectations. Across the world, infrastructure networks, transportation corridors, water supply systems, dams, buildings, energy facilities, and environmental protection systems, are facing increasingly complex hazards whose frequency, magnitude, and interaction patterns are evolving. As a result, modern civil engineering practice has shifted from purely capacity-based design toward risk-informed and resilience-oriented decision-making. The rapid acceleration of climate change, the intensification of extreme weather events, the aging of infrastructure assets, and the expansion of urban environments have introduced new layers of uncertainty that challenge traditional engineering assumptions. Designs once considered conservative may now be exposed to environmental loads beyond historical records. Infrastructure systems are interconnected, and disruptions propagate across sectors, amplifying consequences. In such a context, the ability to systematically analyze hazards, quantify uncertainty, and make defensible decisions under incomplete information has become a central requirement for responsible engineering practice.

Risk analysis provides the conceptual and analytical foundation for addressing these challenges. By linking hazard intensity, exposure, vulnerability, and consequences, it offers a structured framework for understanding how adverse events translate into societal and economic impacts. However, hazard identification alone is insufficient. Engineers must also quantify uncertainty through probabilistic modeling, interpret data using rigorous statistical tools, and integrate these insights into structured decision frameworks capable of balancing cost, safety, resilience, and sustainability.

This book, entitled *Hazard and Risk Assessment in Civil Engineering Design*, presents a systematic and integrated treatment of these interconnected themes. It begins with the fundamental concepts of hazard and risk assessment, clarifying terminology and establishing the foundational risk relationships that underpin modern infrastructure evaluation. It then develops the probabilistic and statistical machinery required for rigorous uncertainty analysis, including descriptive statistics, probability theory, Bayesian updating, conditional probability, joint distributions, and diagnostic performance metrics. Building upon these foundations, the book advances toward structured decision methodologies. Decision trees and expected utility theory are presented as transparent tools for modeling sequential engineering choices under uncertainty. Sensitivity analysis techniques reveal parameter influence and tipping points, while scenario-based decision analysis extends the framework to address deep uncertainty through robustness and regret metrics. The integration of adaptive planning transforms decision-making from a one-time prediction exercise into a dynamic process capable of evolving as new information emerges.

The distinguishing strength of this work lies in its integration of theoretical rigor with practical engineering applications. Through detailed examples, including flood protection design, foundation selection under soil uncertainty, stormwater system sizing, wastewater treatment technology evaluation, seismic retrofitting, pavement material choice, and sustainable building design, the book demonstrates how quantitative risk analysis and structured decision tools translate into resilient, cost-effective, and sustainable infrastructure strategies. The theoretical models, analytical methods, and implementation guidelines presented in this volume are grounded in extensive academic research and engineering practice. They collectively form a coherent framework that bridges hazard science, probability theory, uncertainty modeling, and decision analytics. This integrated perspective reflects the evolving nature of civil engineering, where resilience, sustainability, and accountability are as essential as efficiency and structural adequacy.

It is my strong belief that the publication of this book will contribute significantly to advancing risk-based engineering education and professional practice. By equipping engineers, researchers, and graduate students with systematic tools for managing uncertainty, it supports the development of infrastructure systems capable of performing reliably across diverse and uncertain futures. In doing so, it promotes not only technical excellence but also the long-term resilience and sustainability of the built environment.

Ottawa, Canada Hossein Bonakdair, Ph.D., P.Eng.

Competing Interests The authors have no competing interests to declare that are relevant to the content of this manuscript.

About This Book

This book presents a structured and comprehensive treatment of risk, uncertainty, and decision-making methodologies specifically tailored to civil and environmental engineering. It is designed as both a graduate-level teaching resource and a professional reference for researchers and practitioners seeking analytical rigor combined with practical implementation. The content is organized progressively, moving from foundational principles to advanced decision frameworks. Early chapters establish the analytical groundwork by examining hazard identification, infrastructure risk characterization, and the mathematical structure of uncertainty. The discussion then develops probability theory and Bayesian reasoning in a form directly applicable to engineering diagnostics, reliability assessment, and model updating.

A central feature of the book is its detailed treatment of uncertainty modeling. Descriptive statistics, distribution fitting, parameter estimation techniques, and goodness-of-fit testing are presented with engineering-oriented interpretation. Rather than treating statistical tools abstractly, the book demonstrates how data analysis supports hazard modeling and infrastructure reliability evaluation. The decision-analysis portion of the book provides a complete methodological framework. Decision trees are developed systematically, including construction, probability assignment, utility modeling, and expected value calculation. Sensitivity analysis is presented as a diagnostic extension of decision trees, with one-way, two-way, and threshold approaches illustrated in structured engineering contexts. Scenario-based decision analysis is then introduced to address deep uncertainty, supported by robustness criteria, regret analysis, and adaptive planning strategies. Each analytical framework is supported by fully developed engineering examples. These include applications in flood protection design, foundation system selection, wastewater treatment technology comparison, seismic retrofitting alternatives, pavement material decisions, sustainable building strategies, and climate adaptation planning. Problems are formulated step-by-step, assumptions are stated explicitly, mathematical derivations are shown where necessary, and decision criteria are applied transparently. This systematic approach allows readers to replicate methodologies and adapt them to their own projects.

The book also provides implementation guidance. Workflow structures, performance matrices, and evaluation criteria are presented in formats suitable for classroom use, research applications, and professional practice. The integration of probabilistic modeling tools and decision-support software further enhances practical applicability. What distinguishes this book is its integration of multiple domains that are often treated separately: hazard analysis, statistical inference, Bayesian updating, decision trees, sensitivity diagnostics, scenario evaluation, and adaptive planning. By bringing these components together within a unified engineering framework, the book serves as a coherent reference for structured decision-making under uncertainty.

For students, it offers a clear pathway from fundamental probability concepts to advanced decision analytics. For researchers, it provides a consolidated methodological base for extending risk-informed infrastructure studies. For practitioners, it delivers transparent and defensible tools for evaluating complex engineering alternatives in uncertain environments.

Contents

About the Authors

Dr. Hossein Bonakdari holds a Ph.D. in Civil Engineering and is currently an Associate Professor in the Department of Civil Engineering at the University of Ottawa, Canada. He is a registered Professional Engineer (P.Eng.) in Ontario, Canada. With more than two decades of experience in civil engineering infrastructure, research, and professional practice, Dr. Bonakdari brings extensive expertise in risk-informed infrastructure design and decision-making under uncertainty. Throughout his career, he has been actively involved in a broad range of civil engineering projects, gaining firsthand experience in identifying, evaluating, and mitigating risks associated with natural hazards and environmental variability. This practical engagement forms a strong foundation for this book, ensuring that its methodologies directly address real-world engineering challenges faced by practitioners and decision-makers.

Dr. Bonakdari's research focuses on the development of novel analytical algorithms and AI-driven decision-support tools for solving complex problems in water resources, environmental systems, and infrastructure resilience. His work integrates probabilistic modeling, machine learning, and scenario-based analysis to enhance hazard assessment and infrastructure planning. He has authored 6 textbooks and published more than 360 peer-reviewed journal articles, with an h-index of 59. His research has been widely cited and published in leading international journals and presented at numerous international conferences. He has been recognized among the world's top 2% scientists in Engineering and Technology, as reported by Stanford University, reflecting the global impact and credibility of his scholarly contributions.

Dr. Isa Ebtehaj is a Postdoctoral Researcher specializing in artificial intelligence applications in Civil and Environmental Engineering. Her expertise includes machine learning, hybrid modeling approaches, evolutionary optimization, hydrological time series analysis, and scientific programming. Her research aligns closely with the practical implementation of hazard and risk assessment methodologies in water and environmental systems. Over several years, she has participated in industrial and applied research projects at leading institutions, including the Water and Wastewater Research Center and the Environmental Research Center at Razi University. Her work demonstrates a strong practical orientation toward solving real-world challenges in hydrology and environmental engineering.

Dr. Ebtehaj has published more than 160 peer-reviewed journal articles, with an h-index of 50. Her contributions to data-driven modeling, predictive analytics, and uncertainty quantification have been widely recognized in the scientific community. In addition to her research achievements, she has significant expertise in developing user-friendly computational tools and structured code frameworks that support the implementation of probabilistic and decision-analysis methods. Her programming proficiency ensures that the analytical approaches presented in this book can be translated into practical, replicable tools for engineering applications.

Joseph Ladouceur is a Ph.D. Candidate in Civil Engineering at the University of Ottawa and a registered Engineer-in-Training (EIT) with Professional Engineers Ontario. He brings a strong combination of academic excellence, applied research experience, and professional practice to this book. His research focuses on water engineering, including drinking water treatment technologies, membrane systems, and environmental system performance evaluation. He has collaborated with international research partners on applied projects in water infrastructure and environmental engineering, contributing to peer-reviewed journal publications and presenting his work at international conferences.

In addition to his academic contributions, Mr. Ladouceur has practical experience in heavy civil construction and water infrastructure systems. His industry exposure and domain expertise strengthen the applied relevance of this book, particularly in the context of infrastructure reliability, environmental risk, and lifecycle decision-making. His professional commitment is reflected in his registration as an Engineer-in-Training, demonstrating adherence to engineering standards and continuous professional development.

Chapter 1
Fundamentals of Hazard and Risk Assessment in Civil Engineering

1.1 Introduction

Civil engineering is fundamentally concerned with creating the built environment that sustains modern society, its transportation networks, water systems, buildings, energy infrastructure, and public spaces. These systems must not only function under normal operating conditions but must also endure the wide range of uncertainties inherent in the natural and human-made world. As climate change intensifies extreme weather, urbanization accelerates exposure to hazards, and infrastructure systems age, the capacity to anticipate, evaluate, and manage risk has become a defining element of contemporary engineering practice.

Risk analysis provides the conceptual and analytical foundation for understanding how hazards interact with vulnerable systems to produce consequences. It offers engineers a systematic means of identifying potential threats before they materialize, quantifying their likelihood and severity, and developing strategies to reduce or manage their impacts. In this sense, risk analysis is not a peripheral task but a central pillar of responsible engineering design and decision-making. It supports efforts to improve public safety, allocate resources effectively, enhance infrastructure resilience, and ensure that engineering solutions remain robust in the face of uncertainty.

The increasing complexity of engineered systems further elevates the importance of risk-based approaches. Modern civil infrastructure is deeply interconnected: disruptions in one system can propagate rapidly to others, magnifying consequences in ways that were once difficult to foresee. Understanding these interactions requires tools and frameworks that go beyond traditional design calculations, enabling engineers to evaluate multiple scenarios, balance competing objectives, and integrate quantitative and qualitative risk assessments.

This chapter introduces the fundamental concepts that underpin hazard and risk assessment in civil engineering. It begins by outlining why risk analysis matters in today's engineering landscape, then explores its breadth of applications across

H. Bonakdari et al., *Hazard and Risk Assessment in Civil Engineering Design*,
https://doi.org/10.1007/978-3-032-23600-5_1

subdisciplines, including transportation, hydrology, structural engineering, environmental protection, and urban planning. The chapter also establishes the essential terminology used throughout the field, clarifying distinctions between concepts such as danger, threat, hazard, vulnerability, exposure, and consequences. These foundational ideas form the basis for the basic risk equation and for understanding how engineers quantify and manage risk in practice.

By presenting these principles at the outset, the chapter provides a conceptual framework that supports all subsequent material. As the reader progresses, these ideas will be revisited in various contexts, revealing how risk analysis informs every stage of the engineering process, from early planning and design to inspection, maintenance, and long-term infrastructure management. Through this lens, civil engineering emerges not only as a technical discipline but also as one deeply engaged with uncertainty, societal well-being, and the sustainable stewardship of the built environment.

1.2 Risk Analysis: Why It Matters

Risk assessment and analysis are considered two cornerstones of civil engineering. In this domain, not only are infrastructures designed and constructed, but potential risks affecting these structures and the communities they serve must also be foreseen, understood, and managed. Risk analysis in civil engineering is a systematic process. Potential risks are first identified, then thoroughly evaluated in terms of two key components: the probability of occurrence (likelihood) and the consequences of occurrence (impact). These components are then translated into a risk level or rating. Through this methodological approach, both quantitative and qualitative risk assessments are ensured. The primary goal of risk management in civil engineering applications is to mitigate potential risks to acceptable levels. This proactive approach aims to address risks before they materialize and is not a one-time effort; instead, it is a continuous process that spans the entire lifecycle of the civil engineering project. The safety, reliability, and sustainability of engineering solutions are ensured through continuous risk management from the drawing board to final construction.

Due to the increasing scale and scope of many civil engineering projects, as well as their associated societal impacts, risk analysis and management have taken on prominent roles in the field. A wide range of potential risks, from financial and environmental to societal and life-threatening, are encompassed by projects such as bridges, buildings, dams, or highways. Strategies to effectively mitigate these risks are enabled by the foresight provided by rigorous risk analysis, which anticipates scenarios such as structural failures, cost overruns, and environmental impacts.

As we move forward in an increasingly risk-averse society, incorporating risk analysis into the civil engineering curriculum is indispensable. Risk analysis is no longer viewed as simply a theoretical exercise, but instead has become a practical necessity for any modern civil engineering project. It serves as the foundation for delivering civil engineering projects safely, efficiently, and sustainably.

Future engineers must be equipped with the tools to anticipate and assess risk in civil engineering projects, as well as to respond to the challenges it poses. The capability of engineers to design not only aesthetically pleasing and functional structures but also safe and resilient ones is enhanced by understanding risk analysis. This knowledge is vital in an era where engineering challenges are becoming increasingly complex due to factors such as urbanization, climate change, and technological advancements.

1.3 Risk Assessment Applications

The field of civil engineering is diverse and complex, spanning a number of subdisciplines from transportation to environmental engineering, and from structural to geotechnical engineering. Each branch of the civil engineering field presents its own unique set of challenges and potential hazards, which necessitate risk assessment. In this sub-section, the specific risks associated with each area of civil engineering are discussed, with specific emphasis on how risk assessment is integral in addressing the unique challenges posed by each field. Furthermore, the role of risk assessment strategies in developing more robust, safe, and sustainable solutions across each field is highlighted. Finally, based on these insights, the importance of risk assessment as a fundamental tool in the planning, design, construction, and maintenance of civil engineering projects is underscored. Some applications of risk assessment in civil engineering are presented in Fig. 1.1.

1.3.1 Transportation Engineering

In Transportation Engineering, risk assessments are conducted to ensure the safety and efficiency of transportation systems. Common hazards, such as structural weaknesses in bridges and roadways, are identified, and their potential impacts are carefully evaluated. The focus is often placed on minimizing risks that can lead to accidents, traffic congestion, or environmental degradation. For instance, the risk of a bridge collapse due to structural failure might be assessed by considering factors such as the age of the bridge, material fatigue, susceptibility to flood, scour, or other extreme environmental conditions, and daily traffic loads. The likelihood of the bridge's structural failure, as well as its potential consequences, such as traffic disruption and safety hazards, is estimated. Based on these assessments, engineers can recommend to decision-makers timelines for preventive maintenance and retrofitting, as well as any necessary redesign measures. Such proactive measures are crucial in preventing accidents and ensuring the smooth functioning of transportation infrastructure. Furthermore, the environmental impacts of transportation projects, such as emissions and noise pollution, can be evaluated as part of the risk assessment to reduce their ecological footprint.

Fig. 1.1 Some of the applications of risk assessment

1.3.2 Traffic Engineering

Traffic Engineering prioritizes the safety of road users. Risk assessments are routinely conducted to identify and mitigate hazards in high-risk areas, such as intersections prone to accidents or pedestrian crossings with high foot traffic. The assessment focuses on the probability of accidents and their potential consequences. For example, an intersection known for frequent collisions might be studied. Factors contributing to these collisions, such as poor visibility, inadequate signage, or high-speed limits, are identified. Solutions, such as installing traffic lights, additional signage, or reducing speed limits, are then proposed based on the assessment. These measures are instrumental in enhancing road safety and reducing traffic accidents. Furthermore, traffic flow is analyzed, and strategies are developed to alleviate congestion, ensuring a smoother and safer travel experience for all road users.

1.3.3 Water Resources and Hydrology

In this field, risk assessments are crucial for managing and safeguarding water resources. Risks such as flooding, water scarcity during droughts, and water quality degradation are systematically evaluated. For instance, the risk of flooding in a particular river basin might be assessed. This assessment involves analyzing historical data on rainfall patterns, river flow rates, and the area's topography. Predictions about future flood events, including their frequency and severity, are made based on this data. Mitigation strategies, including the construction of dams, levees, or early flood warning systems, are then proposed. These strategies are vital in protecting communities and agricultural lands from flood damage. Similarly, in drought cases, the risk of water shortages is assessed, and water conservation measures are recommended. The assessment also extends to the quality of water resources, identifying potential sources of contamination and proposing measures to protect water quality. The aim is to ensure a sustainable and safe water supply for various needs, including drinking, agriculture, and industry. A relevant contemporary example is the proposed nuclear waste facility at Ottawa's Chalk River location, where radiological monitoring is required both upstream and downstream to compare against background conditions and assess the risk of radioactive contamination entering the river system.

1.3.4 Environmental Engineering

In Environmental Engineering, the assessment of environmental risks from civil engineering projects is conducted meticulously. Factors such as potential soil contamination, air and water pollution, and disruption to natural habitats and wildlife are evaluated. For example, the risk of contaminating nearby water sources might be assessed in a project to construct a new industrial facility. The types of pollutants that could be released during the facility's operation, their potential pathways into the water system, and the consequent impact on aquatic life and human health are examined. Measures to mitigate these risks, such as installing advanced filtration systems or creating buffer zones, are then proposed. Similarly, the impact of construction activities on air quality is evaluated, and solutions such as dust control measures and the use of low-emission machinery are recommended. These assessments aim to identify and mitigate environmental impacts, ensuring that engineering projects are sustainable and comply with environmental regulations and standards. This proactive approach is essential for preserving the environment and maintaining ecological balance.

1.3.5 Geotechnical Engineering

In Geotechnical Engineering, risk assessments are conducted to understand and mitigate risks associated with earth materials and their interaction with civil engineering projects. Issues such as soil stability, landslide risk, and potential ground subsidence are carefully analyzed. For instance, in the case of constructing a new building, the soil stability on which it is to be built might be evaluated. Factors such as soil composition, moisture content, and load-bearing capacity are taken into account. The likelihood of soil settling or shifting, and the potential impact on the building's structural integrity, are estimated. Based on this assessment, recommendations for foundation design or soil stabilization methods are made. These recommendations are crucial for ensuring the safety and longevity of structures. Additionally, in areas prone to landslides, slope stability is assessed, and measures such as retaining walls or drainage systems can be recommended to prevent soil movement. This risk assessment process is essential for avoiding catastrophic failures and ensuring the resilience of civil engineering structures.

1.3.6 Coastal and Port Engineering

Risk assessments are integral for addressing the unique challenges posed by marine and coastal environments. Issues such as coastal erosion, sea-level rise, and the impact of extreme weather events on coastal infrastructures must be evaluated. For example, in the design of a new port facility, the risk of damage from storm surges and high tides might be assessed. Factors such as historical storm patterns, sea-level trends, and local topography are considered. The potential impacts on the port's infrastructure, including the likelihood of flooding and erosion, and the consequent disruption to port operations, are estimated. Based on these assessments, design adaptations like elevated platforms, sea walls, or breakwaters are recommended to protect the facility. These measures are critical to ensuring the durability and functionality of coastal and port structures in the face of a changing climate and natural hazards. Furthermore, the assessment also considers the environmental impact of port activities, ensuring that marine ecosystems are protected and that sustainable construction and operational practices are implemented.

1.3.7 Structural/Construction Engineering

In Structural/Construction Engineering, risk assessments are crucial for ensuring the safety and integrity of buildings and structures. Hazards such as structural failures, material deficiencies, and construction mishaps are methodically evaluated. For example, in constructing a high-rise building, the risk of structural failure due to load imbalances or seismic activity might be assessed. Factors such as the build and material quality, the specific design, and the potential for natural disasters in the area are considered. The likelihood of structural failure and its potential consequences, ranging from minor damage to catastrophic collapse, is estimated. Based on this assessment, design modifications, material choices, and construction techniques may be recommended to enhance the safety, durability, and reliability of the structure. Additionally, construction site safety is evaluated, with risks to workers and the public being identified. Safety training, protective equipment, and altered construction site protocols are examples of strategies that can be proposed to mitigate these risks. This comprehensive risk assessment process is essential in Structural/Construction Engineering to prevent accidents and ensure the long-term safety of structures.

1.3.8 Municipal or Urban Engineering

In Municipal or Urban Engineering, risk assessments are conducted to ensure the safety, efficiency, and sustainability of urban infrastructure and services. Issues such as the aging of public infrastructure, traffic congestion, challenges in waste management, and the provision of clean water and sanitation services are comprehensively evaluated. For instance, the risk associated with the deterioration of an old water supply system in a city might be assessed. Factors such as the age and material of the pipes, the frequency of maintenance, and the incidence of leaks or failures are examined. The likelihood of system failure and its potential impacts on public health and service disruption are estimated. Recommendations for system upgrades, maintenance schedules, or complete replacements are then proposed based on these assessments. These proactive measures are crucial for maintaining uninterrupted and safe urban services. Additionally, in urban planning, the risks of overcrowding, inadequate public amenities, and environmental degradation are evaluated. Urban engineers use these assessments to design cities that are not only functional but also livable, with a focus on sustainability and residents' quality of life.

1.3.9 Urban Planning

In Urban Planning, risk assessments are carried out to guide the development and management of urban areas in a sustainable and safe manner. Challenges such as urban sprawl, population density, environmental sustainability, and the adequacy of public facilities and services are evaluated. For instance, the risk of inadequate infrastructure to support the growing population might be assessed when planning a new residential area. Factors such as access to public transportation, availability of green spaces, and proximity to essential services, such as schools and hospitals, are considered. The potential impacts of inadequate planning, including traffic congestion, reduced quality of life, and strain on existing services, are estimated. Based on these assessments, urban planners propose comprehensive development plans that balance growth with sustainability and livability. These plans often include recommendations for infrastructure development, zoning regulations, and environmental conservation measures. Additionally, the impact of urban development on local ecosystems is assessed, with strategies being developed to minimize environmental degradation and promote ecological harmony. This risk assessment process is essential to urban planning, creating cities that are not only functional but also provide a high quality of life for their inhabitants.

1.3.10 Surveying

In the field of Surveying, risk assessments are conducted to ensure the accuracy and reliability of survey data. Challenges such as terrain difficulties, inaccuracies due to environmental factors, and technological limitations are systematically evaluated. For example, when conducting a land survey for a construction project, the risk of measurement errors due to uneven terrain or obstructed views might be assessed. Factors such as the type of surveying equipment used, environmental conditions at the site, and the presence of natural or manufactured obstacles are taken into account. The potential impact of these factors on survey data accuracy is estimated. Based on this assessment, appropriate surveying methods and technologies are selected to mitigate the risk of error. This may include the use of advanced GPS (i.e., Global Positioning System) systems, aerial surveying techniques, or additional ground surveys for cross-verification. Ensuring the precision of survey data is crucial as it forms the basis for all subsequent planning and construction activities. Accurate surveying is essential for the successful execution of civil engineering projects, preventing costly errors and delays.

1.3.11 Earthquake Engineering

In Earthquake Engineering, risk assessments are integral to the design of structures that can withstand seismic activity. The potential risks associated with earthquakes, such as structural damage, soil liquefaction, and building collapse, are meticulously evaluated. For instance, in designing a new building in an earthquake-prone area, the risk of structural failure during an earthquake might be assessed. Factors such as the region's seismic history, soil type, and building design are considered. The likelihood of different levels of seismic activity and the building's ability to withstand such forces are estimated. Recommendations for earthquake-resistant design features, such as base isolators, shear walls, and reinforced framing, are then proposed based on these assessments. These design adaptations are critical for ensuring the safety and integrity of buildings during seismic events. Additionally, the potential impact of earthquakes on existing buildings is evaluated, and retrofitting measures are suggested for older structures to improve their resilience. This risk assessment process is essential in Earthquake Engineering to protect lives and minimize property damage in the event of an earthquake.

1.3.12 Forensic Engineering

In Forensic Engineering, risk assessments are performed to analyze failures in engineering projects and to prevent future occurrences. The causes of structural failures, material breakdowns, or accidents are thoroughly investigated. For example, in a building collapse, the factors contributing to the failure might be assessed. Aspects such as design flaws, material defects, construction errors, or inadequate maintenance are examined. The sequence of events leading to the failure and the specific point of origin are determined through detailed analysis. Based on these assessments, recommendations to address identified risks in similar projects are proposed. These may include improvements in design standards, construction practices, or maintenance protocols. Such evaluations are crucial for enhancing the safety and reliability of future engineering projects. Additionally, forensic engineers often play a key role in legal and insurance cases, where their risk assessments help determine liability and guide policy improvements. This risk assessment process is essential in Forensic Engineering to ensure the integrity and safety of engineering practices.

1.3.13 Hydraulic Engineering

In Hydraulic Engineering, risk assessments are carried out to manage and mitigate risks associated with water-based projects and systems. Issues such as dam failures, flood control, and the impact of hydraulic structures on ecosystems are systematically evaluated. For example, in the design of a new dam, the risk of structural failure and its potential impact on downstream communities and ecosystems might be assessed. Factors such as the river's hydrological characteristics, sedimentation patterns, and dam structural design are considered. The likelihood of overtopping, structural breaches, and the consequent risk of flooding are estimated. Based on this assessment, engineering solutions such as spillways, overflow channels, or reinforced structures are proposed to enhance the dam's safety. These measures are vital for protecting lives and property, and for preserving ecological balance. Additionally, the assessment evaluates the impact of hydraulic projects on water flow, aquatic habitats, and water quality, ensuring that these projects are environmentally sustainable.

1.3.14 Railway Engineering

In Railway Engineering, risk assessments are conducted to ensure the safety, reliability, and efficiency of railway systems. Various risks associated with railway construction, maintenance, and operation are evaluated. For example, in developing a new railway line, the risks of derailment, signal failure, and bridge or tunnel collapses might be assessed. Factors such as track alignment, ground stability, signal systems, and the structural integrity of bridges and tunnels are taken into account. The likelihood of these events and their potential impacts on safety, service disruption, and infrastructure damage are estimated. Based on these assessments, engineering solutions like advanced signaling systems, robust track designs, and regular maintenance schedules are proposed to mitigate these risks. These measures are critical for ensuring the continuous and safe operation of railway services. Additionally, the environmental impacts of railway projects, including noise pollution and wildlife disruption, are evaluated. Proposed measures to minimize these environmental impacts ensure that railway projects are not only efficient and safe but also environmentally responsible.

1.4 Terminology & Fundamental Concepts

In this section, the fundamental terminology and concepts of risk analysis in civil engineering are introduced. These terms are the foundation upon which the principles of hazard and risk assessment are built, and are integral to a comprehensive understanding of the subject matter. The significance of these terms cannot be overstated, as they provide the basic language and framework necessary for discussing and analyzing risks in civil engineering. Although these concepts may initially appear abstract or theoretical, they are both practical and critically relevant to real-world applications. The complexities of civil engineering risks are understood, analyzed, and managed through the lens of these fundamental concepts. Accordingly, these terms will be used throughout the remainder of this text.

The terms introduced in this section have been selected based on their importance in the field of risk analysis. They not only offer a starting point for students new to the subject but also serve as a reference point for more advanced discussions. A clear understanding of these terms, as well as their precise and consistent application, is essential for effective communication within the field, especially when collaborating on the development of risk assessment and mitigation strategies.

Moreover, mastering this terminology is not just an academic exercise; it equips future civil engineers with the necessary tools to identify, assess, and address the various risks associated with civil engineering projects. Understanding these concepts is the first step in developing the skills required to navigate the challenges posed by both natural and man-made hazards in the industry.

As this book progresses, these terms will be revisited and applied across a range of contexts, demonstrating their wide-ranging relevance in civil engineering. This approach will not only reinforce the understanding of the terminology but also highlight its practical utility in assessing and managing risks in civil engineering projects. The continual application of these concepts throughout the book is designed to illustrate their importance in real-world scenarios, bridging the gap between theoretical knowledge and practical application in civil engineering.

1.4.1 Danger

The concept of **Danger**, as defined by various authoritative dictionaries, encompasses the potential for harm, injury, or even death. It is a state or condition where there is a likelihood or risk of encountering something harmful or perilous. According to the Cambridge Dictionary, danger is explicitly associated with *the possibility of harm or death to someone*, highlighting the severity and potential fatality it entails. Merriam-Webster emphasizes the aspect *of exposure to, or liability for, injury, pain, harm, or loss, suggesting* vulnerability or susceptibility to negative outcomes. Similarly, Macmillan Dictionary defines danger as *a situation in which harm, death, damage, or destruction is possible, indicating a broader spectrum of harmful outcomes beyond personal injury, extending to damage and destruction that could affect* structures, environments, or systems. In essence, danger implies a situation or condition with a notable risk of negative, harmful, or destructive outcomes.

On the other hand, **Safe**, as defined by these dictionaries, is *a state or condition in which there is no danger or risk*. It denotes a situation where individuals, structures, or environments are secure and protected from harm or injury. Being safe means being free from the risk of harm, with the potential for injury, pain, or damage effectively nullified or minimized. When described as 'not in danger,' it emphasizes the absence of threats or harmful conditions. Additionally, the definition of 'safe' as 'free from risk' or 'harmless' underscores the aspect of security and protection, in which the potential for adverse outcomes is significantly reduced or eliminated. Ensuring that structures, systems, and environments are 'safe' in civil engineering and risk assessment involves rigorous planning, design, and management to protect against potential dangers and minimize risks. A schematic of the differences between the concepts of danger and safety is provided in the following Fig. 1.2.

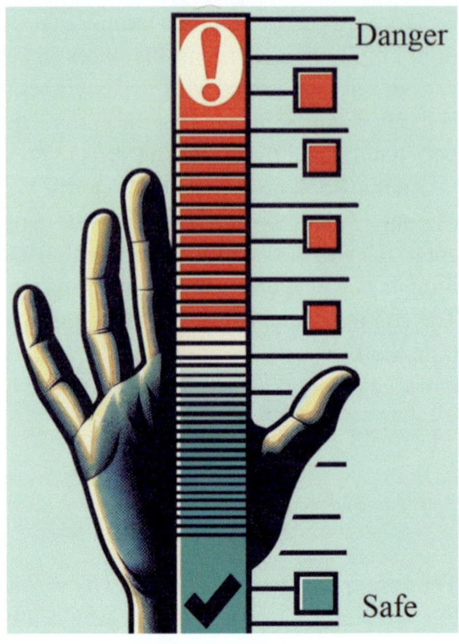

Fig. 1.2 A schematic of the differences between the concepts of danger and safety

1.4.2 Threat

Understanding the term "*threat*" in the context of civil engineering risk analysis requires a thorough examination of its definition and a keen awareness of its application in security and risk management. Cambridge Dictionary defines a threat as *a statement indicating that someone will be harmed, emphasizing the communicative or declaratory aspect of a threat, often expressed* in the form of verbal or written warnings. Merriam-Webster extends this definition to *an expression of intent to inflict evil, injury, or damage, encompassing not only physical harm but also the intent* to cause broader negative impacts.

Macmillan Dictionary offers a broader definition, describing a threat as *a situation or activity that could cause harm or danger*. This interpretation covers a wider range of scenarios, including not only intentional acts but also various conditions or activities that may lead to harmful outcomes.

More specifically, threats, especially in the field of security and risk management, are categorized based on their nature and the type of harm they pose. These can include:

- **Chemical Threats**: Involving the use or release of harmful chemical substances.
- **Biological Threats**: Pertaining to the exposure to harmful biological agents, such as viruses or bacteria.
- **Radiological Threats**: Associated with exposure to harmful radioactive materials.
- **Nuclear Threats**: Involving the use or threat of nuclear weapons or materials.

- **Explosive Threats**: Related to the use of explosive materials to cause harm or damage.
- **Sabotage**: Intentional acts aimed at damaging, destroying, or disrupting vital systems or structures.
- **Cyber Threats**: Involving attacks on or through computer networks and digital systems, potentially leading to data theft, system damage, or disruption of digital services.
- **Environmental Threats**: These include natural disasters such as earthquakes, hurricanes, floods, wildfires, and droughts. Environmental threats can cause significant damage to infrastructure and pose risks to public safety.
- **Technological Threats**: These are associated with the failure or malfunction of technology. Examples include power grid failures, structural failures in engineering works, and accidents in high-risk industries like equipment malfunctions and chemical manufacturing.
- **Economic Threats**: Economic instability or downturns can threaten the viability of engineering projects, lead to reduced investment in infrastructure maintenance, and increase the risk of cost overruns and project delays.
- **Political Threats**: These involve risks arising from political instability, terrorism, wars, and conflicts. Political threats can affect the security and functionality of critical infrastructure and influence policy and regulatory environments.
- **Social Threats**: These include threats from social unrest, public disorder, strikes, and protests. They can disrupt construction projects, affect public safety, and impact existing infrastructure operations.
- **Health Threats**: Epidemics and pandemics, such as the COVID-19 pandemic, pose significant threats not only to public health but also to the continuity of operations in various sectors, including the construction and maintenance of infrastructure, and drinking and wastewater treatment operations.
- **Supply Chain Threats**: These involve risks associated with the disruption of supply chains, impacting the availability of materials and equipment, which can delay construction projects and lead to significant cost increases.
- **Information Security Threats**: Beyond conventional cyber threats, these include risks related to data privacy, intellectual property theft, and misinformation that can impact project planning and execution.
- **Geopolitical Threats**: These encompass risks arising from international relations, such as trade wars, sanctions, and cross-border conflicts, which can impact global projects and international collaborations in civil engineering.

Each of these threat types poses unique challenges for prevention, detection, and mitigation. Understanding and addressing these threats in civil engineering is crucial for designing, constructing, and maintaining safe and resilient structures and systems. It not only involves safeguarding against physical damage but also ensuring the integrity and continuity of services in the face of potential threats.

1.4.3 Hazard

A **Hazard** is broadly characterized as any phenomenon, substance, human activity, or condition that poses a risk of causing significant negative outcomes. These negative outcomes can range from loss of life and injury to broader impacts such as health issues, property damage, and loss of livelihood. Additional negative outcomes associated with hazards include the potential to cause social and economic disruption and environmental degradation. In civil engineering, recognizing and assessing hazards is crucial for designing safe, resilient, and sustainable structures and systems.

1.4.3.1 Types of Hazards

In the context of civil engineering and risk management, identifying and understanding hazards is critical. Hazards can be diverse and encompass a wide array of scenarios, including natural and man-made hazards. Differentiating between natural and man-made hazards is important for developing appropriate risk management strategies. Natural hazards often require an understanding of environmental and geological sciences, whereas man-made hazards might focus more on industrial processes and human error. However, the overlap between these categories is increasingly recognized, especially when human activities exacerbate natural hazards.

1. **Natural Events**: Earthquakes, floods, hurricanes, tornadoes, droughts, tsunamis, and winter storms are all examples of natural events that pose severe hazards to civil engineering infrastructure. These events can also lead to geophysical risks, including landslides, mudslides, and volcanic eruptions. Natural hazards are known to be exacerbated by environmental factors, such as climate change, leading to increased severity and frequency. Due to their inherent power and unpredictability, these natural occurrences can have catastrophic impacts on communities, infrastructure, and the environment. Some examples of catastrophic natural events are listed below for each category:

   - **Earthquakes**: The 2011 Tōhoku earthquake in Japan triggered a massive tsunami and led to the Fukushima Daiichi nuclear disaster.
   - **Floods**: The 1997 Red River flood in Minnesota, North Dakota (US), and southern Manitoba (Canada) became known as the "Flood of the Century.
   - **Hurricanes**: Hurricane Ian (2022) was a unique and challenging storm that had widespread impacts on eastern North Carolina (US).
   - **Tornadoes**: The Selma Tornado in Alabama in 2023 caused significant damage and disruption, marking one of the most devastating natural disasters in the region's recent history.
   - **Droughts**: The 2011–2017 California drought led to significant water shortages and agricultural impacts.

- **Tsunamis**: The 2004 Indian Ocean tsunami, triggered by an undersea earthquake, caused widespread devastation across several countries.
- **Winter Storms**: The 1993 Storm of the Century impacted much of the Eastern U.S. and parts of Canada.

2. **Man-Made Incidents**: These hazards result from human activities and include chemical and nuclear incidents, explosions, and biological attacks. Industrial accidents, structural failures, and pollution are other examples of man-made hazards that can have disastrous effects on public health, safety, and the environment. These events, often resulting from human error, negligence, or malice, can lead to immediate and long-term consequences that are detrimental to the health, safety, and well-being of populations and ecosystems. Some examples of man-made incidents are as follows:

   - **Chemical Accidents**: The 1984 Bhopal gas tragedy in India, where a gas leak from a pesticide plant caused thousands of deaths.
   - **Explosions**: The 2020 Beirut port explosion in Lebanon was caused by the accidental detonation of stored ammonium nitrate.
   - **Biological Attacks**: The 2001 anthrax attacks in the United States through the postal system.
   - **Nuclear Accidents**: The 1986 Chornobyl disaster in the Soviet Union was the worst nuclear power plant accident in history.

The concept of a hazard is central to risk analysis because it represents the potential source of harm or adverse effects. Understanding the nature, scale, and probability of hazards is essential for developing effective strategies to mitigate their impact. This involves not only recognizing the immediate threats they pose but also considering the potential for future events, the vulnerability of populations and structures to these hazards, and the capacity to respond to and recover from them.

1.4.3.2 Magnitude and Impact of Hazards

Understanding the magnitude of a hazard is the essential element of risk assessment, as it directly influences the severity of potential impacts. Magnitude refers to the intensity or scale of a hazardous event or condition, which can be quantified in various ways depending on the type of hazard.

- **Earthquakes**: Ground acceleration is a key measure of an earthquake's magnitude. However, other factors, such as earthquake depth, duration, and distance from urban centers, also play crucial roles in determining potential impact. For instance, the 1995 Kobe earthquake in Japan, despite its moderate magnitude (6.9), caused significant destruction due to its shallow depth and proximity to a densely populated area.
- **Flooding**: The depth of flooding is a basic measure, but the speed of onset, the extent of the inundated area, and the duration of flooding are equally important.

The 2010 Pakistan floods, for example, covered a vast area and lasted for weeks, leading to extensive damage and displacement.

- **Cyclones and Windstorms**: While wind speed is a primary indicator of a cyclone's intensity, other factors such as storm size, rainfall intensity, and storm surge height are critical. Hurricane Katrina (2005) in the U.S. is an example in which the storm surge and subsequent flooding were more devastating than the wind itself.
- **Chemical Spills**: The concentration of the released substance is important, but so are the chemical's toxicity, the area of exposure, and the duration of the release. The Bhopal Gas Tragedy in 1984 was catastrophic not only because of the volume of the toxic gas released, but also because of its high toxicity (methyl isocyanate) and the dense population in the exposed area.
- **Wildfires**: Their magnitude can be measured by burnt area, rate of spread, and fire duration. The 2019–2020 Australian bushfires, known as the "Black Summer," were notable for their vast scale and prolonged duration, leading to extensive ecological and property damage.
- **Droughts**: The intensity of a drought is gauged by its duration, the extent of the affected area, and the degree of moisture deficit. For example, the severity of the California drought from 2011 to 2017 was marked by its multi-year span and widespread impact on water resources.

The impact of these hazards is closely tied to their magnitude. However, it is also influenced by factors such as the vulnerability of the affected population, existing mitigation measures, and response capacity. A high-magnitude hazard may have a relatively low impact in a well-prepared and resilient community, whereas a lower-magnitude event can cause significant damage in a vulnerable area.

1.4.3.3 Identification and Analysis

Identifying potential hazards is pivotal in risk management. This task requires a detailed examination of numerous factors, each of which can significantly influence the types and magnitudes of hazards encountered in a project.

- **Nature of the Project**: Different project types inherently have different hazard profiles. For instance, a high-rise building project in an urban area would need to assess hazards such as structural failures, fire risks, and potential impacts from adjacent construction activities. In contrast, a project in a rural or natural setting might focus more on hazards related to wildlife, weather conditions, or geological instability.
- **Geographical Location**: The location of a project plays a crucial role in determining the types of hazards it may face. A project in a seismically active area, for example, must prioritize earthquake preparedness, while a project in a floodplain must account for flooding risks. Coastal projects are particularly susceptible to hazards such as sea-level rise, storm surges, and coastal erosion, whereas projects

in industrial zones might be more concerned with chemical spills or air quality issues.

- **Environmental Context**: The surrounding environment, including both the natural and built environment, can introduce specific hazards. For example, proximity to rivers or lakes might increase the risk of flooding, while being situated near forests could raise the risk of wildfires. Additionally, the presence of endangered species or sensitive ecosystems might pose environmental compliance risks.

Lifecycle analyses and other structured approaches are often used to systematically identify and assess these hazards. These methodologies allow for a comprehensive evaluation of hazards throughout the entire lifespan of a project, from conception through construction and operation, to decommissioning.

1. **Early-Stage Assessment**: In the early stages of a project, these analyses can help identify potential design and location-related hazards. For instance, site analysis might reveal susceptibility to landslides or flooding, which can influence the design and site preparation.
2. **Construction Phase Assessment**: During construction, hazards related to construction activities, such as accidents, equipment failure, and temporary structures, are assessed. This stage often requires dynamic risk assessments as conditions and activities change frequently.
3. **Operational Phase Assessment**: Once the project is operational, ongoing risk assessments are crucial. These assessments take into account factors such as aging infrastructure, changing usage patterns, and evolving external threats (such as climate change or urban development).
4. **Decommissioning and Post-Operational Assessment**: Finally, decommissioning a project itself can introduce hazards, such as demolition risks, waste disposal issues, and site remediation challenges.

By employing these structured approaches, engineers and planners can not only identify and analyze potential hazards but also develop effective strategies to mitigate them. This proactive stance is key to enhancing the safety, sustainability, and resilience of engineering projects, ensuring they can withstand a range of challenges and hazards throughout their lifespans.

1.4.3.4 Hazard Versus Risk

In the context of civil engineering and risk assessment, it is crucial to distinguish between the terms "Hazard" and "Risk", as they represent different aspects of potential danger.

A "**hazard**" refers to any potential source of harm or adverse effect. In civil engineering, hazards might include natural events such as earthquakes and floods, or man-made occurrences such as chemical spills or structural failures. Hazards are essentially the scenarios or conditions that have the potential to cause damage or injury. For instance, in a construction project, a hazard could be the possibility of a crane collapsing. The hazard exists because there's potential for harm (the crane falling), but it doesn't describe the likelihood of this happening or the potential consequences.

On the other hand, "**risk**" is related to the probability or likelihood that the hazard will cause harm and the severity of the resulting consequences. It is a measure of the chance that a hazard will occur and the extent of its impact if it does. In other words, risk quantifies the potential harm that a hazard could cause. Continuing with our crane example, the risk would involve assessing how likely it is for the crane to collapse (considering factors such as the crane's condition, weather conditions, and operational safety measures, for example) and what the consequences would be if it did (such as damage to property, injury to workers, or delays in the project).

Understanding the distinction between hazard and risk is fundamental in civil engineering. While a hazard represents a potential source of harm, risk measures the likelihood and impact of that harm occurring. Effective risk management in civil engineering involves identifying hazards, evaluating the risks associated with these hazards, and implementing measures to mitigate those risks. This approach ensures the safety and integrity of engineering projects and the well-being of those involved or affected.

1.4.4 Consequences

In the realm of hazard and risk assessment, particularly within civil engineering, **Consequences** refer to the potential outcomes or results that stem from an undesired event. These consequences are pivotal for understanding and quantifying risk. Consequences are estimates of potential losses that may occur due to a specific hazard or threat, encompassing a wide range of values and assets.

Consequences are typically assessed with respect to a number of different categories including:

- **Life and Health**: This pertains to the impact on human life and well-being. In civil engineering, this could mean casualties or injuries resulting from an infrastructure failure or accident, for example.
- **Environment**: The consequences on the natural environment, such as pollution, habitat destruction, or ecological imbalance.
- **Operation, Production, and Service**: This relates to the functionality and continuity of services. For example, the shutdown of a water treatment plant due to contamination would fall under this category.
- **Economic and Material Values**: These are the financial and material losses, such as the cost of repairs, loss of revenue, or replacement of damaged infrastructure.
- **Credibility and Reputation**: The impact on the reputation and trustworthiness of entities involved, which can have long-term effects on their operations and public perception.

Consequences can be either directly or indirectly related to the asset in question. Direct consequences are those that are immediately linked to the asset, such as the costs of repairing or retrofitting a damaged structure. For example, if a bridge is damaged by flooding, the direct consequences would be the costs of repairing it.

Indirect consequences, on the other hand, are outcomes that occur as a result of the event or a decision made in response to the event. Continuing with the bridge example, if the damaged bridge is closed and disrupts transportation routes, the resulting social and economic impacts on the community would be considered indirect consequences.

In practical risk management within civil engineering, modeling both direct and indirect consequences is crucial. This comprehensive approach allows for a more accurate assessment of the total impact of risks. Understanding the interplay between direct and indirect consequences, and how they affect different stakeholders, is key to developing effective risk mitigation strategies. Stakeholders need to recognize these different types of consequences and model them, along with their interactions, to ensure robust and effective risk management practices.

1.4.5 Vulnerability

In the context of hazard and risk assessment, **Vulnerability** refers to the characteristics and circumstances of a community, system, or asset that make it susceptible to the damaging effects of a hazard. This concept is central to understanding how different factors contribute to the risk and impact of hazards on various elements in civil engineering and beyond.

Vulnerability encompasses a wide range of aspects stemming from a diverse array of factors such as:

- **Physical Factors**: This includes elements such as poor design and construction quality of buildings or infrastructure that may not withstand the forces of natural hazards, such as earthquakes or floods.
- **Social Factors**: Social vulnerability arises from factors such as a lack of public information and/or awareness of the risks they face, as well as communities that are not adequately prepared or educated in hazard mitigation.
- **Economic Factors**: Economic vulnerability can be seen in situations where assets are inadequately protected, or in communities that lack the financial resources to effectively respond to or recover from hazardous events.
- **Environmental Factors**: This involves a disregard of appropriate environmental management, which can exacerbate the impacts of natural hazards or can lead to environmental degradation that increases the risk and vulnerability of infrastructure to hazards.

It must be noted that vulnerability is a dynamic concept that varies significantly within a community over time. It is not only dependent on the element's exposure to the hazard, but also on the inherent characteristics of the system, including its physical, technical, organizational, or cultural aspects. These can be exploited by a hazard to cause harm or damage to assets/infrastructure.

Furthermore, the specific understanding of vulnerability and the conceptual frameworks used to address it are influenced by the perspectives of distinct groups, each of which may focus on different aspects of vulnerability. Consequently, vulnerability is:

- **Multi-dimensional**: It encompasses various dimensions, including physical, social, economic, and environmental aspects.
- **Dynamic**: Vulnerability is not static; it can change over time due to a variety of factors, including development, mitigation measures, and changes in the environment.
- **Scale-dependent**: Vulnerability can be assessed at different scales, from individual humans to communities, regions, or even countries.
- **Site-specific**: The characteristics of vulnerability often vary based on the specific location or site and are influenced by the local conditions and contexts that exist.

In the realm of hazard and risk assessment, understanding the different types of vulnerabilities and losses is critical to developing effective management and mitigation strategies. These vulnerabilities can be categorized into a number of classes, including social, physical, economic, and environmental, each of which results in distinct direct and indirect losses (Table 1.1).

1. **Human—Social Vulnerability**: This refers to the potential impacts of events on certain groups within society, particularly those who are more susceptible to harm. These groups include the individuals living in poverty, people experiencing homelessness, single-parent households, pregnant women, people with disabilities, children, and the elderly. Social vulnerability considers factors such as public awareness of risks, the ability of groups to self-cope with disasters, and the effectiveness of the institutional structures designed to help them.
   - Direct Losses: These may include fatalities, injuries, loss of income or employment, and homelessness.
   - Indirect Losses: Longer-term impacts such as diseases, permanent disability, psychological trauma, loss of social cohesion, and political unrest.
2. **Physical Vulnerability**: This is related to the potential for physical impact on the built and natural environment. It can be expressed as elements-at-risk (EaR) and is quantified by the degree of loss to a given EaR resulting from a natural event.
   - Direct Losses: Structural damage or collapse of buildings, nonstructural damage, and damage to infrastructure.
   - Indirect Losses: Progressive deterioration of damaged structures and infrastructure not repaired in time.

Table 1.1 Different types of vulnerabilities and losses

	Human-social	Physical	Economic	Cultural-environmental
Direct losses	• Fatalities • Injuries • Loss of income or employment • Homelessness	• Structural damages or the collapse of a building • Nonstructural damage and damage to contents • Structural damage to infrastructure	• Interruption of business due to damage to buildings and infrastructure • Loss of productive workforce through fatalities, injuries, and relief efforts • Capital costs of response and relief	• Sedimentation • Pollution • Endangered species • Destruction of ecological zones • Destruction of cultural heritage
Indirect losses	• Diseases • Permanent disability • Psychological impact • Loss of social cohesion due to the disruption of the community • Political unrest	• Progressive deterioration of damaged buildings and infrastructure that are not repaired	• Economic losses due to the short-term disruption of activities • Long-term economic losses • Insurance losses weakening the insurance market • Fewer investments • Capital costs of repair • Reduction in tourism	• Loss of biodiversity • Loss of cultural diversity

3. **Economic Vulnerability**: This includes the potential impacts of hazards on economic assets and processes. The particular effect will vary by economic sector and over time.
   - Direct Losses: Interruption of business, loss of productive workforce through fatalities and injuries, and capital costs of response and relief.
   - Indirect Losses: Economic losses due to short-term disruption of activities, long-term economic losses, weakened insurance markets, reduced investments, and repair costs.
4. **Cultural - Environmental Vulnerability**: Concerns about the potential impacts of environmental events, including flora, fauna, ecosystems, and biodiversity.

- Direct Losses: Sedimentation, pollution, endangerment of species, destruction of ecological zones, and cultural heritage.
- Indirect Losses: Loss of biodiversity, loss of cultural diversity, and long-term environmental degradation.

The relationship between vulnerability (V) and capacity (Cap) in the context of civil engineering and risk assessment is an important consideration. These two concepts, though distinct, are intrinsically linked in understanding and managing risks associated with civil infrastructure and systems.

Vulnerability (V): As previously introduced, it refers to a system, community, or asset's susceptibility to the damaging effects of a hazard. It depends on a variety of factors, including physical, social, economic, and environmental factors. Vulnerability is often defined as a system's susceptibility to damage or disruption when exposed to a hazardous event. Different definitions of vulnerability exist, reflecting its multifaceted nature across contexts.

Capacity (Cap): In contrast, capacity indicates a system or component's capability or resilience to perform effectively when exposed to an event. It is a measure of a system's strength, adaptability, and available resources to withstand, respond to, and recover from hazardous events. High capacity often implies better performance and greater resilience in the face of challenges arising from hazardous events.

The relationship between vulnerability and capacity can be understood conceptually as inversely proportional, where $V \propto (Cap)^{-1}$. This implies that as a system's capacity to withstand and adapt to hazards increases, its vulnerability decreases, and vice versa. To illustrate this relationship, consider the following statements:

- "*A particular system (or component) has a low vulnerability.*": This implies the system is less susceptible to damage or disruption from hazards, indicating a high capacity to withstand such events.
- "*A particular system (or component) has high capacity.*": This statement suggests that the system is well-equipped and resilient, thereby having low vulnerability to hazards.

Understanding this inverse relationship is crucial in civil engineering and risk management. It guides the design and implementation of systems and structures, aiming to enhance capacity (thus reducing vulnerability) through robust design, the use of high-quality materials and construction methods, effective planning, and the incorporation of appropriate resilience measures. It is also important to recognize that both vulnerability and capacity are dynamic concepts. They can change over time due to technological advancements, changes in environmental conditions, and socioeconomic developments. The interplay between vulnerability and capacity is a critical consideration in civil engineering and risk assessment. Balancing these two aspects is essential to creating and maintaining resilient, sustainable infrastructure capable of withstanding the diverse challenges posed by natural and artificial hazards.

1.4.6 Example Related to Terminology & Fundamental Concepts

In the field of civil engineering, assessing and managing hazards is a multifaceted endeavor that encompasses a wide range of potential dangers, threats, and vulnerabilities. Each scenario presents unique challenges and requires a tailored approach to ensure safety and provide for effective risk mitigation. This section delves into ten diverse scenarios, each illustrating the intricate relationship between danger, threat, hazard, consequences, and vulnerability.

1.4.6.1 Earthquake

- Danger: Potential structural collapse of buildings.
- Threat: Seismic activity in an earthquake-prone area.
- Hazard: An actual earthquake occurring.
- Consequences: Loss of life, injuries, property damage, disruption of services, and economic activities.
- Vulnerability: Buildings not designed to withstand seismic forces and high population density areas.

1.4.6.2 Chemical Plant Near a Residential Area

- Danger: Exposure to toxic chemicals.
- Threat: Storage and handling of hazardous chemicals in proximity to residential areas.
- Hazard: Chemical spill or leak from the plant.
- Consequences: Health hazards to nearby residents, potential evacuations, and environmental contamination.
- Vulnerability: Residents living close to the plant without adequate protective measures or emergency response plans.

1.4.6.3 Coastal City Facing Rising Sea Levels

- Danger: Flooding and coastal erosion.
- Threat: Climate change leads to rising sea levels.
- Hazard: Increased frequency and severity of coastal flooding.
- Consequences: Damage to coastal infrastructure, loss of property, displacement of communities.
- Vulnerability: Low-lying areas without adequate sea defenses, populations with limited resources to relocate or adapt.

1.4.6.4 Old Bridge Over a River

- Danger: Bridge collapse.
- Threat: Aging infrastructure and increased traffic loads.
- Hazard: Structural failure of the bridge.
- Consequences: Traffic accidents, potential casualties, disruption of transportation and commerce.
- Vulnerability: Lack of regular maintenance, absence of modern safety design features.

1.4.6.5 Farming Community in a Drought-Prone Region

- Danger: Crop failure and water scarcity.
- Threat: Prolonged periods of low rainfall.
- Hazard: Severe drought occurring.
- Consequences: Loss of livelihood for farmers, food shortages, economic stress in the community.
- Vulnerability: Dependence on rain-fed agriculture, lack of alternative water sources, or drought-resistant crops.

1.4.6.6 Urban Area Prone to Air Pollution

- Danger: Health issues related to air quality.
- Threat: High levels of vehicular and industrial emissions.
- Hazard: Persistent smog conditions.
- Consequences: Respiratory illnesses in the population, increased healthcare costs, reduced quality of life.
- Vulnerability: Populations with pre-existing health conditions, areas with poor air circulation.

1.4.6.7 A Dam in a Seismically Active Region

- Danger: Dam failure and subsequent flooding.
- Threat: Earthquake in a region with a large dam.
- Hazard: Structural compromise or failure of the dam due to seismic activity.
- Consequences: Flooding of downstream areas, loss of life, destruction of property, and ecosystems.
- Vulnerability: Communities and ecosystems located downstream, older dams not designed for current seismic standards.

1.4.6.8 Suburban Area Surrounded by Dense Forests

- Danger: Wildfire spreading to inhabited areas.
- Threat: Dry weather conditions in a heavily forested region.
- Hazard: Outbreak of a large wildfire.
- Consequences: Destruction of homes, potential loss of life, and air quality degradation.
- Vulnerability: Homes built close to forest areas, lack of effective firebreaks, or firefighting resources.

1.4.6.9 Industrial Facility Using Heavy Machinery

- Danger: Workplace accidents and injuries.
- Threat: Operation of heavy machinery without proper safety measures.
- Hazard: Machinery malfunction or operator error.
- Consequences: Injuries to workers, potential fatalities, and production delays.
- Vulnerability: Workers without adequate safety training, older machinery lacking modern safety features.

1.4.6.10 Coastal Resort Area

- Danger: Impact from tropical cyclones.
- Threat: Location in a hurricane-prone region.
- Hazard: Landfall of a strong hurricane.
- Consequences: Damage to property, disruption of tourism, economic losses.
- Vulnerability: Buildings not constructed to withstand hurricane-force winds, lack of evacuation plans.

1.5 Basic Risk Equation

In the field of civil engineering, the interplay of threat, hazard, vulnerability, and consequences is central to the understanding and management of risks. Throughout this book, threat and hazard are interchangeably used to refer to degrading events or processes that can negatively impact systems or their components. Hazards, which can range from natural events like earthquakes to man-made incidents such as structural failures, are characterized as situations or conditions capable of impairing the functionality or integrity of civil engineering structures or systems.

Fig. 1.3 The interplay of threat/hazard, vulnerability, and consequences

Vulnerability (V) is the susceptibility or sensitivity of a system or its components to hazards. This metric is crucial for risk assessment, helping quantify a system's exposure or susceptibility to potential damage or degradation. Often expressed subjectively, vulnerability becomes particularly useful in scenarios where quantitative data is limited or the nature of the threat varies significantly.

Consequences are the outcomes or results that occur when a hazard interacts with a vulnerable system. The range of these consequences is diverse, encompassing physical damage, financial losses, environmental impacts, and human casualties, and is influenced by the severity of the hazard and the system's level of vulnerability.

By combining these elements, a comprehensive understanding of risk in civil engineering is established. Risks are seen as arising from the interaction of a threat or hazard (T) with a physical system that possesses certain vulnerabilities (V), resulting in a set of consequences (C) (Fig. 1.3). This conceptual framework, fundamental in risk management, guides engineers and decision-makers in identifying, evaluating, and mitigating potential risks. Understanding this interplay is crucial for ensuring the safety, resilience, and sustainability of civil engineering projects and infrastructure.

In civil engineering, risk (Ri) is a quantifiable metric that can be systematically analyzed and managed. This understanding is based on the fundamental concept that risk is intrinsically linked to three primary components: threats/hazards (T), vulnerability (V) or capacity (Cap), and consequences (C).

It has been established that risk is directly proportional to threats or hazards. This relationship is expressed as $Ri \propto T$. Based on this, as the threat or hazard tends towards zero, the risk also approaches zero ($Ri \rightarrow 0$ as $T \rightarrow 0$).

Similarly, a direct proportionality between risk and vulnerability is observed, formulated as $Ri \propto V$. Consequently, as vulnerability tends to zero, risk also tends to zero ($Ri \rightarrow 0$ as $V \rightarrow 0$). Alternatively, given that capacity and vulnerability are inversely related, it can be shown that the risk diminishes to zero as capacity increases without bound ($Ri \rightarrow 0$ as $Cap \rightarrow \infty$).

The relationship between risk and consequences is also characterized by direct proportionality, denoted as $Ri \propto C{\cdot}Ri \propto C$. Logically, this implies that an increase in consequences accompanies an increase in risk. Mathematically, this is expressed as $Ri \rightarrow 0$ as $C \rightarrow 0$.

Based on these proportional relationships, risk can now be formally expressed as a function of its three fundamental components: threats/hazards, vulnerability or capacity, and consequences, as follows:

The function f(C, T, V), which represents the relationship among consequences (C), threats/hazards (T), and vulnerability (V), is acknowledged as complex. It is often noted that this function, encompassing the three critical components of risk, cannot typically be expressed analytically for the vast majority of situations. The inherent complexity of this function arises not only from the individual characteristics of C, T, and V but also from their potential codependence, further complicating the formulation of f(C, T, V).

The interdependence of these risk components means that changes in one aspect can significantly influence the others. For instance, an increase in vulnerability might amplify the consequences of a given threat, or a heightened threat could escalate the overall risk even if vulnerability remains constant. This codependence adds layers of complexity to the understanding and management of risk in civil engineering projects.

1.6 Risk Subclasses

In the evolving field of civil engineering, the traditional framework of understanding risk through its three basic components—threats/hazards, vulnerability, and consequences—has been significantly augmented by introducing five subclasses: reliability, exposure, likelihood, resilience, and sustainability. These subclasses were introduced to provide a more detailed approach to risk assessment and management in civil infrastructure projects. While the foundational components of risk offer a broad perspective, these subclasses delve deeper into specific aspects of risk, providing a richer, enhanced understanding of the challenges and solutions in civil engineering.

In this section, the intricate relationships between these risk subclasses and the overall concept of risk are explored. Understanding these relationships is vital to the effective use of various risk management tools and processes. This exploration is especially crucial as these emerging concepts are integrated into our risk management strategies. An enhanced, more tailored approach to risk assessment and management will help ensure that our civil infrastructure is not only safe and functional but also resilient and sustainable in the long term, safeguarding the needs of future generations.

1.6.1 Reliability

Reliability (Rel), as a fundamental concept in risk management, plays a pivotal role in ensuring the performance and safety of civil engineering systems. It involves applying scientific knowledge and engineering principles to design and maintain components, products, plants, or processes so they perform as intended, without failure, for a specified duration in a given environment.

Reliability emphasizes a system's dependability throughout its lifecycle. This concept is not just about the initial design and construction of a structure or system; it also encompasses ongoing maintenance and monitoring. The reliability of a component or system is thus seen as its ability to function under stated conditions for a predetermined period. This perspective is crucial in civil engineering, where the longevity and continuous performance of structures like bridges, buildings, and dams are of utmost importance.

The two most significant dimensions of reliability are time and stress. Time refers to the duration over which a system is expected to perform without failure. Stress refers to the various physical, environmental, and operational pressures that the system might encounter. For example, a bridge must be reliable enough to withstand the stress of heavy traffic over many years without significant degradation.

At its core, reliability concerns the relationship between a system's capacity and the demands placed on it. Capacity represents the maximum level of performance a system can sustain, while demand refers to the actual operational or environmental stresses it experiences. The general reliability relationship can be expressed as follows:

$$\mathrm{Rel} = f_{\mathrm{Reliability}}(\mathrm{Capacity},\ \mathrm{Demand}) \tag{1.1}$$

This relationship implies that a system is considered reliable if its capacity consistently meets or exceeds the demands placed upon it.

In practical terms, understanding and enhancing the reliability of a system involves various activities, including robust design, selection of high-quality materials, adherence to construction practices, thorough testing, and regular maintenance. For instance, the reliability of a highway overpass would be ensured by designing it to handle a certain volume and weight of traffic and by regularly inspecting and maintaining its structural components.

Reliability can often be quantified, enabling engineers to predict a system's performance under different scenarios. This quantification is vital for decision-making in design, maintenance, and contingency planning. By calculating the probability of system failure under various conditions, engineers can make informed decisions to enhance reliability.

Reliability engineering is an integral and ongoing process that spans the entire life cycle of a civil engineering facility. From the initial conceptual stage of a project through to its detailed design, construction, operation, and eventual decommissioning, reliability engineering plays a critical role in ensuring the safety, functionality, and longevity of the facility.

The reliability engineering process begins at the very inception of a facility's design. This early-stage integration enables a thorough analysis of potential risks and failure modes, laying the foundation for a robust, resilient design. As the facility progresses through its life cycle, reliability engineering remains a key consideration, with regular assessments and updates to accommodate changes in usage, environmental conditions, and material aging.

Various tools and technologies are employed in reliability engineering. These might include advanced modeling software for stress analysis, simulation tools for predicting failure modes, and monitoring technologies for real-time structural health assessment. Using these tools enables engineers to predict and mitigate potential failures effectively.

One of the primary goals of reliability engineering is to identify and address potential reliability issues as early as possible in the facility's life cycle. This proactive approach is not only beneficial for ensuring safety and performance but also for cost-effectiveness. As indicated in Fig. 1.5, catching reliability problems during the conceptual or detailed engineering design stage can significantly reduce costs compared to addressing them later, such as during construction or operation. The cost of making changes or repairs increases dramatically once the facility is built and operational.

By focusing on reliability in the early stages of design, engineers can make informed decisions that shape the facility's entire life cycle. This might involve choosing more durable materials, designing for ease of maintenance, or incorporating redundancy into critical systems. While potentially increasing initial costs, these decisions can lead to substantial savings in maintenance, repair, and operational efficiency over the facility's lifespan.

The application of reliability engineering principles throughout a facility's life cycle ensures it not only performs its intended functions safely and efficiently but also adapts to changing conditions and requirements over time. This holistic approach is essential for developing sustainable and resilient civil infrastructure.

Figure 1.5 shows the cost of reliability engineering. This figure illustrates two aspects of a facility's lifecycle: the ability to influence cost and the cost of reliability engineering.

Ability to Influence Cost: This line starts high during the conceptual design phase, indicating that the ability to influence the overall cost is greatest at the early stages of a project. As the project moves into the detailed engineering phase and further into procurement, the ability to influence cost decreases sharply. By the time the project reaches the construction phase, the ability to influence costs is quite low, and it remains so through the facility's start-up.

Cost of Reliability Engineering: This line starts low and remains fairly low during the initial phases of conceptual design and detailed engineering. As the lifecycle moves into procurement and construction, the cost of reliability engineering increases significantly, peaking during facility start-up. This suggests that the cost to improve facility reliability is at a minimum during the initial conceptual design stage and reaches a maximum just before the facility goes into operation.

The graph demonstrates a common principle in project management and engineering known as the "Design to Cost" concept. It suggests that early project phases present the best opportunities for cost control and influence over the final project cost. Changes are cheaper and easier to implement during design phases than during construction or after completion. Conversely, the cost of ensuring reliability (such as through testing, quality assurance, and reliability-focused redesigns) is low during design phases, when changes are easier to incorporate. However, these costs rise sharply as the project nears completion, largely because the changes become more complex and expensive to implement once infrastructure is in place and the facility is about to start operations. The implication for managers and engineers is that the overall cost of improved reliability engineering can be reduced through proper planning and investment in thorough project design, rather than attempting to make changes later in the project lifecycle.

Figure 1.6 integrates concepts of reliability and risk in the context of system design and assessment. Recall, the relationship defining reliability is given by Eq. (1.1) and is a function of capacity (the system's ability to withstand load) and demand (the load that is applied to the system). To ensure system safety, the condition *Capacity* ≥ *Demand* must always be met. This means the system should have the capacity to handle or exceed the demand placed on it without failing at any time. The system reaches a failure limit state when *Capacity* = *Demand*. This is a critical condition in which the system is just able to meet demand but has no margin for error. Any additional demand or reduction in capacity could lead to system failure.

Using the general form of the risk equation (Fig. 1.4) and given that the *Capacity* is the inverse of vulnerability and that the *Demand* is another expression of Threat/Hazard, T, a new form of the risk equation can be defined in terms of the capacity (C) and reliability (Rel):

$$Ri = f(C, Rel) \tag{1.2}$$

Civil infrastructure, including bridges, buildings, dams, and other facilities, is constantly exposed to severe operating and environmental conditions throughout its service life. These conditions, ranging from natural weather phenomena to anthropogenic activities, are principal contributors to the deterioration of structural stiffness and stability over time. This gradual weakening can lead to structural failures, potentially causing significant economic and social losses. Given these risks, reliability analysis emerges as a crucial tool in civil engineering.

Reliability analysis is employed to evaluate the safety and serviceability of structures. It involves quantitative assessments to ascertain whether a structure can withstand future extreme forces with a satisfactory level of reliability throughout its

$$\left.\begin{array}{l} Ri \propto T \\ Ri \propto V \\ Ri \propto C \end{array}\right\} \quad Ri = f(C, T, V)$$

f: a complex function

Fig. 1.4 The basic risk equation

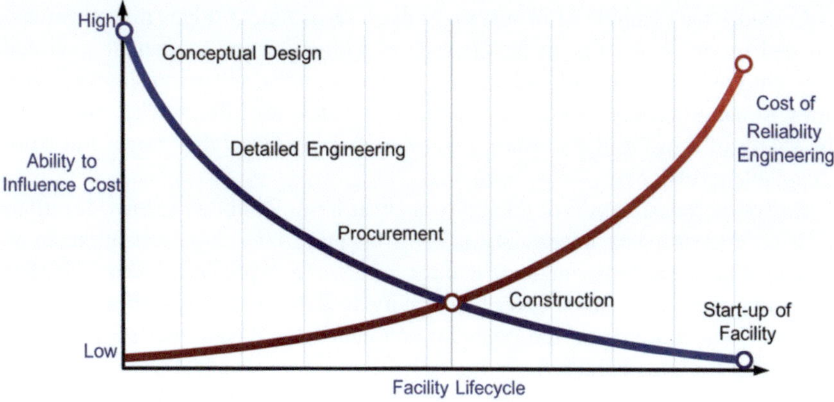

Fig. 1.5 Cost of reliability engineering

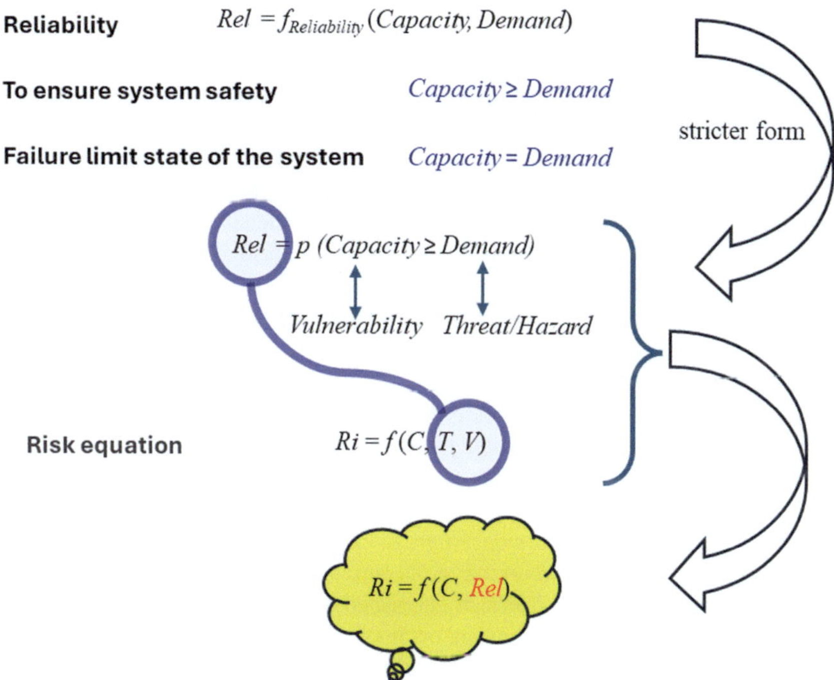

Fig. 1.6 The development of a new risk equation based on reliability

intended service life. This analysis is particularly important in scenarios where structures are subjected to aggressive environmental conditions or increasing operational demands.

Consider the example of in-service bridges. Over time, bridges may experience reduced serviceability due to factors such as aggressive environmental conditions (e.g., chlorides from de-icing salts), rising traffic volumes, and increasing load intensities. Reliability analysis becomes instrumental in making informed decisions about the safety of these bridges, especially when their serviceability might fall below acceptable levels.

One of the key strengths of reliability analysis lies in its ability to provide rational decision criteria when uncertainty is involved in system characterization. In the context of bridge assessment, uncertainties can arise from predicting service loads or estimating the bridge's load-carrying capacity. Reliability analysis offers a systematic approach to compare the potential consequences of various decisions while accounting for uncertainties.

Assessing the reliability of bridges and other structures often involves considering deterioration effects such as corrosion or fatigue. These effects can significantly impact the structural integrity over time. Reliability analysis helps quantify these impacts and plan appropriate maintenance or reinforcement strategies to extend the service life of the structures.

Through reliability analysis, engineers and decision-makers can identify structures at higher risk of failure and prioritize maintenance, rehabilitation, or replacement efforts. This proactive approach not only enhances safety but also optimizes resource allocation, ensuring timely, cost-effective interventions.

In civil infrastructure maintenance and safety, particularly for bridges, the combined effect of service load history and resistance deterioration is pivotal. This dual consideration is essential in accurately assessing a bridge's structural resistance and its ongoing reliability throughout its service life.

Constructing proper probabilistic models for the component variables of reliability is crucial to an effective evaluation of the structural performance of bridges within the reliability analysis framework. These models help quantify the uncertainties associated with both the loads the bridge will bear over its lifespan and the degradation of its materials due to environmental exposure and usage.

Reliability can be more accurately calculated using probabilistic resistance models developed from data representative of known site characteristics, such as service load history. These models provide a more realistic picture of the bridge's ability to withstand future stresses, accounting for its current state and how various loads and environmental factors have historically impacted it.

Over time, as new observations are made or better insights are gained about the bridge's performance and condition, these probabilistic models can and should be updated. Methods such as the Bayesian theorem provide a systematic approach for incorporating new data into existing models, enabling continuous improvement in the accuracy of reliability assessments. Building and maintaining these probabilistic models is not just beneficial but necessary for a comprehensive evaluation of structural performance within the framework of reliability analysis. These models enable engineers to assess the probability of failure or underperformance of bridges across a variety of future scenarios and to make informed decisions regarding maintenance, rehabilitation, or even replacement of bridge structures. This assessment is crucial for prioritizing which structures require immediate attention, planning long-term maintenance schedules, and ensuring that bridges continue to serve their purpose safely and efficiently.

1.6.2 Exposure

As a critical component in risk assessment, exposure plays a crucial role in our ability to understand the potential losses and impacts in hazard-prone areas. Exposure is an essential factor in evaluating the risks faced by people, properties, systems, and other elements located within zones susceptible to various hazards. In its broadest sense, exposure refers to the presence of properties, populations, or ecological elements within areas at risk of hazard events. It encompasses all elements that could potentially suffer losses due to hazards. This includes residential areas, commercial properties, critical infrastructure, and ecological environments that might be affected by natural or man-made hazards.

Exposure measurements often include quantifying the number of people, the types of assets, or the extent of ecological areas within a hazard zone. These measurements are not just about counting the number of elements at risk; they also involve understanding the nature of these elements. For instance, the exposure of a densely populated urban area to flood risks is different from that of a sparsely populated rural area.

To effectively estimate the risks associated with a particular hazard, exposure measures are combined with the specific vulnerability of the exposed elements. In this context, vulnerability refers to the characteristics and circumstances that make a community, system, or asset susceptible to the damaging effects of a hazard. For example, an old building in an earthquake zone represents high exposure (due to its location) and potentially high vulnerability (due to its possible lack of seismic-resistant design).

In civil engineering, understanding exposure is critical for the planning, design, and management of infrastructure projects. It informs decision makers on where to build, what materials to use, and how to design structures to reduce potential losses from hazards. Engineers and planners must not only consider current exposure but also how it might change over time due to factors such as urban development, climate change, and shifting demographic patterns.

In risk assessment in civil engineering, exposure is a key concept that captures the interplay between hazards or threats and vulnerability. It is defined by a specific relationship, represented as $E = f(V, T)$, where exposure, E, is a function of vulnerability, V, and the hazards or threats, T.

Formally, exposure refers to the extent to which a system, community, or asset is subject to potential loss or damage from identified hazards or threats. The function $E = f(V, T)$ seeks to quantify this relationship, combining the susceptibility of the system (vulnerability) with the potential adverse events (hazards/threats) it may face.

By adhering to this definition, exposure can be understood as inversely related to reliability. In essence, higher reliability of a structure or system suggests a lower degree of risk exposure, and conversely, lower reliability implies greater risk exposure. This inverse relationship is pivotal in understanding how improving the reliability of a structure or a system can effectively reduce its overall exposure to risk.

Typically, exposure is expressed more subjectively. This is because exposure often involves qualitative assessments of vulnerability and the potential impact of hazards or threats, which may not always be easily quantifiable. For example, the exposure of a historical building to earthquake hazards might be assessed subjectively, taking into account the building's cultural value and its age- and material-related vulnerability.

Unlike exposure, reliability, and likelihood in risk management are generally computed and expressed in a more objective manner. Reliability often involves quantitative measures based on engineering analyses and probabilistic models, while likelihood deals with the probability of the occurrence of specific hazards or threats.

Understanding the relationship between exposure, reliability, and vulnerability is crucial for civil engineers and risk managers. It guides them in making informed decisions about risk mitigation strategies, focusing on enhancing reliability to reduce exposure. For instance, reinforcing the structural elements of a bridge can increase its reliability and, consequently, decrease its exposure to collapse under heavy traffic or extreme weather conditions.

Figure 1.7 outlines the inspection process for a bridge as a series of steps that assess risk exposure. The bridge inspection process is an integral part of maintaining infrastructure safety and serviceability. The inspection process can be conceptualized as a series of interconnected steps that collectively assess a bridge's exposure to risk, starting from the underlying causes to the potential outcomes of these risks.

1. **Causes**: This initial step identifies the root factors that may contribute to a bridge's vulnerability. Causes could range from environmental conditions to usage patterns that the bridge is subjected to over its lifespan.

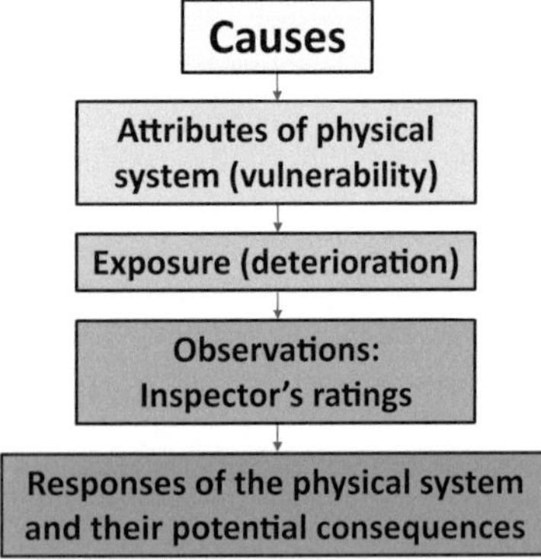

Fig. 1.7 The bridge inspection process

2. **Attributes of Physical System (Vulnerability)**: Here, the focus is on the bridge's inherent attributes that determine its vulnerability. This includes aspects of the bridge's design, materials used, age, and any previous maintenance history that might affect its ability to withstand certain stresses.
3. **Exposure (Deterioration)**: Exposure, in this context, refers to the bridge's deterioration over time. This is influenced by the bridge's vulnerability and the external conditions it faces, such as weather events (freeze–thaw and chlorides from de-icing salts), pollution, and physical wear from traffic.
4. **Observations: Inspector's Ratings**: The actual inspection involves observations and assessments made by inspectors. These are often encapsulated in ratings that reflect the current state of the bridge, noting any signs of deterioration or damage.
5. **Responses of the Physical System and Their Potential Consequences**: The final step in the process involves understanding how the bridge, given its current state as determined by the inspection, will respond to typical and extreme conditions. This also encompasses predicting the potential consequences of these responses, from minor repairs to critical failures that could lead to service interruptions or safety hazards.

By structuring the inspection process in this manner, engineers can systematically evaluate a bridge's risk exposure, which is critical for prioritizing maintenance efforts and ensuring the long-term reliability and safety of the infrastructure.

Different hazards will degrade infrastructure components in different ways depending on their unique composition and form. For instance, in our example of bridges, exposure of steel structural elements to environmental conditions, particularly when combined with de-icing salts, accelerates the corrosion of steel members and reinforcing bars. This chemical reaction not only weakens the structural integrity but also shortens the lifespan of these components. Likewise, roadway pavements are subject to a host of factors that cause deterioration. The constant cyclical loading and unloading from traffic demands exert significant stress on pavement materials, leading to wear and tear over time. Additionally, pavements undergo cycles of freezing and thawing, which can cause the material to expand and contract. These environmental demands lead to cracks, potholes, and other forms of distress in the pavement structure, all of which represent deterioration or exposure that ultimately compromise the serviceability of the roadway. In both cases, understanding the specific hazards and resulting exposures is crucial for effective infrastructure management, enabling the development of targeted maintenance and preservation strategies to extend the service life of these critical components.

The deterioration of infrastructure components commonly leads to observable changes. For example, concrete bridge elements may display discoloration, delamination, spalling, or peeling. Soil erosion results from scouring processes. The corrosion of steel elements first leads to discoloration, then to expansion, before section loss. These observable effects are described in detail in various inspection manuals. Condition rating is the subjective numerical rating that an inspector assigns to the component based on observations. CR is usually measured on a value scale (e.g.,

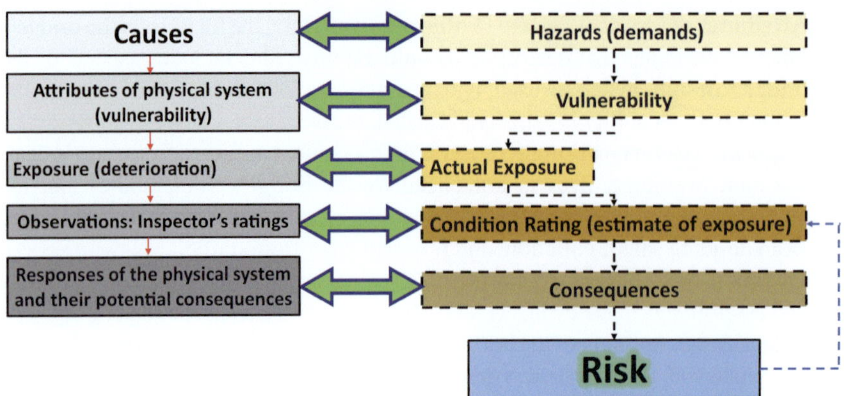

Fig. 1.8 Analogies of inspection processes

0–9), with higher values generally indicating good condition and lower values indicating deteriorated condition. The Analogies of inspection processes are provided in Fig. 1.8.

The bridge inspection process can be conceptualized as a sequence of interrelated steps that collectively assess a bridge's risk profile. This process begins with identifying the causes, akin to the hazards or demands, that the bridge is exposed to over its service life. These could be natural elements, like weather conditions, or anthropogenic factors such as vehicular loads. The hazards or demands are then linked to the physical system's attributes, thereby representing the bridge's vulnerability. This relationship illustrates how specific characteristics of a bridge, such as material strength, design, and age, define its susceptibility to the identified hazards. When the bridge's vulnerabilities interact with external hazards or demands, the result is the bridge's exposure to potential damage or performance degradation. The exposure is, therefore, the cumulative effect of the bridge's inherent vulnerabilities and the external hazards it faces. An inspector evaluates the exposure by providing condition ratings based on observed deterioration or defects. These ratings are informed by the bridge's current physical state, which reflects the impact of the hazards it has been subjected to. The consequences are the outcomes or responses resulting from the bridge's exposure to the identified hazards, taking into account its vulnerabilities. This can include the immediate and long-term effects on the bridge's functionality, safety, and costs associated with potential repairs, rehabilitation, or replacement. The inspection effort is informed by the risk across the entire process, underscoring the need for precise consequence estimation. Accurate estimation of the consequences is critical for prioritizing inspection and maintenance actions, ensuring that resources are allocated effectively to address the most significant risks. By following this idealized inspection process, bridge engineers and inspectors can systematically assess a bridge's condition and risk, ensuring the infrastructure remains safe and functional throughout its intended lifespan.

1.6.3 Likelihood

In the exploration of risk assessment within civil engineering, understanding the concept of **Likelihood** is crucial. Likelihood represents the chance that a particular event will occur. While the terms are often used interchangeably, there is an essential distinction between probability and risk that is essential for accurate risk analysis.

Likelihood concerns the plausibility of a given event. It is a measure of the support that the available data provides for a particular hypothesis or model. When discussing the likelihood of structural failure, for example, engineers assess how well the observed conditions of a structure, such as signs of material fatigue or stress, support the hypothesis that it may fail under certain loads.

Probability, on the other hand, is attached to possible outcomes and is derived from known parameters within a model. It represents the extent to which a particular outcome is expected to occur. In contrast, likelihood is attached to hypotheses and indicates how well a particular set of parameters explains the observed data (Fig. 1.9).

Commonly, the probability of an event is interpreted as the ratio of favorable outcomes to the total number of possible outcomes. If we consider a coin toss, the probability of landing on heads is 0.5, indicating that the outcome is just as likely to occur as not occur. Likelihood pertains to the plausibility of a model or distribution explaining the observed data. In statistical terms, a high likelihood indicates that a particular distribution is a good fit for the data at hand.

Using a Probability Line (Fig. 1.10), the probability is visualized as a value between 0 and 1, where 0 signifies impossibility and 1 signifies certainty. The probability line helps to conceptualize how likely or unlikely an event is based on where its probability lies on the line. For distributions, probabilities are areas under the curve, while likelihoods are the fixed values on the distribution's y-axis corresponding to the observed data points (Fig. 1.11).

Likelihood is employed to assess the probability that a specific event will occur, considering the existing vulnerabilities and the presence of potential hazards. It

Fig. 1.9 The distinction between probability and likelihood

Fig. 1.10 Possibility line

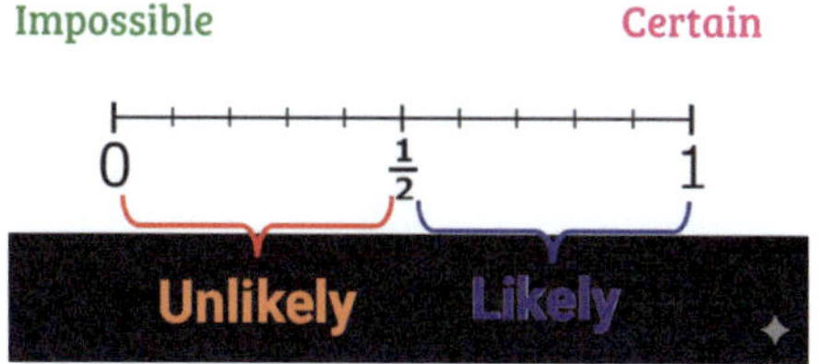

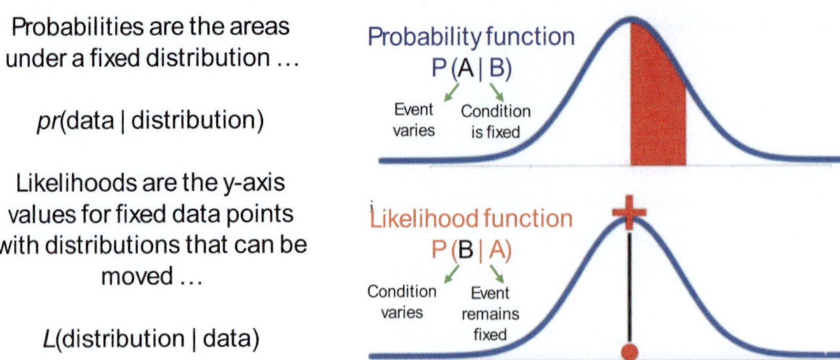

Fig. 1.11 The difference between probabilities and likelihoods based on a distribution

is crucial to develop a comprehensive understanding of the potential impacts that these hazards could have on a system or infrastructure, given its current state of vulnerability.

Mathematically, likelihood is expressed as a function of the vulnerability (V) and hazard (H), which can be represented by the equation $L = f(V, H)$. This functional relationship quantifies the likelihood of an event occurring, factoring in both the system's susceptibility to harm (vulnerability) and the presence of external threats (hazard).

In some cases, the function $f()$ can take on a straightforward form as follows:

$$L = V \times H \tag{1.3}$$

where the likelihood is the product of vulnerability and hazard. This simple multiplicative relationship suggests that the likelihood of an event increases as either the vulnerability or the hazard increases.

1.6.4 Resilience

Resilience is a critical property of systems, communities, or societies subject to various hazards. It embodies the collective ability to withstand adverse events and adapt to changing conditions while maintaining the core structures and functions that are essential to the system's integrity and purpose. Resilience includes not only resistance to the initial effects but also the capacity to absorb and adapt to disruptive events, and most importantly, to recover from them. It signifies a proactive and dynamic approach to managing risk, emphasizing prevention and post-event recovery.

One of the most important characteristics of resilience is a system or community's ability to return to its pre-disaster state or to reach a new, more desirable state after a disaster has occurred. This aspect of resilience acknowledges that simply returning to the status quo may not always be sufficient or desirable, especially if it means reverting to an inherently vulnerable state.

The metaphor of 'springing back' from a shock captures the essence of resilience. It conveys the notion of elasticity and strength, the capacity to rebound from a crisis and to do so with speed and agility. This ability is crucial for communities as it determines how they navigate and emerge from hazard events.

Civil engineers prioritize infrastructure resilience, which involves creating systems that not only withstand disasters but also minimize their impact. Infrastructure resilience deals with enhancing the ability of structures and systems to either avoid, withstand, or recover quickly from various disruptions. The key aim of infrastructure resilience is to reduce the magnitude and duration of disruptive events. This involves ensuring that even when parts of an infrastructure system fail, the overall system can continue to function or can be recovered swiftly. For instance, a bridge built with seismic isolation techniques can remain operational even after a significant earthquake, thereby reducing disruption to transportation networks.

A resilient infrastructure system must be designed with foresight, considering not just the current environmental and operational conditions but also future challenges. This anticipatory approach means designing for climate change, urbanization, and evolving technological landscapes. Infrastructure resilience also refers to a system's ability to withstand the impacts of a hazard without significant loss of function. Following absorption, the ability to adapt— to change or modify components or operations in response to hazards—is crucial. This could mean, for example, using materials in coastal defenses that can adapt to changing sea levels. Post-disaster, the focus shifts to recovery. Resilient infrastructure can be rapidly repaired or restored to full (or improved) functionality. This requires not only robust physical design but also effective emergency planning and logistics.

The National Infrastructure Advisory Council (NIAC) (2009) emphasizes that resilient infrastructure depends on its capacity for anticipation, absorption, adaptation, and/or rapid recovery. This comprehensive approach ensures that infrastructure can maintain critical operations in a crisis and has the resourcefulness to manage and mitigate the damage as it occurs.

The NIAC's (2009) framework delineates resilience in civil infrastructure into three fundamental components—robustness, resourcefulness, and rapid recovery. These components form the cornerstone of resilient infrastructure systems, ensuring they can withstand and recover from crises. Here's an in-depth look at each component:

1. **Robustness**: At the heart of resilience lies robustness—the inherent strength and durability of infrastructure systems. Robustness is the capacity of these systems to maintain critical operations and functions even when subjected to severe stressors. This includes the physical attributes of the structures themselves—how buildings, bridges, dams, and levees are designed to withstand adverse events. It also encompasses systems design strategies like redundancy, where alternative pathways or backups are provided to ensure continued operation in the event

of a failure in one part of the system. For civil engineers, this means incorporating fail-safes and designing beyond minimum code requirements to enhance the endurance of infrastructure.

2. **Resourcefulness**: This component refers to the strategic and efficient management of resources before, during, and after a crisis. Resourcefulness is characterized by the ability to anticipate potential disruptions, respond effectively, and manage crises as they unfold. This involves developing and executing contingency plans, training personnel, effective supply chain management to ensure the availability of critical materials, and prioritizing actions to control and mitigate damage. Effective communication within the organization and with the public is also a crucial aspect of resourcefulness.
3. **Rapid Recovery**: The third pillar of resilience is the ability to recover quickly from a disruptive event, restoring normal operations as efficiently as possible. Rapid recovery entails developing contingency plans that can be activated immediately, conducting competent emergency operations, and swiftly mobilizing the appropriate resources and manpower to the areas where they are needed most. For infrastructure, rapid recovery can involve pre-planned, modular, or easily repairable components that can be quickly replaced or fixed to return the system to operational status.

The addition of redundancy as a fourth key component to resilience is a significant enhancement to the NIAC framework. Redundancy, within the context of civil infrastructure resilience, refers to the inclusion of additional or backup systems that can be called upon in the event of a failure of primary systems. This concept is crucial for maintaining service continuity during unforeseen disruptions.

The concept of redundancy is integral to the resilient design. Engineers incorporate redundancy by ensuring spare capacity or backup systems are in place to handle the full operational load during an emergency. This design principle is applied across various infrastructure systems, from water and wastewater treatment facilities to power grids and transportation networks. An example of redundancy can be seen in the design of a water treatment plant. To ensure continuous operation, designers commonly specify that there should be at least two identical treatment trains, each capable of handling the entire flow independently. This permits the plant to continue operation in the event of an interruption due to equipment malfunction in one of the trains, as well as to perform routine maintenance and upkeep without having to shut down.

Sometimes, these four components, namely robustness, redundancy, rapid recovery, and Resourcefulness (Fig. 1.12), are referred to as the 4Rs of resilience. A closer examination of the 4Rs reveals that resilience is multidisciplinary, requiring cooperation across disciplines to achieve a successful outcome. Without such multidisciplinary cooperation and contributions, achieving infrastructure resiliency is impossible.

Figure 1.13 visually represents the concept and stages of resilience as they pertain to a community or an asset in response to undesirable events such as natural disasters. The stages of resiliency are described below in relation to Fig. 1.13:

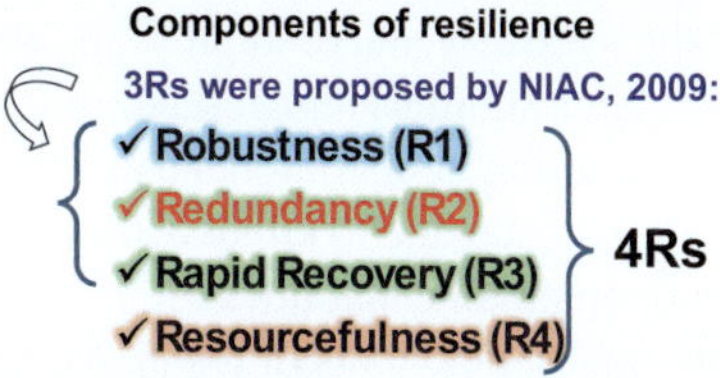

Fig. 1.12 The four components of resilience

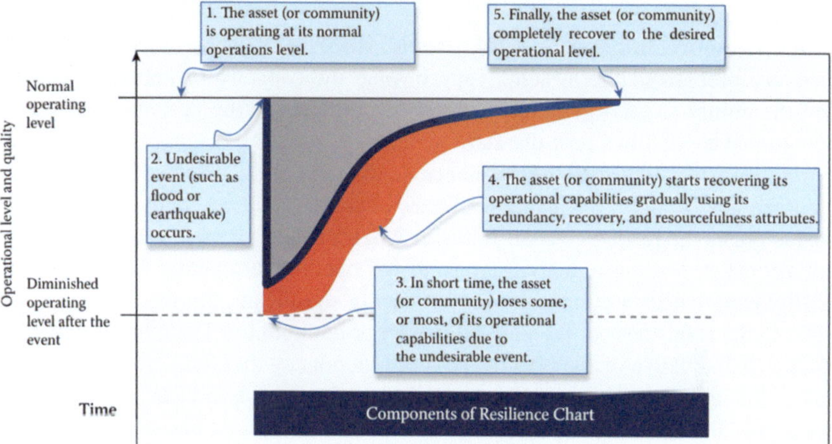

Fig. 1.13 A visual representation of the concept of resilience in the context of a community or an asset

1. **Initial State of Normalcy**: The asset or community functions at its normal operational level. Stable and routine performance is observed in the absence of any stressors or disruptions.
2. **Disruption Event**: An undesirable event occurs, such as a natural disaster (flood, earthquake, etc.), causing immediate and significant disruption. This leads to a drop in the operational level, depicted by the steep decline in the curve.
3. **Immediate Aftermath**: In the short-term, following the event, the asset or community experiences a loss of operational capabilities. This is the point where the operational level is at its lowest, reflecting the maximum impact of the event.
4. **Recovery Phase**: The asset or community begins to recover operational capabilities. This recovery is gradual and leverages the system's inherent redundancy (alternative pathways or backups that allow functionality despite some components being compromised), resourcefulness (the ability to effectively use available resources to restore functionality), and the ability to recover efficiently. The upward slope of the curve indicates improvement over time.
5. **Full Recovery**: The asset or community eventually returns to its original or desired operational level, completing the recovery process.

The shaded area under the recovery curve represents an asset or community's resilience. A smaller shaded area signifies higher resilience, characterized by a rapid recovery and a less severe reduction in operational capabilities. Conversely, a larger shaded area denotes lower resilience due to either a slower recovery process and/or a more significant drop in operational capabilities.

Based on the figure, the four components of resilience can also be defined:

R1 (Robustness) and **R2 (Redundancy)**: These factors determine the extent of damage the system might incur from a hazard. Systems with high robustness and redundancy are likely to sustain less damage and continue functioning close to normal levels after a disaster.

R3 (Resourcefulness) and **R4 (Rapid Recovery)**: These factors drive the recovery process of the system. They involve the capacity to manage resources and the ability to recover quickly. These are critical for the system to restore its operational capabilities post-disaster.

Figure 1.14 compares the resilience of two different assets or communities (denoted as "A" and "B") using resilience charts. A relatively small, shaded area characterizes the resilience chart for asset A. This reflects a rapid recovery from an initial impact, leading to a swift return to full operational quality. The operational quality experiences a noticeable, yet not deeply significant, immediate drop, highlighting the robustness of the particular asset or community. The quick recovery of asset A is indicative of the effectiveness of redundancy measures, resourcefulness, and rapid recovery strategies. This smaller shaded area is a clear sign of higher resilience, allowing asset A to manage disruptions with minimal long-term impact.

For asset B, the resilience chart shows a considerably larger shaded area. This implies a significant initial drop in operational quality that lasts over an extended period, indicating a more substantial impact and a slower recovery to normal operational levels. The prolonged recovery time for asset B may reflect issues with redundancy, resource allocation, or recovery capabilities. The larger shaded area corresponds to lower resilience, suggesting that asset B may experience more prolonged, potentially severe impacts from disruptions.

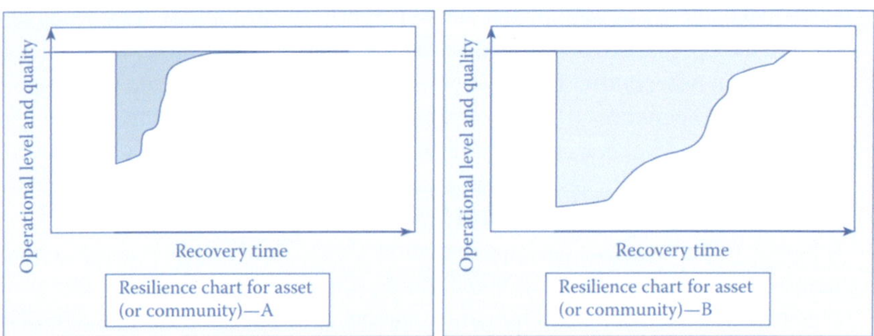

Fig. 1.14 A schematic comparison of the resilience of two different assets or communities (using resilience charts

Comparing assets A and B, it is clear that asset A is more resilient. It is more tolerant to operational impacts and recovers more rapidly from undesirable events. Asset A's recovery curve is notably steeper during the recovery phase, signaling a more robust and efficient response mechanism compared to asset B.

This analysis emphasizes the significance of infrastructure investment and resilience planning. Communities with resilient infrastructure, like that of asset A, recover more quickly from disruptive events compared to those with less resilient assets like asset B. The difference in the shaded areas between the two assets, indicative of time and quality of operations lost, can have considerable economic, social, and environmental consequences. Hence, communities with infrastructure akin to asset A are likely to endure fewer negative effects following disruptive events. In practical terms, resilience translates into maintaining continuity and minimizing downtime, essential for emergency services, economic stability, and community well-being.

Figure 1.15 presents a conceptual framework that links the components of risk with those of resilience, which are often referred to as the "4Rs": robustness (R1), redundancy (R2), rapid recovery (R3), and resourcefulness (R4). The figure illustrates that robustness and redundancy are proactive measures that can mitigate the impact of a hazard and reduce vulnerability. On the other hand, rapid recovery and resourcefulness are reactive measures that deal with the consequences of a hazard by enabling swift restoration and efficient use of resources.

The relationship between threat/hazard and robustness (R1) underscores the importance of recognizing potential threats for developing robust systems that withstand hazards without failing. Similarly, the link between vulnerability and redundancy (R2) emphasizes that redundancy planning depends on identifying vulnerable areas/components to ensure operational continuity during hazardous events. Lastly, the shift from consequences to rapid recovery (R3) and resourcefulness (R4) suggests that by foreseeing potential consequences of hazards, it becomes possible to devise strategies for quick recovery and effective resource management, thereby enhancing overall resilience.

Accurate information on hazards and an asset's vulnerability must be obtained to reduce the magnitude and/or duration of disruptive events. To best manage a hazard's consequences, the ability to return to and/or reconstitute normal operations as quickly

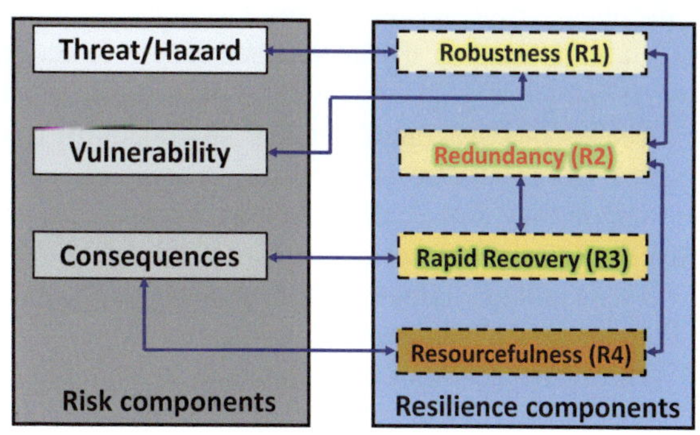

Fig. 1.15 A conceptual model of risk and resilience: Integrating the 4Rs into risk management

and efficiently as possible after a disruption is required. The resourcefulness of the asset and the availability of backup resources are directly related to the consequences and to its ability to respond to and manage a crisis.

1.6.5 Sustainability

Sustainability, especially within the scope of civil engineering, encompasses a commitment to meeting the needs of the present without compromising the ability of future generations to meet their own needs. Though challenging to narrow down into a single definition due to its broad and multi-dimensional nature, certain core objectives are common in all applications of sustainability.

Sustainability is indeed a multifaceted concept that encompasses various aspects of human interaction with the environment and society. The lack of a universal definition for sustainability stems from its broad scope, which integrates environmental, social, economic, and humanitarian considerations. Some of the sustainability objectives include:

- **Operating Within Acceptable and Renewable Limits**

This aspect of sustainability focuses on the idea of using resources at a rate that does not deplete them for future generations. It's about finding a balance between consumption and conservation, ensuring that natural resources are used efficiently and responsibly.

- **Environmental, Societal, Economic, and Humanitarian Concerns Governing Actions**

Here, sustainability is seen as a holistic approach that takes into account not just environmental impact but also the societal, economic, and humanitarian implications of actions. This means making decisions that are not only eco-friendly but also socially responsible, economically viable, and beneficial to humanity as a whole.

- **Respect for Available Natural and Universal Resources**

This objective emphasizes the need to value and preserve our natural resources. It's about understanding the interconnectedness of ecosystems and the importance of maintaining the health and diversity of the environment.

These objectives highlight that sustainability is not just about environmental conservation, but also about ensuring equitable and responsible use of resources to support long-term ecological balance and human well-being. The complexity and diversity of these objectives make it challenging to establish a universal definition or a standardized framework for measuring sustainability, unlike the more defined components of resilience that we saw earlier.

The challenge of not having a universal definition for sustainability components complicates the task of creating a structured table outlining the relationships between sustainability components and risk components. This difficulty stems from several factors:

- **Diverse Interpretations of Sustainability**: Since sustainability encompasses a wide range of areas—environmental, societal, economic, and humanitarian—different organizations, industries, and cultures may have varying interpretations of what constitutes sustainability. This diversity makes it hard to standardize components.
- **Interdisciplinary Nature**: Sustainability cuts across multiple disciplines and sectors. The interplay between environmental, social, and economic factors is complex and often context-specific, which adds layers of complexity to defining universal components.
- **Dynamic and Evolving Concept**: Sustainability is not a static concept. It evolves with technological changes, societal values, environmental conditions, and economic models. What is considered sustainable today might change in the future.
- **Risk Components Association**: Risks in sustainability are multifaceted and can range from immediate environmental risks (like pollution) to long-term societal risks (like social inequality). Mapping these risks into specific sustainability components can be challenging due to their interdependent and often non-linear nature.

Given these challenges, a more flexible and adaptable approach might be required to understand the relationship between sustainability and risk. This could involve case studies, scenario analysis, and stakeholder engagement rather than trying to fit everything into a rigid table format. The goal would be to capture the nuances and complexities of sustainability concerning various risk factors.

Table 1.2 presents a comparative analysis of risk, resilience, and sustainability, focusing on their key components and main emphasis within the context of civil engineering. For risk, the components include vulnerability, threat, and consequences. These elements are crucial in assessing the potential dangers and impacts associated with civil engineering projects. Vulnerability pertains to how susceptible a structure or system is to damage, threat refers to external events that could cause harm, and consequences deal with the potential negative outcomes of these threats.

Table 1.2 Comparisons of different item issues that concern risk, resilience, and sustainability

Item	Risk	Resilience	Sustainability
Components	• Vulnerability	• Robustness	Varies, but can also be related to Vulnerabilities, Threats, and Consequences
	• Threat	• Resourcefulness	
	• Consequences	• Recovery	
		• Redundancy	
Main emphasis	Cost or cost benefits	Continuity of operations	Long-term consequences conservation of resources

Resilience components, on the other hand, consist of robustness, resourcefulness, recovery, and redundancy. These aspects are vital for ensuring that structures and systems can withstand and quickly recover from disruptive events. Robustness relates to the ability to endure shocks without significant loss of functionality, resourcefulness involves effective management of resources in response to threats, recovery focuses on restoring operations post-disruption, and redundancy ensures backup options are in place for continuity.

Sustainability in civil engineering varies more broadly in its components, but often it ties back to vulnerabilities, threats, and consequences, especially in the long term. It emphasizes the conservation of resources and considers the long-term environmental, social, and economic impacts of engineering projects.

The main emphasis of risk is on cost or cost–benefit analysis, assessing potential losses and mitigating costs. Resilience prioritizes operational continuity, maintaining functionality despite challenges in adverse conditions. Sustainability, distinctively focuses on long-term consequences and the conservation of resources, ensuring practices today do not hinder future generations' abilities to meet their needs.

This comprehensive framework underscores the importance of integrating these three concepts in civil engineering. By doing so, it is possible to create structures and systems that are not only safe and efficient in the face of immediate challenges, but also sustainable and beneficial for future generations.

1.7 Elements at Risk

In the context of hazard and risk assessment, the term elements-at-risk (Fig. 1.16) refers collectively to a wide range of entities that are susceptible to damage or harm from hazardous events. This broad category encompasses the human population, which may face risks such as injury, loss of life, health impacts, and psychological trauma. It also includes property, encompassing buildings, homes, and personal items, all of which are vulnerable to destruction or damage. Additionally, bridges, utilities, and communication networks are crucial components of public and private infrastructure elements-at-risk, as damage to these can severely disrupt community functionality and economic stability. The economic domain is also a significant element-at-risk, involving businesses, markets, and overall economic activities that can suffer both direct and indirect losses due to hazardous events. Beyond these, the term extends to cover the disruption of daily life activities, cultural practices, and social functions, which can profoundly affect the social fabric and well-being of communities. Lastly, the natural environment, including ecosystems, wildlife, and natural resources, is also an element-at-risk, as environmental damage can lead to long-term ecological imbalances and diminish the capacity of ecosystems to provide vital services. Overall, understanding and identifying these elements are crucial for effective hazard and risk assessment, enabling the development of targeted strategies for risk mitigation and resilience enhancement.

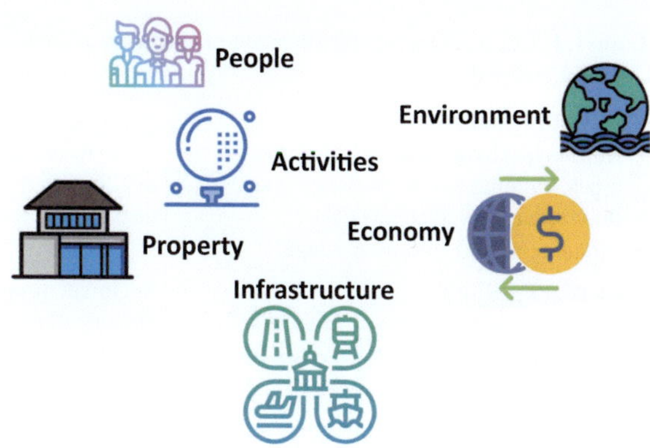

Fig. 1.16 Classes of elements-at-risk

Table 1.3 serves as a demonstrative guide for the segmentation and characterization of elements-at-risk, a step which will influence how the risk associated with these elements is communicated and managed. The complexity in classification arises from the need to cater to different end users—from decision-makers and emergency personnel to the general public. Each group requires the information to be presented in a context relevant to their specific role in disaster preparedness and response. For instance, the number of buildings at risk and their attributes, such as age, construction type, and replacement costs, are critical for urban planners and developers. At the same time, population distribution and demographics are particularly relevant to emergency services, which must plan for evacuations or aid distribution.

Furthermore, the table demonstrates that the characterization of elements-at-risk is not simply a matter of tallying potential losses, but rather involves a deep analysis of many facets that influence their susceptibility or vulnerability to hazards. Physical elements like buildings, with their varying uses, structures, and historical value, are juxtaposed with population metrics that consider the density and the vulnerable segments, such as the disabled or economically disadvantaged.

Essential facilities are critical to the resilience of a community and their ability to rapidly recover from undesirable events, while socio-economic aspects underscore the importance of social structures and support systems that can either mitigate or exacerbate the risks faced by a community. Similarly, the table highlights the need to safeguard transportation and utility lifelines, recognizing that their disruption can paralyze a community's functioning and its ability to respond rapidly to hazards.

The table further expands to consider the economic fabric of a region, including the distribution of activities that contribute to its economic vigor and the potential ripple effects a disaster could have on economic stability and employment. Finally, the table touches upon environmental elements, which are often the most overlooked component in risk assessments, but are crucial for long-term sustainability and recovery.

By delineating elements-at-risk in such detail, the table reflects the complexity of risk assessment in civil engineering. It brings to light the intricate web of variables that need to be considered in risk assessment and underscores the importance of

Table 1.3 Comprehensive Classification of Elements-at-Risk in Hazard and Risk Assessment for Civil Engineering

Physical Elements	Population
• Buildings: Urban land use, construction types, building height, building age, total floor space, replacement costs • Monuments and cultural heritage	Population density, distribution in space, distribution in time, age distribution, gender distribution, handicapped, income distribution
Essential Facilities	**Socio-economic Aspects**
Emergency shelters, schools, hospitals, fire stations, police, ambulance services, libraries, community centers	Organization of population, governance, community organization, government support, and socio-economic levels. Cultural heritage and traditions
Transportation Facilities	**Economic Activities**
Roads, railways, metro, public transportation systems, harbor facilities, and airport facilities	Spatial distribution of economic activities, input–output table, dependency, redundancy, unemployment, economic production in various sectors
Lifelines	**Environmental Elements**
Water supply, electricity supply, gas supply, telecommunications, mobile telephone network, stormwater infrastructure, and sewage system	Ecosystems, protected areas, natural parks, environmentally sensitive areas, forests, wetlands, aquifers, flora, fauna, biodiversity

a comprehensive approach that respects the diversity of elements involved and the various scales at which they operate. This approach is essential in a world where the nature and scope of risks are ever-changing, requiring dynamic and context-sensitive assessments.

This thorough categorization not only aids in formulating effective risk management strategies but also ensures that the diverse needs of different stakeholders are met. It acknowledges that the manner in which elements-at-risk are characterized will shape the risk management policies and actions that follow, making this a foundational aspect of any risk assessment endeavor in civil engineering.

Readers seeking a deeper treatment of the topics introduced in this chapter are encouraged to consult the references cited at the end of the chapter, including foundational and contemporary works such as Amiri et al. (2024, 2025), Ang and Tang (2007), Blaikie et al. (2004), Bruneau et al. (2003), Cutter et al. (2003), Ebtehaj and Bonakdari (2022), Ebtehaj et al. (2024), Holling (1973), Kaplan and Garrick (1981), Melchers and Beck (2018), Nowak and Collins (2013), Qureshi et al. (2025), Soltani et al. (2024), and relevant international guidelines and standards (Intergovernmental Panel on Climate Change, 2022; International Organization for Standardization, 2015, 2019; Joint Committee on Structural Safety, 2001; National Infrastructure Advisory Council, 2009; National Institute of Standards and Technology, 2016; Nations Unies, Stratégie internationale de prévention des catastrophes, 2015; NIAC, 2009; United Nations Office for Disaster Risk Reduction, 2017; World Commission on Environment and Development, 1987) reports. These sources provide additional theoretical background, methodological developments, and practical applications related to the fundamentals of hazard and risk assessment in civil and environmental engineering. Together, they offer complementary perspectives that can support both deeper conceptual understanding and professional practice.

1.8 Summary

This chapter established the conceptual foundation upon which modern hazard and risk assessment practices in civil engineering are built. It began by illustrating why risk analysis is an essential component of engineering decision-making, emphasizing its role in anticipating undesirable events, evaluating their potential impacts, and supporting the design of safer and more resilient infrastructure systems. Across all civil engineering subdisciplines, risk assessment provides a structured means of understanding how hazards interact with vulnerable systems to produce consequences that can affect human life, economic activity, environmental integrity, and societal functioning.

Fundamental terminology was clarified to distinguish between closely related concepts such as danger, threat, hazard, vulnerability, exposure, and consequences. These definitions provide the vocabulary needed to formally analyze and communicate risk. Hazards were shown to include both natural and human-made events, each capable of producing impacts whose magnitude depends on a combination of physical intensity, environmental context, and system susceptibility. The chapter also highlighted the importance of considering both direct and indirect consequences, recognizing that cascading and long-term effects can greatly exceed the initial physical damage.

The basic risk equation was introduced as a conceptual framework linking three core components—hazard or threat (T), vulnerability (V), and consequences (C). Although the exact functional relationships among these components are often complex and not analytically expressible, proportional relationships provide a practical framework for evaluating and comparing risks. This equation also underscores the interdependence of system characteristics and external stressors, reinforcing the idea that effective risk management requires simultaneous attention to hazards, system fragility, and potential impacts.

To deepen this foundation, the chapter expanded the discussion by introducing several important subclasses of risk. Reliability was presented as a quantitative measure of a system's ability to perform as intended under given demands, highlighting the importance of understanding capacity, demand relationships over the life cycle of infrastructure. Exposure was described as the presence of people, assets, or systems within hazard-prone areas, influenced by both vulnerability and the spatial or temporal distribution of hazards. The concept of likelihood was clarified in relation to probability, emphasizing its role in assessing how strongly observed data support particular models of hazard or system behavior.

Resilience was examined as a complementary perspective to risk, focusing not only on avoiding failure but also on absorbing disruptions, adapting to changing conditions, and recovering functionality within acceptable timeframes. The four components of resilience—robustness, redundancy, resourcefulness, and rapid recovery—were articulated as essential attributes of infrastructure and communities seeking to withstand modern hazards. Sustainability was presented as a long-term, multidimensional objective that links environmental, economic, and social considerations to the broader goal of ensuring that today's engineering practices do not compromise the needs of future generations.

Finally, the chapter outlined the concept of elements-at-risk, providing a structured classification of physical, social, economic, infrastructural, and environmental assets that may be affected by hazardous events. Recognizing and characterizing these elements is essential for determining the distribution of potential losses and for tailoring risk management strategies to the needs of diverse stakeholders.

Together, these discussions establish a comprehensive framework for understanding hazard, risk, resilience, and sustainability in civil engineering. The concepts introduced in this chapter will serve as building blocks for subsequent chapters, where they will be applied to more detailed methods, analytical tools, and case studies. By grounding the reader in these essential principles, the chapter prepares the foundation for systematically evaluating and managing the complex risks inherent in the planning, design, operation, and maintenance of civil infrastructure systems.

References

Amiri, A., Soltani, K., Ebtehaj, I., & Bonakdari, H. (2024). A novel machine learning tool for current and future flood susceptibility mapping by integrating remote sensing and geographic information systems. *Journal of Hydrology, 632*, 130936. https://www.sciencedirect.com/science/article/pii/S0022169424003305

Amiri, A., Ebtehaj, I., Soltani, K., Gumiere, S. J., & Bonakdari, H. (2025). Future soil erosion trends in Canadian agricultural lands from runoff and sustainability impacts. *Scientific Reports, 15*(1), 23184.

Ang, A. H.-S., & Tang, W. H. (2007). *Probability concepts in engineering: Emphasis on applications to civil and environmental engineering* (2nd ed.). Wiley. https://books.google.com/books/about/Probability_Concepts_in_Engineering.html?id=G4IoAQAAMAAJ

Blaikie, P., Cannon, T., Davis, I., & Wisner, B. (2004). *At risk: Natural hazards, people's vulnerability and disasters* (2nd ed.). Routledge. https://www.routledge.com/At-Risk-Natural-Hazards-Peoples-Vulnerability-and-Disasters/Blaikie-Cannon-Davis-Wisner/p/book/9780415252164

Bruneau, M., Chang, S. E., Eguchi, R. T., Lee, G. C., O'Rourke, T. D., Reinhorn, A. M., Shinozuka, M., Tierney, K., Wallace, W. A., & von Winterfeldt, D. (2003). A framework to quantitatively assess and enhance the seismic resilience of communities. *Earthquake Spectra, 19*(4), 733–752. https://doi.org/10.1193/1.1623497

Cutter, S. L., Boruff, B. J., & Shirley, W. L. (2003). Social vulnerability to environmental hazards. *Social Science Quarterly, 84*(2), 242–261. https://doi.org/10.1111/1540-6237.8402002

Ebtehaj, I., & Bonakdari, H. (2022). A comprehensive comparison of the fifth and sixth phases of the coupled model intercomparison project based on the Canadian earth system models in spatio-temporal variability of long-term flood susceptibility using remote sensing and flood frequency analysis. *Journal of Hydrology, 617*, 128851. https://www.sciencedirect.com/science/article/pii/S0022169422014214

Ebtehaj, I., Fortin, J., Bonakdari, H., & Grégoire, G. (2024). How climate change will shape pesticide application in quebec's golf courses: insights with deep learning based on assessing CMIP5 and CMIP6. *Applied Sciences, 14*(20), 9209. https://www.mdpi.com/2076-3417/14/20/9209

Holling, C. S. (1973). Resilience and stability of ecological systems. *Annual Review of Ecology and Systematics, 4*, 1–23. https://doi.org/10.1146/annurev.es.04.110173.000245

Intergovernmental Panel on Climate Change. (2022). Summary for policymakers. In H.-O. Pörtner, D. C. Roberts, M. Tignor, E. S. Poloczanska, K. Mintenbeck, A. Alegría, M. Craig, S. Langsdorf, S. Löschke, V. Möller, & A. Okem (Eds.), *Climate change 2022: Impacts, adaptation and vulnerability* (pp. 3–33). Cambridge University Press. https://doi.org/10.1017/9781009325844.001, https://www.ipcc.ch/report/ar6/wg2/downloads/report/IPCC_AR6_WGII_SummaryForPolicymakers.pdf

International Organization for Standardization. (2015). *ISO 2394:2015—General principles on reliability for structures.* ISO. https://www.iso.org/fr/standard/58036.html

International Organization for Standardization. (2019). *IEC 31010:2019—Risk management—Risk assessment techniques.* ISO. https://www.iso.org/fr/standard/72140.html

Joint Committee on Structural Safety. (2001). *JCSS Probabilistic Model Code.* https://www.jcss-lc.org/jcss-probabilistic-model-code/

Kaplan, S., & Garrick, B. J. (1981). On the quantitative definition of risk. *Risk Analysis, 1*(1), 11–27. https://doi.org/10.1111/j.1539-6924.1981.tb01350.x, https://ideas.repec.org/a/wly/riskan/v1y1981i1p11-27.html

Melchers, R. E., & Beck, A. T. (2018). *Structural reliability analysis and prediction* (3rd ed.). Wiley. https://doi.org/10.1002/9781119266105, https://books.google.com/books/about/Structural_Reliability_Analysis_and_Pred.html?id=8yE6DwAAQBAJ

National Infrastructure Advisory Council. (2009). *Critical infrastructure resilience: Final report and recommendations.* U.S. Department of Homeland Security, Cybersecurity and Infrastructure Security Agency (CISA). https://www.cisa.gov/sites/default/files/publications/niac-critical-infrastructure-resilience-final-report-09-08-09-508.pdf

National Institute of Standards and Technology. (2016). *Community resilience planning guide for buildings and infrastructure systems* (Vol. I) (NIST Special Publication 1190). https://doi.org/10.6028/NIST.SP.1190v1, https://nvlpubs.nist.gov/nistpubs/SpecialPublications/NIST.SP.1190v1.pdf

Nations Unies, Stratégie internationale de prévention des catastrophes. (2015). *Cadre d'action de Sendai pour la réduction des risques de catastrophe 2015–2030.* Nations Unies. https://www.unisdr.org/files/43291_frenchsendaiframeworkfordisasterris.pdf

NIAC (2009). National Infrastructure Advisory Council (accessible through the following link: https://www.cisa.gov/sites/default/files/publications/niac-critical-infrastructure-resilience-final-report-09-08-09-508.pdf

Nowak, A. S., & Collins, K. R. (2013). *Reliability of structures* (2nd ed.). CRC Press. https://doi.org/10.1201/b12913

Qureshi, M. U. A., Amiri, A., Ebtehaj, I., Guimere, S. J., Cunderlik, J., & Bonakdari, H. (2025). Coupling HEC-RAS and AI for River morphodynamics assessment under changing flow regimes: Enhancing disaster preparedness for the Ottawa River. *Hydrology, 12*(2), 25. https://www.mdpi.com/2306-5338/12/2/25

Soltani, K., Amiri, A., Ebtehaj, I., Cheshmehghasabani, H., Fazeli, S., Gumiere, S. J., & Bonakdari, H. (2024). Advanced forecasting of drought zones in Canada using deep learning and CMIP6 projections. *Climate, 12*(8), 119. https://www.mdpi.com/2225-1154/12/8/119

United Nations Office for Disaster Risk Reduction. (2017). *Terminology on disaster risk reduction (sendai framework terminology).* https://www.undrr.org/drr-glossary

World Commission on Environment and Development. (1987). *Our common future* (Report transmitted to the UN General Assembly as document A/42/427). United Nations. https://digitallibrary.un.org/record/139811

Chapter 2
Climate Change, Hazards, and Infrastructure Risk

2.1 Introduction

Civil engineering operates at the intersection of natural systems, built environments, and human society. As infrastructure becomes increasingly complex and communities continue to expand into areas exposed to environmental hazards, the need to understand and manage risk has never been more critical. The built environment, comprising transportation networks, water systems, buildings, energy infrastructure, and communication networks, faces a growing range of threats arising from climate change, urbanization, aging assets, and technological interdependencies. These evolving conditions introduce new vulnerabilities and magnify existing ones, underscoring the necessity for systematic approaches to hazard identification, risk assessment, and risk-informed decision-making.

This chapter provides a comprehensive overview of hazard and risk analysis as applied in civil engineering, emphasizing the importance of anticipating adverse events before they manifest as failures or disasters. The discussion begins by examining the consequences of climate change for infrastructure, including how shifts in temperature, precipitation, and extreme weather can compromise structural reliability, community safety, and economic stability. By illustrating how hazards affect different infrastructure sectors, coastal systems, transportation networks, water management structures, and public facilities, the chapter contextualizes risk within real-world engineering practice.

Following this, the chapter introduces the conceptual foundations of hazard analysis and risk assessment, linking them to the broader framework of disaster risk management. The processes of risk identification, consequence evaluation, likelihood estimation, and the development of risk treatment strategies are explained in detail, demonstrating how these tools support evidence-based engineering decisions. Special emphasis is placed on understanding exposure, vulnerability, and system reliability, recognizing that risk emerges not only from the presence of a hazard but also from a system's susceptibility and the value of the elements at risk.

H. Bonakdari et al., *Hazard and Risk Assessment in Civil Engineering Design*, https://doi.org/10.1007/978-3-032-23600-5_2

Finally, the chapter highlights the relevance of risk-based approaches for sustainable civil engineering. By integrating social, environmental, and economic considerations, the three pillars of sustainability, risk assessment enables engineers to design infrastructure that is safer, more resilient, and better suited to withstand uncertainty over its life cycle. As climate change intensifies natural hazards and infrastructure networks age and expand, incorporating rigorous hazard and risk analysis will remain essential to protecting communities, preserving ecosystems, and ensuring long-term economic viability.

2.2 The Consequences of Climate Change on Infrastructure

Climate change poses perhaps the most significant and complex challenge of our generation to civil engineering infrastructure and necessitates a reshaping of the way engineers think about design, construction, maintenance, and resilience. The effects of a changing climate are not a distant threat but an immediate reality, significantly impacting various aspects of civil engineering infrastructure worldwide. These effects, ranging from rising sea levels to increased frequency and intensity of extreme weather events, to shifting precipitation patterns and temperature fluctuations, have profound implications for infrastructure planning and management. These climatic changes threaten the integrity and functionality of vital systems, including transportation networks, water supply systems, urban drainage systems, airports, and coastal defenses.

Additionally, indirect effects of climate change, such as population displacement and the consequent urbanization pressures, also warrant attention. These factors compel civil engineers to reconsider existing infrastructure capabilities and to innovate in the design of resilient, adaptable, and sustainable infrastructure. This subsection will highlight the vulnerabilities of existing infrastructure to a changing climate and discuss emerging strategies and technologies being adopted to mitigate these risks. The aim is to provide a comprehensive understanding of how climate change is reshaping the landscape of civil engineering and how the profession is responding to these unprecedented and emerging challenges (Fig. 2.1).

2.2.1 Coastal Infrastructure

In the realm of coastal infrastructure, significant impacts are driven by a changing climate, including rising sea levels, more frequent severe storms, and coastal erosion. For instance, coastal inundation due to rising sea levels has been observed worldwide. In the United States, the eastern seaboard has been especially impacted by increasing frequency of flood events. One example is in Atlantic City (New Jersey), where the number of flood days per year has increased dramatically from 0.68 during the period from 1950 to 1969 to an average of more than 10 flood days for the period from 2010

Fig. 2.1 Conceptual illustration of the consequences of climate change on infrastructure

to 2022. This leads to the gradual submergence of low-lying areas, with critical infrastructure such as ports, marinas, and seafront promenades being affected. The erosion of coastlines, accelerated by rising sea levels, results in land loss and threatens the stability of structures along the coast.

High-intensity storms and surges, now more commonly seen due to a changing climate, also pose serious risks to coastal structures intended to protect against flooding and waves. These extreme weather events often exceed the design parameters of seawalls, breakwaters, bulkheads, revetments, and other levee-type protective structures, leading to potential failures. For example, a seawall designed for past weather conditions might be inadequate under new climatic patterns, necessitating redesign or reinforcement.

Furthermore, rising sea levels and more intense storm surges in coastal areas have the potential to increase the intrusion of salt water into groundwater aquifers as well as to increase the flow of salt water into estuaries. This affects not only the ecological balance but also the foundation and structural integrity of buildings and infrastructure. The corrosion of materials used in coastal infrastructure, accelerated by increased salinity and humidity, is also a significant concern.

Due to all these challenges, civil and environmental engineers must adapt coastal infrastructure to be more resilient against the effects of climate change. Strategies such as elevating critical infrastructure, redesigning coastal flood protection structures, and using corrosion- and salinity-resistant materials are employed. Efficient project delivery must also balance the stability, resiliency, and integrity of coastal structures under new climatic conditions with considerations of sustainability, ensuring that infrastructure can withstand changing environmental conditions.

2.2.2 Educational Infrastructure

Owing to the effects of climate change, the educational infrastructure sector has experienced profound impacts in recent years, including the increased exposure of schools, universities, and other educational facilities to extreme weather conditions such as heatwaves, floods, and storms. For example, the longer duration and more intense heatwaves resulting from climate change can lead to higher indoor temperatures in educational buildings, which are often not designed for such extremes. Consequently, the learning environment is compromised, with the health and comfort of students and staff being affected.

As with other structures in low-lying and/or flood-prone areas, the vulnerability of educational infrastructure to flooding and storm damage has increased due to a changing climate. For educational facilities, the damages are two-fold. Considering a school located near a flood-prone river, not only would flooding necessitate costly repairs to the school structure, but it would also lead to significant interruptions in the educational calendar. Disruption to educational activities can have significant long-term effects on the learning process and students' educational outcomes.

Additionally, the indirect effects of climate change, such as population displacement due to environmental factors, place additional pressure on educational infrastructure. Schools in areas receiving an influx of displaced populations may face overcrowding and resource strain, impacting the quality of education provided. For example, in schools experiencing such issues, banks of portable classrooms may be installed. These portable classrooms, however, often struggle to control indoor climate, require students to travel outside for access, and may pose a security risk during lockdowns, etc.

In response to these challenges, the adaptation and resiliency of educational infrastructure are being prioritized. Measures such as implementing climate control systems to manage indoor temperatures, reinforcing buildings to withstand extreme weather events, and redesigning facilities to accommodate changing population dynamics are being considered. These adaptations are critical for ensuring that educational facilities remain safe, functional, and conducive to learning in the face of climate change.

2.2.3 Sewer and Drainage Systems

In the context of sewer and drainage systems, significant impacts are experienced due to climate change. Increased frequency and intensity of heavy rainfall events, often associated with climate change, can overload sewer systems and drainage networks. For example, in urban areas with higher impervious-surface coverage, rapid runoff during heavy rains overwhelms the capacity of existing drainage systems, leading to urban flooding.

Furthermore, the volumes of wastewater and stormwater are often subject to significant fluctuations, driven by more extreme weather events, especially during prolonged droughts followed by intense rainfall. These fluctuations challenge the design limits and operational efficiency of sewer systems. The capacity of existing infrastructure, originally designed based on historical weather patterns, is frequently exceeded under these new climatic conditions, posing a risk of system failures and overflows.

Additionally, rising sea levels, which are a direct consequence of climate change, have led to the inundation of low-lying sewer and drainage infrastructure, especially in coastal cities. This not only disrupts the functionality of these systems but also poses risks of contamination as the integrity of sewage systems is compromised.

In response to these challenges, the adaptation and modernization of sewer and drainage systems are being prioritized in many locations. Measures such as expanding system capacity, incorporating green infrastructure elements (rain gardens and permeable pavements, for example), and elevating or relocating critical infrastructure outside of flood-prone areas have been implemented. These measures aim to enhance the resiliency of sewer and drainage systems, ensuring their effective functioning in the face of altered precipitation patterns and rising sea levels due to climate change.

Figure 2.2 shows a schematic of a rain garden cross-section to help the reader understand their structure and purpose in urban and residential landscapes. The cross-sectional view displays the different layers that compose the rain garden and illustrates how they work together. Starting from the top, the mulch layer retains moisture and suppresses weeds while helping to foster water infiltration into lower layers. The plant layer, featuring deep-rooted plants, is crucial for absorbing water and aiding filtration. Below this is the soil layer, which not only supports plant growth but also filters pollutants. The sand layer serves as an additional filtration medium, while the gravel layer helps guide the filtered water toward the drainage system. At the bottom lies the perforated pipe, essential for draining water and preventing waterlogging. It's typically depicted as a cylindrical shape running horizontally across the bottom of the rain garden. The pipe may be shown with small holes or perforations along its length. These perforations allow water to enter the pipe after it has filtered down through the upper layers.

The diagram also shows how rainwater is collected and channeled into the garden, typically from surfaces like rooftops. Surrounding landscaping features, such as grass lawns, pathways, and residential structures, are shown in the schematic, demonstrating the ease with which the rain garden fits into the setting and its role in managing stormwater. Overall, the design encapsulates the effectiveness of rain

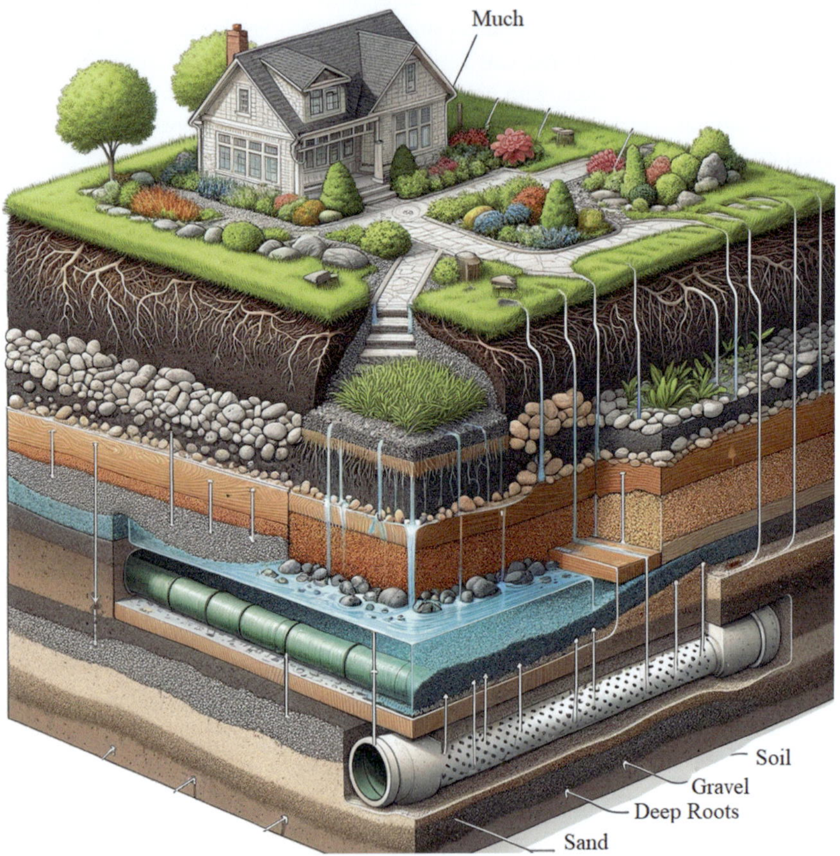

Fig. 2.2 A schematic of the rain garden

gardens in reducing runoff, filtering pollutants, and replenishing groundwater, all while enhancing the aesthetic appeal of the area.

2.2.4 *Railway Systems*

The effects of climate change on railway systems include an increased susceptibility to weather extremes, such as heatwaves, heavy rainfall, and storms. For instance, during heatwaves, railroad tracks and ties experience thermal expansion, which leads to the risk of track distortions and buckling. Such events can disrupt rail services, which would be detrimental for the flow of goods as well as passengers, as well as pose serious environmental and safety hazards from derailment. Moreover, the increased frequency and intensity of heavy rainfall and storms result in challenges such as flooding of tracks, landslides, and debris blocking railway lines. For example, a railway line passing through a mountainous region might experience frequent disruptions due to landslides triggered by heavy rainfall, affecting service reliability and requiring substantial maintenance.

Coastal railway infrastructure is also threatened by rising sea levels resulting from climate change. The inundation of tracks and associated infrastructure in coastal areas is increasingly observed, causing service disruptions and necessitating costly repairs or track relocations. Adaptation measures in railway systems are being implemented to address these challenges. Examples include reinforcing tracks to withstand higher temperatures, improving drainage systems to prevent flooding, and installing early warning systems for landslides and other hazards. The goal of resilient railway infrastructure design is to maintain the safety, reliability, and efficiency of railway systems amid the evolving challenges posed by climate change.

2.2.5 *Airports*

The increased frequency of extreme weather events, such as severe storms, high winds, and flooding, poses challenges for airport operations and infrastructure. For example, runways and airport facilities are at risk of flooding during intense rainfall events, leading to disruptions in airport operations and potential damage to infrastructure. Extreme winds pose safety challenges during takeoff and landing.

Additionally, rising global temperatures associated with climate change lead to higher temperatures at airports. This increase affects aircraft performance, as higher temperatures can reduce engine efficiency and wing lift. As a result, longer runways may be required for safe takeoff and landing under extreme heat conditions.

Furthermore, the issue of rising sea levels poses a specific threat to coastal airports. Seawater encroachment onto airport property is observed, threatening runways, taxiways, and support buildings, especially in low-lying areas. This necessitates implementing protective measures, such as constructing sea barriers and elevating critical infrastructure.

Adaptation strategies for airport infrastructure are being developed to address these challenges. Approaches implemented to date include redesigning and reinforcing runways and buildings to withstand extreme weather, improving drainage systems to prevent flooding, and adjusting operational protocols to account for changing environmental conditions. The aim is to ensure that airports remain safe, functional, and resilient in the face of the evolving impacts of climate change.

2.2.6 *Dams and Reservoirs*

Variations in rainfall patterns and increased frequency of extreme weather events are affecting water storage and management. For example, in regions with more intense and frequent rainfall, dams face greater pressure from higher water inflows, posing challenges for flood control and spillway management. The risk of overtopping and structural failure is heightened under these conditions. Conversely, in areas where climate change leads to reduced rainfall and prolonged droughts, reservoirs experience lower water levels. This reduction impacts their capacity for water supply, hydroelectric power generation, and irrigation. For instance, a reservoir designed based on historical rainfall data might struggle to meet water demand during prolonged dry spells. Furthermore, the altered temperature patterns associated with climate change affect the thermal stratification and water quality in reservoirs. The alteration in water temperature can impact aquatic ecosystems and the operational efficiency of hydroelectric power plants. Impaired water quality can complicate the use of the reservoir for potable water production as well as for recreational activities. For example, under the right conditions (elevated nutrient content, high temperature, and low wind), cyanobacteria can proliferate, producing hepa- and neurotoxins, as well as skin irritants, that can affect humans and animals.

In response to these evolving challenges, dam and reservoir management adaptation measures are being implemented. These include redesigning or retrofitting dam structures to handle increased water flows, implementing advanced water management strategies to optimize storage during variable rainfall conditions, and adopting ecological management practices to preserve water quality and biodiversity. The focus is on enhancing the resilience of dam and reservoir infrastructure to ensure reliable water supply and safety in the face of climate change.

2.2.7 Telecommunication Networks

Extreme weather events such as storms, floods, and heatwaves pose risks to the integrity and functionality of telecommunication networks. For example, the infrastructure of telecommunication networks, including cell towers and ground cables, is vulnerable to damage from severe storms and flooding. Disruptions in services and connectivity are often experienced as a result, affecting communication during critical times. Moreover, many critical infrastructures, like hospitals, rely on access to cloud network solutions in order to operate efficiently. Additionally, higher temperatures associated with climate change are impacting the reliability of telecommunication equipment. Overheating of components in data centers and transmission equipment, which are not designed for such elevated temperatures, has been observed. This results in increased maintenance requirements as well as a higher potential for service interruptions. Furthermore, the threat of sea-level rise poses a specific challenge to telecommunication infrastructure in coastal areas. The submersion and corrosion of underground cables and equipment are potential risks, particularly in low-lying regions.

Adaptation strategies for telecommunication networks include reinforcing infrastructure to withstand extreme weather, elevating critical equipment in flood-prone areas, and implementing cooling systems to manage equipment temperatures. Additionally, deploying more resilient, redundant network architectures is being considered in many localities to ensure continuous connectivity even under adverse climatic conditions. The goal is to maintain robust, reliable telecommunications services in the face of the evolving impacts of climate change.

2.2.8 Agricultural Infrastructure

In the agricultural infrastructure sector, profound impacts are being felt due to climate change. Variabilities in climate patterns, such as changing rainfall regimes, extended droughts, and extreme temperature fluctuations, are affecting agricultural practices and infrastructure. For instance, irrigation systems are increasingly strained due to altered precipitation patterns and water scarcity. The efficiency and capacity of these systems are often challenged, leading to issues in maintaining crop hydration and productivity.

Furthermore, the increased prevalence of extreme weather events, such as floods and storms, poses risks to agricultural buildings, storage facilities, and machinery. Damage and crop loss due to such circumstances are common, impacting food production and security. For example, a grain storage facility in a flood-prone area might face significant water damage, leading to losses of stored produce.

Additionally, the rise in temperatures resulting from climate change has altered the pests and diseases experienced by crops, which can significantly affect crop health and yields. Adapting farming practices and infrastructure to these new challenges is necessary, including developing crop varieties that are resistant to pests and implementing advanced pest management systems.

In response to these evolving challenges, adaptive measures in agricultural infrastructure, such as redesigning or upgrading irrigation systems to optimize water use, building flood-resistant storage facilities, and adopting sustainable farming practices to enhance resilience against climate variability, are being implemented. The focus is on ensuring the sustainability and productivity of agricultural practices in the face of the changing climate, which is crucial for food security and rural economies.

2.2.9 Mining Infrastructure

The increased frequency and severity of extreme weather events, such as heavy rainfall, floods, and storms, pose challenges to mining operations. For instance, open-pit mines and tailing dams are particularly vulnerable to flooding, leading to operational disruptions and potential environmental hazards. The stability of these structures under changing climatic conditions is often compromised, leading to safety concerns.

The effects of drought and water scarcity, exacerbated by climate change, are also affecting water-intensive mining processes. Water availability, essential for various mining operations, is becoming increasingly uncertain. This leads to the need for more efficient water management and conservation strategies within the mining sector.

Furthermore, the thawing of permafrost in arctic and subarctic regions, a direct consequence of rising global temperatures, is affecting mining infrastructure in these areas. The stability of the ground, critical for the structural integrity of mining facilities, is compromised, posing risks to both safety and operations.

In response to these challenges, adaptation measures for the mining infrastructure, such as reinforcing tailing dams and other structures to withstand extreme weather events, adopting water-efficient mining technologies, and redesigning facilities in areas affected by permafrost thaw, are being developed and implemented. These concerns must be appropriately balanced with consideration of the environmental and social impacts of mining operations to reduce the carbon footprint and improve the sustainability of mining activities in the face of climate change.

2.2.10 Healthcare Facilities

The increased incidence of extreme weather events, such as heatwaves, floods, and hurricanes, poses challenges to the resiliency and functionality of healthcare facilities. For example, hospitals and clinics, especially in flood-prone areas, are at risk of inundation during heavy rainfall or storm surges, leading to disruptions in medical services and potential damage to medical equipment.

Furthermore, the rise in temperatures associated with climate change is affecting the indoor environment of healthcare facilities. Increased cooling is required to maintain a safe and comfortable environment for patients and healthcare workers, which not only strains the energy systems of these facilities but also increases operational costs.

Climate change also contributes to the spread of infectious diseases and other health conditions, impacting the demand for healthcare services. The capacity of healthcare facilities to handle these changing health needs is often challenged, requiring additional infrastructure and resource allocations.

In response to these evolving challenges, adaptation strategies for healthcare facilities, such as redesigning and strengthening buildings to withstand extreme weather events, enhancing cooling and ventilation systems to manage higher temperatures, and elevating critical infrastructure in areas susceptible to flooding, are being implemented. Moreover, the increased healthcare needs resulting from the impacts of climate change are now being considered in the development of new healthcare facilities. The aim is to ensure that healthcare facilities remain safe, functional, and able to deliver uninterrupted medical services in the face of climate change.

2.2.11 Historical and Cultural Sites

Significant impacts are being felt in preserving historical and cultural sites due to climate change. These sites, often irreplaceable and of great value, face threats from various climate-related factors. For instance, increased rainfall and flooding are leading to the deterioration of ancient structures, monuments, and archaeological sites, particularly those built with materials susceptible to water damage. As a result, foundation erosion and structural integrity weakening are observed.

Moreover, the rise in temperatures and changes in humidity levels associated with climate change impact delicate artifacts, paintings, and historical documents. Without appropriate climate-controlled preservation conditions, these items experience accelerated degradation, resulting in the loss of historical information.

The rise in sea level also presents a specific threat to coastal historical sites. The submersion and erosion of these sites are being increasingly witnessed, posing risks to their preservation and accessibility. For example, a historic coastal fort or settlement might be at risk of permanent damage or disappearance due to rising sea levels and coastal erosion.

In response to these challenges, adaptation measures to protect historical and cultural sites, such as reinforcing and restoring structures to withstand extreme weather conditions, coastal stabilization techniques, improving climate control systems to preserve artifacts, and relocating at-risk items to safer locations, are being implemented. Furthermore, strategies for using technology to monitor and predict the impacts of climate change on these sites are being developed. The aim is to preserve the historical and cultural heritage for future generations, ensuring that these sites can withstand the changing environmental conditions posed by climate change.

2.2.12 Waste Management Systems

Waste management systems are being significantly impacted by the effects of climate change. Changes in weather patterns and the increased frequency of extreme weather events affect waste collection, processing, and disposal operations. For instance, heavy rainfall and flooding can disrupt waste collection services, leading to overflowing landfills and processing facilities. The contamination of surrounding environments, including water bodies and land areas, often results from such overflows. Severe rainfall events may also affect the efficiency of moisture-sensitive processes, such as composting, where excessive moisture can promote the growth of anaerobic microorganisms. For these facilities, large capital projects may be necessary to weatherproof existing facilities.

Increased temperatures associated with climate change are impacting the decomposition rate of the waste. An acceleration in the generation of methane gas from organic waste in landfills is observed, which can contribute to greenhouse gas emissions if appropriate gas collection infrastructure is not in place. This poses challenges in managing landfill gases and maintaining environmental safety standards.

Climate change is also leading to changes in waste generation patterns. The increase in consumption and use of disposable products during climate-induced events, such as heatwaves and storms, results in higher volumes of waste, straining existing waste management infrastructure.

In response to these challenges, adaptation measures in waste management systems are being implemented. These include enhancing infrastructure to withstand extreme weather events, introducing more efficient and sustainable waste processing methods, and improving landfill gas management techniques. Additionally, waste reduction and recycling initiatives have been widely promoted as a means to reduce the overall volume of waste to be managed. The goal is to ensure that waste management systems are resilient, efficient, and environmentally sustainable in the face of climate change.

2.2.13 Industrial Facilities

The increased frequency and severity of extreme weather events, such as floods, storms, and heatwaves, pose operational challenges and risks to industrial facilities. For example, manufacturing plants and industrial complexes are vulnerable to disruptions caused by flooding or storm damage. The risk of equipment damage and production downtime often results from such events.

Furthermore, the rise in global temperatures associated with climate change is affecting the cooling requirements of industrial processes. An increased demand for energy to maintain optimal operational temperatures in facilities is seen, leading to higher energy consumption and operational costs.

Climate change also impacts the availability and reliability of resources essential for industrial operations. Water scarcity, a result of altered precipitation patterns and prolonged droughts, poses a challenge for industries reliant on water for manufacturing processes. The need for efficient water management and alternative water sources is becoming increasingly important.

In response to these evolving challenges, adaptation measures for industrial facilities include reinforcing and redesigning buildings and infrastructure to withstand extreme weather conditions, implementing energy-efficient cooling systems, and adopting water-saving technologies. Moreover, strategies to reduce carbon footprints and enhance the sustainability of industrial operations are being emphasized, aligning with global efforts to mitigate climate change. The goal is to ensure that industrial facilities remain operational, resilient, and environmentally responsible in the face of climate change.

2.2.14 Public Safety Infrastructure

The increased incidence of extreme weather events, such as hurricanes, floods, wildfires, and heatwaves, poses significant challenges to public safety systems and facilities. For example, emergency response services, including fire stations, police stations, and ambulance facilities, are at risk of being overwhelmed during natural disasters. The capacity of these services to respond effectively is often strained under such extreme conditions.

The infrastructure supporting public safety, including communication systems and transportation networks, is also vulnerable to disruption. Damage to roads, bridges, and communication lines during severe weather events can impede emergency response efforts and hinder access to affected areas.

Furthermore, climate change increases the frequency of health emergencies, such as heat-related illnesses and diseases transmitted through contaminated water or vector-borne pathogens. The demand for emergency medical services is increasing as a result, placing additional pressure on healthcare facilities and first responders.

In response to these challenges, adaptation strategies for public safety infrastructure, such as enhancing emergency response capabilities, fortifying critical infrastructure to withstand extreme weather conditions, and improving communication systems for effective coordination during emergencies, are being developed. Additionally, training and resources are being allocated to better prepare first responders and medical personnel for the unique challenges posed by climate change. The aim is to ensure that public safety infrastructure remains robust, resilient, and capable of protecting communities in the face of evolving climate-related risks.

2.2.15 Green Infrastructure

Green infrastructure encompasses elements like urban green spaces, green roofs, and rain gardens. The functionality and effectiveness of these natural systems in providing environmental benefits, such as stormwater management, air purification, and urban heat reduction, are being altered as a result of climate change. For example, urban green spaces, which are critical for mitigating the urban heat island effect, face challenges due to increased temperatures and changing precipitation patterns. The stress on vegetation and the reduced ability of these spaces to provide cooling are often a result.

Furthermore, the capacity of green infrastructure to manage stormwater is being tested by the increased frequency and intensity of heavy rainfall events. The design parameters of rain gardens, bioswales, and permeable pavements, originally based on historical rainfall data, are frequently exceeded under new weather patterns. This reduces the effectiveness of these systems in preventing urban flooding and managing runoff.

The changing climate conditions are also affecting the biodiversity supported by green infrastructure. The shift in species composition and health of ecosystems within urban green spaces is being observed, impacting their ecological value and the services they provide.

In response to these evolving challenges, adaptation measures for green infrastructure, such as selecting drought-resistant and heat-tolerant plant species, redesigning stormwater management systems to accommodate increased water volumes, and enhancing urban biodiversity through ecological landscaping practices, are being implemented. Moreover, integrating green infrastructure with traditional gray infrastructure (conventional, engineered structures and systems used in urban environments for essential services) is emphasized, creating a more resilient and adaptive urban environment. The aim is to ensure that green infrastructure continues to provide vital environmental and social benefits in the face of changing climate conditions.

2.2.16 Landslides

Changing precipitation patterns, characterized by more intense and frequent rainfall, are major factors contributing to increased landslide frequency and severity. For instance, soil saturation from prolonged or heavy rainfall increases the risk of slope instability and landslides. The destabilization of hillsides and mountainous areas, particularly those already weakened by human activities or natural erosion, is often a consequence.

Wildfires, which are becoming more common due to climate change, have also been linked to increased landslide risk. The loss of vegetation from these fires leads to a concomitant loss of root systems, which stabilize the soil. Without the stabilizing effect of the root systems, subsequent heavy rains on these burnt areas can easily trigger mudslides and debris flows, posing risks to nearby communities and infrastructure.

Ground instability has also been reported as a result of permafrost thaw in colder regions. As a direct consequence of rising global temperatures, the once-frozen ground layers lose their structural integrity upon thawing, increasing the risk of landslides in these areas.

In response to these challenges, proactive measures for landslide risk management, such as monitoring and assessing high-risk areas, implementing early warning systems, and developing slope stabilization strategies, are being put in place. Engineering solutions, such as retaining walls, drainage systems, and reforestation of vulnerable slopes, are being employed to mitigate the risk. Additionally, land-use planning and building regulations are being adapted to reduce the exposure of communities and infrastructure to landslide-prone areas. The goal is to enhance safety and minimize the impacts of landslides, which are becoming more prevalent due to the changing climate.

2.2.17 Bridges

The increased frequency of extreme weather events, such as floods, hurricanes, and heatwaves, poses challenges to the structural integrity and functionality of these vital infrastructures. For instance, the increased intensity and frequency of flooding place greater stress on bridge foundations and supports. A common consequence is the scouring of bridge foundations by strong water flows, which is only exacerbated by increased river flow rates and flood levels.

Additionally, higher temperatures are affecting the expansion and contraction of bridge materials. An increased thermal expansion rate in materials like steel and concrete during heatwaves can lead to structural stress and the need for more frequent maintenance. The risk of expansion joint failures and damage to the bridge deck surface is often observed under such conditions and must be appropriately considered in the design of future climate-resilient bridges.

Coastal bridge infrastructure is also significantly impacted by rising sea levels. Increased exposure to saltwater and higher humidity accelerate corrosion of materials and degrade structural components. This necessitates more rigorous maintenance regimes and may lead to a significant decrease in the bridge lifetime before major retrofit.

In response to these evolving challenges, adaptation measures for bridges, such as redesigning and reinforcing bridge structures to withstand increased loads and extreme weather conditions, using materials and coatings that are more resistant to corrosion and thermal stress, and enhancing monitoring and maintenance practices, are being implemented. Additionally, future climate scenarios are being emphasized in designing new bridges to ensure their resilience and longevity. The aim is to ensure that bridges, as critical components of transportation networks, remain safe and operational in the face of the changing climate.

2.3 Damage of Elements at Risk—Urbanization and Floods

Assessing the potential losses and extent of damage to elements-at-risk is a multifaceted process that requires a detailed analysis of the negative effects that different types of hazards may impose. Classifying these hazards and their specific impacts on various elements-at-risk is crucial for accurate risk assessment in civil engineering.

Different types of hazards, such as undercutting from erosion or landslides, subsidence resulting in loss of support, seismic activity causing shaking, fires, gaseous emissions, structural loading, and inundation from floods (Fig. 2.3), can each affect structures differently. Figure 2.3 illustrates various hazards that can affect buildings. The image captures a cliffside in partial collapse, vividly illustrating the destructive power of erosion or landslides on the structural stability of buildings or roads on cliffs. It also shows a building experiencing a loss of support, with the ground beneath visibly sinking, representing the sudden or gradual phenomenon of subsidence. The figure depicts the effects of an earthquake, where seismic activity leads to immediate destabilization of the building from shaking and to the formation of fissures around its foundation. Another powerful image depicts a city engulfed in flames, highlighting the intense, often uncontrollable nature of urban fires and their potential to cause extensive property damage. The release of hazardous gases is indicated by thick gas plumes emanating from industrial sites or volcanic areas, signaling the risks posed by toxic gas emissions. Additionally, the figure demonstrates the impact of excessive loading through a heavily laden bridge or structure, showing signs of stress and strain. This emphasizes the effect of excessive weight, whether from accumulated snow or human-made structures. Lastly, the figure depicts a coastal city or landscape submerged by floodwaters, conveying the profound effects of flooding caused by storm surges, river overflows, or tsunami events.

The vulnerability of a building to these hazards is determined by a combination of factors, including but not limited to its structural type, the materials used in construction, adherence to building codes, the age and maintenance of the structure, roof type, and physical dimensions, including height, floor space, and volume. Furthermore, the building's location plays a significant role in risk assessment, the proximity to other buildings, the distance from the hazard source, nearby vegetation, and the presence and size of openings in the structure all influence the potential damage a hazard might cause.

Civil engineers and risk assessors can more accurately evaluate the potential degree of damage by understanding the intricate relationships between these factors and the types of hazards. This evaluation is essential for the design and implementation of mitigation strategies, the enhancement of building codes, and the development of emergency response plans to minimize the impact of hazardous events on elements-at-risk. It forms a crucial part of the planning process, ensuring that structures are designed not only for functionality and aesthetics, but also for the specific risks they may face in their particular environments. This comprehensive approach to risk assessment ultimately informs and improves the resilience of infrastructure and communities against the inevitable occurrences of natural and man-made hazards.

Fig. 2.3 Illustration of the various hazards that can affect buildings

Flooding, in terms of property damage, has been recognized as one of the most prevalent and financially impactful disasters in Canada. Accordingly, an extended discussion of the wide-ranging effects of floods is warranted. These natural occurrences are characterized by their ability to inflict significant damage on the built and natural environment, but also have impacts beyond the immediate destruction of property. The repercussions of flooding are far-reaching and influence various sectors, including infrastructure, economic stability, environmental conditions, and the mental health of affected populations. This in-depth examination aims to enhance the readers' understanding of the comprehensive impacts of flooding with an emphasis on the importance of effective response and prevention strategies in the Canadian context.

2.3.1 Hydrograph

Before delving into the economic impacts of flooding, it's essential to understand the underlying fundamental hydrological processes and the influence that urbanization exerts on flood dynamics. The hydrograph depicted in Fig. 2.4 is a visual representation of streamflow, or water volume, over time, highlighting the response of a river to a rainfall event. The shape of the hydrograph is indicative of several key hydrological phenomena.

The initial response of a stream to rainfall is a period of increasing flow known as the rising limb. This part of the hydrograph ascends quickly, reflecting the rapid movement of surface runoff into the stream channels following a rain event. The steepness of the rising limb can be attributed to the relatively swift process of overland

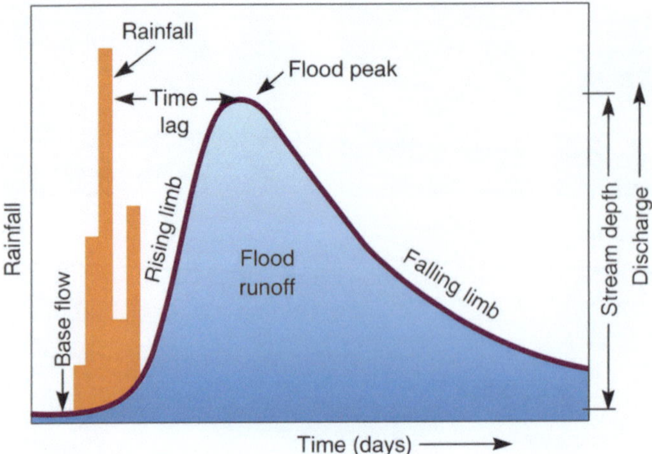

Fig. 2.4 A schematic of a hydrograph

flow, with little to no infiltration into the soil, especially in urban environments where impervious surfaces such as roads and buildings prevent water from soaking into the ground.

Following the apex of the hydrograph, which is known as the flood peak, the water level begins to recede; this is reflected in the falling limb of the hydrograph. The descent is more gradual compared to the ascent because the stream continues to be fed not only by the diminishing surface runoff, but also by base flow, which is the contribution from groundwater sources. Groundwater moves through the soil and rock substrata at a slower rate and continues to feed the stream long after the surface runoff has declined.

Urbanization can significantly impact the hydrograph by increasing both the volume and speed of runoff entering streams and rivers. With their permeable soils and vegetation, natural landscapes typically allow for more significant infiltration and slower runoff, which can mitigate the sharp rise and peak of a flood event. In contrast, with their high concentration of impervious surfaces, urban areas do not absorb rainfall, leading to quicker runoff and a shorter time lag between the rainfall event and the resulting peak flow in the stream. This can cause higher stream levels to occur in a shorter amount of time, increasing both the frequency and intensity of flood events. Moreover, urban infrastructure such as storm drains and sewer systems is often designed to quickly move water away from built-up areas. While this can reduce local flooding, it usually delivers water to streams more rapidly, contributing to the steep rising limb of the urban hydrograph. Additionally, urbanization typically involves altering or removing natural waterways, which can reduce the landscape's natural buffering and water conveyance capacity, further contributing to flood risk.

The hydrograph is a crucial tool for understanding and predicting the behavior of streams during rain events and is essential for effective flood risk management. Hydrologists and civil engineers can assess the impact of land-use changes on flood behavior by analyzing the hydrograph's characteristics, including the rising and falling limbs and the time lag. This knowledge is critical when designing infrastructure to cope with floodwaters, determining flood risk for existing developments, and planning new developments with adequate flood mitigation measures.

For instance, understanding the hydrograph's behavior in urban planning can guide the implementation of sustainable urban drainage systems or green infrastructure, which aim to mimic the natural water cycle disrupted by urbanization. Techniques such as permeable pavements, green roofs, and rain gardens can help to slow down runoff, allowing for infiltration and reducing the volume of water that reaches the stream quickly.

The delayed decrease in stream flow following a peak, as represented by the gradual falling limb, is also a reminder of the prolonged nature of the flood risk. Even after the most visible and dramatic parts of a flood have passed, elevated water levels can persist, posing a continued risk of damage. This underscores the importance of a sustained response to flood events, not just during the immediate crisis but also throughout the recovery period.

Understanding hydrograph dynamics in both natural and urban settings is essential for accurate flood risk assessment and the development of strategies to mitigate the adverse effects of floods. This is particularly vital in light of a changing climate, which is expected to increase the frequency and severity of extreme weather events, including heavy rainfall that can lead to flooding. By carefully analyzing hydrographs and considering urbanization's impact on natural hydrological processes, civil engineers and urban planners can better prepare for and respond to flood events, thereby minimizing potential losses and ensuring community resilience.

2.3.2 What Happens When a Flood Occurs in Urban Areas?

Urbanization significantly alters the hydrological response of a landscape, transforming the way water interacts with the environment during and after rainfall events. When floods occur in urban areas, the consequences of such hydrological changes become particularly evident.

In regions that have undergone urbanization, the natural, permeable soil that once absorbed a portion of rainfall is often covered by impermeable surfaces such as concrete and asphalt. Buildings, roads, and other infrastructure replace natural vegetation, disrupting the natural water cycle. This shift leads to several critical changes in how water behaves during a flood (Fig. 2.5):

- **Reduced Infiltration**: The impervious surfaces created by urban development greatly reduce the area available for rainwater to infiltrate the ground. This lack of infiltration leads to more water running off surfaces directly into storm-sewer systems.
- **Increased Runoff Volumes**: With fewer areas for water to seep into the ground, more rainwater from a storm event becomes surface runoff. This increase in runoff

Fig. 2.5 A simple schematic of what happens when a flood occurs in urban areas

volume can overwhelm stormwater systems and natural watercourses, causing them to exceed their capacity.

- **Accelerated Peak Flows**: The rapid collection and conveyance of rainwater by storm-sewer systems, combined with the speed at which rainwater flows across impervious surfaces, contribute to faster water accumulation in rivers and streams. This acceleration can lead to sudden and more severe peak flows during heavy rainfall, increasing the risk of flooding.
- **Altered River Dynamics**: The changes in runoff patterns and peak flows can also alter the natural dynamics of rivers and streams. The increased volume and speed of water can lead to greater erosion, alter sediment transport, and affect the aquatic ecosystems that depend on natural flow regimes.
- **Increased Flood Severity and Frequency**: Ultimately, altered hydrological responses to urbanization can lead to more frequent and severe flooding. With the soil's natural buffering capacity diminished, urban areas become more susceptible to the immediate, and sometimes catastrophic, impacts of heavy rainfall.

Based on this, urban planners and civil engineers must carefully consider the hydrological impacts of urbanization. Strategies to mitigate these effects include creating green spaces, permeable pavements, rain gardens, and other sustainable urban drainage systems that mimic the natural water cycle and reduce the adverse effects of urban runoff during flood events. These solutions aim to restore some of the natural infiltration capacity and slow down the flow of water, reducing the strain on urban drainage systems and decreasing flood risks.

Therefore, urbanization demands innovative approaches to water management that integrate urban planning with sustainable environmental practices. It is crucial

Fig. 2.6 Urban Expansion in Houston, Texas: A Tale of Two Decades—The 1940s versus the 1960s

to balance development needs with the preservation of natural hydrological functions that protect urban areas from the impacts of flooding.

An example of the impact of rapid urbanization on flooding is found in the city of Houston (Texas), which underwent rapid development over a span of only two decades (Fig. 2.6). In the 1940s, the cityscape was less dense, with fewer high-rise buildings and a greater number of open and green spaces. By the 1960s, the expansion is evident, with the construction of more roadways and a bustling downtown area characterized by numerous high-rise structures. This rapid urban development, particularly on flat floodplains that are easier and more economical to build on due to reduced earthwork and blasting needs, is a common trend in urban growth. However, such development in flood-prone areas can significantly increase the risk of flood damage, as natural land cover is replaced by impervious surfaces, altering the region's hydrological dynamics. The images capture not only the physical change in the urban landscape but also imply the underlying challenges of managing flood risks in rapidly developing cities.

Figure 2.7 illustrates the stark contrast in the hydrological response of Houston, Texas, to similar rainstorms before and after urbanization. In the 1940s, the hydrograph for the October 1949 rainstorm in a rural setting shows a peak discharge of around 1100 cubic feet per second (cfs), taking nearly 8 h to reach its maximum. This suggests that the rural landscape permitted significant infiltration of rainwater into the ground, which tempered the flow into waterways, resulting in a more gradual, lower flood peak. Conversely, following the city's rapid urbanization, the 1960 hydrograph shows a drastically different response. Impervious surfaces such as roofs and pavements prevent rainwater from seeping into the soil, instead causing it to run off quickly into the stormwater system. The result is a much steeper, higher peak discharge, reaching nearly 5000 cfs in a shorter time frame than during the 1949 event. This sharp rise in the hydrograph reflects the reduced lag time between rainfall and runoff peak due to urbanization, which significantly increases the risk of flooding in urban areas from overwhelmed drainage systems and rapidly elevating water levels in streams and rivers.

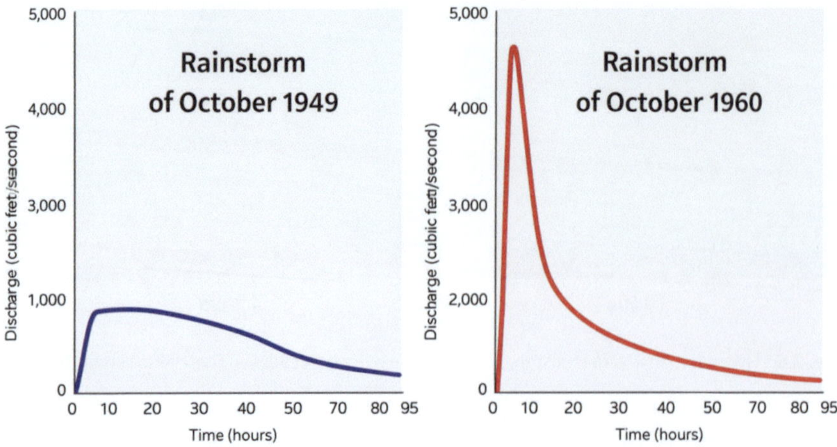

Fig. 2.7 Comparative Hydrographs of October Rainstorms in Houston, Texas: A Tale of Two Decades-1940s versus 1960s (Abbott, 2017)

Figure 2.8 more directly compares the flood hydrographs from the two rainstorm events in Houston, Texas, highlighting the dramatic changes brought by urbanization. In the rural setting of 1949, the hydrograph shows a moderate and slow increase in water discharge due to rainwater infiltration into the soil, which mitigates the flood's intensity. This slow percolation allowed for a gradual release of water into the city's streams, producing a lower peak discharge over an extended period. The contrast with the 1960 urban hydrograph is stark. Post-urbanization, the impermeable surfaces created by development prevent rainwater from entering the soil, leading to rapid surface runoff. The 1960 hydrograph shows a sharp, high peak discharge occurring much more quickly than in 1949, with the peak flow more than quadrupling. This quick, intense response to rainfall, typical of urban environments, means water is delivered to the city's watercourses much more quickly, contributing to the increased frequency and severity of floods. The comparison of these two hydrographs clearly illustrates how urbanization, with its associated increase in impervious surfaces such as roofing and paving, can exacerbate the risk and impact of flooding in metropolitan areas.

2.3.3 Frequency of Flood Events

The frequency of flooding in urban areas is closely related to how human development alters the natural landscape, particularly by adding impervious surfaces like roofs and pavement. This alteration has a profound effect on the hydrological cycle, leading to quicker and more significant runoff during rain events. This can, in turn, cause streams and rivers to rise rapidly, increasing the risk and frequency of flash flooding. Even communities designed with flood mitigation strategies in mind are not

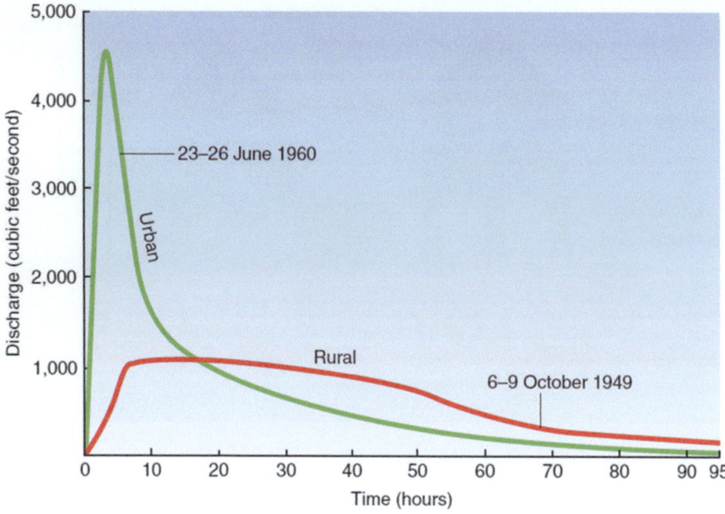

Fig. 2.8 Combined hydrographs of October Rainstorms in Houston, Texas (Abbott, 2017)

immune to the unexpected nature of floods, especially as urban expansion continues. The growth in population and the corresponding extension of urbanization upstream can lead to more frequent floods as impermeable structures replace the pervious or absorbent ground. This means that areas once capable of absorbing rainfall are now contributing to rapid runoff, overwhelming waterways, and rendering existing stormwater management systems inadequate.

The degree of damage that floods inflict on buildings is dependent on the flood type, its velocity, and the height of the water. Fast-flowing water can exert significant lateral pressure on buildings, leading to structural failure or collapse. Furthermore, the velocity of floodwaters can induce severe erosion or scouring, threatening the integrity of embankments, slopes, levees, bridge piers, and building foundations. The depth of the floodwaters determines the extent of a building's submersion, influencing the overall impact force. The duration of a flood is equally critical, as prolonged water exposure can lead to material degradation. Construction materials commonly used in buildings, like drywall or gypsum, can be particularly vulnerable, often requiring replacement after prolonged water exposure.

2.3.4 Effect of Building Characteristics on Flood Impact

The impact of flooding on a building is influenced by a number of factors related to its design, construction, and location (Fig. 2.9). Understanding these characteristics is critical for assessing flood risk and planning appropriate mitigation strategies.

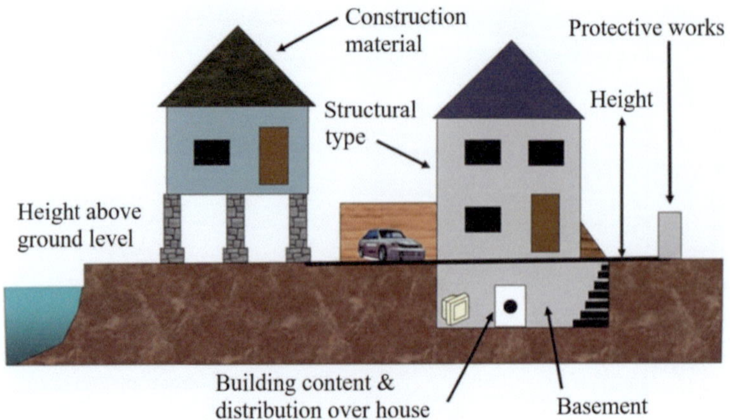

Fig. 2.9 A schematic of the building and its characteristics

- **Building Use**: The function of a building significantly affects the potential human and economic impact of a flood. A building with high occupancy or valuable contents, like a residential complex or a data center, may experience more severe consequences if flooded than, for example, a storage facility.
- **Building Materials**: The reaction of a building to flood exposure depends on its construction materials. Materials like wood or masonry can suffer extensive damage when soaked, while steel or concrete may fare better. The example of gypsum sheeting in Canadian buildings demonstrates the importance of material choice—gypsum must be replaced when wet, increasing repair costs and time.
- **Structural Type**: Different structural designs respond to flooding in various ways. For instance, some structures may be entirely dislodged from their foundation by floodwaters, while others withstand the force due to their foundational design.
- **Height Above Ground**: The elevation of a building's lowest floor above ground level can be the margin between safety and flood damage. For example, following the tsunami events in Japan, some building types were entirely lifted by the flood waters while others remained on their foundations. The height above ground level will undoubtedly play an essential role in determining the impacts of flood events on buildings. In the National Capital Region floods of 2019, many residents were saved from flooding by only one or two inches of height above the water level.
- **Location of Doors and Openings**: The positioning and design of doors and other openings can be critical, particularly in areas like walkout basements that might be prone to water entry. In such a case, a door that may be susceptible to leaks and water infiltration, especially in older homes, is now present instead of a relatively impervious concrete foundation that has been waterproofed.
- **Presence of a Basement**: Basements are typically vulnerable to flooding, often becoming the initial point of water ingress and accumulation. Floodwaters frequently find their way into homes, gravitating toward the basement that lies

below ground level. Regardless of the flood's depth, once the water reaches the structure, it tends to flow preferentially into the basement.

- **Height of the Building**: In cases where a building boasts multiple stories, occupants tend to distribute their belongings across various levels. This practice mitigates the severity of damage when the lower floors are inundated. Additionally, multiple stories facilitate evacuating individuals and valuable possessions to higher floors, enhancing safety and protection.
- **Distance to the Water Body**: Proximity to water channels can significantly increase the risk of structural undermining and collapse due to rapid erosion during floods.
- **Presence of Protective Flood Structures**: Features such as levees, floodwalls, and berms can protect a building by delaying or preventing floodwaters from reaching it or controlling the water's ingress to a manageable rate.

A comprehensive understanding of risk assessment is essential for effective urban planning and emergency response. This necessitates the creation of comprehensive maps that detail not only the intensity of potential hazards but also the distribution of the population exposed to these risks and the expected losses from various disaster scenarios. These maps serve as crucial tools for authorities and responders to visualize risk and allocate resources efficiently.

Moving forward, it is critical to incorporate additional hazards into the risk assessment process, particularly those that are relevant to the specific region or community in question. This could include a broad spectrum of events, from natural disasters like earthquakes and hurricanes to human-made threats such as industrial accidents or infrastructure failures.

Moreover, an often-overlooked but vital risk assessment component is quantifying uncertainty. Future planning and strategy development must account for the inherent uncertainties in predicting hazard events, their impacts, and the effectiveness of planned interventions. This includes understanding the data limitations, potential variability in hazard occurrences, and the dynamic nature of both human populations and built environments.

2.3.5 Costs of Flooding

The data presented by the World Resources Institute's Aqueduct Floods Tool (Ward et al., 2020) paints a concerning picture of the future impact of floods globally and specifically in Canada. By the year 2030, the global population affected by floods is projected to double. This escalation is accompanied by a threefold increase in the damage to urban property due to riverine floods. For Canada, the projections are particularly stark: the number of individuals directly affected by floods is expected to rise from approximately 200,000 in 2010 to over 350,000 by 2030. Looking further ahead to 2050, the situation appears even more dire, with anticipated annual costs of flooding damage forecasted to surpass US$14 billion, impacting around 431,000 people. These figures underscore the urgency for enhanced flood risk management and resilient infrastructure planning to mitigate the impending increase in flood-related impacts.

A recent study conducted by the Canadian Institute of Climate Choices (CICC) (Under Water: The Costs of Climate Change for Canada's Infrastructure) highlights a critical concern regarding the readiness of Canadian homes for the expected increase in extreme weather events due to climate change. With nearly 950,000 buildings located in a 100-year risk zone, areas with a 1% annual chance of flooding, the potential for widespread damage is significant. Additionally, the fact that 550,000 structures are in a 20-year risk zone, which faces a higher probability of flooding (5% annually), underscores the urgency for strategic planning and investment in flood resilience measures.

The CICC points out that, although the probability of flooding in specific areas might appear relatively low on an annual basis, the actual costs when flooding occurs are substantial. Presently, flood damage costs Canada around $1.3 billion each year. This annual cost is projected to escalate dramatically by 2100, with an increase of 4 to 7 times under low (RCP2.6) and high (RCP8.5) emissions scenarios, respectively. The CICC's analysis signals that the cost of inaction is likely to be far greater than the investments required for adaptation and mitigation today.

Preventing construction in flood-prone areas and restoring floodplains are pivotal strategies for reducing flood losses. By allowing rivers the space to accommodate large floods, communities can be shielded more effectively from the dangers of flooding. In Canada, efforts are directed towards discouraging floodplain development and rigorously mapping flood risk areas. However, this is often a complex task, particularly when dealing with pre-existing developments that were established before these risks were fully understood, or such mapping was conducted.

Figure 2.10 visually explains the geological and hydrological changes in a floodplain during normal and flood conditions. The upper section of this figure shows the floodplain during normal conditions. It illustrates a river meandering through the floodplain, which is the flat area on either side of the stream. The subsurface is depicted in layers, with the topmost layer consisting of silt and clay and a lower layer of sand. These materials are typical deposits left by the river as it changes its course and floods over time. The lower section represents the floodplain during a flood event. The area covered by water is significantly larger, submerging the floodplain. Here, the river overflows its banks, spreading across the adjacent land. The water saturation in the soil layers is also depicted, showing potential inundation of the silt, clay, and sand layers. This floodplain schematic shows how the landscape can change dramatically during a flood, emphasizing the importance of these areas for flood management and ecological balance. Floodplains can absorb excess water during such events, reducing the impact of floods on inhabited areas. However, the opposite is true when floodplains are significantly urbanized, as the expanded river banks pose a flood hazard to nearby development.

Figure 2.11 shows the application of flood mapping in predicting the potential extent of a 100-year flood event (which has a 1% chance of occurring in any given year) and highlights structures within the floodplain that are at risk. The purple

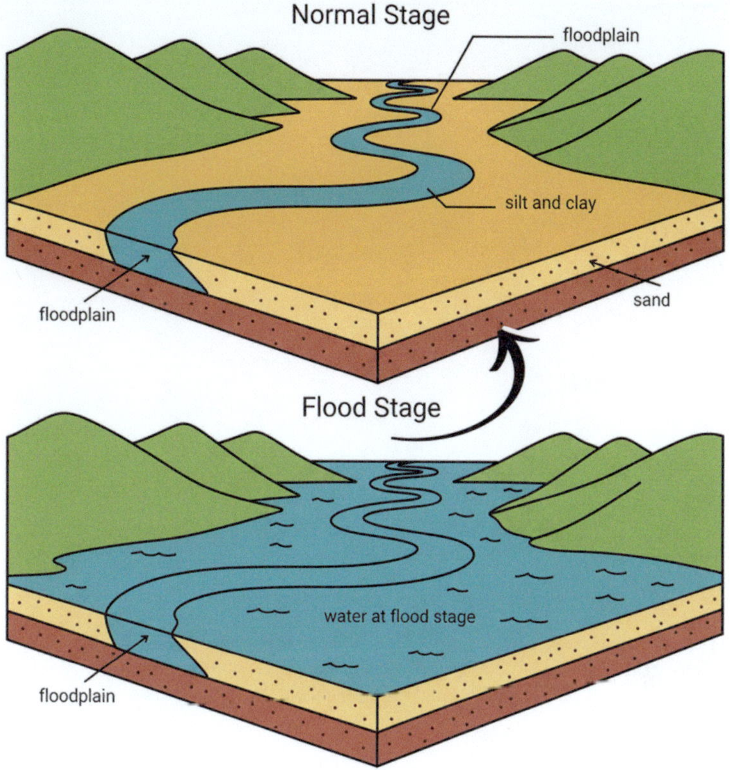

Fig. 2.10 A cross-sectional comparison of a floodplain during normal conditions and during flood events

markings on such maps identify buildings vulnerable to inundation during a major flood event. These maps not only inform current safety and emergency response planning but are also crucial for future land-use planning and development decisions, ensuring that new structures are not placed in harm's way.

2.4 Hazard Analysis

Hazard analysis is a critical tool in risk management. It is essential for identifying and quantifying risks associated with potential hazards so as to prevent negative outcomes like loss of life, injury, health impacts, property damage, environmental degradation, and loss of services. This process is systematic in nature and encompasses several crucial steps:

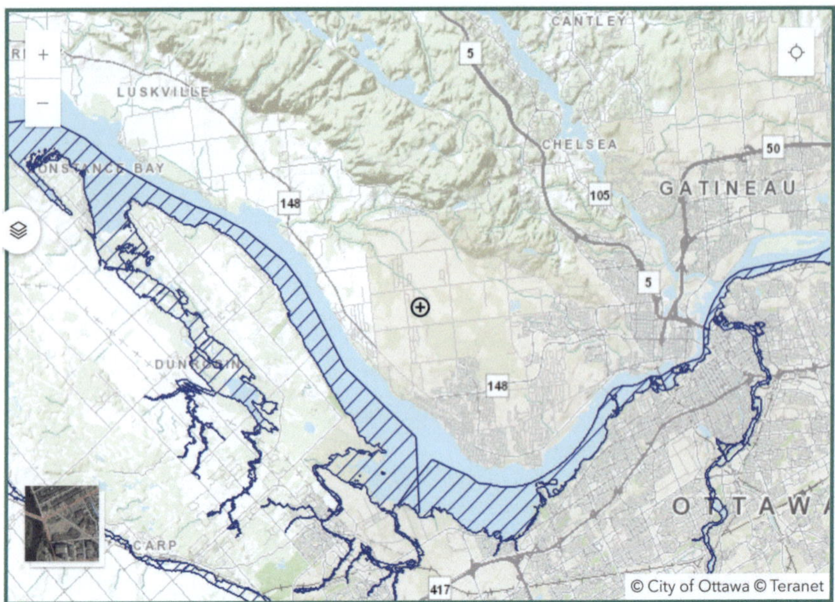

Fig. 2.11 A flood risk assessment for Rideau Gardens, detailing the potential reach of a centennial flood based on current flow data (https://ottawa.ca/en/planning-development-and-construction/maps-and-zoning/flood-plain-mapping-and-climate-change/interactive-map)

- **Frequency Estimation**: This involves assessing historical data, predictive modeling, and expert judgment to determine the likelihood and frequency of a hazardous event's occurrence.
- **Consequence Analysis**: Here, the potential impacts of the hazard on people, the environment, and infrastructure are identified. The severity of outcomes is evaluated on a scale from minor to significant.
- **Risk Reduction Measures**: Utilizing the frequency and consequence assessments, the required level of risk reduction is determined. This may involve implementing safety systems, revising procedures, or introducing new policies to mitigate risks.

The insights from hazard analysis enable organizations and authorities to prioritize safety, allocate resources effectively, and develop tailored emergency response plans. By cataloging different types of hazards, strategies can be developed to either prevent or mitigate these risks. Understanding these hazards is crucial for a proactive approach to safety and risk management, ensuring preparedness for adverse events, and enhancing community and ecosystem resilience against disasters. The ultimate goal is to minimize the likelihood and potential consequences of hazardous events, thereby protecting human life, environmental integrity, and maintaining operational continuity. The various types of hazard analyses are as follows:

Hazard Identification: This is the foundational step in hazard analysis, in which potential problems that could cause harm are identified. For instance, during a site inspection, a construction manager may identify risks such as unguarded heights, or an industrial hygienist might flag a chemical process that releases toxic fumes. A safety analyst could also flag cybersecurity vulnerabilities in a company's software as potential hazards for data breaches.

Exposure analysis: Exposure analysis quantifies the risk by determining how many people are exposed to a hazard, the frequency and duration of their exposure, and their location relative to the hazard source. For example, consider a worker in a sawmill. The time-weighted average limit for softwood sawdust is 5 mg/m^3, which accounts for the number of workers exposed, the duration and frequency of exposure. Other examples include exposure of medical staff to infectious diseases in the healthcare industry and of communities to air pollution from industrial emissions in the environmental sector.

Hazard prevalence: Hazard prevalence assesses the commonality of potential incidents, the severity of each incident, and the likelihood that they lead to loss of life, injury or other health impacts, property damage, loss of livelihoods and services, social and economic disruption, or environmental damage (the severity of consequences of the potential incidents). For example, traffic safety experts might evaluate the prevalence of car accidents at a busy intersection, or a financial auditor may examine the frequency of fraud cases within an organization over a set period. The prevalence of all hazardous events in previous years may be measured to inform the development of safety protocols.

Hazard Ranking: Hazard ranking is a systematic approach to categorize hazards based on their likelihood of occurrence and potential impact. A company may use hazard rankings to determine which activities require additional precautions when their teams perform them. This is commonly used in workplace health and safety plans to identify which activities pose the greatest risk of injury, have the highest severity, and require additional worker protections. For example, in an oil refinery, hazards might be ranked with fire and explosion as high-priority risks due to their potentially catastrophic impacts. In contrast, in a corporate office setting, repetitive strain injuries might be ranked lower but still significant due to their common occurrence.

Exposure Pathways (E-PAT): E-PAT analyzes the routes through which individuals may come into contact with hazards. For instance, in environmental health, this could involve charting the pathways of contamination from a polluted water source to human consumption. In agriculture, it might include tracking the exposure pathways from pesticide application to farmworkers.

Hazard Characteristics (H-C): Hazard characteristics evaluate the intrinsic qualities of a hazard, such as toxicity, flammability, or reactivity, and their potential impact on people and the environment. Chemical engineers might assess the reactivity of a new compound, while urban planners could evaluate the flood risk associated with a particular soil type in a developing area.

2.5 Disaster

Disasters represent a significant rupture in the fabric of society, marked by severe repercussions that extend across human, material, economic, and environmental domains. These events are characterized by their overwhelming scale, which eclipses the capacity of the affected community or society to manage using its own resources.

A disaster is not merely a natural or man-made hazard, but rather is the result of a combination of risk exposure, community vulnerability, and inadequate capacity to mitigate or recover from its effects. The gravity of a disaster is measured not just by immediate damages, but also by the enduring impacts on human well-being, societal functions, economic stability, and environmental integrity.

Examples of disasters and their grave consequences include the Red River Flood in Manitoba in 1997, which caused extensive economic damage and necessitated the displacement of several communities. Hurricane Katrina, the deadly Category 5 hurricane which made landfall just south of New Orleans in 2005, further emphasized the scale of disaster that can occur, highlighting not only the immense physical destruction but also how systemic vulnerabilities and shortcomings in emergency preparedness and response plans can amplify a disaster's impact. A further example is the 2010 earthquake in Haiti, which was a catastrophic seismic event leading to significant loss of life, widespread infrastructure destruction, and persistent socioeconomic upheaval.

Disasters are complex events, often precipitated by a hazard, that result in profound and lasting consequences when they intersect with human systems. They disrupt daily life and can lead to dire long-term outcomes, such as:

- **Loss of Life and Health Impacts**: This includes immediate fatalities and injuries, as well as long-term health issues related to the disaster.
- **Economic Losses**: The destruction of assets and the disruption of economic activities can lead to severe financial strain on individuals and the community.
- **Environmental Degradation**: Disasters can result in long-term damage to ecosystems, which may take years or decades to recover.
- **Social and Psychological Effects**: Beyond physical impacts, disasters can lead to social displacement, psychological trauma, and a breakdown of community networks.

A hazard transitions into a disaster when it causes significant harm to life, property, and the environment. The distinction lies in the extent of the impact and the disruption caused to the societal equilibrium. Addressing disasters requires a holistic approach that encompasses preparedness, mitigation, response, and recovery efforts, tailored to the unique vulnerabilities and capacities of each community or society.

Figure 2.12 illustrates the multifaceted nature of disaster impacts, encompassing a broad spectrum of quantifiable and non-quantifiable effects across various dimensions of society and the environment. The impact of a disaster extends far beyond immediate physical damage. It includes a complex array of consequences that affect

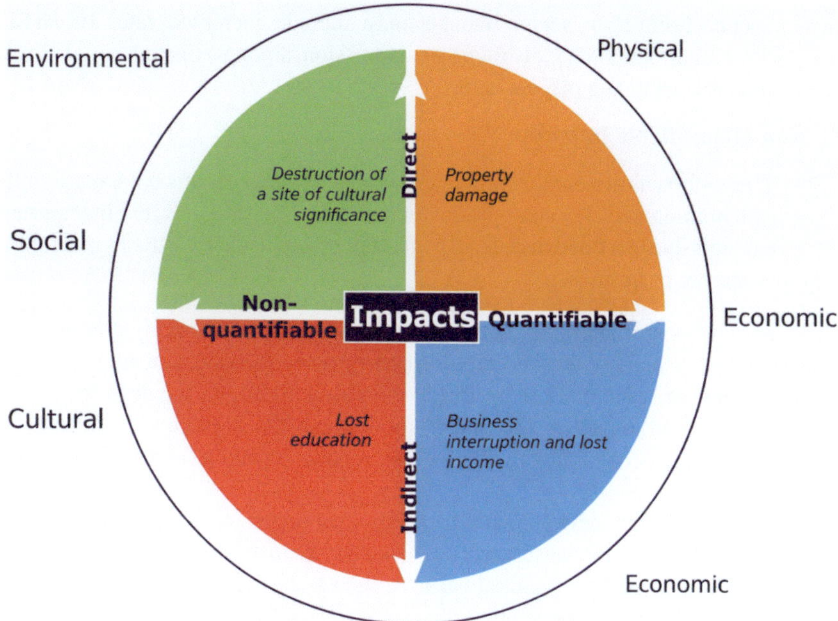

Fig. 2.12 The spectrum of disaster impacts

individuals, communities, and the environment directly and indirectly. These impacts can be categorized as follows:

1. **Direct Impacts**:
   - *Physical*: These are immediate consequences that are often quantifiable, such as property damage or destruction. An example is the cost of repairs to homes and infrastructure.
   - *Environmental*: This includes direct damage to ecosystems, such as pollution of waterways or the destruction of wildlife habitats. An example would be the discharge from an industrial facility of wastewater containing per- and polyfluoroalkyl substances.
   - *Social*: Direct social impacts can be quantified regarding the number of lives lost or persons injured.
2. **Indirect Impacts**:
   - *Economic*: Indirect economic effects include business interruption and lost income, which can ripple through the economy, affecting employment rates and gross domestic product.
   - *Cultural*: Damage to cultural heritage sites or loss of historical artifacts represents a non-quantifiable impact that affects a community's cultural identity.

- *Social*: Long-term social impacts may include increased rates of PTSD (Post-traumatic stress disorder) or depression among affected populations, challenging to quantify monetarily.

3. **Non-Quantifiable Impacts**:

   - These impacts are challenging to express in monetary terms but have significant implications. For example, lost educational opportunities for kindergarten students due to a flood disaster can affect developmental milestones and future academic performance.

The diagram encapsulates the complex nature of disaster impacts, stressing the importance of a comprehensive approach to disaster management and recovery that considers the full breadth of potential effects. It highlights the need for robust data-collection and impact-assessment methodologies that capture the true costs of disasters, including economic costs as well as impacts on human and environmental well-being.

Natural hazards are inevitable and recurring phenomena on our planet, and their occurrence is beyond human control. In Canada, as in many parts of the world, history is full of examples of natural hazards, such as earthquakes, floods, wildfires, and severe weather. Since preventing these natural events is not feasible, engineers and other professionals focus on devising strategies to minimize damage and protect lives and property when they occur.

Risk assessment is an integral component of disaster risk management. It serves as a proactive measure to anticipate the potential damage and prepare accordingly. The process of risk assessment includes:

1. **Risk Identification**: This phase involves recognizing all potential risks and the circumstances under which a natural hazard could lead to a disaster. Analysts must understand what can happen, how, where, and why.
2. **Risk Analysis**: The identified risks are further scrutinized to determine their nature and extent. This involves evaluating the likelihood of occurrence and potential consequences to gain understanding about the potential severity of the impact on communities and infrastructure.
3. **Risk Evaluation**: This step prioritizes the risks based on their analysis, considering factors such as societal values, economic implications, and environmental impacts. The aim is to determine which risks require immediate attention and resources for mitigation efforts.

Disaster risk management is an encompassing term for the systematic approach to identifying, assessing, and reducing disaster risks. It's a holistic discipline that not only considers the immediate measures to mitigate the impacts of disasters, but also the longer-term strategies for prevention and preparedness. At the heart of disaster risk management lies the process of risk assessment. This process is vital as it dissects and categorizes the components of a disaster, allowing for a detailed understanding of various risks and their potential impacts. It serves as a foundational step that informs all subsequent risk management actions. Risk assessment involves:

- **Identification**: Pinpointing the specific hazards that could potentially lead to a disaster, such as earthquakes, floods, or industrial accidents.
- **Analysis**: Evaluating the likelihood of these hazards and the severity of their potential impacts. This includes analyzing historical data, predictive models, and current environmental and social conditions.
- **Evaluation**: Determining the acceptability of risks and identifying which require immediate attention based on their potential to cause harm.

Risk management encompasses a broader scope, including the development and implementation of strategies to address risks identified in the assessment phase. This process involves:

- **Prioritization**: Deciding which risks require urgent attention based on their potential impact and the resources available for mitigation.
- **Strategy Formulation**: Developing comprehensive plans that include mitigation measures, emergency response procedures, and long-term resilience building.
- **Implementation**: Putting into action the planned strategies to manage and mitigate risks, which may involve infrastructure improvement, policy changes, community training, and education.
- **Monitoring and Review**: Continually assessing the effectiveness of risk management strategies and making adjustments as needed, based on new data or evolving circumstances.

These concepts will be explored further in the following sections, focusing on the specific actions, methodologies, and tools used in risk assessment and risk management to safeguard societies against the adverse effects of the disaster.

2.6 Risk

Risk, a term integral to the disciplines of safety, engineering, and management, encapsulates the potential for loss or injury. It is a concept that bridges the likelihood of an event occurring with the spectrum of its negative outcomes. The definition adopted from ISO[1]/IEC[2] Guide 73 highlights two principal facets of risk: probability and consequence.

In everyday language, risk is often associated with the chance of something happening, such as the risk of a car accident while driving. This common understanding primarily focuses on the event's occurrence without necessarily considering the full range of its possible effects. In contrast, in technical and engineering contexts, the term is used more specifically to denote the potential losses and implications of an event. It is not just about the occurrence of the event, but also about quantifying the severity of the outcome. For instance, in civil engineering, the risk

[1] ISO: International Organization for Standardization.

[2] IEC: International Electrotechnical Commission.

of a bridge collapse would be considered not only the probability of the collapse based on structural analysis but also the potential consequences, such as loss of life, economic disruption, and environmental damage.

The perception of risk can vary significantly among individuals and groups, influenced by factors such as knowledge, experience, cultural background, and personal values. Engineers and risk managers must recognize these different perceptions when communicating risk and implementing risk management strategies. Effective risk communication requires conveying both the statistical probability of an event and its potential impacts in a way that is understandable and meaningful to diverse audiences.

Understanding this dual nature of risk is crucial for engineers who are tasked with designing systems and structures that not only minimize the probability of failure but also mitigate the consequences should a failure occur. This comprehensive approach to risk ensures that protective measures are both appropriate and proportional to the potential impacts, thereby safeguarding public well-being and maintaining trust in technological and infrastructural advancements.

Figure 2.13 identifies the core components that constitute risk within the context of disaster risk management. Central to this concept are three elements: hazard, vulnerability, and exposure. A hazard is a potential source of harm or adverse effect, which can be either natural or man-made. Vulnerability represents the susceptibility to harm, influenced by various physical, social, economic, and environmental factors. Exposure highlights the presence of people, property, and other assets in areas that could be negatively affected by hazards. These elements are visually and conceptually interlinked, emphasizing that risk is not a standalone phenomenon but the product of these interacting factors.

At the core of the diagram lies risk, highlighted as the point where hazard, vulnerability, and exposure converge. Risk is conceptualized as the likelihood of experiencing harm or loss and is a function of the three aforementioned elements. Risk becomes actualized when a hazard materializes within the context of vulnerability and exposure, leading to potential damage and loss.

Encircling the central theme of the diagram are various risk factors denoted by black boxes, which act as precursors or contributors to risk. These include environmental degradation, climate change, poor urban planning, weak enforcement of building codes, substandard building quality, social inequity, and inadequately regulated investments. These factors represent the diverse and complex array of elements that can amplify the severity and probability of risks materializing. Each factor can increase a community's susceptibility to disasters or exacerbate the impacts of hazardous events.

The figure implies that the risks represented can be effectively controlled or managed through targeted strategies. These strategies may include countermeasures such as policy changes, infrastructure improvements, and mitigation tactics designed to lessen the impact of hazardous events. Risk management is an ongoing process that requires identifying, assessing, and prioritizing risks followed by coordinated and economical application of resources to minimize, monitor, and control the probability or impact of unfortunate events.

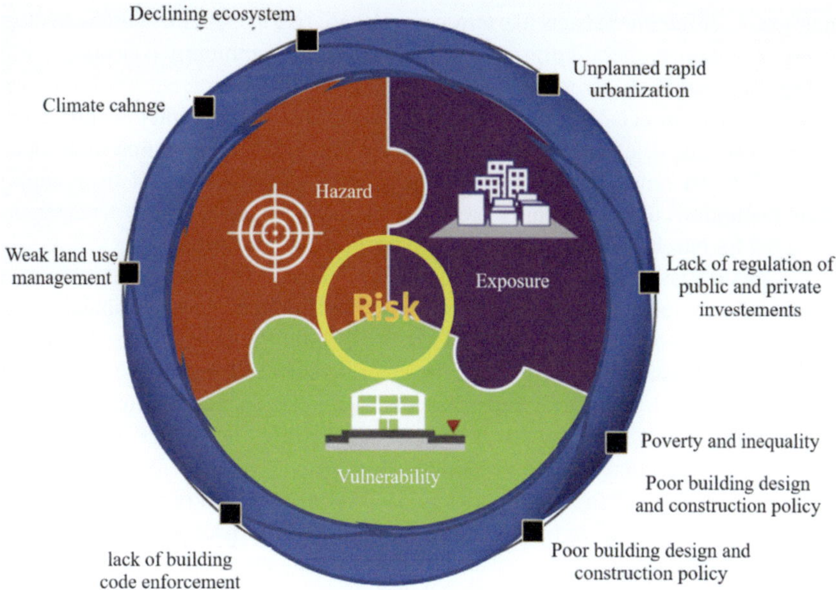

Fig. 2.13 A visual representation of the concept of risk within the context of disaster risk management

An Event or Scenario refers to a specific occurrence or a sequence of events that can trigger the manifestation of risk. This could be a natural disaster like an earthquake, a human-made hazard like a technological failure, or a sequence of events leading to an economic crisis. Such events serve as the catalysts that, when interacting with existing vulnerabilities and exposures, can result in significant consequences with varying degrees of severity.

In practical terms, applying this risk framework involves taking concrete actions to mitigate risk. For instance, reducing vulnerability might involve enhancing the quality of building design and construction to withstand earthquakes. Similarly, reducing exposure could involve relocating communities away from flood-prone areas. The application of risk management principles is critical in planning for, responding to, and recovering from hazardous events, with the ultimate goal of safeguarding lives, property, and the environment.

Risk is an inherent aspect of existence, particularly within the context of engineering, finance, operations, and strategic planning. It encompasses a comprehensive evaluation of threats, vulnerabilities, and potential consequences, aiming to manage and mitigate these elements in a structured and objective way. Several forms of risk exist, including:

Hazard Risk: Risks associated with hazards involve potential dangers that can cause damage or loss. Natural hazards include earthquakes, hurricanes, and floods, which are part of the Earth's dynamic systems. Man-made hazards, on the other hand, include technological or industrial accidents such as chemical spills, nuclear meltdowns, or intentional acts like terrorism. Hazard risk management often involves emergency preparedness, infrastructure resilience, and community education.

Operational Risk: Every operational process carries an inherent level of uncertainty that can affect outcomes. Operational risk arises from internal processes, people, systems, or external events that disrupt the normal functioning of an operation. This can range from equipment failures to human error and from supply chain disruptions to cybersecurity breaches. Effective operational risk management is crucial for ensuring smooth and efficient business operations.

Financial Risk: Financial risk pertains to the possibility of experiencing monetary losses in the conduct of normal business activities or through various financial assets. This might involve market volatility, credit issues, liquidity constraints, or investment underperformance. Financial risks require careful analysis and strategic planning to ensure fiscal health and stability for individuals, businesses, and economies.

Strategic Risk: Strategic risk is associated with high-level decisions that can influence the direction and objectives of an entire organization or sector. These risks can stem from changes in the market environment, regulatory shifts, competition, or the failure to adapt to new technologies. Strategic risk can be considered at numerous scales, ranging from private business to local municipality and from international corporations to national governments. Strategic risk management concerns the alignment of long-term goals with the changing landscape, ensuring adaptability and sustainability.

In each of these domains, risk management plays a pivotal role in guiding decision-making processes. It involves identifying potential risks, assessing their likelihood and potential impact, and implementing strategies to address them effectively. By doing so, engineers, financiers, operators, and strategists can make informed decisions that balance potential benefits against associated risks, ensuring progress and stability in their respective fields.

Risk, as a foundational concept in safety and security disciplines, is designed to holistically address the interplay between threats, vulnerabilities, and their potential consequences. It serves as a measure to evaluate and prioritize various scenarios based on their likelihood and impact, to preserve the integrity of assets, and to ensure the well-being of individuals and communities.

Vulnerability refers to a system, entity, or asset's susceptibility or predisposition to impairment when exposed to a hazard. Vulnerability can arise from various factors, including design flaws, lack of protective measures, or inherent weaknesses. A threat is any potential cause of an unwanted incident that may result in harm to a system or organization. It represents the active potential for a hazard to become an incident, ranging from natural disasters to human-induced actions or events. Consequences are the outcomes or effects of an incident that occurs when a threat exploits a vulnerability. The consequences can be diverse, affecting health, safety, the environment, and the economy, and can be tangible, such as financial loss, or intangible, like reputational damage.

Risk assessment is the process of systematically identifying and evaluating potential risks by analyzing threats, assessing vulnerabilities, and estimating consequences.

It is a proactive step within the broader risk management framework. Risk management, on the other hand, is the overarching process of developing, implementing, and refining strategies/measures to manage and mitigate risks. It involves creating safeguards to protect assets, reducing vulnerabilities to prevent threats from manifesting, and planning responses to minimize the consequences should an incident occur.

Together, risk assessment and risk management form a comprehensive approach to managing potential dangers, grounded in objective analysis and balanced decision-making. They are indispensable tools for informed decision-making and strategic planning across various sectors, from public safety and environmental protection to business continuity and national security.

2.6.1 Risk Assessment

Figure 2.14 illustrates the risk assessment process, outlining the key steps for systematically identifying and managing risks. The steps outlined in this figure are described in more detail below:

1. **Identify the Hazards**: The first step emphasizes the importance of recognizing potential hazards. The nature of hazards is often widely varied, and identifying them is a multidisciplinary effort, often best approached through team brainstorming. This step is crucial because any hazard not identified at this stage will remain unaddressed, potentially posing significant risks during project construction and throughout the asset's lifespan.
2. **Determine Who Might be Harmed and How**: The second step involves detailing who could be affected by the hazards identified in step 1 and how they might be harmed. Understanding the potential impact on different groups or individuals is essential for developing targeted safety measures.
3. **Evaluate the Risks and Take Precautions**: In this crucial step, the assessment considers the likelihood of each hazard occurring and the severity of its potential consequences. This evaluation is vital for prioritizing risks and deciding where best to allocate resources, for example, through engineered measures or other controls to mitigate them.
4. **Record Your Findings and Implement Them**: Once the hazards, potential impacts, and mitigation strategies have been identified, they should be meticulously documented. This documentation step is vital for implementing the safety measures. Moreover, a record must be maintained for future reference and accountability.
5. **Review Assessment and Update if Necessary**: The last step is an ongoing process where the risk assessment is regularly reviewed and updated. This ensures that any new hazards that arise due to changing conditions throughout the asset's life are identified and managed. This allows for the refinement of risk mitigation measures following their implementation.

The emphasis on thoroughness in the risk assessment process underscores that any oversight in identifying hazards can result in unmitigated risks, potentially leading to serious consequences for the design users. Hence, the risk assessment process must be comprehensive, identifying all potential hazards. Furthermore, it is critical to understand that risk assessment is an iterative process and must be continually reviewed and updated to identify new hazards arising from changing asset or environmental conditions, and to develop and implement strategies to mitigate them.

2.6.2 Hazard Analysis Versus Risk Assessment

Hazard analysis and risk assessment are two distinct but interconnected processes in the field of safety and risk management. Understanding their differences and how they complement each other is crucial for effectively managing potential dangers in any domain, from industrial operations to public health.

Hazard analysis focuses primarily on identifying any potentially dangerous conditions that may arise from a product, process, or application. It involves the analyst recognizing all possible scenarios where something could go wrong. This step does not typically involve an in-depth evaluation of the likelihood or consequences of these hazards, but rather ensures a comprehensive catalog of all scenarios that could potentially trigger a dangerous event or condition, referred to as the cause of the hazard.

Risk assessment takes the findings of hazard analysis one step further. It involves evaluating the probability (likelihood) and severity (consequences) of the occurrence of each identified hazard. Risk, in this context, is not just about the presence of a hazard, but also about the likelihood that the hazard will actually cause harm and the potential extent of that harm (exposure). Risk assessment is a more detailed process that builds upon the identified hazards to determine how likely they are to become a reality and how severe the outcomes could be. In essence, hazard analysis is the first step of the risk assessment process.

The process of risk assessment typically includes (1) enumerating the hazards and possible risk scenarios, (2) estimating the likelihood of each hazard occurring and assessing the potential impacts, and (3) comparing the estimated risks against predefined risk criteria to determine their acceptability. These are the first three steps of Fig. 2.14.

While preliminary risk levels might be proposed during the hazard analysis, it is during the risk assessment stage that these are validated and refined. Risk assessment aims to predict the risks more precisely and determine which risks are acceptable and which require intervention.

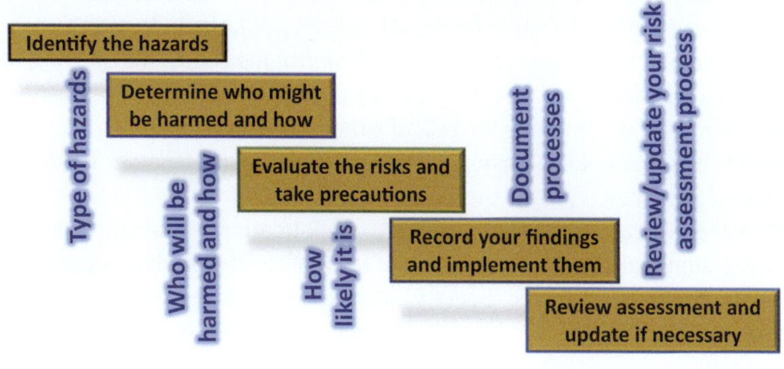

Fig. 2.14 Main steps in the risk assessment process

2.6.3 Risk Management

Figure 2.15 lays out the fundamental components of risk management, the overarching process in the systematic approach to managing uncertainty and minimizing potential harm and loss across various sectors. Each component plays a critical role in the risk management strategy.

Risk Assessment: This initial stage systematically evaluates the potential risks an organization or asset may face. This is not a one-time effort, but instead is an ongoing process that adapts as new risks emerge and existing ones evolve. For civil engineers, risk assessment might involve studying geological survey data to assess the risk of land subsidence for new constructions or analyzing traffic flow patterns when designing roadways to minimize the risk of accidents.

Risk Analysis: The identified risks are then analyzed in terms of their potential frequency and severity. This analysis takes into account both historical data and predictive modeling to estimate the impact of each risk. For example, in the context of water supply, engineers might analyze the risks of contamination, supply disruption, or infrastructure failure due to natural disasters.

Fig. 2.15 The four pillars of risk management

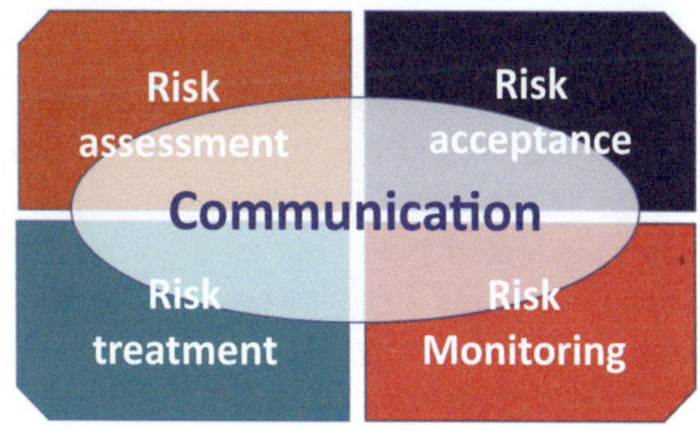

Risk Treatment: This step involves deciding on the best course of action to address the identified risks. Risk treatment options could include avoiding the risk, reducing the risk through various engineered control measures, transferring the risk (e.g., through insurance), or accepting the risk if it falls within the organization's risk tolerance. For example, treating the risk of structural failure could involve additional investments in higher-quality materials, an altered detailed design, or more stringent construction practices.

Risk Acceptance: After implementing the risk treatment strategies, an organization (or decision makers) must decide what level of residual risk is acceptable. This decision is often based on a cost–benefit analysis; for example, the cost of completely eliminating a risk might be so high that it's more reasonable to accept a certain level of risk instead.

Risk Monitoring: Risks are not static; they change as new information becomes available, as external circumstances change, or as the organization/asset itself changes. Regular risk monitoring ensures that the organization remains aware of its risk profile and can respond to changes rapidly. For instance, a civil engineering firm might regularly collect stress/strain and settlement data from a test pile installed near a recently constructed bridge and use this information to update the bridge's risk profile as conditions change.

Communication: Effective communication about risks and risk management strategies is vital to maintaining trust among stakeholders, including the public, investors, and regulatory bodies. This should involve the organization explaining the rationale behind risk assessments, the reasons for accepting certain risks, and the measures taken to mitigate others.

In civil engineering, risk management is essential not only for the success and profitability of individual projects, but also for the safety and reliability of the infrastructure upon which society depends. The process must be dynamic, adapting to new risks as they arise and to changes in the external environment, such as new regulations or changing climate patterns. It requires collaboration among specialists from various fields, from geotechnical engineers to financial analysts, and effective communication with all stakeholders involved.

The process of risk assessment is a structured, methodical approach underpinning the broader discipline of risk management. It constitutes a sequence of steps involving recognizing potential hazards, evaluating their potential impacts, and formulating strategies to address these impacts. Here is an extended discussion of each step in the risk assessment process:

Risk Identification: At the outset, risk identification is a truly multidisciplinary effort where engineers, often working in diverse teams, compile all available data, historical precedents, and experiential knowledge to create a comprehensive picture of the potential risks that an asset or organization may face. This brainstorming session lays the groundwork for identifying, scrutinizing, and cataloging every conceivable source of risk.

Risk Analysis: Following identification, each risk is dissected to understand its constituents, the likelihood of its occurrence, its potential magnitude, and the extent of its impact. This analytical phase is where risks transform from abstract possibilities to quantifiable entities. Engineers may employ statistical models, simulations, or failure mode and effects analysis to gain insights into how each risk could unfold.

Risk Prioritization: With a multitude of risks on the table, not all can be addressed with equal urgency or resources. Prioritization is an exercise in resource allocation, where risks are ranked based on their potential to inflict harm. Those that pose significant threats to project success or safety receive immediate attention, while others may be monitored or accepted as part of the project's risk profile.

Risk Causation Analysis: Understanding the root causes of risks is vital for effective mitigation. Engineers must delve into the 'why' behind each risk, tracing back through the web of project elements to isolate factors that could precipitate a risk event. This investigative work is crucial for developing targeted interventions that address the core, not just the symptoms, of project vulnerabilities.

Risk Acceptance Thresholds: Deciding which risks to accept and which to mitigate is a balancing act, often influenced by subjective judgment informed by experience and industry standards. The thresholds set for risk acceptance are pivotal, as they dictate the actions that will follow. They need to be set in alignment with the organization's risk appetite and the practicalities of the project's context.

The risk treatment component in a risk-based management program then includes three phases:

Phase I—Baseline Mitigation: In this first phase, the risk treatment process begins by establishing basic mitigation measures. These initial actions address identified risks exceeding the organization's acceptance thresholds. Baseline mitigation often involves the application of industry standards, best practices, and regulatory requirements to form a protective groundwork against risks. For example, in civil engineering, this might include the implementation of safety protocols on a construction site, the application of conservative design margins, or the introduction of early warning systems for natural hazard monitoring.

Phase II—Project Plans and Prioritizations: Building upon the groundwork laid in phase I, this phase involves more detailed planning and the prioritization of risk treatment projects. A strategic plan maps out the specific measures needed to mitigate each high-priority risk. This plan considers factors such as resource availability, project timelines, and the interdependencies between various risk treatment activities. Prioritization ensures that the most critical risks are addressed first, with a clear sequence of actions for subsequent risks. During this phase, civil engineers would work closely with project managers to sequence construction activities to minimize exposure to identified risks, such as scheduling work outside of adverse weather windows or ensuring that critical infrastructure components are tested and certified before deployment.

Phase III—Project Execution: The final phase turns plans into action. The project execution phase is where all the risk treatment strategies are implemented. This phase requires diligent oversight and coordination to ensure that each mitigation measure is executed according to plan, within the constraints of time and budget, and with the flexibility to adapt as necessary. For civil engineers, this could mean overseeing the construction of flood defenses, retrofitting buildings to withstand earthquakes, or adjusting project timelines to accommodate unforeseen ground conditions. This is a phase characterized by monitoring, adjusting, and reporting on the effectiveness of the risk treatment measures as well as making course corrections (as required) to achieve the desired reduction in the risk profile.

Risk Communication: The most vital step that holds the entire process together is communication. Clear, transparent, and ongoing dialogue about risks and their management is essential for stakeholder buy-in and support. It involves articulating complex technical issues in accessible language to ensure that decision-makers, funders, and the public understand the rationale behind risk management strategies.

Throughout this journey, risk management is not a static process but a dynamic one that adapts to new information, project progress, and the evolving risk landscape. It requires a commitment to continuous improvement, learning, and adaptation. Each step builds on the last, aiming to create a project environment where risks are known, controlled, and managed to support project success and safety.

2.6.4 Why Hazard and Risk Assessment in Civil Engineering?

The question of why hazard and risk assessment are critical in civil engineering can be addressed by considering the profound implications such assessments have on the three pillars of sustainability: society, the environment, and the economy (Chap. 1, Sect. 1.5.5). Below is an in-depth look at the significance of hazard and risk assessment in civil engineering with respect to these pillars of sustainability:

Societal Implications: Civil engineering projects are not isolated constructs; rather, they are interwoven into the fabric of society. They serve as the backbone of our daily lives, facilitating transportation, housing, water supply, waste management, and more. Ensuring the safety of these projects is paramount. Hazard and risk assessments allow engineers to predict and mitigate potential failures that could result in catastrophic consequences for the community. From the structural integrity of bridges to the safe design of buildings in earthquake-prone areas, risk assessment protects lives and assures public safety.

Environmental Considerations: The environmental impact of civil engineering projects can be significant. It is the responsibility of the engineer to minimize this impact, preserving ecosystems and ensuring the sustainability of natural resources. Hazard and risk assessments help identify and quantify the environmental risks associated with a project, such as soil erosion, water contamination, or habitat destruction. By doing so, engineers can implement design choices and construction practices that protect the environment, such as erosion control measures or preserving green spaces within urban designs (for example).

Economic Efficiency: Civil engineering projects represent substantial financial investments and have long-term economic implications for both the immediate stakeholders and the broader economy. Risk assessments help identify potential cost overruns, delays, and other financial risks that could affect the profitability and economic

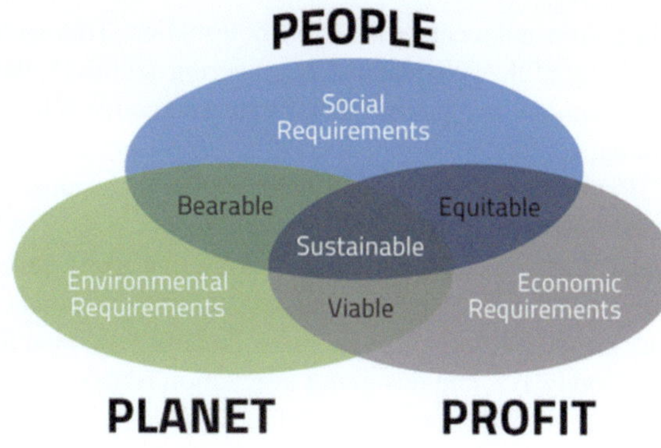

Fig. 2.16 Balancing the Triple Bottom Line: People, planet, and profit as a holistic approach to sustainability

viability of a project. By incorporating risk management into the planning and design phases, civil engineers can optimize resource allocation, reduce waste, and ensure that projects are completed within budget, thereby enhancing their economic efficiency.

Regenerative and Sustainable Future: The "Triple Bottom Line" concept extends beyond the immediate project lifecycle to consider the long-term sustainability of the built environment. By integrating social, environmental, and economic assessments, civil engineers can contribute to the creation of infrastructure that not only serves the present needs but also contributes to the health and well-being of future generations. This includes designing adaptable structures that can be repurposed or easily deconstructed, the use of renewable building materials, and planning for climate change mitigation and resilience.

Figure 2.16 represents the concept of sustainability, often referred to as the "three pillars" or "triple bottom line", which comprises People, Planet, and Profit. These three overlapping ellipses each represent one of the pillars:

- **People**: This represents the social requirements or the human capital in sustainability. It emphasizes the need for social equity, fair labor practices, community development, health and education, and the overall well-being of people.
- **Planet**: This denotes the environmental requirements or the natural capital. It focuses on the importance of environmental stewardship, conservation of natural resources, reduction of pollution, and the overall health of the planet's ecosystems.
- **Profit**: This stands for economic requirements or economic capital. It underscores the necessity of economic growth and development, profitability, and financial viability for businesses and economies.

Where the three ellipses intersect in the middle, the term "Sustainable" indicates that true sustainability is achieved only at the intersection of viable economic practices, equitable social responsibilities, and bearable environmental impact. The goal is to create a balanced approach that considers all these aspects without compromising the ability of future generations to meet their needs.

The 1997 Red River flood in Manitoba, also known as the 'Flood of the Century,' stands as a poignant example of the critical role civil engineering plays in hazard mitigation and meeting the triple bottom line. This event underscores the importance of thoughtful, well-planned engineering solutions that aim to protect people, the environment, and assets from the unpredictable forces of nature and the impacts of human activity.

Protecting People: One of the primary responsibilities of civil engineers is to safeguard human lives. The evacuation of over 27,000 residents during the Red River flood was a massive undertaking that likely saved many lives. Civil engineers are tasked with designing and constructing infrastructure that can withstand such catastrophic events. This includes creating effective flood defenses, such as levees and floodways, and designing evacuation routes and emergency management plans that can be swiftly implemented in the face of imminent danger.

Protecting the Environment: The 1997 flood affected the human population and the surrounding ecosystems. Inundation can lead to significant environmental degradation, disrupting habitats and causing long-term ecological imbalances. Civil engineers contribute to environmental protection by designing sustainable water management systems that mitigate the impact of flooding, preserve natural watercourses, and ensure the continuity of wildlife habitats.

Protecting Assets: The economic toll of the Red River flood was immense, with significant damage to homes, businesses, and infrastructure. Civil engineers strive to protect these assets by incorporating risk assessments into their designs, ensuring that structures are resilient and can either resist floodwaters or minimize damage when flooding occurs. The construction of flood-resistant materials, the elevation of critical infrastructure, and the implementation of waterproofing measures are all examples of engineering interventions aimed at asset protection.

The Red River flood in 1997 serves as a stark reminder of the power of natural events and the need for preparedness. In response to this disaster, civil engineers and policymakers have worked to improve flood prediction models, upgrade existing infrastructure, and implement new mitigation strategies to reduce the risk of future flooding events. This includes the expansion of the Red River Floodway, which was instrumental in preventing further floods of similar magnitude from causing extensive damage to Winnipeg in the years following 1997.

In essence, civil engineering is about more than just building structures; it is about creating a resilient society capable of withstanding the tests of nature and time. Through rigorous training, adherence to stringent standards, and a commitment to innovation, civil engineers are at the forefront of developing the built environment in ways that promote harmony with the natural world, protect human communities, and preserve the wealth of resources and assets that underpin our civilization.

Consider, as an example, how the Red River Floodway project sought to incorporate the 3Ps (people, planet, profit) into project planning and implementation. This initiative involves constructing an inlet gateway, a dike, and a channel outlet. It is imperative that the system's capacity is sufficient to safeguard the City of Winnipeg during major flood events, including the southern and northern suburbs that surround the city. A key focus is also to protect the region south of Winnipeg from artificial flooding caused during the operation of the control gate at the inlet structure. In addition to infrastructure concerns, environmental and social factors are also pivotal. Environmental considerations include protecting endangered species like the peregrine falcon, demonstrating a commitment to ecological stewardship. Socially, the project encompasses initiatives such as training programs for skilled workers and allocating a portion of the work to aboriginal contractors. When exploring different solutions for the floodway, their effectiveness is evaluated based on expected safety for both the city and the environment, as well as the costs associated with the project. These evaluations are crucial as various solutions offer different cost implications and environmental impacts. This comprehensive approach ensures a balanced consideration of all critical factors in establishing the floodway.

The response to the 1997 "Flood of the Century" in Manitoba illustrates the dynamic nature of civil engineering in adapting to and mitigating the risks posed by natural disasters. The critical review by the International Joint Commission (IJC) highlighted a significant vulnerability in Winnipeg's flood defense system, prompting an unprecedented collaboration between local, provincial, and federal entities to bolster the city's resilience against future floods. The historic investment of $628 million by the Federal Government and the province of Manitoba towards the expansion of the Red River Floodway was not just a reaction to a past event, but a forward-looking measure to safeguard against potential future catastrophes. This strategic expansion was designed to significantly exceed the capacity of the watershed during both the historic 1997 flood and even the record-setting flood of 1826. The increase in the channel's capacity from 1,700 cubic meters per second to 4,000 cubic meters per second was not a marginal improvement but a substantial upgrade that would reshape Winnipeg's flood preparedness. The scale of the excavation project echoes the immense importance placed on infrastructure that can withstand the most severe challenges. Ranking just behind the Panama Canal in terms of the scope of earthworks required, the expansion of the Red River Floodway stands as a testament to human ingenuity and determination in the face of natural hazards. Such an undertaking is a vivid example of how civil engineering serves as a guardian against the forces of nature, employing science, design, and collaborative governance to protect communities and maintain the function of society.

Figure 2.17 represents the lifecycle of a civil engineering project, emphasizing the core principles of ensuring the safety of people and the environment alongside the project's economic feasibility. It outlines the sequential stages of project development, from the initial idea and concept, planning and feasibility analysis, to detailed field investigations and tests. These stages precede the critical design phase, where plans are refined and readied for implementation. Following the design, the project moves into the manufacturing or construction phase, where the designs are brought to life. Once operational, the project enters a phase of ongoing operation and maintenance, ensuring continued functionality and safety. Ultimately, the lifecycle concludes with decommissioning, safely retiring the project at the end of its useful life. Each step in this process is interconnected and informed by the central considerations of safety, environmental stewardship, and cost-effectiveness, as exemplified by the Red River Floodway project.

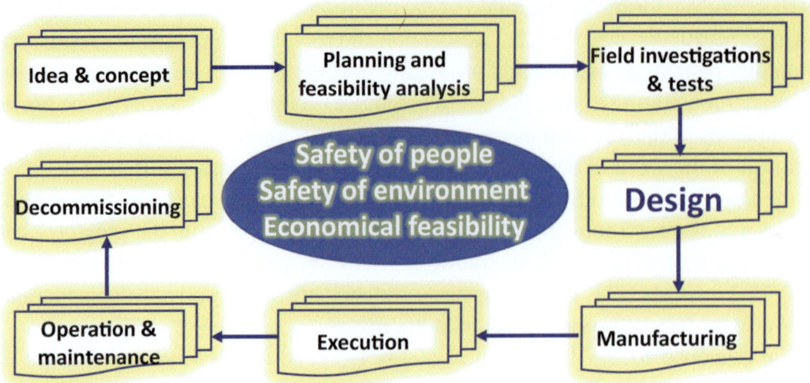

Fig. 2.17 The civil engineering project lifecycle

Sustainability has the role of protecting and maximizing the benefit of the 3Ps. The benefits resulting from the construction of the floodway can be influenced by the uncertainty of various sources, including flood discharge, water level, resistances (materials, soil,…), degradation processes, service, manufacturing costs, executing costs, maintenance, decommissioning costs, etc. These types of concerns should be addressed in the feasibility study of the project.

2.6.5 *Real-World Problems*

When tackling real-world problems, civil engineers employ modeling to simulate scenarios and perform cost–benefit analyses. These models enable decision-makers to evaluate various outcomes and make informed choices. However, it is critical to recognize that models are approximations and inherently possess certain limitations. Data quality and quantity are fundamental to the robustness of a model. Adequate and relevant data are a requirement for accurate predictions and analyses using models. However, data collection is often constrained by financial resources, and expansive, comprehensive datasets may be unattainable, especially for historical data. Consequently, models might be developed with incomplete data, leading to potential inaccuracies. Experience also plays a role in modeling. A model may perform well under conditions for which it was designed, but unforeseen events or variables outside its historical experience could lead to less reliable predictions. This subjectivity and the intrinsic variability of natural processes add layers of complexity to modeling efforts. The civil engineer must be aware of these constraints and recognize that, while models are valuable tools, they are not infallible. Engineers must integrate a margin for uncertainty into their decision-making processes and risk assessments. This involves considering uncertainties throughout the lifecycle of a project—from planning and design, to operation and maintenance, and finally in decommissioning. By acknowledging these uncertainties, engineers can plan for contingencies and develop engineering solutions that have greater resiliency and adaptability.

Understanding and quantifying risk is a multifaceted process in civil engineering that requires assessing various components, such as likelihood, consequences, and the system's reliability. Each component plays a crucial role in determining the overall risk associated with a project or asset. Consequences are the effects of an event on objectives, encompassing immediate to long-term impacts. These can be direct, like repair costs, or indirect, like the socio-economic effects of infrastructure downtime. The interaction between these direct and indirect consequences can amplify an event's total impact. Likelihood, on the other hand, measures the probability of an event occurring. It can be gauged subjectively or objectively using qualitative descriptions or quantitative metrics like probabilities. The art and science of civil engineering involve assessing this likelihood against the backdrop of uncertainty inherent in natural and man-made systems. The interplay between the probability of an event (likelihood) and its potential impacts (consequences) is where risk calculation takes center stage. Risk concerns balancing the expected positive and negative outcomes that stem from an activity, such as constructing or operating an infrastructure asset. In the context of risk assessment, a number of relationships exist which consider the various risk subcomponents:

1. $R_i = f_3$ (E, C)—Here, risk is a function of exposure (E) and consequences (C), capturing how the magnitude of exposure to a hazard translates into potential outcomes.
2. $R_i = f_4$ (Rel, C)—This equation links risk with reliability (Rel) and consequences (C), focusing on how system reliability influences the expected results of potential hazards.
3. $R_i = f_5$ (L, C)—This formulation relates risk to likelihood (L) and consequences (C), indicating how the probability of an event aligns with its potential impacts.

As engineers, the goal is to ensure that a balance is found, allowing infrastructure to provide the expected level of service throughout its lifespan while minimizing the risk of adverse events. This balance hinges on making informed decisions based on both the consequences of potential issues and the likelihood of their occurrence. By considering these fundamental risk components, civil engineers can craft strategies that not only mitigate risk but also optimize the safety, functionality, and sustainability of the built environment.

Figure 2.18 illustrates the complex interplay between various elements that contribute to risk assessment. It starts with the concept of vulnerability and capacity, which are intrinsic qualities that determine how robust a system or structure may be. These qualities contribute to determining the system's overall reliability, or its ability to withstand threats, hazards, and demands, which are external factors that challenge the system. Reliability is further influenced by exposure, which captures the degree to which the system is subjected to potential dangers or stressors. Central to the diagram is risk, which emerges from the combination of exposure to hazards and the potential for negative consequences or impact. The diagram also acknowledges that

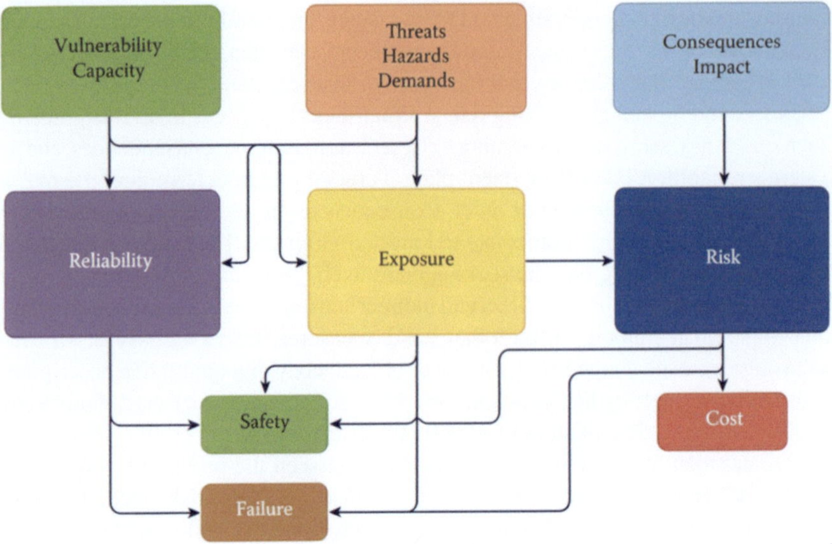

Fig. 2.18 Interconnected framework of risk: Assessing the dynamics of vulnerability, reliability, and exposure

with risk comes the potential for failure, which is the breakdown or malfunction of the system, and this, in turn, threatens safety, a state where risk is controlled or mitigated. Lastly, cost is depicted as an outcome of risk, representing the financial implications associated with both preventive measures to ensure safety and the consequences of failure. The figure encapsulates the notion that risk is not a standalone entity but a result of multiple dynamic factors that interact with one another, requiring careful consideration and management in any system or structure.

2.7 Risk-Based Decision Analysis

Figure 2.19 outlines a six-step workflow for conducting risk-based decision analysis. The process begins with the *Planning* phase, where the scope and objectives of the risk analysis are established, setting the stage for all subsequent activities. This leads to the *Structure* phase, which involves creating a detailed representation of the system to understand its various components and their interrelationships. Next, in the *Identifying Hazards* phase, potential issues within the system are pinpointed, along with how these might occur and methods for their control. This phase is followed by a thorough *Risk Assessment*, which involves two stages: first, an *Analysis of Probability and Consequences* is conducted to determine the likelihood of each hazard and its potential impact; second, *Risk Scenarios* are identified, and *Sensitivity Analyses* are performed to understand how changes in conditions could affect risks. After

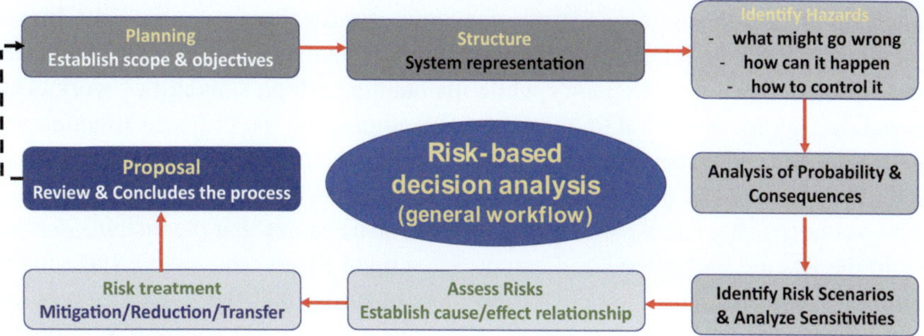

Fig. 2.19 Generalized workflow for risk-based decision analysis: A comprehensive approach to identifying, assessing, and managing risks

assessing the risks, the *Risk Treatment* phase is initiated, focusing on mitigating, reducing, or transferring the identified risks through strategic measures. Finally, the *Proposal* stage involves reviewing the entire process and providing recommendations to manage the risks effectively. This generalized workflow is designed to be iterative, suggesting that the process may loop back to earlier stages as new information emerges or as situations evolve. Each step involves a series of tasks and sub-tasks, ensuring a comprehensive and detailed approach to risk management. The workflow is flexible enough to apply to various contexts, whether documenting acceptable levels of risk or providing a foundation for management decisions. The details of each phase will be presented in the following sub-sections.

2.7.1 Step 1: Planning

Risk analysis (RA) in civil engineering is an essential process that aids in making informed decisions by evaluating the potential risks associated with projects. The process begins with defining the context of the problem, which is accomplished by setting the boundaries and objectives of the RA. This foundational step dictates the direction and scope of the entire analysis and determines the end goal of the RA, as well as how its outcomes will be utilized. Key to this process is the identification of stakeholders, both internal and external, who may contribute valuable data and insights during the consultation process that can shape the outcome of the RA process. Choosing the correct methodology for RA is critical as various approaches exist, each with its own applications, limitations, and levels of precision. The decision regarding which methodology to employ depends on the objectives of the RA, the available information, and the specific project at hand. An appropriately selected RA methodology ensures that the analysis is both accurate and relevant to the project's needs.

Gathering and processing information is a fundamental component of RA. Adequate and relevant information is indispensable, contributing to both an improved understanding of the project and a more accurate assessment of potential risks. Implementing an RA schedule helps to keep the analysis on track, guiding the RA team to meet project deadlines efficiently while maintaining a high standard of work. This structured approach is vital to ensure the RA aligns with the project's timeline and delivers results when needed. Finally, it is essential to communicate the results to all interested parties and confirm that they were engaged throughout the RA process. This not only fosters transparency and trust but also ensures that the insights derived from the RA are understood and can be acted upon by all who have a stake in the project's outcome. This step closes the loop, ensuring that the RA process is not just an academic exercise but instead a practical tool that supports risk-informed decision-making in civil engineering.

Table 2.1 describes some of the potential pitfalls of poor project planning and outlines the expected problems that could arise from each. In the first task, the scope may suffer from being either too narrow, potentially excluding some hazards or subsystems from consideration. The scope could also be too broadly defined, which could result in a superficial understanding of the risks (as well as falling behind on the RA schedule and budget). The objectives of the RA must be clearly defined; otherwise, they may be challenging to understand and implement. Issues with stakeholders and other interested parties involve the risk of poor engagement, leading to resistance and negative attitudes towards the project. Choosing the wrong RA methodology could neglect important aspects such as organizational activities and human factors, or it might be resource-intensive or superficial. Problems with the analysis team, such as non-attendance or dysfunction, could result in a lack of necessary information, making it difficult to reach informed decisions and causing a lack of motivation among team members. Gathering information is crucial, and any gaps or changes in the system information could lead to serious mistakes in the risk analysis, including the under- or overestimation of risks. Finally, a poorly managed schedule can lead to stressful situations characterized by a lack of time, a poor team atmosphere, and a decrease in creativity, all of which are detrimental to the planning process.

2.7.2 Step 2: System Definition (Structure)

The second step in risk analysis is to meticulously define the system that is under scrutiny. This involves breaking down the system into functional blocks, which is a method of compartmentalizing a complex system into more manageable sub-systems or processes. This division is crucial as it allows for a more focused analysis of each component's role and potential risks in the context of the entire system by the RA team.

To ensure a comprehensive RA, it is essential to include all critical elements that could influence the system's future performance and outcomes. This inclusion should be exhaustive to prevent the oversight of potential risks associated with any component or function of the system.

Table 2.1 Anticipating challenges in risk analysis: potential deviations and their consequences during the planning stage

Task	Deviation	Expected problems
Scope	Too narrow	Some hazards or subsystems could not be included
	Too broad	Could lead to superficial results
Objectives	Unclear aim	Could it be challenging to understand and implement
Interested parties	Poorly rooted	Active or passive resistance (internal and/or external)
		Negative attitudes (e. g. lack of collaboration)
Choose methodology	Inappropriate	Could neglect certain aspects (e.g., organizational activities, human factors)
		Requires excessive resources or time
		Could it be too superficial?
Choose analysis team	Failure to attend	Missing information makes it difficult to reach decisions
	Team does not function	Lack of motivation
Gathering information	Information Unavailable	Certain parts of the object cannot be analyzed
	Incomplete information	Risk for serious mistakes in the analysis. Under- or overestimation of risk
	Changes in the system	Does not help hazard identification. Risk of serious mistakes in the analysis
Schedule	Stressful situations	Lack of time, poor team atmosphere, poor creativity

Identifying primary areas of concern is also pivotal in this step. The analysis can prioritize where to allocate resources and attention by pinpointing these areas. It guides the risk assessment process toward the most significant aspects that could impact the system's safety, functionality, and reliability.

The division of the system into functional blocks should be logical and systematic, aligning with the context and scope defined in the first step. Each block should be clearly defined, with boundaries that help to understand its interactions with other blocks and the system as a whole. To validate this division, the following questions should be considered:

- Was the system appropriately divided into functional blocks that reflect its operational reality?
- Have all important elements, including those that may seem peripheral, but could impact the system under certain conditions, been included in the analysis?
- Have the primary areas of concern been correctly identified and defined, based on their potential impact on the system's performance and risk profile?

Table 2.2 Striking the right balance: challenges in system division for effective risk assessment

Task	Deviation	Expected problems
Analysis division	Object too narrowly defined	Incomplete analysis (serious risks can be missed)
	Parts excluded	Incomplete analysis (serious risks can be missed)
	Division into sections is too crude	Incomplete and superficial analysis
	Too detailed	Could the RA process become complex and take too much time?

By considering these questions during the division of the project, it can be ensured that the system definition is robust and all-encompassing, laying a solid foundation for the subsequent steps in the risk analysis process.

Table 2.2 outlines potential issues in the division stage of risk assessment (RA), focusing on how the definition and division of the project can affect the completeness and efficiency of the RA process. If the object of the analysis is defined too narrowly or certain parts are excluded, there's a risk of an incomplete analysis, leading to potentially serious risks being missed. Conversely, the analysis may be incomplete and superficial if the division is too crude. On the other hand, an incredibly detailed division could overcomplicate the RA process, resulting in it becoming time-consuming and challenging to manage. This highlights the delicate balance needed in the system definition stage to ensure a thorough yet efficient RA process. The division must be detailed enough to capture all relevant parts and potential risks without becoming so complex that it hampers the analysis.

2.7.3 Step 3: Hazard Identification

In the third step of the RA process, the focus shifts to hazard identification. Hazards are the inherent properties or conditions within a risk source that have the potential to cause harm or adverse effects. This step is pivotal as it directly informs the scope and focus of the risk management strategies that will be employed later. The identification of hazards is a detailed process that typically involves three key actions:

1. **Decomposition of the System**: This action entails breaking down the system into its constituent components and subsystems. By doing so, each element can be scrutinized individually, which is crucial for a thorough understanding of the system's complexity and for uncovering hidden risks that might not be apparent when considering the system as a whole.
2. **Identification of Failure States**: Once the system is decomposed, the potential failure states for each component and subsystem should be pinpointed. This involves considering the various ways in which each component could fail and

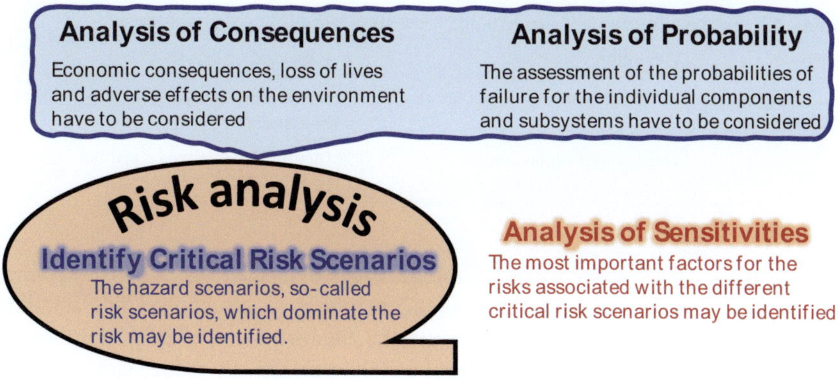

Fig. 2.20 Comprehensive risk analysis process: From consequence and probability assessment to critical risk identification and sensitivity analysis

the hazards associated with these failures. Historical data, failure databases, and expert knowledge play a crucial role in identifying these failure states.

3. **Realization of Hazards**: After identifying potential failure states, the analysis seeks to understand how these failures might manifest as hazards. This involves developing failure scenarios that map out the chain of events that could lead from a component failure to system-level consequences. It is essential to consider common-cause failures during this step—situations where a single cause could lead to multiple component failures, significantly increasing the risk.

Throughout this process, it is critical to ensure that the identification of hazards is comprehensive, considering all possible sources of risk, including technical, human, and external factors. The thoroughness of this step sets the stage for effective risk assessment and management, allowing for the development of strategies that are both preventive and responsive to the full risk profile of the system in question.

Figure 2.20 illustrates the integral components of RA within a project or system assessment framework. It encapsulates four key aspects:

1. **Analysis of Consequences**: This step involves evaluating the potential outcomes of risk scenarios, focusing on economic impacts, the potential loss of life, and environmental damage. These consequences align with those outlined in the acceptance criteria of the project, ensuring that the risk analysis is consistent with the project's defined tolerances.
2. **Analysis of Probability**: This part of the process assesses how likely it is that any given component or subsystem will fail. By understanding these probabilities, the risk analysis can account for the likelihood of different scenarios occurring.
3. **Identify Critical Risk Scenarios**: The most significant risk scenarios are identified by analyzing consequences and probabilities. These scenarios present the highest risk due to their potential impact and probability of occurrence. Ranking these scenarios by their contribution to the overall risk helps prioritize them for further action.

Table 2.3 The consequences of overlooking or omitting hazards in risk assessment

Task	Deviation	Expected problems
Identify hazards	Some parts of the object are neglected	Incomplete analysis (serious risks can be missed)
	Deliberated omissions	Incomplete analysis (serious risks can be missed)

4. **Analysis of Sensitivities**: The final step is a sensitivity analysis, which determines the most influential factors to the identified critical risk scenarios. Understanding these factors helps in planning effective risk treatment strategies by highlighting where control or mitigation measures can most effectively reduce risk.

The process depicted is cyclic, implying that risk analysis is an iterative process. Once critical risk scenarios and their sensitivities are identified, this information feeds back into refining the analyses of consequences and probabilities, thus improving the understanding of risks and enhancing the decision-making process for risk management.

Table 2.3 discusses some of the potential shortcomings that may occur during the hazard identification step of RA, emphasizing the impact of such deviations. Two types of deviations are highlighted: these are the neglect and omission of some of the project components, both of which can lead to an incomplete analysis and the possibility of missing serious risks. Both deviations underscore the critical nature of thoroughness in the hazard identification process. Incomplete analysis, whether due to oversight or deliberate omissions, can have significant implications for the safety and integrity of the system being evaluated. This step is foundational, as undetected hazards cannot be mitigated and may lead to unforeseen consequences.

2.7.4 Step 4. Risk Assessment

In this critical phase of risk assessment, where risk acceptability is gauged, the analysis must pivot from identification to characterization and judgment. This involves a multi-faceted approach that seeks not only to map out potential scenarios and their likelihood, but also to understand the depth of their impact and the effectiveness of existing safety measures. An overview of the process is described below:

1. **Risk Characterization**: This involves a detailed understanding of the risk scenarios under consideration. Each hazard is paired with its possible outcomes to characterize the risk associated its potential to cause harm. This step often requires a thorough review of historical data, theoretical models, and expert judgment to outline the risk landscape accurately.
2. **Cause/Effect Relationships**: The next step is delineating the causal pathways linking each identified hazard to its potential consequences. This task is vital

for risk mitigation as it allows for the identification of intervention points where safety measures can be most effectively applied.

3. **Risk Estimation**: Evaluating risk estimations involves scrutinizing the probability of occurrence and the severity of consequences. It often includes quantitative methods such as fault tree analysis or qualitative assessments based on expert opinions. The reliability and validity of the data used for these estimations are critical for assuring the accuracy of the risk assessment.
4. **Identification of Safety Barriers**: Safety barriers are safeguards designed to either prevent a hazardous event or mitigate its effects, should they occur. Identifying these barriers includes cataloging existing preventative, mitigative, and emergency response measures as well as evaluating their adequacy and effectiveness.
5. **Decision on Risk Tolerability**: Ultimately, the process culminates in a decision on whether the level of risk is tolerable. This decision is informed by comparing the risk level against established acceptance criteria, which may be based on legal standards, industry best practices, or specific organizational risk thresholds.

The overarching goal is to ensure that the risk level is within a range that is deemed acceptable by the stakeholders and aligns with both regulatory requirements and societal expectations. This phase underscores the need for a strategic approach to risk management, one that balances technical feasibility with ethical considerations and public acceptability. The decisions made here will direct subsequent actions in risk treatment and influence the long-term safety and sustainability of the engineering project.

Table 2.4 shows potential issues that may arise during the risk assessment phase, highlighting how subjective judgments can lead to significant deviations. Overestimation may result in minor risks being given undue attention, obscuring more serious risks within a mass of less critical issues. Underestimation, on the other hand, can lead to significant risks being overlooked or dismissed as inconsequential. Additionally, failing to define clear assessment criteria can yield results that are too vague to be actionable, preventing the effective prioritization and treatment of risks. The risk of subjective assessment underscores the importance of objectivity and the establishment of clear, standardized criteria in the risk assessment process. Without this, there is a danger that the assessment will not accurately reflect the true risk landscape, potentially leading to serious oversights or misallocation of resources.

Table 2.4 Subjectivity in risk assessment: the perils of overestimation, underestimation, and vague criteria

Task	Deviation	Expected problems
Assessment of risks	Overestimation	Serious risks could get lost in a mass of minor risks
	Underestimation	Important risks could "disappear."
	Criteria not clearly defined	Results too vague

2.7.5 Step 5. Risk Treatment

Formulating safety barriers is an essential approach in risk management for ensuring the integrity of civil engineering projects. Four approaches to developing and proposing safety barriers are described below:

1. **Risk Mitigation**: This approach aims to eliminate the possibility of a hazard occurring by altering the system design or operational parameters. For example, structural failure risk due to concrete corrosion can be mitigated using corrosion-resistant materials like fiber-reinforced polymers (FRP) instead of traditional steel reinforcement. This proactive stance on risk ensures the long-term durability of structures and minimizes the need for frequent maintenance.
2. **Risk Reduction**: Unlike mitigation, which seeks to nullify the hazard, risk reduction strategies aim to lessen the likelihood of occurrence or minimize the potential impact. This could involve incorporating redundant systems, enhancing existing safety features, or adopting newer, safer technologies. For instance, the risk of fatigue failure in welded joints can be reduced by employing advanced welding techniques or opting for bolted connections that distribute stress more evenly and allow for easier inspection and maintenance.
3. **Risk Transfer**: Here, the responsibility and potential financial burden of a risk are shifted to a third party, typically through insurance or contractual agreements. This strategy is often employed when the risk cannot be fully mitigated or reduced due to technological or financial constraints. Transferring risk requires a comprehensive understanding of the legal and financial aspects of the agreements to ensure that the third party adequately covers the scope of the potential hazard.
4. **Risk Acceptance**: Sometimes, after evaluating all other options, a decision may be made to accept certain risks. This is often the case with residual risks that are deemed low in probability or impact. However, acceptance is rarely an option when the safety of the public is at stake. When risks are accepted, they are closely monitored, and contingency plans are put in place to respond effectively should the risk materialize.

Each of these approaches requires thorough analysis and careful consideration of the project's context, the potential hazards, and the acceptability of risks. Engineers must balance the cost of implementing these safety barriers against the potential costs of inaction, including damage to infrastructure, environmental impact, and harm to human life. The selection of the most appropriate risk treatment method hinges on a combination of factors, including, but not limited to, the nature of the hazard, available technologies, regulatory requirements, societal expectations, and the overall risk management strategy of the organization.

Table 2.5 presents issues that can occur during the risk treatment phase, specifically with regard to the tasks of making and revising proposals for safety measures. Deviations in making proposals, such as lacking substance or being inadequate, lead to proposals with an insufficient basis or ineffective safety barriers. When revising proposals, excluding good proposals could result in choosing cost-prohibitive

Table 2.5 Challenges in risk treatment

Task	Deviation	Expected problems
Making proposals	Lacking substance	Insufficient basis for proposals
	Inadequate	Ineffectual safety barriers proposed
Revising proposals	Good proposals excluded	Too expensive implies too much exchange
	Poor recommendations	Take resources from better safety barriers

options or unnecessary complexity. Meanwhile, poor recommendations might divert resources away from more effective safety barriers. These problems underscore the need for a thorough and balanced approach to proposing and revising risk treatment strategies.

2.7.6 Step 6. Summary (Proposal)

The culmination of the risk analysis process is encapsulated in the summary or proposal step, which not only wraps up the findings, but also sets the stage for future actions and reviews. This conclusive phase involves several key tasks:

1. **Report Risky Processes**: A comprehensive report must be compiled detailing all identified risky processes. This report should be structured to clarify the risks, their origins, and the circumstances under which they could potentially lead to adverse outcomes. The documented processes should be transparent and traceable for future reference and audits.
2. **Propose Safety Barriers**: The report should clearly outline recommendations for safety barriers designed to mitigate identified risks. The proposal should indicate not only the types of safety barriers required, but also the parties responsible for their implementation and oversight.
3. **Inform Stakeholders**: Disseminating the findings of the risk analysis is crucial. All stakeholders, including project team members, clients, regulatory bodies, and potentially affected community members, should be informed of the risks and the proposed measures to address them.
4. **Emergency Response Evaluation**: The proposal should assess the current state of emergency response plans. It must scrutinize the adequacy and effectiveness of these plans in addressing potential disasters, ensuring that quick and efficient action can be taken if necessary.
5. **Identify Areas for Further Investigation**: The risk analysis might reveal areas that need more in-depth study. These could include environmental assessments, ergonomic evaluations, or detailed safety inspections. These areas should be highlighted for future research and investigation.
6. **Future Planning**: The report should address current concerns and include provisions for continuous monitoring and updating of the risk analysis. This ensures

that the system or project remains within the accepted risk thresholds throughout its lifecycle, adapting to new information or environmental changes.

The report generated in this final step presents a record of the entirety of the RA process as well as a blueprint for ongoing risk management. It must be thorough, actionable, and adaptable, reflecting the dynamic nature of risk and the necessity for vigilance in civil engineering projects.

Table 2.6 delineates several pitfalls that can occur during the various stages of risk management, particularly in reporting, decision-making on safety barriers, their implementation, and information dissemination.

1. **Reporting**: Incompleteness or poor readability can lead to misunderstandings or a negative perception of the project, potentially resulting in proposals being rejected.

Table 2.6 Critical fault lines in risk management: from reporting to safety barrier implementation and team dynamics

Task	Deviation	Expected problems
Reporting	Incomplete	A poor impression is given. Proposals rejected
	Hard to read	Misunderstanding. Proposals rejected
Safety barriers decision	Proposals rejected	Hazard remains
	Poor proposals accepted	Take resources from better safety barriers
	Decision unclear	Uncertainty in implementation. Delays
Implementation of safety barriers	Inadequate follow-up	Not carried through
	Lack of resources	Lack of skilled people or inadequate financial resources
	Unclear responsibility	Was not defined or due to an unclear decision
	In the branch of decision	Inadequate time
	Implementation failure	Inadequate description of safety barriers
		Inadequate skills
		System changed, but safety barriers were not adopted
Information	Insufficient spread	Ignorance of the risks
		Delay in implementation
Analysis team	Failure to attend	Information is lacking at meetings, and it is difficult to reach decisions
	Team does not function	Inappropriate composition, lack of motivation

2. **Safety Barrier Decision**: This phase is fraught with risks such as the rejection of crucial proposals, acceptance of subpar safety measures, indecisiveness, and insufficient follow-through on decisions made.
3. **Safety Barrier Implementation**: Challenges here include a lack of resources, unclear responsibilities, inadequate time allocation, and failures in the implementation process itself.
4. **Information**: Insufficient distribution of critical information can lead to delayed implementation and a lack of awareness among stakeholders.
5. **Analysis Team**: Non-attendance or dysfunctional team dynamics can result in a lack of necessary information, making decision-making difficult and potentially jeopardizing the project's success.

Each of these potential issues underscores the need for a clear, well-structured approach to risk management, with detailed, readable reporting, decisive and well-supported safety barrier decisions, effective implementation plans, thorough information dissemination, and a functional analysis team.

2.8 Sources of Risk in Engineering

In engineering, risks arise from various sources, each with its own set of unique challenges and implications. Understanding these sources is critical for effective risk management and mitigation strategies. The primary sources of risk in engineering projects include:

1. **Natural Hazards**: These are risks stemming from natural events like floods, earthquakes, hurricanes, and other weather-related phenomena. The unpredictability and often immense power of these natural occurrences make them significant risk factors in engineering projects. For instance, designing a floodway requires accounting for the potential of extreme weather events and their impact on the structure's integrity and functionality.
2. **Technical Failures**: These risks are associated with the failure of technical systems or components. They can be due to design flaws, material defects, or inadequate maintenance practices. In engineering, ensuring that designs adhere to the highest standards, use quality materials, and implement rigorous testing and maintenance routines are vital practices that can help mitigate these risks.
3. **Operational Errors**: Human factors play a crucial role in operational risks. These can include errors in judgment, lapses in safety protocols, or mistakes in machine or system operations. Training, effective operational procedures, and a strong safety culture are essential to minimize these risks.
4. **Malevolence**: This category encompasses risks due to intentional harmful actions, such as sabotage, terrorism, or vandalism. While less frequent, the potential impact of such actions can be severe. Security measures, surveillance, and contingency planning are critical to address these risks.

Each of these risk sources requires a tailored approach to risk assessment and management. For instance, managing natural hazards may involve incorporating resilience and adaptability into designs, while mitigating technical failures might focus on rigorous quality control and testing. A discussion of the approaches commonly followed for each source of risk is provided in the following subsections.

2.8.1 Risks Due to Natural Hazards

As the second-largest country in the world by geography, Canada's vast and varied landscape exposes it to a wide array of natural hazards. The country's diverse geography and climate make it susceptible to numerous natural disasters, including floods, droughts, ice storms, tornadoes, hail, wildfires, freeze–thaw cycles, hurricanes, earthquakes, tsunamis, and landslides. This wide spectrum of hazards presents a significant challenge in terms of disaster preparedness and risk management.

The impact of climate change has been particularly evident in the increasing frequency and severity of these weather-related events. This trend is not only a significant concern for the safety and well-being of the Canadian population but also has substantial financial implications. The growing cost of severe weather events is a testament to the changing climate's tangible impacts. In 2021, the financial toll of insured damage from severe weather events in Canada reached $2.1 billion, as reported by Catastrophe Indices and Quantification Inc. (CatIQ). This figure highlights the economic strain that natural disasters can impose, affecting both insurers and taxpayers.

Notable events contributing to this substantial cost include the devastating November floods in British Columbia, which caused widespread destruction and displacement, and the destructive summer hailstorms in Calgary. These events are just a few examples of the diverse and impactful weather phenomena that Canada faces.

On a global scale, the financial impact of natural disasters is even more staggering. Munich Reestimated (Munich Re. 2022) that global insured losses from natural disasters in 2021 reached approximately USD $120 billion, making it one of the costliest years on record for the insurance sector. This global perspective underscores the widespread and significant impact of natural disasters, driven partly by climate change, emphasizing the need for robust disaster preparedness and effective risk management strategies in Canada and worldwide.

2.8.2 Risks Due to Technical Failures

Risks due to technical failures are a significant concern in various fields, including engineering, manufacturing, and technology. These failures can have far-reaching consequences, from financial losses to safety hazards. In the field of civil engineering, technical failures can lead to catastrophic events with widespread consequences. Some notable examples specifically related to civil engineering are described below:

1. **Johnstown Flood (1889)**: A catastrophic failure of the South Fork Dam led to the Johnstown Flood in Pennsylvania, causing the death of over 2,200 people. Poor maintenance and alterations to the dam's structure were identified as key factors contributing to its failure.
2. **Quebec Bridge Collapse (1907 and 1916)**: The Quebec Bridge suffered two collapses during its construction due to design flaws and poor project management. The initial collapse in 1907, caused by an underestimation of the load and inadequate design, resulted in 75 deaths. The second collapse in 1916, caused by a hoisting device failure, killed 13 workers.
3. **Vajont Dam Disaster (1963)**: In Italy, a massive landslide caused a reservoir to overtop the Vajont Dam, resulting in a mega-tsunami that killed around 2,000 people. While the dam itself remained intact, the disaster was due to a failure to recognize and address the geological instability of the region during the dam's construction and operation.
4. **Silver Bridge Collapse (1967)**: The collapse of the Silver Bridge over the Ohio River, which resulted in 46 deaths, was attributed to the failure of an eye-bar in the suspension chain resulting from a small defect. This incident highlighted the importance of regular and thorough inspection of aging infrastructure.
5. **Hyatt Regency Walkway Collapse (1981)**: In one of the deadliest structural failures in U.S. history, two walkways collapsed at the Hyatt Regency Hotel in Kansas City, Missouri, killing 114 people and injuring over 200. The collapse was due to a design flaw in the walkways' suspension rods, which could not bear the weight of the crowded walkways.
6. **Genoa Bridge Collapse (Morandi Bridge, 2018)**: The collapse of the Morandi Bridge in Genoa, Italy, resulted in 43 deaths. The failure was attributed to the deterioration of concrete and the corrosion of the steel cables within the stays due to environmental factors. This incident raised questions about the durability of similar bridge structures worldwide.

These incidents in civil engineering demonstrate the critical importance of meticulous design, rigorous safety standards, continuous maintenance, and the consideration of environmental and contextual factors in construction and infrastructure projects. They serve as stark reminders of the potential consequences of technical failures in this field.

2.8.3 Risks Due to Operational Errors

The Algo Centre Mall roof collapse in Elliot Lake, Ontario, is indeed a tragic example of a civil engineering failure with devastating consequences. The incident highlights critical issues in structural integrity, maintenance, and safety standards in civil engineering.

Elliot Lake, once a thriving mining community due to the discovery of uranium in the area, experienced significant growth during the 1960s as Canada invested heavily in the development of its nuclear industry. However, the closure of the uranium mine in 1990 led to a decline in the town's population, transitioning it largely into a retirement community. The Algo Centre Mall was a central hub for this small community.

On June 23, 2012, a significant portion of the mall's rooftop parking deck, measuring approximately 12 m by 24 m, catastrophically collapsed. This collapse caused metal and concrete debris to fall through two floors of the shopping center. The disaster tragically resulted in the loss of two lives and injured more than 20 individuals. The collapse not only brought immense grief and disruption to the community but also raised serious questions about the structural safety and maintenance of public buildings.

Investigations into the collapse revealed a history of water infiltration and corrosion issues that were inadequately addressed. The failure pointed to lapses in maintenance, inspection, and accountability, underscoring the importance of rigorous structural health monitoring, especially in aging infrastructure. This event served as a somber reminder of the critical need for ongoing maintenance and safety checks in civil engineering structures to prevent such tragedies.

The Algo Centre Mall collapse is a case study of the potential consequences of neglecting structural integrity and maintenance, serving as a cautionary tale for civil engineers, architects, and builders about their vital responsibility in ensuring their structures' safety and durability.

2.8.4 Risks Due to Malevolence

Risks due to malevolence in civil engineering involve intentional acts that aim to cause harm or destruction. These can include terrorism, sabotage, vandalism, or other criminal activities. Such risks pose significant challenges, as they are often unpredictable and can have devastating effects on infrastructure and public safety. Some examples highlighting risks due to malevolence in the context of civil engineering include:

1. **Oklahoma City Bombing (1995)**: The bombing of the Alfred P. Murrah Federal Building in Oklahoma City was a result of a domestic terrorist attack. The explosion caused a significant portion of the building to collapse, killing 168 people. This incident highlighted the vulnerability of public buildings to malevolent acts and led to increased security measures in government buildings.
2. **World Trade Center Attacks (2001)**: Perhaps the most famous example of malevolence impacting civil engineering structures is the terrorist attacks on the World Trade Center in New York City. The deliberate crashing of airliners into the Twin Towers led to their collapse, causing massive loss of life and property. This

tragic event reshaped the approach to building design, particularly concerning structural robustness and emergency evacuation procedures.

3. **London Underground Bombings (2005)**: The coordinated terrorist bombings on London's public transport system, including the Underground trains, resulted in 52 deaths and highlighted the vulnerabilities in public transport infrastructure to malevolent acts. This led to increased security measures and emergency response planning in public transport systems worldwide.
4. **Cyberattacks on Infrastructure**: In recent years, cyberattacks have emerged as a significant malevolent risk in civil engineering, particularly concerning critical infrastructure like power grids, water treatment facilities, and transportation systems. For instance, the Stuxnet computer worm, discovered in 2010, was designed to damage Iran's nuclear program and highlighted the potential for cyberattacks to cause physical damage to infrastructure.
5. **Destruction of Cultural Heritage Sites**: Malevolent destruction of cultural heritage sites, often during armed conflicts, is another form of risk. An example is the destruction of the ancient city of Palmyra in Syria by ISIS, which not only had historical and cultural implications but also highlighted the need for protecting such sites from intentional destruction.

These examples underscore the importance of considering security and resilience in civil engineering projects, particularly for high-profile or critical infrastructure. The potential for malevolent acts requires that engineers and planners incorporate robust security measures, including physical barriers, surveillance systems, and cybersecurity protocols, into their designs. Additionally, emergency response and recovery plans are crucial for mitigating the impacts of such malevolent actions.

The concepts and approaches presented in this chapter build upon a broad body of international research and standards addressing resilience, hazard characterization, and risk-informed decision making. Together, these works provide a unified foundation for understanding how infrastructure systems respond to natural and anthropogenic hazards, how uncertainty should be incorporated into engineering analysis, and how risk-based methodologies support sustainable and resilient design across the infrastructure life cycle. They collectively inform the definitions of hazard, vulnerability, exposure, and resilience, guide the development of probabilistic assessment frameworks, and support the integration of climate change considerations into planning, design, operation, and policy formulation in civil engineering practice (Abbott, 2023; Bruneau et al., 2003; Canadian Institute for Climate Choices, 2021; Ebtehaj et al., (2020a, 2020b, 2021; Ebtehaj & Bonakdari, 2022a, 2022b, 2024; Gholami et al., 2020; Hallegatte et al., 2019; Intergovernmental Panel on Climate Change, 2019, 2022; International Electrotechnical Commission and International Organization for Standardization, 2019; International Organization for Standardization, 2009, 2015, 2015, 2018; Kaplan & Garrick, 1981; Kramer, 2014; Leopold, 1968; National Institute of Standards & Technology, 2016, 2024; U.S. Department of Agriculture, Natural Resources Conservation Service, 2021; U.S. Geological Survey, 2016; U.S. Global Change Research Program, 2023; United Nations General Assembly, 2016;

United Nations Office for Disaster Risk Reduction, 2015; Ward et al., 2020; Walton et al., 2019).

2.9 Summary

Civil engineering functions at the interface between natural processes, the built environment, and human society, which makes infrastructure inherently exposed to environmental uncertainty. As urbanization expands and infrastructure systems become more interconnected, the consequences of disturbances are no longer isolated but propagate across multiple sectors. Climate change, aging assets, and technological interdependencies are therefore transforming traditional engineering challenges into risk management problems. Instead of focusing solely on structural capacity, modern engineering must anticipate failure scenarios through systematic hazard identification and probabilistic assessment. The chapter introduces the idea that infrastructure reliability depends not only on design strength but also on exposure, vulnerability, and the value of the elements at risk.

A major portion of the chapter examines how climate change alters environmental loading conditions. Rising temperatures, changing precipitation patterns, and increasing extreme weather events reduce safety margins that were assumed during original design. Coastal infrastructure faces sea-level rise, erosion, and storm surge intrusion, while water and sewer systems experience capacity exceedance due to intense rainfall. Transportation systems such as railways, bridges, and airports suffer from thermal stresses, flooding, and scour, leading to operational disruptions. Dams and reservoirs must operate under altered hydrological regimes, and telecommunication networks are increasingly vulnerable due to dependence on power and exposure to storms. Public facilities including healthcare and educational buildings may lose functionality during disasters, while industrial and waste management facilities can release contaminants during floods. Agricultural and mining infrastructure encounter soil instability and water variability, and slopes become more susceptible to landslides. Overall, climate change primarily amplifies existing hazards by increasing their frequency, magnitude, and spatial reach.

Urbanization further intensifies risk, particularly through flooding. The expansion of impervious surfaces accelerates runoff and modifies watershed hydrographs, producing higher peak flows and shorter response times. The chapter explains how floods develop in urban environments, how building characteristics influence damage levels, and how both direct and indirect economic losses accumulate. Flood risk therefore emerges from the interaction between meteorological forcing, land-use change, and infrastructure vulnerability rather than from rainfall alone.

To manage such challenges, the chapter clarifies the distinction between hazards, disasters, and risk. A hazard is a potentially damaging event, a disaster is the realized disruption, and risk represents the combination of likelihood and consequence. Risk is conceptualized as the interaction of hazard, exposure, and vulnerability. Hazard analysis focuses on identifying threatening events, while risk assessment evaluates the probability of occurrence and associated impacts on systems and communities. These concepts form the basis of disaster risk management in civil engineering.

The chapter then presents a structured risk-based decision framework. Engineers begin with planning and system definition to identify objectives, stakeholders, and infrastructure boundaries. Hazard identification determines credible adverse events, followed by risk assessment to estimate probabilities and consequences. Risk treatment options—such as mitigation measures, adaptation strategies, or acceptance—are then evaluated, and the results are communicated to support engineering and policy decisions. This workflow distinguishes hazard analysis, which studies events, from risk assessment, which evaluates impacts and supports action.

Infrastructure risk originates from multiple sources. Natural hazards such as floods and landslides interact with technical failures including design deficiencies and material deterioration. Operational errors related to human factors and maintenance practices can trigger failures, while intentional actions may also threaten system functionality. Because infrastructure networks are interconnected, disruptions can cascade, transforming localized failures into widespread societal impacts.

The chapter concludes by emphasizing the importance of risk-based engineering for sustainability. By integrating environmental, social, and economic considerations, risk assessment enables infrastructure to be designed and managed for resilience throughout its life cycle. Rather than relying on deterministic safety factors alone, engineers evaluate acceptable risk levels and allocate resources efficiently. In the context of intensifying climate variability and expanding urban systems, systematic hazard and risk analysis becomes essential for protecting communities, preserving ecosystems, and ensuring long-term economic stability.

References

Abbott, P. L. (2017). *Natural disasters* (10th ed.). New York, US: McGraw-Hill Education.

Abbott, P. L. (2023). *Natural disasters* (12th Edn.). McGraw-Hill Education. https://www.mheducation.ca/product/natural-disasters-ise-9781265125554-can-group

Bruneau, M., Chang, S. E., Eguchi, R. T., Lee, G. C., O'Rourke, T. D., Reinhorn, A. M., Shinozuka, M., Tierney, K., Wallace, W. A., & von Winterfeldt, D. (2003). A framework to quantitatively assess and enhance the seismic resilience of communities. *Earthquake Spectra, 19*(4), 733–752. https://doi.org/10.1193/1.1623497

Canadian Institute for Climate Choices. (2021). *Under Water: The Costs of Climate Change for Canada's Infrastructure*. https://climatechoices.ca/wp-content/uploads/2021/09/Infrastructure-English-FINAL-jan17-2022.pdf

Ebtehaj, I., Bonakdari, H., Safari, M. J. S., Gharabaghi, B., Zaji, A. H., Madavar, H. R., & Mehr, A. D. (2020a). Combination of sensitivity and uncertainty analyses for sediment transport modeling in sewer pipes. *International Journal of Sediment Research, 35*(2), 157–170. https://www.sciencedirect.com/science/article/pii/S1001627918303810

Ebtehaj, I., Bonakdari, H., Zeynoddin, M., Gharabaghi, B., & Azari, A. (2020b). Evaluation of preprocessing techniques for improving the accuracy of stochastic rainfall forecast models. *International Journal of Environmental Science and Technology, 17*(1), 505–524. https://doi.org/10.1007/s13762-019-02361-z

Ebtehaj, I., Sammen, S. S., Sidek, L. M., Malik, A., Sihag, P., Al-Janabi, A. M. S., Chau, K. W., & Bonakdari, H. (2021). Daily water level prediction by using new hybrid machine learning models. *Engineering Applications of Computational Fluid Mechanics, 15*(1), 1343–1361. https://doi.org/10.1080/19942060.2021.1966837

Ebtehaj, I., & Bonakdari, H. (2022a). A reliable hybrid outlier robust non-tuned rapid machine learning model for multi-step ahead flood forecasting in Quebec, Canada. *Journal of Hydrology, 614*, 128592. https://www.sciencedirect.com/science/article/pii/S0022169422011623

Ebtehaj, I., & Bonakdari, H. (2022b). A comprehensive comparison of the fifth and sixth phases of the coupled model intercomparison project based on the Canadian earth system models in spatio-temporal variability of long-term flood susceptibility using remote sensing and flood frequency analysis. *Journal of Hydrology, 617*, 128851. https://www.sciencedirect.com/science/article/pii/S0022169422014214

Ebtehaj, I., & Bonakdari, H. (2024). Generalized structure of group method of data handling: Novel technique for flash flood forecasting. *Water Resources Management, 38*(9), 3235–3253. https://doi.org/10.1007/s11269-024-03811-1

Gholami, A., Bonakdari, H., Ebtehaj, I., & Khodashenas, S. R. (2020). Reliability and sensitivity analysis of robust learning machine in prediction of bank profile morphology of threshold sand rivers. *Measurement, 153*, 107411. https://www.sciencedirect.com/science/article/pii/S0263224119312783

Hallegatte, S., Rentschler, J., & Rozenberg, J. (2019). Lifelines: The resilient infrastructure opportunity. *World Bank*. https://doi.org/10.1596/978-1-4648-1430-3

Intergovernmental Panel on Climate Change. (2019). *Summary for Policymakers.* In H.-O. Pörtner, D. C. Roberts, V. Masson-Delmotte, P. Zhai, M. Tignor, E. Poloczanska, K. Mintenbeck, A. Alegría, M. Nicolai, A. Okem, J. Petzold, B. Rama, & N. M. Weyer (Eds.), *IPCC special report on the ocean and cryosphere in a changing climate.* https://www.ipcc.ch/site/assets/uploads/sites/3/2019/11/03_SROCC_SPM_FINAL.pdf

Intergovernmental Panel on Climate Change. (2022). *Summary for Policymakers.* In H.-O. Pörtner, D. C. Roberts, E. S. Poloczanska, K. Mintenbeck, M. Tignor, A. Alegría, M. Craig, S. Langsdorf, S. Löschke, V. Möller, & A. Okem (Eds.), *Climate change 2022: Impacts, adaptation and vulnerability.* Cambridge University Press. https://doi.org/10.1017/9781009325844.001

International Electrotechnical Commission and International Organization for Standardization. (2019). *Risk management—Risk assessment techniques* (IEC/ISO 31010:2019). https://www.iso.org/standard/72140.html

International Organization for Standardization. (2009). *Risk management—Vocabulary* (ISO Guide 73:2009). https://www.iso.org/standard/44651.html

International Organization for Standardization. (2015). *General principles on reliability for structures* (ISO Standard No. 2394:2015). https://www.iso.org/standard/58036.html

International Organization for Standardization. (2018). *Risk management—Guidelines* (ISO Standard No. 31000:2018). https://www.iso.org/standard/65694.html

Kaplan, S., & Garrick, B. J. (1981). On the quantitative definition of risk. *Risk Analysis, 1*(1), 11–27. https://doi.org/10.1111/j.1539-6924.1981.tb01350.x

Kramer, M. G. (2014, October). *Enhancing Sustainable Communities with Green Infrastructure: A Guide to Help Communities Better Manage Stormwater While Achieving Other Environmental, Public Health, Social, and Economic Benefits* (EPA 100-R-14-006). U.S. Environmental Protection Agency. https://www.epa.gov/system/files/documents/2024-03/green-infrastructure.pdf

Leopold, L. B. (1968). *Hydrology for urban land planning: A guidebook on the hydrologic effects of urban land use* (U.S. Geological Survey Circular 554). U.S. Geological Survey. https://pubs.usgs.gov/circ/1968/0554/report.pdf

National Institute of Standards and Technology. (2016). *Community resilience planning guide for buildings and infrastructure systems* (Vol. I) (NIST Special Publication 1190). https://doi.org/10.6028/NIST.SP.1190v1

National Institute of Standards and Technology. (2024, February 26). *The NIST cybersecurity framework (CSF) 2.0* (NIST CSWP 29). https://doi.org/10.6028/NIST.CSWP.29

U.S. Department of Agriculture, Natural Resources Conservation Service. (2021, February). *Engineering field handbook: Chapter 2—Estimating runoff volume and peak discharge* (2nd ed.). https://directives.nrcs.usda.gov/sites/default/files2/1712930818/31754.pdf

U.S. Geological Survey. (2016, December 15). *Flood inundation mapping*. https://www.usgs.gov/science/science-topics/flood-inundation-mapping

U.S. Global Change Research Program. (2023). *Fifth National Climate Assessment*. https://www.nrcc.cornell.edu/workshops/webinars/2023/12/present_1.pdf

United Nations General Assembly. (2016). *Report of the Open-ended Intergovernmental Expert Working Group on Indicators and Terminology Relating to Disaster Risk Reduction: Note by the Secretary-General* (A/71/644). United Nations. https://digitallibrary.un.org/record/852089/files/A_71_644-EN.pdf

United Nations Office for Disaster Risk Reduction. (2015). *Sendai Framework for Disaster Risk Reduction 2015–2030*. United Nations. https://www.undrr.org/publication/sendai-framework-disaster-risk-reduction-2015-2030

Walton, R., Binns, A., Bonakdari, H., Ebtehaj, I., & Gharabaghi, B. (2019). Estimating 2-year flood flows using the generalized structure of the Group Method of Data Handling. *Journal of Hydrology, 575*, 671–689. https://www.sciencedirect.com/science/article/pii/S002216941930513X

Ward, P. J., Winsemius, H. C., Kuzma, S., Bierkens, M. F. P., Bouwman, A., de Moel, H., Díaz Loaiza, A., et al. (2020). *Aqueduct Floods Methodology* (Technical Note). World Resources Institute. https://www.wri.org/research/aqueduct-floods-methodology

Chapter 3
Probability and Bayesian Methods for Hazard and Risk Assessment

3.1 Introduction

In this chapter, essential concepts and methods needed for the analysis and interpretation of statistical data within the realm of civil engineering are presented. This includes strategies for understanding data representation, descriptive measures of statistical data, and graphical techniques crucial for effective data analysis.

Data representation is the most basic foundation that is needed in order to accurately describe and summarize data sets. In this chapter, several methods for data representation, which permit succinct and clear conveyance of complex data sets, will be explored in detail. Descriptive measures of statistical data, including the mean, median, mode, and standard deviation (amongst others), are vital for summarizing and understanding the central tendencies and variability of data. A review of these basic statistical descriptors will be presented in this chapter.

The majority of this chapter will be devoted to the applications of conditional probability in civil engineering. Conditional probability, which is defined as the likelihood of an event occurring given that another event has already occurred, is a fundamental concept in risk assessment and decision-making in engineering projects. How the concept of conditional probability is applied in real-world scenarios will be illustrated, enhancing the reader's ability to make informed decisions based on probabilistic outcomes.

Bayes' theorem, a powerful tool for updating probabilities based on new evidence, will also be introduced. Known as the formula for the probability of "causes", Bayes' theorem is instrumental in refining hypotheses and improving predictive models. Its theoretical basis and practical applications will be covered, providing the skills necessary to utilize this theorem effectively.

H. Bonakdari et al., *Hazard and Risk Assessment in Civil Engineering Design*,
https://doi.org/10.1007/978-3-032-23600-5_3

3.2 Risk Analysis

In civil engineering, understanding and quantifying risk is vital for effective decision-making. The integration of probability theory provides a robust framework for addressing the uncertainties inherent in engineering projects. By systematically assessing both the likelihood of various events as well as their potential consequences, engineers can make informed choices, prioritizing resources and efforts where they are most needed to mitigate risks.

Figure 3.1 illustrates the conceptual definition of risk as the product of two critical components: probability and consequence. This relationship underscores the idea that risk is not merely a function of how likely an event is to occur (probability), but also of the potential impact or severity of the event's outcome (consequence).

Probability theory offers the mathematical tools and principles necessary for this rigorous analysis. It allows for the consistent treatment of uncertainties, ensuring that all possible outcomes are considered and appropriately weighted in the decision-making process. This approach not only enhances the accuracy of risk assessments but also improves the reliability and safety of engineering designs.

When modeling risk, it is crucial to quantify the probability of events accurately. To do so, models must be updated to reflect the latest data and insights as new information becomes available. Figure 3.2 illustrates the conceptual steps involved in developing a probabilistic model, beginning with data collection and culminating in the calculation of the probability of events. This figure outlines a structured modeling approach that ensures a systematic and rigorous assessment of uncertainties.

The process begins with the collection of data, which serves as the foundation for model estimation. Accurate and comprehensive data collection is crucial, as the

Fig. 3.1 A conceptual definition of the risk based on the probability and consequence

Data

Model estimation

Probabilistic model

Probability of events

Fig. 3.2 Conceptual steps from data collection to probability calculation

quality of the data directly impacts the reliability of the developed model. This data can include historical records, experimental results, or observational studies relevant to the specific civil engineering context.

Next, the collected data is used for model estimation. This step involves selecting and calibrating a model that best represents the underlying processes or phenomena being studied. Statistical techniques and computational methods are employed to fit the model to the data, ensuring that it accurately captures the underlying relationships and patterns. The estimated model is considered to be probabilistic as it incorporates the inherent uncertainties and variabilities present in the data, allowing for a more realistic representation of real-world conditions.

The final step in the process is the calculation of the probability of events. Using the probabilistic model, the likelihood of various events occurring can be determined. This involves analyzing the model's outputs to quantify the probabilities of different outcomes, considering both the frequency and impact of potential events.

These calculated probabilities provide valuable insights for decision-making in civil engineering. By understanding the likelihood of various risks and their potential consequences, engineers can develop more robust designs, implement effective mitigation strategies, and enhance overall project resilience.

3.3 Probability Interpretation Principles

3.3.1 Defining Interpretation Principles

The purpose of probability theory is to facilitate the quantitative assessment of the likelihood of different events. In principle, there are three distinct interpretations of probability: Frequentistic, Classical, and Bayesian. Each interpretation offers a unique perspective on how probabilities are understood and applied in various contexts.

The frequentistic interpretation, also known as the frequentist approach, defines probability as the long-run relative frequency of an event occurring. This interpretation is grounded in the idea that if an experiment is repeated many times under identical conditions, the proportion of times the event occurs will converge to a stable value. For example, in civil engineering:

I. The probability of a certain type of structural defect appearing in a batch of concrete beams can be determined by inspecting a large number of beams and calculating the proportion of defects found.
II. The likelihood of traffic congestion at a specific intersection during peak hours can be estimated by recording traffic flow data over several months and analyzing the frequency of congestion events.
III. The probability of soil failure under load can be assessed by conducting numerous soil tests under controlled conditions and observing the frequency of failure occurrences.

The classical interpretation, sometimes referred to as the axiomatic or theoretical approach, defines probability based on the principle of equally likely outcomes. According to this interpretation, if an experiment has a finite number of equally likely outcomes, the probability of an event is the ratio of the number of favorable outcomes to the total number of possible outcomes. For instance:

I. The probability of selecting a defective bolt from a batch of bolts, where the number of defective and non-defective bolts are known, can be calculated by dividing the number of defective bolts by the total number of bolts.
II. In structural design, the probability of a particular mode of failure occurring (such as bending versus shear failure) can be determined if the modes are equally likely and the total number of possible failure modes is known.
III. When considering the failure of a component due to different possible causes (such as corrosion, fatigue, or overload), the probability of each cause can be calculated if the causes are equally likely and their total number is known.

The Bayesian interpretation, also known as the subjective or degree-of-belief approach, views probability as a measure of an individual's belief, or confidence, in the occurrence of an event, given the available information. In this framework, probabilities are updated as new evidence is obtained, using Bayes' theorem. This interpretation is particularly useful in situations where prior knowledge or expert judgment is available and can be combined with new data to refine the assessment of probabilities. For example:

I. The probability of a dam failing can be updated as new inspection data or monitoring results are obtained, taking into account the initial design specifications and historical performance data.
II. The likelihood of meeting a project completion deadline can be updated by considering both initial estimates as well as ongoing observations as new progress reports and performance metrics become available.
III. The probability of a bridge sustaining damage during an earthquake can be revised by combining new seismic data and updated structural health monitoring information with prior knowledge of the bridge's design and construction quality.

Table 3.1 compares the three interpretations of probability in civil engineering. By understanding and applying these three interpretations of probability, engineers and decision-makers can select the most appropriate method for their specific needs and contexts. The Bayesian approach is capable of incorporating prior information into the analysis and can compute the probability that a particular hypothesis is true, while the frequentist approach relies solely on the data and long-run probabilities. Whether focusing on long-term frequencies, theoretical outcomes, or subjective beliefs, probability theory provides a versatile and powerful toolset for assessing and managing uncertainty.

Table 3.1 An overview of probability theories in civil engineering

	Frequentistic interpretation	Classical interpretation	Bayesian interpretation
Defining principle	The long-run relative frequency of an event occurring	Probability based on equally likely outcomes	Probability as a measure of belief, updated with new evidence
Basis	Repeated trials or experiments under identical conditions	Theoretical analysis of all possible outcomes	Prior knowledge combined with new data
Key concept	Relative frequency over many trials	Ratio of favorable outcomes to total outcomes	Updating probabilities with new information using Bayes' theorem
Common applications	– Statistical inference – Hypothesis testing	– Games of chance – Combinatorial problems	– Machine learning – Decision analysis – Risk assessment
Advantages	Provides objective probabilities based on empirical data	Simple and intuitive for problems with clearly defined, equally likely outcomes	Incorporates prior knowledge and allows for updating with new data
Disadvantages	Requires a large number of trials; may not be applicable to unique events	Assumes equally likely outcomes, which may not always be realistic	Subjective nature of initial probabilities; computationally intensive

3.3.1.1 Frequentistic Interpretation

In the Frequentistic interpretation, the probability of an event A occurring (denoted as P(A)) is simply the relative frequency of occurrence of event A in an experiment with n trials. This means that the probability of an event A is defined as the number of times that Event A occurs, divided by the total number of experiments carried out. Mathematically, this is expressed as shown in Eq. 3.1:

$$P(A) = \lim \frac{N_A}{n_{\exp}} \left(for\ \ n_{\exp} \to \infty\right) = \frac{Number\ of\ times\ event\ A\ occurs}{Total\ Number\ of\ Trials} \tag{3.1}$$

where N_A is the number of times event A occurred, and n_{exp} is the total number of experiments.

As the number of trials increases to infinity, the empirical frequency converges to some value, which is the true probability of the event. This interpretation relies on the Law of Large Numbers, which states that as the number of trials grows, the relative frequency of an event's occurrence will approach the true probability.

For example, imagine you are flipping a coin 1000 times. If the head side is observed 479 times, you would estimate the probability of achieving a head when flipping the coin to be $479\,heads/1000\,flips = 0.479$. This estimation improves as more trials are conducted, ultimately converging to 0.5 for a fair coin (since both heads and tails are equally likely outcomes). Similarly, in civil engineering, we can use the frequentistic interpretation of probability to estimate the likelihood of various events based on observed data from real-world experiments. Some examples are presented below.

Structural Defects in Concrete Beams: In a quality control test, 500 concrete beams are inspected for structural defects. Suppose that 25 beams are found to have defects, then, using the frequentistic interpretation, the probability of a beam having a structural defect is found to be:

$$P(Defect) = \frac{25}{500} = 0.05$$

As more beams are inspected over time, this probability will stabilize around the true defect rate for the production process.

Traffic Congestion at an Intersection: To estimate the probability of traffic congestion at a particular intersection during peak hours, traffic flow data is recorded over 200 weekdays. If congestion is observed on 120 of those days, then the probability of its occurrence is:

$$P(Congestion) = \frac{120}{200} = 0.6$$

Recording more data over additional days will help refine this estimate, providing a more accurate probability of congestion.

Soil Failure under Load: In a series of 300 soil tests conducted under identical conditions, soil failure was observed in 45 tests. The probability of soil failure under these conditions is:

$$P(Soil\ Failure) = \frac{45}{300} = 0.15$$

As more tests are conducted, this probability will converge to the true likelihood of soil failure.

Water Quality Analysis: The farthest location in a distribution system from the water treatment facility is sampled and analyzed for disinfection by-products (DBP) during a period of stable water quality. If 1 test was found to exceed the maximum contaminant level out of a series of 50 samples, then the probability of exceedance can be calculated as:

$$P(PDB\ Exceedance) = \frac{1}{50} = 0.02$$

These examples illustrate how the frequentistic interpretation of probability is applied in civil engineering contexts. By conducting a large number of trials and observing the frequency of specific events, engineers can estimate the probabilities of various outcomes. These estimates are for the design and implementation of safe and efficient civil and environmental infrastructure.

3.3.1.2 Classical Interpretation

According to the Classical Interpretation of probability, for equally likely outcomes, the probability of event A is the number of times A is observed (n_A), divided by the total number of equally likely outcomes in the experiment (n_{tot}). Mathematically, this is expressed as:

$$P(A) = \frac{n_A}{n_{tot}} \tag{3.2}$$

The classical interpretation of probability can be illustrated using the example of a coin flip. Here, the probability of achieving a "head" would be 0.5 as there is only one possible way to achieve a "head" from two equally likely outcomes (head and tail). In this interpretation, the experiment does not need to be carried out because it must be determined a priori that all possible outcomes are equally likely and the probabilities are determined theoretically by logical reasoning. Similarly, in practical civil engineering applications, we can use the classical interpretation of probability to estimate the likelihood of various events based on observed data from real-world experiments.

Probability of Selecting a Defective Bolt: Suppose a batch of 100 bolts contains 10 defective bolts and 90 non-defective bolts. The probability of randomly selecting a defective bolt from this batch is given by:

$$P(Defective\ Bolt) = \frac{10}{100} = 0.1$$

Since each bolt has an equal chance of being selected, this probability is derived without conducting the experiment.

Probability of Bridge Failure Modes: Consider a simple bridge design that can fail in three equally likely ways: bending failure, shear failure, and fatigue failure. The probability of the bridge failing due to bending is:

$$P(Bending\ Failure) = \frac{1}{3} = 0.33$$

Each failure mode is considered equally likely, so this probability is known without performing any tests.

Probability of Overload versus Corrosion in Pipeline: If a pipeline has two equally likely failure modes, namely overload and corrosion, the probability of the pipeline failing due to overload is:

$$P(Overload\ Failure) = \frac{1}{2} = 0.5$$

Here, both failure modes are equally probable, and this probability is determined by logical deduction.

To summarize, the classical interpretation of probability relies on this inherent symmetry and fairness of the situation, with all possible outcomes being equally likely to occur. Calculation of the classical probability can be determined theoretically as the proportion of possible outcomes that include the event in question to all possible equally likely outcomes. This approach is particularly useful when dealing with theoretical models as well as problems where all outcomes are equally likely and can be logically enumerated. By understanding and applying this interpretation, civil engineers can make informed decisions about the likelihood of various outcomes based on the known characteristics and configurations of their systems.

3.3.1.3 Bayesian Interpretation

In the Bayesian interpretation, the probability of the event A is formulated as a degree of belief that A will occur. This degree of belief, also referred to as prior belief or prior probability, is assigned before obtaining any further knowledge about the event. The Bayesian approach is inherently subjective, meaning it depends on the individual's state of mind, experience, expertise, and preferences.

For instance, consider the coin-flipping problem. An engineer following the Bayesian interpretation would argue that there are two possible outcomes—heads or tails. Given no preference or additional information favoring either outcome, they would judge the probability of achieving a head to be 0.5. This belief is based on the assumption of two equally likely outcomes.

The Bayesian interpretation allows for the incorporation of new evidence to update this prior probability. As new data or information becomes available, the initial degree of belief is revised using Bayes' theorem. This continuous updating process refines the probability estimates, making them more accurate over time. Some examples of the Bayesian interpretation of probability in civil engineering are given below:

Dam Failure Probability: An engineer initially believes there is a 5% chance (this is the prior probability) of a dam failing based on historical data and expert judgment. After conducting an inspection and discovering signs of significant wear, the engineer updates this probability to reflect the new evidence. Using Bayesian methods, the updated probability might increase to 15%, reflecting the heightened risk due to the newly available information concerning the dams' condition.

Project Completion Time: At the beginning of a construction project, an engineer estimates there is a 70% chance of completing the project within six months based on initial plans and typical project timelines. As the project progresses and new information about delays and resource availability is gathered, the engineer uses Bayesian updating to revise this probability. If unexpected delays occur, the updated probability might decrease to 50%, providing a more realistic assessment of the project's completion time.

Earthquake Damage to a Bridge: An engineer initially assesses a 10% probability of a bridge sustaining significant damage during an earthquake based on regional seismic activity and structural design. After installing sensors and collecting real-time monitoring data over several months, the engineer uses Bayesian analysis to update this probability. If the new data indicates increased seismic activity, the probability might be updated to 20%, enhancing the engineer's ability to plan for necessary reinforcements or emergency measures.

Failure of Water Infrastructure: A water main pipe is entering its 50th year of service with a design life of 100 years. Based on the age of a section of water main pipe, an engineer assesses the possibility of rupture to be 15% (prior probability). Water flow to a section of the pipe was turned off, and a robotic inspection revealed pronounced cracking in the pre-stressed concrete despite no abnormalities in pressure loss within the section. Using this information, the engineer revises the estimate for the rupture probability to 40%. In this example, the revised rupture probability, as determined using Bayes' theorem, enables decision making authorities to prioritize resources for the repair of the section of water main and prevent critical infrastructure failures.

The Bayesian interpretation of probability is instrumental in civil engineering because it allows for the integration of both prior knowledge and new data as it becomes available. This flexibility makes it an invaluable tool for decision-making in the face of uncertainty. It enables engineers to continuously refine their risk assessments and improve the reliability and safety of their designs and projects.

3.3.2 Types of Information

Engineers and scientists commonly need to solve problems and make decisions based on only limited information concerning one or more of the parameters of the problem. The types of information available to them can be either objective or subjective, which are conceptually defined in Fig. 3.3.

Objective information is based on experimental results or direct observations. This type of information is measurable, verifiable, reproducible and not influenced

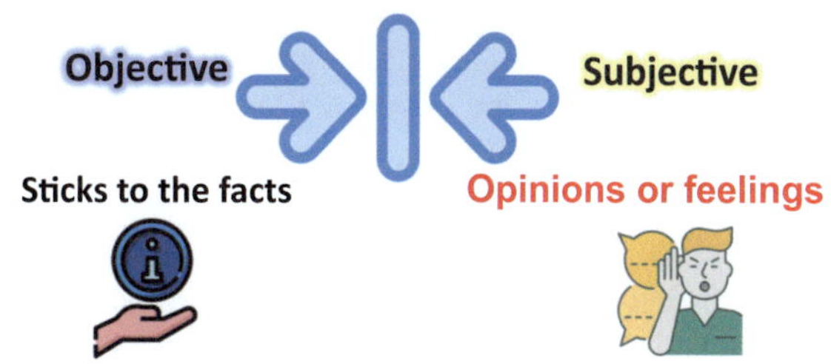

Fig. 3.3 The conceptual definition of objective and subjective information

by personal feelings or opinions. Common sources of objective information include experiments, measurements, and/or historical records. For example, in a concrete strength test, the compressive strength of different concrete samples is measured and recorded: the results are readily measurable/obtainable, are reproducible, and are not prejudiced. On the other hand, **Subjective information** is based on experience and intuition stemming from previous problems similar to the one under consideration or the physics of the problem. Personal beliefs, interpretations, and prior knowledge influence this type of information. For example, an engineer estimates the potential failure rate of a new design based on their intuition and experience with similar projects in the past.

It is common in engineering to encounter problems where both subjective and objective types of information must be used to obtain solutions or make decisions. In these cases, the subjective probabilities are assumed to result from prior knowledge regarding a parameter, which has been gained using objective information (or probabilities). By combining these two types of information, we can develop posterior knowledge through a process called Bayesian Inference. This process is described by the Oxford Dictionary of Statistics as "an approach concerned with the consequences of modifying our previous beliefs as a result of receiving new data" (Upton & Cook, 2014).

Combining the two types of information produces what is termed posterior knowledge in the field of statistics. This process leads us to Bayesian inference, which is the process of confronting alternative hypotheses with new data and using Bayes' Theorem to update our beliefs in each hypothesis.

The Oxford Dictionary of Statistics (Upton & Cook, 2014) describes Bayesian inference as "an approach concerned with the consequences of modifying our previous beliefs as a result of receiving new data."

3.3.3 *Bayesian Interpretation of Probability*

The Bayesian interpretation of probability has become increasingly important in the field of civil and environmental engineering, largely due to the more complex and variable problems encountered by practitioners. Consider the construction of a new building—as project stakeholders demand more complex and unconventional structures, the similarity between existing and new structures decreases, with the number of identical structures being very small or even none. Furthermore, the operational and loading conditions are often significantly different from one structure to another. Differing locations, the effect of time, and a changing climate also contribute to varying loading conditions across structures.

In such cases, the Bayesian interpretation of probability is the most appropriate strategy for handling risk analysis because it can accommodate the unique and evolving nature of each project. By incorporating prior knowledge and continuously updating probabilities with new data, engineers can make more informed decisions, effectively manage risks, and enhance the reliability and safety of their designs.

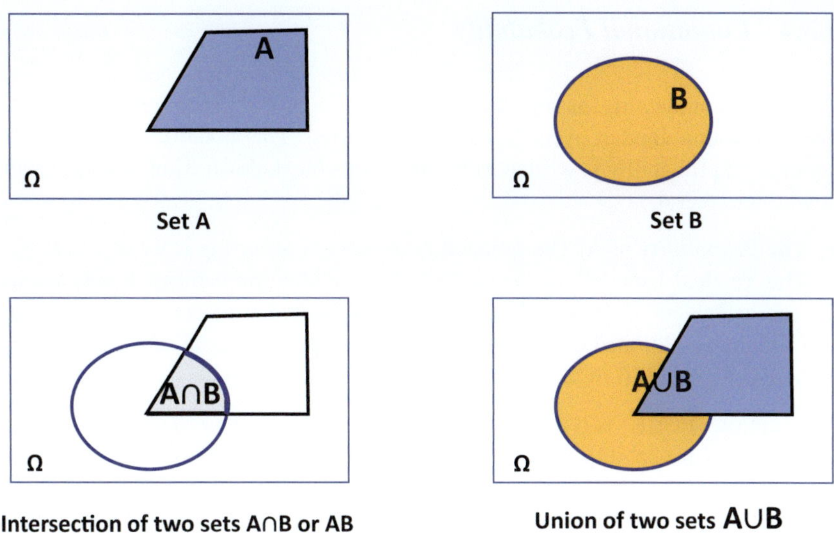

Fig. 3.4 An example of the fundamentals of conditional probability (Venn diagram)

Figure 3.4 demonstrates the fundamental principles of conditional probability by using Venn diagrams (Venn diagram). At the top of the figure, there are two sets of events, namely event A and event B. Both sets are within the total space of possible events, denoted by the Greek letter omega (Ω). These sets help us visualize the relationships between different events.

On the bottom left of the figure, the intersection of sets A and B is shown. This intersection is represented as A ∩ B, which includes all events common to *both* sets A and B. In other words, these are the events that belong to both sets simultaneously. For example, if set A represents the event of having a red car and set B represents the event of having a convertible, then A ∩ B represents having a red convertible.

On the bottom right of the figure, the union of sets A and B is illustrated. This union is represented as A ∪ B, which includes all events that belong to either set A, set B, or both. In logical terms, this means that the union of event A and event B includes any event that is in A or B, or both. For instance, if set A represents the event of having a red car and set B represents the event of having a convertible, then A ∪ B represents having a red car, a convertible, or a red convertible.

To reiterate, the intersection (A ∩ B) means both A and B occur together, while the union (A ∪ B) means either A or B, or both occur. These concepts are essential for understanding how different events relate to each other in probability. This foundation is crucial for more advanced topics like conditional probability, where the probability of one event depends on the occurrence of another. By visualizing these relationships with Venn diagrams, we can better grasp the fundamentals of probability and apply them to real-world scenarios in civil engineering and beyond.

3.3.4 Conditional Probability

Conditional probability, as introduced earlier, is the probability of one event occurring, given that another event has already occurred. This notation used to denote conditional probability is written as P(A|K). Let's break down what this means and how to interpret it:

- **The Probability of A**: The probability of event A occurring is written as P(A).
- **The Vertical Line "|"**: In the context of conditional probability, this symbol is read as "given" or "given that". It separates the event we are interested in (A) from the conditioned event (K). Note that this line is vertical, not slanted like a division symbol (/) or set difference symbol (\).

Therefore, P(A|K) is read as "the probability of A given K." This means we are looking at the probability of event A occurring under the condition that event K has occurred.

To illustrate, consider an example where we want to find the probability that a structure will fail given that it is subjected to a specific load—we can denote this as P(Failure|Load). This represents the probability of structural failure under the condition of a specific load being applied. Similarly, to determine the likelihood of traffic congestion during peak hours, we use P(Congestion|Peak Hours). This indicates the probability of congestion, given that it is peak commuting hours. Moreover, in a construction project, if we are interested in the probability of concrete cracking when the temperature is high, we write P(Cracking|High Temperature). This shows the probability of cracking conditioned on the temperature being high.

The notation P(A|K) is essential in understanding and calculating conditional probabilities. It allows us to consider how the occurrence of one event (K) affects the likelihood of another event (A), a situation that is frequently of interest in the field of civil engineering.

The conditional probability of event A, given that event B has occurred, can be calculated using Eq. 3.3:

$$P(A|B) = \frac{P(A \cap B)}{P(B)} \tag{3.3}$$

This formula expresses the probability of A occurring, given that B has already occurred. It is important to note that P(A|B) is only defined if P(B) $\neq$ 0. If the conditioning event B is an empty set (meaning P(B) = 0), the conditional probability P(A|B) is not defined (since we cannot divide by zero).

The conditional probability P(A|B) represents the likelihood of A occurring under the condition that B has occurred. It is calculated by taking the probability of the intersection of events A and B (P(A $\cap$ B)) and dividing it by the probability of event B (P(B)). If two events, A and B, are independent, the occurrence of one does not affect the probability of the other. Mathematically, this is expressed as:

$$P(A \cap B) = P(A) \times P(B) \tag{3.4}$$

In this case, the conditional probability P(A|B) simplifies to:

$$P(A|B) = \frac{P(A) \times P(B)}{P(B)} = P(A) \tag{3.5}$$

Therefore, if the two events are mutually independent, then knowledge regarding the occurrence of event B does not change the probability of event A occurring.

We can further understand conditional probability by looking at the case of complementary events. Consider two events, namely E_1 and E_2, which form the space of events, Ω.

The conditional probability of event E_1, given that event E_2 has occurred, is written as:

$$P(E_1|E_2) = \frac{P(E_1 \cap E_2)}{P(E_2)} \tag{3.6}$$

Now, let's consider the complementary event, which is the event $\overline{E_1}$ conditioned on the event E_2. According to the definition of conditional probability, we have:

$$P\left(\overline{E_1}|E_2\right) = \frac{P\left(\overline{E_1} \cap E_2\right)}{P(E_2)} \tag{3.7}$$

Since the total probability of all possible outcomes in the sample space must equal one, the sum of the probabilities of E_1 and $\overline{E_1}$ conditioned on E_2 must also equal one:

$$P(E_1|E_2) + P\left(\overline{E_1}|E_2\right) = 1 \tag{3.8}$$

Substituting Eqs. 3.6 and 3.7 into Eq. 3.8, we arrive at:

$$\frac{P(E_1 \cap E_2)}{P(E_2)} + \frac{P\left(\overline{E_1} \cap E_2\right)}{P(E_2)} = 1 \tag{3.9}$$

Since $P(E_1 \cap E_2) + P\left(\overline{E_1} \cap E_2\right) = P(E_2)$, the above equation simplifies to:

$$\frac{P(E_2)}{P(E_2)} = 1 \tag{3.10}$$

This derivation shows that the sum of the conditional probabilities of an event and its complement, conditioned on another event sums to one.

From the rearrangement of Eq. 3.8, we can also see that the probability of the complement of E_1 conditioned on E_2 is simply one minus the probability of E_1 conditioned on E_2. This is represented mathematically as:

$$P(\overline{E_1}|E_2) = 1 - P(E_1|E_2) \tag{3.11}$$

It is important to recognize that in the equation $P(\overline{E_1}|E_2) = 1 - P(E_1|E_2)$, the conditioning event E_2 is the same sample space on both sides of the equation. This ensures consistency in the application of conditional probability.

As an example, suppose the probability of structural failure (E_1) given a specific load (E_2) is 0.3. The probability of no structural failure, which is the complement of the event E_1 ($\overline{E_1}$), given the same load, is:

$$P(\overline{E_1}|E_2) = 1 - P(E_1|E_2) = 1 - 0.3 = 0.7$$

In the following subsections, we will provide several real-world examples to illustrate the concept of conditional probability. These examples will help to demonstrate how conditional probability is applied in various practical scenarios, enhancing the readers' understanding of this fundamental statistical concept.

3.3.4.1 Probability Analysis of Serious Personal Accident

The probability of a serious personal accident at the ABC construction company on a given day is 0.001. The probability that construction will be stopped for the rest of the day after a serious personal accident is 0.73. What, then, is the probability of a personal accident followed by a stop in construction for the rest of the day?

Solution

In this problem, the reader should first identify what information is provided. Firstly, we are told the probability of a serious personal injury occurring, denoted as event A and P(A) = 0.001. We are also told the probability of a work stoppage *given* that there is a serious personal injury. This is the conditional probability of a work stoppage, let us call it event B, conditioned on there being a serious personal injury(P(B|A) = 0.73). We are asked to find the probability of either a serious personal injury or a work stoppage a work stoppage (i.e., the intersection of these two events).

$$P(B|A) = 0.73$$

$$P(A \cap B) = ?$$

From the definition of conditional probability (see Eq. 3.3), we know the following relationship, for which we can directly enter the values and calculate the probability of the intersection, which is found to be 0.00073.

$$P(A \cap B) = P(B|A) \times P(A) = 0.73 \times 0.001 = 0.00073$$

3.3.4.2 Probability Analysis of Cracking Beams in Mini Platforms

ABC Marine Solutions constructs mini platforms with two load-carrying beams, referred to as beam A and beam B. An analysis of 468 mini platforms reveals that 23 of them have a crack in beam A, 18 have a crack in beam B, and 15 have cracks in both beams. Your company owns a mini platform constructed by this company, and you have just discovered a crack in beam B. What is the probability that beam A also has a crack?

Solution

Based on the problem description, we want to find the probability of a crack in beam A *given* that there is a crack in beam B—this is the conditional probability.

So, if we designate the probabilities of cracks in beams A and B as event A and event B, respectively, then using the frequentist approach the probability of a crack in beam A is 23/468, the probability of a crack in beam B is 18/468, and the probability of a crack in both beams A and B is 15/468. Note that the intersection is represented in its convenient form as P(AB). It is important to be careful here—we are not asked to find P(AB); we are asked to find the probability of A given B, or P(A|B).

We let sets A and B designate cracks in beams A and B, respectively. Our probabilities are:

$$\begin{aligned} P(A) &= \tfrac{23}{468} \\ P(B) &= \tfrac{18}{468} \\ P(AB) &= \tfrac{15}{468} \\ P(A|B) &= ? \end{aligned}$$

Using our previously introduced relationship for conditional probability (Eq. 3.3), we can replace the respective probabilities with their fractions and compute the result.

$$P(A|B) = \frac{P(AB)}{P(B)} = \frac{\frac{15}{468}}{\frac{18}{468}} = \frac{15}{18} \approx 0.83$$

Therefore, there is an 83% probability that, given beam B in your mini platform has a crack, beam A will also have a crack.

3.3.4.3 Probability Analysis of Winter Road Closures

There are two highways from City A to City B, as shown in Fig. 3.5. Route 1 is on flat terrain, whereas Route 2 is a scenic route that passes through mountainous terrain. During adverse winter conditions, one or both routes may be closed to traffic due to heavy snowfalls, with Route 2 being more likely to be closed than Route 1.

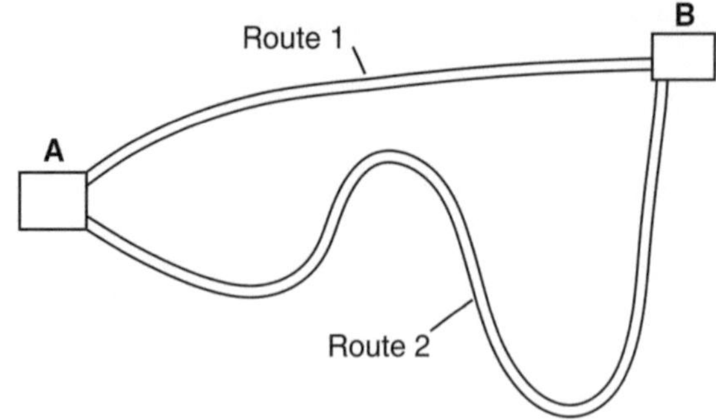

Fig. 3.5 The schematic of two highways from City A to City B

Furthermore, the condition of Route 1 during a severe snowstorm may depend on whether or not Route 2 is open to traffic.

Suppose that during a severe snowstorm, the probability that Route 1 is open to traffic is 0.75, the probability that Route 2 is open to traffic is 0.5, and the probability that both routes will be open is 0.4. What is the probability that Route 1 is open, given that Route 2 is closed?

Solution

Now, the first step in approaching any type of conditional probability problem is to outline the information you have—specifically, the given probabilities and the probability that the question is asking for. Here, we let $P(E_1)$ be the probability that Route 1 is open to traffic, while $P(E_2)$ is the probability that Route 2 is open to traffic. The probability that both routes are open is represented by their intersection, $P(E_1 \cap E_2)$. Thus, we are given:

$$P(E_1) = 0.75$$

$$P(E_2) = 0.5$$

$$P(E_1E_2) = 0.4$$

Based on the problem statement, we are looking for the probability that Route 1 is open while Route 2 is closed. Therefore, this can be denoted as the probability of E_1 conditioned on the complement of E_2. Recall that the complement of Route 2 being open (represented as E_2) is $\overline{E_2}$.

Firstly, we know from the fundamental relationships of the conditional probability that the probability of E_1 conditioned on E_2 is given by the intersection of E_1 and E_2 divided by the probability of E_2. Based on the information from the problem statement, we can simply plug in the probabilities that we have, and we find the result to be 0.8.

$$P(E_1|E_2) = \frac{P(E_1E_2)}{P(E_2)} = \frac{0.4}{0.5} = 0.8$$

On the other hand, the probability that Route 1 is open while Route 2 is closed (i.e., the conditional probability of E_1 given $\overline{E_2}$) is given by the intersection of E_1 and $\overline{E_2}$ divided by the probability of $\overline{E_2}$. From fundamental principles, the intersection of the complements of E_1 and E_2 is equal to 1 minus the union of E_1 and E_2. Recall that the union of E_1 and E_2 is defined as the probability of E_1 plus the probability of E_2 minus the intersection of E_1 and E_2. From the problem statement, we have all the values and can simply plug them in to find the result, which is 0.15.

$$P\left(\overline{E_1}|\overline{E_2}\right) = \frac{P\left(\overline{E_1}\,\overline{E_2}\right)}{P\left(\overline{E_2}\right)}$$

$$\begin{aligned} P\left(\overline{E_1}\,\overline{E_2}\right) = 1 - P\left(\overline{\overline{E_1}\,\overline{E_2}}\right) &= 1 - P(E_1 \cup E_2) \\ &= 1 - [P(E_1) + P(E_2) - P(E_1E_2)] \\ &= 1 - [0.75 + 0.5 - 0.4] = 0.15 \end{aligned}$$

Since the complement of E_2 is simply 1 minus the probability of E_2, we know that the probability of the complement of E_2 is 0.5. Therefore, all values needed to find the probability of the complement of E_1 conditioned on the complement of E_2 have been calculated. Through substitution, we can arrive at a final value of 0.3.

Finally, to find the probability of E_1 conditioned on the complement of E_2, we simply subtract the probability of the complement of E_1 conditioned on the complement of E_2 from 1, yielding an answer of 0.7.

$$P\left(\overline{E_1}|\overline{E_2}\right) = \frac{P\left(\overline{E_1}\,\overline{E_2}\right)}{P\left(\overline{E_2}\right)} = \frac{0.15}{0.5} = 0.3$$

$$P\left(E_1|\overline{E_2}\right) = 1 - P\left(\overline{E_1}|\overline{E_2}\right) = 1 - 0.3 = 0.7$$

3.3.4.4 Probability Analysis of Buildings Earthquake

A region is prone to earthquakes, and there are two key buildings, Building A and Building B, which are critical for disaster response. Building A is newer and built with modern earthquake-resistant techniques, while Building B is older and considered more vulnerable. During an earthquake, either (or both) buildings may be damaged and become non-operational.

Suppose that during an earthquake, the probability that Building A remains operational is 0.8, the probability that Building B remains operational is 0.6, and the probability that both buildings remain operational is 0.5. What is the probability that Building A is operational, given that Building B is not operational?

Solution

Define the events:

Let E_A be the event that Building A is operational ($P(E_A) = 0.8$).

Let E_B be the event that Building B is operational ($P(E_A) = 0.6$).

Let $E_A \cap E_B$ be the event that both Building A and Building B are operational ($P(E_A \cap E_B) = 0.5$).

We are interested in finding the probability that Building A is operational, given that Building B is not operational, $P(E_A|\overline{E_B})$.

First, we need to find the probability that Building B is not operational:

$$P(\overline{E_B}) = 1 - P(E_B) = 1 - 0.6 = 0.4$$

Next, we use the definition of conditional probability. We need to find

$$P(E_A \cap \overline{E_B}) = P(E_A) - P(E_A \cap E_B) = 0.8 - 0.5 = 0.3$$

Now, we can find the conditional probability $P(E_A|\overline{E_B})$:

$$P(E_A|\overline{E_B}) = \frac{P(E_A \cap \overline{E_B})}{P(\overline{E_B})} = \frac{0.3}{0.4} = 0.75$$

So, the probability that Building A is operational, given that Building B is not operational, is 0.75.

3.3.4.5 Probability Analysis of Flood Risk Assessment with Multiple Conditions

A river basin is protected by two levees, Levee A and Levee B. Both levees are crucial in preventing flooding during heavy rainfalls. During a severe storm, multiple factors determine the integrity of these levees, including the probability of failure under specific conditions.

Suppose that during a severe storm:

- The probability that Levee A fails is 0.2.
- The probability that Levee B fails is 0.3.
- The probability that both Levee A and Levee B fail simultaneously is 0.1.
- The probability that the storm will lead to significant downstream flooding if either or both levees fail is 0.6.
- The probability of significant downstream flooding if neither levee fails is 0.05.

What is the probability that significant downstream flooding occurs during the storm?

Solution

Define the events:

Let F_A be the event that Levee A fails ($P(F_A) = 0.2$).

Let F_B be the event that Levee B fails ($P(F_B) = 0.3$).

Let $F_A \cap F_B$ be the event that both Levee A and Levee B fail ($P(F_A \cap F_B) = 0.1$).

Let D be the event of significant downstream flooding.

We are interested in finding the probability of significant downstream flooding, P(D).

First, we need to calculate the probability of different combinations of levee failures:

1. Probability that neither levee fails:

$$P\left(\overline{F_A} \cap \overline{F_B}\right) = 1 - P(F_A) - P(F_B) + P(F_A \cap F_B) = 1 - 0.2 - 0.3 + 0.1 = 0.6$$

2. Probability that only Levee A fails:

$$P\left(F_A \cap \overline{F_B}\right) = P(F_A) - P(F_A \cap F_B) = 0.2 - 0.1 = 0.1$$

3. Probability that only Levee B fails:

$$P\left(F_B \cap \overline{F_A}\right) = P(F_B) - P(F_A \cap F_B) = 0.3 - 0.1 = 0.2$$

4. Probability that both Levee A and Levee B fail:

$$P(F_A \cap F_B) = 0.1$$

Next, we use the law of total probability to find P(D):

$$\begin{aligned} P(D) = {} & P(D|F_A \cap F_B)P(F_A \cap F_B) \\ & + P\left(D|F_A \cap \overline{F_B}\right)P\left(F_A \cap \overline{F_B}\right) \\ & + P\left(D|F_B \cap \overline{F_A}\right)P\left(F_B \cap \overline{F_A}\right) \\ & + P\left(D|\overline{F_A} \cap \overline{F_B}\right)P\left(\overline{F_A} \cap \overline{F_B}\right) \end{aligned}$$

Given

$$\begin{aligned} P(D|F_A \cap F_B) &= 0.6 \\ P\left(D|F_A \cap \overline{F_B}\right) &= 0.6 \\ P\left(D|F_B \cap \overline{F_A}\right) &= 0.6 \\ P\left(D|\overline{F_A} \cap \overline{F_B}\right) &= 0.05 \end{aligned}$$

Substitute the values:

$$\begin{aligned} P(D) &= 0.6 \times 0.1 + 0.6 \times 0.1 + 0.6 \times 0.2 + 0.05 \times 0.6 \\ &= 0.06 + 0.06 + 0.12 + 0.03 = 0.27 \end{aligned}$$

So, the probability that significant downstream flooding occurs during the storm is 0.27.

3.3.5 *Law of Total Probability*

Figure 3.6XEXE illustrates the concept of conditional probability using two sets, A and B. The top row shows individual sets A and B, their intersection (A ∩ B), and their union (A ∪ B). The bottom row demonstrates the fundamentals of conditional probability: P(A|B) and P(B|A). For P(A|B), the probability is determined by dividing the intersection of A and B by the probability of B, visually represented as the wedge of A within B. Similarly, P(B|A) is found by dividing the intersection of A and B by the probability of A, shown as the wedge of B within A.

The bottom section of this figure provides a graphical shortcut for understanding conditional probability. The key is to identify the condition, which is always in the denominator. First, determine the denominator space, then see what fraction of it is occupied by the numerator. For example, for P(AlB) shown on the bottom left of the figure, the probability is conditioned on B, making the denominator space the probability of B. We examine this space to see what fraction is occupied by A, which represents the intersection, and serves as the numerator. Following the same approach, this method can also be applied to find the probability of B conditioned on A, as shown in the bottom right-hand side of Fig. 3.6.

Figure 3.7 graphically illustrates the multiplication rule of conditional probability. In this visualization, the various elements have been rearranged to isolate the intersection of A and B. The diagram demonstrates that the joint probability P(A ∩ B) can be expressed as P(A|B) × P(B) or as P(B|A) × P(A).

Now, let's look at the Law of Total Probability or Bayes' Rule. Often, we have a sample space that is partitioned into n events that are mutually exclusive and collectively exhaustive. We can denote each of these partitions as E_1, E_2, …, E_n. Therefore, the union of E_1, E_2, …, E_n forms the sample space S. If we are interested in the probability of another event A within this sample space, we can derive the theorem of total probability.

From the Venn diagram shown in Fig. 3.8, we see that event A, which is shaded, is composed of the intersections of A with E_1, E_2, …, E_n. Therefore, we can express the probability of event A in terms of the marginal probabilities of events E_n and the conditional probability of B given E_n.

Mathematically, we express A as:

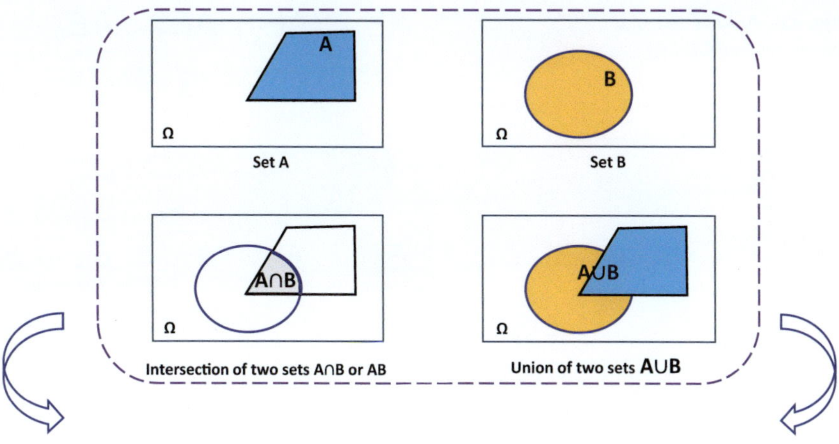

$$P(A|B) = \frac{P(A \cap B)}{P(B)} \qquad P(B|A) = \frac{P(A \cap B)}{P(A)}$$

B = B A = A

Fig. 3.6 Graphical representation of conditional probability along with an example of the P(A|B) and P(B|A)

Joint probability

$$P(A \cap B) = P(A|B) \times P(B) \qquad P(A \cap B) = P(B|A) \times P(A)$$

= × B = × A

Fig. 3.7 Pictorial representation of the multiplication rule of conditional probability

$$A = A \cap S = A \cap (E_1 \cup E_2 \cup \dots E_n) = (A \cap E_1) \cup (A \cap E_2) \cup \dots (A \cap E_n) \quad (3.12)$$

Since $A \cap E_1, A \cap E_2, \dots, A \cap E_n$ are mutually exclusive, we have the following:

$$P(A) = P(A \cap E_1) + P(A \cap E_2) + \dots + P(A \cap E_n) \quad (3.13)$$

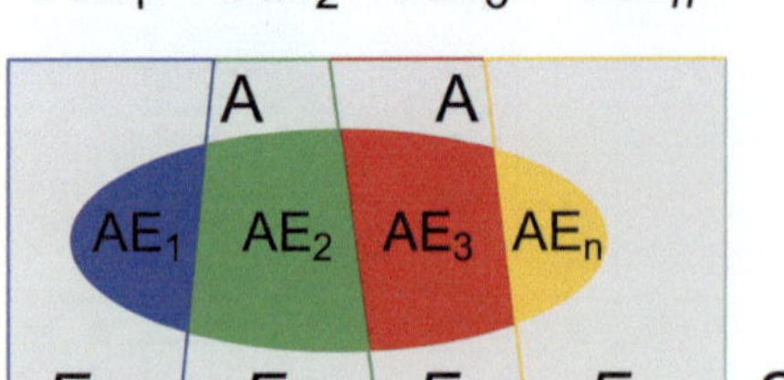

Fig. 3.8 A schematic of the law of total probability

Thus, the theorem of total probability can be defined as:

$$P(A) = \sum_{i=1}^{n} P(A \cap E_i) = \sum_{i=1}^{n} P(A|E_i) \times P(E_i) \tag{3.14}$$

By using this theorem, we can determine the probability of event A based on the probabilities of the partitions E_i and the conditional probabilities of A given each E_i.

The multiplication rule is a fundamental principle in probability theory. It can be applied whether the probability is conditioned on event E_2 or E_1. In both cases, the joint probability of the two events, or their intersection, is on the left-hand side of the equation. The rule is expressed as follows:

$$\begin{aligned} P(E_1E_2) &= P(E_1|E_2)P(E_2) \\ P(E_1E_2) &= P(E_2|E_1)P(E_1) \end{aligned} \tag{3.15}$$

The marginal probability used depends on the event the probability is conditioned on. If the probability is conditioned on event E_2, then the marginal probability of E_2 is used, and vice versa for E_1.

If two events, E_1 and E_2, are statistically independent, the probability of their joint occurrence is equal to the product of their individual probabilities:

$$P(E_1E_2) = P(E_1)P(E_2) \tag{3.16}$$

Moreover, if E_1 and E_2 are statistically independent, their complements $\overline{E_1}$ and $\overline{E_2}$ will also be statistically independent.

$$P(\overline{E}_1\overline{E}_2) = P(\overline{E}_1)P(\overline{E}_2) \tag{3.17}$$

When applying the multiplication rule to three events, which is common in the study of various processes, it can be expressed as follows:

$$\begin{aligned} P(E_1E_2E_3) &= P(E_1|E_2E_3)P(E_2E_3) \\ &= P(E_1|E_2E_3)P(E_2|E_3)P(E_3) \end{aligned} \tag{3.18}$$

This shows the joint probability of three events as a product of conditional probabilities and the marginal probability.

If these three events are statistically independent, the probability of their joint occurrence simplifies to the product of their individual probabilities:

$$P(E_1E_2E_3) = P(E_1)P(E_2)P(E_3) \tag{3.19}$$

This simplification is due to the fact that, for independent events, the occurrence of one event does not affect the probabilities of the others. Therefore, the joint probability is simply the product of their marginal probabilities.

We might emphasize that all the mathematical rules pertaining to the probabilities of events apply equally to the conditional probabilities of events that are conditioned on the same reconstituted sample space, including the addition rule and the multiplication rule. In particular, we may observe the following:

For two events E_1 and E_2, the conditional probability rules are:

$$P(E_1 \cup E_2|A) = P(E_1|A) + P(E_2|A) - P(E_1E_2|A) \tag{3.20}$$

$$P(E_1E_2|A) = P(E_1|A) + P(E_2|A) \tag{3.21}$$

If E_1 and E_2 are statistically independent events, given event A:

$$P(E_1E_2|A) = P(E_1|A)P(E_2|A) \tag{3.22}$$

This shows that the same rules for probabilities of events hold when conditioned on another event, maintaining the principles of addition and multiplication.

Using the chain rule, the joint probability for n events can be broken down into $n - 1$ conditional probabilities and one marginal probability. The derivation of the joint probability for n events is as follows:

$$\begin{aligned} P(E_1E_2...E_n) &= P(E_1|E_2...E_n)P(E_2...E_n) \\ &= P(E_1|E_2...E_n)P(E_2|E_3...E_n)P(E_3...E_n) \\ &= P(E_1|E_2...E_n)P(E_2|E_3...E_n)...P(E_{n-1}|E_n)P(E_n) \end{aligned} \tag{3.23}$$

From this formula, we can see that if we had, for example, 3 events, the joint probability would be the marginal probability of event 3, the conditional probability of event 2 given event 3, and the conditional probability of event 1 given the intersection of events 2 and 3. This approach helps in understanding the dependencies and relationships between multiple events.

3.3.5.1 Probability Analysis of the Failure Modes in Tall Building Foundations

The foundation of a tall building may fail due to inadequate bearing capacity or excessive settlement. Let B and S represent these respective modes of foundation failure, with P(B) = 0.001, P(S) = 0.008, and P(B|S) = 0.10, which is the conditional probability of bearing capacity failure given that there is excessive settlement.

What is the probability of failure of the foundation? What is the probability that the building experiences excessive settlement without bearing capacity failure?

Solution

The first step in solving the problem is to identify the given probabilities. We are told that the probability of bear capacity failure is 0.001, the probability of settlement failure is 0.008, and the probability of bearing capacity failure conditioned on excessive settlement is 0.1.

$$\mathrm{P(B)} = 0.001$$

$$\mathrm{P(S)} = 0.008$$

$$\mathrm{P(B|S)} = 0.1$$

Next, we must identify which probabilities are being asked of us in the problem. Here, we want to know the probability of the foundation's failure. Of course, this probability is denoted as the union of the events B and S because either B or S or both will lead to foundation failure (P(B ∪ S)). In the second part of the question, we are asked to find the probability that the building will experience excessive settlement but not bearing capacity failure. This can be represented by the intersection of the settlement with the complement of B, $P\left(S \cap \overline{B}\right)$.

We know from fundamentals that the union of two events can be computed using the rule shown on this slide. However, we do not have the probability of the joint events B and S. So, the question becomes, how can we calculate this?

$$P(B \cup S) = P(B) + P(S) - P(B \cap S)$$

Here, we can see that we have been given the two marginal probabilities from the problem statement, but that the joint probability is lacking. To determine it, we must use some of the joint probability rules that we have discussed up to this point.

Using the joint probability rule, we know that the probability of the joint event is simply the product of the probability of B conditioned on S and the probability of S. Both of these values were provided in the problem statement. So we are now able to calculate P(B ∪ S) by replacing the value of P(B) = 0.001, P(S) = 0.008, and P(B|S) = 0.1. This yields an answer of 0.0082.

$$\begin{aligned} P(B \cup S) &= P(B) + P(S) - P(B \cap S) \\ &= P(B) + P(S) - P(B|S) \times P(S) \\ &= 0.001 + 0.008 - 0.1 \times 0.008 = 0.0082 \end{aligned}$$

Now, we want to find the probability that the building will experience excessive settlement, but without bearing capacity failure. This suggests that we should find the $P(S \cap \overline{B})$. Again, we can use the joint probability rule $P(S \cap \overline{B}) = P(\overline{B}|S) \times P(S)$. However, we do not know the probability of the complement of B conditioned on S. To find this, we can relate the conditional probability of a complementary event using the formula $P(\overline{B}|S) = 1 - P(B|S)$. Once we have defined this, we can simply plug in the values provided in the problem statement to arrive at the final solution, which is 0.0072.

$$\begin{aligned} P(S \cap \overline{B}) &= P(\overline{B}|S) \times P(S) \\ &= [1 - P(B|S)] \times P(S) \\ &= [1 - 0.1] \times 0.008 = 0.0072 \end{aligned}$$

3.3.5.2 Probability Analysis of Air Quality Compliance in Urban Areas

Air quality in an urban area is monitored to ensure compliance with environmental regulations. The primary concerns are high levels of particulate matter (PM) and nitrogen dioxide (NO_2). Let A and B represent these respective modes of air quality non-compliance, with P(A) = 0.02, P(B) = 0.015, and P(A|B) = 0.25, which is the conditional probability of PM non-compliance, given that there is NO_2 non-compliance.

1. What is the probability of air quality non-compliance in the urban area?
2. What is the probability that the urban area experiences NO_2 non-compliance without PM non-compliance?

Solution

1. Probability of Air Quality Non-Compliance

We need to find the probability that the air quality is non-compliant, which can be due to either high levels of PM, NO_2, or both.

Using the law of total probability:

$$\mathrm{P(Non-compliance) = P(A \cup B)}$$

Using the formula for the union of two events:

$$\mathrm{P(A \cup B) = P(A) + P(B) - P(A \cap B)}$$

We can calculate P(A ∩ B) using the conditional probability:

$$P(A \cap B) = P(A|B) \times P(B)$$

$$P(A \cap B) = 0.25 \times 0.015 = 0.00375$$

Now, substitute the values:

$$P(A \cup B) = 0.02 + 0.015 - 0.00375 = 0.03125$$

So, the probability of air quality non-compliance is:

$$P(\text{Non - compliance}) = 0.03125$$

2. Probability of NO_2 Non-Compliance without PM Non-Compliance

We need to find the probability that there is NO_2 non-compliance but not PM non-compliance, denoted as $P(B \cap \overline{A})$.

Using the definition of conditional probability and the complement rule:

$$P(B \cap \overline{A}) = P(B) - P(A \cap B)$$

We already calculated P(A ∩ B) = 0.00375.

So, substitute the values:

$$P(B \cap \overline{A}) = 0.015 - 0.00375 = 0.01125$$

So, the probability that the urban area experiences NO_2 non-compliance without PM non-compliance is: $P(B \cap \overline{A}) = 0.01125$.

3.3.5.3 Probability Analysis of the Spring Snowmelt

The likelihood of a river flooding in the spring season depends on the amount of snow accumulation in the mountains during the preceding winter season. Snow accumulation can be categorized as heavy, normal, or light. If the snow accumulation in the mountains is heavy, the probability of flooding in the following spring will be high. Conversely, if the snow accumulation is light, the probability will be low. Additionally, flooding may also result from spring rainfalls. Consider the following notations:

- F: occurrence of flooding in the river
- H: heavy accumulation of snow
- N: normal accumulation of snow
- L: light (including no) accumulation of snow.

Assume the following probabilities:

- P(F|H) = 0.9
- P(F|N) = 0.4
- P(F|L) = 0.1.

In a given winter season, the probabilities of different snow accumulations are:

- P(H) = 0.2
- P(N) = 0.5
- P(L) = 0.3.

What is the probability of flooding in the river during the following spring season?

Solution

As usual, the first step in analyzing any problem is to identify which quantities are known from the problem statement. The accumulations of snow in the past winter—namely heavy, normal, and light—are mutually exclusive. Furthermore, the probabilities of the three levels of snow accumulation add up to 1.0. This makes sense because, logically, there is a 100% chance that there will be either heavy, normal, light, or no snowfall. Therefore, the three events, H, N, and L, are mutually exclusive and collectively exhaustive. We are seeking to identify the probability of flooding in the river, which is P(F).

Since we have mutually exclusive and exhaustive events and we are interested in the probability of another event, F, occurring, it is clear that we need to employ the theorem of total probability to find the probability of flooding in the river during the following spring season. The theorem of total probability relates the probability of event F to the product of the marginal and conditional probabilities of each partition. At this stage, all the required information was provided in the problem statement, and the probability of flooding is determined to be 0.41, as shown mathematically below (Fig. 3.9).

$$\begin{aligned} P(F) &= P(F|H)P(H) + P(F|N)P(N) + P(F|L)P(L) \\ &= 0.9 \times 0.2 + 0.4 \times 0.5 + 0.1 \times 0.3 = 0.41 \end{aligned}$$

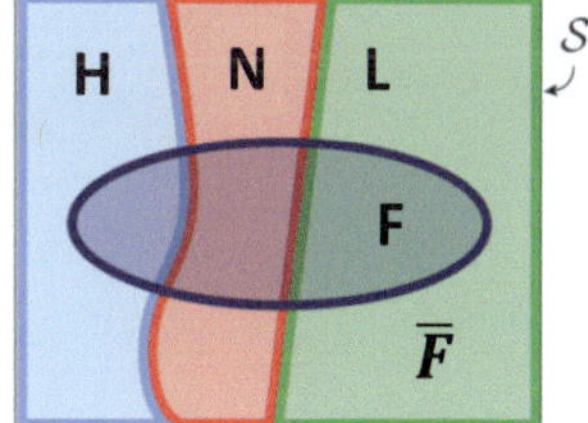

Fig. 3.9 The sample space of the spring snowmelt problem

3.3.5.4 Probability Analysis of Soil Contamination in Urban Redevelopment

The likelihood of soil contamination in an urban redevelopment area depends on the history of industrial activity in the area. The industrial history can be categorized as heavy, moderate, or light. If the industrial activity in the area is heavy, the probability of soil contamination will be high. Conversely, the probability will be low if the industrial activity is light. Additionally, soil contamination may also result from other environmental factors. Consider the following notations:

- C: occurrence of soil contamination
- H: heavy industrial activity
- M: moderate industrial activity
- L: light industrial activity.

Assume the following probabilities:

- $P(C|H) = 0.85$
- $P(C|M) = 0.45$
- $P(C|L) = 0.10$.

In a given area, the probabilities of different levels of industrial activity are:

- $P(H) = 0.25$
- $P(M) = 0.50$
- $P(L) = 0.25$.

What is the probability of soil contamination in the urban redevelopment area?

Solution

We need to find the probability of soil contamination, P(C).

Using the law of total probability:

$$P(C) = P(C|H) \cdot P(H) + P(C|M) \cdot P(M) + P(C|L) \cdot P(L)$$

Substitute the given values:

$$P(C) = (0.85 \times 0.25) + (0.45 \times 0.50) + (0.10 \times 0.25)$$

$$P(C) = 0.2125 + 0.225 + 0.025 = 0.4625$$

So, the probability of soil contamination in the urban redevelopment area is: $P(C) = 0.4625$.

3.3.5.5 Probability Analysis of Highway Pavement Acceptance

Before a section (1/10 km long) of a newly constructed highway pavement is accepted by the Ministry of Transportation, the thickness of a 25-cm pavement is inspected for specification compliance by taking ultrasonic readings. Each 1/10-km section will be accepted if the measured thickness is at least 23 cm; otherwise, the entire section will be rejected, or a penalty will be imposed.

From past inspection data, it is known that in 90% of cases the ultrasonic test indicates a thickness of at least 23 cm (that is, the test result is "accept"). Furthermore, the performance of the ultrasonic device is such that:

- When the test result is "accept", the pavement is actually in compliance (true thickness $\geq$23 cm) 80% of the time, and
- When the test result is "reject", the pavement is actually not in compliance (true thickness <23 cm) 80% of the time.

Let G denote the event that the pavement is well constructed (true thickness $\geq$23 cm), and let A denote the event that the ultrasonic test result is "accept" (measured thickness $\geq$23 cm).

1. What is the probability that a well-constructed section of the pavement will be accepted on the basis of the ultrasonic test?
2. What is the probability that a well-constructed section will be rejected on the basis of the ultrasonic test?
3. What is the probability that a poorly constructed section will be accepted on the basis of the ultrasonic test?
4. What is the probability that a poorly constructed section will be rejected on the basis of the ultrasonic test?

Solution

As usual, the first step in solving the problem is to identify the givens from the problem statement, as well as what is required from us. Allow us to consider the notations of G = constructed thickness of pavement is at least 23 cm, and A = ultrasonic measured thickness is $\geq$23 cm. The provided statement "the ultrasonic test is 80% reliable" means P(G|A) = 0.8 and also, $P\left(\overline{G}|\overline{A}\right) = 0.8$. This may also be interpreted to mean that an ultrasonic reading could have an error of 20%, which means $P\left(\overline{G}|A\right) = P\left(G|\overline{A}\right) = 0.2$. Based on the contractor's past construction record (test results), we may assume that 90% of his constructed pavement will have acceptable ultrasonic readings; hence, P(A) = 0.90.

The four conditional probabilities that we must find can be mathematically represented as:

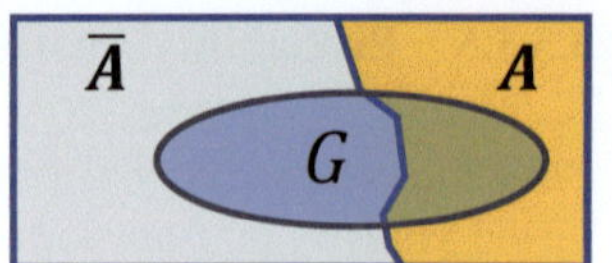

Fig. 3.10 The Venn diagram for the example of the highway pavement acceptance

$$P(A|G)$$
$$P(\overline{A}|G)$$
$$P(A|\overline{G})$$
$$P(\overline{A}|\overline{G})$$

The probability that a particular 1/10-km section of the pavement is well constructed and will be accepted by the Ministry is the probability of A conditioned on G. Therefore, from the multiplication rules, we have the following relationship.

$$P(A|G) = \frac{P(GA)}{P(G)}$$

We can rearrange and solve for the joint probability in terms of the probability of G conditioned on A, which are two values that were provided in the problem statement. Now, the main difficulty with this problem is the calculation of the P(G):

$$P(GA) = P(A|G)P(G) = 0.8 \times 0.9 = 0.72$$

To calculate P(G), consider the Venn diagram shown in Fig. 3.10. As shown in the Figure:

$$P(G) = P(GA) + P\left(G\overline{A}\right)$$

However, if we want to detail it based on the base rules of probability, we can say:

The union of a set and its complement is the universe $A \cup \overline{A} = U$, and the intersection of a set with the universe is the set itself, $G \cap U = G$.

Now, if we combine these two rules, we have $G \cap \left(A \cup \overline{A}\right) = G \cap U$ by simplifying this, we have $P(G) = P(GA) + P\left(G\overline{A}\right)$.

Now, we are able to go back to our expression, which was derived from the multiplication rule, and calculate the probability of A conditioned on G.

$$\begin{aligned} P(G) &= P(GA) + P\left(G\overline{A}\right) \\ &= P(G|A)P(A) + P\left(G|\overline{A}\right)P\left(\overline{A}\right) \\ &= 0.8 \times 0.9 + 0.2 \times 0.1 = 0.74 \end{aligned}$$

$$P(A|G) = \frac{P(GA)}{P(G)} = \frac{0.72}{0.74} = 0.973$$

The probability that a well-constructed section of the pavement will be rejected is $P(\overline{A}|G)$. This means that the ultrasonic device did not read a value of 23 cm. However, the constructed thickness of the pavement was 23 cm. This can be simply calculated by considering the complement, which is the probability of A conditioned on G. The value of this conditional probability was found to be 0.973, and therefore, $P(\overline{A}|G) = 1 - P(A|G) = 1 - 0.973 = 0.027$. This value can also be found in the same manner by using the value of P(G) that we previously found.

$$P(\overline{A}|G) = 1 - P(A|G) = 1 - 0.973 = 0.027$$

$$P(\overline{A}|G) = \frac{P(G\overline{A})}{P(G)}$$
$$P(G\overline{A}) = P(G|\overline{A})P(\overline{A}) = 0.2 \times 0.1 = 0.02$$
$$P(G) = 0.74$$
$$P(\overline{A}|G) = \frac{0.02}{0.74} = 0.027$$

The probability that a poorly constructed section is accepted is given by the probability of A conditioned on the complement of G. From the multiplication rule, we can compute the joint probability of A and the complement of G directly. As we have already calculated the P(G) value, its complement can be easily computed by subtracting P(G) from a value of 1. Finally, we can directly calculate the required quantity to be 0.69.

$$P(A|\overline{G}) = \frac{P(\overline{G}A)}{P(\overline{G})}$$
$$P(\overline{G}A) = P(\overline{G}|A)P(A) = 0.2 \times 0.9 = 0.18$$
$$P(\overline{G}) = 1 - 0.74 = 0.26$$
$$P(A|\overline{G}) = \frac{P(\overline{G}A)}{P(\overline{G})} = \frac{0.18}{0.26} = 0.69$$

Lastly, the probability that a poorly constructed pavement will be rejected is given by the probability of the complement of A conditioned on the complement of G. There are two approaches to arrive at this value. The first is to subtract the complement from 1, the complement being the probability of A conditioned on the complement of G. Alternatively, we could define the conditional probability using the multiplication rule and then simply plug in all the known values to arrive at the answer.

Method 1:

$$P(\overline{A}|\overline{G}) = 1 - P(A|\overline{G}) = 1 - 0.69 = 0.31$$

Method 2:

$$P(\overline{A}|\overline{G}) = \frac{P(\overline{G}\overline{A})}{P(\overline{G})} = \frac{P(\overline{G}|\overline{A})P(\overline{A})}{P(\overline{G})} = \frac{0.8 \times 0.1}{0.26} = 0.31$$

3.3.5.6 Probability Analysis of Wastewater Treatment Plant Compliance

Before a newly constructed section (1/10 km^2) of a wastewater treatment plant is approved by the Environmental Protection Agency (EPA), the quality of the treated water is inspected for compliance with environmental standards by taking sample readings. Each 1/10 km^2 section will be approved if the measured pollutant concentration is below the regulatory threshold; otherwise, the entire section will be rejected, or additional treatment will be required. Suppose, from past experience, that 85% of all wastewater treatment plants constructed by the contractor were found to be in compliance with standards. However, the sample reading determination is only 75% reliable; i.e., there is a 25% chance that a conclusion based on the sample test may be erroneous.

- What is the probability that a compliant section of the wastewater treatment plant will be approved by the EPA based on the sample test?
- What is the probability that a compliant section will be rejected based on the sample test?
- What is the probability that a non-compliant section will be approved based on the sample test?
- What is the probability that a non-compliant section will be rejected based on the sample test?

Solution

1. **Probability that a compliant section will be approved**:

Let CCC denote a compliant section, and A denote approval based on the sample test.

Given:

- P(C) = 0.85
- Reliability of the test: 75% accurate (i.e., P(A|C) = 0.75).

  Therefore, the probability that a compliant section will be approved is:

$$P(A|C) = 0.75$$

2. **Probability that a compliant section will be rejected**:

The probability that a compliant section will be rejected is the complement of the probability that it will be approved:

$$P(R|C) = 1 - P(A|C)$$

where R denotes rejection.

So,

$$P(R|C) = 1 - 0.75 = 0.25.$$

3. **Probability that a non-compliant section will be approved**:

Let NC denote a non-compliant section.

Given:

- $P(NC) = 1 - P(C) = 0.15$
- Reliability of the test: 25% erroneous (i.e., $P(A|NC) = 0.25$).

Therefore, the probability that a non-compliant section will be approved is:

$$P(A|NC) = 0.25.$$

4. **Probability that a non-compliant section will be rejected**:

The probability that a non-compliant section will be rejected is the complement of the probability that it will be approved:

$$P(R|NC) = 1 - P(A|NC)$$

So,

$$P(R|NC) = 1 - 0.25 = 0.75$$

Summary

- The probability that the EPA will approve a compliant section of the wastewater treatment plant based on the sample test is 0.75.
- The probability that a compliant section will be rejected based on the sample test is 0.25.
- The probability that a non-compliant section will be approved based on the sample test is 0.25.
- The probability that a non-compliant section will be rejected based on the sample test is 0.75.

3.4 Bayes' Theorem

The mathematical definition of the Bayes' Theorem, which plays a central role in risk analysis and decision-making under uncertainty, can be expressed in several equivalent forms. In the context of this chapter, it is presented in terms of conditional probability, and its mathematical formulation is shown.

To derive Bayes' Theorem, let's first write out the expressions for conditional probability:

$$\begin{cases} P(A|B) = \dfrac{P(A \cap B)}{P(B)} \\ P(B|A) = \dfrac{P(A \cap B)}{P(A)} \end{cases} \tag{3.24}$$

Both of these expressions contain the intersection of events A and B, P(A ∩ B). By equating these two expressions, we can derive Bayes' Theorem.

Starting with:

$$\begin{cases} P(A|B) \times P(B) = P(A \cap B) \\ P(B|A) \times P(A) = P(A \cap B) \end{cases} \tag{3.25}$$

Since both expressions yield P(A ∩ B), they can be set equal to one another:

$$P(A|B) \times P(B) = P(B|A) \times P(A) \tag{3.26}$$

To isolate for P(A|B), divide both sides by P(B):

$$P(A|B) = \frac{P(B|A) \times P(A)}{P(B)} \tag{3.27}$$

This is Bayes' Theorem. It shows how to update the probability of event A occurring given that event B has occurred, based on prior knowledge of P(A), P(B), and P(B|A) (Fig. 3.11).

Bayes' rule provides the basis for something called Bayesian Inference, a concept that we will be discussing further in this chapter. Bayesian Inference is a method of statistical inference in which Bayes' Theorem is used to update the probability of a hypothesis as more evidence or information becomes available. In other applications, such as modeling, Bayes' rule allows us to make parallel comparisons. Learning from the data can be seen as an indirect problem: we have a known observation or quantity, denoted by B, and we want to use this information to infer knowledge about an unknown variable A and model parameters (Fig. 3.12). This is often seen in scenarios where we have data (B), and we want to infer the underlying cause or parameter (A).

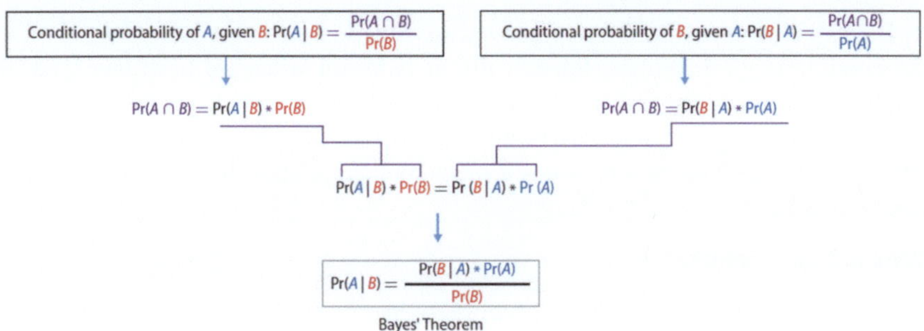

Fig. 3.11 The mathematical definition of Bayes' theorem

Fig. 3.12 Definition of observations and unknown (hypothesis) in Bayes' Theorem

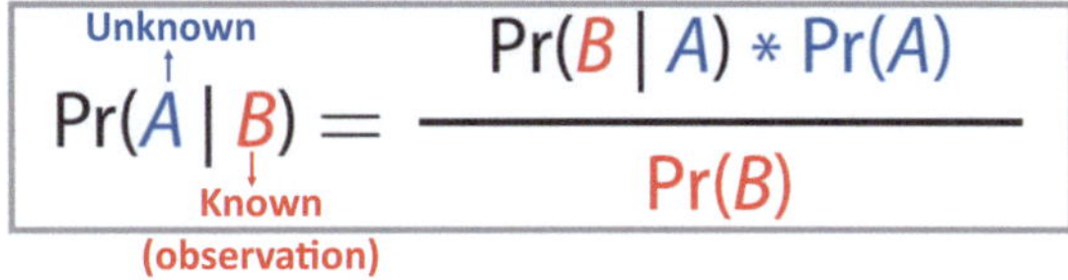

So, what is remarkable about this figure? Bayes' Theorem expresses how a subjective degree of belief should rationally change to account for evidence. This is important for several reasons:

- **Dynamic Updating**: Bayes' Theorem allows us to continuously update our beliefs and knowledge based on new data. This dynamic process of updating probabilities in light of new evidence is crucial in fields where conditions change rapidly, such as finance, medicine, and engineering.
- **Combining Prior Knowledge with New Evidence**: One of the most significant aspects of Bayesian Inference is that it seamlessly combines prior knowledge (prior probabilities) with new data (likelihood). This integration helps to make more informed and accurate decisions, especially when prior knowledge is robust but needs refinement with new evidence.
- **Rational Decision Making**: The theorem provides a rational framework for decision-making under uncertainty. It prescribes a logical way to update our degree of belief, ensuring that decisions are based on the most current and relevant information.
- **Handling Uncertainty**: In many real-world problems, uncertainty is a significant factor. Bayes' Theorem provides a structured approach for quantifying and managing this uncertainty, enabling better risk assessment and management.
- **Foundation for Advanced Techniques**: Bayes' Theorem is the foundation for many advanced statistical methods and machine learning algorithms. Techniques such as Bayesian Networks, Markov Chain Monte Carlo methods, and Bayesian optimization are built upon the principles of Bayesian Inference, providing powerful tools for data analysis and prediction.

Bayes' Theorem should be employed in situations where we need to reverse the conditional probabilities. Specifically, it is useful when we have one type of conditional probability and wish to find its inverse. For instance, if we know the probability of A given B (P(A|B)) but need to find the probability of B given A (P(B|A)), Bayes' Theorem provides a mathematical relationship that connects the two (Fig. 3.12).

Why is This Useful?

- **Ease of Calculation**: In many real-world scenarios, it might be easier to calculate one conditional probability over the other. For example, it might be straightforward to collect data on P(A|B) but challenging to measure P(B|A) directly. Bayes' Theorem allows us to leverage the easier-to-calculate probability to determine the more challenging one.
- **Data Collection Constraints**: Sometimes, due to practical constraints, we can only gather data for one of the conditional probabilities. For instance, in civil engineering, it might be easier to determine the probability of a bridge failing given a specific load condition (P(Failure|Load)) than to determine the probability of that load condition given that the bridge has failed (P(Load|Failure)). Bayes' Theorem helps bridge this gap by relating the two probabilities.
- **Connecting Prior and Posterior Probabilities**: Bayes' Theorem is fundamental in Bayesian Inference, where we update our prior beliefs (prior probabilities) with new evidence (likelihood) to develop updated beliefs (posterior probabilities). This process is crucial for iterative learning and decision-making in uncertain environments.
- **Rational Decision-Making**: By providing a structured way to update beliefs and probabilities based on new evidence, Bayes' Theorem supports rational and informed decision-making. It ensures that our decisions are based on the most current and comprehensive information available.

Figure 3.13 shows the terminology associated with Bayes' Rule.

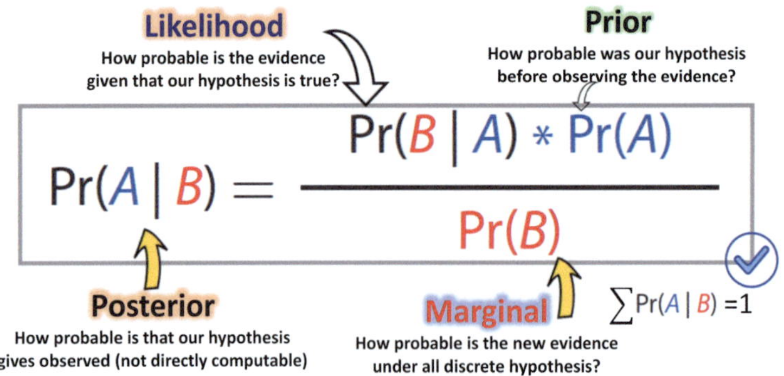

Fig. 3.13 Terminology associated with Bayes' rule

The prior distribution, denoted by the marginal probability of A, expresses our belief about the hypothesis before observing the evidence. The prior distribution describes how probable our test hypothesis is without having taken into account any of the data. Prior knowledge can be expressed in a number of ways and can, for example, be based on available data, direct measurement, or expert opinion. It is considered prior information because it is typically based on relevant information that exists before the test.

The posterior probability distribution describes how probable it is that our hypothesis occurs after considering new information. For this reason, the posterior probability is an updated or revised probability based on the information contained within the likelihood function. The posterior probability combines the prior distribution and the likelihood of the observed data to provide a new probability assessment.

The likelihood function describes the conditional probability of observing event B, given the values that A can take. It plays a crucial role in Bayes' formula as it represents the information provided by the data B that modifies our prior knowledge of A. The likelihood function essentially tells us how probable the observed data is under different possible values of A, thus impacting our posterior belief about A.

The term in the denominator is referred to as the evidence and acts as a normalization constant, ensuring that the sum of all posterior probabilities equals 1. The marginal probability (Evidence) P(B) is the total probability of the observed data under all possible hypotheses. It is computed by integrating or summing over all possible values of A, weighted by their prior probabilities and the likelihood of the observed data given those values.

Detailed Breakdown

- **Prior** (P(A)): This term reflects our initial belief about hypothesis A before considering the new data B. It is the baseline probability of A based on prior knowledge, which could come from historical data, expert opinion, or other sources of information available before the current observation.
- **Likelihood** (P(B|A)): This term quantifies how likely it is to observe the data B, assuming that A is true. It measures the compatibility of the observed evidence B with hypothesis A. The likelihood function is crucial because it updates our prior belief in light of the new evidence.
- **Posterior** (P(A|B)): This is the updated probability of hypothesis A after taking the observed data B into account. It combines the prior probability and the likelihood of the observed data to form a revised belief. The posterior probability gives us a new, informed perspective on the hypothesis after considering the evidence.
- **Marginal** (P(B)): This term serves as a normalizing constant to ensure that the total probability distribution sums to one. It is calculated by summing or integrating the product of the prior probability and the likelihood of all possible A values. The marginal probability represents the total probability of the evidence B across all hypotheses.

In summary, Bayes' rule incorporates new evidence from a test into the prior belief or hypothesis about the test result, which we then use to update the posterior

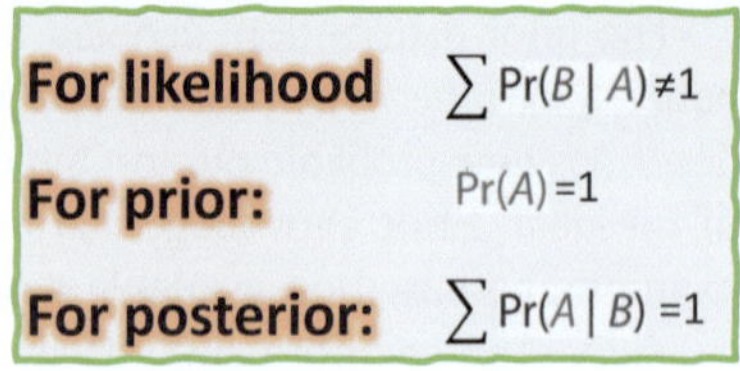

Fig. 3.14 Key properties of the likelihood, prior, and posterior distributions in Bayesian analysis

probability. This process can be repeated: our posterior distribution becomes the new prior distribution whenever we continue to update with new evidence. This iterative process is fundamental to Bayesian analysis, allowing us to refine and update our beliefs about a parameter based on the results of ongoing experiments.

So now, you may be wondering how exactly we compute the likelihood. Computing the likelihood of the observed data is a critical part of Bayesian analysis. Throughout this book, we will learn about several probability distributions that will enable you to estimate the likelihood of the observed data under each hypothesis.

The key takeaway is that Bayes' Theorem tells us that the posterior distribution is proportional to the product of the prior distribution and the standardized likelihood function. In other words, Bayes' Theorem informs us that the probability distribution for A after observing data B (the posterior distribution) is proportional to the product of the distribution for A before observing the data (the prior distribution) and the likelihood of observing B given A (Fig. 3.14).

- **Likelihood**: The sum of the likelihood $\sum P(B|A)$ is not equal to 1. This indicates that the likelihood function represents the conditional probability of the observed data B given different values of hypothesis A, and it is not a probability distribution over A.
- **Prior Probability**: The prior probability P(A) must sum to 1. This ensures that the prior distribution is a valid probability distribution over hypothesis A, reflecting our initial beliefs before observing the data.
- **Posterior Probability**: The sum of the posterior probabilities $\sum P(A|B)$ equals 1. This ensures that the posterior distribution is a valid probability distribution over hypothesis A, reflecting our updated beliefs after incorporating the observed data B.

3.4.1 Marginalization

Earlier in this chapter, the concept of conditional probability was introduced, where the probability of A given B is discussed. It is noted that this probability is proportional to the likelihood of B under the hypothesis of A_1.

Marginalization is a concept whereby a marginal probability is obtained from a joint probability and is an especially useful concept in the application of Bayes' Theorem, where the normalization constant is often quite difficult to estimate in many real-world applications. Consider a sample space S which is formed by a series of

$$= P(A_1|B) \times P(B)$$
$$= P(A_2|B) \times P(B)$$
$$= P(A_3|B) \times P(B)$$
$$\sum_{j=1}^{n} P(A_j|B) \times P(B) = P(A)$$

Fig. 3.15 Graphical representation of the concept of marginalization

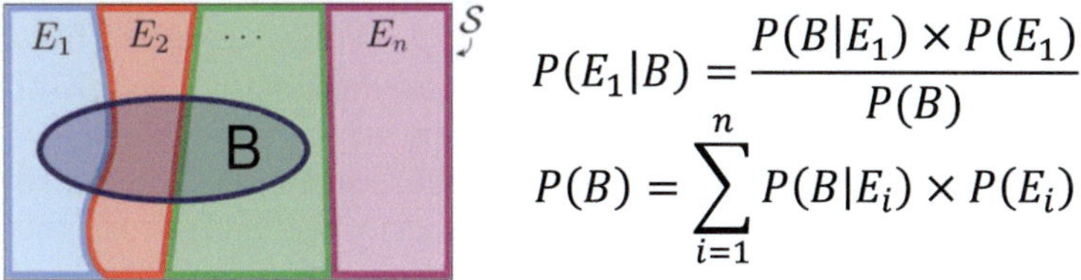

Fig. 3.16 A graphical representation of the marginalization process

into n mutually exclusive and collectively exhaustive events, denoted as $\{E_1, E_2, \ldots, E_n\}$. Consider that we are interested in another event, B, within the sample space S. Then, from Fig. 3.16, it can be observed that event B, which is shaded, consists of the intersections of B with $E_1, E_2, \ldots, E_n$. The probability of event B can be determined by summing the product of the joint probability of B and each event E_i, along with the probability of event E_i. The marginalization principle provides a means to derive probabilities of new propositions based on existing probabilities.

In Fig. 3.15, the concept of marginalization is represented graphically. It is fundamental knowledge that the probability of A can be equated to the sum of the intersections of A and B. Through the rule of conditional probability, it is possible to express the intersection as the probability of A_j given B, multiplied by the probability of B. In the example provided, three different A_j have been considered, and it is demonstrated pictorially how the sum of their conditional probabilities adds up to the probability of A.

3.4.2 *Joint Probability Distribution*

Understanding the relationships between multiple variables is crucial for modeling and predicting complex phenomena in probability and statistics. In this regard, joint probability distributions serve as a fundamental tool, providing a comprehensive framework to capture and analyze the interdependencies between variables. Whether dealing with discrete or continuous variables, joint probability enables us to comprehend the simultaneous occurrence of events and make informed predictions based on partial information. In the field of civil engineering, the ability to accurately model and predict various phenomena is essential for designing safe and efficient structures and systems. The significant role that joint probability distributions play in civil and environmental engineering applications will be better understood with some real-world examples. Consider the case of performing a flood risk assessment, which involves multiple variables such as precipitation, river flow, and temperature. Understanding the joint probability distribution of these variables allows engineers to estimate the likelihood of flooding under different scenarios. For instance, the probability of high precipitation combined with high-temperature conditions can inform the design of flood control measures and emergency response plans. In structural engineering, the safety and reliability of buildings and bridges depend on factors such as material strength, load conditions, and environmental impacts. By analyzing the joint probability distribution of these variables, engineers can assess the probability of structural failure under different combinations of loads and environmental conditions. This is crucial for ensuring that structures can withstand extreme events such as earthquakes and hurricanes. In transportation engineering, traffic flow and safety are influenced by variables such as vehicle speed, traffic density, and weather conditions. Joint probability distributions help engineers to model the interactions between these factors to better predict traffic congestion and accident risk. Moreover, joint probability distributions are invaluable in environmental engineering, where they aid in assessing pollution levels, resource management, and the impact of climatic changes. By comprehensively understanding the interdependencies between environmental factors, engineers can develop more effective strategies for pollution control, water resource management, and sustainable development.

3.4.2.1 Discrete Variables

Discrete variables are defined as those that can take on a finite number of distinct values. In civil engineering, these variables are often used to represent categorical data or measurements that can be classified into distinct groups. For example, the levels of damage to a structure can be categorized as no damage, minor damage, or major damage. Similarly, soil types can be grouped into classes, such as clay, sand, or gravel. Another example involves traffic volume, which can be classified into low, medium, or high categories. These discrete variables allow for the clear classification and analysis of data, which is essential for making informed decisions in civil engineering projects.

After understanding discrete variables and their applications in civil engineering, it becomes essential to explore how these variables can be analyzed together to understand their relationships. One effective way to do this is through the concept of joint probability distributions, which allows us to examine how two discrete variables interact. The process for handling discrete variables in joint probability calculations is

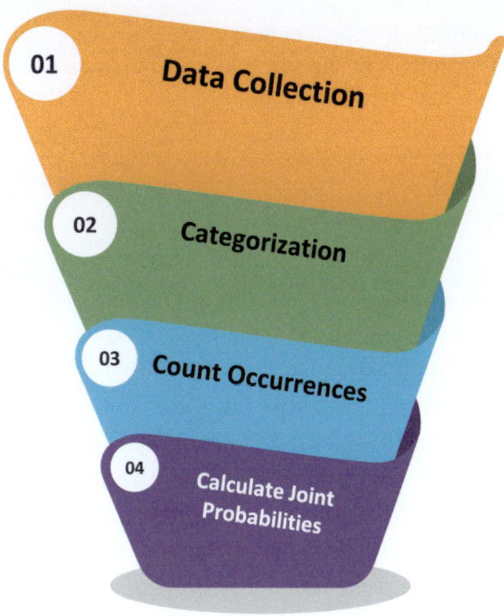

Fig. 3.17 Steps for handling discrete variables in joint probability calculations

outlined in Fig. 3.17. The procedure begins with data collection, where observations are gathered, each including values for the two discrete variables of interest. These data may be obtained from various sources such as field measurements, surveys, or historical records. Once the data is collected, the next step is categorization, which involves defining the possible categories for each variable. For instance, precipitation levels might be categorized as low, medium, or high, while soil types could be classified into categories such as clay, sand, and gravel. Following categorization, the occurrences of each possible combination of the variable categories are counted, creating a contingency table. Finally, the joint probabilities for each combination are calculated by dividing the count of each category combination by the total number of observations. This results in the development of a joint probability distribution table, which provides valuable insights into the relationship between the two discrete variables.

To illustrate this concept, let's consider the example of a flood risk assessment where we want to evaluate the joint probability distribution of precipitation levels and river flow categories. This joint distribution provides insight into how often different precipitation levels coincide with specific river flow rates, which is crucial for designing effective flood management strategies. Let's apply the previously outlined steps to this example:

In this example, the variables of interest are the precipitation level, denoted as X, and the river flow rate, denoted as Y. The precipitation level X is categorized into three distinct levels: Low (x_1), Medium (x_2), and High (x_3). Similarly, the river flow rate Y is also categorized into three levels: Low (y_1), Medium (y_2), and High (y_3).

Suppose we have collected data on precipitation levels and river flow rates over 30 days. Each day, the precipitation level is recorded as either low, medium, or high, and the corresponding river flow rate is noted as low, medium, or high. These observations

Table 3.2 Example of data collection for precipitation levels and river flow rates

Day	Precipitation level (X)	River flow rate (Y)
1	Low (x_1)	Low (y_1)
2	Medium (x_2)	Medium (y_2)
3	High (x_3)	High (y_3)
…	…	…
30	Medium (x_2)	High (y_3)

are categorized to help analyze the relationship between these two variables. Table 3.2 is an example of how the data might be structured, showing the precipitation levels and river flow rates for each day.

After categorizing the data, the next step is to count how often each combination of precipitation level and river flow rate occurs (Table 3.3). This involves reviewing the dataset and tallying the frequency of each pair of categories (e.g., Low precipitation with Low river flow, Medium precipitation with High river flow, etc.). These counts form the basis for constructing a contingency table, which helps in understanding the relationship between the variables. For example, if there are five instances where both precipitation and river flow are low, this count is recorded as part of the joint probability analysis. By counting the occurrences for all possible combinations, we can quantify how often certain conditions happen together.

Once the occurrences for each combination of precipitation level and river flow rate have been counted, the next step is to calculate the joint probabilities. This is done by dividing the count of each specific combination by the total number of observations, which in this case is 30. The resulting value represents the probability of that particular combination occurring. For instance, if there are five days where both precipitation and river flow are low, the joint probability for this combination would be calculated as 5/30, or approximately 0.167 (Table 3.4). These joint probabilities provide a quantitative measure of how likely each combination of conditions is to occur.

Table 3.3 Occurrence table for different combinations of precipitation levels and river flow rates

Precipitation (X)	River flow (Y)	Count
Low (x_1)	Low (y_1)	5
Low (x_1)	Medium (y_2)	3
Low (x_1)	High (y_3)	2
Medium (x_2)	Low (y_1)	4
Medium (x_2)	Medium (y_2)	6
Medium (x_2)	High (y_3)	5
High (x_3)	Low (y_1)	2
High (x_3)	Medium (y_2)	2
High (x_3)	High (y_3)	1

Table 3.4 Computing the joint probability distribution for precipitation levels and river flow rates

Precipitation (X)	River flow (Y)	Count	Joint probability P(X,Y)
Low (x_1)	Low (y_1)	5	5/30 = 0.167
Low (x_1)	Medium (y_2)	3	3/30 = 0.100
Low (x_1)	High (y_3)	2	2/30 = 0.067
Medium (x_2)	Low (y_1)	4	4/30 = 0.133
Medium (x_2)	Medium (y_2)	6	6/30 = 0.200
Medium (x_2)	High (y_3)	5	5/30 = 0.167
High (x_3)	Low (y_1)	2	2/30 = 0.067
High (x_3)	Medium (y_2)	2	2/30 = 0.067
High (x_3)	High (y_3)	1	1/30 = 0.033

Table 3.5 Joint probability distribution for precipitation levels and river flow rates

P(X,Y)		River flow rate		
		Low (y_1)	Medium (y_2)	High (y_3)
Precipitation	Low (x_1)	0.167	0.1	0.067
	Medium (x_2)	0.133	0.2	0.167
	High (x_3)	0.067	0.067	0.033

This result can be more conveniently represented with only the computed joint probability values, as shown in Table 3.5.

3.4.2.2 Continuous Variables

Continuous variables are defined as those that can take any value within a specified range, without being restricted to distinct, separate categories. Unlike discrete variables, which are limited to finite (countable) values, continuous variables can represent any value along a continuum, making them ideal for modeling quantities that require precision and granularity. Common examples of continuous variables include measurements such as temperature, load, stress, time, and distance—quantities that can be expressed as a continuous spectrum rather than distinct units.

In civil engineering, continuous variables play a crucial role in accurately modeling and predicting complex, real-world phenomena. For instance, the amount of rainfall over a region can vary continuously and needs to be measured with a high degree of accuracy to forecast potential flooding. Similarly, the strength of construction materials, such as concrete or steel, is not fixed but can vary depending on several factors, including the quality of materials and environmental conditions. Accurately measuring these variables is essential for ensuring the safety and reliability of engineering designs. Furthermore, continuous variables are often used to describe traffic speeds, which can vary continuously across different highway segments or at different

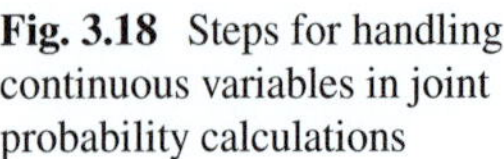

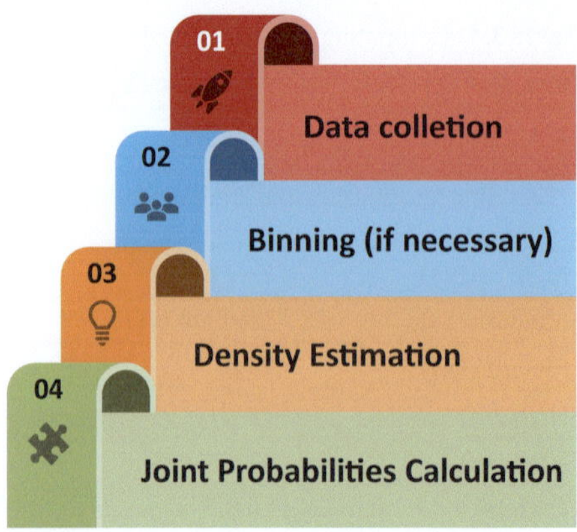

Fig. 3.18 Steps for handling continuous variables in joint probability calculations

times of the day. Engineers rely on continuous data to analyze traffic flow, optimize transportation infrastructure, and enhance road safety. By employing continuous variables in models and simulations, engineers can better understand and manage the inherent variability in natural and man-made systems, leading to more effective and efficient design, construction, and maintenance practices in civil engineering.

To analyze the relationship between continuous variables, the concept of the joint probability distribution can be applied. This process involves several steps (see Fig. 3.18 for a summary) that are crucial for understanding how two continuous variables interact.

The first step in this process is data collection. Here, data must be gathered, with each observation including values for the two continuous variables of interest. These data points can be obtained from various sources such as sensors, experiments, or simulations. For example, in civil engineering, sensors might be used to collect continuous data on environmental factors like temperature and humidity over time, or experiments might be conducted to measure the relationship between material stress and strain. The quality and accuracy of this data are critical, as they form the foundation for all subsequent analyses.

Next, binning might be considered, though this step is optional. Binning involves dividing the range of values for each continuous variable into intervals or bins, which can simplify the analysis process. This approach is particularly useful when dealing with large datasets or when the goal is to create a more manageable data representation. However, binning can also result in a loss of detail due to the grouping of numerous observations into a single category, so it should be used judiciously depending on the specific requirements of the analysis and the method chosen for calculating joint probabilities.

Following data collection and binning, the next step is density estimation. The joint probability distribution is typically represented by a probability density function (PDF) for continuous variables. Estimating the PDF is crucial because it provides a continuous representation of the probability distribution across the entire range of the variables. Several methods are available for density estimation, such as Kernel density estimation, which is a non-parametric method that smooths the data to estimate the PDF, or fitting a parametric model (e.g., a multivariate normal distribution), which involves assuming a specific form for the distribution and estimating its parameters. The choice of method depends on the characteristics of the data and the desired level of accuracy.

Finally, joint probability calculation involves integrating the joint PDF over the specific ranges or intervals of interest. The integration process allows for determining the probability that the variables fall within certain ranges, providing valuable insights into their relationship. Typically, this calculation is performed using numerical methods or specialized software tools, as manual integration can be challenging, especially for complex distributions.

By following these steps, engineers and researchers can effectively analyze the joint distribution of continuous variables, leading to more informed decisions and better predictions in their respective fields.

Scenario 1: Structural Load and Material Strength

In structural engineering, understanding the joint probability distribution of load and material strength is essential for assessing the probability of structural failure under various conditions. By analyzing this relationship, engineers can determine the likelihood of different combinations of load levels and material strength, which permits the design of structures that can withstand these stresses and environmental conditions, minimizing the risk of failure and maximizing safety.

In this example scenario, we focus on two continuous variables: X, which represents the load on a bridge measured in kilonewtons (kN), and Y, which denotes the material strength measured in megapascals (MPa). These two variables are critical for understanding the relationship between the forces applied to a structure and the strength of the materials used to build it. By analyzing the joint probability distribution of X and Y, engineers can gain insights into the likelihood of various load levels and material strength combinations, which is essential for assessing the risk of structural failure and ensuring the safety and reliability of the bridge.

Continuous data has been collected for 50 different bridge components in this scenario. Both the load applied to the bridge and the material strength have been recorded for each component. These data points are typically gathered through experiments or by using field mounted stress–strain gauges that monitor the performance of the bridge components under different stress conditions. Accurate data collection is crucial for ensuring that the joint probability distribution reflects real-world conditions and helps engineers make informed decisions about the structure's safety and durability. The collected data for all 50 components are provided in Table 3.6 for further analysis.

Following data collection, the next step in creating the joint probability distribution is binning. Although this step is optional, here we will divide the continuous range for each of the two variables into discrete intervals to facilitate the analysis of the joint probability distribution. This step helps manage the complexity of continuous

Table 3.6 Field data for 50 bridge components

Component	Load (X) (kN)	Material strength (Y) (MPa)
1	100	250
2	150	275
3	200	300
…	…	…
50	180	290

Table 3.7 Frequency of different combinations of load and material strength

Load (X) (kN)	Material strength (Y) (MPa)	Count
100–150	250–275	10
100–150	275–300	5
100–150	300–325	0
150–200	250–275	8
150–200	275–300	12
150–200	300–325	2
200–250	250–275	3
200–250	275–300	7
200–250	300–325	3

data by grouping values into manageable categories. For this scenario, the load (X) on the bridge is divided into three intervals: 100–150, 150–200, and 200–250 kN. Similarly, the material strength (Y) is divided into three intervals: 250–275, 275–300, and 300–325 MPa. These bins are chosen to cover the entire range of observed data for both variables. By creating this grid, it becomes easier to count the occurrences of each combination of load and material strength. This discretization is particularly useful in cases where the dataset is large or a simplified representation of the joint distribution is needed.

After creating the grid, the next step is to count how many observations fall into each combination of the bins for load and material strength. This involves tallying the number of bridge components with load values and corresponding material strength values within each respective interval. For example, we count the number of components with a load between 100 and 150 kN and material strength between 250 and 275 MPa. These counts represent the frequency of each bin combination Table 3.7.

With the counts of each combination of load and material strength bins determined, the next step is calculating the joint probabilities for these combinations. This is done by dividing the count of each specific bin combination by the total number of observations, which, in this case, is 50. The resulting values represent the probability that a randomly selected bridge component will have load and material strength values falling within the specified ranges.

For example, if there are 10 observations where the load is between 100 and 150 kN and the material strength is between 250 and 275 MPa, the joint probability for

Table 3.8 Computing the joint probability distribution for the combination of load and material strength

Load (X)	Material strength (Y)	Count	Joint probability $P(X,Y)$
100–150	250–275	10	$10/50 = 0.20$
100–150	275–300	5	$5/50 = 0.10$
100–150	300–325	0	$0/50 = 0.00$
150–200	250–275	8	$8/50 = 0.16$
150–200	275–300	12	$12/50 = 0.24$
150–200	300–325	2	$2/50 = 0.04$
200–250	250–275	3	$3/50 = 0.06$
200–250	275–300	7	$7/50 = 0.14$
200–250	300–325	3	$3/50 = 0.06$

Table 3.9 Joint probability distribution table

$P(X,Y)$	250–275 MPa	275–300 MPa	300–325 MPa
100–150	0.20	0.10	0.00
150–200	0.16	0.24	0.04
200–250	0.06	0.14	0.06

this combination is calculated as $10/50 = 0.200$. This means there is a 20% chance that a randomly selected component will have these characteristics.

By repeating this process for all combinations, we create a joint probability distribution that provides a detailed understanding of the relationship between load and material strength. This distribution is essential for assessing structural reliability, as it helps identify which combinations are most likely to occur and whether they might pose a risk of failure. The calculated joint probabilities are then organized into a table. Table 3.8 displays the count of occurrences for each bin combination along with the calculated joint probabilities, which are derived by dividing each count by the total number of observations (50).

After calculating the joint probabilities for each combination of binned load and material strength, we can organize these probabilities into a comprehensive joint probability distribution table. This table offers a clearer view of how the load and material strength variables interact, highlighting the likelihood of each specific combination occurring. The joint probability distribution table is provided in Table 3.9, where the joint probabilities are presented alongside the corresponding counts, offering a valuable resource for further analysis and decision-making in structural engineering.

Scenario 2: Earthquake Intensity and Structural Damage

In this scenario, the goal is to understand the relationship between earthquake intensity and the likelihood of structural damage. Earthquake intensity is a crucial factor in determining the extent of damage a structure might endure during a seismic event. To analyze this relationship, earthquake intensities are categorized into three distinct levels: Light (L), Moderate (M), and Important (I). Structures subjected to these earthquakes can either remain undamaged $\overline{D}$ or sustain damage (D). By examining the conditional probabilities of structural damage given the intensity of the earthquake, we can better predict and mitigate the risks associated with seismic activity.

Based on prior experience and historical data, it is known that the majority of earthquakes fall into the light category, with an estimated 90% of all events being light in intensity. Moderate earthquakes are less common, accounting for about 8% of occurrences, while important or severe earthquakes are rare, making up only about 2% of all events. These probabilities provide a baseline understanding of earthquake occurrences and allow us to assess the potential risks more accurately.

We are interested in determining the conditional probabilities, or likelihood, of structural damage occurring given each level of earthquake intensity. This involves calculating the probability that a structure will be in a damaged state (D), given that a light, moderate, or important earthquake has occurred. Understanding these conditional probabilities is crucial for engineers and decision-makers, as it helps in designing structures that can better withstand seismic forces, implementing effective emergency response plans, and improving overall earthquake preparedness.

By analyzing the conditional probabilities of damage for each intensity level, we can develop strategies to minimize the impact of earthquakes on structures. For example, suppose it is found that structures have a high likelihood of damage during moderate or important earthquakes. In that case, building codes can be updated to enforce stricter design standards in earthquake-prone areas. Additionally, this analysis can inform the retrofitting of existing structures to enhance their resilience against seismic events.

Step 1. Data Collection

In this step, we focus on gathering historical data related to earthquake events, which serves as the foundation for our analysis. The primary data of interest includes the intensity of each recorded earthquake and the resulting condition of structures, classified as either damaged (D) or undamaged ($\overline{D}$). This historical data is crucial because it provides empirical evidence of how different intensities of earthquakes impact structural integrity.

The data collection process involves compiling records from various sources, such as seismic monitoring agencies, geological surveys, and historical archives that document past earthquakes and their effects on structures. By systematically collecting this information, we aim to capture a wide range of earthquake events across different intensities—light (L), moderate (M), and important (I)—and the corresponding outcomes for structures. This comprehensive dataset allows us to identify patterns and correlations between earthquake intensity and the likelihood of structural damage.

Additionally, the dataset should include a significant number of events to ensure statistical reliability and generalizability. The quality and accuracy of this data are paramount, as it forms the basis for calculating the conditional probabilities that will be used to predict the likelihood of damage given a specific earthquake intensity.

Step 2: Categorization

For this analysis, suppose we have collected a dataset consisting of 1000 samples. Each sample includes two key pieces of information: the earthquake intensity and the structure state. The earthquake intensity is categorized into three levels: light (L), moderate (M), or important (I), reflecting the severity of each seismic event. The structure state indicates whether the structure was damaged (D) or remained undamaged ($\overline{D}$) as a result of the earthquake.

To facilitate further analysis, this data is organized into a structured table (Table 3.10), which helps in visualizing the distribution of structural states across the different categories of earthquake intensity. With this organized data, we can proceed to explore how different intensities of earthquakes influence the likelihood of structural damage. This information can be used in the improvement of building earthquake resilience and safety.

Summary of the Data Collection

In this hypothetical dataset, we have gathered a total of 1000 samples of earthquake events, meticulously categorized by the intensity of the earthquakes and the resulting state of the structures—whether they were damaged or undamaged. This data serves as the foundation for the analysis and understanding of the relationship between earthquake intensity and the likelihood of structural damage. By examining this relationship, we can gain insights into how different intensities of earthquakes impact structural integrity, which is vital for designing safer buildings and infrastructure.

1. **Light (L)**:

   - Total Samples: 900
   - Damaged Structures: 9
   - Undamaged Structures: 891.

   The majority of the earthquake events fall into the light category, with 900 samples recorded. Within this category, it is observed that only 9 structures sustained damage, while 891 structures remained undamaged. This indicates that light earthquakes generally have a low likelihood of causing structural damage.
2. **Moderate (M)**:

   - Total Samples: 80
   - Damaged Structures: 8
   - Undamaged Structures: 72.

Table 3.10 Sample data of earthquake events and structural states. Each row represents an individual sample, detailing the earthquake intensity categorized as light (L), moderate (M), or important (I), and the corresponding structure state, indicating whether the structure was damaged (D) or undamaged ($\overline{D}$)

Sample	Earthquake intensity	Structure state
1	Light (L)	Undamaged ($\overline{D}$)
2	Light (L)	Undamaged ($\overline{D}$)
3	Light (L)	Undamaged ($\overline{D}$)
4	Light (L)	Undamaged ($\overline{D}$)
5	Light (L)	Undamaged ($\overline{D}$)
6	Light (L)	Undamaged ($\overline{D}$)
7	Light (L)	Undamaged ($\overline{D}$)
...	...	...
900	Light (L)	Undamaged ($\overline{D}$)
901	Moderate (M)	Undamaged ($\overline{D}$)
902	Moderate (M)	Undamaged ($\overline{D}$)
903	Moderate (M)	Undamaged ($\overline{D}$)
904	Moderate (M)	Undamaged ($\overline{D}$)
905	Moderate (M)	Undamaged ($\overline{D}$)
906	Moderate (M)	Undamaged ($\overline{D}$)
907	Moderate (M)	Undamaged ($\overline{D}$)
908	Moderate (M)	Damage (d)
909	Moderate (M)	Damage (D)
...	...	...
980	Moderate (M)	Damage (D)
981	Important (I)	Undamaged ($\overline{D}$)
982	Important (I)	Undamaged ($\overline{D}$)
983	Important (I)	Undamaged ($\overline{D}$)
984	Important (I)	Undamaged ($\overline{D}$)
985	Important (I)	Undamaged ($\overline{D}$)
986	Important (I)	Undamaged ($\overline{D}$)
987	Important (I)	Undamaged ($\overline{D}$)
988	Important (I)	Damage (D)
989	Important (I)	Damage (D)
990	Important (I)	Damage (D)
991	Important (I)	Damage (D)
992	Important (I)	Damage (D)
993	Important (I)	Damage (D)
994	Important (I)	Damage (D)
995	Important (I)	Damage (D)
996	Important (I)	Damage (D)
997	Important (I)	Damage (D)
998	Important (I)	Damage (D)

(continued)

Table 3.10 (continued)

Sample	Earthquake intensity	Structure state
999	Important (I)	Damage (D)
1000	Important (I)	Damage (D)

For moderate earthquakes, 80 samples have been collected. Of these, 8 structures were reported as damaged, while 72 remained undamaged. This data suggests that as the intensity of the earthquake increases to a moderate level, the likelihood of structural damage also increases. However, a significant number of structures still manage to withstand the seismic forces and remain undamaged.

3. **Important (I)**:

- Total Samples: 20
- Damaged Structures: 12
- Undamaged Structures: 8.

The dataset also includes 20 samples of important (or severe) earthquakes. In this category, a higher proportion of structures experienced damage, with 12 out of the 20 structures being damaged and only 8 remaining undamaged. This indicates a much higher risk of structural damage during significant seismic events, underscoring the importance of designing structures to withstand such forces.

This comprehensive dataset not only provides the raw numbers necessary for calculating probabilities but also highlights the varying levels of risk associated with different earthquake intensities. By understanding the frequency and impact of each intensity level, we can use this data to calculate the conditional probabilities of damage given an earthquake's intensity. This, in turn, informs the development of strategies for mitigating damage and improving the resilience of structures in earthquake-prone areas.

Step 3: Count Occurrences

In this step, we focus on calculating the prior probabilities by counting the number of earthquakes that fall into each intensity category. Additionally, we aim to calculate the conditional probabilities of damage given each intensity level. These calculations are crucial for understanding the overall likelihood of different earthquake intensities occurring and their potential impact on structures.

Count the Number of Earthquakes of Each Intensity

The first part of this step involves counting the total number of earthquake events categorized into light, moderate, and important intensities. As identified in the dataset, 900 of the 1000 recorded earthquakes are classified as light, indicating that the majority of seismic events tend to be less severe. Moderate earthquakes account for 80 samples, while important or severe earthquakes are relatively rare, with only 20 samples.

- **Light (L)**: 900 samples

- **Moderate (M)**: 80 samples
- **Important (I)**: 20 samples.

Calculating the Prior Probabilities

We can calculate the prior probabilities for each earthquake intensity category based on these counts. The prior probability represents the likelihood of an earthquake falling into each intensity category before considering any additional information. This is calculated by dividing the number of events in each category by the total number of samples (1000):

Light (L)

$$P(L) = \frac{900}{1000} = 0.9$$

This indicates that there is a 90% probability that any given earthquake will be light in intensity.

Moderate (M)

$$P(M) = \frac{80}{1000} = 0.08$$

This shows that moderate earthquakes have an 8% probability of occurring.

Important (I)

$$P(I) = \frac{20}{1000} = 0.02$$

There is a 2% probability that an earthquake will be of important intensity, indicating that such severe events are relatively rare.

Count the Number of Damaged Structures for Each Intensity

Next, we should analyze the dataset to determine the number of damaged structures associated with each earthquake intensity. This involves counting how many structures were damaged or remained undamaged for light, moderate, and important earthquakes.

For Light (L):

- Total samples: 900
- Damaged: 9
- Undamaged: 891.

Despite the high occurrence of light category earthquakes, only a small fraction of structures—9 out of 900—sustained damage during light earthquakes, suggesting that such events typically pose a lower risk of structural damage.

For Moderate (M):

- Total samples: 80
- Damaged: 8
- Undamaged: 72.

While less frequent, moderate earthquakes show a higher likelihood of causing structural damage. Out of 80 recorded moderate earthquakes, 8 resulted in structural damage. This indicates a greater potential risk for damage as the intensity increases from light to moderate.

For Importance (I):

- Total samples: 20
- Damaged: 12
- Undamaged: 8.

Important or severe earthquakes are rare, with only 2% occurrence, but have a significant impact on structures. In this category, 12 out of 20 events resulted in damage, highlighting the considerable risk posed by high-intensity earthquakes. This underscores the need for robust structural design and earthquake-resistant construction practices in areas prone to severe seismic activity.

Calculate Conditional Probabilities

With the counts of damaged structures identified for each intensity level, we can now calculate the conditional probabilities of damage given each intensity. These conditional probabilities indicate the likelihood that a structure will be damaged if an earthquake of a specific intensity occurs. They are calculated by dividing the number of damaged structures by the total number of earthquakes for each intensity category:

- Conditional Probability of Damage Given Light (L) Intensity Earthquake:

$$P(D|L) = \frac{\text{Number of damaged structures given Light}}{\text{Total number of Light earthquakes}} = \frac{9}{900} = 0.01$$

- Conditional Probability of Damage Given Moderate (M) Intensity Earthquake:

$$P(D|L) = \frac{\text{Number of damaged structures given Moderate}}{\text{Total number of Moderate earthquakes}} = \frac{8}{80} = 0.1$$

- Conditional Probability of Damage Given Important (I) Intensity Earthquake:

$$P(D|L) = \frac{\text{Number of damaged structures given Important}}{\text{Total number of Important earthquakes}} = \frac{12}{20} = 0.6$$

Table 3.11 Joint probability distribution table for earthquake intensity and structure state

Earthquake intensity	Structure state	Count	Joint probability *P*(Intensity, state)
Light (L)	Damage (D)	9	9/1000 = 0.009
Light (L)	Undamaged ($\overline{D}$)	891	891/1000 = 0.891
Moderate (M)	Damage (D)	8	8/1000 = 0.008
Moderate (M)	Undamaged ($\overline{D}$)	72	72/1000 = 0.072
Importance (I)	Damage (D)	12	12/1000 = 0.012
Importance (I)	Undamaged ($\overline{D}$)	8	8/1000 = 0.008

Step 4: Calculate Joint Probabilities

In this step, we calculate the joint probabilities for each combination of earthquake intensity and structure state (damaged or undamaged). Joint probabilities provide a comprehensive view of the likelihood that both an earthquake of a certain intensity and a specific structural state will occur simultaneously.

To calculate these joint probabilities, we take the count of each category combination (e.g., light earthquakes resulting in undamaged structures) and divide it by the total number of observations in the dataset, which is 1000 in this case. This process yields a joint probability distribution that quantifies how often each combination occurs relative to the entire dataset. Table 3.11 shows the calculated joint probabilities for each combination of earthquake intensity (light, moderate, and important) and structural outcome (damaged or undamaged). The joint probabilities are derived by dividing the count of each category combination by the total number of observations (1000).

Calculating Conditional Probabilities from Joint Probability

To determine the conditional probabilities, or the likelihood of structural damage given each earthquake intensity, we make use of the joint probabilities and the prior probabilities that we have previously calculated. Conditional probabilities provide a deeper understanding of how the occurrence of one event (in this case, an earthquake of a certain intensity) influences the probability of another event (structural damage). This is crucial for risk assessment and preparedness in earthquake-prone areas.

The conditional probability of damage given a specific earthquake intensity is calculated using the formula:

$$P(D|\text{Intensity}) = \frac{P(D \cap \text{Intensity})}{P(\text{Intensity})}$$

where $P(D|\text{Intensity})$ is the conditional probability of damage given the intensity, P(Intensity ∩ D) is the joint probability of the specific intensity and damage occurring together, and P(Intensity) is the prior probability of the earthquake occurring at that specific intensity.

Let's apply this formula to calculate the conditional probabilities for each intensity category:

- **Conditional Probability of Damage Given Light (L)**:

$$P(D|\mathrm{L}) = \frac{P(L \cap D)}{P(\mathrm{L})} = \frac{0.009}{0.9} = 0.01$$

  This result indicates that, given a light earthquake, there is a 1% probability that structural damage will occur.
- **Conditional Probability of Damage Given Moderate (M)**:

$$P(D|\mathrm{M}) = \frac{P(M \cap D)}{P(\mathrm{M})} = \frac{0.008}{0.08} = 0.01$$

  This calculation shows that the likelihood of damage occurring given a moderate earthquake is 10%.
- **Conditional Probability of Damage Given Important (I)**:

$$P(D|\mathrm{I}) = \frac{P(I \cap D)}{P(\mathrm{I})} = \frac{0.012}{0.02} = 0.6$$

  For important earthquakes, the probability of structural damage occurring rises significantly to 60%.

As highlighted in this example, conditional probabilities can be calculated using either the counts directly from Step 3 or by utilizing the joint probabilities from Step 4. Both approaches are valid and will yield the same results, offering flexibility in how we analyze and interpret the data. Understanding these two methods is important as it allows us to choose the most appropriate approach depending on the context and available data.

- **Direct Method (Step 3)**: This method involves using the counts directly to find the conditional probabilities. By simply dividing the number of occurrences where both events happen (e.g., the number of times a structure is damaged given a specific earthquake intensity) by the total count of the event of interest (e.g., the total number of earthquakes of that intensity), we can obtain the conditional probability. This approach is straightforward and effective when dealing with raw data, as it requires minimal calculations.
- **Joint Probability Method (Step 4)**: In this method, we use the joint probabilities along with the prior probabilities to calculate the conditional probabilities. This involves dividing the joint probability of the two events occurring together by the prior probability of the conditioning event. Although this method involves an additional step of calculating the joint probabilities first, it is particularly useful when a more refined analysis is needed or when the dataset is large and complex. It also provides a deeper understanding of the relationship between the events by

explicitly considering both the joint occurrence and the overall likelihood of the conditioning event.

Both methods ultimately lead to the same results, as demonstrated in the calculations. Whether we use the counts from Step 3 directly or the joint probabilities from Step 4, we arrive at the same conditional probabilities, confirming the robustness of our analysis. This dual approach allows for flexibility in data analysis. It reinforces the accuracy of our conclusions, making it an invaluable tool in assessing risks and impacts, such as those associated with earthquake-induced structural damage.

By understanding and applying both methods, we can enhance our ability to analyze complex relationships between variables, leading to better-informed decisions and more effective strategies for mitigating risks in various fields, including structural engineering and disaster preparedness.

Scenario 3: Quality Assurance of Aggregates for Construction

In this scenario, we wish to ensure the quality of aggregates used to construct a reinforced concrete building. Aggregates are supplied by two companies, Company A and Company B. Orders are for Company A to deliver 600 truckloads per day and for Company B to deliver 400 truckloads per day. From prior experience, 3% of the material from Company A is expected to be substandard, whereas 1% of the material from Company B is substandard. What is the probability that a load of aggregates is from Company A, given that it is found to be substandard?

A = Aggregates supplied by Company A.

B = Aggregates supplied by Company B.

E = Aggregates are substandard.

Step 1: Data Collection

In this quality assurance scenario, the first step involves collecting data on the daily supply of aggregates to understand the distribution between the two suppliers and assess the potential risk of receiving substandard materials. This step is crucial for establishing a clear picture of the supply chain and identifying any patterns that might influence the quality of the aggregates used in construction.

We have a total of 1000 truckloads of aggregates delivered to the construction site each day, sourced from two different companies:

- **Company A** supplies a significant portion of the aggregates, delivering 600 truckloads per day. This means that Company A is responsible for 60% of the total aggregate supply. Given this high volume, any issues with the quality of aggregates from Company A could substantially impact the overall quality of the concrete used in the construction project.
- **Company B** supplies the remaining 400 truckloads per day, accounting for 40% of the total aggregate supply. Although Company B delivers a smaller quantity,

ensuring the quality of these aggregates is equally important for maintaining the structural integrity of the building.

By understanding this supply distribution, we can gain insights into the likelihood of encountering substandard aggregates from each company. Knowing that 3% of Company A's aggregates and 1% of Company B's aggregates are substandard, we can use this data to calculate the overall probability of receiving substandard materials and to identify which supplier is more likely to be the source of these materials. This knowledge is essential for implementing quality control measures, such as more rigorous testing or selective sourcing, to ensure that only high-quality aggregates are used in the construction process.

Collecting and analyzing this data allows us to make informed decisions about supplier selection and quality assurance strategies. It provides a foundation for calculating the conditional probability of receiving a substandard load from a specific supplier, which will be further explored in the following steps.

Step 2: Categorization

In this step, we categorize the collected data to understand better the quality distribution of aggregates supplied by the two companies. Recall that categorization helps organize the data into meaningful groups, which, in this example, will be essential for conducting further analysis and drawing accurate conclusions about the likelihood of receiving substandard materials from each supplier.

For this scenario, suppose we have a total of 1000 samples, each representing a single truckload of aggregates. Each sample contains the following information:

1. **Supplier**:
   - **Company A**: Out of the 1000 samples, 600 are supplied by Company A. Thus, Company A is the primary supplier, delivering the majority of the aggregates used in the construction project. Given the volume of aggregates provided by this supplier, it is important to closely monitor the quality to prevent any potential issues that could affect the integrity of the construction.
   - **Company B**: The remaining 400 samples are supplied by Company B. While Company B supplies a smaller portion of the aggregates, ensuring the quality of these materials is equally crucial. The categorization allows us to compare the quality of aggregates from both suppliers and identify any disparities that may exist.
2. **Substandard Loads**:
   - **Yes (Substandard)**: This category identifies the samples that were found to be substandard, meaning they did not meet the required quality standards for use in concrete production. Substandard aggregates could include materials that are too fine or coarse, have impurities, display alkali-aggregate reactivity, or lack the necessary strength characteristics.
   - **No (Non-substandard)**: This category includes the samples that meet the quality requirements and are suitable for use in construction. Ensuring a high

Table 3.12 Categorization of aggregate samples from Company A and Company B

Sample	Supplier	Substandard
1	A	No
2	A	No
…	…	…
582	A	No
583	A	Yes
…	…	…
600	B	Yes
601	B	No
602	B	No
…	…	…
996	B	No
997	B	Yes
998	B	No
999	B	No
1000	B	Yes

proportion of non-substandard aggregates is key to maintaining the structural integrity and durability of the concrete building.

By organizing this data into a structured table, we can clearly see the distribution of samples between the two suppliers and identify the proportion of substandard and non-substandard loads. This structured dataset serves as a foundation for calculating the probabilities associated with each supplier and the likelihood of receiving substandard materials. By analyzing this categorized data, we can make informed decisions about supplier quality and implement targeted quality control measures to ensure that only high-quality aggregates are used in the construction process.

Table 3.12 displays each sample along with its supplier and whether it was identified as substandard or non-substandard. The organization of collected field data in such a manner permits an easier analysis of the probabilities associated with receiving substandard materials in the steps that follow.

Step 3: Count Occurrences

In this step, we determine the prior probabilities of selecting a truckload from either supplier, the joint probabilities of receiving a substandard or non-substandard load from each company, and the conditional probabilities of a substandard sample, given its supplier.

Prior Probabilities

The prior probabilities represent the likelihood of selecting a truckload from either Company A or Company B without considering the quality of the aggregates. These probabilities give us a baseline understanding of the distribution of the supply:

Table 3.13 Joint probability distribution of aggregate quality from Company A and Company B

Supplier	Substandard?	Count	Joint probability *P*(Supplier, substandard)
A	Yes	18	18/1000 = 0.018
A	No	582	582/1000 = 0.582
B	Yes	4	4/1000 = 0.004
B	No	396	396/1000 = 0.396

- Probability of aggregates being supplied by Company A:

$$P(A) = \frac{600}{1000} = 0.6$$

  This indicates that there is a 60% chance of selecting a truckload from Company A.
- Probability of aggregates being supplied by Company B:

$$P(B) = \frac{400}{1000} = 0.4$$

  This shows a 40% probability of selecting a truckload from Company B.

These prior probabilities serve as the foundation for understanding the distribution of the supply before considering any quality factors.

Joint Probabilities

The joint probabilities represent the probability of selecting a truckload that is both from a specific company and either substandard or non-substandard. This calculation gives us a more detailed picture of the likelihood of receiving a substandard load from each supplier. As with the previous examples, we establish our occurrence table by listing all possible combinations of supplier and aggregate quality. The count of samples for each occurrence is divided by the total number of samples in the population to develop the joint probability distribution (Table 3.13).

Calculate Conditional Probabilities

Conditional probabilities allow us to assess the likelihood that a truckload is substandard given that it is from a specific company. If we denote a substandard aggregate load by the event E, then:

- Conditional Probability of Substandard Aggregates from Company A

$$P(E|A) = \frac{\text{Number of substandard aggregates from A}}{\text{Total number of aggregates from A}} = \frac{18}{600} = 0.03$$

  This indicates that there is a 3% chance that a truckload from Company A will be substandard.

- Conditional Probability of Substandard Aggregates from Company B

$$P(E|B) = \frac{\text{Number of substandard aggregates from B}}{\text{Total number of aggregates from B}} = \frac{4}{400} = 0.01$$

For Company B, the probability of a truckload being substandard is 1%.

These conditional probabilities provide insight into the quality of aggregates from each supplier, allowing us to assess and compare their performance. By understanding these probabilities, construction managers and quality assurance teams can make informed decisions on how to address quality issues, whether by increasing inspections, adjusting the sourcing strategy, or implementing more stringent acceptance criteria for the aggregates supplied by each company.

Step 4: Joint Probabilities

In this step, we focus on determining the joint probabilities, which represent the likelihood of selecting a truckload that is both from a specific company and either substandard or non-substandard. Joint probabilities give us a comprehensive understanding of how the quality of aggregates is distributed across the suppliers, helping to identify patterns or trends in the quality assurance process.

To calculate the joint probabilities, we divide the count of each category combination (e.g., the number of substandard loads from Company A) by the total number of observations (1000). This allows us to determine the likelihood of each specific combination occurring within the entire dataset.

- **Joint Probability of Aggregates from Company A Being Substandard *P*(*A* ∩ *E*):**

$$P(A \cap E) = \frac{\text{Number of substandard aggregates from A}}{\text{Total number of aggregates}} = \frac{18}{1000} = 0.018$$

- **Joint Probability of Aggregates from Company A Being Non-substandard *P*(*A* ∩ $\overline{E}$):**

$$P(A \cap \overline{E}) = \frac{\text{Number of non-substandard aggregates from A}}{\text{Total number of aggregates}} = \frac{582}{1000} = 0.582$$

- **Joint Probability of Aggregates from Company B Being Substandard *P*(*B* ∩ *E*):**

$$P(B \cap E) = \frac{\text{Number of substandard aggregates from B}}{\text{Total number of aggregates}} = \frac{4}{1000} = 0.004$$

- **Joint Probability of Aggregates from Company B Being Non-substandard *P*(*B* ∩ $\overline{E}$):**

Table 3.14 Joint probability distribution of aggregate quality from Company A and Company B

Supplier	Substandard	Count	Joint probability P(Supplier, substandard)
A	Yes	18	0.018
A	No	582	0.582
B	Yes	4	0.004
B	No	396	0.396

$$P(B \cap \overline{E}) = \frac{\text{Number of non-substandard aggregates from B}}{\text{Total number of aggregates}} = \frac{396}{1000} = 0.396$$

These calculations can be summarized in a Joint Probability Distribution Table, as shown in Table 3.14, for easier viewing and interpretation.

These joint probabilities give us insight into the distribution of substandard and non-substandard aggregates between the two suppliers. By knowing these probabilities, we can better assess the quality of aggregates and take appropriate quality control measures.

Calculate Conditional Probabilities

After calculating the joint probabilities, we proceed to determine the conditional probabilities, which represent the likelihood of a truckload being substandard, given that it is from a specific company. This is done using the joint probabilities along with the prior probabilities calculated in Step 3.

- Conditional Probability of Substandard Aggregates from Company A

$$P(E|\text{A}) = \frac{P(A \cap E)}{P(\text{A})} = \frac{0.018}{0.6} = 0.03$$

This indicates that there is a 3% probability that a truckload from Company A will be substandard.

- Conditional Probability of Substandard Aggregates from Company B

$$P(E|\text{B}) = \frac{P(B \cap E)}{P(\text{B})} = \frac{0.004}{0.4} = 0.01$$

This shows that there is a 1% probability that a truckload from Company B will be substandard.

Understanding these conditional probabilities is crucial for making informed decisions about quality control and supplier management. By identifying the likelihood of receiving substandard aggregates from each supplier, we can implement targeted measures to improve overall material quality, such as increasing inspection rates for certain suppliers or negotiating quality standards with them.

Scenario 4: Quality Control of Concrete for Reinforced Concrete Construction

In this scenario, we aim to ensure the quality of concrete material used in reinforced concrete construction. Quality control is a critical aspect of construction projects, as the strength and durability of concrete significantly affect the structural integrity and safety of the completed building. To monitor and ensure the quality of the concrete, compressive strength cylinder samples are collected randomly (in accordance with the contract documents) from the concrete mixes delivered to the construction site.

Past records indicate that 80% of the concrete cylinders cast from mixes delivered to the site are of satisfactory quality. This high percentage suggests a generally reliable mixing process, but quality assurance is still necessary to identify and mitigate potential issues. To achieve this, the engineer requires that at least one cylinder from each batch delivered daily be tested for minimum compressive strength after a 7-day curing period. The compressive strength test is a standard procedure in construction quality control and provides an indication of the concrete's ability to bear loads without cracking or failing.

However, it is important to note that the test method employed is not perfect. Its reliability is 90%, meaning there is a 90% chance that a good-quality concrete cylinder will pass the test. Conversely, there is still a 10% chance that a good-quality cylinder could fail the test due to inherent imperfections in the testing process. Additionally, the test has a 10% probability of incorrectly passing a poor-quality cylinder.

The objective in this scenario is to determine the probability that a concrete mix is of good quality, given that the cylinder sample passes the compressive strength test. To do so, we must calculate the conditional probability, considering both the prior probability of receiving good-quality concrete and the reliability of the testing method. Understanding this probability is crucial for making informed decisions about the quality and safety of the concrete used in construction, ensuring that the structure meets the necessary standards for strength and durability.

Step 1: Data Collection

To illustrate this example, suppose that throughout the construction project, 1000 compressive strength cylinders are cast and collected. This sample set will be used to demonstrate the quality control process and the testing outcomes. By gathering these samples, we can better understand the distribution of good and poor-quality concrete, as well as the effectiveness of the test method in identifying the quality of the material. If we denote a cylinder being good quality concrete by the event G, then:

- **Good quality concrete (G)**: 80% of 1000 = 800 samples
- **Poor quality concrete ($\overline{G}$)**: 20% of 1000 = 200 samples.

Table 3.15 shows a subset of the 1000 samples collected for quality control of concrete used in reinforced concrete construction. Each sample is categorized based on its quality (good or poor) and the test result (pass or fail).

Table 3.15 Sample data for concrete quality and test results

Sample	Quality (G)	Test result (T)
1	Good	Pass
2	Good	Pass
3	Good	Pass
4	Good	Fail
5	Poor	Pass
…	…	…
800	Good	Pass
801	Poor	Fail
…	…	…
1000	Poor	Fail

Step 2: Categorization

For each of our 1000 samples in this example, the information collected includes:

1. **Concrete Quality**:
    - Good (800 samples)
    - Poor (200 samples)
2. **Compressive Strength Test Result**:
    - Pass
    - Fail.

By categorizing the samples into these groups, we can systematically evaluate the performance of the test and understand how often good and poor-quality concrete is correctly identified.

Step 3: Count Occurrences and Calculate Prior Probabilities

For each of the categories identified in step 2, we should count the number of occurrences from the data set.

- **Good quality concrete (G)**:
    - Pass: $0.9 \times 800 = 720$
    - Fail: $0.1 \times 800 = 80$
- **Poor quality concrete ($\overline{G}$)**:
    - Pass: $0.1 \times 200 = 20$
    - Fail: $0.9 \times 200 = 180$.

These counts help us understand the distribution of test outcomes for good and poor-quality concrete.

Prior Probabilities

The prior probability data were provided to the reader in the problem statement. This information is formulated based on our past experience with the concrete supplier and their track record of performance. Based on these past records, these prior probabilities represent the likelihood of the concrete being of good or poor quality and can be used in the calculation of the conditional probabilities related to the test outcomes.

- **Probability of good quality concrete P(G):**

$$P(\mathrm{G}) = 0.8$$

- **Probability of poor quality concrete P($\overline{G}$):**

$$P(\overline{G}) = 0.2$$

Conditional Probabilities Calculated using Direct Counts

If we denote passing the compressive strength test by the event E, then:

- Conditional Probability of Passing the Test Given Good Quality

$$P(T|G) = \frac{\text{Number of passes compressive tests}}{\text{Number of accepted compressive tests}} = \frac{720}{800} = 0.9$$

This means there is a 90% probability that a good-quality concrete sample will pass the test, reflecting the reliability of the test method.

- Conditional Probability of Passing the Test Given Good Quality

$$P(T|\overline{G}) = \frac{\text{Number of passes compressive tests}}{\text{Number of accepted compressive tests}} = \frac{20}{200} = 0.1$$

This implies a 10% probability that a poor-quality concrete sample will pass the test, indicating a limitation in the test method's ability to detect substandard material.

These conditional probabilities are derived directly from the count data that details the number of compressive strength tests that passed and the total samples for each quality category. They play a crucial role in assessing the test's effectiveness in quality control, allowing engineers to understand the likelihood of both true positives (correctly identifying good quality) and false positives (incorrectly passing poor quality) in the testing process. This understanding is essential for making informed decisions about the quality and safety of the concrete used in construction.

Step 4: Calculate Joint Probabilities

In this step, we focus on calculating the joint probabilities, representing the likelihood of each combination of concrete quality and test result occurring together. These joint probabilities provide a more detailed understanding of the relationship

Table 3.16 Joint probability distribution of concrete quality and compressive strength test results

Quality	Test results	Count	Joint probability P(Quality, Test)
Good	Pass	720	$720/1000 = 0.72$
Good	Fail	80	$80/1000 = 0.08$
Poor	Pass	20	$20/1000 = 0.02$
Poor	Fail	180	$180/1000 = 0.18$

between the quality of the concrete and the outcome of the compressive strength test. By examining these joint probabilities, we can assess the test's effectiveness in distinguishing between good and poor-quality concrete.

Table 3.16 shows the joint probabilities for each combination of concrete quality (good or poor) and test result (pass or fail). The joint probabilities are calculated by dividing the count of each category combination by the total number of samples (1000). This distribution helps in understanding the effectiveness of the testing method in distinguishing between good and poor-quality concrete.

Conditional Probabilities Calculated Using Joint Probabilities

We can now determine the conditional probabilities using the joint probabilities and the prior probabilities calculated in previous steps.

- **Conditional Probability of Passing the Test Given Good Quality** $P(T|G)$: Using Bayes' theorem:

$$P(T|G) = \frac{P(G \cap T)}{P(G)} = \frac{0.72}{0.8} = 0.9$$

  This confirms that there is a 90% probability of a good-quality concrete sample passing the test.
- **Conditional Probability of Passing the Test Given Poor Quality** $P(T|\overline{G})$): Using Bayes' theorem:

$$P(T|\overline{G}) = \frac{P(\overline{G} \cap T)}{P(\overline{G})} = \frac{0.02}{0.2} = 0.1$$

This indicates a 10% chance that a poor-quality concrete sample will incorrectly pass the test.

3.4.3 Sensitivity and Specificity

Sensitivity and specificity are key performance metrics used to evaluate the effectiveness of diagnostic tests. These metrics are particularly important in fields like civil engineering, where accurate detection of issues such as structural defects, soil contamination, or material weaknesses can have significant safety and economic implications.

3.4.3.1 Sensitivity in Diagnostic Testing

Sensitivity, also known as the true positive rate, is the measure of the number of positive results measured during a diagnostic test, compared to the actual number of positive results. It is calculated as:

$$\text{Sensitivity} = \frac{\text{True Positives } (TP)}{\text{True Positives } (TP) + \text{False Negatives } (FN)} \tag{3.28}$$

A test with high sensitivity effectively detects the presence of a condition of interest when it actually exists. In this case, there would be very few false negatives, minimizing the risk of missing cases that require attention. Consider a test designed to detect cracks in concrete structures. A sensitivity of 0.95 implies that the test correctly identifies 95% of actual cracks, while 5% might be missed by the test and not recorded. High sensitivity is vital in this context because undetected cracks (false negatives) could be significant and compromise structural integrity, and lead to potential failures.

3.4.3.2 Specificity in Diagnostic Testing

Specificity, also known as the true negative rate, measures the proportion of actual negatives that are correctly identified by the test. It indicates how well the test can identify those without the condition of interest. The formula for specificity is:

$$\text{Speciticity} = \frac{\text{True Negatives } (TN)}{\text{True Negatives } (TN) + \text{False Positives } (FP)} \tag{3.29}$$

A test with high specificity is effective at correctly identifying cases that satisfy the negative condition, meaning it has a low rate of false positives. This is crucial when the cost or consequences of a false positive are high. Consider a non-destructive test designed to detect structural defects in bridges. If this test has a specificity of 0.90, it means that 90% of the bridges that are actually defect-free are correctly identified as such. The remaining 10% might be incorrectly flagged as having defects (false positives). High specificity is important because false positives can lead to unnecessary repairs or project delay, causing undue costs and resource allocation.

In the context of civil engineering, high specificity ensures that tests do not incorrectly classify safe structures as defective. This can save on unnecessary inspections, repairs, and costs. For example, if a bridge without any critical structural defects is wrongly identified as having one, this could result in unwarranted closures, as well as performing costly (and potentially time-consuming) additional testing on the structure. In the extreme case, it could result in unnecessary repair work. Hence, specificity is a key metric in ensuring that resources are focused on truly defective structures, improving overall safety and efficiency in infrastructure management.

3.4.3.3 Balancing Sensitivity and Specificity

While sensitivity and specificity are both critical to ensuring precise and accurate non-destructive tests, there is often a trade-off between the two. A highly sensitive test might have lower specificity, resulting in more false positives, and vice versa. The ideal balance depends on the context and the consequences of false negatives and false positives. For instance, in safety–critical applications like bridge inspection, ensuring high sensitivity might be prioritized to catch all possible defects, even if it means accepting a greater incidence of false positives.

3.4.3.4 Practical Examples of Balancing Sensitivity and Specificity in Civil Engineering

Bridge Safety Inspections

- **Sensitivity**: In bridge safety inspections, high-sensitivity tests are critical for identifying major structural defects such as cracks, reinforcement corrosion, or material fatigue. A highly sensitive test ensures that nearly all existing defects are detected. Missing a defect (false negative) carries severe consequences, including the risk of structural failure, accelerated deterioration, and costly emergency repairs. Because the consequences of overlooking a true defect are far more severe than those associated with false alarms, bridge inspection programs deliberately prioritize sensitivity, even if doing so results in more follow-up assessments. For instance, if a non-destructive crack detection test has a sensitivity of 0.95, it means that 95% of actual defects are correctly identified, providing strong protection against missing issues that could compromise the bridge's integrity.
- **Specificity**: High specificity in bridge safety inspections ensures that structures without defects are not incorrectly classified as problematic (false positives). This accuracy is important for avoiding unnecessary repairs, secondary inspections, or temporary closures, all of which entail financial costs and disrupt transportation networks. However, in the context of critical infrastructure, specificity is typically balanced against the need for high sensitivity. While engineers aim for tests that correctly identify defect-free structures, they often accept somewhat lower specificity if it allows sensitivity to remain high. For example, a non-destructive test with a specificity of 0.90 correctly classifies 90% of non-defective bridges as safe. Any false positives can be addressed through secondary testing, ensuring that resources are managed efficiently without compromising public safety.

Flood Risk Assessment

- **Sensitivity**: In flood risk assessment models, high sensitivity is critical for accurately identifying areas at risk of flooding. A highly sensitive model ensures that regions susceptible to flooding are flagged, allowing for timely evacuation, the construction of flood defenses, and the implementation of preventive measures. For example, if a flood prediction model has a sensitivity of 0.92, this would suggest that 92% of the areas that are genuinely at risk of flooding are correctly identified. This reduces the chances of these regions being caught unprepared for a flood event, which would otherwise have devastating impacts on the communities and critical infrastructure.
- **Specificity**: High specificity in flood risk assessment ensures that areas not at risk of flooding are not incorrectly identified as flood-prone (false positives). This is important to avoid unnecessary panic, economic disruption, and the misallocation of resources to areas that are not actually threatened. For instance, if the model specificity is 0.88, it means that 88% of the areas not at risk are correctly identified as safe. High specificity helps maintain public trust and ensures that emergency services and resources are directed efficiently to the regions that need them the most.

3.4.3.5 Visualizing Sensitivity and Specificity

A confusion matrix is a valuable tool that provides a summary of a diagnostic test's performance. This matrix helps to visualize and understand the relationship between the actual conditions (e.g., presence or absence of structural defects) and the test results. The matrix allows us to calculate key metrics such as sensitivity, specificity, and predictive values. Table 3.17 shows the format of a confusion matrix.

Where:

- **True Positive (TP)**: The number of cases where the test correctly identifies the presence of a condition (e.g., a bridge with defects is correctly identified as defective).
- **False Positive (FP)**: The number of cases where the test incorrectly identifies the presence of a condition (e.g., a non-defective bridge is incorrectly identified as defective).
- **False Negative (FN)**: The number of cases where the test fails to identify the condition (e.g., a bridge with defects is incorrectly identified as non-defective).

Table 3.17 Structure of a confusion matrix

	Predicted positive	Predicted negative
Actual positive	True positive (TP)	True negative (TN)
Actual negative	False positive (FP)	False negative (FN)

- **True Negative (TN)**: The number of cases where the test correctly identifies the absence of a condition (e.g., a non-defective bridge is correctly identified as non-defective).

3.4.3.6 Importance of Sensitivity and Specificity in Civil Engineering

In civil engineering, the importance of sensitivity and specificity in diagnostic testing extends across several key areas, including safety, cost-effectiveness, and regulatory compliance, each of which plays a crucial role in the successful execution and maintenance of engineering projects.

1. **Safety**
   - **Critical Importance**: Civil engineers play a vital role in protecting the health and safety of the public. Accurate detection of structural issues, soil contamination, and other hazards is essential to protect public safety and ensure the longevity of engineering projects. High sensitivity in diagnostic tests ensures that potential hazards, such as bridge cracks, contaminated soil, or polluted water, are identified early. This early detection allows for timely interventions, preventing catastrophic failures that could lead to injuries, fatalities, or extensive property damage.
   - **Real-World Impact**: An example of the importance of sensitivity to safety in engineering is the work of performing a bridge inspection. Here, a highly sensitive test can identify even minor defects that might develop into major issues if left unaddressed. By accurately detecting these problems, engineers can take corrective actions to maintain structural integrity and prevent accidents.
2. **Cost-effectiveness**
   - **Resource Management**: High specificity in diagnostic tests helps avoid unnecessary interventions, which can be both time-consuming and expensive. When a test accurately identifies structures that are free of defects, it prevents the misallocation of resources toward unnecessary repairs, inspections, or remediation efforts. This is crucial in large-scale projects where budget constraints and resource allocation are significant considerations. It is also of critical importance in time-constrained projects, such as in the deployment of temporary flood mitigation measures (for example), as responders cannot afford to misallocate these resources into regions that are not prone to flooding.
   - **Efficiency**: By reducing false positives, engineers can focus their attention and resources on addressing true issues, leading to more efficient project management. This not only saves costs but also ensures that maintenance efforts are directed where they are truly needed, optimizing the use of available funds and labor.
3. **Regulatory Compliance**

- **Meeting Standards**: Civil engineering projects are often subject to stringent regulations and standards that require precise and reliable testing. Regulatory bodies demand high sensitivity and specificity in tests to ensure that all safety and environmental guidelines are met. Failure to comply with these regulations can result in legal consequences, project delays, or even shutdowns.
- **Quality Assurance**: Reliable testing contributes to quality assurance and control in construction and maintenance. High sensitivity and specificity in diagnostic tests provide the data needed to demonstrate that a project complies with all relevant safety and environmental standards, which is critical for obtaining permits and approvals.
- **Risk Mitigation**: Adhering to regulatory standards through accurate testing also helps mitigate risks associated with liability and public trust. Ensuring that structures are safe and comply with regulations builds confidence among stakeholders, including government agencies, clients, and the public.

3.4.4 Bayes Theorem Example

3.4.4.1 Example 1: Assessing the Probability of Actual Soil Contamination

In a region in Ontario, soil contamination by a specific hazardous chemical has a prevalence of 0.5%. When the local authority issues building permits, a mandatory soil contamination test is conducted, which has the following characteristics:

- **Sensitivity** (Probability that the test correctly identifies a contaminated site): 99%
- **Specificity** (Probability that the test correctly identifies a non-contaminated site): 98%.

Given these parameters, we obtain a positive test result indicating contamination. However, considering the relatively low prevalence of contamination, this raises an important question: *What is the actual probability that the site is genuinely contaminated given a positive test result?*

Solution

To answer this question, let's break down the process using Bayes' Theorem to find the probability that a site is actually contaminated, given a positive test result.

Provided Information:

If we denote that the site is contaminated by the event A and the test correctly identifies a contaminated site by the event B, then we have:

- **Prevalence of Contamination (P(A))**: The probability that a site is contaminated in the region is 0.005 (or 0.5%). This value reflects how common contamination is within this specific area.

- **Sensitivity (P(B|A))**: The test's ability to correctly identify a contaminated site is 0.99. This means that if we have 1,000 truly contaminated sites, 990 will test positive (true positives), while 10 will test negative (false negatives).
- **Specificity** (P(B′|A′)): The probability that the test correctly identifies a non-contaminated site is 0.98. If there are 1,000 non-contaminated sites, 980 will correctly test negative (true negatives), while 20 will incorrectly test positive (false positives).

Step-by-Step Calculation

1. **Calculate the probability of a Positive Test Result (P(B))**:

   We need to determine the overall probability that a test will come back positive, whether or not the site is actually contaminated. This is found using the law of total probability:

$$P(B) = P(B|A) \times P(A) + P(B|A') \times P(A')$$

   where:

   - $P(B|A')$: The probability of testing positive given that the site is not contaminated $P(B|A') = 1 - \text{specificity} = 1 - 0.98 = 0.02$.
   - $P(A')$: The probability that the site is not contaminated

$$P(A') = 1 - \mathrm{P(A)} = 1 - 0.005 = 0.995$$

   Plugging in the values:

$$\begin{aligned} P(B) &= (0.99 \times 0.005) + (0.02 \times 0.995) \\ &= 0.00495 + 0.0199 \\ &= 0.02485 \end{aligned}$$

2. **Apply Bayes' Theorem** to find the probability that a site is contaminated given a positive test result (P(A|B)):

$$P(A|B) = \frac{P(B|A) \times P(A)}{P(B)}$$

   Plugging in the numbers:

$$P(A|B) = \frac{0.99 \times 0.005}{0.02485} \approx 0.1992$$

Therefore, the probability that a site is actually contaminated, given that it tested positive, is approximately 19.92%. This result shows that even with a highly sensitive and specific test, the low prevalence of contamination in the region significantly affects the positive predictive value of the test. It highlights the importance of considering the base rate of an event when interpreting diagnostic test results, a key insight provided by Bayes' Theorem.

3.4.4.2 Example 2: Determining the Probability of Actual Structural Defects in Bridges

In this example, we evaluate the probability that a bridge has critical structural defects after receiving a positive result from a non-destructive test used during a safety inspection. This is another scenario where Bayes' Theorem can be applied to update the probability of an event (the presence of structural defects) based on new evidence (the test result).

- **Prevalence of Critical Structural Defects (P(A))**: The probability that a bridge in a given region has critical structural defects is 0.02, or 2%. This represents the base rate or prior probability of bridge defects before testing.
- **Sensitivity of the non-destructive test** (P(B|A)): The probability that the test correctly identifies a bridge with defects is 0.95. This means that if there are 1,000 bridges with actual structural defects, the test will correctly identify 950 of them (true positives). The remaining 50 will be false negatives, meaning the test incorrectly identifies them as defect-free.
- **Specificity of the non-destructive test (P(B′|A′))**: The probability that the test correctly identifies a bridge without defects is 0.90. This implies that out of 1,000 defect-free bridges, 900 will correctly test negative (true negatives), while 100 will incorrectly test positive (false positives).

Given these parameters, what is the probability that a bridge, which tested positive for structural defects, actually has critical structural defects?

Solution Using Bayes' Theorem

Given the data for this scenario:

- **Prevalence (P(A))**: 0.02 (2% of bridges have critical structural defects)
- **Sensitivity (P(B|A))**: 0.95 (95% of bridges with defects are correctly identified)
- **Specificity** (P(B′|A′)): 0.90 (90% of defect-free bridges are correctly identified)

Understanding Sensitivity and Specificity

- **Sensitivity**: This is the probability that the test will correctly identify a bridge with defects when truly defective. In this example, a sensitivity of 0.95 means that if there are 100 defective bridges, the test will correctly identify 95 of them (true positives). The remaining 5 will be false negatives, where the test fails to detect the defects.
- **Specificity**: This is the probability that the test will correctly identify a bridge without defects when it is indeed not defective. With a specificity of 0.90, out

of 100 non-defective bridges, 90 will be correctly identified as defect-free (true negatives), while 10 will be false positives, incorrectly flagged as having defects.

Step 1: Calculating the Probability of a Positive Test Result (P(B))

Using the law of total probability, we calculate the overall probability that the test returns a positive result:

$$P(B) = P(B|A) \times P(A) + P(B|A') \times P(A')$$

where:

- $P(B|A')$: The probability of testing positive given that the bridge does not have defects $P(B|A') = 1 - \text{specificity} = 1 - 0.9 = 0.1$
- $P(A')$: The probability of not having defects

$$P(A') = 1 - \text{P(A)} = 1 - 0.02 = 0.98.$$

Plugging in the values:

$$\begin{aligned} P(B) &= (0.95 \times 0.02) + (0.10 \times 0.98) \\ &= 0.019 + 0.098 \\ &= 0.117 \end{aligned}$$

Therefore, the overall probability of obtaining a positive test result from the NDT is 11.7%.

Step 2: Applying Bayes' Theorem to Find P(A|B)

Now, we can use Bayes' theorem to determine the probability that a bridge has defects given a positive test result:

$$P(A|B) = \frac{P(B|A) \times P(A)}{P(B)}$$

$$P(A|B) = \frac{0.95 \times 0.02}{0.117}$$

$$P(A|B) = \frac{0.019}{0.117} \approx 0.1624$$

Therefore, the probability that a bridge actually has critical structural defects, given that it tested positive, is approximately 16.24%. Despite the high sensitivity and reasonable specificity of the test, the relatively low prevalence of defects results in a moderate probability that a bridge with a positive test result is truly defective. This example highlights how even effective tests can yield a significant number of false positives in scenarios where the condition being tested for is relatively rare.

3.4.4.3 Example 3: Evaluating the Probability of Reinforcement Corrosion in a Concrete Beam

Scenario

From experience, a design engineer expects that the probability of corrosion in the steel reinforcement of a reinforced concrete beam after 5 years exposed to the elements is 1%. A 5-year inspection is performed using a half-cell potential test to determine whether the steel reinforcing is corroded or not. Based on this, the following information is available:

- **Prevalence (P(CI))**: The probability that corrosion of the reinforcement has initiated is 0.01, or 1%. This represents the base rate or prior probability of corrosion before any inspection is performed.
- **Sensitivity (P(Test Positive|CI)**: The probability that the inspection method will indicate corrosion, given that corrosion has been initiated, is 0.8. This means that if corrosion is present, the inspection method will correctly identify it 80% of the time.
- **False Positive Rate (P(Test Positive|CI)**: The probability that the inspection method will indicate corrosion even when no corrosion has been initiated is 0.1, or 10%. This is the rate at which the test incorrectly identifies corrosion where none exists.

Question

What is the probability that corrosion of the reinforcement has been initiated, given that the half-cell potential test during inspection indicated corrosion has been initiated?

Solution using Bayes' Theorem

From the provided information in the problem statement, we have the following:

- **Prevalence (P(CI)**: The probability that corrosion has initiated in a reinforced concrete beam is 0.01, or 1%. This means that out of every 100 reinforced concrete beams, we expect that, on average, 1 will have initiated corrosion after 5 years.
- **Sensitivity** (P(I|CI)): Let I be the event where the inspection method indicates corrosion has initiated. Then, the probability that the inspection method correctly identifies corrosion when it has actually initiated in the beam is 0.8. This implies that the inspection method will correctly identify 80% of the corroded beams.
- **False Positive Rate (P(I|CI′))**: The probability that the inspection method incorrectly indicates corrosion when it has not actually initiated in the beam is 0.1. This means that 10% of the beams that have not initiated corrosion are incorrectly identified as having corroded by the inspection method.

  Applying Bayes' Theorem:

1. **Define the Probabilities**

- Probability that corrosion has initiated (P(CI)) = 0.01
- Probability that corrosion has not initiated (P(CI′)) = 1 − P(CI) = 1 − 0.01 = 0.991
- Sensitivity (P(I|CI)) = 0.8
- False positive rate (P(I|CI′)) = 0.1

2. **Calculate the Probability of a Positive Inspection Result (P(I))**
Using the law of total probability:

$$P(I) = P(I|CI) \times P(CI) + P(I|CI') \times P(CI')$$

By entering the values available from the problem statement:

$$P(I) = (0.8 \times 0.01) + (0.1 \times 0.99)$$
$$= 0.008 + 0.099 = 0.107$$

3. **Apply Bayes' Theorem to find P(CI|I)**:

$$P(CI|I) = \frac{P(I|CI) \times P(CI)}{P(I)} = \frac{0.8 \times 0.01}{0.107} = \frac{0.008}{0.107} \approx 0.0748.$$

Therefore, the probability that corrosion of the reinforcement has actually been initiated, given that the inspection indicated corrosion, is approximately 7.48%. Despite the inspection method indicating the presence of corrosion, the relatively low prevalence of corrosion initiation and false positives lead to a lower probability that corrosion is truly present.

This analysis underscores the importance of considering both the test's sensitivity and the base rate of the condition when interpreting inspection results. By using Bayes' Theorem, engineers can make more informed decisions about the likelihood of structural issues, ensuring that interventions are targeted and effective while avoiding unnecessary repairs when they are not needed.

3.4.4.4 Example 4: Inferring Earthquake Intensity from Observed Structural Damage

In this scenario, we want to infer the intensity of an earthquake based on the observed damage to structures. The earthquake can be categorized into three intensities: light (L), moderate (M), and important (I). The structure can either be damaged (D) or undamaged ($\overline{D}$). We are given the likelihood of damage for each intensity, as well as the prior probabilities of each earthquake intensity occurring.

Given Information:

- **Likelihood of Damage Given Earthquake Intensity (Sensitivity for each intensity)**:

  - Pr(D|L) = 0.01 (Sensitivity for light earthquake)
  - Pr(D|M) = 0.1 (Sensitivity for moderate earthquake)
  - Pr(D|I) = 0.6 (Sensitivity for important earthquake)

- **Prior Probabilities of Each Earthquake Intensity (Prevalence)**:
  - Pr(L) = 0.9 (Prevalence of light earthquake)
  - Pr(M) = 0.08 (Prevalence of moderate earthquake)
  - Pr(I) = 0.02 (Prevalence of important earthquake).

Question

What is the probability of each earthquake intensity (L, M, I), given that damage has been observed following an earthquake event?

Solution Using Bayes' Theorem

To find the probability of each earthquake intensity given that damage has been observed, we need to use Bayes' Theorem to calculate the posterior probabilities: Pr(L|D), Pr(M|D), and Pr(I|D).

Step 1: Calculate the Total Probability of Damage (Pr(D))

Using the law of total probability, the total probability of damage is given by:

$$\Pr(D) = \Pr(D|L) \times \Pr(L) + \Pr(D|M) \times \Pr(M) + \Pr(D|I) \times \Pr(I)$$

The prevalence and likelihood data were provided in the problem statement, and so we can simply plug in the given values to find the probability of structural damage:

$$\Pr(D) = (0.01 \times 0.9) + (0.1 \times 0.08) + (0.6 \times 0.02)$$

$$\Pr(D) = 0.009 + 0.008 + 0.012 = 0.029$$

Step 2: Apply Bayes' Theorem for Each Intensity:

- Probability of Light Earthquake Given Damage (Pr(L|D))

$$\Pr(L|D) = \frac{\Pr(D|L) \times \Pr(L)}{\Pr(D)} = \frac{0.01 \times 0.9}{0.029} = \frac{0.009}{0.029} \approx 0.31$$

- Probability of Moderate Earthquake Given Damage (Pr(L|D))

$$\Pr(M|D) = \frac{\Pr(D|M) \times \Pr(M)}{\Pr(D)} = \frac{0.1 \times 0.08}{0.029} = \frac{0.008}{0.029} \approx 0.28$$

- Probability of Importance Earthquake Given Damage (Pr(L|D))

$$\Pr(I|D) = \frac{\Pr(D|I) \times \Pr(I)}{\Pr(D)} = \frac{0.6 \times 0.02}{0.029} = \frac{0.0128}{0.029} \approx 0.41$$

Given that damage has been observed following an earthquake, the probabilities of each earthquake intensity are approximated as:

- Light earthquake (Pr(L|D)) ≈ **31%**
- Moderate earthquake (Pr(M|D)) ≈ **28%**
- Important earthquake (Pr(I|D)) ≈ **41%**.

From this example, we can see that, despite the low prior probability of an important earthquake, the high likelihood of damage resulting from high importance earthquakes significantly increases the probability that a damaged structure was the result of an important intensity event. This example demonstrated how Bayes' Theorem can be used to update our understanding of an event based on new evidence, in this case, using data on observed structural damage to infer the earthquake's intensity. This approach is essential in earthquake engineering and risk assessment, guiding emergency response and structural evaluation efforts.

3.4.4.5 Example 5: Inferring Groundwater Contamination Given a Positive Test Detection

Scenario

In this situation, we are concerned with groundwater contamination by a specific chemical. The contamination can be classified into three levels of severity, namely, low (L), moderate (M), and high (H). The groundwater can either be contaminated (C) or not contaminated ($\overline{C}$). We are given the likelihood of detecting the contaminant for each contamination level and the prior probabilities of each contamination level occurring. Using Bayes' Theorem, we want to infer the contamination level given that a test has detected the contaminant.

Given Information

- **Likelihood of Detection Given Contamination Level (Sensitivity for each level)**:
  - Pr(D|L) = 0.05: Probability of detecting the contaminant given low contamination.
  - Pr(D|M) = 0.4: Probability of detecting the contaminant given moderate contamination.
  - Pr(D|H) = 0.9: Probability of detecting the contaminant given high contamination.
- **Prior Probabilities of Each Contamination Level (Prevalence)**:
  - Pr(L) = 0.7: Probability of low contamination in the groundwater.
  - Pr(M) = 0.2: Probability of moderate contamination in the groundwater.

– Pr(H) = 0.1: Probability of high contamination in the groundwater.

Question

What is the probability of each contamination level (low, moderate, high) given that a test has detected the contaminant in the groundwater?

Solution using Bayes' Theorem

Step 1: Calculate the Probability of Detection (Pr(D))

Using the law of total probability, we calculate the overall probability of detecting the contaminant in the groundwater:

$$\Pr(D) = \Pr(D|L) \times \Pr(L) + \Pr(D|M) \times \Pr(M) + \Pr(D|H) \times \Pr(H)$$

Plugging in the values:

$$\begin{aligned}\Pr(D) &= (0.05 \times 0.7) + (0.4 \times 0.2) + (0.9 \times 0.1)\\ &= 0.035 + 0.08 + 0.09 = 0.205\end{aligned}$$

Step 2: Calculate the Posterior Probabilities

Using Bayes' theorem, we can calculate the probability of each contamination level given that a detection has been observed (Pr(D)):

- **Probability of Low Contamination Given Detection (Pr(L|D))**:

$$\Pr(L|D) = \frac{\Pr(D|L) \times \Pr(L)}{\Pr(D)} = \frac{0.05 \times 0.7}{0.205} = \frac{0.035}{0.205} \approx 0.1707$$

- **Probability of Moderate Contamination Given Detection (Pr(M|D))**:

$$\Pr(M|D) = \frac{\Pr(D|M) \times \Pr(M)}{\Pr(D)} = \frac{0.4 \times 0.2}{0.205} = \frac{0.08}{0.205} \approx 0.3902$$

- **Probability of High Contamination Given Detection (Pr(H|D))**:

$$\Pr(H|D) = \frac{\Pr(D|H) \times \Pr(H)}{\Pr(D)} = \frac{0.9 \times 0.1}{0.205} = \frac{0.09}{0.205} \approx 0.439$$

Given that detection (D) has been observed, the posterior probabilities for the contamination level of groundwater are:

- Probability of Low Contamination Given Detection (Pr(L|D)) $\approx$ 17.07%
- Probability of Moderate Contamination Given Detection (Pr(M|D)) $\approx$ 39.02%
- Probability of High Contamination Given Detection (Pr(H|D)) $\approx$ 43.90%.

This analysis shows that even though high contamination has the lowest prior probability, the high sensitivity of detecting contaminants at this level significantly increases its posterior probability when a detection is made. Bayes' Theorem allows us to update our beliefs about the contamination level, providing a clearer understanding of the situation based on observed evidence.

3.4.4.6 Example 6: Inferring Groundwater Contamination Level Given a Positive Test Detection

Scenario

In this situation, we are concerned about groundwater contamination by a specific chemical, and we are given the likelihood of correctly identifying the absence of contamination (specificity) for different levels of contamination. The groundwater can be either contaminated (C) or not contaminated ($\overline{C}$). Using the provided specificity values and prior probabilities, we want to infer the contamination level given that a test has detected the contaminant.

Given Information

- **Specificity for Different Contamination Levels** (Probability of correctly identifying the absence of contamination):
  - $\Pr(\overline{D}|\text{L}) = 0.95$: Specificity for low contamination (95% of low contamination cases are correctly identified as not contaminated).
  - $\Pr(\overline{D}|\text{M}) = 0.6$: Specificity for moderate contamination (60% of moderate contamination cases are correctly identified as not contaminated).
  - $\Pr(\overline{D}|\text{H}) = 0.1$: Specificity for high contamination (10% of high contamination cases are correctly identified as not contaminated).
- **Prior Probabilities of Each Contamination Level**:
  - $\Pr(L) = 0.7$: Prevalence of low contamination.
  - $\Pr(M) = 0.2$: Prevalence of moderate contamination.
  - $\Pr(H) = 0.1$: Prevalence of high contamination.

Question

What is the probability of each contamination level (low, moderate, high) given that a test has detected the contaminant (positive test result)?

Step-by-Step Solution

1. **Calculate the Probability of a Positive Test Detection (Pr(D))**:

   Since we have the specificity, we need to convert it to sensitivity to find the probability of a positive test result.

$$\Pr(D|L) = 1 - \Pr(\overline{D}|L) = 1 - 0.95 = 0.05$$

$$\Pr(D|M) = 1 - \Pr(\overline{D}|M) = 1 - 0.6 = 0.4$$

$$\Pr(D|H) = 1 - \Pr(\overline{D}|H) = 1 - 0.1 = 0.9$$

Using the law of total probability to find the overall probability of detecting the contaminant:

$$\Pr(D) = \Pr(D|L) \times \Pr(L) + \Pr(D|M) \times \Pr(M) + \Pr(D|H) \times \Pr(H)$$

Plugging in the values:

$$\begin{aligned} \Pr(D) &= (0.05 \times 0.7) + (0.4 \times 0.2) + (0.9 \times 0.1) \\ &= 0.035 + 0.08 + 0.09 = 0.205 \end{aligned}$$

2. **Apply Bayes' Theorem for Each Contamination Level**:
   - **For Low Contamination (Pr**(L|D)):

$$\Pr(L|D) = \frac{\Pr(D|L) \times \Pr(L)}{\Pr(D)} = \frac{0.05 \times 0.7}{0.205} = \frac{0.035}{0.205} \approx 0.1707$$

   - **For Moderate Contamination (Pr**(M|D)):

$$\Pr(M|D) = \frac{\Pr(D|M) \times \Pr(M)}{\Pr(D)} = \frac{0.4 \times 0.2}{0.205} = \frac{0.08}{0.205} \approx 0.3902$$

   - **For High Contamination (Pr**(H|D)):
   - $\Pr(H|D) = \frac{\Pr(D|H) \times \Pr(H)}{\Pr(D)} = \frac{0.9 \times 0.1}{0.205} = \frac{0.09}{0.205} \approx 0.439$

Given that the test has detected the contaminant, the inferred probabilities of each contamination level in the groundwater are approximately:

- Probability of Low Contamination Given Detection (**Pr(L|D)**) $\approx$ 17.07%
- Probability of Moderate Contamination Given Detection (**Pr(M|D)**) $\approx$ 39.02%
- Probability of High Contamination Given Detection (**Pr(H|D)**) $\approx$ 43.90%.

This outcome shows how Bayes' Theorem integrates the test specificity and prevalence data to provide a more informed assessment of the contamination level when a positive detection is made. Even though low contamination has the highest prior probability, a positive test result significantly increases the likelihood of high contamination due to the test's higher sensitivity in detecting higher contamination levels. This insight is vital for environmental monitoring and effective groundwater management.

3.4.4.7 Example 7: Quality Control for Aggregate Loads at a Work Site

Scenario

In this example, aggregates for the construction of a reinforced concrete building are supplied by two companies: Company A and Company B. Company A delivers 600 truckloads per day, and Company B delivers 400 truckloads per day. From prior experience, we know that 3% of the material from Company A is substandard, while 1% of the material from Company B is substandard. We want to determine the probability that a substandard load came from Company A, given that a load of aggregates is found to be substandard.

Given Information

- **Number of Truckloads Delivered per Day**:
  - Company A delivers 600 truckloads.
  - Company B delivers 400 truckloads.
- **Proportion of Substandard Loads**:
  - Company A: 3% of the material is substandard ($\Pr(S|A) = 0.03$).
  - Company B: 1% of the material is substandard ($\Pr(S|B) = 0.01$).
- **Prior Probabilities (Proportion of Total Truckloads)**:
  - The total number of truckloads per day is $600 + 400 = 1000$.
  - $\Pr(A) = 600/1000 = 0.6$: The probability that a truckload is from Company A.
  - $\Pr(B) = 400/1000 = 0.4$: The probability that a truckload is from Company B.

Question

What is the probability that a substandard load of aggregates came from Company A, given that a load is found to be substandard?

Solution

Step 1: Calculate the Prior Probabilities (Prevalence)

First, we calculate the prior probabilities for each company delivering a truckload of aggregates. This is based on the number of truckloads delivered per day:

- **Pr**(A): The probability that a load is from Company A:

$$\Pr(A) = \frac{600}{600 + 400} = 0.6$$

- **Pr**(B): The probability that a load is from Company B:

$$\Pr(B) = \frac{400}{600 + 400} = 0.4$$

Thus, the prior probabilities are:

- Pr(A) = 0.6
- Pr(B) = 0.4.

Step 2: Calculate the Probability of a Substandard Load Given Each Company (Sensitivity)

In this case, the **sensitivity** measures the probability that a load is substandard, given that it is from a particular company.

- **Pr(**S|A): The probability that a load from Company A is substandard:

$$\Pr(S|A) = 0.03$$

- **Pr(**S|B): The probability that a load from Company B is substandard:

$$\Pr(S|B) = 0.01$$

Step 3: Calculate the Overall Probability of a Substandard Load (Pr(S))

We now calculate the overall probability of a substandard load using the law of total probability. This takes into account the proportion of loads delivered by each company and their respective substandard rates:

$$\Pr(S) = \Pr(S|A) \times \Pr(A) + \Pr(S|B) \times \Pr(B)$$

Plugging in the values:

$$\Pr(S) = (0.03 \times 0.6) + (0.01 \times 0.4) = 0.018 + 0.004 = 0.022$$

The overall probability of a substandard load is Pr(S) = 0.022 or 2.2%.

Step 4: Calculate the Posterior Probability (Pr(A|S))

Finally, we use Bayes' Theorem to calculate the probability that the substandard load came from Company A:

$$\Pr(A|S) = \frac{\Pr(S|A) \times \Pr(A)}{\Pr(S)}$$

Substituting the values:

$$\Pr(A|S) = \frac{0.03 \times 0.6}{0.022} = \frac{0.018}{0.022} \approx 0.8182$$

Given that a load of aggregates is found to be substandard, the probability that it came from Company A is approximately 81.82%. This high probability indicates that Company A is more likely to deliver a substandard load than Company B. This makes logical sense, as they deliver more loads per day and have a higher incidence rate of substandard loads. This understanding can assist in quality control, supplier evaluation, and decision-making in procurement for construction projects.

3.4.4.8 Example 8: Nitrate Contamination in Watersheds

Scenario

In a water resource management scenario, two watersheds, Watershed A and Watershed B, supply water to a region. Watershed A supplies 700 cubic meters of water per day, while Watershed B supplies 300 cubic meters of water per day. From prior experience, it is known that 5% of the water from Watershed A has high nitrate levels (above the safety threshold), while 2% of the water from Watershed B has high nitrate levels. Given that a sample of water is found to have high nitrate levels, we want to determine the probability that it came from Watershed A.

Given Information

- **Daily Water Supply**:
  - Watershed A: 700 cubic meters.
  - Watershed B: 300 cubic meters.
- **Proportion of Water with High Nitrate Levels**:
  - Watershed A: 5% of the water has high nitrate levels ($\Pr(H|A) = 0.05$).
  - Watershed B: 2% of the water has high nitrate levels ($\Pr(H|B) = 0.02$).
- **Prior Probabilities (Proportion of Total Water Supply)**:
  - The total water supplied per day is $700 + 300 = 1000$ cubic meters.
  - $\Pr(A) = 700/1000 = 0.7$: Probability that the water is from Watershed A.
  - $\Pr(B) = 300/1000 = 0.3$: Probability that the water is from Watershed B.

Question

What is the probability that a sample of water with high nitrate levels came from Watershed A?

Solution using Bayes' Theorem

Step 1: Calculate the Prior Probabilities (Prevalence)

The prior probabilities represent the chances that a sample is from each watershed based on the volume of water supplied by each.

- **Pr(A)**: Probability that a sample is from Watershed A:

$$\Pr(A) = \frac{700}{700 + 300} = 0.7$$

- **Pr(B)**: Probability that a sample is from Watershed B:

$$\Pr(B) = \frac{300}{700 + 300} = 0.3$$

Thus, the prior probabilities are:

- Pr(A) = 0.7
- Pr(B) = 0.3.

Step 2: Calculate the Probability of High Nitrate Levels Given Each Watershed (Sensitivity)

Sensitivity measures the probability of detecting high nitrate levels, given that the sample is from a specific watershed. In this case, we are given the following sensitivities:

- **Pr(H|A):** Probability of high nitrate levels given the sample is from Watershed A:

$$\Pr(H|A) = 0.05$$

- **Pr**(H|B): Probability of high nitrate levels given the sample is from Watershed B:

$$\Pr(H|B) = 0.02$$

Step 3: Calculate the Overall Probability of a Sample Having High Nitrate Levels (Pr(H))

We apply the law of total probability to calculate the overall probability of a sample having high nitrate levels:

$$\Pr(H) = \Pr(H|A) \times \Pr(A) + \Pr(H|B) \times \Pr(B)$$

Plugging in the values:

$$\begin{aligned} \Pr(H) &= (0.05 \times 0.7) + (0.02 \times 0.3) \\ &= 0.035 + 0.006 \\ &= 0.041 \end{aligned}$$

The overall probability of a sample having high nitrate levels is Pr(H) = 0.041 or 4.1%.

Step 4: Calculate the Posterior Probability (Pr(A|H))

Using Bayes' Theorem, we can calculate the probability that a sample with high nitrate levels came from Watershed A:

$$\Pr(A|H) = \frac{\Pr(H|A) \times \Pr(A)}{\Pr(H)}$$

Substituting the values:

$$\Pr(A|H) = \frac{0.05 \times 0.7}{0.041} = \frac{0.035}{0.041} \approx 0.8537$$

Given that a water sample has high nitrate levels, the probability that it came from Watershed A is approximately **85.37%**. This high probability reflects the larger volume of water supplied by Watershed A and its higher rate of nitrate contamination, making it much more likely that a sample with high nitrate levels originated from Watershed A. This result is valuable for water quality management, as it helps focus mitigation and monitoring efforts on the watershed that is more likely to contribute to nitrate contamination.

3.4.4.9 Example 9: High Nitrate Levels in Watersheds—Part II

Scenario

Consider a water resource management scenario where water is supplied by four watersheds, Watershed A, Watershed B, Watershed C, and Watershed D, to a region. Watershed A supplies 500 m^3 of water per day, Watershed B supplies 300 m^3, Watershed C supplies 150 m^3, and Watershed D supplies 50 m^3 per day. From prior experience, it is known that 5% of the water from Watershed A has high nitrate levels (above the safety threshold), 2% of the water from Watershed B has high nitrate levels, 8% of the water from Watershed C has high nitrate levels, and 10% of the water from Watershed D has high nitrate levels. Given that a sample of water is found to have high nitrate levels, what is the probability that it is from Watershed A?

Solution using Bayes' Theorem

Step 1: Calculate the Prior Probabilities (Prevalence)

- Probability that a sample is from Watershed A: $\Pr(A) = \frac{500}{500+300+150+50} = 0.5$
- Probability that a sample is from Watershed B: $\Pr(B) = \frac{300}{500+300+150+50} = 0.3$
- Probability that a sample is from Watershed C: $\Pr(C) = \frac{150}{500+300+150+50} = 0.15$
- Probability that a sample is from Watershed D: $\Pr(D) = \frac{50}{500+300+150+50} = 0.05$.

Prevalence represents the likelihood that a randomly selected individual (or event, in this case) from the population has the condition. It is a measure of how common the condition is within the population. In this example, the prevalence of water from each watershed is given by their respective prior probabilities: Pr(A) = 0.5, Pr(B) = 0.3, Pr(C) = 0.15, and Pr(D) = 0.05.

Step 2: Calculate the Probability of High Nitrate Levels Given Each Watershed (Sensitivity)

In this example, the sensitivity for each watershed is given by the probability of a sample having high nitrate levels, given that it is from that watershed:

- Pr(H|A) = Probability of high nitrate levels given the sample is from Watershed A = 0.05
- Pr(H|B) = Probability of high nitrate levels given the sample is from Watershed B = 0.02
- Pr(H|C) = Probability of high nitrate levels given the sample is from Watershed C = 0.08
- Pr(H|D) = Probability of high nitrate levels given the sample is from Watershed D = 0.10.

Step 3: Calculate the Overall Probability of a Sample Having High Nitrate Levels (Pr(H))

Using the law of total probability, we calculate the overall probability of a sample having high nitrate levels. This takes into account the contributions from each watershed:

$$\begin{aligned} \Pr(H) &= \Pr(H|A) \times \Pr(A) + \Pr(H|B) \times \Pr(B) \\ &+ \Pr(H|C) \times \Pr(C) + \Pr(H|D) \times \Pr(D) \end{aligned}$$

Plugging in the values:

$$\Pr(H) = (0.05 \times 0.5) + (0.02 \times 0.3) + (0.08 \times 0.15) + (0.10 \times 0.05) = 0.048$$

Step 4: Calculate the Posterior Probability (Pr(A|H))

Using Bayes' Theorem, we can now calculate the probability that a sample with high nitrate levels came from Watershed A:

$$\Pr(A|H) = \frac{\Pr(H|A) \times \Pr(A)}{\Pr(H)}$$

Substituting the values:

$$\Pr(A|H) = \frac{0.05 \times 0.5}{0.048} = \frac{0.025}{0.048} \approx 0.5208$$

Given that a water sample has high nitrate levels, the probability that it came from Watershed A is approximately 52.08%. This result is significant and reflects both the volume of water supplied by Watershed A and the nitrate contamination levels, as well as contributions from other watersheds. This probability helps guide water quality management, monitoring efforts, and decision-making regarding potential sources of contamination.

3.4.4.10 Example 10: Evaluating the Probability of Reliable Concrete Compressive Strength Testing

In order to ensure the quality of concrete material used in a reinforced concrete construction, concrete cylinder samples are collected at random from concrete mixes delivered to the construction site by a mixing plant. Past records of concrete from the same plant show that 80% of concrete mixes are of satisfactory quality, meaning that they meet the specified minimum compressive strength. To further ensure that the concrete delivered on site is of good quality, the engineer requires that one cylinder among those collected each day be tested (after 7 days of curing) for a minimum compressive strength. The test method is not perfect—its reliability is only 90%—meaning the probability that a good-quality concrete cylinder will pass the test is 0.90, or that a poor-quality cylinder can pass the test is 0.10. If a concrete cylinder passes the test, what is the probability of good-quality concrete being delivered on site?

Now, suppose also that the engineer is not satisfied with just testing one cylinder and would prefer that a second cylinder be tested. If the second cylinder tested also gave a positive result, what is the probability that the concrete is of good quality? If the test result of either one of the two cylinders tested was negative, what is the probability of good-quality concrete?

Solution of Part 1: Probability of Good-Quality Concrete Given One Cylinder Passes the Test

Step 1: Calculate the Prior Probabilities (Prevalence)

- Pr(G) = Probability that a concrete mix is good quality = 0.80
- Pr(P) = Probability that a concrete mix is poor quality = 0.20.

Step 2: Calculate the Probability of Passing the Test Given Each Quality (Sensitivity)

- Pr(Pass|G) = Probability that a good-quality cylinder passes the test = 0.90
- Pr(Pass|P) = Probability that a poor-quality cylinder passes the test = 0.10.

Step 3: Calculate the Overall Probability of a Cylinder Passing the Test (Pr(Pass))

We use the law of total probability to calculate the overall probability of passing the test:

$$\Pr(Pass) = \Pr(Pass|G) \times \Pr(G) + \Pr(Pass|P) \times \Pr(P)$$

Plugging in the values:

$$\Pr(Pass) = (0.90 \times 0.80) + (0.10 \times 0.20) = 0.72 + 0.02 = 0.74$$

Step 4: Calculate the Posterior Probability (Pr(G|Pass))

We use Bayes' theorem to find the probability that the concrete is of good quality given that a cylinder passes the test:

$$\Pr(G|Pass) = \frac{\Pr(Pass|G) \times \Pr(G)}{\Pr(Pass)}$$

Substituting the values:

$$\Pr(G|Pass) = \frac{0.90 \times 0.80}{0.74} = \frac{0.72}{0.74} = 0.97297$$

So, the probability of good-quality concrete delivered on site, given that one cylinder passes the test, is approximately 97.30%.

Solution of Part 2: Probability of Good-Quality Concrete Given Two Cylinders Pass the Test

Step 1: Calculate the Probability of Both Cylinders Passing the Test

When two independent tests are conducted, and both cylinders pass, the probability that the concrete is of good quality can be calculated as:

$$\Pr(\text{Good}|\text{Pass, Pass}) = \frac{\Pr(\text{Pass, Pass}|\text{Good}) \times \Pr(\text{Good})}{\Pr(\text{Pass, Pass})}$$

where:

- $\Pr(\text{Pass,Pass}|\text{Good}) = (0.90) \times (0.90) = 0.81$: The probability that both tests pass given that the concrete is of good quality.
- $\Pr(\text{Pass,Pass}|\text{Poor}) = (0.10) \times (0.10) = 0.01$: The probability that both tests pass given the concrete is of poor quality.

Step 2: Calculate the Overall Probability of Both Cylinders Passing the Test

$$\mathrm{Pr}(\mathrm{Pass, Pass}) = \mathrm{Pr}(\mathrm{Pass, Pass|G}) \times \mathrm{Pr}(\mathrm{G}) + \mathrm{Pr}(\mathrm{Pass, Pass|P}) \times \mathrm{Pr}(\mathrm{P})$$

Plugging in the values:

$$\mathrm{Pr}(\mathrm{Pass, Pass}) = (0.81 \times 0.80) + (0.01 \times 0.20) = 0.648 + 0.02 = 0.65$$

Step 3: Calculate the Posterior Probability (Pr(G|Pass,Pass))

We use Bayes' theorem to find the probability that the concrete is of good quality, given that both cylinders pass the test:

$$\mathrm{Pr}(\mathrm{G|Pass, Pass}) = \frac{\mathrm{Pr}(\mathrm{Pass, Pass|Good}) \times \mathrm{Pr}(\mathrm{G})}{\mathrm{Pr}(\mathrm{Pass, Pass})}$$

Plugging in the values:

$$\mathrm{Pr}(\mathrm{G|Pass, Pass}) = \frac{0.81 \times 0.80}{0.65} = 0.9969$$

So, the probability of good-quality concrete delivered on site, given that both cylinders pass the test, is approximately 99.69%.

Solution of Part 3: Probability of Good-Quality Concrete Given One Cylinder Fails the Test

If either one of the two cylinders fails the test, we consider two scenarios: one cylinder passes, and one fails, or both fail.

Step 1: Calculate the Probabilities for Each Scenario

For one cylinder passing and one failing:

$$\mathrm{Pr}(Pass\_one|G) = 2 \times (\mathrm{Pr}(Pass|G) \times Pr(Fail|G)) = 2 \times (0.90 \times 0.10) = 0.18$$

$$\mathrm{Pr}(Pass\_one|G) = 2 \times (\mathrm{Pr}(Pass|P) \times \mathrm{Pr}(Fail|P)) = 2 \times (0.10 \times 0.90) = 0.18$$

For both cylinders failing:

$$\mathrm{Pr}(Fail\_both|G) = \mathrm{Pr}(Fail|G) \times \mathrm{Pr}(Fail|G) = 0.10 \times 0.10 = 0.01$$

$$\mathrm{Pr}(Fail\_both|P) = \mathrm{Pr}(Fail|P) \times \mathrm{Pr}(Fail|P) = 0.90 \times 0.90 = 0.81$$

Step 2: Calculate the Overall Probabilities

$$\begin{aligned}\Pr(Fail\_one) &= \Pr(Pass\_one|G) \times \Pr(G) + Pr(Pass\_one|P) \times \Pr(P) \\ &= (0.18 \times 0.80) + (0.18 \times 0.20) = 0.18\end{aligned}$$

$$\begin{aligned}\Pr(Fail\_both) &= \Pr(Fail\_both|G) \times \Pr(G) + Pr(Fail\_both|P) \times \Pr(P) \\ &= (0.01 \times 0.80) + (0.81 \times 0.20) = 0.17\end{aligned}$$

Step 3: Calculate the Posterior Probability (Pr(G|Fail_one)) and Pr(G|Fail_both))

For one cylinder failing:

$$\Pr(G|Fail\_one) = \frac{\Pr(Pass\_one|G) \times \Pr(G)}{\Pr(Fail\_one)} = \frac{0.18 \times 0.80}{0.18} = 0.80$$

For both cylinders failing:

$$\Pr(G|Fail\_both) = \frac{\Pr(Fail\_both|G) \times \Pr(G)}{\Pr(Fail\_both)} = \frac{0.01 \times 0.80}{0.17} = 0.047$$

So, the probability of good-quality concrete delivered on site given that one of the two cylinders fails the test, is approximately 80%. The probability of good-quality concrete if both cylinders fail the test is approximately 4.7%.

3.4.4.11 Example 11: Assessing Landslide Risk in a Mountainous Region

In a mountainous region, a government agency is tasked with assessing the risk of landslides. The region is divided into four zones: Zone A, Zone B, Zone C, and Zone D. Each zone has different geological characteristics and varying levels of precipitation, which influence the likelihood of landslides. The agency collects data on past landslide occurrences to estimate the prior probabilities of landslides in each zone:

- **Zone A**: Accounts for 25% of the total area.
- **Zone B**: Accounts for 30% of the total area.
- **Zone C**: Accounts for 20% of the total area.
- **Zone D**: Accounts for 25% of the total area.

From historical data, it is known that:

- 15% of landslides occur in Zone A.
- 20% of landslides occur in Zone B.

- 40% of landslides occur in Zone C.
- 25% of landslides occur in Zone D.

Additionally, the agency has developed a detection system that uses satellite imagery to identify areas at risk of landslides. The system's reliability is as follows:

- In Zone A, the system correctly identifies a landslide-prone area 70% of the time.
- In Zone B, the system correctly identifies a landslide-prone area 80% of the time.
- In Zone C, the system correctly identifies a landslide-prone area 90% of the time.
- In Zone D, the system correctly identifies a landslide-prone area 85% of the time.

However, the system also has a false positive rate:

- In Zone A, the system incorrectly identifies a non-landslide-prone area as at risk 10% of the time.
- In Zone B, the system incorrectly identifies a non-landslide-prone area as at risk 15% of the time.
- In Zone C, the system incorrectly identifies a non-landslide-prone area as at risk 20% of the time.
- In Zone D, the system incorrectly identifies a non-landslide-prone area as at risk 5% of the time.

Given that the system has identified an area as at risk of landslide, the agency wants to know:

1. What is the probability that the area identified as at risk is in Zone C?
2. What is the probability that the area identified as at risk is not in Zone C?
3. If the agency decides to conduct a second assessment using ground sensors, which have a 95% accuracy rate in detecting landslide-prone areas in Zone C and no false positives, what is the probability that the area is indeed at risk if both the satellite system and the ground sensors indicate a risk?

Solution of Part 1

Step 1: Calculate the Prior Probabilities (Prevalence)

The prevalence of landslides in each zone of the region was provided directly from the problem statement using historical data.

- $\Pr(A) = 0.25$
- $\Pr(B) = 0.30$
- $\Pr(C) = 0.20$
- $\Pr(D) = 0.25$.

Step 2: Calculate the Probability of Detection (Sensitivity) and False Positives

Again, this information was directly provided in the problem statement and needed only to be extracted. Let R denote a reliable detection event by the satellite sensor, and FP denote a false positive detection event.

- Pr(R|A) = 0.70
- Pr(R|B) = 0.80
- Pr(R|C) = 0.90
- Pr(R|D) = 0.85
- Pr(FP|A) = 0.10
- Pr(FP|B) = 0.15
- Pr(FP|C) = 0.20
- Pr(FP|D) = 0.05.

Step 3: Calculate the Overall Probability of Detection (Pr(R))

We use the law of total probability to calculate the overall probability of detection:

$$\Pr(R) = \Pr(R|A) \times \Pr(A) + \Pr(R|B) \times \Pr(B) + \Pr(R|C) \times \Pr(C) + \Pr(R|D) \times \Pr(D)$$

Plugging in the values:

$$\Pr(R) = (0.70 \times 0.25) + (0.80 \times 0.30) + (0.90 \times 0.20) + (0.85 \times 0.25) = 0.8075$$

Solution of Part 2: Calculate the Posterior Probabilities

1. **Probability of Zone C Given Detection (Pr(C|R))**:

$$\Pr(C|R) = \frac{\Pr(R|C) \times \Pr(C)}{\Pr(R)} = \frac{0.90 \times 0.20}{0.8075} = 0.2228$$

2. **Probability of Not Zone C Given Detection (Pr($\overline{\text{C}}$|R))**:

$$Pr(\overline{C}|R) = 1 - Pr(C|R) = 1 - 0.2228 = 0.7772$$

Solution of Part 3: Second Assessment Using Ground Sensors

Step 1: Calculate the Probability of Detection by Ground Sensors

For Zone C with ground sensors:

$$\Pr(GS|C) = 0.95$$

Since ground sensors have no false positives, then

$$\Pr(FP|GS) = 0$$

Step 2: Calculate the Posterior Probability After Two Positive Results

Using the results from the satellite system and ground sensors:

$$\Pr(C|R, GS) = \frac{\Pr(GS|C) \times \Pr(C|R)}{\Pr(GS)}$$

First, calculate Pr(GS):

$$\Pr(GS) = \Pr(GS|C) \times \Pr(C|R) + \Pr(GS|\overline{C}) \times \Pr(\overline{C}|R)$$

Since $\Pr(GS|\overline{C}) = 0$, then

$$\Pr(GS) = 0.95 \times 0.2228 + 0 = 0.21166$$

Now, calculate Pr(C|R,GS):

$$\Pr(C|R, GS) = \frac{\Pr(GS|C) \times \Pr(C|R)}{\Pr(GS)} = \frac{0.95 \times 0.2228}{0.21166} \approx 1$$

Therefore, when both the ground sensor and satellite methods are used to identify landslide risk, and both methods indicate that Zone C is at risk, there is almost a 100% chance that zone C is reliably at risk.

3.4.4.12 Example 12: Assessing Bridge Structural Deficiencies Using Drones and Ground Sensors

Scenario

A transportation agency is responsible for assessing the structural integrity of bridges in a metropolitan area. The area has four major bridges: Bridge A, Bridge B, Bridge C, and Bridge D. Each bridge has different traffic loads, age, and maintenance history, influencing the likelihood of structural issues. The agency collects data on past structural inspections to estimate the prior probabilities of structural issues in each bridge.

Given Information

- **Bridge A**: 30% of the total inspections.
- **Bridge B**: 25% of the total inspections.
- **Bridge C**: 20% of the total inspections.
- **Bridge D**: 25% of the total inspections.

From historical data, it is known that:

- 10% of the structural issues are found in **Bridge A**.

- 20% of the structural issues are found in **Bridge B**.
- 50% of the structural issues are found in **Bridge C**.
- 20% of the structural issues are found in **Bridge D**.

The agency uses a detection system that employs drones equipped with sensors to identify areas with potential structural issues. The system's reliability is as follows:

- In **Bridge A**, the system correctly identifies structural issues 85% of the time.
- In **Bridge B**, the system correctly identifies structural issues 80% of the time.
- In **Bridge C**, the system correctly identifies structural issues 90% of the time.
- In **Bridge D**, the system correctly identifies structural issues 75% of the time.

However, the system also has a false positive rate:

- In **Bridge A**, the system incorrectly identifies bridges having no issues as having issues 5% of the time.
- In **Bridge B**, the system incorrectly identifies bridges having no issues as having issues 10% of the time.
- In **Bridge C**, the system incorrectly identifies bridges having no issues as having issues 15% of the time.
- In **Bridge D**, the system incorrectly identifies bridges having no issues as having issues 20% of the time.

Questions

Given that the system has identified an area as having potential structural issues, the agency wants to know:

1. What is the probability that the area identified as having issues is in Bridge C?
2. What is the probability that the area identified as having issues is not in Bridge C?
3. If the agency decides to conduct a second assessment using ground sensors, which have a 95% accuracy rate in detecting structural issues in Bridge C and no false positives, what is the probability that the area is indeed at risk if both the drone system and the ground sensors indicate a risk?

Solution of Part 1

Solution using Bayes' Theorem

Step 1: Calculate the Prior Probabilities (Prevalence)

The prevalence was directly given in the problem statement as the frequency of inspection by the agency.

- $\Pr(A) = 0.30$
- $\Pr(B) = 0.25$
- $\Pr(C) = 0.20$
- $\Pr(D) = 0.25$.

Step 2: Calculate the Probability of Detection (Sensitivity) and False Positives

The sensitivity and false positive probabilities were directly provided in the problem statement and are represented mathematically below.

- Pr(R|A) = 0.85
- Pr(R|B) = 0.80
- Pr(R|C) = 0.90
- Pr(R|D) = 0.75
- Pr(FP|A) = 0.05
- Pr(FP|B) = 0.10
- Pr(FP|C) = 0.15
- Pr(FP|D) = 0.20.

Step 3: Calculate the Overall Probability of Detection (Pr(R))

We use the law of total probability to calculate the overall probability of detection:

$$\begin{aligned}\Pr(R) &= \Pr(R|A) \times \Pr(A) + \Pr(R|B) \times \Pr(B) \\ &\quad + \Pr(R|C) \times \Pr(C) + \Pr(R|D) \times \Pr(D)\end{aligned}$$

All of the values were extracted from the problem statement and can be directly input to the expression:

$$\begin{aligned}\Pr(R) &= (0.85 \times 0.30) + (0.80 \times 0.25) + (0.90 \times 0.20) \\ &\quad + (0.75 \times 0.25) = 0.8225\end{aligned}$$

Step 4: Probability of Bridge C Given Detection (Pr(C|R)):

$$\Pr(C|R) = \frac{\Pr(R|C) \times \Pr(C)}{\Pr(R)} = \frac{0.90 \times 0.20}{0.8225} = 0.2187$$

Therefore, given that the sensor detects a structural deficiency, the probability that it is from Bridge C is **21.87%**.

Solution of Part 2: Probability of not Bridge C Given Detection ($\Pr(\overline{C}|R)$)

$$\begin{aligned}\Pr(\overline{C}|R) &= 1 - Pr(C|R) \\ &= 1 - 0.2187 = 0.7813\end{aligned}$$

Conversely, given a detection of deficiency, the probability that it is from any bridge other than Bridge C is **78.13%**.

Solution of Part 3: Second Assessment Using Ground Sensors

Step 1: Calculate the Probability of Detection by Ground Sensors

For Bridge C with ground sensors:

$$\Pr(GS|C) = 0.95$$

Since ground sensors have no false positives,

$$\Pr(FP|GS) = 0$$

Step 2: Calculate the Posterior Probability After Two Positive Results

Using the results from the drone system and ground sensors:

$$\Pr(C|R, GS) = \frac{\Pr(GS|C) \times \Pr(C|R)}{\Pr(GS)}$$

First, calculate Pr(GS):

$$\Pr(GS) = \Pr(GS|C) \times \Pr(C|R) + \Pr(GS|\overline{C}) \times \Pr(\overline{C}|R)$$

Since $\Pr(GS|\overline{C}) = 0$

$$\Pr(GS) = 0.95 \times 0.2187 + 0 = 0.207765$$

Now, calculate Pr(C|R,GS):

$$\Pr(C|R, GS) = \frac{\Pr(GS|C) \times \Pr(C|R)}{\Pr(GS)} = \frac{0.95 \times 0.2187}{0.207765} \approx 1$$

The additional use of ground sensors significantly increases the confidence in the assessment, highlighting the importance of multiple layers of verification in managing the structural integrity of bridges.

3.4.4.13 Example 13: Assessing Pothole Risk on City Roads Using Aerial Drones and Ground Sensors

A transportation agency is responsible for assessing the risk of potholes on various roads in a city. The city has four major roads: Road A, Road B, Road C, and Road D. Each road has different traffic volumes, age, and maintenance history, influencing the likelihood of pothole formation. The agency collects data on past pothole occurrences to estimate the prior probabilities of potholes in each road:

- **Road A:** 35% of the total road network.
- **Road B:** 25% of the total road network.
- **Road C:** 20% of the total road network.
- **Road D:** 20% of the total road network.

From historical data, it is known that:

- 10% of the potholes are found on Road A.
- 30% of the potholes are found on Road B.
- 50% of the potholes are found on Road C.
- 10% of the potholes are found on Road D.

The agency uses a detection system that employs aerial drones equipped with cameras to identify areas with potential potholes. The system's reliability is as follows:

- On Road A, the system correctly identifies potholes 80% of the time.
- On Road B, the system correctly identifies potholes 85% of the time.
- On Road C, the system correctly identifies potholes 90% of the time.
- On Road D, the system correctly identifies potholes 75% of the time.

However, the system also has a false positive rate:

- On Road A, the system incorrectly identifies non-pothole areas as having potholes 5% of the time.
- On Road B, the system incorrectly identifies non-pothole areas as having potholes 10% of the time.
- On Road C, the system incorrectly identifies non-pothole areas as having potholes 15% of the time.
- On Road D, the system incorrectly identifies non-pothole areas as having potholes 20% of the time.

Given that the system has identified an area as having potential potholes, the agency wants to know:

1. What is the probability that the area identified as having potholes is on Road C?
2. What is the probability that the area identified as having potholes is not on Road C?
3. If the agency decides to conduct a second assessment using ground sensors, which have a 95% accuracy rate in detecting potholes on Road C and no false positives, what is the probability that the area is indeed at risk if both the aerial system and the ground sensors indicate a risk?

Solution of Part 1: Probability of Area Identified as Having Potholes is on Road C

Step 1: Calculate the Prior Probabilities (Prevalence)

- Pr(A) = 0.35
- Pr(B) = 0.25
- Pr(C) = 0.20
- Pr(D) = 0.20.

Step 2: Calculate the Probability of Detection (Sensitivity) and False Positives

- Pr(R|A) = 0.80
- Pr(R|B) = 0.85
- Pr(R|C) = 0.90
- Pr(R|D) = 0.75
- Pr(FP|A) = 0.05
- Pr(FP|B) = 0.10
- Pr(FP|C) = 0.15
- Pr(FP|D) = 0.20.

Step 3: Calculate the Overall Probability of Detection (Pr(R))

We use the law of total probability to calculate the overall probability of detection:

$$\Pr(R) = \Pr(R|A) \times \Pr(A) + \Pr(R|B) \times \Pr(B) + \Pr(R|C) \times \Pr(C) + \Pr(R|D) \times \Pr(D)$$

Plugging in the values:

$$\Pr(R) = (0.80 \times 0.35) + (0.85 \times 0.25) + (0.90 \times 0.20) + (0.75 \times 0.20) = 0.8225$$

Step 4: Calculate the Posterior Probabilities

1. **Probability of Road C Given Detection (Pr(C|R))**:

$$\begin{aligned}\Pr(R) &= \frac{\Pr(R|C) \times \Pr(C)}{\Pr(R)}\\ &= \frac{0.90 \times 0.20}{0.8225} = 0.2187\end{aligned}$$

Solution of Part 2: Probability of not Road C Given Detection (Pr($\overline{C}$|R))

$$\begin{aligned}\Pr(\overline{C}|R) &= 1 - \Pr(C|R)\\ &= 1 - 0.2187\end{aligned}$$

$$= 0.7813$$

Solution of Part 3: Second Assessment Using Ground Sensors

Step 1: Calculate the Probability of Detection by Ground Sensors

For Road C with ground sensors:

$$\Pr(GS|C) = 0.95$$

Since ground sensors have no false positives,

$$\Pr(FP|GS) = 0$$

Step 2: Calculate the Posterior Probability After Two Positive Results

Using the results from the aerial system and ground sensors:

$$\Pr(C|R, GS) = \frac{\Pr(GS|C) \cdot \Pr(C|R)}{\Pr(GS)}$$

First, calculate Pr(GS):

$$Pr(GS) = \Pr(GS|C) \times \Pr(C|R) + \Pr(GS|\overline{C}) \times \Pr(\overline{C}|R)$$

Since $\Pr(GS|\overline{C}) = 0$

$$\Pr(GS) = 0.95 \times 0.2187 + 0 = 0.207765$$

Now, calculate $\Pr(C|R, GS)$

$$\Pr(C|R, GS) = \frac{\Pr(GS|C) \cdot \Pr(C|R)}{\Pr(GS)} = \frac{0.95 \times 0.2187}{0.207765} \approx 1$$

3.4.4.14 Example 14

A structural engineering firm is responsible for assessing the risk of cracks in beams within a large commercial building. The building has four types of beams: Type A, Type B, Type C, and Type D. Each beam type has different load capacities, ages, and maintenance histories, influencing the likelihood of cracks. The firm collects data on past crack occurrences to estimate the prior probabilities of cracks in each type of beam:

- **Type A**: 40% of the total beams.
- **Type B**: 30% of the total beams.
- **Type C**: 20% of the total beams.
- **Type D**: 10% of the total beams.

From historical data, it is known that:

- 5% of the cracks are found in Type A beams.
- 25% of the cracks are found in Type B beams.
- 50% of the cracks are found in Type C beams.
- 20% of the cracks are found in Type D beams.

The firm uses a detection system that employs ultrasonic testing to identify potential cracks in the beams. The system's reliability is as follows:

- For Type A beams, the system correctly identifies cracks 85% of the time.
- For Type B beams, the system correctly identifies cracks 80% of the time.
- For Type C beams, the system correctly identifies cracks 90% of the time.
- For Type D beams, the system correctly identifies cracks 75% of the time.

However, the system also has a false positive rate:

- For Type A beams, the system incorrectly identifies non-cracked beams as cracked 5% of the time.
- For Type B beams, the system incorrectly identifies non-cracked beams as cracked 10% of the time.
- For Type C beams, the system incorrectly identifies non-cracked beams as cracked 15% of the time.
- For Type D beams, the system incorrectly identifies non-cracked beams as cracked 20% of the time.

Given that the system has identified a beam as having potential cracks, the firm wants to know:

1. What is the probability that the beam identified as having cracks is a Type C beam?
2. What is the probability that the beam identified as having cracks is not a Type C beam?
3. Suppose the firm decides to conduct a second assessment using a more accurate detection method, such as X-ray testing, which has a 95% accuracy rate in detecting cracks in Type C beams and no false positives. What is the probability that the beam is indeed at risk if both the ultrasonic system and the X-ray testing indicate a risk?

Solution of Part 1: Probability of Type C Beam Given Detection

Solution using Bayes' Theorem

Step 1: Calculate the Prior Probabilities (Prevalence)

- Pr(A) = 0.40
- Pr(B) = 0.30
- Pr(C) = 0.20
- Pr(D) = 0.10.

Step 2: Calculate the Probability of Detection (Sensitivity) and False Positives

- Pr(R|A) = 0.85
- Pr(R|B) = 0.80
- Pr(R|C) = 0.90
- Pr(R|D) = 0.75
- Pr(FP|A) = 0.05
- Pr(FP|B) = 0.10
- Pr(FP|C) = 0.15
- Pr(FP|D) = 0.20.

Step 3: Calculate the Overall Probability of Detection (Pr(R))

We use the law of total probability to calculate the overall probability of detection:

$$\Pr(R) = \Pr(R|A) \times \Pr(A) + \Pr(R|B) \times \Pr(B) + \Pr(R|C) \times \Pr(C) + \Pr(R|D) \times \Pr(D)$$

Plugging in the values:

$$\Pr(R) = (0.85 \times 0.40) + (0.80 \times 0.30) + (0.90 \times 0.20) + (0.75 \times 0.10) = 0.835$$

Step 4: Calculate the Posterior Probabilities

Probability of Type C Beam Given Detection (Pr(C|R)):

$$\begin{aligned}\Pr(C|R) &= \frac{\Pr(R|C) \cdot \Pr(C)}{\Pr(R)} \\ &= \frac{0.90 \times 0.20}{0.835} = 0.2156\end{aligned}$$

Solution of Part 2: Probability of not Type C Beam Given Detection (Pr($\overline{\mathbf{C}}$|R))

$$\Pr(\overline{C}|R) = 1 - Pr(C|R)$$

$$= 1 - 0.2156 = 0.7844$$

Solution of Part 3: Second Assessment Using X-Ray Testing

Step 1: Calculate the Probability of Detection by X-Ray Testing

For Type C beams with X-ray testing:

$$\Pr(X\,|C) = 0.95$$

$$\Pr(\mathrm{X}|\mathrm{C}) = 0.95$$

Since X-ray testing has no false positives,

$$\Pr(FP|X) = 0$$

Step 2: Calculate the Posterior Probability After Two Positive Results

Using the results from the ultrasonic system and X-ray testing:

$$\Pr(C|R, X) = \frac{\Pr(X\,|C) \times \Pr(C|R)}{\Pr(X)}$$

First, calculate Pr(X):

$$\Pr(X) = \Pr(X\,|C) \times \Pr(C|R) + \Pr(X\,|\overline{C}) \times \Pr(\overline{C}|R)$$

Since $\Pr(X\,|\overline{C}) = 0$:

$$\Pr(X) = 0.95 \times 0.2156 + 0 = 0.20482$$

Now, calculate $\Pr(C|R, X)$:

$$\begin{aligned}\Pr(C|R, X) &= \frac{\Pr(X\,|C) \times \Pr(C|R)}{\Pr(X)}\\ &= \frac{0.95 \times 0.2156}{0.20482} \approx 1\end{aligned}$$

3.4.4.15 Example 15

A water management agency is responsible for assessing the risk of cracks in different sections of a dam. The dam is divided into six sections: Section A, Section B, Section C, Section D, Section E, and Section F. Each section has different exposure levels to water pressure, age, and maintenance histories, influencing the likelihood of cracks. The agency collects data on past crack occurrences to estimate the prior probabilities of cracks in each section:

- **Section A**: 15% of the total dam area.
- **Section B**: 20% of the total dam area.
- **Section C**: 25% of the total dam area.
- **Section D**: 10% of the total dam area.
- **Section E**: 15% of the total dam area.
- **Section F**: 15% of the total dam area.

From historical data, it is known that:

- 10% of the cracks are found in Section A.
- 20% of the cracks are found in Section B.
- 30% of the cracks are found in Section C.
- 15% of the cracks are found in Section D.
- 15% of the cracks are found in Section E.
- 10% of the cracks are found in Section F.

The agency uses a detection system that employs sonar technology to identify potential cracks in the dam. The system's reliability is as follows:

- In Section A, the system correctly identifies cracks 85% of the time.
- In Section B, the system correctly identifies cracks 80% of the time.
- In Section C, the system correctly identifies cracks 90% of the time.
- In Section D, the system correctly identifies cracks 75% of the time.
- In Section E, the system correctly identifies cracks 70% of the time.
- In Section F, the system correctly identifies cracks 80% of the time.

However, the system also has a false positive rate:

- In Section A, the system incorrectly identifies non-cracked sections as having cracks 5% of the time.
- In Section B, the system incorrectly identifies non-cracked sections as having cracks 10% of the time.
- In Section C, the system incorrectly identifies non-cracked sections as having cracks 15% of the time.
- In Section D, the system incorrectly identifies non-cracked sections as having cracks 20% of the time.
- In Section E, the system incorrectly identifies non-cracked sections as having cracks 25% of the time.

- In Section F, the system incorrectly identifies non-cracked sections as having cracks 10% of the time.

Given that the system has identified an area as having potential cracks, the agency wants to know:

1. What is the probability that the area identified as having cracks is in Section C?
2. What is the probability that the area identified as having cracks is not in Section C?
3. Suppose the agency decides to conduct a second assessment using ground-penetrating radar (GPR), which has a 95% accuracy rate in detecting cracks in Section C and no false positives. What is the probability that the area is indeed at risk if both the sonar system and the GPR indicate a risk?

3.4.4.16 Solution of Part 1: Probability of Section C Given Detection

Step 1: Calculate the Prior Probabilities (Prevalence)

- Pr(A) = 0.15
- Pr(B) = 0.20
- Pr(C) = 0.25
- Pr(D) = 0.10
- Pr(E) = 0.15
- Pr(F) = 0.15.

Step 2: Calculate the Probability of Detection (Sensitivity) and False Positives

- Pr(R|A) = 0.85
- Pr(R|B) = 0.80
- Pr(R|C) = 0.90
- Pr(R|D) = 0.75
- Pr(R|E) = 0.70
- Pr(R|F) = 0.80
- Pr(FP|A) = 0.05
- Pr(FP|B) = 0.10
- Pr(FP|C) = 0.15
- Pr(FP|D) = 0.20
- Pr(FP|E) = 0.25
- Pr(FP|F) = 0.10.

Step 3: Calculate the Overall Probability of Detection (Pr(R))

We use the law of total probability to calculate the overall probability of detection:

$$
\begin{aligned}
\Pr(R) = {} & \Pr(R|A) \times \Pr(A) + \Pr(R|B) \times \Pr(B) \\
& + \Pr(R|C) \times \Pr(C) + \Pr(R|D) \times \Pr(D) \\
& + \Pr(R|E) \times \Pr(E) + \Pr(R|F) \times \Pr(F)
\end{aligned}
$$

Plugging in the values:

$$\begin{aligned}\Pr(R) &= (0.85 \times 0.15) + (0.80 \times 0.20) \\ &\quad + (0.90 \times 0.25) + (0.75 \times 0.10) \\ &\quad + (0.70 \times 0.15) + (0.80 \times 0.15) \\ &= 0.8125\end{aligned}$$

Step 4: Calculate the Probability of Section C Given Detection:

$$\begin{aligned}\Pr(C|R) &= \frac{\Pr(R|C) \times \Pr(C)}{\Pr(R)} \\ &= \frac{0.90 \times 0.25}{0.8125} = 0.2769\end{aligned}$$

Solution of Part 2: Probability of not Section C Given Detection

$$\begin{aligned}\Pr(\overline{C}|R) &= 1 - \Pr(C|R) \\ &= 1 - 0.2769 = 0.7231\end{aligned}$$

Solution of Part 3: Second Assessment Using Ground-Penetrating Radar (GPR)

Step 1: Calculate the Probability of Detection by GPR

For Section C with GPR:

$$\Pr(GPR|C) = 0.95$$

Since GPR has no false positives,

$$\Pr(FP|GPR) = 0$$

Step 2: Calculate the Posterior Probability After Two Positive Results

Using the results from the sonar system and GPR:

$$\Pr(C|R, GPR) = \frac{\Pr(GPR|C) \cdot \Pr(C|R)}{\Pr(GPR)}$$

First, calculate Pr(GPR):

$$\Pr(GPR) = \Pr(GPR|C) \times \Pr(C|R) + \Pr(GPR|\overline{C}) \times \Pr\,(\overline{C}|R)$$

Since $\Pr(GPR|\overline{C}) = 0$

$$\Pr(GPR) = 0.95 \times 0.2769 + 0 = 0.263055$$

Now, calculate Pr(C|R,GPR):

$$\begin{aligned}\Pr(C|R, GPR) &= \frac{\Pr(GPR|C) \cdot \Pr(C|R)}{\Pr(GPR)} \\ &= \frac{0.95 \times 0.2769}{0.263055} \approx 1\end{aligned}$$

The statistical and probabilistic framework presented in this chapter builds upon classical probability theory, Bayesian inference, and modern statistical learning approaches widely adopted in engineering risk analysis, integrating foundational probability concepts and their engineering applications (Ang & Tang, 2007; Jaynes, 2003; Ross, 2023), originating from the formulation of conditional probability and inference introduced by Bayes (1763). Bayesian updating and probabilistic model calibration are essential for structural reliability and hazard assessment (Beck & Au, 2002; Gelman et al., 2013; Kass & Raftery, 1995; Robert & Casella, 2004), while graphical probabilistic models provide a rigorous framework for representing dependencies and uncertainty (Koller & Friedman, 2009). In engineering decision making, uncertainty characterization and risk quantification rely on established reliability and risk definitions and standardized statistical practices (Kaplan & Garrick, 1981; ISO, 2006, 2018, 2023; NIST & SEMATECH, 2003–2022), supported by descriptive statistics and exploratory data analysis techniques for understanding data behavior (Montgomery & Runger, 2018; Tufte, 2001; Tukey, 1977). Diagnostic performance metrics and classification evaluation methods such as sensitivity, specificity, and ROC analysis further support predictive modeling and hazard detection (Altman & Bland, 1994; Fawcett, 2006; Pepe, 2003). These statistical foundations enable informed interpretation of engineering datasets and decision making under uncertainty and are actively applied in contemporary civil and environmental engineering studies including flood susceptibility mapping, drought forecasting, erosion assessment, and AI assisted hydrological modeling (Amiri et al., 2024, 2025; Bontemps et al., 2025; Ebtehaj & Bonakdari, 2022; Ebtehaj et al., 2024; Soltani et al., 2021, 2024).

3.5 Summary

This chapter develops the mathematical and conceptual foundations needed to analyze uncertainty in civil and environmental engineering. It begins by outlining how data are represented and summarized using descriptive statistics, then moves toward the deeper probabilistic structures that govern risk assessment and decision-making.

The first major theme is risk analysis, where risk is framed as a function of both the probability of an event and its consequences. The chapter explains how engineers convert raw observations into probabilistic models, following a sequence from data collection to model estimation and finally to probability calculation. These models allow engineers to quantify uncertainty in events such as structural failure, soil instability, flooding, or traffic disruption.

Next, the chapter introduces three interpretations of probability, frequentistic, classical, and Bayesian, each offering a different perspective on how probability statements should be understood. The frequentist view relies on long-run frequencies observed in repeated trials; the classical view interprets probability as ratios of equally likely outcomes; and the Bayesian view interprets probability as a degree of belief updated with new evidence. Throughout, civil engineering examples illustrate how each interpretation informs practical decision-making.

Building on this, the chapter develops the machinery of conditional probability, explaining intersections, unions, complements, and independence with the aid of Venn diagrams. Numerous engineering-based examples, accident analysis, cracking in structural beams, winter road closures, flood risk scenarios, and building performance during earthquakes, show how conditioning reshapes expectations when partial information is known.

A key result is the Law of Total Probability, which describes how the probability of an event can be decomposed across mutually exclusive scenarios. This leads naturally to Bayes' Theorem, presented both symbolically and conceptually. Bayes' Theorem is shown to be a powerful method for inverting conditional probabilities, integrating prior information with data, and forming posterior beliefs. It is demonstrated through examples involving soil contamination, bridge defects detected by non-destructive testing, corrosion detection in reinforced concrete, inference of earthquake intensity based on observed damage, and groundwater contamination assessments.

The chapter then introduces marginalization, the process of deriving marginal probabilities from joint distributions. This sets the stage for working with joint probability distributions for both discrete and continuous variables. Through examples such as precipitation–river flow interactions, structural load–material strength relationships, and earthquake intensity–damage relationships, the chapter shows how joint distributions capture dependence between variables and allow for more sophisticated risk assessments.

Several extended scenarios highlight how probabilistic tools support engineering judgment: evaluating suppliers whose materials have different defect rates, determining concrete quality using imperfect compression tests, and assessing contamination risks using reliability metrics from diagnostic testing.

The final portion of the chapter focuses on sensitivity and specificity, key measures of test performance widely used in engineering diagnostics. The chapter explains true positives, false positives, true negatives, and false negatives, and discusses the trade-offs between sensitivity and specificity in contexts such as bridge inspections and flood forecasting. Confusion matrices are introduced as a structured way to visualize diagnostic accuracy.

Overall, the chapter provides a conceptual bridge between simple descriptive statistics and advanced probabilistic reasoning. Through extensive examples rooted in civil engineering practice, it shows how probability theory underpins risk quantification, model updating, data interpretation, and decision-making under uncertainty.

References

Altman, D. G., & Bland, J. M. (1994). Diagnostic tests 1: Sensitivity and specificity. *BMJ, 308*(6943), 1552. https://doi.org/10.1136/bmj.308.6943.1552

Amiri, A., Soltani, K., Ebtehaj, I., & Bonakdari, H. (2024). A novel machine learning tool for current and future flood susceptibility mapping by integrating remote sensing and geographic information systems. *Journal of Hydrology, 632*, 130936. https://www.sciencedirect.com/science/article/pii/S0022169424003305

Amiri, A., Ebtehaj, I., Soltani, K., Gumiere, S. J., & Bonakdari, H. (2025). Future soil erosion trends in Canadian agricultural lands from runoff and sustainability impacts. *Scientific Reports, 15*(1), 23184. https://www.nature.com/articles/s41598-025-05947-5

Ang, A. H.-S., & Tang, W. H. (2007). *Probability concepts in engineering: Emphasis on applications to civil and environmental engineering* (2nd ed.). Wiley. https://books.google.com/books/about/Probability_Concepts_in_Engineering.html?id=G4IoAQAAMAAJ

Bayes, T. (1763). An essay towards solving a problem in the doctrine of chances. *Philosophical Transactions of the Royal Society of London, 53*, 370–418. https://doi.org/10.1098/rstl.1763.0053

Beck, J. L., & Au, S.-K. (2002). Bayesian updating of structural models and reliability using Markov chain Monte Carlo simulation. *Journal of Engineering Mechanics, 128*(4), 380–391. https://doi.org/10.1061/(ASCE)0733-9399(2002)128:4(380)

Bontemps, J., Ebtehaj, I., Deslauriers, G., Rousseau, A. N., Bonakdari, H., & Dessureault-Rompré, J. (2025). Advanced efficient feature selection integrating augmented extreme learning machine and particle swarm optimization for predicting nitrogen use efficiency and yield in corn. *Agronomy, 15*(1), 244. https://www.mdpi.com/2073-4395/15/1/244

Ebtehaj, I., & Bonakdari, H. (2022). A comprehensive comparison of the fifth and sixth phases of the coupled model intercomparison project based on the Canadian earth system models in spatio-temporal variability of long-term flood susceptibility using remote sensing and flood frequency analysis. *Journal of Hydrology, 617*, 128851. https://www.sciencedirect.com/science/article/pii/S0022169422014214

Ebtehaj, I., Fortin, J., Bonakdari, H., & Grégoire, G. (2024). How climate change will shape pesticide application in Quebec's golf courses: Insights with deep learning based on assessing CMIP5 and CMIP6. *Applied Sciences, 14*(20), 9209. https://www.mdpi.com/2076-3417/14/20/9209

Fawcett, T. (2006). An introduction to ROC analysis. *Pattern Recognition Letters, 27*(8), 861–874. https://doi.org/10.1016/j.patrec.2005.10.010

Gelman, A., Carlin, J. B., Stern, H. S., Dunson, D. B., Vehtari, A., & Rubin, D. B. (2013). *Bayesian data analysis* (3rd ed.). Chapman and Hall/CRC. https://doi.org/10.1201/b16018

International Organization for Standardization. (2006). *Statistics—Vocabulary and symbols—Part 1: General statistical terms and terms used in probability (ISO Standard No. 3534-1:2006).* https://www.iso.org/standard/40145.html

International Organization for Standardization. (2018). *Risk management—Guidelines (ISO Standard No. 31000:2018).* https://www.iso.org/standard/65694.html

International Organization for Standardization. (2023). *Accuracy (trueness and precision) of measurement methods and results—Part 1: General principles and definitions (ISO Standard No. 5725-1:2023).* https://www.iso.org/standard/69418.html

Jaynes, E. T. (2003). Probability theory: The logic of science. *Cambridge University Press.* https://doi.org/10.1017/CBO9780511790423

Kaplan, S., & Garrick, B. J. (1981). On the quantitative definition of risk. *Risk Analysis, 1*(1), 11–27. https://doi.org/10.1111/j.1539-6924.1981.tb01350.x

Kass, R. E., & Raftery, A. E. (1995). Bayes factors. *Journal of the American Statistical Association, 90*(430), 773–795. https://doi.org/10.1080/01621459.1995.10476572

Koller, D., & Friedman, N. (2009). *Probabilistic graphical models: Principles and techniques.* MIT Press. https://mitpress.mit.edu/9780262013192/probabilistic-graphical-models/

Montgomery, D. C., & Runger, G. C. (2018). *Applied statistics and probability for engineers* (7th ed.). Wiley. https://www.wiley.com/en-ca/Applied%2BStatistics%2Band%2BProbability%2Bfor%2BEngineers%2C%2B7th%2BEdition-p-9781119400363

National Institute of Standards and Technology, & SEMATECH. (2003–2022). *NIST/SEMATECH e-handbook of statistical methods (NIST Handbook 151).* https://doi.org/10.18434/M32189

Pepe, M. S. (2003). *The statistical evaluation of medical tests for classification and prediction.* Oxford University Press. https://academic.oup.com/book/52788

Robert, C. P., & Casella, G. (2004). *Monte Carlo statistical methods* (2nd ed.). Springer. https://doi.org/10.1007/978-1-4757-4145-2

Ross, S. M. (2023). *A first course in probability* (10th ed.). Pearson. https://www.pearson.com/en-ca/subject-catalog/p/first-course-in-probability-a/P200000006334/9780137504589

Soltani, K., Ebtehaj, I., Amiri, A., Azari, A., Gharabaghi, B., & Bonakdari, H. (2021). Mapping the spatial and temporal variability of flood susceptibility using remotely sensed normalized difference vegetation index and the forecasted changes in the future. *Science of The Total Environment, 770*, 145288. https://www.sciencedirect.com/science/article/pii/S0048969721003545

Soltani, K., Amiri, A., Ebtehaj, I., Cheshmehghasabani, H., Fazeli, S., Gumiere, S. J., & Bonakdari, H. (2024). Advanced forecasting of drought zones in Canada using deep learning and CMIP6 projections. *Climate, 12*(8), 119. https://www.mdpi.com/2225-1154/12/8/119

Tufte, E. R. (2001). *The visual display of quantitative information* (2nd ed.). Graphics Press. https://unesdoc.unesco.org/ark:/48223/pf0000145210

Tukey, J. W. (1977). *Exploratory data analysis.* Addison-Wesley. https://books.google.com/books/about/Exploratory_Data_Analysis.html?id=UT9dAAAAIAAJ

Upton, G., & Cook, I. (2014). *Oxford dictionary of statistics* (2nd ed.). Oxford, UK: Oxford University Press.

Chapter 4
Descriptive Statistics and Uncertainty Analysis in Civil Engineering

4.1 Introduction

Understanding and quantifying uncertainty is fundamental to engineering design. This is especially true in the field of civil and environmental engineering, where variability in materials, measurements, and natural phenomena can significantly affect outcomes. This chapter provides a structured and practical introduction to descriptive statistics and their role in data analysis and uncertainty modeling within engineering contexts. It serves as a foundation for interpreting raw data and preparing it for probabilistic modeling and risk assessment.

The chapter begins by defining the concept and purpose of descriptive statistics, emphasizing their value in transforming complex datasets into simplified, interpretable formats. It covers both quantitative approaches (e.g., mean, median, mode, variance, standard deviation, percentiles, coefficient of variation) and graphical methods (e.g., histograms, box plots, violin plots) for data description. These tools are essential for characterizing data distribution, identifying trends and outliers, as well as summarizing key statistical properties in both a visually and analytically meaningful way.

Building on this foundation, the chapter transitions into a discussion of uncertainty in engineering. It introduces and contrasts aleatory and epistemic uncertainty, explaining their sources by providing real-world examples encountered in civil and environmental engineering practice. The chapter highlights how these uncertainties manifest themselves in material testing and design decisions, for example, and how they can be represented and managed through the use of random variables.

H. Bonakdari et al., *Hazard and Risk Assessment in Civil Engineering Design*,
https://doi.org/10.1007/978-3-032-23600-5_4

Following this, the focus of the chapter shifts to discuss probability distributions and the role of probability density functions and cumulative distribution functions in civil engineering applications. It uses some of the most frequently measured variables in civil and environmental engineering, for example, rainfall, wind speed, and structural loads, to explore how these probability distributions can be used to help describe the likelihood of different outcomes. The chapter further elaborates on discrete and continuous random variables, statistical moments (mean, variance, skewness, and kurtosis), and their relevance to uncertainty modeling and hazard prediction.

In closing this chapter, the final sections are dedicated to guiding readers through methods used for distribution selection and validation, including the method of moments, least-squares estimation, and maximum likelihood estimation. It concludes with a practical discussion on goodness-of-fit tests such as Chi-square, Kolmogorov–Smirnov, and Anderson–Darling tests, supported by code and visual tools to evaluate the adequacy of different distribution models.

Each section is explained with engineering-focused examples and implemented through both MATLAB and Python codes, enabling practical application and computational insight.

4.2 Descriptive Statistics

The level of risk associated with a given event is typically assessed by first investigating the available data, which includes observations and test results, by using descriptive statistics. Descriptive statistics are employed without making assumptions about the degree or nature of the randomness inherent in the data being analyzed. Instead, they are viewed as a convenient tool to reduce the data to a more manageable form. This makes the data suitable for further analysis and allows it to be communicated in a standardized format to other professionals.

Figure 4.1 illustrates the sequence of steps to be used when analyzing data, beginning with the collection of data and its summary using descriptive statistics. Following this, model estimation is conducted to develop a probabilistic model that can quantify the likelihood of specific events. Probabilistic models are helpful when estimating the probability of different events, providing valuable insights into the potential risks associated with the various outcomes. Based on these first steps, the reader can clearly see how raw data is handled to provide actionable insights through the use of statistical methods. It is important to note that before any model estimation or probability calculations are performed, a thorough understanding of the descriptive statistics is essential. Descriptive statistics provide the foundation by which the data is structured, interpreted, and prepared for subsequent analysis, ensuring that the insights are based on well-organized and comprehensible data.

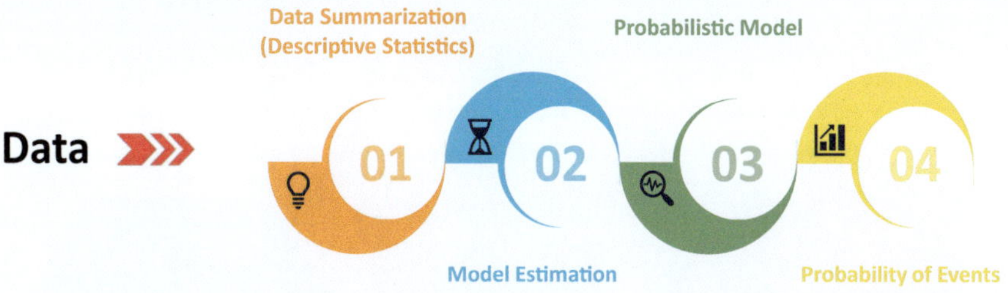

Fig. 4.1 An overview of the data analysis process by descriptive statistics, including model estimation, probabilistic model generation, and estimating the probability of events

In order to provide the most accurate assessment of risk in engineering practice, it is critical to first have a solid understanding of the data and its distribution. In this first step, the goal is simply to describe the data, which can be done by two different methods:

- **Numerically**: By using statistical measures such as mean, median, standard deviation, variance, and range, which help summarize the central tendency, dispersion, and overall distribution of the data.
- **Graphically**: By employing visual representations such as histograms, box plots, and scatter plots, which allow for a quick and intuitive understanding of any trends in the data, potential outliers, as well as any possible relationships between variables.

It should be stressed that, prior to moving forward with more complex analyses such as model estimation and probability calculations, it is essential to gain a clear understanding of the data distributions by assessing them both numerically and graphically.

4.3 Data Description and Treatment

As mentioned, a detailed analysis of the data is a requirement in the process of risk assessment; therefore, it is important to have a strong foundation in the application of summary statistics and their use in data description. Summary statistics can be divided into several key categories (Fig. 4.2):

- Measures of central tendency: These include the average (mean), median, and mode, all of which provide insight into where the center or typical value of the data lies.

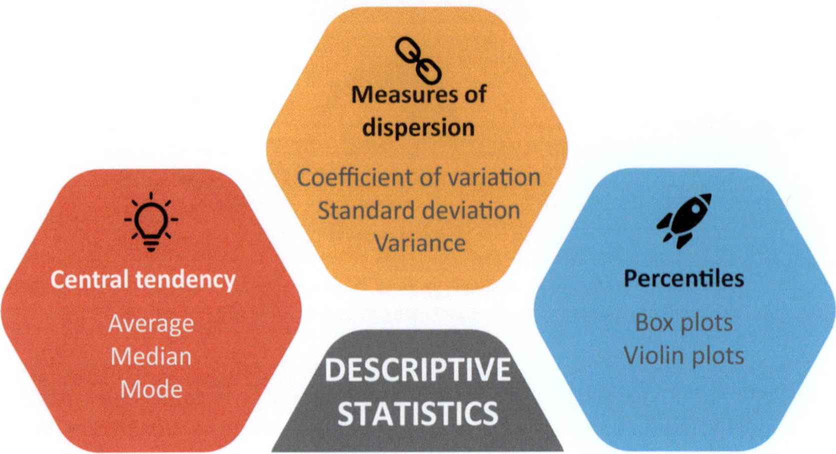

Fig. 4.2 Key categories of descriptive statistics

- Measures of dispersion: These include the variance, standard deviation, and the coefficient of variation, which help to describe the spread or variability of the data from the central values.
- Percentiles: These offer information about the distribution of the data across different ranks. Graphical methods such as box plots and violin plots can be used to visualize these percentiles, highlighting the range and concentration of the data.

4.3.1 *Measures of Central Tendency*

Measures of central tendency are statistical metrics used to summarize a set of data by identifying the center or expected value within the dataset. These measures provide insights into the data's overall trend or the most common values, helping to understand where the data points are clustered. The three most commonly used measures of central tendency are.

4.3.1.1 Mean (Average Value)

The arithmetic mean, or average value, is the most commonly used descriptor of central tendency. The mean, denoted by $\overline{X}$ is calculated for a set of n observations by taking the sum of all individual values, x_i, and dividing this total by the number of observations. This can be expressed using Eq. 4.1. below:

$$\overline{X} = \frac{1}{n}\sum_{i=1}^{n} x_i \tag{4.1}$$

Here, $\sum_{i=1}^{n} x_i$, represents the sum of all values in the dataset, and n denotes the total number of observations. Inherent in this calculation is the assignment of equal weight to each observation, which makes the mean a useful measure for understanding the overall trend of the dataset. While the arithmetic mean is the most widely applied measure of central tendency across various fields of study, it is susceptible to the presence of extreme outlier values.

4.3.1.2 Median (Middle Value)

The median is defined as the value that separates the upper and lower halves of the dataset. It is the value above which 50% of the data lies and below which the other 50% of the data is located. To find the median, the data values must first be arranged in either increasing or decreasing order. If the total number of observations, n, is odd, the median is the value located at the rank position $\frac{n+1}{2}$. If n is even, the median is calculated by taking the average of the two middle values—those positioned at ranks $\frac{n}{2}$ and $\frac{n+1}{2}$. This can be expressed mathematically as:

$$x_{50} = \begin{cases} x_{\frac{n+1}{2}} & \text{if } n \text{ is odd} \\ x_{\frac{n}{2}} & \text{if } n \text{ is even} \end{cases} \tag{4.2}$$

It should be noted that the median is less sensitive to outliers and skewed data. This is because, unlike the mean, the median depends solely on the ordered position of values rather than their magnitude. Extreme observations do not alter the location of the middle value in the dataset, and therefore have little to no influence on its calculation.

4.3.1.3 Mode (Most Frequent Value)

The mode value, denoted as x_d, is defined as the point or points within a dataset where the highest frequency or relative frequency of occurrence is observed and can be determined with the aid of a frequency histogram. If no value is repeated, then the mode does not exist for that particular data set.

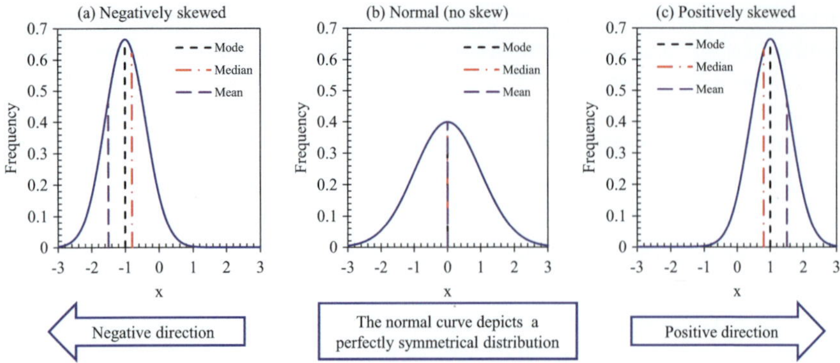

Fig. 4.3 Examples of different data distributions: **a** Negatively skewed distribution where the mode is greater than the median and mean; **b** Symmetrical distribution where the mean, median, and mode coincide at the center; and **c** Positively skewed distribution where the mode is less than the median and mean. The direction of the tail in the distribution indicates the skewness

Examples of normal, positive, and negatively skewed data distributions can be seen in Fig. 4.3. In a symmetrical distribution, the mean, median, and mode all coincide at the center of the distribution, indicating no skewness. In a negatively skewed distribution, the tail extends toward the negative direction of the x-axis, and the three measures follow their characteristic order: mean < median < mode, with the mean pulled furthest in the direction of the tail. Conversely, in a positively skewed distribution, where the tail extends toward the positive direction, the ordering reverses to mode < median < mean, again with the mean displaced most strongly toward the tail.

4.3.1.4 Limitations of Central Tendency Measures and the Importance of Data Distribution

Although measures of central tendency, such as the mean, median, and mode, provide important insights into a dataset, they do not fully characterize the underlying data. These measures only summarize certain aspects of the data, specifically its central point or the most frequent value. However, they fail to convey the shape, spread, or variability of the data distribution. For instance, two variables can have the same mean but differ significantly in their distribution, as illustrated by the two histograms shown in Fig. 4.4. This discrepancy shows that central tendency measures alone are insufficient to fully describe the dataset. The shape and spread of the data, which can be captured by histograms or other visual tools, provide additional insights that are necessary to gain a complete understanding of the dataset.

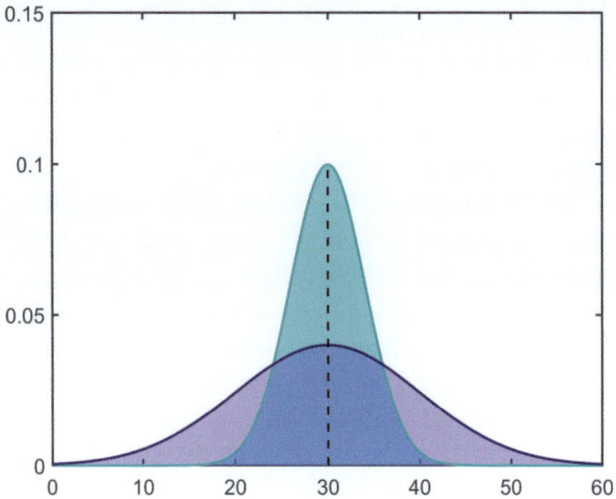

Fig. 4.4 Histograms showing two datasets with the same mean, but drastically different distributions

4.3.1.5 MATLAB Implementation of Central Tendency Measures

The three measures of central tendency discussed up to this point, namely the mean, median, and mode, can be conveniently calculated in MATLAB using Code 4.1. The use of this code will be demonstrated for a dataset containing concrete compressive strength measurements in megapascals (MPa). The script is clearly structured and annotated with explanations, making it suitable for beginners in both MATLAB and statistics.

The script begins with the initialization section on Lines 1 to 3, where the `clc`, `clear`, and `close all` commands are used. `clc` clears the command window to provide a clean display, `clear` removes all previously defined variables from memory, and `close all` shuts down any open figures. This ensures that the workspace is free from remnants of earlier code, avoiding potential interference.

In Lines 5 to 14, the "`%`" operator is used to provide code comments. In the case of Code 4.1, the comments contain a block of text explaining the purpose of the script to help any future users. It describes that the code calculates the mean (average), median (middle value), and mode (most frequent value), and then presents them in a neatly formatted table. These comments are not executed by MATLAB but are helpful guides for future users' reference and understanding.

The data definition section on Lines 16 to 18 creates the dataset. A vector named `data` is defined, which includes 20 numerical values representing the compressive strength of concrete samples in MPa as measured in the field. These values are stored as a row vector, allowing for vectorized operations in MATLAB, which simplifies the analysis.

Next, the data display section on Lines 20 to 23 prints the original dataset to the command window. The `fprintf` function is used with the format specifier `'%.1f '` to show each value rounded to one decimal place, separated by spaces. This clean visual format helps the user verify that the dataset is loaded and formatted correctly before performing any calculations.

The calculation of the dataset mean spans Lines 26 to 29. Here, the mean (or average) value of the dataset is computed using MATLAB's built-in `mean` function and stored in the variable `mean_value`. The mean provides a simple measure of central tendency by summing all values and dividing by the number of observations.

In the median calculation section, which can be seen on Lines 32 to 35, MATLAB's `median` function is used to find the central value of the sorted dataset. This value, stored in the variable `median_value`, divides the data into two equal halves. The median is particularly useful when the data contains outliers, as it is not affected by extreme values.

Following this, the `mode` function is used in Lines 38–40 to determine the most frequently occurring value in the dataset. The result is stored in the variable `mode_value`. If multiple values appear with equal highest frequency, MATLAB returns the smallest one. If all values are unique, the mode may not be informative.

Lines 43 to 46 contain the code used to display the results in a tabular format. Here, the MATLAB function `char` is used to define the statistic names (`'Mean (MPa)'`, `'Median (MPa)'`, and `'Mode (MPa)'`) as a character array in `StatNames`, and stores the corresponding calculated values in `StatValues`. These are combined into a table using the `table` function, which organizes the names and values into a readable structure.

Finally, the table display section on Lines 48 to 53 handles the presentation of the output. For spacing purposes, the code begins with two blank lines using the `disp(' ')` function, then prints a header line for context, and converts the `StatNames` to string format so they are displayed correctly as labels. The command `disp(resultsTable)` then outputs the final statistics table to the command window in a readable format.

Code 4.1 MATLAB script for computing and displaying central tendency measures (mean, median, and mode) of a concrete compressive strength dataset in metric units (MPa)

Code 4.1. MATLAB script for computing and displaying central tendency measures (mean, median, and mode) of a concrete compressive strength dataset in metric units (MPa)

```
1   clc
2   clear
3   close all
4
5   % ======================================================================
```

```
Central tendency measures
% ============================================================================
% This script calculates and explains the following descriptive statistics:
% 1. Mean (Average)
% 2. Median (Middle Value)
% 3. Mode (Most Frequent Value)
% Results are presented in a table with two columns: Name and Value.
% Units: Megapascals (MPa)
% ============================================================================

% Sample dataset: Concrete compressive strength (in MPa)
data = [28.5, 30.2, 27.8, 32.0, 29.6, 28.5, 31.1, 33.5, 29.0, 28.9, ...
30.0, 29.4, 30.8, 27.5, 29.8, 30.2, 31.0, 29.2, 28.5, 32.4];

% Display the dataset
fprintf('Original Data (MPa):\n');
fprintf('%.1f ', data);
fprintf('\n');

%% 1. Mean:
% The mean is the average value of the dataset.
% It is calculated as the sum of all values divided by the number of values.
mean_value = mean(data);

%% 2. Median:
% The median is the middle value when the data is sorted.
% It separates the higher half from the lower half of the data.
median_value = median(data);

%% 3. Mode:
% The mode is the value that appears most frequently in the dataset.
% If all values are unique, the mode is not meaningful.
mode_value = mode(data);

%% Create a results table
StatNames = char('Mean (MPa)', 'Median (MPa)', 'Mode (MPa)');
StatValues = [mean_value; median_value; mode_value];
resultsTable = table(StatNames, StatValues);

%% Display the table
disp(' ');
disp(' ');
disp('Descriptive Statistics Results (Metric Units - MPa)');
resultsTable.StatNames = string(resultsTable.StatNames);
```

The output of Code 4.1, as it appears in the MATLAB Command Window, is shown in Fig. 4.5. Here we can see that the code computes the descriptive statistics for the dataset of interest and then outputs the results in a labeled table.

```
Command Window
Original Data (MPa):
28.5  30.2  27.8  32.0  29.6  28.5  31.1  33.5  29.0  28.9  30.0  29.4  30.8  27.5  29.8  30.2  31.0  29.2  28.5  32.4

Descriptive Statistics Results (Metric Units - MPa)
        StatNames          StatValues
      _____________        __________

      "Mean (MPa)   "        29.895
      "Median (MPa)"         29.7
      "Mode (MPa)   "        28.5
```

Fig. 4.5 Command Window output of Code 4.1

4.3.2 Measures of Dispersion

While measures of central tendency describe the middle or typical values in a dataset, they do not provide information about the spread or variability of the data. Measures of dispersion are used to describe the level of scatter or variability in the data around the central tendency. These measures give a clearer picture of how much the data points deviate from the center, helping to assess the distribution's spread.

4.3.2.1 Variance

The variance, one of the most commonly used measures of dispersion, quantifies the extent to which data points in a dataset differ from the mean value. For a set of n observations, variance S^2 is calculated by summing the squared differences between each data point and the mean, and then dividing this sum by $n-1$. The formula for variance is given by Eq. 4.3:

$$S^2 = \frac{1}{n-1}\sum_{i=1}^{n}\left(x_i - \overline{X}\right) = \frac{1}{n-1}\left[\sum_{i=1}^{n}x_i^2 - \frac{1}{n}\left(\sum_{i=1}^{n}x_i\right)^2\right]. \tag{4.3}$$

Here, x_i represents the individual data points, and $\overline{X}$ is the mean of the data. The units of the variance are the square of the units of the variable x, which reflects the squared differences from the mean.

In these equations, the denominator $n-1$, representing the number of degrees of freedom, is used instead of n to compute the average squared deviation from the mean. This adjustment ensures an unbiased estimate of the population variance when using a sample. The use of $n-1$, also known as Bessel's correction, corrects for the fact that the sample mean is likely to be slightly closer to the individual data points than the true population mean, thus preventing the underestimation of the true variance. The units of the variance are the square of the units of the variable x.

Variance serves as a foundation for other important measures, such as the standard deviation, which is simply the square root of the variance. The standard deviation is useful as it returns the dispersion to the same units as the data, making it more interpretable in practical terms. Both variance and standard deviation are critical for understanding the spread and reliability of the data. The standard deviation S is then defined as the square root of the variance (Eq. 4.4):

$$S = \sqrt{S^2} \tag{4.4}$$

The standard deviation measures the average distance of the data points from the mean, providing an understanding of how spread out the numbers are. When more samples are closer to the mean, the standard deviation (or variance) decreases. Conversely, the corresponding standard deviation (or variance) increases if the samples are more widely spread out.

Figure 4.6 illustrates the effect of standard deviation on the shape of a probability distribution by using two sample datasets, one with a small standard deviation (represented by the red curve) and one with a large standard deviation (represented by the blue curve). In the case of the red curve, a high degree of precision is observed as most of the measurements are clustered closely around the most probable value, leading to a narrower and taller distribution with a low standard deviation.

On the other hand, the blue curve represents a distribution with low precision, where the measurements are more spread out from the most probable value. This results in a wider and shorter distribution, indicating a higher standard deviation. The visual distinction between these two distributions highlights the effect of variability in data, as a higher spread results in a larger standard deviation, while a lower spread results in a smaller standard deviation.

Understanding both the variance and standard deviation is crucial for accurately interpreting data variability. These measures, alongside visual representations such as that shown in Fig. 4.6, allow researchers and analysts to gain insights into the consistency, precision, and reliability of their data.

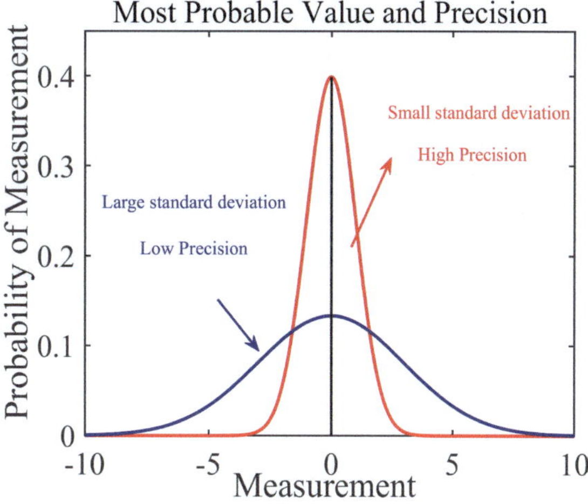

Fig. 4.6 Illustration of the effect of standard deviation on the shape of a probability distribution

4.3.2.2 Coefficient of Variation

A second commonly used measure derived from the variance/standard deviation is the coefficient of variation (COV). As we saw in the previous section, the standard deviation (denoted as S) is by definition the square root of the variance (Eq. 4.4). It provides an indication of the spread of the data about the mean value and is expressed in the same units as the original data.

While the standard deviation provides an absolute measure of dispersion, the coefficient of variation (COV) is a relative measure of variability. It expresses the standard deviation as a proportion of the mean, making it a dimensionless quantity that is useful for comparing variability across different datasets or units of measurement. The COV is defined using Eq. 4.5:

$$COV = \frac{S}{\overline{X}} \tag{4.5}$$

where S is the sample standard deviation and $\overline{X}$ is the sample mean. A higher COV indicates greater variability relative to the mean, which suggests more spread in the data around the central value. Conversely, a lower coefficient of variation indicates less variability, signifying a more precise estimate of the mean.

The use of the COV is particularly beneficial when comparing datasets with different scales or units of measurement. Since it is dimensionless, it allows for direct comparisons of variability between datasets.

To better understand the application of the statistical measures introduced, we will present an example involving the determination of the compressive strength of concrete.

Example

As part of their quality control protocol, ABC Contracting cast five samples to determine the compressive strength of their concrete, measured in kilopounds per square inch (ksi). After being broken in the lab, the results were as follows: 2.5, 3.5, 2.2, 3.2, and 2.9 ksi. To verify that the concrete meets specifications, the mean, variance, standard deviation, and coefficient of variation (COV) of the concrete strength must be determined.

Solution

The descriptive statistics, as specified by the problem statement, are computed using the previously introduced equations. By directly applying these formulas to the dataset, each statistical measure can be calculated:

Mean Value: The mean value of the concrete strength $\overline{X}$ is calculated using Eq. 4.1 as shown below:

$$\overline{X} = \frac{2.5 + 3.5 + 2.2 + 3.2 + 2.9}{5} = 2.86 \text{ ksi}$$

Variance: The variance S^2 is computed using Eq. 4.3 as shown below:

$$S^2 = \frac{2.5^2 + 3.5^2 + 2.2^2 + 3.2^2 + 2.9^2 - \left(\frac{2.5+3.5+2.2+3.2+2.9}{5}\right)^2}{5-1} = 0.273\,\text{ksi}^2$$

Standard Deviation: The standard deviation, S, is given by the square root of the sample variance (Eq. 4.4):

$$S = \sqrt{0.273} = 0.52249\,\text{ksi}$$

Coefficient of Variation: The coefficient of variation (COV) can be computed using Eq. 4.5 and the sample mean and standard deviation calculated above:

$$COV(X) = \frac{S}{\overline{X}} = \frac{0.52249}{2.86} = 0.183$$

The relatively large COV of 18.3% suggests that the average compressive strength value might not be reliable, as there is a considerable level of variability in the test results. This high variability suggests the need for additional measurements to obtain a more precise estimate of the concrete's compressive strength and could be the result of several factors. If the COV remains large with further testing, relatively large factors of safety should be considered for projects utilizing this specific concrete mix to account for the uncertainty.

The computed descriptive statistics provide insight into both the central tendency and the variability of the dataset, enabling more informed decisions regarding the reliability of the concrete's compressive strength.

4.3.2.3 MATLAB Implementation of Dispersion Measures

Code 4.2 presents a MATLAB script that calculates the three important measures of dispersion introduced earlier, namely the variance, standard deviation, and coefficient of variation (COV). These measures help quantify how much variability exists in a dataset. The execution of this code will be demonstrated through the use of a practical example with a dataset consisting of slump test results in millimeters (mm), a test commonly employed in civil engineering practice to assess the consistency and workability of fresh concrete. The slump test is conducted by filling a standard metal cone with freshly mixed concrete, then lifting the cone and measuring the vertical drop—or subsidence—of the concrete. This drop, known as the slump value, reflects how easily the concrete flows. A higher slump value indicates more workable, fluid concrete, while a lower value suggests a stiffer, drier mix. The dataset used in this script includes 15 measurements of slump collected on-site during a concrete pour, providing a realistic sample for analyzing variability in concrete workability.

The script begins with Lines 1 to 3, where `clc` is used to clear the Command Window, `clear` removes all variables currently stored in memory, and `close` `all` shuts any open figure windows. This ensures a fresh and interference-free environment for running the new analysis.

In Lines 5 to 15, the script includes header comments that describe the purpose of the code. It is explained that the program calculates three key statistical metrics: variance (how data spreads around the mean), standard deviation (the average distance from the mean), and COV (a relative measure of dispersion that compares the standard deviation to the mean). The comments also mention that the script uses sample slump data and outputs the results in a table format.

Lines 17 and 18 define the input data. The variable `slump_data` stores a list of 15 values representing measured slump heights in millimeters. As noted earlier, these values correspond to real-world data from a slump test, a standard procedure used during the construction of civil structures where the workability of fresh concrete is measured by how much it "slumps" after the removal of a cone-shaped mold.

Lines 20 to 23 are used to display the dataset to the user in the Command Window. The `fprintf` function is used to print each value with no decimal places, as in practice, slump test results are usually recorded in whole millimeters. This display step helps the user visually confirm that the correct data has been loaded.

The script moves on to the first dispersion measure in Lines 25 to 28, where variance is calculated. Variance measures the average of the squared differences between each value and the dataset's mean. It tells us how far the data points tend to be from the average value. MATLAB's built-in `var()` function is used here, and it automatically applies Bessel's correction by dividing by $(n-1)$, which is appropriate for sample data.

In Lines 30 to 33, the script computes the standard deviation using MATLAB's `std()` function. This metric is simply the square root of the variance and is expressed in the same units as the original data (in this case, millimeters). In practice, the standard deviation is easier to interpret than variance (being in the same units as the parameter under consideration) and tells us how tightly the data points cluster around the mean.

Lines 35 to 39 focus on calculating the coefficient of variation (COV). COV is a dimensionless measure that relates the standard deviation to the mean, making it useful for comparing variability across datasets with different units or scales. First, the script calculates the mean (`mean_slump`), then computes the COV (as a percentage) using the formula `(standard deviation / mean) * 100`.

The results are compiled into a table using Lines 41 to 44 of the code. The `StatNames` variable holds the names of the three statistics, while `StatValues` contains their calculated values. The `table()` function is used to assemble these into a structured and readable table stored in `resultsTable`.

Finally, in Lines 46 to 50, the table is displayed to the user in the Command Window. Two blank lines (`disp(' ')`, shown in Lines 47–48) create visual spacing before the table header is printed. Then, `disp(resultsTable)` outputs the formatted table, allowing the user to easily read the variance, standard deviation, and COV results for the concrete slump data.

Code 4.2 MATLAB script for computing and presenting dispersion measures (variance, standard deviation, and coefficient of variation) of a concrete slump dataset in metric units (mm)

```
clc
clear
close all

% ==========================================================================
%                              Dispersion measures
% ==========================================================================
% This script calculates key statistical measures of dispersion:
% 1. Variance - quantifies how far data points spread from the mean.
% 2. Standard Deviation - measures the average distance from the mean.
% 3. Coefficient of Variation (COV) - expresses the standard deviation
% relative to the mean (dimensionless), useful for comparing across datasets.
% The script uses assumed slump test data in mm and outputs a table with
% the calculated values. The data represent a single sample group.
% ==========================================================================

% Sample Dataset Explanation:
slump_data = [65, 70, 72, 64, 69, 75, 78, 71, 67, 66, 76, 80, 73, 68, 74];

% Display the dataset
fprintf('Original Slump Data (mm):\n');
fprintf('%.0f ', slump_data);
fprintf('\n');

%% 1. Variance:
% Variance quantifies the spread of the data from the mean.
% It is computed as the average of the squared differences from the mean.
variance_value = var(slump_data); % MATLAB uses n-1 for sample variance

%% 2. Standard Deviation:
% The standard deviation is the square root of the variance,
% expressed in the same units as the original data.
std_dev = std(slump_data);

%% 3. Coefficient of Variation:
% The coefficient of variation (COV) is a normalized measure of dispersion.
% It is calculated as (standard deviation / mean) * 100.
mean_slump = mean(slump_data);
cov_value = (std_dev / mean_slump) * 100;

%% Create results table
```

```
Command Window
Original Slump Data (mm):
65 70 72 64 69 75 78 71 67 66 76 80 73 68 74


Dispersion Statistics for Concrete Slump (Metric Units - mm)
                  StatNames                     StatValues
    ________________________________         __________

    {'Variance (mm^2)'                }        23.171
    {'Standard Deviation (mm)'        }        4.8137
    {'Coefficient of Variation (%)'}           6.7608
```

Fig. 4.7 Command window output of Code 4.2

```
StatNames = {'Variance (mm^2)'; 'Standard Deviation (mm)'; 'Coefficient of Variation (%)'};
StatValues = [variance_value; std_dev; cov_value];
resultsTable = table(StatNames, StatValues);

%% Display the table
disp(' ');
disp(' ');
disp('Dispersion Statistics for Concrete Slump (Metric Units - mm)');
disp(resultsTable);
```

The output of Code 4.2, which displays the calculated dispersion measures (variance, standard deviation, and coefficient of variation) for the dataset, is shown in Fig. 4.7 as it appears in the MATLAB Command Window.

4.3.3 Percentile

Percentiles are statistical measures that provide information about the relative standing of values within a dataset. They indicate the position of a particular value relative to the rest of the data by dividing the dataset into 100 equal parts. The most common percentiles include the 25th percentile (first quartile, Q1), the 50th percentile (median, Q2), and the 75th percentile (third quartile, Q3). Percentiles are widely used in data analysis to understand the distribution and spread of values in a dataset, particularly when dealing with skewed data or in identifying outliers.

Two powerful graphical tools for visualizing percentiles, and more generally the data distribution as a whole, are the box plot and the violin plot. Both plots are similar in that they offer valuable insights into the distribution, variability, and presence of outliers in the data, but are different in how they emphasize different aspects of the dataset. A box plot, also known as a whisker plot or box and whisker plot, displays a summary of the central tendency and variability of a dataset, showing the median, quartiles, and potential outliers. A violin plot provides a more detailed view of the data distribution by combining elements of a box plot with a density plot, giving a clearer picture of the data's underlying distribution and its spread. In the following sub-sections, each plot will be described in greater detail to highlight its unique features and applications.

4.3.3.1 Box Plot

A box plot is a graphical representation that conveys information about the central tendency and spread of the data. It visualizes the median (or mean) value along with the first quartile (Q1) and third quartile (Q3).

The first quartile, denoted by Q1, is the median of the lower half of the data set. This indicates that approximately 25% of the numbers in the data set lie below Q1, while about 75% of the numbers lie above it. Identifying Q1 helps in understanding the spread of the lower portion of the dataset and provides insight into the distribution of values closer to the minimum.

The third quartile, denoted by Q3, is the median of the upper half of the data set. It marks the value below which approximately 75% of the observations fall, meaning that roughly the upper 25% of the data lie at or above Q3. Identifying Q3 is crucial for understanding the spread of the higher portion of the dataset and offers insights into the range of values closer to the maximum.

Figure 4.8 presents the format of a box-and-whisker plot. This plot is especially useful in civil engineering when analyzing large sets of measurements—such as rainfall records, material strengths, or environmental data—where understanding data variability and outliers is critical for design and risk assessments.

The box-and-whisker plot displays several important statistical components:

- **Median (Q2)**: This is the middle value of the dataset, which divides the data into two equal halves. It is represented by the horizontal line inside the box.
- **First Quartile (Q1)**: Also known as the 25th percentile, this marks the value below which 25% of the data fall.
- **Third Quartile (Q3)**: The 75th percentile, indicating the value below which 75% of the data are located.
- **Interquartile Range (IQR)**: The range between Q1 and Q3 (i.e., IQR = Q3–Q1). It represents the spread of the middle 50% of the data and is a key indicator of statistical dispersion.
- **Minimum and Maximum (Whiskers)**: These represent the smallest and largest values within a defined range, often excluding outliers. The whiskers extend from the quartiles to these extreme values.

Using this plot, engineers can quickly assess the symmetry, spread, and potential outliers in the data. For instance, if the median line is closer to the bottom or top

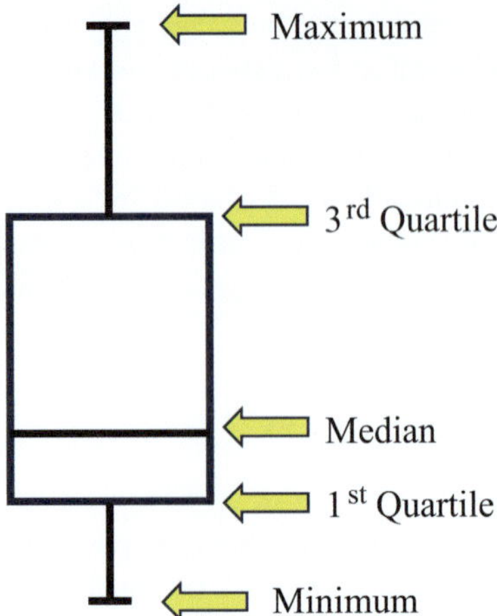

Fig. 4.8 A box plot (box-and-whisker plot) showing the key components of a data distribution: minimum, first quartile (Q1), median, third quartile (Q3), maximum, and the interquartile range (IQR)

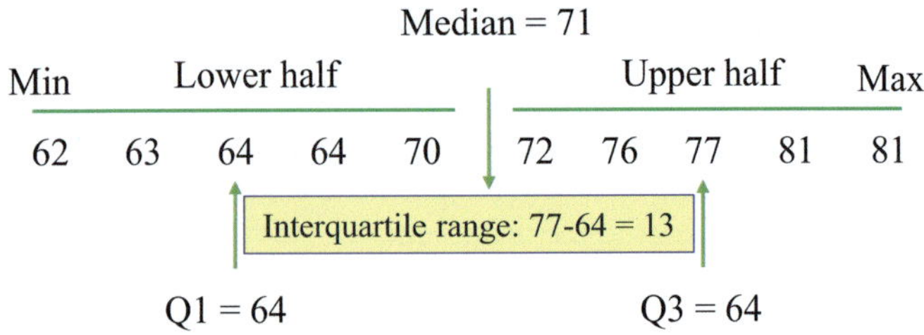

Fig. 4.9 Step-by-step computation of quartiles from an ordered dataset

of the box (as we can see in Fig. 4.8), it may indicate skewness. A longer whisker on one side suggests a longer tail of data, which may impact decisions such as factor-of-safety selection or drainage design in flood risk zones. Other examples of boxplot applications in civil engineering include screening for non-uniformity in material properties or environmental inputs. For example, if a boxplot of soil strength shows a wide interquartile range, engineers might consider more conservative design assumptions or call for additional site investigations.

Figure 4.9 illustrates a practical example of how quartiles are computed from a dataset arranged in ascending order. This visual representation helps us grasp the concepts of median, quartiles, and the interquartile range (IQR), all of which are key descriptive statistics used in data analysis.

The dataset consists of ten numerical values:

$$62, 63, 64, 64, 70, 72, 76, 77, 81, 81$$

Since the dataset has an even number of values (n = 10), the median (Q2) is calculated as the average of the two middle values, which are 70 and 72. In the event that there was an uneven number of samples, the median value could be taken directly as the middle value:

$$\text{Median} = \frac{70 + 72}{2} = 71$$

Next, we divide the dataset into two halves:

- **Lower half**: 62, 63, 64, 64, 70
- **Upper half**: 72, 76, 77, 81, 81.

Each of these halves contains an odd number of values (n = 5), so the first quartile (Q1) and third quartile (Q3) can be identified directly as the middle values of these subsets:

- Q1 = 64
- Q3 = 77.

The Interquartile Range (IQR) is then calculated by subtracting Q1 from Q3:

$$IQR = Q3 - Q1 = 77 - 64 = 13$$

The IQR is particularly helpful because it focuses on the middle 50% of the data, ignoring extreme values or potential outliers. This makes it a reliable measure of data spread, especially when evaluating resilience in design, such as determining stormwater storage volume based on rainfall variability or analyzing material properties to ensure structures meet required tolerances.

To assist with larger datasets, the table below provides a summary of essential Excel® formulas used to compute descriptive statistics, including the minimum, maximum, median, and the first and third quartiles. These functions are particularly useful for automating calculations in large spreadsheets. Each formula operates directly on a selected range of cells, and when computing quartiles, the second argument is set to either 1 or 3 to return the first (25%) or third (75%) quartile, respectively. The complete set of formulas is provided in Table 4.1 of the attached Excel file. It should be noted that Excel provides two quartile functions: QUARTILE.INC and QUARTILE.EXC. The INC version ("inclusive") includes both the minimum and maximum values in the calculation, which can shift the quartile positions toward the extremes, especially when the dataset is small. To avoid this issue, we use QUARTILE.EXC. This function follows the exclusive quartile definition, which excludes the endpoints and aligns more closely with the classical statistical definition of Q1 and Q3. Using EXC prevents Q1 or Q3 from being pulled artificially toward the minimum or maximum values and provides a more accurate representation of the distribution.

Table 4.1 Summary of built-in excel functions for calculating key statistical measures: minimum, median, maximum, and the First (Q1) and Third (Q3) quartiles from a selected range of data

Statistic	Excel formula
Minimum	= MIN(selected cells)
First quartile (Q1)	= QUARTILE.EXC (selected cells, 1)
Median	= MEDIAN(selected cells)
Third quartile (Q3)	= QUARTILE.EXC (selected cells, 3)
Maximum	= MAX (selected cells)

Code 4.3 shows how MATLAB can be used to visualize the spread and variability of a sample hydrological dataset using box plots. The script is divided into logical sections, making it suitable for students and beginners learning about statistical graphics. The goal of this script is to help users explore how rainfall varies over time or across locations through the use of a visual statistical representation.

In hydrology, rainfall data are fundamental for understanding water cycle processes such as runoff, infiltration, and flood generation. Annual rainfall data helps assess long-term climatic conditions and water availability, while monthly rainfall data provides insights into seasonal variations and supports decision making in reservoir management, agriculture, and urban drainage, for example. Box plots are especially valuable for hydrologists because they provide a compact view of data spread, central tendency, and potential anomalies, helping to compare conditions between years or among locations.

As usual, we begin the script by clearing the MATLAB environment in Lines 1 to 3. The command `clc` clears the command window, `clear` removes all variables from the memory, and `close all` closes any previously opened figure windows. This prevents any interference from earlier scripts or analyses and ensures that the output will reflect only the results of the current run.

The purpose and structure of the script are described in Lines 5 to 11 by using `'%'` to add a comment. The comments indicate that the script will create two different visualizations: a single box plot (for example, annual rainfall at one station) and multiple box plots (e.g., rainfall comparison across three stations).

Lines 13 to 19 provide a dedicated customization section where users can change visual settings. These include the color of the box (`boxColor`), the marker used to indicate outliers (`markerStyle`), the color of the median line (`medianColor`), the font size used in labels and titles (`fontSize`), and the thickness of box plot outlines (`boxLineWidth`). This section is useful for tailoring the output to specific presentation or publication standards. All options for `markerStyle` and `medianColor` are provided in Tables 4.2 and 4.3, respectively. Several of these marker symbols (such as `o`, `s`, `d`, `^`, `v`, `>`, `<`, `p`, and `h`) support filling. In MATLAB, users can specify both the marker edge color and the marker face color using the properties `MarkerEdgeColor` and `MarkerFaceColor`. This option is especially useful for producing publication-quality figures, where filled markers improve visibility and help distinguish overlapping data points or outliers.

Table 4.2 List of valid `markerStyle` options in MATLAB, including their symbols and corresponding descriptions used for plotting outliers and data points

Symbol	Description
o	Circle
+	Plus sign
*	Asterisk
.	Point (Very small)
x	Cross
s	Square
d	Diamond
^	Upward-pointing triangle
v	Downward-pointing triangle
>	Right-pointing triangle
<	Left-pointing triangle
p	Pentagon
h	Hexagon
None	No marker shown

Table 4.3 Standard color codes accepted by MATLAB for customizing plot elements such as the Median line in boxplots

Code	Color name
r	Red
g	Green
b	Blue
c	Cyan
m	Magenta
y	Yellow
k	Black
w	White

In Lines 21 to 30, sample data is defined, which in this case is the measured monthly rainfall depths. Two types of datasets are created. First, Line 24 defines a vector named `annual_rainfall` representing the annual rainfall depth (in millimeters) recorded at a single station over the course of ten years. This simulates long-term monitoring data used to study trends or inter-annual variability. Second, Lines 27 to 29 create three sets of monthly rainfall values (each with 12 entries) collected from three different stations. These are combined into a 12×3 matrix named `monthly_rainfall` in Line 30, which will later be used for comparison in a group boxplot.

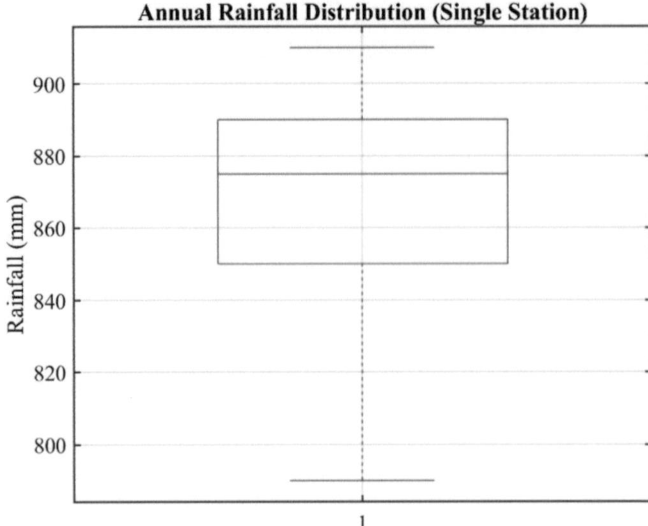

Fig. 4.10 Single-station box plot displaying the distribution of annual rainfall data, including median, quartiles, and potential outliers for one observation site

Lines 32 to 39 generate the first plot—a single box plot of the annual rainfall dataset. The MATLAB built-in `boxplot()` function is used to create the chart, with customization options such as the color for the median line and outlier markers shown in the function input. A title and y-axis label are added using `title()` and `ylabel()`, while visual clarity is improved by using `grid on` and `set(gca)` to control axis properties. The '`grid on`' simply displays the background grid, whereas `set(gca,...)` modifies the properties of the current axes. In MATLAB, `gca` refers to the "current axes," and `set()` allows us to adjust visual elements such as line widths, font sizes, or tick direction to improve the clarity and appearance of the figure.

Next, Lines 41 to 49 create the second figure, which combines multiple boxplots into a single figure for the monthly rainfall data from the three stations. Again, the `boxplot()` function is used, but this time the matrix input results in three separate boxes, one for each station. Custom labels (`xticklabels`) are used to identify which station each boxplot corresponds to, and the grid and font size settings enhance readability. After running this code, two figures will be plotted, shown in Figs. 4.10 and 4.11, representing the single-station and multi-station box plots, respectively.

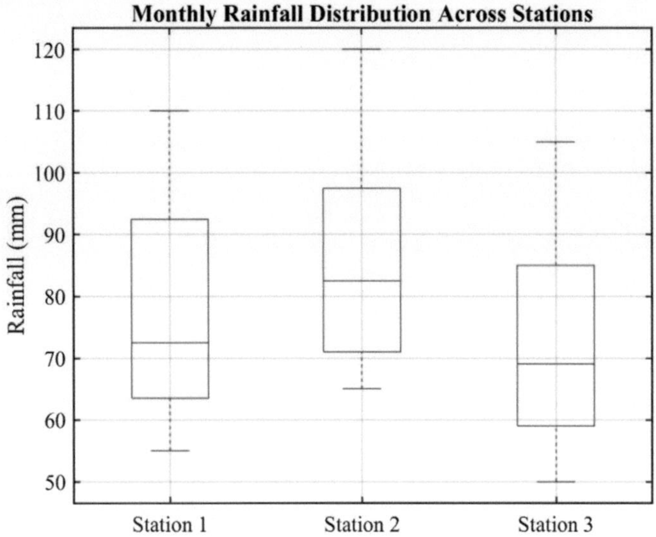

Fig. 4.11 Multi-station box plots comparing monthly rainfall distributions across three different observation stations

Code 4.3 MATLAB script for generating and customizing box plots to visualize dispersion in hydrological data

```
clc;
clear;
close all;

% =========================================================================
%        MATLAB Script for Box Plot Visualization in Hydrology
% =========================================================================
% This script generates box plots for assumed hydrological data.
% The first plot shows a single box plot (e.g., annual rainfall).
```

```
% The second plot compares multiple categories (e.g., rainfall across stations).
% ==========================================================================

%% ------------------------ User Customization Options --------------------
% General appearance settings
boxColor = [0.2, 0.6, 0.8]; % RGB for box color
markerStyle = 'o'; % Marker type for outliers ('o', '+', '*', etc.)
medianColor = 'r'; % Color for the median line
fontSize = 12; % Font size for labels and title
boxLineWidth = 1.5; % Line width for boxes

%% ------------------------ Generate Sample Data -------------------------
% Concept: Rainfall depth measurements (in mm) from hydrology
% Case 1: Annual rainfall data from a single station
annual_rainfall = [820, 790, 860, 900, 880, 850, 870, 890, 910, 885];

% Case 2: Monthly rainfall from 3 different stations (synthetic)
station1 = [65, 70, 68, 90, 105, 110, 95, 88, 75, 60, 55, 62];
station2 = [72, 78, 85, 95, 120, 115, 100, 85, 80, 70, 65, 68];
station3 = [60, 65, 70, 85, 98, 105, 85, 75, 68, 58, 50, 55];
monthly_rainfall = [station1; station2; station3]'; % 12x3 matrix

%% ------------------------ Single Box Plot -----------------------------
figure;
boxplot(annual_rainfall, 'Colors', medianColor, 'Symbol', markerStyle, ...
'Widths', 0.5, 'BoxStyle', 'outline', 'Whisker', 1.5);
title('Annual Rainfall Distribution (Single Station)', 'FontSize', fontSize);
ylabel('Rainfall (mm)', 'FontSize', fontSize);
grid on;
set(gca, 'FontSize', fontSize, 'LineWidth', 1);

%% ------------------------ Multiple Box Plots -----------------------
figure;
boxplot(monthly_rainfall, 'Colors', medianColor, 'Symbol', markerStyle, ...
'Widths', 0.4, 'BoxStyle', 'outline', 'Whisker', 1.5);
title('Monthly Rainfall Distribution Across Stations', 'FontSize', fontSize);
ylabel('Rainfall (mm)', 'FontSize', fontSize);
xticklabels({'Station 1', 'Station 2', 'Station 3'});
grid on;
set(gca, 'FontSize', fontSize, 'LineWidth', 1);
```

It should also be noted that box and whisker plots can be conveniently plotted directly from Excel. To do so, you must first navigate to the *Insert* tab, then the *Charts* ribbon, and open the Histogram dropdown menu by clicking on the downward pointing arrow (see Fig. 4.12), where you will find the option to insert a Box and Whisker Plot. This chart type automatically visualizes the five-number summary—minimum, first quartile (Q1), median, third quartile (Q3), and maximum—based on your dataset. Excel will also detect and display outliers if they exist in your data. This

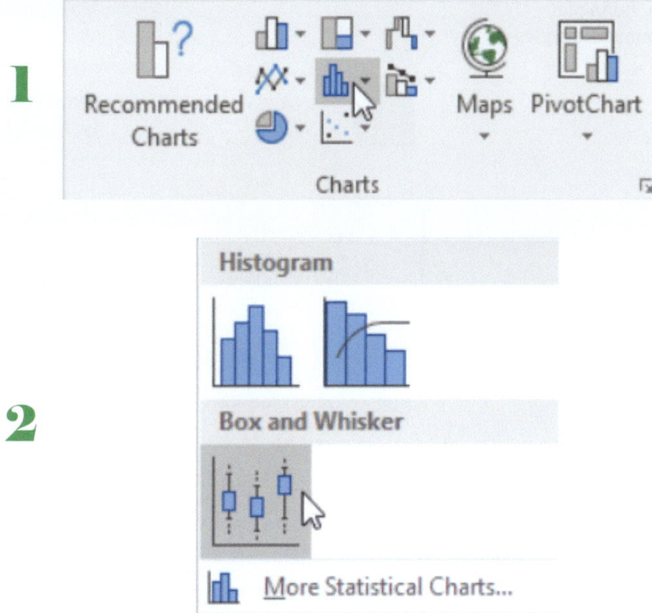

Fig. 4.12 Steps to insert a Box and Whisker Plot in Excel®: (1) Navigate to the "Insert" tab and select the "Charts" group; (2) From the histogram dropdown menu, choose "Box and Whisker" to generate the plot based on the selected dataset

built-in functionality is especially useful for quick exploratory data analysis without the need for manual calculations.

4.3.3.2 Violin Plot

Figure 4.13 presents a violin plot, a powerful data visualization tool that builds upon the boxplot by also displaying the data distribution (or density). This type of plot is especially helpful to gain deeper insight into the spread of the data and where it is most concentrated.

While a violin plot might appear unusual to those encountering it for the first time, it is, in the most basic sense, a combination of a boxplot and a density plot. This means that in addition to the summary statistic information provided by a boxplot (minimum, first quartile, median, third quartile, maximum, and interquartile range), the violin plot also includes a mirrored kernel density plot, which resembles the shape of a violin—hence the name.

The components of a violin plot are broken down and discussed individually below to further the reader's comprehension.

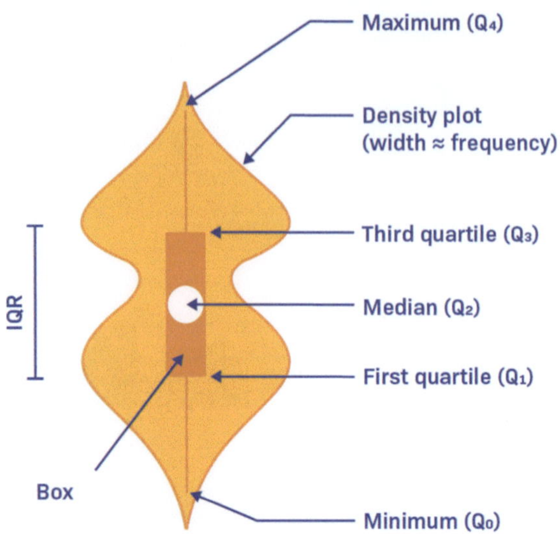

Fig. 4.13 A violin plot showing summary statistics

- **Vertical Axis (Values)**: This shows the range of values contained within the response variable of your dataset (e.g., rainfall depth, concrete compressive strength, soil moisture content, or daily temperature).
- **Horizontal Width (Density)**: Compared to the singular boxplot, this is the most important visual addition of the violin plot. The wider the plot at a given value in the vertical axis, the more frequent that value is in the dataset. Narrow parts indicate values that occur less often.
- **Inner Box Plot**: Located within the boundaries of the kernel density plot is a vertical box, whose lower, middle, and upper lines represent the first quartile (Q_1), median (Q_2), and third quartile (Q_3), respectively, just as is the case for a regular boxplot.
- **Ends of the Violin**: These represent the minimum and maximum values, completing the full range of data.

Figure 4.13 features of a violin plot.

While a boxplot is helpful for summarizing the spread of the data, it does not show where the data is most concentrated. A violin plot, however, does show the data concentration, which is extremely useful when the data is not symmetrically distributed or has multiple peaks (called a multimodal distribution). The violin plot uses the width of the kernel density distribution to communicate how often values occur:

- The wider the plot at a certain point, the higher the frequency of those values.
- The narrower the plot, the less frequent those values are.

Fig. 4.14 Annotated violin plot showing how plot width reflects data frequency

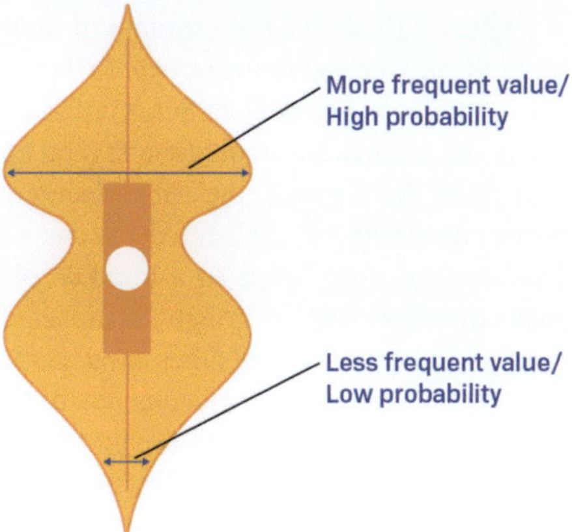

This characteristic helps users quickly identify which ranges in the dataset are most common and which are outliers or rare values.

In Fig. 4.14, for example:

- The top half of the violin plot is wider, which means that more data points fall in this upper range—this indicates a high probability of values occurring there.
- The lower tail of the plot is narrow and pointed, suggesting that data values in that region are less frequent and possibly outliers.

For civil engineering applications, this visualization technique is particularly valuable when analyzing environmental or material data:

- When evaluating wind speed distributions for bridge design, engineers may want to know not just the maximum wind speed, but also how frequently certain speeds occur.
- In hydrology, a violin plot can highlight typical daily flows of a river while still capturing the potential for flood extremes, giving a more comprehensive view of variability than a boxplot alone.
- In traffic engineering, this can be used to show the typical congestion patterns across multiple times of day or road segments.
- In wastewater treatment, the plot could be used to show the distribution of effluent concentrations for a contaminant of interest, such as biochemical oxygen demand or nutrients such as nitrogen and phosphorus.

By visually emphasizing which values are more likely, violin plots support better decision-making under uncertainty, a core aspect of hazard and risk assessment in civil engineering design.

Figure 4.15 shows the example of using two violin plots to compare soil shear strength measurements from two different construction sides, represented by **Data Set A** and **Data Set B**. Data Set A represents a site located in a reclaimed urban area where the ground has undergone several episodes of disturbance and re-compaction over time. As a result, the subsurface conditions are highly variable, with some zones consisting of engineered fill and others still containing weak, native soils. This heterogeneity leads to a bimodal distribution of shear strength values—one peak corresponding to stronger, compacted materials, and another to weaker, undisturbed soil layers. On the other hand, Data Set B corresponds to a site that has been recently prepared under strict quality control using a uniform soil type and consistent compaction practices. The violin plot for this site exhibits a narrow, unimodal distribution, indicating low variability and consistent shear strength throughout the site. This example illustrates how violin plots can be especially useful when comparing the center, spread, and distribution shapes of multiple datasets in a single view.

When interpreting violin plots, there are a number of insights that the engineer should be looking for. These will be illustrated below using Data Set A and Data Set B, shown in Fig. 4.15:

- Data Set A has a wider shape overall, indicating that its data values are more dispersed (i.e., it has a larger spread or variability).

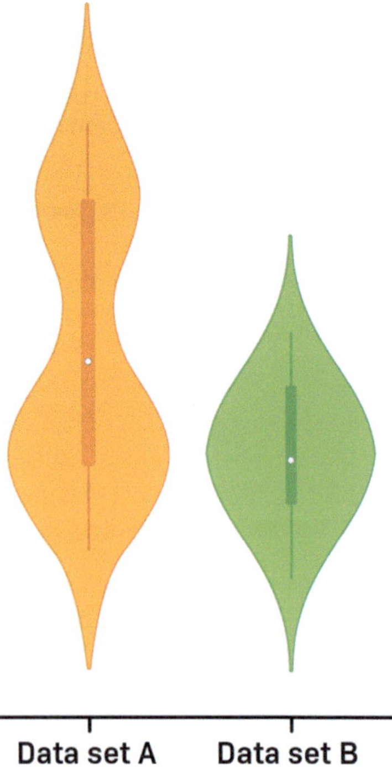

Fig. 4.15 Violin plots comparing two datasets

- Data Set B appears more compact, with a smaller interquartile range (IQR) and a lower median, suggesting that its values are more concentrated near the center.
- Median Comparison: The white dot in each plot represents the median. Data Set B's median is clearly lower than that of Data Set A.
- Spread and IQR: The inner box is shorter in Data Set B, reflecting a tighter range between the first and third quartiles compared to Data Set A.
- Density and Distribution Shape: Data Set B exhibits an unimodal distribution (a single peak), whereas Data Set A shows signs of a bimodal distribution—there are two prominent bulges. This suggests that Data Set A may contain two distinct subpopulations.

By combining Codes 4.4 to 4.8 shown below, a custom function named `ViolinPlotF` is obtained, which can be used to create violin plots in MATLAB. In violin plots, MATLAB's built-in `ksdensity` function is used to provide a detailed visual representation of the data distribution by kernel density estimation. As noted earlier, this helps in analyzing both the central tendency (mean, median) and spread (distribution shape) of grouped data. The `ViolinPlotF` function is especially helpful for engineering students and researchers who want to visually inspect how test results (e.g., compressive strength or rainfall) vary across conditions or groups.

Code 4.4 shows the beginning of the custom MATLAB function `ViolinPlotF`, which is designed to create violin plots. The function is defined in Line 1, where five outputs are specified: `h` (plot handle), `L` (legend object), `MX` (mean values), `MED` (median values), and `bw` (bandwidths used in the plot). The main input, `Y`, represents the dataset to be visualized, and `varargin` allows users to optionally provide additional arguments for customizing the plot.

Lines 3 to 14 set the default plotting parameters. These include the default x-axis labels `xL` (Line 4), face color `fc` (Line 5, set to orange as `[1 0.5 0]`), edge color `lc` (Line 6, set to black), and transparency `alp` (Line 7, set to 0.5 for semi-transparency). In this function, the face color corresponds to the fill color of the violin shape, while the edge color specifies the outline color. The transparency parameter controls how opaque the filled region appears, which is useful when multiple violins overlap. The line colors for plotting the mean and median are set by `mc` and `medc` in Lines 8–9, respectively. The variable `b` (Line 10) is used to control the kernel bandwidth for density estimation and is initially left empty, meaning MATLAB will use its default method. When the bandwidth is empty, MATLAB's `ksdensity` function automatically selects the bandwidth using its default rule, which is based on a normal-reference (Gaussian) estimation method. This method chooses the bandwidth by minimizing the mean integrated squared error under the assumption that the data follow a normal distribution. The Boolean flags `plotlegend`, `plotmean`, and `plotmedian` are set to values of 1 in Lines 11 to 13, which enables these features by default. Finally, `x` is initialized as an empty array in Line 14, allowing users to specify x-axis positions manually, if desired.

Lines 16 to 19 are added to ensure that the input dataset is handled in a uniform format. If the input `Y` is not already a cell array (checked using `iscell(Y)` in Line 17), it is converted into one using `num2cell(Y,1)` (Line 18). In MATLAB, a matrix stores all values in a single numeric array, whereas a cell array allows each column (or group of values) to be stored and processed independently. This format is useful here because each cell can represent a separate data distribution, such as one violin in the plot. This conversion splits the input matrix column-wise into individual cells, which allows the function to handle both matrix and cell-array formats consistently during plotting. This part of the code is important because it makes the function more flexible and robust, ensuring that the following sections can process the data without additional checks or errors.

Code 4.4 Initialization and input processing

```
1   function[h,L,MX,MED,bw]=ViolinPlotF(Y,varargin)
2
3   % Defaults:
4   xL=[];
5   fc=[1 0.5 0];
6   lc='k';
7   alp=0.5;
8   mc='k';
9   medc='r';
10  b=[];
11  plotlegend=1;
12  plotmean=1;
13  plotmedian=1;
14  x = [];
15
16  % Convert single columns to cells:
17  if iscell(Y)==0
18  Y = num2cell(Y,1);
19  end
```

Code 4.5 covers Lines 21 to 65 of the combined violin plot function, where the useris permitted to customize the display of the violin plot. This section parses the optional inputs passed through `varargin` and updates the default settings accordingly. In Lines 21 to 24, the code checks if the user has specified x-axis labels using the keyword `'xlabel'`. If found, it assigns the provided label array (`{'Group A', 'Group B'},` for example) to the variable `xL`, replacing the default empty setting. Next, Lines 25 to 27 check for a user-specified face color using the keyword `'facecolor'`. If present, this value replaces the default orange (`[1 0.5 0]`) used for filling the violin shapes. In MATLAB, colors can be specified using RGB triplets, where the three values represent the relative intensities of red, green, and blue, each ranging from 0 to 1. The combination `[1 0.5 0]` corresponds to an orange color, with full red, moderate green, and no blue. Different RGB combinations produce different colors.

Lines 28 to 30 handle the edge color of the violin shapes, allowing the user to customize the outline (The default is set to black, denoted by 'k' – other color options are presented earlier in Table 4.3). In Lines 31 to 33, the code lets the user adjust the transparency (opacity) of the violins using the 'facealpha' option. Low values of this input parameter, such as 0.2, would make the plot more transparent, while a value of 1 would make it fully opaque. Lines 34 to 41 allow for customization of the mean line color. The code checks if 'mc' (mean color) is provided and, if a value is given, it updates the variable mc and turns on plotmean. If the user explicitly sets the color to empty, the code will disable the mean plotting by setting plotmean = 0. A similar structure follows for the median line color in Lines 42 to 48, using the 'medc' keyword. If a valid color is provided, it's assigned to medc, and plotmedian is activated. If the color is set to empty, median plotting is disabled.

In Lines 50 to 59, the function processes the 'bw' option, which controls the bandwidth used in the kernel density estimation (a key element in violin plots). If a single bandwidth value is provided, it applies the same value to all columns (Line 53). If an array is given, it checks that its size matches the number of data groups. If this size check fails (Line 55), a warning and error are triggered to prevent mismatched inputs. Finally, lines 60 to 65 handle two additional options: 'plotlegend' and 'x'. If 'plotlegend' is passed, the value is stored to determine whether the plot should include a legend. When plotlegend is set to 1, a legend is generated; when it is set to 0, the legend is suppressed. The legend labels themselves are not defined in this section but are created later in the code, typically based on the plotted data series or the corresponding x-axis labels. Similarly, if 'x' is specified, the code will plot the violins at the specified locations, something that can be helpful when overlaying plots or using custom x-positions for clarity.

Code 4.5 enhances the flexibility of the violin plot function and makes it highly customizable for different datasets, visual preferences, or publication requirements. It ensures that users—especially beginners—can easily tailor the appearance of their plots using clear, optional arguments.

Code 4.5 User-defined customization options for violin plots

```
21  % Get additional input parameters (varargin)
22  if isempty(find(strcmp(varargin,'xlabel')))==0
23      xL = varargin{find(strcmp(varargin,'xlabel'))+1};
24  end
25  if isempty(find(strcmp(varargin,'facecolor')))==0
26      fc = varargin{find(strcmp(varargin,'facecolor'))+1};
```

```
27 end
28 if isempty(find(strcmp(varargin,'edgecolor')))==0
29     lc = varargin{find(strcmp(varargin,'edgecolor'))+1};
30 end
31 if isempty(find(strcmp(varargin,'facealpha')))==0
32     alp = varargin{find(strcmp(varargin,'facealpha'))+1};
33 end
34 if isempty(find(strcmp(varargin,'mc')))==0
35     if isempty(varargin{find(strcmp(varargin,'mc'))+1})==0
36         mc = varargin{find(strcmp(varargin,'mc'))+1};
37         plotmean = 1;
38     else
39     plotmean = 0;
40     end
41 end
42 if isempty(find(strcmp(varargin,'medc')))==0
43     if isempty(varargin{find(strcmp(varargin,'medc'))+1})==0
44         medc = varargin{find(strcmp(varargin,'medc'))+1};
45         plotmedian = 1;
46     else
47         plotmedian = 0;
48     end
49 end
50 if isempty(find(strcmp(varargin,'bw')))==0
51     b = varargin{find(strcmp(varargin,'bw'))+1}
52 if length(b)==1
53     disp(['same bandwidth bw = ',num2str(b),' used for all cols'])
54     b=repmat(b,size(Y,2),1);
55 elseif length(b)~=size(Y,2)
56     warning('length(b)~=size(Y,2)')
57     error(Please enter either a single bandwidth value or an array whose
       length matches the number of columns in the dataset)
58 end
59 end
60 if isempty(find(strcmp(varargin,'plotlegend')))==0
61     plotlegend = varargin{find(strcmp(varargin,'plotlegend'))+1};
62 end
63 if isempty(find(strcmp(varargin,'x')))==0
64     x = varargin{find(strcmp(varargin,'x'))+1};
65 end
```

Code 4.6 (Lines 67 to 84) is a critical part of the violin plot function. This section of the code focuses on preparing the color settings and estimating the probability distribution of each data group using kernel density estimation (KDE). Recall, the KDE forms the shape of each violin. Line 67 begins with a check: if only one face color has been specified for all violins (i.e., if the face color array has only one row), MATLAB replicates that single color for each group using the `repmat` command in Line 68. This ensures that each group has a color assigned, even when the user provides only a single input.

The next major block begins with Line 71, where the "%" operator is used to provide a comment that this section calculates the kernel density for each data group. The loop in Line 72 runs through each column of data (each group) individually.

Within the loop:

- Line 73 checks if a specific bandwidth was provided for KDE. If yes, the function `ksdensity()` uses that value in Line 74. Bandwidth controls how smooth or sharp the violin's density curve will appear. A small bandwidth shows more detail but might look noisy, while a large one smooths the curve.
- If no bandwidth is given, the `elseif` command is used in Lines 75–76 to specify MATLAB's default bandwidth settings.

Once the density estimate `f` and its associated values `u` are computed, Line 78 normalizes the density values using a scaling factor of `*0.3` so that they all fit within a consistent horizontal width of the violin. This ensures that violins have comparable widths even if density magnitudes differ greatly.

Next, the code stores:

- The normalized density values `f` into the matrix `F` (Line 79).
- The variable `u`, which contains the associated vertical axis values/variables such as rainfall or strength into `U` (Line 80).
- The median and mean of the dataset into `MED` and `MX` (Lines 81–82). Note that the calculation of the median and mean values ignores any NaN values present in the dataset.
- The bandwidth into `bw` (Line 83), which may be useful for legend or analysis purposes.

By the end of Code 4.6, the function has collected all the statistical and plotting information necessary to draw the violin shapes that represent the distribution of each data group.

Code 4.6 Estimating distributions using kernel density

```
67  if size(fc,1)==1
68  fc=repmat(fc,size(Y,2),1);
69  end
70
71  %% Calculate the kernel density
72  for i=1:size(Y,2)
73      if isempty(b)==0
74          [f, u, bb]=ksdensity(Y{i},'bandwidth',b(i));
75      elseif isempty(b)
76          [f, u, bb]=ksdensity(Y{i});
77      end
78      f=f/max(f)*0.3; %normalize
79      F(:,i)=f;
80      U(:,i)=u;
81      MED(:,i)=nanmedian(Y{i});
82      MX(:,i)=nanmean(Y{i});
83      bw(:,i)=bb;
84  end
```

Code 4.7 (Lines 85 to 131) handles the rendering of the violin plots on the figure. In this code, the horizontal positions of the violins are determined, the smooth distribution shapes are plotted, and the mean and median lines for each group are overlaid.

The section begins with Lines 85–97, which check how the x-axis positions should be handled. If the user did not specify custom positions (i.e., `x` is empty, verified in Line 86), MATLAB assigns default horizontal positions using a vector of zeros, one for each group (Line 87), and sets the flag `setX` to 0 (Line 88) to indicate default spacing should be used. However, if `x` is specified (Line 89), then `setX` is marked as 1 (Line 90), and the function checks for a possible conflict. If x-positions and x-labels (`xL`) are both provided, MATLAB raises an error (Lines 91–95) because these options cannot be used together. When only text labels (`xL`) are specified, the function automatically places the violins at default integer positions (1, 2, …, N) along the x-axis and assigns the labels to these positions. In contrast, when custom x-positions are provided, the user is responsible for setting any axis labels manually after plotting.

The coding necessary to render the violin plots begins in Line 99, with a loop from Lines 101–131 that processes each data group one by one.

Within this loop:

- Lines 102–114 use the `fill()` function to draw the violin shapes. This function creates the symmetrical curved body of each violin using the previously computed density (`F`) and value axis (`U`). The plot is mirrored and flipped vertically to produce the full violin. Custom x-axis positions (`x`) for the violin plot may be specified by `setX`. The face and edge colors (`fc`, `lc`) as well as the transparency (`alp`), as defined by the user in Code 4.5, are applied accordingly.
- The `hold on` command in Line 115 ensures that all violin plots and overlays appear on the same figure. If this command was not included, the violin plot for each new data column/variable would overwrite the previous entry.
- Lines 116 to 130 overlay the mean and median lines onto each violin plot. These lines show horizontal bars across the violin at the correct vertical height corresponding to the computed mean (`MX`) or median (`MED`).
  - For default x-spacing (`setX == 0`), the plotting is done using `interp1` (Lines 118 and 121) to map the appropriate widths for the left and right sides of the violin.
  - For custom spacing (`setX == 1`), similar plotting is performed; however, the position must be adjusted using the user-input x-position value (Lines 125 and 128).

The lines representing the mean or median use the color specified in Code 4.4 (`mc` or `medc`) and a line width of 2 for improved visibility. The interpolation ensures the lines span the full width of the violin at the appropriate vertical level.

In summary, Code 4.7 is used to render the violin plot. It uses the kernel density estimates from earlier sections to build smooth curves, place them properly on the plot, and then overlay meaningful statistics, including the mean and median. The user has full control over the color, positioning, and styling; however, the code specifies default values throughout to ensure that even the most novice user can generate clean and informative visuals.

Code 4.7 Plotting the violin shapes and central tendency lines

```
%% Check x-value options
if isempty(x)
    x = zeros(size(Y,2));
    setX = 0;
else
    setX = 1;
    if isempty(xL)==0
        disp('___________________________________________________')
        warning(This function does not support using x-axis positions together
        with text labels.')
        warning('If you provide custom x-axis positions, you should apply x-
        axis labels separately afterwards.)
        error('You can only use one option—either specify custom x-axis
        positions or provide x-axis labels, but not both')
    end
end

%% Plot the violins
i=1
for i=i:size(Y,2)
    if isempty(lc) == 1
        if setX == 0
            h(i)=fill([F(:,i)+i;flipud(i-
            F(:,i))],[U(:,i);flipud(U(:,i))],fc(i,:),'FaceAlpha',
            alp,'EdgeColor','none');
        else
            h(i)=fill([F(:,i)+x(i);flipud(x(i)-
            F(:,i))],[U(:,i);flipud(U(:,i))],fc(i,:),'FaceAlpha',
            alp,'EdgeColor','none');
        end
    else
        if setX == 0
            h(i)=fill([F(:,i)+i;flipud(i-
            F(:,i))],[U(:,i);flipud(U(:,i))],fc(i,:),'FaceAlpha',
            alp,'EdgeColor',lc);
```

```
        else
            h(i)=fill([F(:,i)+x(i);flipud(x(i)-
            F(:,i))],[U(:,i);flipud(U(:,i))],fc(i,:),'FaceAlpha',
            alp,'EdgeColor',lc);
        end
    end
hold on
    if setX == 0
        if plotmean == 1
            p(1)=plot([interp1(U(:,i),F(:,i)+i,MX(:,i)),
            interp1(flipud(U(:,i)),flipud(i-F(:,i)),MX(:,i)) ],[MX(:,i)
            MX(:,i)],mc,'LineWidth',2);
        end
        if plotmedian == 1
            p(2)=plot([interp1(U(:,i),F(:,i)+i,MED(:,i)),
            interp1(flipud(U(:,i)),flipud(i-F(:,i)),MED(:,i)) ],[MED(:,i)
            MED(:,i)],medc,'LineWidth',2);
        end
    elseif setX == 1
        if plotmean == 1
            p(1)=plot([interp1(U(:,i),F(:,i)+i,MX(:,i))+x(i)-i,
            interp1(flipud(U(:,i)),flipud(i-F(:,i)),MX(:,i))+x(i)-i],[MX(:,i)
            MX(:,i)],mc,'LineWidth',2);
        end
        if plotmedian == 1
          p(2)=plot([interp1(U(:,i),F(:,i)+i,MED(:,i))+x(i)-i,
            interp1(flipud(U(:,i)),flipud(i-F(:,i)),MED(:,i))+x(i)-
            i],[MED(:,i) MED(:,i)],medc,'LineWidth',2);
        end
    end
end
```

Code 4.8 (Lines 133 to 166) finalizes the violin plot's appearance. In this code, the rendered violin plots from Code 4.7 are modified with the addition of a legend, the axis limits are set, and the x-axis labels are provided. These final adjustments are essential for improving the presentation and readability of the plot, especially for use in reports, presentations, or publications.

This block of code begins in Line 133 by checking whether the user has specified the inclusion of a legend. The code checks whether the condition `plotlegend = = 1` is met and if at least one of the mean (`plotmean`) or median (`plotmedian`) values are being shown. If these conditions are met, then the code proceeds to construct a legend.

- In Lines 135–141, the legend is configured to include only those lines that were actually included within the violin plot:
  - If both the mean and median values are shown (Line 136), the legend includes both.
  - If only one is shown, the legend is configured to only display the value that is included within the plot (Lines 138 or 140).
- The `legend()` function returns a legend object, saved as `L`.
- Line 142 customizes the legend's appearance by turning off the border (`'box','off'`) and setting the font size to 14 for readability. These entries

are customizable by the user. These properties can be modified by the user if a different legend style is preferred, such as displaying the border or adjusting the font size to suit specific formatting or publication requirements.

- If no legend is needed, Line 144 sets `L` to the empty set, `L = []`.

The axis limits are adjusted depending on whether custom x-positions (`x`) were used.

- If default positioning is used (`setX == 0`), Line 149 sets the horizontal axis to span from `0.5` to the number of groups + `0.5`, and the vertical axis from the minimum to the maximum of the values stored in `U` (the vertical range of the data).
- If custom `x` values are provided (`setX == 1`), **Line 151** calculates the axis limits based on the minimum and maximum `x` values, adding a small margin (5%) on either side using `range(x)`.

These adjustments are used to ensure that all violins are fully visible and spaced appropriately.

In the last section, depending on whether the user specified `xlabel` values earlier in Code 4.5, x-axis labels are added to the plot.

- Line 155 initializes an empty cell array `xL2` with a blank entry to begin the label string.
- A for loop from Lines 156 to 158 builds the cell array `xL2` by inserting each x-axis label (such as `'Water Cured'`, `'Air Cured'`, etc.) and placing an empty entry between successive labels. These empty entries are required because each violin occupies two x-tick positions (left and right halves of the violin shape). The blank labels prevent text from being displayed at intermediate tick positions and ensure that each label appears centered beneath its corresponding violin plot.
- Line 160 disables tick marks and sets the font size to 12 for the axis.
- Line 161 enables a box around the entire plot for clarity.
- If an `xlabel` was specified (Line 163), then Line 164 sets the x-axis labels to `xL2`.

Finally, Line 166 marks the end of the `ViolinPlotF` function definition.

In summary, this last section of the code ensures that the plot is clearly labeled and formatted for presentation. The legend identifies for the reader what values the lines overlain on the plot represent (i.e., mean or median value), while the axis adjustments and labels make the chart easy to interpret and identify which dataset each violin plot represents. These steps polish the visualization and make it suitable for academic or professional use.

Code 4.8 Final display settings: legend, axis, and labels

```
%% Add legend if requested
if plotlegend==1 & plotmean==1 | plotlegend==1 & plotmedian==1
if plotmean==1 & plotmedian==1
L=legend([p(1) p(2)],'Mean','Median');
elseif plotmean==0 & plotmedian==1
L=legend([p(2)],'Median');
elseif plotmean==1 & plotmedian==0
L=legend([p(1)],'Mean');
end
set(L,'box','off','FontSize',14)
else
L=[];
end

%% Set axis
if setX == 0
axis([0.5 size(Y,2)+0.5, min(U(:)) max(U(:))]);
elseif setX == 1
axis([min(x)-0.05*range(x) max(x)+0.05*range(x), min(U(:)) max(U(:))]);
end

%% Set x-labels
xL2={''};
i=1;
for i=1:size(xL,2)
xL2=[xL2,xL{i},{''}];
end
set(gca,'TickLength',[0 0],'FontSize',12)
box on

if isempty(xL)==0
set(gca,'XtickLabel',xL2)
end
end % End of function
```

By combining Code 4.4 to Code 4.8, a complete and fully functional MATLAB function named `ViolinPlotF` is obtained. This function enables users to create violin plots that display both the distribution and central tendency of one or more datasets using kernel density estimation. Once this function is saved in a `.m` file (e.g., `ViolinPlotF.m`), it can be called like any other MATLAB function. Users can simply pass their data into this function—along with optional customization inputs such as labels, colors, or bandwidth—and generate detailed violin plots for visual analysis.

An example showing the use of the `ViolinPlotF` function to visualize real or simulated civil engineering data is shown in Code 4.9. This example will help the reader to understand how the code plots the distribution and central tendency measures of the data, and will also show them how the user-defined customizations are manifested in the final violin plot.

The script begins by clearing the MATLAB environment to avoid interference from previous sessions. Lines 1 to 3 use `clc`, `clear`, and `close all` to clear the command window, remove all variables, and close any open figure windows, respectively. This step ensures that the script starts with a clean slate for accurate and reproducible results.

Lines 5 to 11 include comments that explain the purpose of the code. The script is intended to visualize the distribution of data collected from concrete compressive strength measurements. Two violin plots will be generated: one for a single dataset collected from a class experiment, and another comparing three types of curing methods (Water, Air, and Steam) using simulated lab data. The violin plot not only shows the density of the distribution but also overlays the mean and median values, which are essential for understanding the data's spread and central tendency.

In Lines 13 to 16, a dataset called `strength_standard` is defined. This dataset contains the compressive strength values (in megapascals, MPa) collected from standard cured concrete cylinders tested after 28 days. This kind of test is commonly performed in undergraduate civil engineering labs and is crucial for evaluating material performance.

Next, Lines 18 to 25 create the first violin plot using the `ViolinPlotF` function. A figure window is opened, and the data is wrapped in a cell array as required by the function. Customization options are provided: the violin shape is colored blue (`[0.2 0.6 0.8]`), the edge is set to black (`'k'`), and the transparency is set to 0.6. The mean and median lines are colored black and red, respectively, and a label is added to the x-axis to identify the dataset as "Standard Curing."

Lines 27 to 31 finalize the appearance of the first figure. A title is added to the plot, and the y-axis is labeled in megapascals (MPa). The command `grid on` enables background grid lines aligned with the axis ticks, improving visual guidance and readability. The font size is increased using `set(gca, 'FontSize', 12)`, which modifies the font size of the axes elements, including tick labels and axis labels, but does not affect the title or legend. The legend is added separately in the lower-left corner ("southwest") and retains its default font size of 14, as specified earlier, to ensure clear identification of the mean and median lines.

Lines 34 to 41 define three datasets representing different concrete curing techniques: `water_cured`, `air_cured`, and `steam_cured`. This comparison is crucial in civil engineering education to explore how curing methods influence the final strength of concrete—a key factor in the durability and safety of structures.

Lines 43 to 51 create a violin plot for these three curing groups using the `ViolinPlotF ()` function. The input is structured as a cell array (`multi_data`) to accommodate multiple datasets. The x-axis labels are set using `'xlabel'` to clearly identify each group of data associated with the different curing techniques. The three curing techniques are differentiated visually by assigning unique face colors to each dataset using a 3×3 RGB matrix with the `'facecolor'` option. The `'edgecolor'` is set to black for clear borders, and `'facealpha'`, which controls the transparency, is set to 0.6 for visual clarity. The `'mc'` and `'medc'` options define the colors of the mean and median lines, respectively. Here, the mean line is specified as black (`'k'`) while the median line is specified as a red dashed line (`'r--'`). Additionally, the `'bw'` parameter is specified as `0.4` to adjust the bandwidth of the kernel density estimate, affecting the smoothness of the violin shapes.

Lines 53 and 54 explicitly define the x-axis tick positions and labels to ensure full control over the final figure layout. While the `ViolinPlotF` function supports internal labeling, manually setting the tick positions (Line 53) and corresponding labels (Line 54) allows the user to override the default behavior and guarantee consistent alignment and labeling when producing publication-quality figures. This approach is particularly useful when precise formatting is required or when additional customization is applied after plotting.

Lines 56 to 60 refine the overall appearance of the figure. A descriptive title is added to convey the chart's purpose—comparing concrete strength by curing method. The y-axis is labeled with units (MPa), and `grid on` adds horizontal grid lines for better readability. Font size is increased using `set(gca, 'FontSize', 12)` for consistency with the earlier plot. Finally, a legend is placed in the lower-left corner (`'southwest'`) to distinguish between the plotted mean and median lines, enhancing interpretability.

Code 4.9 MATLAB script for visualizing data distributions using violin plots, including both single and multi-group comparisons with customization for color, transparency, and statistical overlays (mean and median).

```
clc;
clear;
close all;

%% ==========================================================================
% This script visualizes the distribution of concrete compressive strength
% using violin plots. The first figure shows a single dataset, while the
```

```
% second compares three curing methods: Water Curing, Air Curing, and Steam Curing.
% The violin plots display distribution density, mean, and median values.
% Units: Megapascals (MPa)
%% ==========================================================================

%% Single Dataset - Concrete Compressive Strength after 28 Days (MPa)
% Collected from a class experiment under standard conditions
strength_standard = [29.5, 30.2, 28.9, 30.8, 31.0, 30.5, 29.8, 31.2, 30.0, 30.7, ...
29.7, 31.1, 30.3, 30.6, 30.1];

figure;
violin({strength_standard}, ... % Wrap in cell array!
'facecolor', [0.2 0.6 0.8], ...
'edgecolor', 'k', ...
'facealpha', 0.6, ...
'mc', 'k', ...
'medc', 'r', ...
'xlabel', {'Standard Curing'});

title('Compressive Strength Distribution - Single Dataset');
ylabel('Compressive Strength (MPa)');
grid on;
set(gca, 'FontSize', 12);
legend({'Mean', 'Median'}, 'Location', 'southwest', 'Box', 'off');


%% Multi Dataset - Comparing Different Curing Methods
% Simulated lab data to demonstrate the effect of curing on concrete strength

water_cured = [31.2, 32.0, 30.5, 31.5, 32.1, 30.9, 31.8, 31.0, 30.7, 32.4];
air_cured = [26.1, 27.5, 26.8, 27.0, 27.3, 26.5, 27.1, 27.2, 26.4, 27.0];
steam_cured = [33.5, 34.0, 33.8, 34.2, 33.6, 34.1, 33.9, 34.3, 33.7, 34.0];

multi_data = {water_cured, air_cured, steam_cured};

figure;
violin(multi_data, ...
'xlabel', {'Water Cured', 'Air Cured', 'Steam Cured'}, ...
'facecolor', [0.3 0.7 0.5; 0.9 0.6 0.2; 0.4 0.4 0.9], ...
'edgecolor', 'k', ...
'facealpha', 0.6, ...
'mc', 'k', ...
```

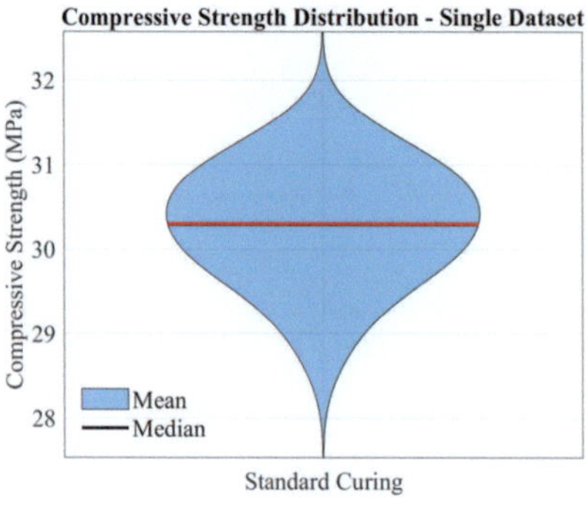

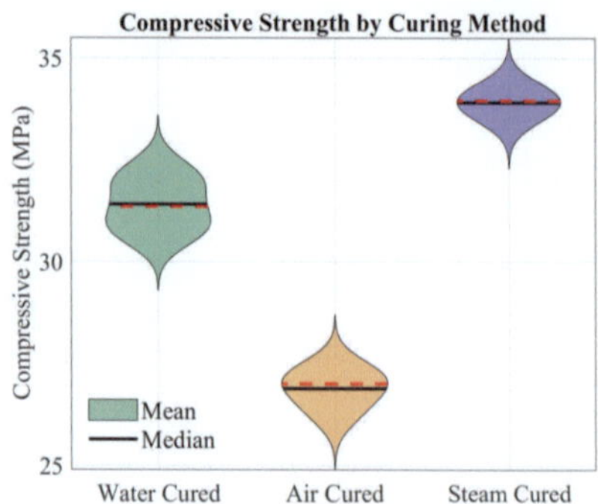

Fig. 4.16 Visualization of distribution characteristics using violin plots: the first plot represents a single dataset, while the second compares multiple groups based on different conditions

```
'medc', 'r--', ...
'bw', 0.4);

set(gca, 'XTick', 1:3); % Ensure there are ticks at each group
set(gca, 'XTickLabel', {'Water Cured', 'Air Cured', 'Steam Cured'});

title('Compressive Strength by Curing Method');
ylabel('Compressive Strength (MPa)');
grid on;
set(gca, 'FontSize', 12);
legend({'Mean', 'Median'}, 'Location', 'southwest', 'Box', 'off');
```

After implementing Code 4.9, two figures will be generated automatically. These figures illustrate the distribution of the provided data using violin plots and are presented in Fig. 4.16.

4.4 Understanding Uncertainty in Civil Engineering Design

Engineering problems in the real world often involve variable or incomplete data, limiting our ability to make decisions with complete confidence—this is what we call uncertainty. Acknowledging uncertainty is a fundamental part of risk assessment, especially when designing safe, reliable, and cost-effective infrastructure. Uncertainty stems from how we collect, process, and interpret data, and how we use this data to make decisions—such as predicting the probability of future events (e.g., a flood, structural failure, or extreme traffic demand).

4.4.1 Sources of Uncertainty

Uncertainty in engineering problems can arise from a number of different sources, which can broadly be grouped into three categories (Fig. 4.17).

4.4.1.1 Inherent Natural Variability

This type of uncertainty is caused by natural, unpredictable changes in the system. It is irreducible, meaning no matter how many measurements we take, the variability will still exist because the physical processes themselves are not fully deterministic. For example, variations in wind loads, earthquakes, and snow loads are driven by complex atmospheric and geological processes that exhibit significant randomness. Because these forces originate from natural systems, their exact intensity, duration, and distribution cannot be precisely predicted. As a result, engineers must account for this variability when designing structures, often by incorporating safety factors and relying on statistical models. Similarly, variations in soil strength, material stiffness, and moisture content, even within a single construction site, are common manifestations of inherent natural variability. Soils and construction materials are naturally heterogeneous, and their properties can differ significantly over short distances due to differences in mineral composition, compaction, and local environmental conditions. These inconsistencies make it difficult to define exact values for engineering parameters, necessitating probabilistic approaches and robust design strategies to ensure safety and performance.

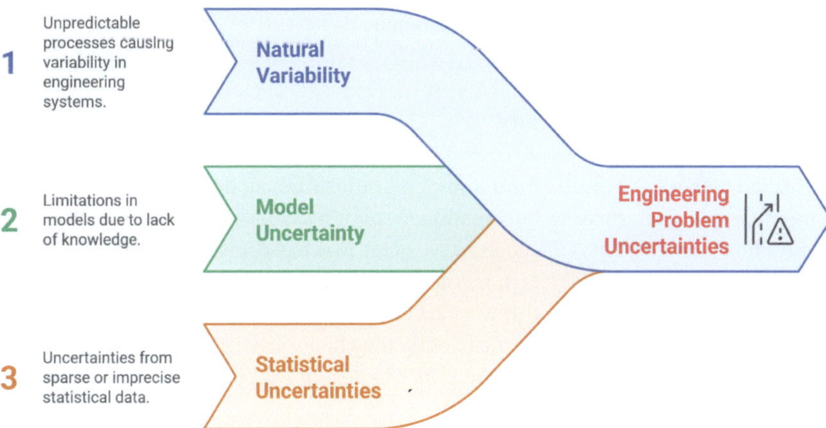

Fig. 4.17 Sources of uncertainty in engineering

4.4.1.2 Model Uncertainty

This class of uncertainty arises due to limitations in the mathematical or computational models that are used in engineering practice.

In civil engineering, model uncertainty arises from the limitations and assumptions embedded in the mathematical and computational models used to represent complex physical systems. One common example is the use of simplified hydrological models for flood estimation. These models often rely on generalized assumptions about rainfall-runoff processes, land surface characteristics, and watershed responses, which may not fully capture the complexity and variability of real-world flood events. Consequently, the estimated flood magnitudes and timings can deviate from actual observations, introducing uncertainty into flood risk assessments and design decisions.

Another source of model uncertainty is found in structural analysis, where idealized boundary conditions are often assumed, or certain load types may be neglected. In practice, boundary conditions can be complex and difficult to define precisely, yet models frequently simplify them to reduce computational complexity. Similarly, loads such as dynamic effects, temperature changes, or long-term creep may be omitted, potentially leading to inaccurate predictions of structural behavior and performance.

Model uncertainty is also a key factor when predicting infrastructure deterioration, as there is often limited data available on long-term effects. For instance, forecasting the degradation of materials like concrete or steel over decades often involves empirical models calibrated on data collected over the short-term. Due to the lack of comprehensive long-term observations and the influence of site-specific factors, these models may not reliably represent future conditions, leading to uncertainty in maintenance planning and life-cycle cost analysis.

4.4.1.3 Statistical Uncertainty

Statistical uncertainty results from imprecise or insufficient data. It often stems from limited sample sizes, missing information, or poor measurement accuracy. Engineers commonly face this type of uncertainty when making decisions about design and safety based on limited datasets or incomplete observations. For example, estimating a 100 year flood using only a few years of rainfall records introduces significant statistical uncertainty. Without a sufficiently long historical dataset, the frequency and magnitude of extreme events may be either under- or overestimated, compromising the reliability of flood risk assessments.

Another typical case arises when estimating traffic loads on a bridge based on only one week's worth of sensor data. Such a short monitoring period may fail to capture the full range of traffic patterns, seasonal variations, or rare heavy-load events. For example, if the data collection coincides with a national or bank holiday, regular commuter traffic may be significantly reduced, leading to a severe underestimation of the typical loading conditions. As a result, load estimates derived from this limited data may not be representative of long-term operational demands, potentially resulting in either overly conservative or unsafe design decisions.

The use of outdated or uncalibrated instruments to measure material strength (for example) also contributes to statistical uncertainty. Measurement errors and inconsistencies reduce data quality, reducing the accuracy of material property estimations. This uncertainty can propagate through structural analyses and design calculations, potentially compromising the overall safety and performance of infrastructure systems. Engineers must have a clear understanding of the variables being monitored within their system, as well as the method used during data collection. Such an awareness may allow for the early detection of potentially erroneous measurements arising from faulty equipment and or improperly maintained devices.

4.4.2 *Types of Uncertainty: Aleatory Versus Epistemic*

Uncertainty can also be classified based on whether it can be reduced (Fig. 4.18).

4.4.2.1 Aleatory Uncertainty

This type of uncertainty is random and cannot be eliminated, only modeled probabilistically. For example, no model can predict the exact moment and magnitude of an earthquake, but probabilistic models can estimate the likelihood of one occurring within a given time frame. This type of uncertainty is closely associated with the inherent variability of natural processes.

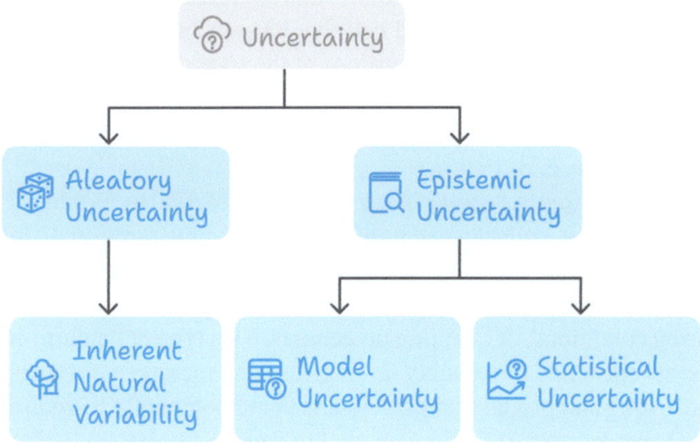

Fig. 4.18 Type of uncertainty in engineering

4.4.2.2 Epistemic Uncertainty

This uncertainty arises due to a lack of knowledge. It can often be reduced by improving models, collecting more data, or refining measurement techniques. For instance, uncertainty in the compressive strength of a concrete mix design can be reduced by increasing the number of cylinders cast and tested in the laboratory. Epistemic uncertainty is primarily linked to two sources:

- **Model uncertainty**: from simplified or imperfect modeling assumptions.
- **Statistical uncertainty**: from sparse data, limited sampling, or imprecise measurements.

4.4.2.3 Distinguishing Aleatory and Epistemic Uncertainty Through Material Testing

To better understand how different types of uncertainty manifest in civil engineering practice, let us consider a common laboratory procedure: the compression testing of concrete cylinders. This test is widely used to determine the compressive strength of concrete—a critical parameter in structural design and quality assurance.

Imagine a scenario where a single batch of concrete cylinders is produced and tested using the same loading equipment, under strictly controlled laboratory conditions. Despite the uniform procedure, you may still observe variations in the measured compressive strength values. This type of variability—arising from natural randomness in the material properties, micro-cracks, air voids, or aggregate distribution—is referred to as aleatory uncertainty. It is an inherent property and is considered to be irreducible, meaning it must be accounted for through probabilistic methods rather than eliminated.

However, if we suspect that some test results are affected by bias or errors in the measurement process, such as a load cell that has not been properly calibrated or inconsistencies in measuring the dimensions of the specimens, this would introduce epistemic uncertainty. This kind of uncertainty originates from a lack of knowledge or precision and is often reducible through improvements in testing procedures, equipment maintenance, or training.

Now consider testing a second batch of concrete cylinders produced using the same mix design, but collected at a different site or by a different inspector. If the average compressive strength of this batch differs significantly from the first, the observed variability may be attributable to differences in cylinder casting techniques, curing conditions, or handling procedures. This type of batch-to-batch variation represents epistemic uncertainty, as it arises from incomplete control or knowledge of the testing and production process and could potentially be reduced through improved standardization, training, and quality control.

Furthermore, a common mistake in engineering design is to use only the average value of compressive strength (from a few specimens or batches) in calculations, without considering the spread or distribution of the values. Doing so effectively ignores the natural variability and could lead to underestimating the actual risk of failure. This oversight also contributes to epistemic uncertainty, as it stems from simplifying assumptions or the omission of critical information about data variability.

Figure 4.19 summarizes the key categories of uncertainty commonly encountered in concrete testing, including aleatory variability, epistemic errors, batch-to-batch differences, and the risk of oversimplifying data in engineering decisions. In summary, by analyzing the sources of variability in a seemingly straightforward experiment, one can begin to distinguish between uncertainty that must be accepted (aleatory) and uncertainty that can be reduced (epistemic). Making this distinction helps engineers make better-informed decisions, design more reliable structures, and manage risk more effectively throughout the lifecycle of a project.

As engineers, we are often asked to predict the performance, safety, or reliability of structures and systems over time. However, as the prediction horizon extends further into the future, uncertainty inevitably increases. This relationship between time and uncertainty is commonly illustrated using the Future Cone concept, adopted from the futures and foresight literature. In this framework, the horizontal axis represents increasing time into the future, while the vertical axis reflects growing uncertainty and decreasing predictability.

The near-term region, labeled "Will," represents outcomes that are highly constrained and can be predicted with relatively high confidence. The medium-term region, labeled "Can," encompasses multiple plausible futures that are technically feasible but increasingly uncertain. The distant future, labeled "Ought," reflects normative or value-driven considerations, where predictions shift from what is likely

Fig. 4.19 Summary of uncertainties involved in concrete strength testing

to what is desirable or intended, often informed by policy, ethics, or long-term societal goals.

In the near future (F–*Will*), uncertainty is low, predictability is high, and system behavior is less likely to deviate from the norm. This is where forecasting models are most applicable, as we usually have access to real-time or recently collected data. For example, predicting next week's rainfall or the estimation of short-term traffic flow in urban infrastructure can be done with relatively high accuracy.

In the medium-term (S–*Can*), uncertainty increases, and predictability decreases. This is the realm where scenario planning and simulation models become necessary. Rather than giving precise predictions, engineers generate multiple plausible scenarios to anticipate a range of outcomes. For example, planning climate-resilient infrastructure over the next 10–20 years requires scenario-based design due to the range of possible temperature and precipitation changes.

In the distant future (H–*Ought*), predictability is very low, and uncertainty dominates. Here, long-term projections are heavily influenced by assumptions and value judgments. This zone reflects the ethical and strategic vision of what should happen, not necessarily what will or can happen. For instance, envisioning the future of carbon-neutral cities or the use of autonomous construction equipment decades from now involves deep uncertainty.

This conceptual framework has practical implications in civil engineering, especially in risk-based design, sustainability planning, and resilience assessment:

- Engineers must select the appropriate tools (forecasting, simulation, or exploratory design) based on the time scale and nature of uncertainty.
- The longer the time horizon, the less we can rely on deterministic methods, and the more we need probabilistic, adaptive, and scenario-driven approaches.
- Acknowledging uncertainty is essential: It is not a weakness of our models, but instead a realistic representation of the complexity and dynamism of real-world systems (Fig. 4.20).

4.4.3 The Role of Random Variables in Representing Uncertainty

A random variable can be a variable whose value is unknown or a function that assigns values to each of an experiment's possible outcomes. The use of random variables is most common in probability and statistics, where they are employed to quantify the outcomes of random events. In hazard modeling and risk assessment, random variables are essential tools for estimating the likelihood of adverse scenarios, such as structural failure, natural hazards, or system overloads.

From an engineering standpoint, random variables encapsulate all forms of uncertainty that influence the outcome of a model. These uncertain inputs can arise from various sources and must be carefully considered during model development to

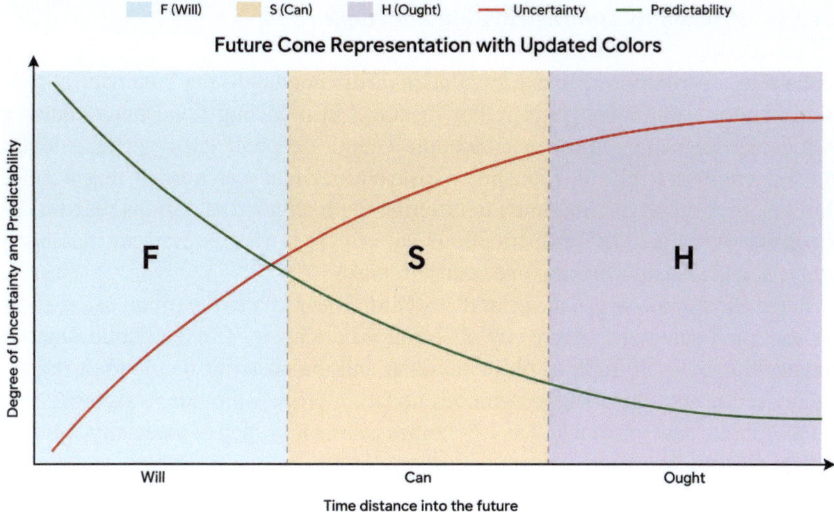

Fig. 4.20 The Future Cone: A visual representation of how uncertainty and predictability evolve over time. Near-term predictions are more reliable, whereas long-term visions are shaped by deeper uncertainty and aspirational thinking

ensure realistic and robust predictions. In the previous section, uncertainty was classified based on its nature, distinguishing between epistemic uncertainty (stemming from limited knowledge) and aleatory uncertainty (arising from inherent randomness). In practice, engineers also classify uncertainty based on its source within the modeling and analysis process. From this perspective, the primary sources of uncertainty are commonly grouped as physical, statistical, and model uncertainties.

- Physical uncertainty stems from the variability in material properties, the geometry of the structure, and the external loading environment. These uncertainties are often inherent and difficult to eliminate entirely.
- Statistical uncertainty arises due to incomplete data or a limited number of experimental observations, such as when only a small number of material tests are available to characterize variability.
- Model uncertainty results from idealizations and simplifications in mathematical models used to represent complex real-world behavior. These may include assumptions made during model formulation or the selection of probability distributions.

Random variables are the fundamental carriers of all these uncertain components into the model. Every form of variability affecting system performance must be encoded probabilistically through appropriate random variables to facilitate a meaningful hazard or risk assessment. By recognizing the distinct origins of uncertainty and correctly modeling them using random variables, engineers can perform more realistic simulations and evaluations. This practice supports the development of safer and more resilient infrastructure systems under uncertain future conditions.

4.4.3.1 Probability and Distribution Functions

Probability distributions serve a foundational role in quantifying both natural variability and epistemic uncertainty. For instance, in modeling flood magnitudes or earthquake intensities, the exact outcome cannot be predicted deterministically. Instead, engineers rely on probability distributions (such as normal, log-normal, Gumbel, or Weibull distributions) to describe the likelihood of various magnitudes occurring over time. These distributions are critical in risk assessment, reliability analysis, and decision-making under uncertainty.

In the Bayesian context, the prior distribution encapsulates our initial belief about the uncertain parameter before any observations are made. The likelihood function updates this belief in light of observed data, and the posterior distribution reflects the revised uncertainty. This continuous updating process provides a rigorous way to incorporate new information as it becomes available, which is especially valuable in hazard scenarios where data may be sparse or evolving over time.

Moreover, the choice of an appropriate probability distribution is not merely a technical detail—it significantly influences the predictions and the associated confidence levels. Misrepresenting uncertainty, either by oversimplifying the distribution or by ignoring dependencies among variables, may lead to flawed conclusions and an underestimation of potential risks.

Therefore, proficiency in interpreting and applying probability density functions, understanding the characteristics of different distribution families, and recognizing their implications in hazard models is a prerequisite for robust and credible analysis in engineering and environmental sciences.

Cumulative Distribution Function (CDF)

The Cumulative Distribution Function (CDF) is a core concept used to define a probability distribution. For a random variable X, the CDF, denoted by $F_X(x)$, gives the probability that X will take a value less than or equal to x:

$$F_X(x) = P(X \leq x) \tag{4.6}$$

In hazard-related applications, the CDF is used to estimate the probability that a hazard will not exceed a certain critical threshold. For instance, engineers may use CDFs to determine the probability that the wind speed in a given location will remain below a certain value during a storm event.

Probability Density Function (PDF)

For continuous random variables, the Probability Density Function (PDF) provides a more granular view of the distribution. The PDF, denoted as $f_X(x)$, is the derivative of the CDF:

$$f_X(x) = \frac{\partial F(x)}{\partial x} \tag{4.7}$$

It represents the likelihood that the random variable is exactly equal to some value x. While the probability of a continuous variable taking on an exact value is technically zero, the PDF allows us to compute probabilities over intervals using integration.

Engineering Interpretation

In civil engineering applications, PDFs and CDFs are used in reliability analysis, load and resistance factor design (LRFD), and risk assessments. Consider the following examples:

- Flood risk modeling: The CDF of annual maximum river discharge helps determine the return period of floods.
- Wind engineering: The PDF of wind speeds is used to estimate the most likely wind speeds affecting a structure.
- Seismic design: Ground acceleration data are modeled using PDFs to derive seismic hazard curves.

The graphical relationship between a PDF and a CDF is insightful. The CDF is always a non-decreasing function ranging from 0 to 1, while the PDF typically appears as a bell-shaped curve (for normal distributions) or skewed depending on the underlying uncertainty. These graphical insights help civil engineers visually understand and communicate risk.

4.4.3.2 Continuous and Discrete Random Variables

Figure 4.21 illustrates the fundamental relationship between the PDF and the CDF of a standard normal distribution, which is commonly used in engineering risk analysis. The PDF is represented as a smooth, bell-shaped blue curve centered around the mean. This function quantifies the relative likelihood of a random variable taking on a specific value. Its shape depends on the distribution's mean and standard deviation—for the normal distribution, it is symmetric around the mean with decreasing probability as one moves away from the center.

In contrast, the CDF is shown as an orange dashed curve that increases monotonically from 0 to 1. This function represents the cumulative probability of the variable being less than or equal to a specific value. The steepest part of the CDF corresponds to the peak of the PDF, highlighting the range where the most probable outcomes are concentrated. As one moves farther from the mean in either direction, the CDF flattens, reflecting the decreasing likelihood of extreme events.

Understanding this relationship is particularly important for civil engineers and decision-makers who often assess uncertainty and risk in environmental, hydraulic,

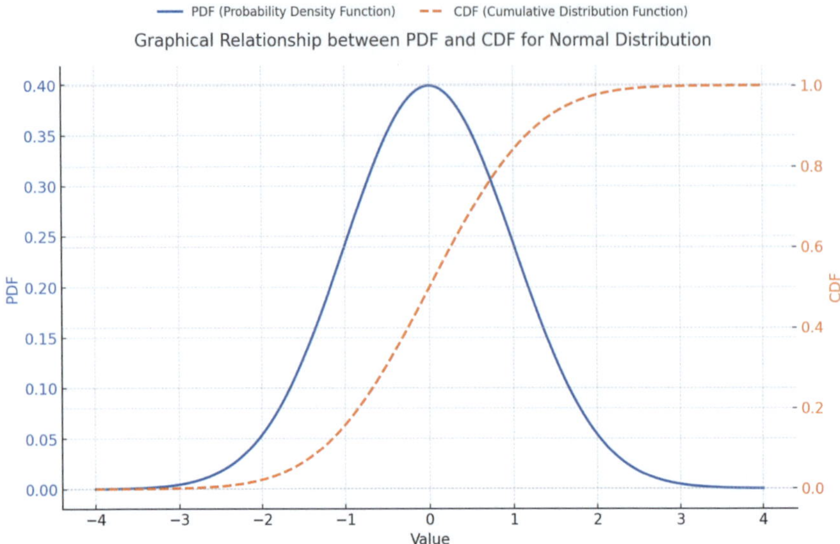

Fig. 4.21 Graphical relationship between the PDF (blue curve) and the CDF (orange dashed curve) for a standard normal distribution

and structural systems. The PDF provides insights into where outcomes are most likely to occur, while the CDF allows for straightforward calculation of probabilities, such as the chance that a flood level will not exceed a certain threshold. Together, these functions offer a comprehensive graphical toolkit for visualizing and communicating uncertainty in engineering applications.

If X is a continuous random variable, probabilities are not associated with specific, individual values of x, but rather with intervals of x. This distinction arises because continuous variables can take on an uncountably infinite number of possible outcomes within a specified range of possibilities. Events are therefore defined over ranges on the real number line—such as (a, b] or (−∞, x]—not at single points.

As a result, the probability that X takes on any exact value, represented by P(X=x), is always equal to zero. This may seem counterintuitive at first, but it is a foundational concept in probability theory. Instead, what we can meaningfully define is the density of the probability at that point, represented by the PDF, $f_X(x)$. The PDF describes how densely packed the probability mass is around a certain point. However, it does not directly give probabilities; instead, it must be integrated over an interval to determine the probability that X falls within that range.

Figure 4.22 provides a graphical representation of the PDF and the CDF for a continuous random variable X. The left-hand panel shows the CDF, which represents the cumulative probability $F(x) = P(X \leq x)$. For any given value of x, the corresponding point on the CDF curve gives the probability that the random variable takes a value less than or equal to x. The right-hand panel displays the PDF, where the blue curve represents the probability density function $f(x)$. The probability that X is less than a given value x, denoted $P(X < x)$, is obtained by integrating the PDF

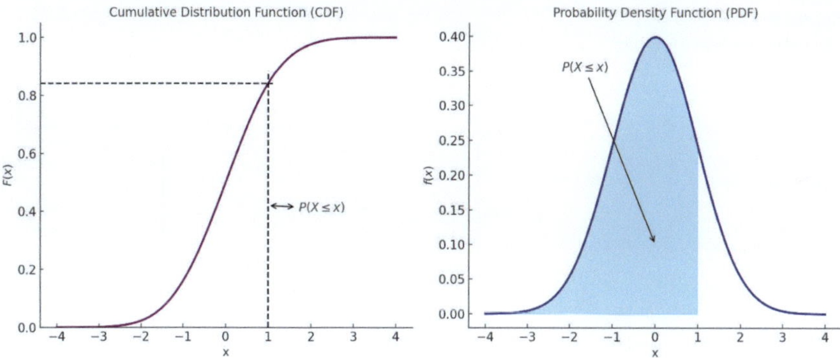

Fig. 4.22 Graphical representation of the PDF and the CDF for a continuous random variable X

from $-\infty$ to x. This probability is illustrated by the shaded area under the curve. Thus, the value of the CDF at a given point x is equal to the area under the PDF curve accumulated up to that same value. Note that for continuous random variables, the probability at any single point is zero (i.e., $P(X = x) = 0$); only probabilities over intervals are non-zero.

In the context of civil engineering design and hazard assessment, this understanding becomes particularly important. For example, when analyzing load exceedance probabilities, flood frequency, or material failure probabilities, engineers use these statistical functions to evaluate risk. Rather than considering the chance of a single, exact measurement (which is zero), we calculate the likelihood that a value lies within a critical range—such as a flood discharge exceeding 3000 m^3/s or a stress level surpassing a material's yield strength. Thus, the interpretation of the PDF and CDF is not merely theoretical but is foundational in developing reliable, probabilistically informed designs in engineering.

Figure 4.23 illustrates the CDF and the PDF for a discrete random variable. Discrete random variables are those that can take on only a finite or countably infinite number of distinct values, such as integers or categorized outcomes. In the upper plot, the CDF is shown as a step function, which increases only at the possible values of the random variable X. Each jump in the CDF corresponds to the addition of the probability associated with that specific value of X, as determined by the PDF. For example, the jump at $x = 1$ reflects the probability that $X = 1$, and the total height of the CDF at any point x gives $P(X \leq x)$. This visual clearly represents how probabilities accumulate as we move across the domain of the variable. In the lower plot, the PDF (also called the Probability Mass Function (PMF) for discrete variables) displays individual probabilities for each specific value that X can take. These are shown as vertical bars—each bar's height reflects the probability of that outcome. For instance, a bar height of 0.5 at $x = 1$ means there is a 50% chance that the variable X equals 1.

Importantly, if a certain value does not belong to the domain of X (e.g., $x = 1.5$), its probability is zero, and no bar appears in the PDF. This reinforces a key

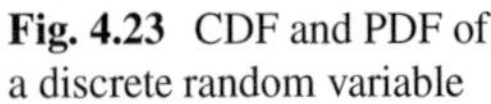

Fig. 4.23 CDF and PDF of a discrete random variable

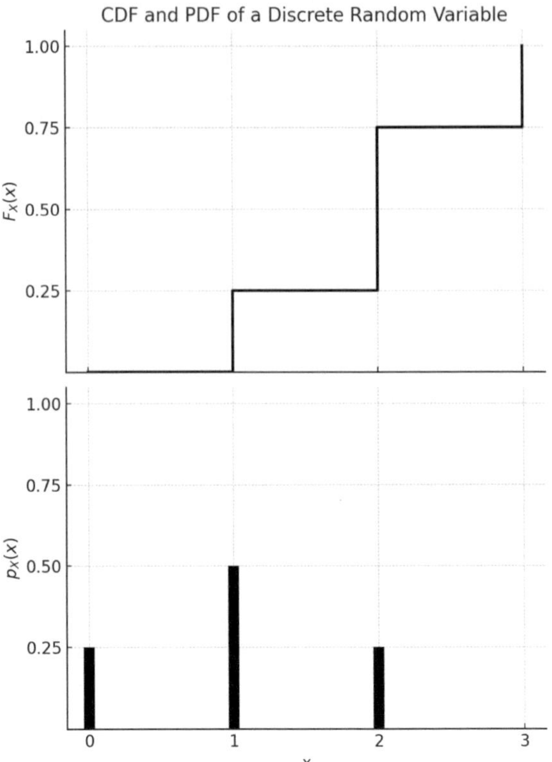

principle in probability: only defined values of the random variable can have non-zero probabilities.

Now, let's consider the following example to solidify our understanding of discrete probability functions. Table 4.4 presents the annual number of hurricanes in the United States of America for each year from 1970–1971 to 1984–1985. These values form a finite and countable sample space, which means they are best modeled using discrete random variables.

To construct the PDF for this discrete dataset, we apply the classical approach: divide the number of hurricanes in a particular year by the total number of hurricanes that occurred over the entire time period. For example, during the year 1970–1971, there were 5 hurricanes. Since the total number of hurricanes over the entire 15 year span is 34, the probability assigned to the year 1970–1971 is calculated as: $5 \div 34 \approx 0.15$. This calculation is repeated for each year to generate the entire PDF, as shown in the upper panel of Fig. 4.24.

Once the PDF is available, the CDF is constructed by successively adding the PDF values year by year. For instance, the CDF for the year 1971–1972 is obtained by summing the PDF values from 1970–1971 to 1971–1972. This process accumulates the probabilities, providing the likelihood that the number of hurricanes is less than or equal to a given year's total.

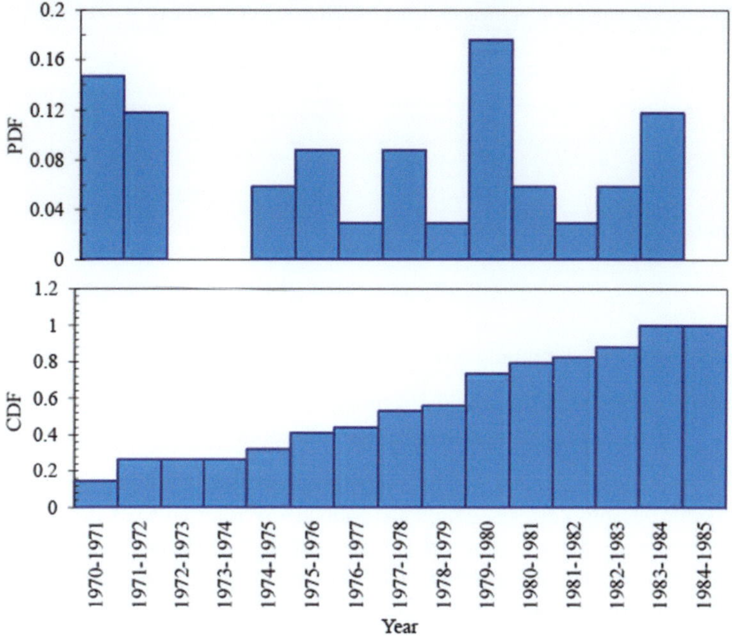

Fig. 4.24 Bar charts displaying the discrete PDF and CDF for hurricane occurrences in the United States from 1970–1971 to 1984–1985

An important feature to observe is that the CDF curve exhibits plateaus for years with zero hurricane counts. For example, the years 1972–1973 and 1973–1974 had no hurricanes. As a result, the PDF for those years is zero, and the CDF curve remains flat, indicating no increase in cumulative probability.

This example clearly demonstrates how discrete PDFs and CDFs work in practical, real-world civil engineering and environmental hazard scenarios. Understanding such distributions is essential when assessing risks associated with natural phenomena, such as hurricanes, floods, and earthquakes.

Describing Probability Distributions Using Parameters and Moments

The probabilistic properties of a random variable can be fully characterized when its probability distribution function is known. This includes its CDF or PDF, along with the associated parameters that define the distribution's shape, location, and scale. These parameters are often denoted by p in mathematical notation.

Table 4.4 Annual number of hurricanes in the United States of America

Year	NH	Year	NH	Year	NH
1970–1971	5	1975–1976	3	1980–1981	2
1971–1972	4	1976–1977	1	1981–1982	1
1972–1973	0	1977–1978	3	1982–1983	2
1973-74	0	1978–1979	1	1983–1984	4
1974–1975	2	1979–1980	6	1984–1985	0

NH: Number of Hurricanes

For example, we commonly write:

- $F_X(x, p)$: to represent the cumulative distribution function with parameter set p
- $f_X(x, p)$: to represent the probability density function with the same parameter set.

The parameters contained within the set p can include the mean, standard deviation, skewness, and kurtosis, which are collectively referred to as the moments of the distribution. Table 4.4 shows the main statistical moments of a random variable. These moments are essential as they provide insight into the central tendency, variability, and shape of the distribution.

Even when the underlying probability distribution is fully specified (either analytically or through an assumed model), moments remain critical because they summarize the distribution's key properties without requiring the full function. They are often easier to interpret and apply in practice—especially when analyzing the behavior of variables related to civil engineering phenomena such as rainfall intensity, wind speed, structural loads, or earthquake magnitudes.

In many real-world engineering scenarios, particularly in hazard assessment, limited data availability or complex system behavior often prevents the reliable identification of a closed-form probability distribution for the variable of interest. In such cases, a practical approach is to develop an approximate description of the random variable through its main descriptors (e.g., sample mean, sample variance).

This approach is beneficial when:

- Historical data that is available is sparse or noisy
- Models rely on simulation-based approaches rather than closed-form solutions
- Probabilistic analysis must be performed quickly for preliminary engineering design decisions.

Even when the full distribution is known, these descriptors remain useful for preliminary assessments, comparative studies, or sensitivity analyses.

Fundamental Moments of Random Variables and Their Role in Probability Analysis

The *i*th moment of a random variable, denoted as m_i, provides critical information about the behavior and characteristics of the variable's probability distribution. In civil engineering and other applied sciences, moments are frequently used to summarize and simplify the probabilistic characteristics of uncertain quantities, such as material strengths, loads, or environmental hazards.

For a continuous random variable, the *i*th moment is mathematically defined by the following integral:

$$m_i = \int_{-\infty}^{\infty} x^i f_X(x) dx \tag{4.8}$$

According to Eq. 4.8, we multiply each possible value of the variable x raised to the power of i (the desired moment) by the probability density at that value, integrating this product over the entire range of possible x values. The result yields a weighted average of x^i that emphasizes how values far from the center contribute, depending on the order i.

For a discrete random variable, where the variable takes on specific isolated values x_j, the moment is computed as:

$$m_i = \sum_{j=1}^{n} x^i p_X(x_j) \tag{4.9}$$

Here, $p_X(x_j)$ is the probability that the variable takes on the value x_j, and the summation is taken over all possible values of x_j.

To better characterize the behavior of random variables in engineering applications, we now explore the meaning and importance of statistical moments.

- The first moment, m_1, is known as the expected value or mean. It represents the average value of the random variable if the experiment or measurement were repeated many times. In simple terms, it gives us a central value around which the outcomes are distributed.
- The second moment, m_2, provides a measure of how widely the values are spread about the mean, or central value. When this moment is centered by subtracting the mean from the random variable prior to squaring, it yields the classical definition of the variance, which quantifies the dispersion of the data around the mean. The square root of the variance is known as the standard deviation, a commonly used descriptor because it expresses variability in the same units as the original data.
- Higher-order moments, such as the third moment (skewness) and fourth moment (kurtosis), provide insights into the asymmetry and peakedness of the distribution, respectively. These become especially important when the data or underlying

variable is not symmetrically distributed, as may happen with extreme weather events or irregular load occurrences.

The mean, also referred to as the expected value, is one of the most fundamental descriptors of a random variable. Whether dealing with a continuous or discrete random variable, the expected value provides a concise measure of central tendency. In formal terms, it is defined as the first moment of the probability distribution of the variable. The expected value is commonly denoted as E[X], and we often use the Greek letter μ (mu) to represent it.

For a continuous random variable, the expected value is computed by integrating the product of the variable x and its PDF $f_X(x)$ over the entire range of possible values:

$$\mu_X = E[X] = \int_{-\infty}^{\infty} x f_X(x) dx. \tag{4.10}$$

This formulation implies that the expected value is a weighted average of all possible values of the random variable, where the weights are given by their corresponding probabilities.

In the case of a discrete random variable, the form of the relationship is identical; however, it is employed through summation rather than integration. The expected value is calculated by summing the product of each possible value x_j and its probability mass function (PMF) $p_X(x_j)$:

$$\mu_X = E[X] = \sum_{j=1}^{n} x_j \cdot p_X(x_j) \tag{4.11}$$

This approach captures the average outcome over a series of repeated observations or experiments. If an experiment were repeated a large number of times under identical conditions, the average of the observed values would converge toward this expected value.

The variance, denoted as $\sigma^2{}_X$, is one of the most widely used descriptors in probability and statistics. It represents the second central moment of a random variable and quantifies the extent to which values of the variable deviate from the mean. In other words, it measures the dispersion or spread of the distribution around its average value.

Mathematically, variance is defined as the expected value of the squared deviation of the variable from its mean. The greater the variance, the more spread out the values are. This is critically important in engineering applications—especially in risk and hazard assessments—where uncertainty and variability must be quantified and accounted for in the design process.

For a continuous random variable, the variance is calculated by integrating the square of the deviation of x from its mean μ_X, multiplied by the probability density function $f_X(x)$, over the entire domain of x:

$$\sigma_X^2 = Var[X] = E\left[(X - \mu_X)^2\right] = \int_{-\infty}^{\infty} (x - \mu_X)^2 f_X(x)dx \tag{4.12}$$

This formulation captures the average squared deviation, weighted by the probability of each value. Since it involves squaring the deviation, variance gives greater weight to larger differences, making it sensitive to outliers and variations in the distribution.

For a discrete random variable, the same concept is applied, but via summation rather than integration. For a discrete variable, we compute the variance by summing the product of squared deviations from the mean and their corresponding probabilities $p_X(x_j)$:

$$\sigma_X^2 = Var[X] = \sum_{j=1}^{n} (x_j - \mu_X)^2 \cdot p_X(x_j) \tag{4.13}$$

Each term in the summation reflects the contribution of a specific value x_j to the overall variability. If a value is far from the mean and has a high probability, it will significantly influence the variance.

An essential aspect of interpreting data in engineering lies in understanding its variability. This variability is commonly quantified through the variance and its square root, the standard deviation, both of which play central roles in design reliability, risk assessment, and system performance evaluation. While variance is fundamental from a theoretical standpoint, the standard deviation is often more practical in engineering applications because it is expressed in the same units as the original data and is directly used in the construction of confidence intervals and design criteria.

- **Design margins**: Understanding variability allows engineers to design with safety buffers that accommodate for deviation in input variables like rainfall, wind speed, or material strength.
- **Risk analysis**: High variance in environmental loads (e.g., flood discharges, seismic activity) may signal a greater risk, requiring robust hazard mitigation strategies.
- **Data reliability**: Low variance implies consistent behavior or measurements, which is crucial when modeling hydrological or structural systems.

Properties of the Moments Operators

In the previous section, it was shown that the expected value E[X] of a discrete random variable X is defined as the first moment, which represents the average or central tendency of a probability distribution. In this section, we will explore some important properties of the expectation operator, which make it a powerful and flexible tool in probability theory, particularly for modeling uncertainties in engineering.

It is useful to note that the expectation operator, denoted as $E[\cdot]$, exhibits several mathematical properties that make it resemble other familiar operations like integration and differentiation. These properties help to simplify expressions and enable analytical manipulation of complex random variables and functions of those variables. Suppose that *a*, *b*, and *c* are constants and X is a random variable. Then:

Constant Rule:

$$E[c] = c \tag{4.14}$$

This property means the expected value of a constant is just the constant itself.

Multiplication by a Constant:

$$E[cX] = cE[X] \tag{4.15}$$

When the random variable X is multiplied by a constant, the expectation of the product is simply the constant multiplied by the expectation of X.

Linearity of Expectation:

$$E[a + bX] = a + bE[X] \tag{4.16}$$

This tells us that expectation is linear, i.e., it distributes over addition and scalar multiplication.

Additivity for Functions of a Random Variable:

$$E\left[g_1X + g_2X\right] = E\left[g_1X\right] + E\left[g_2X\right] \tag{4.17}$$

This rule implies that the expectation operator can be distributed across sums, even for non-linear functions of X.

This linearity property of expectation is particularly useful because it enables us to derive other important statistical relationships. This can be seen in the derivation of the formula for the variance of a random variable X, expressed in terms of its first and second moments:

$$Var[X] = E\left[(X - \mu_X)^2\right] \tag{4.18}$$

Expanding the square inside the expectation using algebraic identities yields:

$$E\left[X^2 + \mu_X^2 - 2\mu_X X\right] = \mu_X^2 + E\left[X^2\right] - 2\mu_X E[X] \tag{4.19}$$

Since E[X]=μ_X, we can simply it to:

$$\begin{aligned} &E\left[X^2 + \mu_X^2 - 2\mu_X X\right] = \mu_X^2 + E\left[X^2\right] - 2\mu_X E[X] \\ &= \mu_X^2 + E\left[X^2\right] - 2\mu_X^2 = E\left[X^2\right] - \mu_X^2 \end{aligned} \tag{4.20}$$

This formula expresses the variance as the difference between the second moment and the square of the first moment of the random variable. Additional properties of the variance operator can be explored by considering *a*, *b*, and *c* as constants and X as a random variable.

By definition, the variance of a constant is zero since its value cannot vary.

$$Var[c] = 0 \tag{4.21}$$

When a random variable, X, is scaled by a constant, c, the expansion of Eq. 4.18 reveals that its variance is scaled by c^2.

$$Var[cX] = c^2 Var[X] \tag{4.22}$$

Using the linearity rule as well as Eqs. 4.21 and 4.23 below demonstrates that a linear combination of a constant with a random variable X scaled by some factor c simply scales the variance by b^2.

$$Var[a + bX] = b^2 Var[X] \tag{4.23}$$

These properties allow for efficient manipulation of variances in more complex expressions, especially when dealing with linear transformations or multiple random variables.

Conditional Moments

The conditional probability density function (CPDF) of a continuous random variable X_1, given that another random variable X_2 has taken a specific value, is denoted as $f_{X_1|X_2}(x_1, x_2)$. This function is derived from the joint probability density function $f_{X_1,X_2}(x_1, x_2)$ and the marginal probability density function of X_2 and $f_{X_2}(x_2)$, using the definition of conditional probability:

$$f_{X_1|X_2}(x_1, x_2) = \frac{f_{X_1|X_2}(x_1, x_2)}{f_{X_2}(x_2)} \tag{4.24}$$

This formulation ensures that we are analyzing the behavior of X_1 under the condition that X_2=x_2, mirroring the logic of conditional probability for events $P(A|B) = \frac{P(A \cap B)}{P(B)}$.

The concept of conditional moments builds on this conditional PDF. When X_1 and X_2 are jointly distributed continuous random variables, the conditional expected value of X_1 given X_2=x_2 is defined as:

$$\mu_{X_1|X_2} = E[X_1|X_2 = x_2] = \int_{-\infty}^{\infty} x f_{X_1|X_2}(x_1, x_2) dx \tag{4.25}$$

This expression represents the first moment of the conditional distribution and provides critical insight into the expected behavior of X_1 when information about X_2 is known. Conditional expectations and higher-order conditional moments are widely used in statistical inference, machine learning, and stochastic modeling.

4.4.3.3 Examples of Random Variables

Analyzing Severe Snowstorms Using Probability Concepts

In City A, a severe snowstorm is defined as one in which the snowfall exceeds 10 inches. Let the random variable X represent the amount of snowfall (in inches) during a storm in City A. The cumulative distribution function (CDF) for X is given by:

$$F_X(x) = \begin{cases} 1 - \left(\frac{10}{x}\right)^4 & x \geq 10 \\ 0 & x < 10 \end{cases}$$

We are asked to:

a. Determine the median snowfall.
b. Find the expected amount of snowfall.
c. Calculate the percentage of severe snowstorms that are disastrous, defined as those with more than 15 inches of snow.

Solution:

(a) **Determining the Median of X**:

The median of a random variable X is defined as the value that divides the probability distribution into two equal halves. In other words, it is the point where the probability of the variable taking a value less than or equal to it is exactly 50%. Mathematically, this means:

$$P(X \leq x_m) = 0.5$$

This definition aligns with the CDF, which is given by:

$$F_X(X) = P(X \leq x)$$

So, when we say that $P(X \leq x_m) = 0.5$, we are directly referring to the CDF evaluated at x_m. Therefore, the condition for the median can be equivalently written as:

$$F_X(x_m) = 0.5$$

Substitute the expression for $F_X(x)$:

$$1 - \left(\frac{10}{x_m}\right)^4 = 0.5$$

$$\left(\frac{10}{x_m}\right)^4 = 0.5 \Rightarrow \frac{10}{x_m} = \sqrt[4]{.5} \Rightarrow x_m = \frac{10}{\sqrt[4]{.5}} \approx 11.9$$

Conclusion: The median snowfall in a severe snowstorm is approximately 11.9 inches.

(b) **Expected Snowfall in a Severe Storm**:

To calculate the expected value, E[X], we first need the PDF, obtained by differentiating the CDF:

$$f_X(x) = \frac{d}{dx}\left[1 - \left(\frac{10}{x}\right)^4\right] = 4 \cdot 10^4 \cdot x^{-5}, \quad x \geq 10$$

Then the expected value is:

$$E[X] = \int_{10}^{\infty} x \cdot 4 \cdot 10^4 \cdot x^{-5} dx = \int_{10}^{\infty} 4 \cdot 10^4 \cdot x^{-4} dx$$

$$= 4 \cdot 10^4 \left[\frac{x^{-3}}{-3}\right]_{10}^{\infty} = \frac{4 \cdot 10^4}{3} \cdot \left(0 - (-10^{-3})\right) = \frac{4}{3} \cdot 10^4 \cdot 10^{-3} = \frac{4}{3} \cdot 10 = 13.3$$

Conclusion: The expected snowfall is approximately 13.3 inches.

(c) **Probability of a Disastrous Storm (X > 15)**

We define a disastrous storm as one with snowfall exceeding 15 inches. Thus, we want:

$$P(X > 15) = 1 - P(X \leq 15) = 1 - F_X(15)$$

Using the CDF:

$$F_X(15) = 1 - \left(\frac{10}{15}\right)^4 = 1 - \left(\frac{2}{3}\right)^4 = 1 - \frac{16}{81} = \frac{65}{81} \approx 0.8$$

$$P(X > 15) = 1 - 0.8 = 0.2$$

Conclusion: About 20% of severe snowstorms are disastrous.

Roof Load and Probability of Collapse

The total load, X (in tons), on the roof of a building has the following PDF:

$$f_x(x) = \begin{cases} \frac{24}{x^3} & 3 \leq x \leq 6 \\ 0 & elsewhere \end{cases}$$

(a) Determine the CDF of X, i.e ., $F_X(x)$.
(b) What is the expected total load?
(c) Determine the coefficient of variation of the total load.
(d) Suppose the roof can carry only 5.5 tons before collapse. What is the probability that the roof will collapse?

Solution

(a) **Determine the CDF $F_X(x)$**

To compute the CDF $F_X(x)$, we integrate the PDF from the lower bound (which is 3 in this case, not $-\infty$ because the PDF is zero outside the interval [3, 6]) up to a general value x. That is:

$$F_X(x) = \int_3^x \frac{24}{t^3} dt$$

Using standard integration rules:

$$F_X(x) = \left[-\frac{12}{t^2}\right]_3^x = \frac{4}{3} - \frac{12}{x^2}, \text{ for } 3 \leq x \leq 6$$

Outside this range:

$$F_X(x) = \begin{cases} 0 & 3 < x \\ 1 & 6 > x \end{cases}$$

This function gives the probability that the total load on the roof is less than or equal to any value x between 3 and 6.

(b) **Find the Expected Total Load E[X]**

$$E[X] = \int_3^6 x \cdot f_X(x) dx = \int_3^6 x \cdot \frac{24}{x^3} dx = \int_3^6 \frac{24}{x^2} dx$$

Solving the integral:

$$E[X] = 24\left(\frac{1}{3} - \frac{1}{6}\right) = 24 \cdot \frac{1}{6} = 4$$

So, the average roof load is 4 tons.

(c) **Determine the Coefficient of Variation (CoV)**

The Coefficient of Variation is a standardized measure of dispersion, defined as:

$$CoV = \frac{\sigma}{\mu}$$

where:

$$\mu = E[X] = 4$$

$$\sigma^2 = Var(X) = E\left[X^2\right] - (E[X])^2$$

We compute E[X²]] as:

$$E\left[X^2\right] = \int_3^6 x^2 \cdot f_X(x)dx = \int_3^6 x^2 \cdot \frac{24}{x^3}dx = \int_3^6 \frac{24}{x}dx$$

This integral evaluates to:

$$E\left[X^2\right] = 24 \cdot \ln\left(\frac{6}{3}\right) = 24 \cdot \ln(2)$$

Then the variance:

$$Var(x) = 24 \cdot \ln(2) - (4)^2$$

And finally:

$$CoV = \frac{\sqrt{24 \cdot \ln(2) - (4)^2}}{4} \approx 0.199.$$

This result tells us that the variation in roof loads is about 19.9% of the mean value, which can help evaluate safety margins.

(d) **Probability the Roof Will Collapse**

In the problem statement, we are told that the roof is only designed to handle 5.5 tons before collapsing. We must find:

$$P(\text{collapse}) = P(X > 5.5) = 1 - P(X \leq 5.5) = 1 - F_X(5.5)$$

Substituting into the CDF:

$$F_X(5.5) = \frac{4}{3} - \frac{12}{(5.5)^2} \approx 0.937$$

$$P(\text{collapse}) = 1 - 0.937 = 0.063$$

Therefore, there is approximately a 6.3% chance that the roof will collapse under these conditions.

4.4.4 Histograms and Probability Distributions

Understanding the relationship between raw data and probability distributions is fundamental in statistics and probability theory. The process begins with data collection, where raw samples are gathered from experiments, observations, or simulations. Because these data points often lack immediate clarity, the next step in the process should be histogram creation, where the data is organized into bins that represent ranges of values. This allows for an initial view of the data's structure.

Once a histogram is constructed, histogram analysis can be performed to identify the presence of outliers as well as any underlying patterns such as symmetry, skewness, and modality. At this stage, the histogram reflects the empirical properties of the sample and gives insight into how the data may be distributed. However, it is important to recognize that histograms are only approximations of the true population distribution—they depend on the specific sample and can vary due to randomness.

To achieve a more accurate and generalizable understanding, we proceed to distribution fitting, where a smooth, mathematically defined curve is fitted to the histogram. This fitted curve, referred to as a probability distribution function, provides a complete and continuous representation of the data's behavior. Practically speaking, it simplifies statistical computations and enables analytical modeling.

With the distribution in place, we can perform distribution analysis to compute probabilities, expectations, variances, and other statistical measures. This forms the basis for statistical inference, the final step where predictions and decisions are made about the underlying population based on the observed sample.

Figure 4.25 visually summarizes this flow, beginning with the collection of raw data to making statistical inferences using probability distributions. In Fig. 4.26, the key differences between histograms and probability distributions are highlighted. Histograms are sample-based, use discrete bins, and are typically used for initial data exploration. In contrast, probability distributions are smooth, continuous mathematical functions that facilitate deeper statistical analysis. The histogram reveals data features like spread and outliers, whereas the distribution quantifies probabilities and moments. Together, these figures and explanations underscore the importance of transitioning from raw sample observations to formalized distributions for effective statistical reasoning and modeling.

Let us consider a real-world application of histograms through a brief case study of the Confederation Bridge, a 12.9 km long multi-span structure that connects

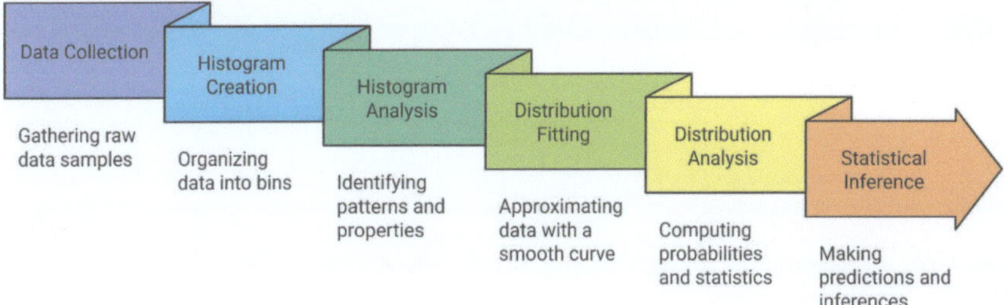

Fig. 4.25 Process of developing and implementing probability distributions from raw data collection

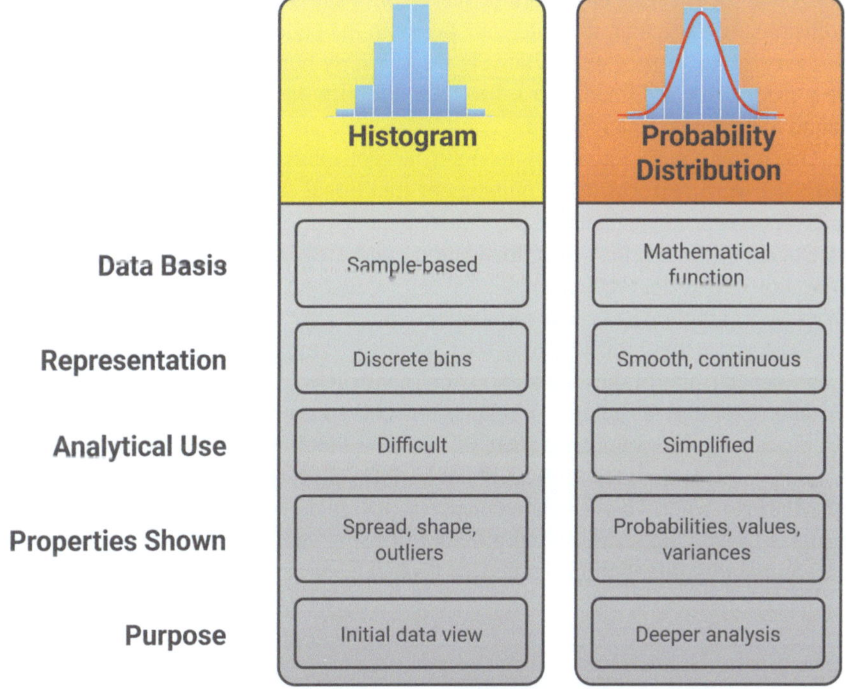

Fig. 4.26 A comparison of histograms and probability distributions

the provinces of Prince Edward Island and New Brunswick on Canada's east coast (Fig. 4.27). Notably, it holds the distinction of being the *longest continuous marine span bridge over ice-covered water* in the world. The project was undertaken as a design-build initiative by the Government of Canada, with construction commencing in October 1993 and the bridge officially opening to traffic on June 1, 1997.

The construction of the bridge posed numerous geotechnical engineering challenges due to a combination of deep water, high lateral and eccentric loads, complex

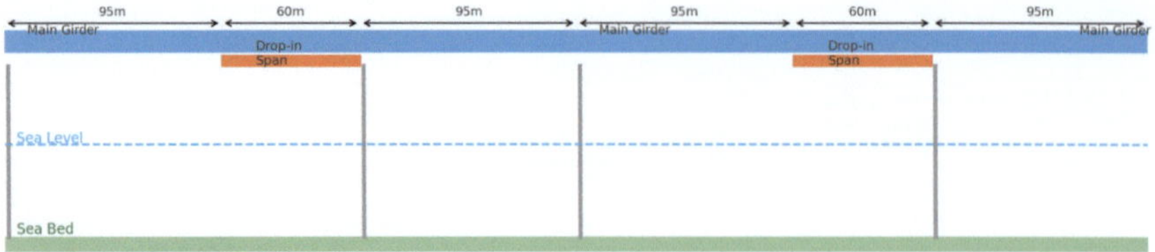

Fig. 4.27 A schematic of the Confederation Bridge, which connects Prince Edward Island and New Brunswick on the east coast of Canada

geological conditions, variable bedrock strength, and a limited construction window restricted by ice and adverse weather. One of the main engineering concerns was the horizontal loads acting on the bridge piers from wind, ice, and the potential for ship collisions. In particular, the mudstone layers in the bedrock foundation presented a risk for sliding, as they were relatively weak, nearly horizontal, and varied in thickness from a few millimeters to 1.5 m. The majority of these layers, however, were found to be thinner than 50 mm.

Given the natural variability inherent in geotechnical parameters, probabilistic analyses were performed to complement traditional deterministic models. These analyses aimed to characterize the statistical distribution of the bearing and sliding resistance of the bridge's pier foundations under diverse loading and foundation conditions. This approach also allowed for a quantitative estimation of failure risk from environmental loads and ship impact.

To illustrate this, a histogram of the unconfined compressive strength of the sandstone beneath the bridge foundation was constructed. This histogram (Fig. 4.28) visually captures the relative frequency of compressive strength values observed during testing. From the histogram, it is evident that the most frequently occurring strength value is close to 20 MPa, although values lower than 10 MPa and higher than 35 MPa also occur. This variation reflects the natural heterogeneity of the subsurface materials and underscores the importance of using statistical tools like histograms and probability distributions in modern geotechnical analysis and design.

4.5 Probability Distributions

For practical applications, both discrete and continuous probability distribution functions play a vital role in data analysis and modeling. These functions are especially valuable for several reasons (Fig. 4.29). First, a distribution function may arise from an underlying physical process and can often be derived based on well-established and physically reasonable assumptions. Second, some distributions emerge as the result of limiting processes, capturing the behavior of random variables under specific theoretical or empirical conditions. Third, many distributions are widely recognized

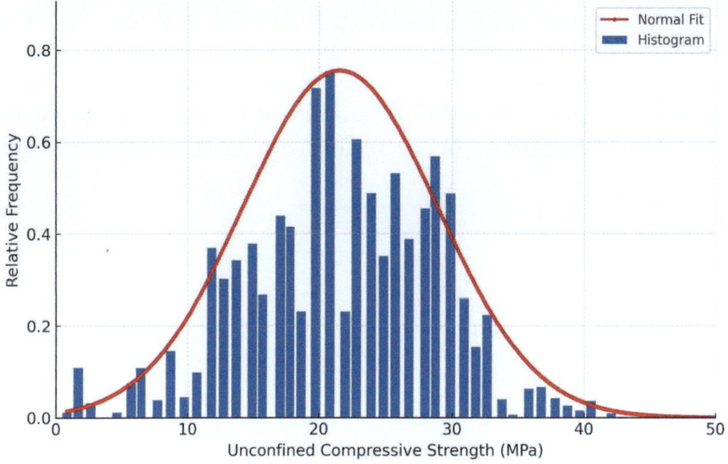

Fig. 4.28 Compressive strength distribution of sandstone beneath the confederation bridge

and studied, making it easier to access the necessary statistical and probability-related information—such as probability tables, moment equations, and analytical tools—which are commonly available in textbooks and software libraries.

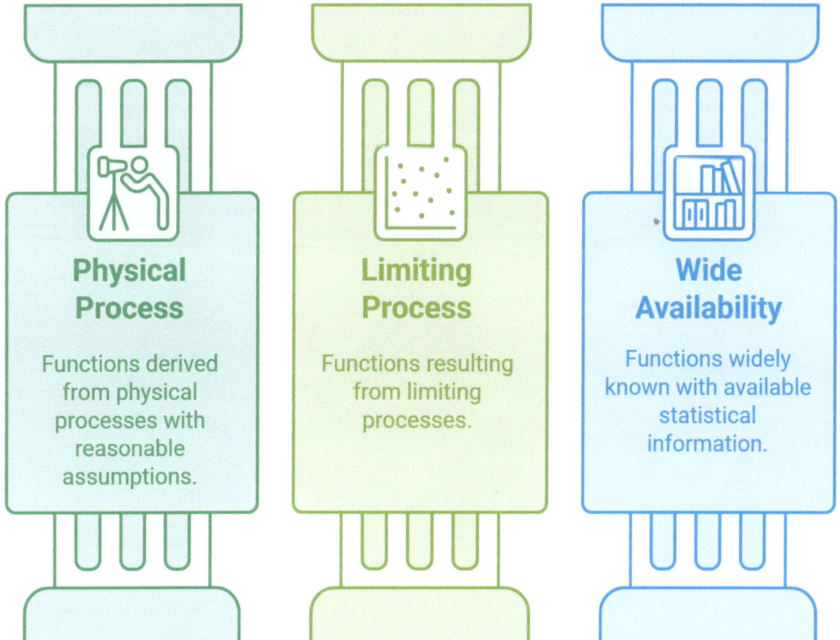

Fig. 4.29 Motivation for using specific probability distribution functions

There exists a diverse array of discrete probability distributions, each suited to particular types of data or experimental scenarios. In Fig. 4.30, a number of discrete and continuous probability distributions are presented. The selection of an appropriate distribution depends heavily on the nature of the data, such as whether the outcomes are countable, limited, or governed by chance events with specific patterns.

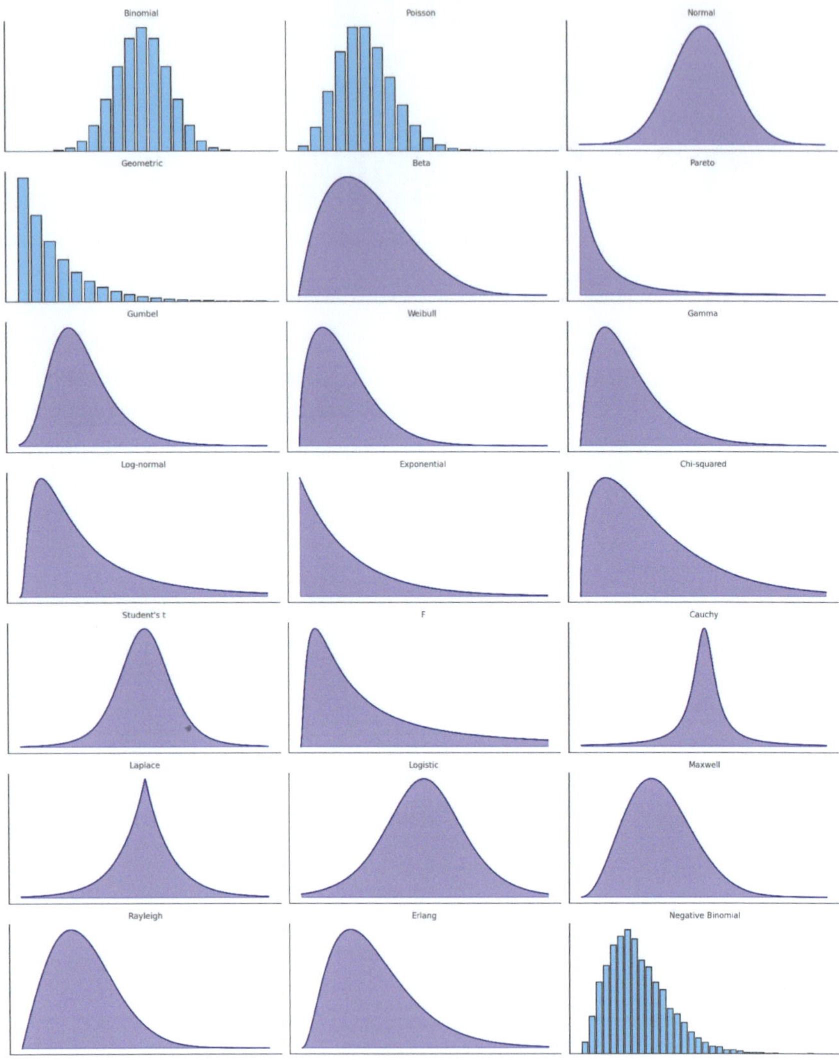

Fig. 4.30 Examples of discrete and continuous probability distributions

Discrete and continuous probability distributions are foundational in modeling a variety of real-world phenomena. For discrete random variables, the Binomial distribution is frequently used to evaluate the probability of events with two outcomes—typically referred to as success or failure. It finds practical applications in areas such as quality control, where products are either accepted or rejected based on test criteria. Similarly, the Poisson distribution is applied when estimating the likelihood of a given number of events occurring within a fixed time or space interval. This distribution is especially useful in traffic engineering to analyze vehicle arrival rates, study spacing between cars, assess parking availability, investigate accident types, and design features such as left-turn lanes.

In contrast, for continuous random variables, several distribution functions exist. Among them, the Normal distribution is the most commonly encountered. It is characterized by a symmetric, bell-shaped curve and is particularly valuable due to its versatility in modeling a wide range of natural and social phenomena. The Normal distribution provides a good approximation for the distribution of many population characteristics, making it a key topic of discussion in subsequent materials.

4.5.1 Normal Probability Distribution

The Normal distribution, also commonly referred to as the Gaussian distribution, is the best known and most widely applied probability distribution in statistics and engineering disciplines. It represents a family of continuous probability distributions that can effectively model a wide range of real-world phenomena, particularly datasets where the distribution of observations forms a symmetric, bell-shaped distribution centered around the mean. The importance of the Normal distribution stems from its capacity to accurately describe natural variability across numerous contexts, making it a foundational tool in probabilistic analysis and statistical modeling.

Every normal distribution is essentially a version of the standard normal distribution, modified by applying a number of transformations to shift its horizontal position (i.e., translation) or its width (i.e., scaling). Specifically, the mean (denoted as μ) and the standard deviation (denoted as σ) govern the exact form of any given normal distribution. The mean, referred to as the location parameter, determines the center of the distribution along the horizontal axis. A shift in the mean results in a corresponding shift of the entire curve along the x-axis: increasing μ moves the curve to the right, whereas decreasing μ shifts it to the left. Thus, the mean directly controls the horizontal positioning of the distribution without altering its overall shape.

The standard deviation σ, on the other hand, serves as the scale parameter and dictates the spread or dispersion of the distribution around the mean. A small value of σ results in a narrow, sharply peaked curve, indicating that the data points are closely concentrated around the mean value. Conversely, a larger standard deviation produces a flatter, wider distribution, reflecting greater variability among the observed values. In essence, the standard deviation adjusts the width of the distribution while maintaining its characteristic symmetry about the mean.

Due to its mathematical tractability and the prevalence of naturally occurring processes that approximate a normal distribution, it occupies a central role in statistical theory, including hypothesis testing, confidence interval estimation, and regression analysis. Its bell-shaped and symmetric nature simplifies the calculation of probabilities and supports inferential conclusions about broader populations based on sample data.

In the context of practical applications, it is essential to recognize that adjusting the mean μ and standard deviation σ provides a simple but powerful mechanism to tailor the normal distribution to fit observed data precisely. Consequently, understanding the influence of these two parameters is crucial for the correct interpretation and application of the Normal distribution in both theoretical analyses and engineering practice.

Figure 4.31 shows the behavior of different normal distributions based on varying mean (μ) and standard deviation (σ) values. Each curve corresponds to a different set of parameters, demonstrating independently the effects of changing μ or σ on the distribution's shape and position. Curve A ($\mu = 0$, $\sigma = 1$) represents the standard normal distribution, serving as a baseline for comparison. It occupies a central position, both horizontally and in terms of spread. Curves B and C illustrate vertical scaling effects due to the standard deviation, while curves D and E illustrate horizontal shifting effects due to the mean.

The impact of changing the mean (μ) is evident when comparing the distributions A, D, and E. Curves A ($\mu = 0$, $\sigma = 1$) and D ($\mu = 1$, $\sigma = 1$) have identical spreads (equal σ) but are centered at different locations along the x-axis. Curve D, with a higher mean, is shifted to the right relative to curve A. Curve E ($\mu = -1$, $\sigma = 1$) is similarly shifted to the left. This confirms that varying the mean results in a horizontal translation of the distribution without altering its spread or symmetry.

The effect of changing the standard deviation (σ) while keeping the mean constant is highlighted by comparing curves A, B, and C. All three distributions have $\mu = 0$. Curve B ($\sigma = 0.5$) is narrower and taller than curve A ($\sigma = 1$), showing how a smaller standard deviation concentrates more probability density around the mean.

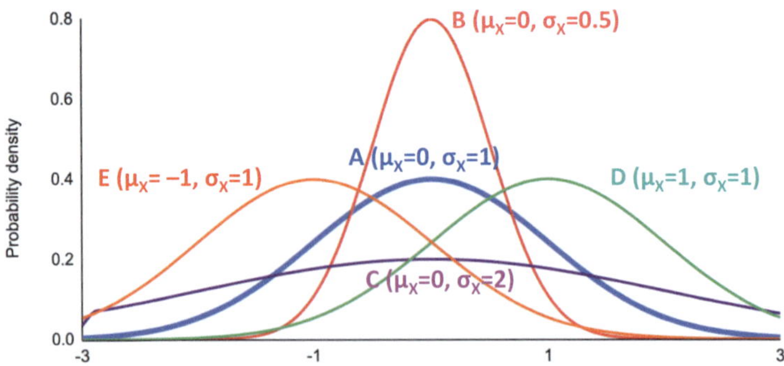

Fig. 4.31 Impact of mean (μ) and standard deviation (σ) on the standard normal distribution

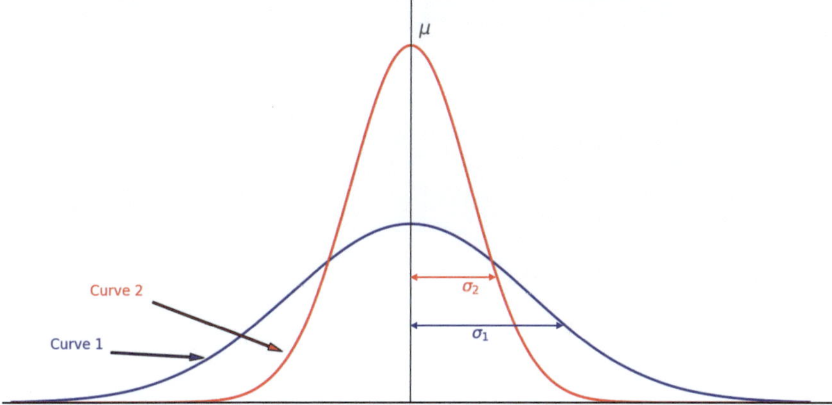

Fig. 4.32 Effect of standard deviation on the spread of a normal distribution. Curve 1 (larger σ_1) shows a wider, flatter distribution, while Curve 2 (smaller σ_2) shows a narrower, steeper distribution, both centered at the same mean (μ)

Conversely, curve C ($\sigma = 2$) is flatter and wider, showing that a larger standard deviation disperses the probability density over a broader range of values.

Figure 4.32 illustrates the effect of the standard deviation (σ) on the spread of a normal distribution curve. In this figure, two curves, labeled Curve 1 and Curve 2, with the same mean (μ) value are plotted. The horizontal position of the curves remains centered at μ, but their shapes differ distinctly due to variations in their respective standard deviations, σ_1 and σ_2.

Curve 1 represents a normal distribution with a larger standard deviation (i.e., $\sigma_1 > \sigma_2$). The curve appears wider and flatter, indicating that the data is more dispersed around the mean. A larger standard deviation spreads the values further from the mean, resulting in a curve with gentle slopes and a lower peak height. The points of inflection, where the curve changes concavity, are located farther from the mean, at distances corresponding to $\pm\sigma_1$.

In contrast, Curve 2 corresponds to a normal distribution with a smaller standard deviation (σ_2). The curve is narrower and taller compared to Curve 1. A smaller standard deviation indicates that the values are more tightly clustered around the mean, producing a steeper, sharper peak. The points of inflection for Curve 2 are closer to the mean, reflecting the reduced variability in the dataset.

Both curves clearly show that the inflection points—where the curve transitions from concave upward to concave downward—occur exactly one standard deviation away from the mean ($\mu \pm \sigma$). This property provides a convenient visual method for identifying the dispersion of data in a normal distribution. By locating these points, one can quickly estimate the standard deviation relative to the mean value. This comparison between Curve 1 and Curve 2 highlights the critical role that σ plays in determining the spread and shape of a normal distribution. While the mean dictates

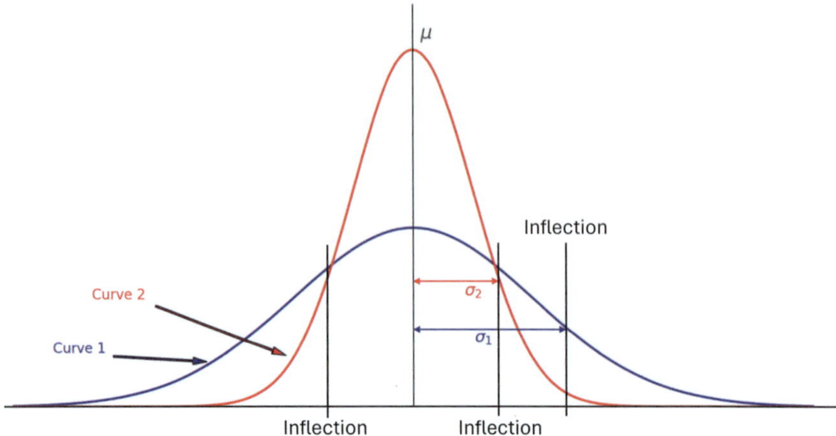

Fig. 4.33 Normal distributions with different standard deviations emphasizing the position of the inflection points

the central location of the curve along the horizontal axis, the standard deviation defines how concentrated or dispersed the distribution is around the mean.

Figure 4.33 builds upon the previous illustration by explicitly marking the inflection points of each normal distribution. While both Curve 1 and Curve 2 remain centered at the same mean (μ) and differ in standard deviation ($\sigma_1 > \sigma_2$), this figure emphasizes that the points of inflection are located precisely one standard deviation away from the mean ($\mu \pm \sigma$).

For Curve 1, with the larger standard deviation σ_1, the inflection points are farther from the mean, resulting in a broader curve. Conversely, for Curve 2, with the smaller standard deviation σ_2, the inflection points are closer to the mean, leading to a steeper and narrower distribution. The visual highlighting of these inflection points underscores how the standard deviation controls not only the overall spread but also the location of the curvature change in a normal distribution.

Figure 4.34 illustrates the application of the 68–95–99.7 rule for the normal distribution. This empirical rule describes how the area under the normal curve relates to the standard deviation from the mean (μ). In a normal distribution, approximately 68% of the total area under the curve is concentrated within one standard deviation of the mean, that is, between $\mu - \sigma$ and $\mu + \sigma$. This central region captures the majority of typical observations and is critical in evaluating variability near the mean. Expanding further, about 95% of the total area lies within two standard deviations from the mean, extending from $\mu - 2\sigma$ to $\mu + 2\sigma$. This range includes nearly all observations in many natural and engineering datasets, providing a broader envelope of expected variability. Finally, nearly 99.7% of the total area under the curve falls within three standard deviations from the mean, spanning from $\mu - 3\sigma$ to $\mu + 3\sigma$. This interval essentially covers almost the entire distribution, leaving only 0.3% of the probability in the tails beyond $\pm 3\sigma$. Importantly, while the specific values of μ and σ

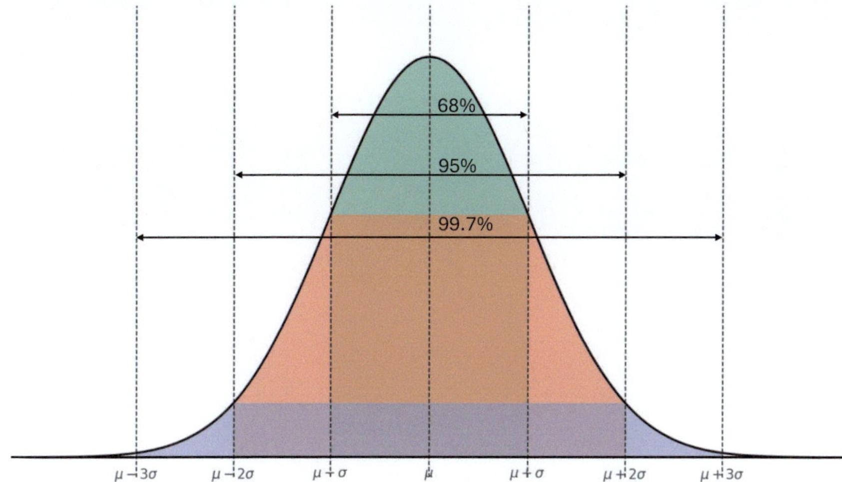

Fig. 4.34 Application of the 68–95–99.7 rule to the normal distribution, illustrating the percentages of the area under the curve that fall within one, two, and three standard deviations from the mean

may differ among normal distributions, the proportions described by the 68–95–99.7 rule remain invariant. This property allows consistent interpretation of the spread and concentration of values across different normal random variables, facilitating both analysis and prediction in practical applications.

An example of the practical application of the 68–95–99.7 rule can be observed in the analysis of annual rainfall for the Ottawa region. Assuming that the rainfall amounts are approximately normally distributed and given that the mean annual rainfall is 920 mm with a standard deviation of 75 mm, the rule can be used to interpret the distribution of expected rainfall values.

Applying the 68–95–99.7 rule:

- Approximately 68% of the annual rainfall values are expected to lie within one standard deviation from the mean. Therefore, in Ottawa, around 68% of the years will have rainfall amounts between 920 ± 75 mm (a range of 845 mm to 995 mm).
- Extending to two standard deviations, about 95% of the annual rainfall values are expected to be contained within the interval $920 \pm (2 \times 75)$ mm, or between 770 mm and 1070 mm. This broader range captures almost all typical annual variations in rainfall.
- Considering three standard deviations, approximately 99.7% of the annual rainfall measurements are expected to fall within $920 \pm (3 \times 75)$ mm, corresponding to the range of 695 mm to 1145 mm. Rainfall values outside this interval would be considered extremely rare.

This application demonstrates how the 68–95–99.7 rule provides an immediate and quantitative understanding of the distribution of natural phenomena, allowing

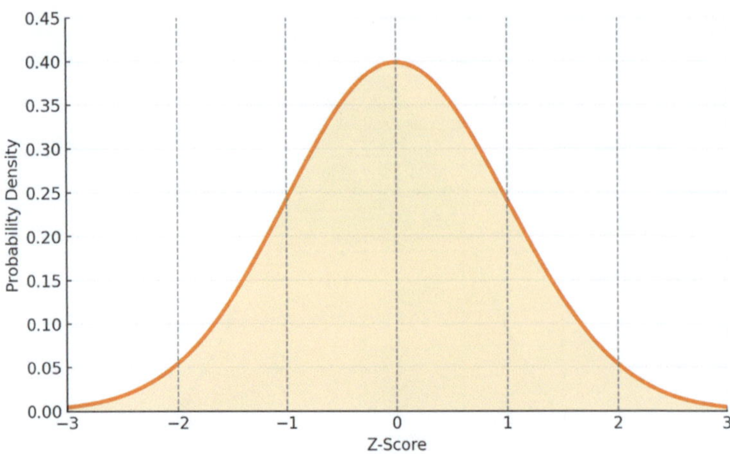

Fig. 4.35 Standard normal distribution curve ($\mu = 0$, $\sigma = 1$) showing probability density relative to standardized z-scores

for effective interpretation of the variability and extremity of annual rainfall in the Ottawa region.

Figure 4.35 illustrates the Standard Normal Distribution, which is a special case of the normal distribution where the mean (μ) is 0 and the standard deviation (σ) is 1. Standardizing a normal distribution involves converting the original variable into corresponding z-scores, thereby transforming the original distribution into this standard form.

The process of standardization is crucial when dealing with normal variables because not all real-world datasets have a mean of 0 and a standard deviation of 1. Different datasets can have widely varying means and spreads. To meaningfully compare or compute probabilities for such variables, we must transform the original normal distribution into the standard normal form.

The z-score is a measure of how many standard deviations an individual data point is from the mean. It is calculated using the formula:

$$z = \frac{X - \mu}{\sigma} \tag{4.26}$$

where X is the original value, μ is the mean, and σ is the standard deviation of the original normal distribution.

By converting raw scores into z-scores, we can reference a Standard Normal Table (also called the z-table), which lists the cumulative probabilities associated with specific z-values. This approach allows for the determination of probabilities even when the values do not fall exactly at $\pm 1\sigma$, $\pm 2\sigma$, or $\pm 3\sigma$.

The standardized curve shown in Fig. 4.35 demonstrates the symmetry of the standard normal distribution, centered precisely at a z-score of 0, with inflection points at z = ±1. The horizontal axis is marked in units of standard deviations from the mean, enabling clear interpretation of probability regions under the curve.

The PDF and CDF are fundamental to understanding the behavior of the standard normal distribution, which plays a central role in statistics and probability theory.

The PDF of the standard normal distribution is defined as:

$$f_Z(z) = \varphi(z) = \frac{1}{\sqrt{2\pi}} \exp\left(-\frac{1}{2}z^2\right) \tag{4.27}$$

This equation describes the precise shape of the standard normal curve, a smooth, symmetric, bell-shaped curve centered at zero. The PDF gives the relative likelihood of the random variable Z taking on a particular value. Although the PDF itself does not directly give probabilities, it is essential because the area under the curve between two points represents the probability of the variable falling within that interval.

Recall, the CDF is simply the integral of the PDF, and for the standard normal distribution is expressed as:

$$F_Z(z) = \Phi(z) = \frac{1}{\sqrt{2\pi}} \int_{-\infty}^{z} \exp\left(-\frac{1}{2}x^2\right) dx \tag{4.28}$$

The CDF provides the probability that the random variable Z is less than or equal to a specific value z. Unlike the PDF, the CDF accumulates probabilities from $-\infty$ up to z, offering a cumulative view of the probability distribution. The CDF is particularly important for finding probabilities associated with ranges of values and is widely used in hypothesis testing, confidence intervals, and other inferential procedures.

Understanding the PDF and CDF for the standard normal distribution is critical for several reasons:

- **Standardization**: Many real-world data sets follow approximately normal distributions, but with different means and standard deviations. By standardizing these variables into z-scores, we can universally apply standard normal tables to compute probabilities.
- **Simplicity and Universality**: Since the standard normal distribution has fixed parameters (μ=0, σ=1), it serves as a reference model, simplifying complex probability calculations and comparisons across different problems.
- **Foundation for Advanced Concepts**: Mastery of the PDF and CDF for the standard normal distribution forms the basis for understanding many statistical techniques, including regression analysis, statistical inference, hypothesis testing, and quality control processes.

After the Normal random variable has been standardized, its cumulative probability can be directly obtained using a Standard Normal table (z-table). Figure 4.36

illustrates how cumulative probability corresponds to the shaded area under the standard normal distribution curve, extending to the left of a given z-score. This cumulative probability is equivalent to the probability that a standard normal random variable is less than or equal to a given value.

To accurately use the z-table, it is important to understand its structure. The leftmost column in Table 4.5 displays the z-score rounded to the nearest tenth (e.g., −1.6, −1.5, etc.), while the top row provides the hundredths digit (from 0.00 to 0.09). The intersection of these two components yields the cumulative probability. For example, to find the cumulative probability associated with a z-score of −1.75, one would locate −1.7 in the left column and move across to the column labeled 0.05. The table entry at this intersection, approximately 0.0401, represents the area under the standard normal curve to the left of $z = -1.75$. The table simplifies the computational process, allowing practitioners to assess probabilities without direct integration of the probability density function.

Notably, values less than or equal to $z = -3.50$ yield extremely small cumulative probabilities, typically less than 0.0002, signifying extremely rare occurrences in a normal distribution. These tail probabilities are often used in the context of detecting outliers or evaluating events in the extreme lower tail.

In a similar fashion, cumulative probabilities for positive z-scores are listed in Table 4.6. Figure 4.37 illustrates the cumulative probability to the right of a specified z-score, which complements the interpretation shown in Fig. 4.36. To find this right-tail probability, one simply subtracts the cumulative probability from 1. For instance, for $z = 1.5$, Table 4.6 provides a cumulative probability of approximately 0.9332 to the left of this value. Therefore, the area to the right of $z = 1.5$ is given by 1–0.9332 = 0.0668.

This approach is especially useful when computing upper-tail probabilities in one-tailed hypothesis tests or when determining critical values in right-tailed regions. As shown in the table, cumulative probabilities increase with increasing z-values and

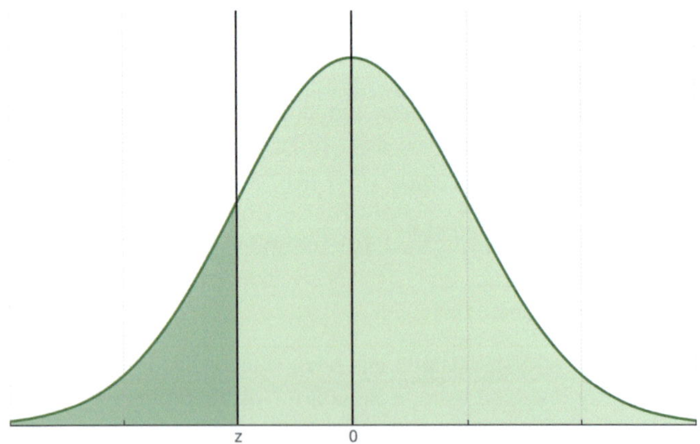

Fig. 4.36 Shaded area under the standard normal distribution curve to the left of a specified z-score

Table 4.5 Principal statistical moments of a random variable

Moment	Symbol	Description	Interpretation in engineering context
Mean	μ or E[X]	The average or expected value of a random variable	Indicates the central tendency (e.g., average rainfall, load)
Variance	σ^2	The expected squared deviation from the mean	Measures the spread or dispersion around the mean
Standard deviation	σ	The square root of variance	Measure of spread in identical units to the variable being measured: Easier-to-interpret
Skewness	γ_1 or Sk(X)	The degree of asymmetry in the distribution	Positive skew = longer right tail (e.g., extreme flood events)
Kurtosis	γ_2 or $K_u(X)$	The "peakedness" or tail heaviness of a distribution	High kurtosis = more extreme outliers (important in hazard models)
Coefficient of variation	$C_V = \sigma/\mu$	Relative measure of dispersion normalized by the mean	Useful for comparing variability across variables or units

approach values very close to 1 for large positive z-scores. For example, for z = 3.50 and beyond, the cumulative probability exceeds 0.9998, indicating a very high likelihood that the observed value is less than this z-score.

Together, Figs. 4.36 and 4.37 and Tables 4.5 and 4.6 provide a complete framework for interpreting and computing cumulative probabilities associated with both negative and positive z-scores. These tools are indispensable for evaluating normal probabilities and conducting a wide range of statistical analyses involving the standard normal distribution.

The symmetry property of the standard normal distribution, which is fundamental when working with cumulative probabilities of z-scores. Due to the perfect symmetry of the bell-shaped standard normal curve about the mean (μ = 0), the cumulative probability of a negative z-score is equal to one minus the cumulative probability of its corresponding positive z-score. Mathematically, this relationship is expressed as:

$$\Phi(-z) = 1 - \Phi(z) \tag{4.29}$$

This property is illustrated visually in Fig. 4.38, where the left panel shows the cumulative probability Φ(−z), corresponding to the area under the curve to the left of the negative z-score. The right panel shows the area to the left of a positive z-score z, represented as Φ(z). Because these two areas are symmetric about the mean, we can easily compute probabilities for negative z-scores using the more commonly available positive z-score cumulative tables by applying this identity.

Table 4.6 Cumulative probabilities for negative z-scores from the standard normal distribution

z	0	0.01	0.02	0.03	0.04	0.05	0.06	0.07	0.08	0.09
–3.4	0.0003	0.0003	0.0003	0.0003	0.0003	0.0003	0.0003	0.0003	0.0003	0.0002
–3.3	0.0005	0.0005	0.0005	0.0004	0.0004	0.0004	0.0004	0.0004	0.0004	0.0003
–3.2	0.0007	0.0007	0.0006	0.0006	0.0006	0.0006	0.0006	0.0005	0.0005	0.0005
–3.1	0.0010	0.0009	0.0009	0.0009	0.0008	0.0008	0.0008	0.0008	0.0007	0.0007
–3.0	0.0013	0.0013	0.0013	0.0012	0.0012	0.0011	0.0011	0.0011	0.0010	0.0010
–2.9	0.0019	0.0018	0.0018	0.0017	0.0016	0.0016	0.0015	0.0015	0.0014	0.0014
−2.8	0.0026	0.0025	0.0024	0.0023	0.0023	0.0022	0.0021	0.0021	0.0020	0.0019
–2.7	0.0035	0.0034	0.0033	0.0032	0.0031	0.0030	0.0029	0.0028	0.0027	0.0026
–2.6	0.0047	0.0045	0.0044	0.0043	0.0041	0.0040	0.0039	0.0038	0.0037	0.0036
–2.5	0.0062	0.0060	0.0059	0.0057	0.0055	0.0054	0.0052	0.0051	0.0049	0.0048
–2.4	0.0082	0.0080	0.0078	0.0075	0.0073	0.0071	0.0069	0.0068	0.0066	0.0064
–2.3	0.0107	0.0104	0.0102	0.0099	0.0096	0.0094	0.0091	0.0089	0.0087	0.0084
–2.2	0.0139	0.0136	0.0132	0.0129	0.0125	0.0122	0.0119	0.0116	0.0113	0.0110
–2.1	0.0179	0.0174	0.0170	0.0166	0.0162	0.0158	0.0154	0.0150	0.0146	0.0143
–2.0	0.0228	0.0222	0.0217	0.0212	0.0207	0.0202	0.0197	0.0192	0.0188	0.0183
–1.9	0.0287	0.0281	0.0274	0.0268	0.0262	0.0256	0.0250	0.0244	0.0239	0.0233
–1.8	0.0359	0.0351	0.0344	0.0336	0.0329	0.0322	0.0314	0.0307	0.0301	0.0294
–1.7	0.0446	0.0436	0.0427	0.0418	0.0409	0.0401	0.0392	0.0384	0.0375	0.0367
–1.6	0.0548	0.0537	0.0526	0.0516	0.0505	0.0495	0.0485	0.0475	0.0465	0.0455
–1.5	0.0668	0.0655	0.0643	0.0630	0.0618	0.0606	0.0594	0.0582	0.0571	0.0559
–1.4	0.0808	0.0793	0.0778	0.0764	0.0749	0.0735	0.0721	0.0708	0.0694	0.0681
–1.3	0.0968	0.0951	0.0934	0.0918	0.0901	0.0885	0.0869	0.0853	0.0838	0.0823
–1.2	0.1151	0.1131	0.1112	0.1093	0.1075	0.1056	0.1038	0.1020	0.1003	0.0985
–1.1	0.1357	0.1335	0.1314	0.1292	0.1271	0.1251	0.1230	0.1210	0.1190	0.1170
–1.0	0.1587	0.1562	0.1539	0.1515	0.1492	0.1469	0.1446	0.1423	0.1401	0.1379
–0.9	0.1841	0.1814	0.1788	0.1762	0.1736	0.1711	0.1685	0.1660	0.1635	0.1611
–0.8	0.2119	0.2090	0.2061	0.2033	0.2005	0.1977	0.1949	0.1922	0.1894	0.1867
–0.7	0.2420	0.2389	0.2358	0.2327	0.2296	0.2266	0.2236	0.2206	0.2177	0.2148
–0.6	0.2743	0.2709	0.2676	0.2643	0.2611	0.2578	0.2546	0.2514	0.2483	0.2451
–0.5	0.3085	0.3050	0.3015	0.2981	0.2946	0.2912	0.2877	0.2843	0.2810	0.2776
–0.4	0.3446	0.3409	0.3372	0.3336	0.3300	0.3264	0.3228	0.3192	0.3156	0.3121
–0.3	0.3821	0.3783	0.3745	0.3707	0.3669	0.3632	0.3594	0.3557	0.3520	0.3483
–0.2	0.4207	0.4168	0.4129	0.4090	0.4052	0.4013	0.3974	0.3936	0.3897	0.3859
–0.1	0.4602	0.4562	0.4522	0.4483	0.4443	0.4404	0.4364	0.4325	0.4286	0.4247
0.0	0.5000	0.4960	0.4920	0.4880	0.4840	0.4801	0.4761	0.4721	0.4681	0.4641

Table 4.7 Cumulative probabilities for positive z-scores from the standard normal distribution

z	0	0.01	0.02	0.03	0.04	0.05	0.06	0.07	0.08	0.09
0	0.5000	0.5040	0.5080	0.5120	0.5160	0.5199	0.5239	0.5279	0.5319	0.5359
0.1	0.5398	0.5438	0.5478	0.5517	0.5557	0.5596	0.5636	0.5675	0.5714	0.5753
0.2	0.5793	0.5832	0.5871	0.5910	0.5948	0.5987	0.6026	0.6064	0.6103	0.6141
0.3	0.6179	0.6217	0.6255	0.6293	0.6331	0.6368	0.6406	0.6443	0.6480	0.6517
0.4	0.6554	0.6591	0.6628	0.6664	0.6700	0.6736	0.6772	0.6808	0.6844	0.6879
0.5	0.6915	0.6950	0.6985	0.7019	0.7054	0.7088	0.7123	0.7157	0.7190	0.7224
0.6	0.7257	0.7291	0.7324	0.7357	0.7389	0.7422	0.7454	0.7486	0.7517	0.7549
0.7	0.7580	0.7611	0.7642	0.7673	0.7704	0.7734	0.7764	0.7794	0.7823	0.7852
0.8	0.7881	0.7910	0.7939	0.7967	0.7995	0.8023	0.8051	0.8078	0.8106	0.8133
0.9	0.8159	0.8186	0.8212	0.8238	0.8264	0.8289	0.8315	0.8340	0.8365	0.8389
1	0.8413	0.8438	0.8461	0.8485	0.8508	0.8531	0.8554	0.8577	0.8599	0.8621
1.1	0.8643	0.8665	0.8686	0.8708	0.8729	0.8749	0.8770	0.8790	0.8810	0.8830
1.2	0.8849	0.8869	0.8888	0.8907	0.8925	0.8944	0.8962	0.8980	0.8997	0.9015
1.3	0.9032	0.9049	0.9066	0.9082	0.9099	0.9115	0.9131	0.9147	0.9162	0.9177
1.4	0.9192	0.9207	0.9222	0.9236	0.9251	0.9265	0.9279	0.9292	0.9306	0.9319
1.5	0.9332	0.9345	0.9357	0.9370	0.9382	0.9394	0.9406	0.9418	0.9429	0.9441
1.6	0.9452	0.9463	0.9474	0.9484	0.9495	0.9505	0.9515	0.9525	0.9535	0.9545
1.7	0.9554	0.9564	0.9573	0.9582	0.9591	0.9599	0.9608	0.9616	0.9625	0.9633
1.8	0.9641	0.9649	0.9656	0.9664	0.9671	0.9678	0.9686	0.9693	0.9699	0.9706
1.9	0.9713	0.9719	0.9726	0.9732	0.9738	0.9744	0.9750	0.9756	0.9761	0.9767
2	0.9772	0.9778	0.9783	0.9788	0.9793	0.9798	0.9803	0.9808	0.9812	0.9817
2.1	0.9821	0.9826	0.9830	0.9834	0.9838	0.9842	0.9846	0.9850	0.9854	0.9857
2.2	0.9861	0.9864	0.9868	0.9871	0.9875	0.9878	0.9881	0.9884	0.9887	0.9890
2.3	0.9893	0.9896	0.9898	0.9901	0.9904	0.9906	0.9909	0.9911	0.9913	0.9916
2.4	0.9918	0.9920	0.9922	0.9925	0.9927	0.9929	0.9931	0.9932	0.9934	0.9936
2.5	0.9938	0.9940	0.9941	0.9943	0.9945	0.9946	0.9948	0.9949	0.9951	0.9952
2.6	0.9953	0.9955	0.9956	0.9957	0.9959	0.9960	0.9961	0.9962	0.9963	0.9964
2.7	0.9965	0.9966	0.9967	0.9968	0.9969	0.9970	0.9971	0.9972	0.9973	0.9974
2.8	0.9974	0.9975	0.9976	0.9977	0.9977	0.9978	0.9979	0.9979	0.9980	0.9981
2.9	0.9981	0.9982	0.9982	0.9983	0.9984	0.9984	0.9985	0.9985	0.9986	0.9986
3	0.9987	0.9987	0.9987	0.9988	0.9988	0.9989	0.9989	0.9989	0.9990	0.9990
3.1	0.9990	0.9991	0.9991	0.9991	0.9992	0.9992	0.9992	0.9992	0.9993	0.9993
3.2	0.9993	0.9993	0.9994	0.9994	0.9994	0.9994	0.9994	0.9995	0.9995	0.9995
3.3	0.9995	0.9995	0.9995	0.9996	0.9996	0.9996	0.9996	0.9996	0.9996	0.9997
3.4	0.9997	0.9997	0.9997	0.9997	0.9997	0.9997	0.9997	0.9997	0.9997	0.9998

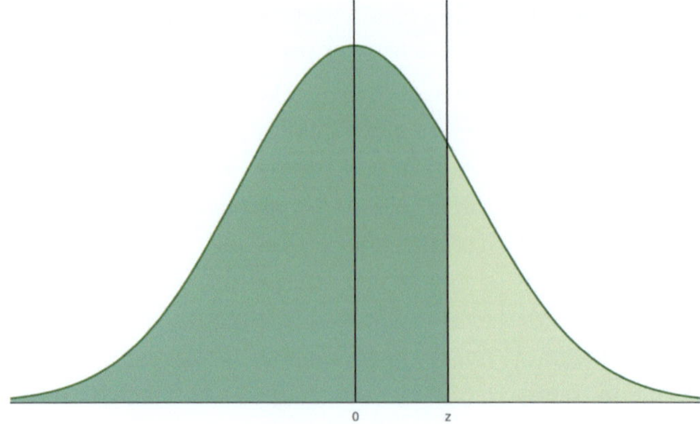

Fig. 4.37 The shaded area under the standard normal distribution curve to the right of a specified z-score

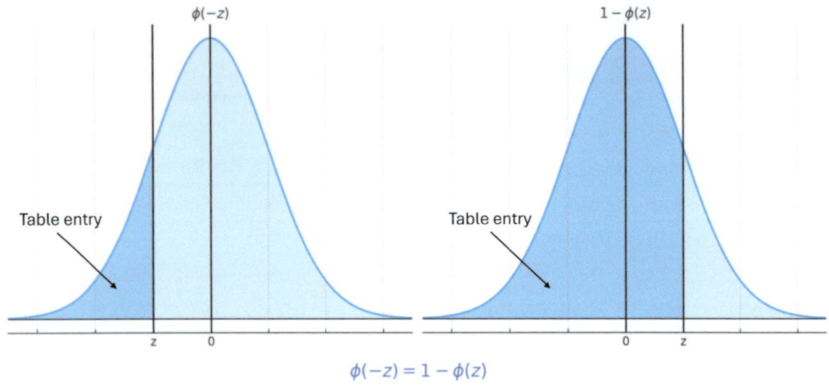

Fig. 4.38 Symmetry property of the standard normal distribution

This identity simplifies calculations by reducing the need to tabulate both tails of the standard normal distribution. Instead, it allows us to derive left-tail probabilities for negative z-values directly from known right-tail values for positive z-values.

To determine the probability that a value lies between two points *a* and *b* on a standard normal distribution, we use the difference between their cumulative probabilities. As illustrated in Fig. 4.39, the shaded region under the curve from point *a* to point *b* represents the probability $P(a \leq Z \leq b)$. This area can be equivalently determined by subtracting the area under the curve up to point *a* (shown in the rightmost plot) from the area under the curve up to point *b*. Mathematically, this probability is expressed as follows:

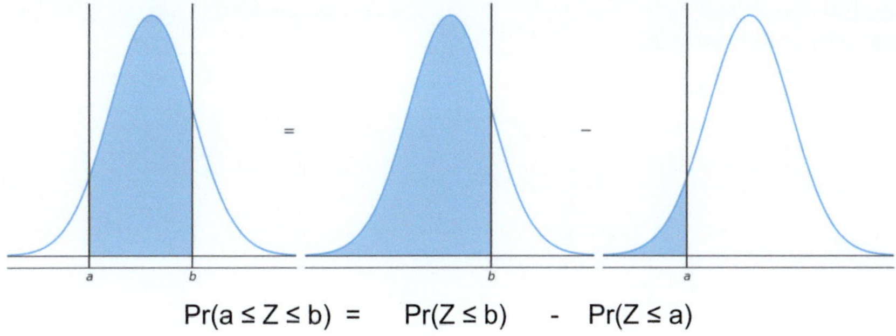

Fig. 4.39 Graphical interpretation of calculating the probability between two z-scores, P($a \leq Z \leq b$)

$$P(a \leq Z \leq b) = \Phi(b) - \Phi(a) \tag{4.30}$$

where Φ(z) is the CDF of the standard normal distribution.

This approach is crucial for evaluating the likelihood of a variable falling within a specific range and is widely used in hypothesis testing, confidence interval construction, and probabilistic modeling in statistics and engineering applications.

4.5.1.1 Code-Based Exploration of the Normal Distribution

Probability Functions

In this section, Code 4.10 will be used to demonstrate how the PDF of a normal distribution can be plotted in MATLAB using two different built-in functions: `normpdf` and `pdf`. As was the case in all previous coding examples, the script begins by clearing the command window, workspace variables, and closing all figure windows (Lines 1–3). This ensures a clean environment before executing the script.

In Lines 5–7, the user defines the key parameters of the normal distribution. Setting `mu = 0` and `sigma = 1` specifies the standard normal distribution, which is commonly used as a reference case in probability and statistics. More generally, the code is flexible and can be applied to generate the probability density function of any normal distribution by assigning different values to the mean (`mu`) and standard deviation (`sigma`).

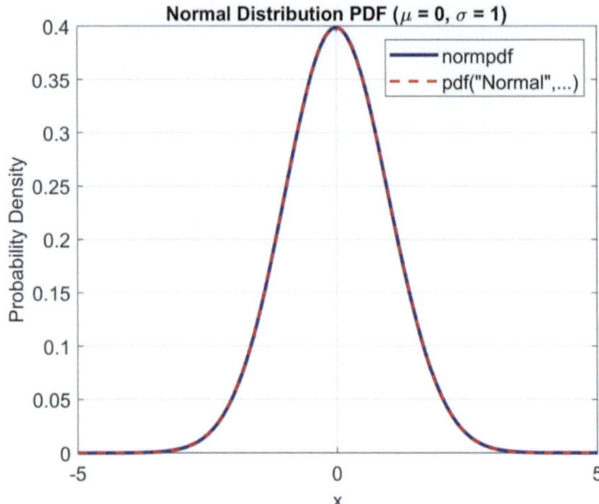

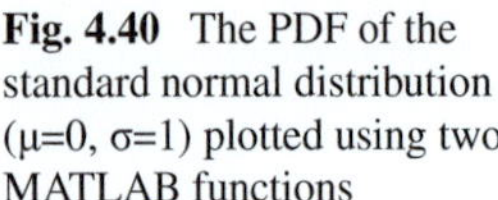
Fig. 4.40 The PDF of the standard normal distribution (μ=0, σ=1) plotted using two MATLAB functions

Next, in Line 10, a vector termed `x` is created with values ranging from –5 to 5 based on a step size of 0.1. This range is sufficient to capture the shape of the standard normal distribution, whose values, as we have seen earlier, are negligible beyond ±3 standard deviations.

In Line 13, the function `normpdf` is used to compute the PDF values at each point in the vector `x`. An equivalent approach is shown in Line 16 using the more general `pdf` function, where the distribution type `'Normal'` is explicitly stated. Both functions return the same values. In the `pdf` function, the function arguments include the vector of values at which to evaluate the PDF, the distribution mean, and the standard deviation. It should be noted that the `pdf` function can return the probability density for a large number of commonly employed distribution types, including binomial, chi-square, F, Gamma, Geometric, Lognormal, and Weibull distributions, for example.

The plotting begins in Line 19 with the opening of a new figure window. Line 20 plots the PDF using `normpdf` as a solid blue line, and Line 22 overlays the results from the `pdf` function as a dashed red line (for illustrative purposes). The `hold on` command in Line 21 ensures that both plots are shown together, as can be seen in Fig. 4.40.

Lines 25–29 are used to annotate the plot. A title is added to describe the purpose of the figure and to indicate the parameters of the distribution. The horizontal axis is labeled as z, reflecting that the standard normal distribution is shown, while the vertical axis represents the probability density. A legend is included to distinguish between the two methods used to generate the PDF, and a background grid is enabled to improve visual reference. The spacing of the grid lines is controlled by the axis tick locations, which can be refined by specifying custom x-axis ticks when greater visual resolution is desired.

Code 4.10 MATLAB implementation of PDF for a normal distribution using both `normpdf` and `pdf` functions.

```
1  clc
2  clear
3  close all
4
5  %% Step 1: Define parameters of the normal distribution
6  mu = 0; % Mean of the distribution
7  sigma = 1; % Standard deviation of the distribution
8
9  %% Step 2: Define the range of x-values for which to compute the PDF
10 x = -5:0.1:5; % Create a vector of x values from -5 to 5 with 0.1 intervals
11
12 %% Step 3a: Calculate the PDF using the 'normpdf' function
13 pdf_values_1 = normpdf(x, mu, sigma); % Computes the normal PDF at each x
14
15 %% Step 3b: Alternatively, calculate the PDF using 'pdf' function syntax
16 pdf_values_2 = pdf('Normal', x, mu, sigma); % Equivalent result to normpdf
17
18 %% Step 4: Plot the PDF using the first method
19 figure; % Open a new figure window
20 plot(x, pdf_values_1, 'b-', 'LineWidth', 2); % Plot PDF as a blue line
21 hold on; % Keep current plot so we can overlay
22 plot(x, pdf_values_2, 'r--', 'LineWidth', 1.5); % Overlay PDF using 'pdf' function in red dashed line
23
24 %% Step 5: Label the plot
25 title('Normal Distribution PDF (\mu = 0, \sigma = 1)');
26 xlabel('x'); % Label x-axis
27 ylabel('Probability Density'); % Label y-axis
28 legend('normpdf', 'pdf("Normal",...)'); % Legend to distinguish the two methods
29 grid on; % Add grid for better readability
```

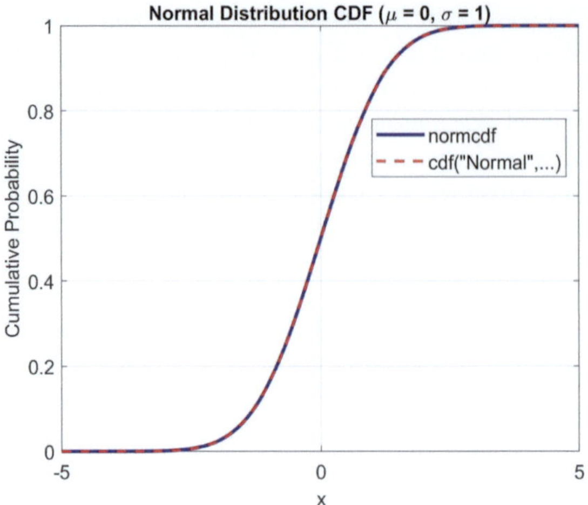

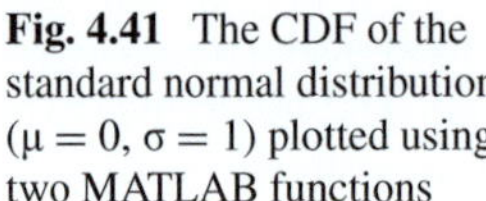

Fig. 4.41 The CDF of the standard normal distribution ($\mu = 0$, $\sigma = 1$) plotted using two MATLAB functions

Code 4.11 demonstrates how to compute and visualize the CDF of a normal distribution using MATLAB. As in the previous example, the script starts by resetting the environment using `clc`, `clear`, and `close all` (Lines 1–3). In Lines 5–7, the user sets the mean (`mu`) and standard deviation (`sigma`) of the distribution, which in this example are set to zero and one, respectively, for the standard normal distribution. A range of values from −5 to 5 is created in Line 10 and stored within the vector `x` to evaluate the CDF over a broad interval of the distribution.

The CDF values are computed in two ways:

- In Line 13, the `normcdf` function directly computes the cumulative probability for each `x`.
- Line 16 shows an alternative approach using the generic `cdf` function, specifying `'Normal'` as the distribution type. As was the case for the `pdf` function used in Code 4.11, this approach can be extended to a wide number of distributions depending on the given application.

Plotting begins in Line 19, where a blue solid line is used to display the result from `normcdf`, while Line 22 specifies the use of a dashed red line to overlay the result from `cdf`. The `hold on` command ensures both graphs are shown together (Fig. 4.41). In the same manner as described earlier for Code 4.10, the plot is labeled in Lines 25–29 with a title, axis labels, and a legend for clarity. A grid is also added to enhance readability.

Code 4.11 MATLAB code to plot the CDF of a Normal Distribution

```
clc
clear
close all

%% Step 1: Define parameters of the normal distribution
mu = 0; % Mean of the distribution
sigma = 1; % Standard deviation of the distribution

%% Step 2: Define the range of x-values for which to compute the CDF
x = -5:0.1:5; % Create a vector of x values from -5 to 5 with 0.1 intervals

%% Step 3a: Calculate the CDF using the 'normcdf' function
cdf_values_1 = normcdf(x, mu, sigma); % Computes the normal CDF at each x

%% Step 3b: Alternatively, calculate the CDF using the 'cdf' function syntax
cdf_values_2 = cdf('Normal', x, mu, sigma); % Equivalent result to normcdf

%% Step 4: Plot the CDF using both methods
figure; % Open a new figure window
plot(x, cdf_values_1, 'b-', 'LineWidth', 2); % Plot CDF as a blue line
hold on; % Keep current plot so we can overlay
plot(x, cdf_values_2, 'r--', 'LineWidth', 1.5); % Overlay CDF using 'cdf' function in red dashed line

%% Step 5: Label the plot
title('Normal Distribution CDF (\mu = 0, \sigma = 1)');
xlabel('x'); % Label x-axis
ylabel('Cumulative Probability'); % Label y-axis
legend('normcdf', 'cdf("Normal",...)'); % Legend to distinguish the two methods
grid on; % Add grid for better readability
```

The Inverse Cumulative Distribution Function (Inverse CDF) is a fundamental concept in probability and statistics. It determines the value of a random variable, denoted as x, such that the probability of the variable being less than or equal to x equals a specified cumulative probability. In simpler terms, if you are given a probability and need to find the corresponding value within a given distribution, you can use the Inverse CDF tool. This is especially useful in probabilistic modeling, statistical analysis, and simulations where specific probability thresholds need to be translated into variable values.

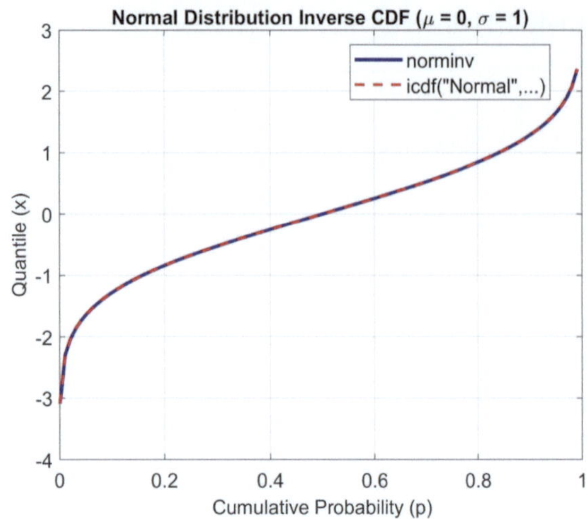

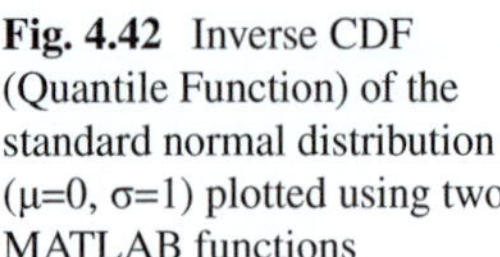

Fig. 4.42 Inverse CDF (Quantile Function) of the standard normal distribution (μ=0, σ=1) plotted using two MATLAB functions

The Inverse CDF is also known as the Quantile Function because it identifies the quantile, or threshold, that matches a particular probability level. For instance, the 0.5 quantile (also known as the 50th percentile) represents the median—the value below which 50% of the data lies. Similarly, the 0.95 quantile indicates the value below which 95% of the data is expected to fall.

Code 4.12 illustrates the use of MATLAB to compute and plot the Inverse CDF—also known as the Quantile Function—for a normal distribution. The environment is reset with `clc`, `clear`, and `close all` (Lines 1–3), after which the mean and standard deviation of the distribution are defined in Lines 5–7. In Line 10, a range of cumulative probabilities `p` is defined from 0.001 to 0.999. This avoids the extremes of 0 and 1, which would yield $\pm\infty$ in a theoretical normal distribution.

The inverse CDF (i.e., the value of `x` for which $P(X \leq x) = p$ is calculated in two ways:

- Using `norminv(p, mu, sigma)` in Line 13.
- Using the general form `icdf('Normal', p, mu, sigma)` in Line 16.

Both methods produce equivalent results and return the quantile values corresponding to each probability in `p`. For both functions, the input arguments include the probability at which you want to find the quantiles as well as the distribution descriptors, including the mean and standard deviation. For the more general `icdf` function, the input argument `'Normal'` is also included to specify the distribution type to consider.

Lines 19–22 are used to generate the plot of the inverse CDF, which is shown in Fig. 4.42. The plot shows how increasing the cumulative probability maps to increasing quantiles of the normal distribution, and how probabilities closer to 0 or 1 correspond to more extreme negative or positive quantile values, respectively. A title, axis labels, and a legend are added for clarity in Lines 25–29.

Code 4.12 MATLAB Code to plot the Inverse CDF of a Normal Distribution

```
clc
clear
close all

%% Step 1: Define parameters of the normal distribution
mu = 0; % Mean of the distribution
sigma = 1; % Standard deviation of the distribution

%% Step 2: Define a range of cumulative probabilities
p = 0.001:0.01:0.999; % Probability values from 0.001 to 0.999

%% Step 3a: Calculate inverse CDF using 'norminv' function
inv_values_1 = norminv(p, mu, sigma); % Returns the x-values for each probability

%% Step 3b: Alternatively, calculate using the 'icdf' function
inv_values_2 = icdf('Normal', p, mu, sigma); % Equivalent result to norminv

%% Step 4: Plot the inverse CDF using both methods
figure;
plot(p, inv_values_1, 'b-', 'LineWidth', 2); % Blue solid line
hold on;
plot(p, inv_values_2, 'r--', 'LineWidth', 1.5); % Red dashed line

%% Step 5: Label the plot
title('Normal Distribution Inverse CDF (\mu = 0, \sigma = 1)');
xlabel('Cumulative Probability (p)');
ylabel('Quantile (x)');
legend('norminv', 'icdf("Normal",...)');
grid on;
```

To illustrate how the inverse CDF function can be used to find the quantile value for a given probability, consider the standard normal distribution, which has a mean of 0 and a standard deviation of 1. Suppose you want to find the value x such that the probability $P(X \leq x) = 0.975$. Using the Inverse CDF (or `norminv` function in MATLAB), the calculation is:

```
x = norminv(0.975, 0, 1); % Output: x ≈ 1.96
```

This means that approximately 97.5% of the values in a standard normal distribution lie below 1.96. This kind of computation is widely used in hypothesis testing, confidence interval estimation, and risk analysis.

Code 4.13 presents a complete example where the use of the three core functions associated with the normal distribution, namely the PDF, CDF, and Inverse CDF (also known as the quantile function), is demonstrated. This implementation uses Python's scientific computing libraries—NumPy, SciPy, and Matplotlib—to replicate functionality commonly used in MATLAB.

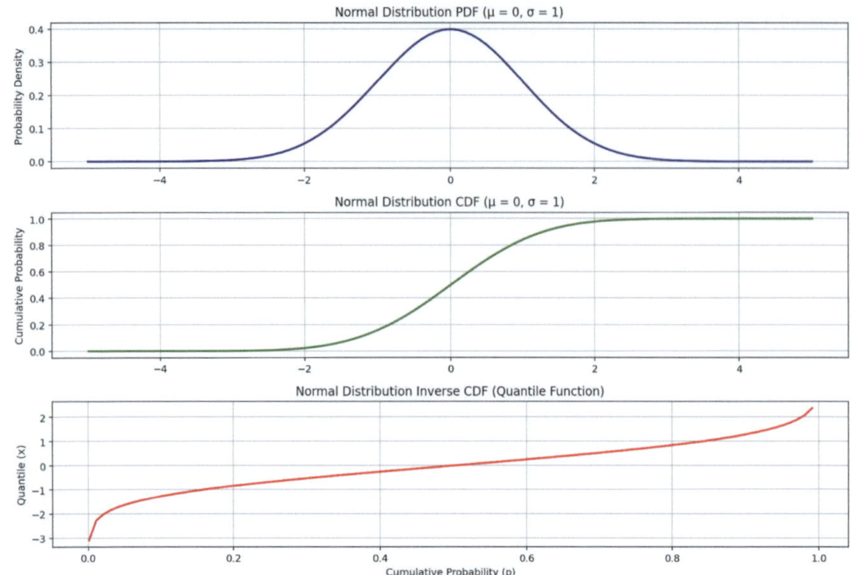

Fig. 4.43 Visual representation of key functions of the standard normal distribution

The script begins by importing the required libraries in Lines 1–3. `NumPy` is used for creating numerical arrays and ranges; `Matplotlib.pyplot` provides plotting capabilities, and `scipy.stats.norm` offers direct access to the normal distribution's functions such as `pdf()`, `cdf()`, and `ppf()` (percent point function, which is equivalent to `icdf` in MATLAB).

Next, the distribution parameters are defined (Lines 5–7). The mean (`mu`) is set to 0 and the standard deviation (`sigma`) to 1, which together define a standard normal distribution. Then, in Line 10, a range of `x` values spanning from –5 to 5 in increments of 0.1 is created using `np.arange()`. This range will be used for evaluating both the PDF and the CDF, so that both the central shape and tails of the distribution are sufficiently captured.

Lines 13 and 14 compute the PDF and CDF using the functions `norm.pdf()` and `norm.cdf()` respectively. These functions return the corresponding probability values for each point in the `x` range specified in Line 10. Following that, in Lines 17 and 18, a new array `p` is defined to include cumulative probability values between 0.001 and 0.999. This range avoids the extremes of 0 and 1, which would correspond to infinite values in a theoretical normal distribution. The `norm.ppf()` function is then used to compute the quantiles (or inverse CDF values) that correspond to each probability in `p`.

Plotting of the results begins in Line 21 with the creation of a new figure. The plotting is structured into three vertically stacked subplots. The first subplot (Lines 24–28) shows the PDF as a blue line, representing the relative likelihood of different values occurring in the distribution. The second subplot (Lines 31–35) visualizes the

CDF in green, showing the accumulation of probability from left to right. Finally, the third subplot (Lines 38–43) displays the inverse CDF (quantile function) in red, indicating the value of the variable associated with a given cumulative probability.

To ensure clarity and prevent overlapping elements, `plt.tight_layout()` (Line 45) adjusts the spacing between subplots. Lastly, `plt.show()` (Line 46) renders the complete figure. The result is a concise, efficient, and educational visualization that encapsulates the foundational functions of the normal distribution in one script. The output of the code for the mentioned subplots are provide in Fig. 4.43.

Code 4.13 Plotting the PDF, CDF, and Inverse CDF of a Normal Distribution in Python

```
1   import numpy as np
2   import matplotlib.pyplot as plt
3   from scipy.stats import norm
4
5   # Step 1: Define parameters of the normal distribution
6   mu = 0
7   sigma = 1
8
9   # Step 2: Define x values for PDF and CDF
10  x = np.arange(-5, 5.1, 0.1)  # From -5 to 5 with 0.1 step
11
12  # Step 3: Compute PDF and CDF
13  pdf_values = norm.pdf(x, mu, sigma)
14  cdf_values = norm.cdf(x, mu, sigma)
15
16  # Step 4: Define probabilities for inverse CDF (quantiles)
17  p = np.arange(0.001, 0.9991, 0.01) # Avoid 0 and 1 due to infinite tails
18  inv_values = norm.ppf(p, mu, sigma)
19
20  # Step 5: Plot PDF, CDF, and Inverse CDF in subplots
21  plt.figure(figsize=(12, 8))
22
23  # Subplot 1: PDF
24  plt.subplot(3, 1, 1)
25  plt.plot(x, pdf_values, 'b-', linewidth=2)
26  plt.title('Normal Distribution PDF (μ = 0, σ = 1)')
27  plt.ylabel('Probability Density')
28  plt.grid(True)
29
30  # Subplot 2: CDF
31  plt.subplot(3, 1, 2)
32  plt.plot(x, cdf_values, 'g-', linewidth=2)
```

```
plt.title('Normal Distribution CDF (μ = 0, σ = 1)')
plt.ylabel('Cumulative Probability')
plt.grid(True)

# Subplot 3: Inverse CDF
plt.subplot(3, 1, 3)
plt.plot(p, inv_values, 'r-', linewidth=2)
plt.title('Normal Distribution Inverse CDF (Quantile Function)')
plt.xlabel('Cumulative Probability (p)')
plt.ylabel('Quantile (x)')
plt.grid(True)

plt.tight_layout()
plt.show()
```

Fitting Data to the Normal Distribution

Beyond possessing an understanding of the various theoretical distributions that exist, it is necessary in many practical applications to assess how well empirical data conforms to a known distribution. This is particularly useful in statistical modeling, quality control, and natural phenomena analysis, where the assumption of normality can justify the use of parametric methods.

When a normal distribution is assumed, fitting the data involves estimating the mean and standard deviation of the Gaussian distribution that best represents the observed sample. In this context, "best-fit" refers to maximum likelihood estimation (MLE), which identifies the parameter values that maximize the probability of observing the given data under the assumed distribution. MATLAB provides the function `fitdist()` for this purpose, which applies MLE to fit a user-specified probability distribution and returns a probability distribution object containing the estimated parameters. While least-squares approaches are commonly used in regression problems to fit deterministic models, MLE is the standard and most widely used method for estimating parameters of probability distributions. For the normal distribution in particular, the MLE estimates have a direct relationship with classical sample statistics, such as the sample mean and variance.

The MLE is a fundamental statistical method used to estimate the parameters of a probability distribution based on observed data. The core idea behind MLE is to find the values of the distribution parameters that maximize the likelihood function, which represents the probability of observing the given data under a specific model. For example, when fitting a normal distribution, MLE identifies the values of the mean (μ) and standard deviation (σ) that make the observed dataset most probable under the assumption of normality.

In the example below, the MATLAB environment will be used to demonstrate how we can generate synthetic data from a normal distribution, fit a normal distribution model to the data using `fitdist`, and visualize the fit against the histogram of the data.

Lines 1–3 reset the MATLAB environment to ensure a clean slate before executing the main part of the script. `clc` clears the command window, `clear` removes all existing variables from the workspace, and `close all` closes any open figure windows. This practice helps prevent unexpected results due to residual data or plots from earlier runs.

Lines 6–8 generate synthetic data that follows a normal (Gaussian) distribution. The true mean (`mu_true`) is set to 5 and the true standard deviation (`sigma_true`) to 2. Using the `normrnd` function, the code creates a 1-by-1000 array of random samples drawn from this normal distribution and stores them within the `data` variable in Line 8. This synthetic dataset simulates observations that might resemble real-world data with an approximately normal pattern.

Line 11 fits a normal distribution model to the generated data using MATLAB's `fitdist` function. Since `fitdist` requires a column vector, the data is transposed by applying the `" ' "` operator to the `data` variable. The resulting object, stored in `pd`, is a probability distribution object that holds the parameters of the fitted normal distribution that were estimated using the MLE criterion. In addition to storing the estimated mean and standard deviation, this object allows the user to directly evaluate probabilistic quantities of interest—such as computing the probability density at a given value, determining cumulative probabilities, or extracting quantiles—using built-in functions, rather than manually deriving or coding the corresponding equations.

Lines 14–15 display the estimated parameters—the mean (`mu`) and standard deviation (`sigma`)—of the fitted distribution in the MATLAB Command Window using the `disp` function. These values are the result of the fitting process and give insight into how well the fitted distribution matches the original data.

Lines 18–27 create a visual comparison between the empirical data and the fitted normal distribution. A new figure is created in Line 18, while Line 19 is used to generate the histogram of the data. Note that the `'Normalization'` input argument is specified, which sets the sum of all bar heights equal to one. Therefore, the histogram represents the empirical probability density, with the height of each bar equal to the probability of selecting an observation within that bin interval. The script then overlays a smooth curve representing the fitted normal probability density function by evaluating the PDF over a linearly spaced range of x-values (lines 21–23). The vector of linearly spaced values is created using the `linspace` function in Line 21, whose input arguments represent the range of the vector values (i.e., `min(data), max(data)`) and the number of values to be contained within the vector (i.e., `100`). Additional commands set the title, labels, and legend for the plot (lines 24–27), and `grid on` (line 28) enhances readability. `grid on` only *displays* the grid; the *spacing* of the grid is determined by `XTick` and `YTick`. The output of this code is shown in Fig. 4.44, where it can be seen that this visual comparison helps assess the goodness-of-fit between the data and the model.

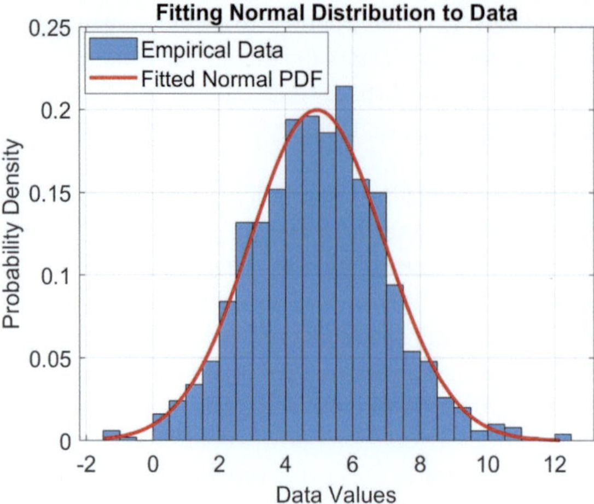

Fig. 4.44 Fitting a normal distribution to empirical data using maximum likelihood estimation (MLE)

Code 4.14 Fit Normal Distribution to Data Using MATLAB

```
clc
clear
close all

%% Step 1: Generate synthetic data (normally distributed)
mu_true = 5;
sigma_true = 2;
data = normrnd(mu_true, sigma_true, [1, 1000]); % Generate 1000 random
samples

%% Step 2: Fit a normal distribution to the data
pd = fitdist(data', 'Normal'); % Fit normal distribution using MLE

%% Step 3: Display estimated parameters
disp(['Estimated Mean: ', num2str(pd.mu)]);
disp(['Estimated Standard Deviation: ', num2str(pd.sigma)]);

%% Step 4: Plot histogram and fitted normal curve
figure;
histogram(data, 'Normalization', 'pdf'); % Plot normalized histogram
hold on;
x = linspace(min(data), max(data), 100); % Generate x range for plotting
y = pdf(pd, x); % Evaluate fitted PDF
plot(x, y, 'r-', 'LineWidth', 2); % Plot fitted normal curve
title('Fitting Normal Distribution to Data');
xlabel('Data Values');
ylabel('Probability Density');
legend('Empirical Data', 'Fitted Normal PDF');
grid on;
```

This same exercise is repeated in Code 4.15 below for the Python environment. The code begins by importing the essential libraries (Lines 1–3): `NumPy` for numerical operations, `Matplotlib` for plotting, and `scipy.stats.norm` to work with the normal distribution. In Lines 6–8, a synthetic dataset is created using `np.random.normal`, where 1,000 samples are drawn from a normal distribution with a true mean (`mu_true`) of 5 and a standard deviation (`sigma_true`) of 2. This dataset simulates empirical observations that are approximately normally distributed.

In Line 11, the script estimates the parameters of the best-fitting normal distribution using the `norm.fit()` function, which applies MLE to determine the most likely values of the mean and standard deviation based on the data. The resulting estimates, `mu_est` and `sigma_est`, represent the parameters of the fitted distribution.

Lines 14–15 create a histogram of the empirical data with 30 bins, normalized so that its area equals 1 (a requirement for overlaying with a probability density function since the probability densities must sum to unit). The histogram uses a semi-transparent steel-blue color and is labeled for clarity.

Next, Lines 18–20 compute and plot the fitted normal PDF. One hundred evenly spaced values spanning between the minimum and maximum of the data are generated in Line 18 using the `np.linspace` function and stored within the variable `x`. Then, `norm.pdf` is used to evaluate the fitted normal curve at each `x` using the MLE-based estimates. This PDF is plotted as a red line over the histogram.

Finally, the plot is annotated (Lines 23–27) with a title, axis labels, legend, and grid lines. `plt.tight_layout()` (Line 28) ensures a clean layout without overlapping elements, and `plt.show()` (Line 29) displays the resulting figure. The outcome is a clear visual demonstration of how well a normal model fits the observed data. The output of this code resembles Fig. 4.44. However, the distribution of the data may vary slightly due to the use of randomly generated samples.

Code 4.15 Fitting a Normal Distribution to Empirical Data Using Python

```
1   import numpy as np
2   import matplotlib.pyplot as plt
3   from scipy.stats import norm
4
5   # Step 1: Generate synthetic data (normally distributed)
6   mu_true = 5
7   sigma_true = 2
8   data = np.random.normal(mu_true, sigma_true, size=1000)  # 1000 samples
9
10  # Step 2: Fit a normal distribution to the data using MLE
11  mu_est, sigma_est = norm.fit(data)  # Estimate mean and std dev
12
```

```
# Step 3: Create histogram of the data
plt.figure(figsize=(8, 6))
count, bins, ignored = plt.hist(data, bins=30, density=True, alpha=0.6, color='steelblue', label='Empirical Data')

# Step 4: Plot the fitted normal PDF
x = np.linspace(min(data), max(data), 100)
pdf_fitted = norm.pdf(x, mu_est, sigma_est)
plt.plot(x, pdf_fitted, 'r-', linewidth=2, label='Fitted Normal PDF')

# Step 5: Annotate the plot
plt.title('Fitting Normal Distribution to Data')
plt.xlabel('Data Values')
plt.ylabel('Probability Density')
plt.legend()
plt.grid(True)
plt.tight_layout()
plt.show()
```

4.5.1.2 Evaluating Goodness-of-Fit Using Statistical Tests

After fitting empirical data to a theoretical probability distribution, such as the normal or exponential distribution, it is important to evaluate how well the fitted model represents the observed data. This process is known as goodness-of-fit analysis, and it ensures that the statistical assumptions underlying a model are valid for the given dataset. Several formal statistical tests are available to assess the quality of fit. The most commonly used ones include the Kolmogorov–Smirnov test (KST), Anderson–Darling test (ADT), and the Chi-square test (CST). Each of these tests provides a numerical and statistical basis for determining whether the observed data significantly deviates from the hypothesized distribution.

Kolmogorov–Smirnov Test (KST)

The Kolmogorov–Smirnov test (KST) is a non-parametric statistical test used to compare a sample's empirical distribution function with a specified theoretical CDF. The test is widely used to determine whether a sample comes from a particular distribution, such as normal, exponential, or uniform (for example).

Let $F_n(x)$ be the empirical distribution function of a sample of size n and F(x) be the CDF of the theoretical distribution to which the data is being compared. The KST statistic is calculated using Eq. 4.31 below:

$$D = \sup_{x} |F_n(x) - F(x)| \tag{4.31}$$

Here, 'sup' refers to the supremum (maximum) of the absolute differences between the empirical and theoretical CDFs over the domain of the data. In simple terms, D represents the largest vertical distance between the two curves.

The KST is widely appreciated for its simplicity and flexibility. A significant advantage of this test is its non-parametric nature, meaning that it does not assume any specific parameters for the distribution, making it applicable to a wide range of continuous distributions. Another key strength is that it provides an intuitive visual interpretation, as it measures the maximum vertical distance between the empirical and theoretical cumulative distribution functions. There are, however, a number of important limitations of the KST that should be noted. Firstly, it is less sensitive in the tails of the distribution compared to the Anderson–Darling test (ADT), which can be problematic when assessing fit in datasets with extreme values. Second, the test is designed for continuous data only and may not be valid for discrete distributions. Finally, the KST test assumes the theoretical distribution's parameters are known in advance; when parameters are estimated from the data (as with fitted distributions), the test results may be biased or overly optimistic unless adjustments are applied.

To interpret the KST results, two main outputs are considered: the test statistic D and the p-value. The test statistic D quantifies the largest difference between the empirical and theoretical cumulative distribution functions. A small D value suggests a close match, while a large D indicates a poor fit. The p-value helps determine statistical significance: if the p-value is greater than a chosen significance level (commonly 0.05), there is insufficient evidence to reject the null hypothesis, meaning the data plausibly follow the assumed distribution. Conversely, a p-value less than 0.05 implies a significant deviation, and the distribution is likely not a good fit. Alternatively, in some software outputs, the test statistic is compared to a critical value from statistical tables. In hypothesis testing, a critical value is a threshold for the test statistic that separates outcomes considered consistent with the assumed model from those considered too extreme to be plausibly explained by it. If $D > D_{critical}$, the null hypothesis is rejected. Both approaches—p-value and critical value—serve the same purpose: to judge whether the difference between the empirical and theoretical distributions is statistically meaningful.

Anderson–Darling test (ADT)

The Anderson–Darling test (ADT) is a statistical goodness-of-fit test used to evaluate whether a sample of data comes from a specific continuous probability distribution, such as the normal, exponential, or Weibull distribution. It can be viewed as a refinement of the Kolmogorov–Smirnov test, as it evaluates discrepancies between empirical and theoretical cumulative distribution functions while assigning greater weight to deviations occurring in the tails of the distribution.

The ADT belongs to the family of empirical distribution function tests, which compare the empirical distribution of observed data with the theoretical CDF of the assumed distribution. The central idea is to measure how much the observed data deviates from what we would expect under the fitted distribution, with more weight placed on discrepancies in the tails.

For a sample of size n sorted in ascending order, let $X_1, X_2, \ldots, X_n$ be the ordered observations, and F(x) be the CDF of the hypothesized distribution. The ADT statistic A^2 is computed using the following equation:

$$A^2 = -n - \frac{1}{n}\sum_{i=1}^{n}\left[(2i-1)\cdot(\ln F(X_i) + \ln(1 - F(X_{n+1-i})))\right] \tag{4.32}$$

This formula involves computing the log-likelihood of the CDF at both ends of the data (i.e., the lower and upper tails), which is the component that gives this test enhanced sensitivity to deviations at the extremes.

The ADT works by first sorting the data in ascending order and then computing the CDF, F(x), of the hypothesized distribution at each sorted value X_i. Using these values, the ADT statistic A^2 is calculated using its specific formula. This computed statistic is then compared against critical values that are distribution-specific and precomputed using simulations that vary depending on the target distribution (such as normal or exponential). If the A^2 value exceeds the critical threshold at a chosen significance level (e.g., 0.05), the null hypothesis is rejected, indicating that the sample data likely does not follow the assumed distribution. Unlike the KST, which gives equal weight to all deviations, the ADT places more emphasis on the tails of the distribution, making it particularly sensitive to discrepancies that occur at the extremes of the data range.

The ADT is a powerful goodness-of-fit test used to assess whether a dataset follows a specific probability distribution. A major advantage of the ADT lies in its emphasis on tail behavior, which makes it particularly valuable in applications where extreme values govern system performance or risk. This includes fields such as hydrology, reliability engineering, and finance, where rare but severe events can have disproportionate consequences. It also provides distribution-specific critical values, allowing for more accurate assessments when comparing to known distributions like the normal, exponential, or Weibull. However, the ADT does come with limitations. It is distribution-dependent, meaning that the test statistic and critical values are calculated differently depending on the distribution being tested. This makes it less generalizable than the KST. Furthermore, while some implementations return p-values, others only provide test statistics and critical values, requiring users to perform a manual comparison to interpret the results. Also, similar to KST, the ADT is designed for continuous distributions and assumes that the data does not contain ties or missing values.

When interpreting the ADT, the key output is the test statistic, which measures the weighted squared differences between the observed and theoretical cumulative distribution functions, with higher weights in the tails. A larger test statistic indicates a greater discrepancy between the data and the assumed distribution. Unlike the Kolmogorov–Smirnov test, the ADT does not always return a p-value; instead, it often provides a list of critical values for several significance levels (e.g., 15, 10,

5, 2.5, and 1%). These critical values are derived from the sampling distribution of the test statistic under the null hypothesis and are typically obtained using analytical approximations or Monte Carlo simulations. Importantly, the critical values are distribution-specific and depend on whether the parameters of the assumed distribution are known or estimated from the data. As a result, separate critical value tables are used for different target distributions and significance levels. To draw conclusions, one compares the calculated test statistic to these critical values. If the test statistic is greater than the critical value at a given significance level (e.g., 5%), the null hypothesis—that the data follow the specified distribution—is rejected at that level. This means the observed data deviate significantly from the expected distribution. However, if the test statistic returned is less than all critical values, there is insufficient evidence to reject the null hypothesis, and the fit is considered acceptable. When p-values are reported (as in Python's `scipy.stats.anderson()` for the normal distribution), they should be interpreted similarly: a small p-value suggests poor fit, while a large one indicates consistency with the target distribution.

Chi-Square (CST)

The Chi-square test (CST) is one of the oldest and most widely used goodness-of-fit tests in statistics. It is designed to assess how well an observed frequency distribution matches the expected frequency distribution derived from a theoretical probability model. The CST is particularly suitable for categorical, or grouped data, and can be used with both discrete and continuous distributions (with the latter requiring binning). The test evaluates whether the observed discrepancies between the data and the theoretical model are due to random variation or if they indicate a poor model fit.

The test is based on the principle that, if the observed data follow the hypothesized distribution, then the observed and expected frequencies in each class (bin) should be reasonably close. Significant deviations suggest that the data may not conform to the assumed distribution.

For a dataset grouped into k intervals (bins), the Chi-square test statistic is calculated using the following equation:

$$\chi^2 = \sum_{i=1}^{k} \frac{(O_i - E_i)^2}{E_i} \tag{4.33}$$

where O_i is the observed frequency in bin i, E_i is the expected frequency in bin i, based on the theoretical distribution, and k is the number of bins. The result is a single value—the Chi-square statistic—which quantifies the total discrepancy between observed and expected values across all bins.

The CST works by first dividing the data into k mutually exclusive intervals or bins, where the choice of the number and size of bins plays a crucial role in the accuracy of the results. Observed frequencies O_i are then obtained by counting the number of data points in each bin. Next, expected frequencies E_i are calculated based on the theoretical distribution being tested, making sure the total expected count equals the sample size. Using these values, the test statistic is computed by Eq. (4.33). The degrees of freedom are determined using the formula $df = k - \mathrm{p} - 1$, where p represents the number of parameters estimated from the data (such as mean and variance). This statistic is then compared to a critical value from the Chi-square distribution at a chosen significance level (e.g., 0.05), or a p-value is calculated using statistical software. If the test statistic exceeds the critical value, or if the p-value is less than the significance level, the null hypothesis is rejected, indicating that the observed data significantly deviates from the expected distribution. It is also important to note that each expected frequency should typically be at least 5 (i.e., $E_i \geq 5$) to ensure the test's validity.

The CST is one of the most widely used statistical tests for assessing the goodness-of-fit between observed frequencies and expected frequencies from a hypothesized probability distribution. One of its main advantages is that it is conceptually simple and highly versatile—it can be applied to both discrete and continuous data (the latter typically grouped into intervals). It is also useful when dealing with categorical data, making it valuable in survey analysis, genetics, quality control, and more. However, the CST has several limitations. First, it requires that the data be grouped into bins, and the test results can be sensitive to how those intervals are chosen; poor binning can reduce accuracy. Second, the test relies on large-sample approximations, requiring sufficiently large expected frequencies in each bin (commonly at least five). As a result, the CST may perform poorly for small sample sizes or when data are sparsely distributed across bins. Additionally, the CST assumes independence of observations, which may not always hold in practice. Finally, unlike the KST and ADT, the CST is not based on cumulative distributions and may miss subtle deviations in distribution shape, particularly in the tails.

Goodness of Fit Coding

Code 4.16 illustrates the MATLAB script for the evaluation and displaying of the goodness-of-fit tests (KST, ADT, CST) with the rejection criteria based on the test statistic and p-value at multiple significance levels. The script begins with a standard cleanup phase using `clc`, `clear`, and `close all` (Lines 1–3) to clear the command window, remove all existing variables from the workspace, and close any open figures, respectively. This ensures that the script runs in a clean environment without residual data or figures from previous sessions, causing interference.

Next, in Lines 5–7, synthetic data are generated using MATLAB's `normrnd` function. This function draws 15 random values from a normal distribution with a mean of 5 and a standard deviation of 2. The number of samples, denoted by `n`, is also calculated and stored for use in subsequent statistical analyses.

Following data generation, the script fits a normal distribution to the data using the `fitdist` function (Lines 9–10) whose input arguments include the data vector as well as the desired distribution type (i.e., `normal`). This step is essential as it provides the parameters (mean and standard deviation) of the fitted distribution, which are then used in the goodness-of-fit tests that follow.

To prepare for statistical decision-making, a vector of common significance levels is defined in line 13. These include α-values of 0.2, 0.1, 0.05, 0.02, and 0.01 so that the script may test the null hypothesis at varying levels of confidence.

The KST is implemented first, beginning at Line 15. The `kstest` function is applied to the data using the cumulative distribution function (`CDF`) of the fitted normal distribution (`pd`). The output of this function includes the test statistic (`h_ks`), p-value (`p_ks`), and a binary decision flag (`ks_stat`) as shown in Line 16. Next, Line 17 provides the critical values for the KST at each predefined alpha level. When the distribution parameters are estimated from the data, as in this example, some goodness-of-fit tests require adjusted critical values; the standard Kolmogorov–Smirnov test assumes that the underlying distribution is fully specified. To determine whether the null hypothesis should be rejected, Lines 18–19 compare the test statistic and p-value against the critical thresholds. Recall, if the p-value is low, the null hypothesis is rejected, meaning that the sample data deviates significantly from the fitted distribution. The outcomes are converted from logical (true/false) to user-friendly strings ("Yes" or "No") for clarity. The results are then organized into a summary table (`T1`) in Lines 21–27, listing the test name, significance level, test statistic, and rejection decisions based on both the statistic and the p-value.

After completing the KST, the script moves on to the ADT, which is executed starting from Line 29 using the `adtest` function. The syntax for this function is of the same format as for the `kstest` function introduced earlier, with outputs including the test statistic (`h_ad`), p-value (`p_ad`), and a binary decision flag (`ad_stat`). In Line 31, the corresponding critical values for different alpha levels are defined. Lines 32–33 then compare the test results against these critical values and determine whether the null hypothesis should be rejected. Again, the logical outcomes are formatted as "Yes" or "No". The results are stored in table `T2` (lines 35–41), following the same structure used for the KST.

The final goodness-of-fit test in the script is the CST, which begins at Line 43. The script first defines bin edges for histogramming the data (Line 44) and counts the observed frequencies in each bin using `histcounts` (Line 45). Then, using the fitted distribution's cumulative function, it computes the expected probabilities per bin and scales them by the sample size to obtain expected frequencies (Lines 46–47). The chi-squared statistic is calculated in Line 49 using the standard formula for comparing observed and expected frequencies. The degrees of freedom for the test are determined in Line 50 as the number of bins minus the number of estimated parameters (mean and standard deviation) minus one. The p-value is then derived

```
Command Window
=================== Kolmogorov-Smirnov Test (KST) ===================
    Test     Alpha    Statistic    Reject_Stat    p_value    Reject_P
    _____    _____    _________    ___________    _______    ________

    "KST"     0.2     0.13255         "No"        0.92357      "No"
    "KST"     0.1     0.13255         "No"        0.92357      "No"
    "KST"    0.05     0.13255         "No"        0.92357      "No"
    "KST"    0.02     0.13255         "No"        0.92357      "No"
    "KST"    0.01     0.13255         "No"        0.92357      "No"

==================== Anderson-Darling Test (ADT) ======================
    Test     Alpha    Statistic    Reject_Stat    p_value    Reject_P
    _____    _____    _________    ___________    _______    ________

    "ADT"     0.2     0.29251         "No"        0.94297      "No"
    "ADT"     0.1     0.29251         "No"        0.94297      "No"
    "ADT"    0.05     0.29251         "No"        0.94297      "No"
    "ADT"    0.02     0.29251         "No"        0.94297      "No"
    "ADT"    0.01     0.29251         "No"        0.94297      "No"

======================== Chi-Squared Test (CST) =======================
    Test     Alpha    Statistic    Reject_Stat    p_value    Reject_P
    _____    _____    _________    ___________    _______    ________

    "CST"     0.2     0.29989         "No"        0.86076      "No"
    "CST"     0.1     0.29989         "No"        0.86076      "No"
    "CST"    0.05     0.29989         "No"        0.86076      "No"
    "CST"    0.02     0.29989         "No"        0.86076      "No"
    "CST"    0.01     0.29989         "No"        0.86076      "No"
```

Fig. 4.45 Output of Code 4.16 displaying the results of the Kolmogorov–Smirnov Test (KST), Anderson–Darling Test (ADT), and Chi-Squared Test (CST) for goodness-of-fit analysis

from the chi-squared cumulative distribution function (Line 51), and the critical values for the test are calculated in Line 52 based on the specified significance levels. The rejection decisions are evaluated and formatted in Lines 53–54, and the final results are compiled into table `T3` (Lines 56–62).

Finally, the script presents the test results to the user in Lines 64–74. Each test is displayed with a clear, labeled header using the `disp` function. To enhance readability, the script separates the outputs of the KST, ADT, and CST using spacing and formatting. Each table presents the alpha levels, test statistic, p-value, and whether the null hypothesis should be rejected based on both the test statistic and the p-value. This structure makes it easy for users—especially students or researchers evaluating statistical assumptions—to understand and compare the outcomes across different tests and significance levels. The output of this code is presented in Fig. 4.45.

Code 4.16 MATLAB script for performing and displaying goodness-of-fit tests (KST, ADT, CST) with rejection decision based on test statistic and p-value at multiple significance levels

```
clc;
clear;
close all;

%% Step 1: Generate or load data
data = normrnd(5, 2, [15, 1]); % Replace with your dataset
n = length(data);

%% Step 2: Fit a normal distribution
pd = fitdist(data, 'Normal');

%% Step 3: Define significance levels
alpha = [0.2, 0.1, 0.05, 0.02, 0.01];

%% ========== Kolmogorov-Smirnov Test (KST) ==========
[h_ks, p_ks, ks_stat] = kstest(data, 'CDF', pd);
cv_table_ks = [0.26588, 0.30397, 0.3376, 0.37713, 0.4042];
reject_ks_stat = replace(string(ks_stat > cv_table_ks), ["true", "false"], ["Yes", "No"]);
reject_ks_pval = replace(string(p_ks < alpha), ["true", "false"], ["Yes", "No"]);

T1 = table();
T1.Test = repmat("KST", 5, 1);
T1.Alpha = alpha';
T1.Statistic = repmat(ks_stat, 5, 1);
T1.Reject_Stat = reject_ks_stat';
T1.p_value = repmat(p_ks, 5, 1);
T1.Reject_P = reject_ks_pval';

%% ========== Anderson-Darling Test (ADT) ==========
[h_ad, p_ad, ad_stat] = adtest(data, 'Distribution', pd);
cv_table_ad = [1.3749, 1.9286, 2.5018, 3.2892, 3.9074];
reject_ad_stat = replace(string(ad_stat > cv_table_ad), ["true", "false"], ["Yes", "No"]);
reject_ad_pval = replace(string(p_ad < alpha), ["true", "false"], ["Yes", "No"]);

T2 = table();
T2.Test = repmat("ADT", 5, 1);
T2.Alpha = alpha';
T2.Statistic = repmat(ad_stat, 5, 1);
T2.Reject_Stat = reject_ad_stat';
T2.p_value = repmat(p_ad, 5, 1);
T2.Reject_P = reject_ad_pval';

%% ========== Chi-Squared Test (CST) ==========
```

```
edges = linspace(min(data), max(data), 6); % 5 bins
obs = histcounts(data, edges);
expected_probs = diff(cdf(pd, edges));
exp = expected_probs * n;

chi2_stat = sum((obs - exp).^2 ./ exp);
df = length(obs) - 1 - 2; % k - p - 1
p_chi2 = 1 - chi2cdf(chi2_stat, df);
cv_table_chi = chi2inv(1 - alpha, df);
reject_chi_stat = replace(string(chi2_stat > cv_table_chi), ["true", "false"], ["Yes", "No"]);
reject_chi_pval = replace(string(p_chi2 < alpha), ["true", "false"], ["Yes", "No"]);

T3 = table();
T3.Test = repmat("CST", 5, 1);
T3.Alpha = alpha';
T3.Statistic = repmat(chi2_stat, 5, 1);
T3.Reject_Stat = reject_chi_stat';
T3.p_value = repmat(p_chi2, 5, 1);
T3.Reject_P = reject_chi_pval';

%% ========== Display Clean Output ==========
disp('=================== Kolmogorov-Smirnov Test (KST) =================== ');
disp(T1);

disp('==================== Anderson-Darling Test (ADT) ====================== ');
fprintf(' ');
disp(T2);

disp('======================== Chi-Squared Test (CST) ======================== ');
fprintf(' ');
disp(T3);
```

Code 4.17 shows the Python script for performing and displaying goodness-of-fit tests (KST, ADT, CST) with rejection decision based on test statistic and p-value at multiple significance levels. The script starts by importing essential Python libraries in Lines 1 to 4. `NumPy` is imported as `np` for numerical operations like array handling and random number generation. `Pandas` (`pd`) is used to create structured tables for displaying results. Then, from `scipy.stats`, general statistical tools (`stats`), the normal distribution model (`norm`), the Anderson–Darling test (`anderson`), and the chi-squared distribution (`chi2`) are loaded for later use in testing.

In Lines 6 to 9, the script begins with data generation. The random seed is set in line 7 using `np.random.seed(42)` to ensure reproducibility—that is, the same data will be generated each time the script runs. In Line 8, 15 data points are drawn from a normal distribution with a mean of 5 and a standard deviation of 2 using `np.random.normal(5, 2, 15)`. The number of data points is stored in the variable `n` on Line 9 using the `len()` function.

Next, the script fits a normal distribution to the data in Lines 11 to 13. Line 12 uses `norm.fit(data)` to estimate the mean and standard deviation of the sample using the MLE criterion. These parameters are then used to create a normal distribution object in Line 13 (`fitted_dist`), which will be referenced in the goodness-of-fit tests.

In Line 16, the script defines an array of five commonly used significance levels (alpha values): 0.2, 0.1, 0.05, 0.02, and 0.01. These are stored in `np.array` and will be used to determine whether to reject the null hypothesis in each statistical test.

The KST is then implemented in Lines 18 to 31. Line 19 computes the KST statistic (`ks_stat`) and its corresponding p-value (`p_ks`) by comparing the sample to the fitted distribution's cumulative distribution function. In Line 20, the critical values for KST at each alpha level are provided manually in an array (`cv_ks`). Line 21 then checks whether the test statistic exceeds the critical values, returning "Yes" if so and "No" if not. Similarly, Line 22 checks if the p-value is less than each alpha level to make a second rejection decision. The results are organized into a table `T1` in Lines 24 to 31, showing the test name, significance levels, test statistic, p-value, and decisions based on both the statistic and p-value.

After the KST, Lines 33 to 48 present the coding to perform the ADT. Line 34 runs the test with `anderson(data, dist='norm')` and stores the full result. Line 35 extracts the test statistic (`ad_stat`), and Line 36 retrieves the critical values from the result. Line 37 checks if the statistic exceeds each critical value, producing the "Yes"/"No" flags. Since `anderson()` does not provide a p-value, Line 38 interpolates an approximate p-value using the inverse of the critical values, and Line 39 uses this to generate the rejection decisions. These results are compiled into a table `T2` in Lines 41 to 48, following the same format as `T1`.

The CST is conducted in Lines 50 to 69. First, Line 51 defines 5 bins using `np.linspace` to divide the data range equally. Line 52 counts how many data points fall into each bin (observed frequencies), and Line 53 calculates the expected probabilities for each bin based on the fitted normal distribution. Line 54 then converts these probabilities to expected counts by multiplying by the sample size. Line 55 calculates the chi-squared test statistic using the standard formula. The degrees of freedom are determined in Line 56 using the earlier defined relationship with the number of bins as k and the number of parameters estimated (mean and standard deviation) as p. In Line 57, the chi-squared p-value is calculated while Line 58 computes the critical values of the chi-squared distribution for the given degrees of freedom at each alpha level. Lines 59 and 60 determine rejection decisions based on both the test statistic as well as the p-value. The full set of CST results is compiled into a table `T3` in Lines 62 to 69.

Finally, the script prints the results of all three tests in a structured and readable format in Lines 71 to 79. Each block is clearly labeled with the name of the test (e.g., KST, ADT, CST), and the corresponding table is displayed using `print()`. This approach ensures that users can easily interpret whether their data fits a normal distribution at each significance level using multiple classical goodness-of-fit methods.

Code 4.17 Python script for performing and displaying goodness-of-fit tests (kst, adt, cst) with rejection decision based on test statistic and p-value at multiple significance levels

```
1   import numpy as np
2   import pandas as pd
3   from scipy import stats
4   from scipy.stats import norm, anderson, chi2
5
6   # Step 1: Generate or load data
7   np.random.seed(42)
8   data = np.random.normal(5, 2, 15)
9   n = len(data)
10
11  # Step 2: Fit normal distribution
12  mu, sigma = norm.fit(data)
13  fitted_dist = norm(loc=mu, scale=sigma)
14
15  # Step 3: Define significance levels
16  alpha = np.array([0.2, 0.1, 0.05, 0.02, 0.01])
17
18  # ================== KST ===================
19  ks_stat, p_ks = stats.kstest(data, fitted_dist.cdf)
20  cv_ks = np.array([0.26588, 0.30397, 0.3376, 0.37713, 0.4042])
21  reject_ks_stat = np.where(ks_stat > cv_ks, "Yes", "No")
```

```
reject_ks_pval = np.where(p_ks < alpha, "Yes", "No")

T1 = pd.DataFrame({
    "Test": ["KST"] * 5,
    "Alpha": alpha,
    "Statistic": [ks_stat] * 5,
    "Reject_Stat": reject_ks_stat,
    "p-value": [p_ks] * 5,
    "Reject_P": reject_ks_pval
})

# =================== ADT ===================
ad_result = anderson(data, dist='norm')
ad_stat = ad_result.statistic
cv_ad = np.array(ad_result.critical_values)
reject_ad_stat = np.where(ad_stat > cv_ad, "Yes", "No")
p_ad = np.interp(ad_stat, cv_ad[::-1], np.array(alpha[::-1]))  # approximate p-value
reject_ad_pval = np.where(p_ad < alpha, "Yes", "No")

T2 = pd.DataFrame({
    "Test": ["ADT"] * 5,
    "Alpha": alpha,
    "Statistic": [ad_stat] * 5,
    "Reject_Stat": reject_ad_stat,
    "p-value": [p_ad] * 5,
    "Reject_P": reject_ad_pval
})

# =================== CST ===================
bins = np.linspace(np.min(data), np.max(data), 6)
obs, _ = np.histogram(data, bins)
expected_probs = np.diff(fitted_dist.cdf(bins))
exp = expected_probs * n
chi2_stat = np.sum((obs - exp) ** 2 / exp)
df = len(obs) - 1 - 2  # k - p - 1
p_chi2 = 1 - chi2.cdf(chi2_stat, df)
cv_chi = chi2.ppf(1 - alpha, df)
reject_chi_stat = np.where(chi2_stat > cv_chi, "Yes", "No")
reject_chi_pval = np.where(p_chi2 < alpha, "Yes", "No")

T3 = pd.DataFrame({
    "Test": ["CST"] * 5,
    "Alpha": alpha,
    "Statistic": [chi2_stat] * 5,
```

4.5.1.3 Confidence Intervals in Normal Distribution Fitting

When fitting a probability distribution—such as the normal distribution—to a set of empirical data, we obtain estimates of its parameters, namely the mean (μ) and the standard deviation (σ). However, these estimates are based on sample data, which naturally introduces uncertainty. This is where confidence intervals (CIs) come into play.

A CI provides a range of values within which the true population parameter is likely to fall, with a specified level of confidence. In many domains of science and engineering, the level of significance is often chosen as $\alpha = 0.05$, a value which was popularized by Sir Ronald Fisher because the z-score associated with this CI could be rounded conveniently to 2. Practically, a 95% confidence interval for the mean implies that if we were to repeat the sampling and fitting process many times, approximately 95% of the resulting intervals would contain the true mean of the distribution.

In the context of fitting a normal distribution, we obtain:

- A point estimate for the mean and standard deviation (via MLE),
- A 95% confidence interval for each parameter

Visually, these intervals can be used not only to report statistical uncertainty in the point estimators, but also to plot a band or range of plausible normal curves that reflect the uncertainty in the fitted model. This approach provides a more complete picture than a single fitted curve alone, especially when communicating variability and reliability in data-driven modeling.

Code 4.18 shows how to visually represent the uncertainty associated with a fitted normal distribution by incorporating confidence bounds directly into the plot. Rather than displaying confidence intervals as static numeric ranges, this example illustrates how the entire shape of the normal curve can vary based on the lower and upper confidence bounds of the distribution's parameters.

The script begins by initializing the environment with `clc`, `clear`, and `close all` (Lines 1–3), which ensures a clean workspace for reproducible analysis. In Lines 6–8, synthetic data is generated by drawing 1,000 random samples from a known normal distribution with a true mean (`mu_true`) of 10 and a standard deviation (`sigma_true`) of 3. This provides a simulated empirical dataset on which fitting and evaluation will be performed.

In Line 11, a normal distribution is fitted to the data using MATLAB's `fitdist()` function with the MLE criterion. This yields the MLE estimate for the mean and standard deviation based on the sample. Following the fitting, Lines 14–16 compute the 95% confidence intervals for the estimated parameters using the `paramci()` function. The returned matrix contains the lower and upper bounds for both the mean and the standard deviation. These bounds are extracted into separate variables (`mu_lower`, `mu_upper`, `sigma_lower`, `sigma_upper`) to facilitate plotting.

Visualization begins in Line 19 by creating a new figure. A normalized histogram of the data is plotted in Line 20 with reduced opacity (`'FaceAlpha', 0.3`) to allow overlaying of the fitted curves. Line 23 generates a smooth range of `x` values for evaluating the probability density functions.

The main fitted curve—corresponding to the MLE estimates—is computed in Line 26 using the `pdf()` function and plotted as a solid red line in Line 27. Two additional curves are then constructed to represent the confidence bounds. The lower bound PDF, based on the lower values of both mean and standard deviation, is computed in Line 30 and plotted as a blue dashed line in Line 31. Similarly, the upper bound PDF, using the upper confidence estimates, is generated in Line 34 and plotted in black dashed style in Line 35.

Finally, Lines 38–43 handle plot annotations. The title and axis labels are added for clarity, and the legend distinguishes between the empirical data, the fitted PDF, and the two confidence-bound PDFs. The use of `grid on` enhances visual readability. The resulting plot clearly demonstrates how uncertainty in the estimated parameters translates into a range of plausible normal distributions, offering a visual representation of model reliability. The output of this code is presented in Fig. 4.46, which visually illustrates the estimated normal distribution along with its 95% confidence bounds derived from the sample data.

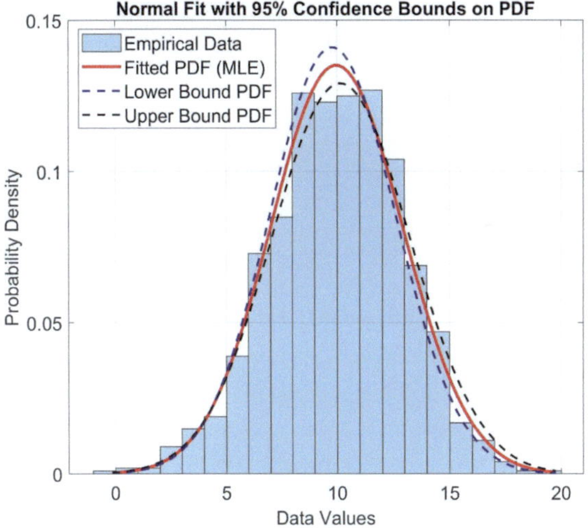

Fig. 4.46 Fitting the PDF of simulated normal data with 95% CIs

Code 4.18 Visualizing confidence intervals on the fitted normal distribution in MATLAB

```
        "Reject_Stat": reject_chi_stat,
        "p-value": [p_chi2] * 5,
        "Reject_P": reject_chi_pval
})

# ================== Output ==================
print("="*10 + " Kolmogorov-Smirnov Test (KST) " + "="*10)
print(T1.to_string(index=False))

print("\n" + "="*10 + " Anderson-Darling Test (ADT) " + "="*10)
print(T2.to_string(index=False))

print("\n" + "="*10 + " Chi-Squared Test (CST) " + "="*10)
print(T3.to_string(index=False))
```

```
clc
clear
close all

%% Step 1: Generate synthetic data
mu_true = 10;
sigma_true = 3;
data = normrnd(mu_true, sigma_true, [1, 1000]);

%% Step 2: Fit normal distribution
pd = fitdist(data', 'Normal');

%% Step 3: Get 95% confidence intervals
ci = paramci(pd); % Rows: [lower; upper], Columns: [mu, sigma]
mu_lower = ci(1,1); mu_upper = ci(2,1);
sigma_lower = ci(1,2); sigma_upper = ci(2,2);

%% Step 4: Plot histogram and fitted PDF
figure;
histogram(data, 'Normalization', 'pdf', 'FaceAlpha', 0.3); % Transparent histogram
hold on;

x = linspace(min(data), max(data), 100);

% Fitted PDF (MLE)
y_fit = pdf(pd, x);
plot(x, y_fit, 'r-', 'LineWidth', 2);

% Lower bound PDF
```

```
y_lower = normpdf(x, mu_lower, sigma_lower);
plot(x, y_lower, 'b--', 'LineWidth', 1.5);

% Upper bound PDF (now in black)
y_upper = normpdf(x, mu_upper, sigma_upper);
plot(x, y_upper, 'k--', 'LineWidth', 1.5); % 'k--' for black dashed line

%% Step 5: Annotate the plot
title('Normal Fit with 95% Confidence Bounds on PDF');
xlabel('Data Values');
ylabel('Probability Density');
legend('Empirical Data', 'Fitted PDF (MLE)', ...
'Lower Bound PDF', 'Upper Bound PDF');
grid on;
```

Code 4.18 demonstrates how to fit a normal distribution to a synthetic dataset in Python, compute the 95% confidence intervals for the estimated parameters, and visualize the resulting distribution with associated uncertainty bounds. This approach mirrors the analysis presented previously in MATLAB, offering a Python-based workflow that is equally accessible for statistical modeling and data visualization.

The script begins by importing essential Python libraries (Lines 1–3). These include `NumPy` for numerical operations, `Matplotlib` for visualization, and `SciPy`'s statistical functions from `scipy.stats`, particularly the `norm` distribution used for probability density calculations.

In Lines 6–8, synthetic data is generated by drawing 1,000 random samples from a normal distribution with a true mean (`mu_true`) of 10 and a standard deviation (`sigma_true`) of 3. This dataset simulates real-world measurements that follow a Gaussian pattern. Once the data is generated, Lines 11–12 compute the MLE of the mean and standard deviation using the sample's arithmetic mean and unbiased standard deviation (`ddof=1` ensures the correct degrees of freedom for the estimate).

To assess the reliability of the estimated parameters, the code proceeds in Lines 15–17 to set up the parameters needed for computing confidence intervals. The sample size `n`, significance level `alpha = 0.05` (corresponding to 95% confidence), and the critical z-score are defined using the inverse cumulative distribution function (quantile function) of the standard normal distribution.

Lines 20–22 calculate the confidence interval for the mean. This is done by first computing the standard error of the mean (SEM) and then using the formula $\mu \pm$ z SEM to obtain the lower and upper bounds for the mean, denoted `mu_lower` and `mu_upper`. For smaller sample sizes, the critical value z should be replaced by the corresponding Student's t value.

The confidence interval for the standard deviation is addressed in Lines 25–29. Because standard deviation estimates follow a chi-squared distribution rather than a normal one, the script uses the `chi2.ppf` function to find the appropriate quantiles and then derives the lower and upper bounds (`sigma_lower` and `sigma_upper`) for the standard deviation using the chi-squared formula.

Visualization begins in Line 32, where a new figure window is created with sufficient space. A normalized histogram of the data is plotted in Line 33 using 30 bins with partial transparency (`alpha=0.3`), allowing overlaid probability curves to be clearly visible. In Line 35, a range of `x` values is generated across the minimum and maximum of the data for evaluating the fitted and boundary PDFs.

The fitted normal PDF using the estimated mean and standard deviation is computed in Line 38 and plotted as a red solid line in Line 39. To visualize the uncertainty associated with the fit, the script also evaluates and plots two additional PDFs: the lower bound curve (Lines 42–43), using `mu_lower` and `sigma_lower`, is plotted as a blue dashed line, while the upper bound curve (Lines 46–47), using `mu_upper` and `sigma_upper`, appears as a black dashed line. These two curves enclose the likely shape of the true distribution given the variability in parameter estimation.

Finally, Lines 50–56 add necessary plot annotations, including a title, axis labels, and a legend to distinguish each element in the figure. The `grid(True)` function improves readability, and `tight_layout()` ensures that all labels and annotations fit cleanly within the figure boundaries. The visualization is then displayed using `plt.show()`.

It is important to note that although the structure and purpose of this code mirror those of the MATLAB implementation presented earlier in Code 4.17, the numerical values and resulting curves may differ slightly. This discrepancy is expected because the samples are randomly generated in each execution, and unless both environments use the same random seed and algorithm, the datasets will not match exactly.

Code 4.19 Fitting and plotting a normal distribution with 95% confidence bounds in Python

```
1  import numpy as np
2  import matplotlib.pyplot as plt
3  from scipy.stats import norm
4
5  # Step 1: Generate synthetic data
6  mu_true = 10
7  sigma_true = 3
8  data = np.random.normal(mu_true, sigma_true, 1000)
```

```
# Step 2: Estimate parameters (MLE)
mu_hat = np.mean(data)
sigma_hat = np.std(data, ddof=1)  # Sample std dev (unbiased)

# Step 3: Compute 95% confidence intervals
n = len(data)
alpha = 0.05
z = norm.ppf(1 - alpha / 2)  # 1.96 for 95%

# Confidence interval for the mean
sem = sigma_hat / np.sqrt(n)
mu_lower = mu_hat - z * sem
mu_upper = mu_hat + z * sem

# Confidence interval for the standard deviation using chi-square approximation
from scipy.stats import chi2
chi2_lower = chi2.ppf(alpha / 2, df=n-1)
chi2_upper = chi2.ppf(1 - alpha / 2, df=n-1)
sigma_lower = np.sqrt((n - 1) * sigma_hat**2 / chi2_upper)
sigma_upper = np.sqrt((n - 1) * sigma_hat**2 / chi2_lower)

# Step 4: Plot histogram and PDFs
plt.figure(figsize=(10, 6))
plt.hist(data, bins=30, density=True, alpha=0.3, label='Empirical Data')

x = np.linspace(min(data), max(data), 100)

# Fitted PDF
y_fit = norm.pdf(x, mu_hat, sigma_hat)
plt.plot(x, y_fit, 'r-', linewidth=2, label='Fitted PDF (MLE)')

# Lower bound PDF
y_lower = norm.pdf(x, mu_lower, sigma_lower)
plt.plot(x, y_lower, 'b--', linewidth=1.5, label='Lower Bound PDF')

# Upper bound PDF
y_upper = norm.pdf(x, mu_upper, sigma_upper)
plt.plot(x, y_upper, 'k--', linewidth=1.5, label='Upper Bound PDF')

# Step 5: Annotate
plt.title('Normal Fit with 95% Confidence Bounds on PDF')
plt.xlabel('Data Values')
```

4.5.1.4 Applications of the Normal Distribution in Real-World Contexts

Probability of Exceeding the Average Annual Rainfall in Quebec

Given that the average annual rainfall in Quebec is 1150 mm with a standard deviation of 150 mm, what is the probability that a year will have more than 1380 mm of rainfall?

Solution

To determine the probability that the average annual rainfall in Quebec exceeds 1380 mm, we assume that annual rainfall totals can be reasonably approximated by a normal distribution with a mean $\mu = 1150$ mm and a standard deviation $\sigma = 150$ mm. This assumption is commonly adopted for annual or seasonal rainfall totals, as they represent the aggregation of many individual precipitation events over time, which often results in an approximately symmetric distribution. In practice, this assumption should be verified using historical rainfall data before being applied in design or risk analysis.

Step 1: Standardize the Observation

Since the value of interest (1380 mm) is not at a clear multiple of σ from the mean, we must first standardize it using the *z-score formula*:

$$z = \frac{X - \mu}{\sigma} = \frac{1380 - 1150}{150} = 1.53$$

This means that 1380 mm is 1.53 standard deviations above the mean.

Step 2: Use the Standard Normal Table

Once we have the z-score, we use the Standard Normal Table (Z-table) (Table 4.6) to find the cumulative probability corresponding to z=1.53. Using the Z-table, the area under the standard normal curve to the left of $z = 1.53$ is:

$$P(Z < 1.53) = 0.9370$$

This indicates that there is a 93.7% chance that rainfall will be less than or equal to 1380 mm.

Step 3: Compute the Desired Probability

However, since the problem asks for the probability of rainfall exceeding 1380 mm, we want the area in the upper tail (i.e., to the right of the z-score). Using the complement rule:

$$P(X > 1380) = 1 - P(Z < 1.53) = 1 - 0.9370 = 0.0630$$

Thus, there is approximately a 6.3% probability that the average annual rainfall in Quebec will exceed 1380 mm.

Determining the Required Strength for a Concrete Mix Design

Calculate the required average strength for a concrete mix design if the specified compressive strength (characteristic strength) is 25 MPa. Assume sufficient data exist for which a standard deviation of 5.61 MPa has been calculated, and the required degree of confidence is 90% (i.e., 10% of results are allowed to fall below the target average strength).

Solution

In this problem, we are provided with the specified compressive strength (often referred to as the characteristic strength) $f_{ck} = 25$ MPa, a standard deviation $\sigma = 5.61$ MPa, and a confidence level of 90%, which implies that only 10% of the compressive strengths can fall below our average target value. This level of confidence corresponds to a cumulative probability of 0.90 on the standard normal distribution curve.

From the standard normal (z) table (Table 4.6), we look up the z-score associated with a cumulative probability of 0.9000, which is found to be a value of 1.28.

Now, using the standardization (z-score) formula, we can calculate the required average strength X:

$$z = \frac{X - \mu}{\sigma} \quad \text{(where } \mu = \text{target strength)}$$

$$1.28 = \frac{X - 25}{5.61} \Rightarrow X = 1.28 \times 5.61 + 25 = 32.2\,\text{MPa}$$

Therefore, for a mix in which we have a known variability (reported as the standard deviation) of 5.61 MPa and for which we must have a compressive strength of 25 MPA with 90% confidence, the mix design should have a specified strength of 32.2 MPa.

Introduction to the NORMDIST Function in Excel

In Excel, the 'NORMDIST' function is designed to return the normal distribution for a specified mean and standard deviation. That is, it calculates either the normal PDF or the CDF for a given set of parameters. This function is highly useful in statistical analysis for modeling data that follows a bell-shaped curve, such as heights, test scores, or measurement errors.

Syntax: '=NORMDIST(x, mean, standard_dev, cumulative)'

- **x**: The value for which you want the distribution.
- **mean**: The arithmetic mean of the distribution.
- **standard_dev**: The standard deviation of the distribution, which must be positive.

- **cumulative**: A required argument that is a logical value. It specifies the type of distribution to be used:

  TRUE: Indicates a CDF is considered. This returns the probability that a random variable is less than or equal to 'x'.
  FALSE: Indicates a Normal PDF is considered. This returns the value of the probability density at 'x'.

Example To find the cumulative probability for a value of 1380 with a mean of 1150 and a standard deviation of 150 (As we saw in our rainfall example 4.5.1.4.1), the syntax you would use is '=NORMDIST(1380, 1150, 150, TRUE)', which aligns with the z-score calculation ($z = (1380 - 1150)/150) = 1.53$) and a corresponding probability (e.g., 0.9370 as seen earlier).

NORMINV Function in Excel

The 'NORMINV' function is categorized under Excel's Statistical functions and is used to calculate the inverse of the normal cumulative distribution. Given a probability, mean, and standard deviation, it returns the value x such that the specified probability P follows $P(X \leq x)$. This is particularly useful for determining critical values or thresholds in hypothesis testing and confidence interval construction.

It is important to note that the mean and standard deviation supplied to NORMINV are treated as population parameters. If these values are estimated from sample data, the resulting quantiles should be interpreted as approximate, since the uncertainty associated with parameter estimation is not accounted for. In such cases, users should ensure that the appropriate form of the standard deviation is used and recognize the distinction between population-based calculations and sample-based inference.

Syntax: '=NORMINV(probability, mean, standard_dev)'

- **probability**: The probability corresponding to the normal distribution, a value between 0 and 1.
- **mean**: The arithmetic mean of the distribution.
- **standard_dev**: The standard deviation of the distribution, which must be positive. When working with sample data, the standard deviation is often estimated using Excel's `STDEV.S` function; however, this estimate is then treated as a population parameter when used in `NORMINV`.

Example To find the value X for which the cumulative probability is 0.9370 with a mean of 25 and a standard deviation of 5.61 (as in our concrete compressive strength example 4.5.1.4.2), you could use '=NORMINV(0.9370, 25, 5.61)', which would yield a value close to 32.2 when considering the z-score of 1.28 (as derived earlier).

Reliability Analysis of Traffic Lights

A city installs 2000 electric traffic lights. The functional life of these lights is normally distributed with a mean (μ) of 1000 h and a standard deviation (σ) of 200 h. Based on this information, answer the following questions:

(a) How many lights are expected to fail in the first 700 functional hours?
(b) How many lights are expected to fail between 900 and 1300 functional hours?
(c) What is the probability that a light will function between 899 and 901 h before failing?
(d) After how many hours would we expect 90% of the lights to still be functional?
(e) After how many hours would we expect only 10% of the lights to still be functional?

Solution

In this analysis, we apply the normal distribution to predict the reliability of a large number of electric traffic lights. Given a sample of 2000 lights with a normally distributed lifespan (mean = 1000 h, standard deviation = 200 h), we can calculate failure probabilities and expected counts using z-scores and cumulative distribution functions.

(a) **Expected failures before 700 h**

To determine how many lights will fail prior to 700 h, we calculate the z-score:

$$z = \frac{700 - 1000}{200} = -1.5$$

Using the standard normal table, we find (Table 4.5):

$$P(Z < -1.5) = 0.0668$$

Or using Excel:

$$\text{Pr(x} < 700) = \text{NORMDIST}(700, 1000, 200, \text{ TRUE}) = 0.0668$$

Thus, 6.68% of the lights will fail prior to 700 h. To find the number of failures, we multiply this probability by the population number:

$$2000 \times 0.0668 = 133.6 \approx 134 \quad \text{lights}$$

So, approximately 134 lights are expected to fail before reaching 700 h.

(b) **Expected failures between 900 and 1300 h**

First, compute z-scores:

$$z_1 = \frac{900 - 1000}{200} = -0.5, \quad z_2 = \frac{1300 - 1000}{200} = 1.5$$

From the table (Tables 4.5 and 4.6):

$$P(Z < 1.5) = 0.9332, \quad P(Z < -0.5) = 0.3085$$

So the probability:

$$P(900 < X < 1300) = 0.9332 - 0.3085 = 0.6247$$

Or using Excel:

$$\begin{aligned} &= \text{NORMDIST}(1300, 1000, 200, \ \text{TRUE}) \\ &- \text{NORMDIST}(900, 1000, 200, \ \text{TRUE}) = \underline{0.6247} \end{aligned}$$

Multiplying by 2000 lights:

$$2000 \times 0.6247 = 1249.4 \approx 1249 \quad \text{lights}$$

Thus, approximately 1249 lights will fail in that interval.

(c) **Probability of failing between 899 and 901 h**

Calculate the z-scores

$$z_1 = \frac{899 - 1000}{200} = -0.505, \quad z_2 = \frac{901 - 1000}{200} = -0.495$$

From the normal table (Table 4.5):

$$\begin{aligned} &P(Z < -0.495) = 0.3103, \\ &P(Z < -0.495) = 0.3103, \quad P(Z < -0.505) = 0.3067 \\ &P(899 < X < 901) = 0.3103 - 0.3067 = 0.0035 \end{aligned}$$

Or using Excel:

$$\begin{aligned} &= \text{NORMDIST}(901, 1000, 200, \ \text{TRUE}) \\ &- \text{NORMDIST}(899, 1000, 200, \ \text{TRUE}) = 0.0035 \end{aligned}$$

So,

$$2000 \times 0.0035 = 7 \quad \text{lights}$$

Only about 7 lights will fail within that 2-hour window.

(d) **Time when 90% of the lights still work**

We look for the time X such that:

$$P(X > x) = 0.9 \Rightarrow P(X < x) = 0.1$$

From the z-table (Table 4.5), the z-score for 0.1 is:

$$z = -1.28$$

Now solve for X:

$$-1.28 = \frac{X - 1000}{200} \Rightarrow X = -1.28 \times 200 + 1000 = 744\,\text{h}$$

Or using Excel:

$$= \text{NORMINV}(0.10, 1000, 200) = 743.6897$$

So, after about 744 hours, 90% of the lights will still be functional.

(e) **Time when only 10% of the lights still work**

Here, P(X < x) = 0.9 and we find the corresponding z-score:

$$z = 1.28$$

Solving for X:

$$1.28 = \frac{X - 1000}{200} \Rightarrow X = 1.28 \times 200 + 1000 = 1256\,\text{h}$$

Or using Excel:

$$= \text{NORMINV}(0.90, 1000, 200) = 1256.31$$

So, after 1256 h, only 10% of the lights will still be operational.

Vehicular Accident Probability Analysis for the City of Ottawa

Statistical data of vehicular accidents in Ottawa show that the annual vehicle miles (i.e., miles per vehicle per year) driven between traffic accidents (of all severities) can be represented by a normal random variable with a mean of 15,000 miles per year and a coefficient of variation of 25%.

(a) For a typical driver who drives 10,000 miles per year, what is the probability of having an accident in a year?
(b) If the driver has driven 8000 miles in a given year without encountering any accident, what is the probability of having an accident for the remainder of that year?

Solution

First, we should calculate the standard deviation:

$$\sigma = \text{COV} \times \mu = 0.25 \times 15,000 = 3,750 \text{ miles}$$

So the distribution becomes:

$$X \sim N(15,000,\ 3,750^2)$$

(a) **Probability of an accident for a 10,000-mile driver**

We want:

$$P(X < 10,000)$$

Convert to Z-score:

$$z = \frac{10,000 - 15,000}{3,750} = -1.33$$

Using Z-table:

$$P(X < 10,000) = P(Z < -1.33) \approx 0.092$$

This means there's a 9.2% chance the driver will have an accident within 10,000 miles.

The probability of not having an accident = 1−0.092=0.9081 or 90.8%.

(b) **The probability that the driver has an accident after reaching 8,000 miles accident-free**

We want:

$$P(X < 10,000 | X > 8,000)$$

This problem can be resolved using the relationships for conditional probability introduced in Chapter:

$$P(A|B) = \frac{P(A \cap B)}{P(B)}$$

Which becomes:

$$P(8,000 < X < 10,000|X > 8,000) = \frac{P(8,000 < X < 10,000)}{P(X > 8,000)}$$

Now calculate Z-scores:

$$z_1 = \frac{10,000 - 15,000}{3,750} = -1.33, \quad z_2 = \frac{8,000 - 15,000}{3,750} = -1.87$$

From Z-table:

$$P(X < 10,000) = 0.092, \quad P(X < 8,000) = \Phi(-1.87) = 0.0308$$

Thus:

$$P(8,000 < X < 10,000) = 0.092 - 0.0308 = 0.0612$$
$$P(X > 8,000) = 1 - 0.0308 = 0.9692$$

Final conditional probability:

$$P(X < 10,000|X > 8,000) = \frac{0.0612}{0.9692} \approx 0.063.$$

So there's a 6.3% chance the driver will have an accident in the remainder of the year. Conversely, there's a 93.7% chance of driving accident-free beyond 8,000 miles.

4.5.2 Log-Normal Probability Distribution

The log-normal distribution is a commonly used probability distribution in engineering, particularly in situations where the values of the variable of interest are strictly positive. It arises naturally when a normally distributed variable is exponentiated, and it is especially helpful when modeling phenomena with right-skewed behavior.

Figure 4.47 visually compares the Normal distribution and the Lognormal distribution. The Normal distribution is symmetric and defined by the well-known probability density function:

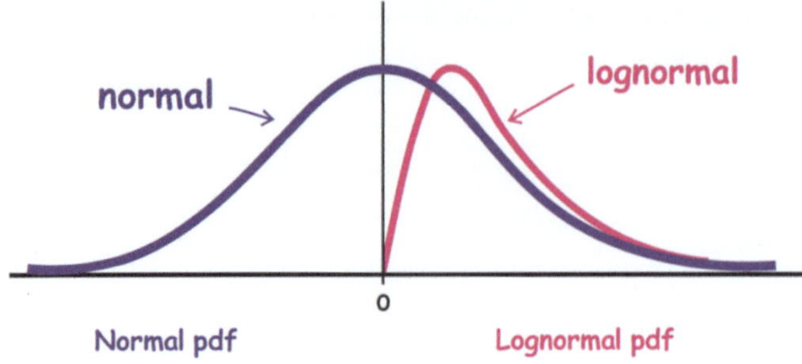

Fig. 4.47 Comparison of normal and lognormal probability density functions

$$f(x) = \frac{1}{\sigma\sqrt{2\pi}} e^{-\frac{1}{2}\left(\frac{x-\mu}{\sigma}\right)^2} \tag{4.34}$$

where μ and σ are the mean and standard deviation, respectively. In contrast to the symmetry seen with the normal distribution, the Lognormal distribution is asymmetric and positively skewed (i.e., tails to the right). Its probability density function (PDF) is defined for x>0 as:

$$f(x) = \frac{1}{x\sigma_{\ln(x)}\sqrt{2\pi}} e^{-\frac{1}{2}\left(\frac{\ln(x)-\mu_{\ln(x)}}{\sigma_{\ln(x)}}\right)^2} \tag{4.35}$$

where $\mu_{\ln(x)}$ and $\sigma_{\ln(x)}$ are the mean and standard deviation of the population after having been transformed using the natural logarithm. These formulas can be seen in Fig. 4.47, where the purple curve represents the normal distribution and the magenta curve represents the lognormal distribution.

The lognormal distribution possesses several key characteristics that distinguish it from other probability distributions. First and foremost, its domain is strictly positive, meaning the distribution is only defined for values where x > 0. This makes it particularly useful in modeling phenomena where negative values are not physically meaningful, such as income, rainfall, or fatigue life, which represents the time or number of stress cycles until a material or component fails under repeated loading.

The distribution is characterized by two parameters: $\mu_{\ln(x)}$ and $\sigma_{\ln(x)}$. These represent the mean and standard deviation of the natural logarithm of the variable, not the variable itself. In other words, if a random variable X is lognormally distributed, then ln(X) is normally distributed with mean $\mu_{\ln(x)}$ and standard deviation $\sigma_{\ln(x)}$.

In terms of shape, the log-normal distribution is right-skewed, with a long tail stretching towards higher values. Unlike the symmetric bell curve of the normal distribution, the log-normal distribution displays a steep ascent on the left side and a gradual decline on the right. This asymmetry makes it especially suitable for modeling real-world data that cluster near a lower bound but occasionally produces very large values.

The expression for the lognormal distribution of a random variable X (Eq. 4.35) can be reformulated using two key parameters: $\xi = \sigma_{\ln(x)}$, which represents the standard deviation of the natural logarithm of X, and $\lambda = \mu_{\ln(x)}$, the mean of the natural logarithm of X. These parameters help simplify the representation of the lognormal distribution and facilitate its integration over a defined interval.

To compute the probability that the random variable X falls within an interval [a, b], we perform an integration of the probability density function from a to b. The integral is expressed as:

$$P(a \leq X \leq b) = \int_a^b \frac{1}{x\xi\sqrt{2\pi}} \exp\left(-\frac{1}{2}\left(\frac{\ln(x) - \lambda}{\xi}\right)^2\right) dx \tag{4.36}$$

Resolving this integral leads to a simplified result in terms of the CDF, denoted as φ, of the standard normal distribution. The closed-form solution for the probability becomes:

$$P(a \leq X \leq b) = \varphi\left(\frac{\ln(b) - \lambda}{\xi}\right) - \varphi\left(\frac{\ln(a) - \lambda}{\xi}\right) \tag{4.37}$$

One of the most accessible ways to evaluate the lognormal distribution is through Excel's built-in function:

= **LOGNORM.DIST**(**x**, **mean**, **standard_dev**, **cumulative**)

This function computes either the PDF or the CDF of the lognormal distribution. The parameter `x` refers to the value at which the function is evaluated. The `mean` and `standard_dev` refer to the mean and standard deviation of the natural logarithm of the variable (i.e., ln(x)), not of x itself. The fourth parameter `cumulative` is a Boolean value—set to `TRUE` to calculate the CDF, or `FALSE` to return the PDF.

Excel evaluates the lognormal distribution assuming that ln(x) is normally distributed with the given parameters. This makes it an efficient tool for quickly computing probabilities and visualizing the behavior of positively skewed data, thereby supporting more informed engineering and design decisions.

In the context of the lognormal distribution, the parameters λ and ξ are derived from the mean μ_X and standard deviation σ_X of the original random variable X, not from the log-transformed values. The probability that X falls within a specific interval [a, b] is computed using the standard normal cumulative distribution function φ (Eq. 4.37).

To determine the shape of the lognormal distribution, we must express its parameters in terms of the mean and standard deviation of the original variable X. The parameter ξ^2 is calculated as:

$$\xi^2 = \ln\left(1 + \left(\frac{\sigma_X}{\mu_X}\right)^2\right) = \ln\left[1 + (\text{CoV})^2\right] \tag{4.38}$$

This tells us that ξ^2 is equal to the natural logarithm of one plus the square of the standard deviation divided by the mean, which is equivalent to the natural logarithm of one plus the coefficient of variation squared.

Once ξ^2 is known, the parameter λ can be found using:

$$\lambda = \ln \mu_X - \frac{1}{2}\xi^2 \tag{4.39}$$

That is, λ is equal to the natural logarithm of the mean of X minus one-half of ξ squared. These transformations allow us to fully define the lognormal distribution from basic descriptive statistics.

Since the lognormal distribution is not symmetric like the normal distribution, the median is often preferred over the mean to describe its central value. By definition, the median x_m of a random variable X following a lognormal distribution is the value at which there is a 50% chance that X is less than or equal to x_m. Mathematically, this is expressed as $P(X \leq x_m) = 0.5$.

To relate this to the parameters of the lognormal distribution, we use the CDF, φ, of the standard normal distribution. This gives us:

$$\varphi\left(\frac{\ln(x_m) - \lambda}{\xi}\right) = 0.5. \tag{4.40}$$

Taking the inverse of the standard normal CDF on both sides isolates the expression:

$$\frac{\ln(x_m) - \lambda}{\xi} = \varphi^{-1}(0.5) = 0. \tag{4.41}$$

Since the inverse CDF of 0.5 is zero (the mean of the standard normal distribution), it follows that:

$$\ln(x_m) - \lambda = 0 \tag{4.42}$$

Therefore, we conclude that the parameter λ, which represents the mean of the natural logarithm of the variable X, is equal to the natural logarithm of the median:

$$\lambda = \ln(x_m) \tag{4.43}$$

This result provides a convenient way to estimate λ directly from the median value, which is especially helpful in real-world scenarios where the median can be readily calculated from the dataset.

4.5.2.1 Code-Based Explanation of the Lognormal Distribution

Lognormal Distribution: PDF, CDF, and Inverse CDF Visualization

Code 4.20 demonstrates how to compute and visualize the PDF, CDF, and inverse CDF (also called the quantile function) of a log-normal distribution using MATLAB. All three plots are presented in a single figure using subplots, making this script a valuable educational tool for students learning about probability distributions and statistical modeling.

The script begins in Line 1 by resetting the MATLAB environment. The `clc` command clears the command window, `clear` removes all variables from memory, and `close all` closes any previously opened figure windows. Including these commands at the beginning of every MATLAB code is good practice as it ensures that the script starts anew without leftover data or figures that could affect the output.

In Lines 3 to 5, the parameters of the lognormal distribution are defined. The variables `mu` and `sigma` represent the mean and standard deviation of the underlying normal distribution from which the lognormal distribution is derived. These values are used to shape the distribution: in this case, `mu = 1` and `sigma = 0.5` create a moderately skewed distribution that only applies to positive values, as the log-normal distribution is not defined for values ≤ 0.

Lines 7 to 10 define the range of values needed to compute and plot the functions. The vector `x` is created using `linspace` to generate 500 evenly spaced points between 0.01 and 10. The starting value is set to 0.01 to avoid the undefined logarithm of zero. This range of values will be used for calculating the PDF and CDF. A second vector `p` is defined to represent probabilities from 0.01 to 0.99, which are used to compute the inverse CDF—i.e., to determine the data value associated with a given probability.

The calculations for the three functions are performed in Lines 12 to 15. The function `lognpdf` (Line 13) is used to calculate the PDF values, which represent the likelihood of different outcomes in the distribution. The `logncdf` function (Line 14) computes the CDF, which accumulates the total probability up to each point in `x`. The `logninv` function (Line 15) calculates the inverse CDF, which maps each probability value from `p` to a corresponding quantile in the lognormal distribution. These MATLAB functions are built-in and specifically designed for this distribution.

The visual representation of the three distributions begins in Line 17 with the creation of a new figure window. It should be first mentioned that the figures will be constructed in the form of three stacked subplots. To do so, the built-in `subplot(m,n,p)` command is used (see Lines 21,28,35). The `subplot` command builds a `mxn` grid to display the figures and then places the plot in question in position `p`. Lines 20 to 25 draw the first subplot, which displays the PDF using a smooth blue curve. The input arguments for this function include the x and y values, which are the values that define the PDF (`x` and `pdf_vals`, respectively), as well as a sequence of string scalar values that are used to customize the style of the plot. Here, `'b'` is used to specify the color of the curve, `'-'` is added after the `'b'` to indicate a solid/smooth line style for the curve, and `'LineWidth'` is used to specify the line thickness. The title (Line 23) and y-axis label (Line 24) help interpret the graph as a density function. The grid lines assist in reading the values more accurately.

Lines 27 to 32 create the second subplot to display the CDF. Again, MATLAB's `plot` function is used; however, this time the color of the curve is set to green using '`g`' as the input argument. This plot shows how probability accumulates across the range of `x`. For instance, the curve shows that a large portion of the probability mass lies within a certain interval—important for risk assessment and threshold analysis.

Finally, Lines 34 to 40 generate the third subplot, which visualizes the inverse CDF. To differentiate from the other two curves, the color is set to red by modifying the `plot` input argument to '`r`'. This function allows users to determine the value of the random variable associated with a given probability. For example, if a user wants to find the 95th percentile of the distribution, they can use this curve to read off the corresponding quantile.

Overall, this script offers a comprehensive and interactive way to understand how data behaves under a lognormal distribution, helping beginners build both coding skills and statistical intuition through MATLAB's visualization tools. The output of this code is shown in Fig. 4.48.

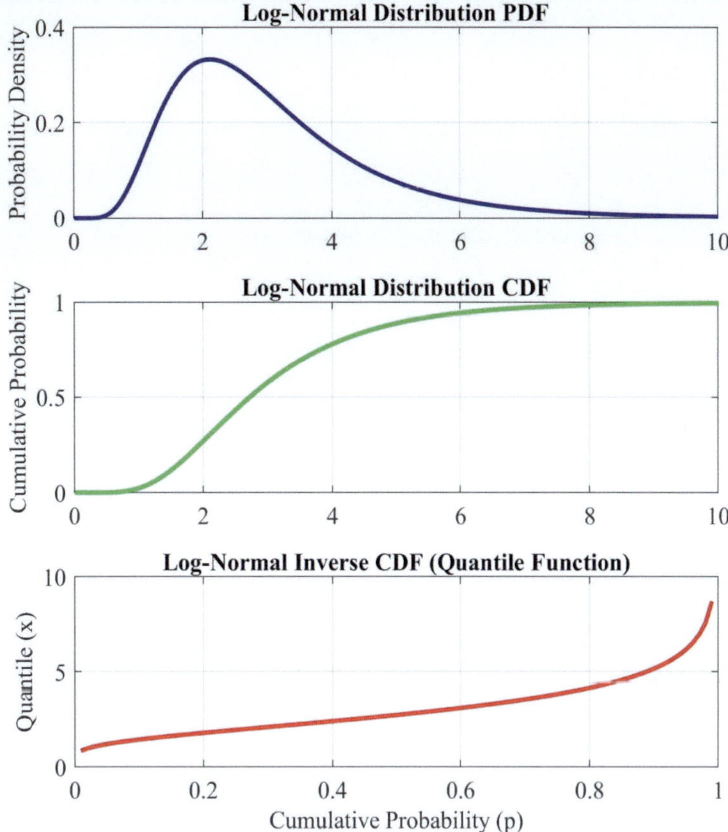

Fig. 4.48 Visualization of a lognormal distribution (PDF, CDF, & Inverse CDF)

Code 4.20 MATLAB script for calculating and plotting the PDF, CDF, and inverse CDF (quantile function) of a lognormal distribution using subplots in a single figure

```
plt.ylabel('Probability Density')
plt.legend()
plt.grid(True)
plt.tight_layout()
plt.show()
```

```
clc; clear; close all;

%% Step 1: Define parameters for the log-normal distribution
mu = 1; % Mean of the underlying normal distribution (log-space)
sigma = 0.5; % Standard deviation of the underlying normal distribution

% Range of x-values for PDF and CDF
x = linspace(0.01, 10, 500); % Avoid 0 to prevent log(0)
% Range of probability values for inverse CDF
p = linspace(0.01, 0.99, 100);

%% Step 2: Calculate PDF, CDF, and Inverse CDF values
pdf_vals = lognpdf(x, mu, sigma); % PDF
cdf_vals = logncdf(x, mu, sigma); % CDF
inv_vals = logninv(p, mu, sigma); % Inverse CDF (Quantile Function)

%% Step 3: Plot all three in one figure using subplots
figure;

% Subplot 1: PDF
subplot(3,1,1);
plot(x, pdf_vals, 'b-', 'LineWidth', 2);
title('Log-Normal Distribution PDF');
ylabel('Probability Density');
grid on;

% Subplot 2: CDF
subplot(3,1,2);
plot(x, cdf_vals, 'g-', 'LineWidth', 2);
title('Log-Normal Distribution CDF');
ylabel('Cumulative Probability');
grid on;

% Subplot 3: Inverse CDF (Quantile Function)
subplot(3,1,3);
plot(p, inv_vals, 'r-', 'LineWidth', 2);
title('Log-Normal Inverse CDF (Quantile Function)');
xlabel('Cumulative Probability (p)');
ylabel('Quantile (x)');
grid on;
```

```
1 import numpy as np
2 import matplotlib.pyplot as plt
3 from scipy.stats import lognorm
4
5 # Step 1: Define parameters for the log-normal distribution
6 mu = 1          # Mean of the underlying normal distribution (log-space)
7 sigma = 0.5      # Standard deviation of the underlying normal distribution
8
9 # In SciPy, lognorm takes 's' (shape) as sigma, and scale = exp(mu)
10 s = sigma
11 scale = np.exp(mu)
12
13 # Range of x-values for PDF and CDF
14 x = np.linspace(0.01, 10, 500)  # Avoid 0 to prevent log(0)
15 # Range of cumulative probabilities for inverse CDF
16 p = np.linspace(0.01, 0.99, 100)
17
18 # Step 2: Calculate PDF, CDF, and Inverse CDF
19 pdf_vals = lognorm.pdf(x, s, scale=scale)
20 cdf_vals = lognorm.cdf(x, s, scale=scale)
21 inv_vals = lognorm.ppf(p, s, scale=scale)  # Percent point function = inverse CDF
22
23 # Step 3: Plot all three in one figure using subplots
24 fig, axs = plt.subplots(3, 1, figsize=(8, 10))
25
26 # Subplot 1: PDF
27 axs[0].plot(x, pdf_vals, 'b-', linewidth=2)
28 axs[0].set_title('Log-Normal Distribution PDF')
29 axs[0].set_ylabel('Probability Density')
30 axs[0].grid(True)
31
```

Code 4.21 repeats this exercise; however, this time uses the Python environment to perform the required operations and figure generation. The Python script begins in Lines 1 to 3 by importing the necessary libraries. `NumPy` is imported as `np` to handle numerical operations and create sequences of data. `Matplotlib.pyplot` is imported as `plt` for plotting graphs—very similar to how plotting works in MATLAB. `lognorm` from `scipy.stats` is imported to access built-in functions for the lognormal distribution, including its PDF, CDF, and inverse CDF.

In Lines 5 to 7, the script defines the key parameters of the lognormal distribution. The variable `mu` represents the mean of the underlying normal distribution (in log-space), and `sigma` represents its standard deviation. These two values define the shape of the lognormal distribution, which is a continuous probability distribution for a random variable whose logarithm is normally distributed.

Lines 9 to 11 explain a crucial detail for using the `lognorm` functions in SciPy. Unlike MATLAB, SciPy parameterizes the lognormal distribution by its shape parameters, that is, `sigma` and a scale parameter, which must be set to `exp(mu)`. This transformation ensures that the distribution matches the intended mean and spread in log-space.

Next, in Lines 13 to 16, the code sets up the range of values over which the distribution functions will be evaluated. The array `x` includes 500 values ranging from 0.01 to 10, carefully excluding 0 to avoid issues with `log(0)`. These `x` values are used for plotting both the PDF and CDF. A separate array `p` is created to represent cumulative probabilities ranging from 0.01 to 0.99; this is used for evaluating the inverse CDF.

Lines 18 to 21 carry out the actual computations. The script calculates the PDF using `lognorm.pdf()`, which describes the likelihood of different outcomes. The CDF is computed using `lognorm.cdf()`, which accumulates probabilities from the lower tail of the distribution up to each point in `x`. Finally, the inverse CDF (also called the quantile function) is computed using `lognorm.ppf()`, which determines the value `x` that corresponds to a given probability `p`.

The results are plotted in Lines 23 to 43 using `plt.subplots()` to create a figure with three vertical subplots. In Lines 27 to 30, the first subplot visualizes the PDF as a blue curve, titled "Log-Normal Distribution PDF" and labeled on the y-axis as "Probability Density." Grid lines are added to enhance readability. Lines 33 to 36 plot the CDF as a green curve in the second subplot, labeled "Cumulative Probability." This graph shows how the probability accumulates from left to right along the distribution. The third subplot, defined in Lines 39 to 43, shows the inverse CDF as a red curve. It maps each cumulative probability `p` to its corresponding quantile value `x`, which is particularly useful for percentile analysis.

Finally, Line 45 calls `plt.tight_layout()` to automatically adjust subplot spacing for a cleaner layout. Line 46 calls `plt.show()` to display the complete figure on the screen. It should be mentioned that the output of this Python script is visually identical to the plots generated by the MATLAB-based code (Fig. 4.48).

Code 4.21 Python script for calculating and plotting the PDF, CDF, and inverse CDF (quantile function) of a log-normal distribution using subplots in a single figure.

```
# Subplot 2: CDF
axs[1].plot(x, cdf_vals, 'g-', linewidth=2)
axs[1].set_title('Log-Normal Distribution CDF')
axs[1].set_ylabel('Cumulative Probability')
axs[1].grid(True)

# Subplot 3: Inverse CDF
axs[2].plot(p, inv_vals, 'r-', linewidth=2)
axs[2].set_title('Log-Normal Inverse CDF (Quantile Function)')
axs[2].set_xlabel('Cumulative Probability (p)')
axs[2].set_ylabel('Quantile (x)')
axs[2].grid(True)

plt.tight_layout()
plt.show()
```

Fitting a Lognormal Distribution with Confidence Bounds

Code 4.22 shows how to fit a lognormal distribution to synthetic data and compute 95% confidence bounds on the estimated parameters and corresponding PDFs. The script is structured into five logical steps, each clearly marked in the code. The script begins in Line 1 by clearing the workspace, command window, and any open figure windows using `clc`, `clear`, and `close all`. This ensures a clean working environment.

In Lines 3 to 6, synthetic data is generated using the `lognrnd()` function, which draws random values from a lognormal distribution. The variables `mu_true` = 2 and `sigma_true` = `0.5` (Lines 4 and 5) define the mean and standard deviation of the underlying normal distribution of the log-transformed data. Then, Line 6 generates a row vector of 1000 samples, simulating a realistic dataset (e.g., environmental measurements or material properties) for statistical analysis.

Lines 8 to 13 perform the model fitting. In Line 9, the function `fitdist()` is used to fit a lognormal distribution to the data using MLE in the same way as was seen earlier for normally distributed data (Code 4.18). The data is transposed (`data'`) to ensure it is a column vector, as required by the function. Lines 12 and 13 compute the estimate for the mean, `mu_hat`, and standard deviation, `sigma_hat`, of the log-transformed data, `log(data)`.

In Lines 15 to 18, the script calculates the 95% confidence intervals for the fitted parameters. Line 16 uses the `paramci()` function, which returns the lower and upper bounds of the confidence intervals for each parameter of the distribution under consideration. These values are separated into `mu_lower`, `mu_upper`, `sigma_lower`, and `sigma_upper` in Lines 17 and 18. These intervals help quantify the

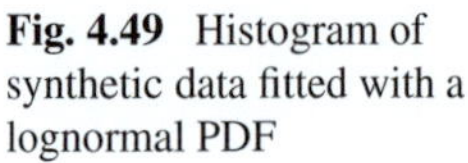

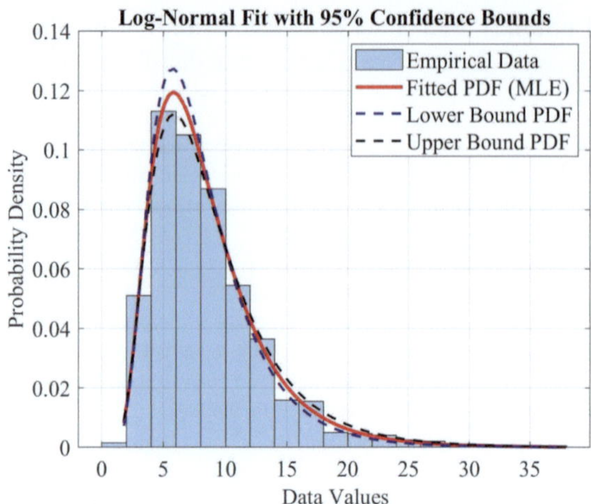

Fig. 4.49 Histogram of synthetic data fitted with a lognormal PDF

uncertainty of the estimated parameters, providing a statistical range where the true value is likely to lie.

Lines 20 to 22 display the estimated parameters and their confidence intervals using `fprintf()`, which displays the CI results in the command window for reporting or validation purposes. Note that in the syntax for the input argument, "`%4.4f`" is used, which specifies that the output is a floating-point number with 4 digits which are all after the decimal point. The visual representation of the fit and bounds is constructed in lines 24 to 41. Line 25 starts a new figure. Line 26 plots a normalized histogram of the data using `histogram()` with `'Normalization', 'pdf'`, so it can be overlaid with the fitted probability density function. The `FaceAlpha` property is set to 0.3 to make the histogram semi-transparent. Line 27 enables multiple plotting commands on the same figure with `hold on`. In Line 29, a linearly spaced vector `x` is generated over the range of the data to use as the input for evaluating the PDF curves.

Lines 32 to 33 plot the fitted lognormal PDF using `lognpdf()` with the estimated parameters. The resulting curve is shown in solid red using the '`r-`' input argument in the `plot` function in Line 33. In Lines 36 to 37, the lower bound PDF is computed using the lower confidence interval values as input to `lognpdf()` and plotted as a blue dashed line ('`b- -`'). Similarly, Lines 40 to 41 plot the upper bound PDF using the upper confidence interval values, shown in a black dashed line.

Finally, Lines 43 to 50 are used to annotate the figure. The title and axis labels are added in Lines 44 to 46 using MATLAB's built-in `title`, `xlabel`, and `ylabel` functions to describe what is being visualized in the figure. Lines 47 to 49 use `legend()` to identify each plot element (empirical data, MLE fit, lower and upper bounds), and place the legend box in the northeast corner. Line 50 activates grid lines to enhance readability. The output of this code is provided as Fig. 4.49.

Code 4.22 Fitting a Log-normal distribution with confidence bounds in MATLAB

```
clc; clear; close all;

%% Step 1: Generate synthetic data from a log-normal distribution
mu_true = 2; % True mean of log(data)
sigma_true = 0.5; % True standard deviation of log(data)
data = lognrnd(mu_true, sigma_true, [1, 1000]); % Generate 1000 samples

%% Step 2: Fit a log-normal distribution using Maximum Likelihood Estimation (MLE)
pd = fitdist(data', 'Lognormal'); % Fit requires column vector

% Extract estimated parameters
mu_hat = pd.mu; % Mean of log(data)
sigma_hat = pd.sigma; % Std dev of log(data)

%% Step 3: Compute 95% confidence intervals for mu and sigma
ci = paramci(pd); % 95% confidence intervals from MLE
mu_lower = ci(1,1); mu_upper = ci(2,1);
sigma_lower = ci(1,2); sigma_upper = ci(2,2);

% Display results
fprintf('Estimated mu: %.4f (95%% CI: [%.4f, %.4f])\n', mu_hat, mu_lower, mu_upper);
fprintf('Estimated sigma: %.4f (95%% CI: [%.4f, %.4f])\n' sigma_lower, sigma_upper);

%% Step 4: Plot histogram and fitted PDF with confidence bounds
figure;
histogram(data, 'Normalization', 'pdf', 'FaceAlpha', 0.3); % Transparent histogram
hold on;

x = linspace(min(data), max(data), 100);

% Fitted PDF (MLE)
y_fit = lognpdf(x, mu_hat, sigma_hat);
plot(x, y_fit, 'r-', 'LineWidth', 2);

% Lower bound PDF
y_lower = lognpdf(x, mu_lower, sigma_lower);
plot(x, y_lower, 'b--', 'LineWidth', 1.5);
```

```
% Upper bound PDF
y_upper = lognpdf(x, mu_upper, sigma_upper);
plot(x, y_upper, 'k--', 'LineWidth', 1.5);

%% Step 5: Annotate the plot
title('Log-Normal Fit with 95% Confidence Bounds');
xlabel('Data Values');
ylabel('Probability Density');
legend('Empirical Data', 'Fitted PDF (MLE)', ...
'Lower Bound PDF', 'Upper Bound PDF', ...
'Location', 'northeast');
grid on;
```

Code 4.23 provides a complete workflow for fitting a lognormal distribution to synthetic data in Python and estimating confidence intervals using bootstrapping. Lines 1–3 import the necessary Python libraries. `numpy` is used for generating random numbers and numerical operations. `matplotlib.pyplot` enables plotting, while `scipy.stats.lognorm` provides functions to compute the log-normal PDF, CDF, and inverse CDF, along with fitting tools.

Lines 5–8 generate synthetic data that follows a log-normal distribution. The parameters `mu_true = 2` and `sigma_true = 0.5` define the mean and standard deviation of the underlying normal distribution. `np.random.lognormal()` is used to create 1000 data points. This simulates a scenario in which we observe data and wish to understand its distribution and confidence bounds.

Lines 10–13 perform MLE to fit a lognormal distribution to the generated data. The `lognorm.fit()` function is called with `floc=0` to fix the location parameter at zero (standard assumption for lognormal distributions). The function returns `shape`, `loc`, and `scale`, which are used to compute the estimated mean (`mu_hat = log(scale)`) and standard deviation (`sigma_hat = shape`) of the underlying normal distribution.

Lines 15–26 define and implement a custom `bootstrap_params()` function to estimate the 95% confidence intervals for the parameters using bootstrapping. Inside this function, Line 17 starts the function definition. Lines 18–25 resample the data 1000 times (with replacement), fit the lognormal distribution for each sample, and store the estimated `(mu, sigma)` pairs. Line 26 returns the 2.5th and 97.5th percentiles of these bootstrapped estimates, providing the lower and upper bounds for both parameters.

Lines 28–30 extract the confidence intervals by calling the bootstrap function. The resulting matrix `ci_bounds` is split to assign meaningful names to the values: `mu_lower`, `mu_upper`, `sigma_lower`, and `sigma_upper`.

Lines 32–33 print out the final estimated parameters (`mu_hat` and `sigma_hat`) along with their 95% confidence intervals using formatted strings. This helps verify how closely the estimated parameters match the true values used to generate the data.

Lines 35–39 prepare the x-values and the corresponding PDF curves to visualize the fitted log-normal distribution and its confidence bounds. `x` is a linear space covering the data range. Then, three PDF curves are computed: one using the MLE estimates, and two using the lower and upper confidence bounds for the parameters.

Lines 41–45 plot the results. A histogram of the empirical data is created with semi-transparency in Line 42. The fitted PDF is plotted in red on Line 43. The lower and upper confidence bound curves are added with dashed lines (blue and black, respectively) in Lines 44 and 45.

Lines 47–54 add annotations and formatting to enhance the clarity of the plot. Titles and axis labels are added (Lines 48–50), the legend is placed in the upper-right corner (Line 51), a grid is enabled (Line 52), layout spacing is optimized (Line 53), and finally, the plot is rendered with `plt.show()` (Line 54).

This code provides a solid and practical example for students to learn distribution fitting, parameter estimation, bootstrapping, and data visualization in Python. It should be noted that the output of this code is exactly the same as the output provided in MATLAB.

Code 4.23 Fitting a Lognormal distribution with confidence bounds in Python

```
1   import numpy as np
2   import matplotlib.pyplot as plt
3   from scipy.stats import lognorm
4
5   # Step 1: Generate synthetic data from a log-normal distribution
6   mu_true = 2
7   sigma_true = 0.5
8   data = np.random.lognormal(mean=mu_true, sigma=sigma_true, size=1000)
9
10  # Step 2: Fit a log-normal distribution (MLE)
11  shape, loc, scale = lognorm.fit(data, floc=0)  # loc fixed to 0
12  mu_hat = np.log(scale)
13  sigma_hat = shape
14
15  # Step 3: Estimate 95% confidence intervals using bootstrap (manual
    method)
16  # Bootstrapping log-space estimates
```

```
def bootstrap_params(data, n_resamples=1000):
    estimates = []
    rng = np.random.default_rng()
    for _ in range(n_resamples):
        sample = rng.choice(data, size=len(data), replace=True)
        s, l, sc = lognorm.fit(sample, floc=0)
        mu = np.log(sc)
        sigma = s
        estimates.append((mu, sigma))
    return np.percentile(estimates, [2.5, 97.5], axis=0)

ci_bounds = bootstrap_params(data)
mu_lower, mu_upper = ci_bounds[:, 0]
sigma_lower, sigma_upper = ci_bounds[:, 1]

print(f"Estimated mu: {mu_hat:.4f} (95% CI: [{mu_lower:.4f},
{mu_upper:.4f}])")
{sigma_upper:.4f}])")

# Step 4: Plot histogram and fitted PDF with confidence bounds
x = np.linspace(min(data), max(data), 100)
y_fit = lognorm.pdf(x, s=sigma_hat, scale=np.exp(mu_hat))
y_lower = lognorm.pdf(x, s=sigma_lower, scale=np.exp(mu_lower))
y_upper = lognorm.pdf(x, s=sigma_upper, scale=np.exp(mu_upper))

plt.figure(figsize=(8, 6))
plt.hist(data, bins=30, density=True, alpha=0.3, label='Empirical Data')
plt.plot(x, y_fit, 'r-', linewidth=2, label='Fitted PDF (MLE)')
plt.plot(x, y_lower, 'b--', linewidth=1.5, label='Lower Bound PDF')
plt.plot(x, y_upper, 'k--', linewidth=1.5, label='Upper Bound PDF')

# Step 5: Annotate plot
plt.title('Log-Normal Fit with 95% Confidence Bounds')
plt.xlabel('Data Values')
plt.ylabel('Probability Density')
plt.legend(loc='upper right')
plt.grid(True)
plt.tight_layout()
plt.show()
```

4.5.2.2 Applications of the Lognormal Distribution in Real-World Contexts

Probability of Operating Equipment Past and its Useful Life Threshold

Suppose the lifetime of an electric motor in a centrifugal pump follows a lognormal distribution. What is the probability that the motor lasts more than 12,000 hours, given that the mean and standard deviation of the natural logarithm of the lifetime (i.e., $\mu_{\ln(x)}$ and $\sigma_{\ln(x)}$, respectively) are 11 and 1.3 h, respectively?

Solution

Consider X to be a lognormally distributed random variable representing the operating hours of a centrifugal pump motor. We are asked to determine P(X > 12,000) where: $\mu_{\ln(x)} = 11$ and $\sigma_{\ln(x)} = 1.3$.

Step 1: Use the Excel built-in function

Apply the following in Excel:

```
=LOGNORM.DIST(12000, 11, 1.3, TRUE)
```

This computes P(X<12000) based on the CDF of the lognormal distribution. Excel returns a value of 0.1082. Then, P(X>12000) is calculated as follows:

$$P(X > 12000) = 1 - P(X < 12000) = 1 - 0.1082 = 0.8918$$

Therefore, the probability that the motor lasts for greater than 12,000 hours is 89.18%.

Step 2: Use the Z-table via transformation

The Z-score is computed using the lognormal transformation:

$$z = \frac{\ln(12000) - \mu_{\ln(x)}}{\sigma_{\ln(x)}} = \frac{\ln(12000) - 11}{1.3} \approx \frac{9.3927 - 11}{1.3} = -1.236$$

Since standard normal tables list cumulative probabilities at discrete z-values, interpolation is required when the desired z-score falls between tabulated values. In this example, the z-score -1.236 lies between -1.23 and -1.24. The cumulative probabilities corresponding to these values are obtained from Table 4.5, and a linear interpolation is performed between them to estimate the probability at $z = -1.236$. This interpolated value is approximately 0.1082. Finally, the P(X>12000) is calculated as follows:

$$P(X > 12000) = 1 - 0.1082 = 0.8918$$

Again, we find a probability of 89.18%, consistent with Excel.

Since standard normal tables list cumulative probabilities at discrete z-values, interpolation is required when the desired z-score falls between tabulated values. In this example, the z-score -1.236 lies between -1.23 and -1.24. The cumulative probabilities corresponding to these values are obtained from Table 4.5, and a linear interpolation is performed between them to estimate the probability at $z = -1.236$. This interpolated value is approximately 0.1082.

Inspection Interval for Oil Platform Equipment

The time T between breakdowns of a major equipment piece on an oil platform is defined by a lognormal distribution with a median of 6 months and a CoV of 0.30.

What is the required interval between inspections and repairs to ensure that the equipment will be operational at any time with a 95% certainty?

If the platform is in operational condition at the time it is scheduled for regular maintenance and inspection (with a 95% probability), what is the probability that it will remain operational without breakdowns for another 2 months?

Solution:

Part 1: Determine Inspection Interval to Ensure 95% Reliability

Step 1: Use the relationship between CoV and the lognormal parameters.

According to the problem statement, we are given the following:

- $X_m = 6$ months (median)
- CoV=0.30

Using this information, we can calculate the lognormal parameters:

$$\lambda = \ln(X_m) = \ln(6) = 1.792$$

$$\xi^2 = \ln(1 + \text{CoV}^2) = \ln(1 + 0.3^2) = \ln(1.09) = 0.293$$

$$\xi \approx \sqrt{0.293} \approx 0.54 \text{ (though the example approximates to 0.3 for simplicity)}$$

Step 2: Solve for t_0 such that $P(T > t_0) = 0.95$

Let t_0 be the inspection time (or service time) such that there is a 95% probability the component has not failed before that time. Rewriting:

$$P(T > t_0) = 0.95 \Rightarrow P(T \leq t_0) = 0.05 \Rightarrow \varphi\left(\frac{\ln(t_0) - \lambda}{\xi}\right) = 0.05$$

From z-tables:

$$\varphi(z) = 0.05 \Rightarrow z = -1.65$$

Substitute values:

$$\frac{\ln(t_0) - 1.792}{0.3} = -1.65 \Rightarrow \ln(t_0) = 1.297 \Rightarrow t_0 = e^{1.297} = 3.66 \text{ months}$$

So, the required inspection interval is approximately 3.66 months.

Part 2: Conditional probability of no failure for 2 more months after inspection

Again, let T be the time between breakdowns of the oil machinery. We are asked:

$$P(T > 5.66|T > 3.66) = ?$$

Since the event $T > 5.66$ is a subset of the event $T > 3.66$, their intersection reduces to $T > 5.66$. This is a conditional probability:

$$P(T > 5.66|T > 3.66) = \frac{P(T > 5.66)}{P(T > 3.66)}$$

We already know:

$$\text{P(T >3.66)} = 0.95$$

Now we should compute $P(T > 5.66)$:

Step 1: Compute the Z-score for 5.66 months

$$z = \frac{\ln(5.66) - 1.792}{0.3} = \frac{1.7346 - 1.792}{0.3} = -0.195$$

From z tables:

$$\varphi(-0.195) \approx \frac{0.4247 + 0.4207}{2} = 0.423 \Rightarrow P(T > 5.66) = 1 - 0.423 = 0.577$$

Step 2: Compute Conditional Probability

$$P(T > 5.66 \middle| T > 3.66) = \frac{0.577}{0.95} \approx 0.61$$

There is a 61% probability that the system remains operational without failure for 2 more months following an inspection where the system was still working.

Earthquake Occurrence

The time between severe earthquakes at a given region in western Canada follows a lognormal distribution with a coefficient of variation of 40%. The expected time between severe earthquakes is 80 years.

(a) Determine the parameters of the lognormally distributed recurrence time, T.
(b) Determine the probability that a severe earthquake will occur within a 20 yr window following the previous one.
(c) Suppose the last severe earthquake in the region took place 100 yr ago. What is the probability that a severe earthquake will occur over the next year?

Solution

Let T be a lognormally distributed random variable representing the recurrence time between severe earthquakes.

(a) **Determine the parameters λ and ξ**

To compute the parameters of the lognormal distribution, we calculate the standard deviation of ln(T) (denoted ξ) using:

$$\xi^2 = \ln(1 + \mathrm{CoV}^2) = \ln(1 + 0.4^2) = \ln(1.16) = 0.148 \Rightarrow \xi = 0.385$$

Then, compute the mean of ln(T) (denoted λ) using:

$$\lambda = \ln(\mu) - \frac{1}{2}\xi^2 = \ln(80) - \frac{1}{2}(0.148) = 4.382 - 0.074 = 4.308$$

Thus, we now know the parameters governing the lognormal distribution of earthquake recurrence time:

- $\lambda = 4.308$
- $\xi = 0.385$

(b) **Probability that a Severe Earthquake Occurs Within 20 Years**

We are asked to compute:

$$P(T \leq 20)$$

Apply the standard lognormal transformation:

$$P(T \leq 20) = \Phi\left(\frac{\ln(20) - 4.308}{0.385}\right) = \Phi(-3.406)$$

From the z-table:

$$\Phi(-3.406) = 0.0003$$

So, the probability of a severe earthquake occurring within 20 years is 0.03%, or very rare.

(c) **Conditional probability of an earthquake in the next year given 100 years have passed**

We now compute:

$$P(100 < T < 101|T > 100)$$

Since the event $100 < T < 101$ is entirely contained within the event $T > 101$ their intersection is simply $100 < T < 101$. As a result, the conditional probability simplifies to:

$$P(100 < T < 101|T > 100) = \frac{P(100 < T < 101)}{P(T > 100)}$$

This is evaluated as:

$$\begin{aligned} P(100 < T < 101) &= \Phi\left(\frac{\ln(101) - 4.308}{0.385}\right) - \Phi\left(\frac{\ln(100) - 4.308}{0.385}\right) \\ &= \Phi(0.432) - \Phi(0.391) \end{aligned}$$

From the z-table:

$$\Phi(0.432) = 0.787 \quad \text{and} \quad \Phi(0.391) = 0.780 \Rightarrow P(100 < T < 101) = 0.007$$

Also:

$$P(T > 100) = 1 - \Phi(0.391) = 1 - 0.780 = 0.220$$

Thus:

$$P(100 < T < 101|T > 100) = \frac{0.007}{0.220} \approx 0.034$$

So, the probability of a severe earthquake occurring in the next year, given that it's already been 100 years since the last one, is approximately 3.4%.

4.6 Random Processes and Extremes

In many practical situations, especially in engineering and environmental science, we are interested not in a single observation, but in how a variable behaves over time. For instance, consider the water demand (WD) in a city such as Guelph (Ontario, Canada).

When observed over time, WD exhibits clear and often predictable patterns, such as diurnal cycles, weekday–weekend differences, and seasonal variations between summer and winter. At the same time, daily demand is influenced by numerous random factors—including weather fluctuations, human behavior, and unexpected events such as fire demand—introducing variability that cannot be predicted exactly. As a result, water demand is best described as a time-varying process with both structured (deterministic) components and random uncertainty.

A random process is a mathematical model that represents this kind of time-varying uncertainty. More precisely, a random process is a collection of random variables, each associated with a specific point in continuous time (or sometimes space). That means:

- At each moment in time, the process gives you a random variable.
- If you pick different times (e.g., different days, months, or years), you are essentially looking at different random variables from the same process.

Imagine measuring the daily water demand for over 1,000 days. Each day you record a value, but due to randomness (e.g., weather, usage habits), these values fluctuate. This creates a time-indexed collection of values—each one a realization of a random variable from the random process.

A continuous-time random process assumes that observations could be made at any point in time, no matter how small the time step (e.g., measuring temperature changes every second). In contrast, a discrete-time random process, also referred to as a **random sequence**, is defined only at specific, separated time points, such as yearly population data or monthly rainfall amounts. In Fig. 4.50, the top graph shows daily water demand (continuous time) as it evolves over time, reflecting a random process. In contrast, Fig. 4.50b shows the average yearly water demand and population data, recorded once yearly—a typical random sequence.

Understanding random processes can help us to model and predict complex systems such as climate trends, traffic flow, or energy usage with greater efficiency. It also enables us to quantify uncertainties in engineering design and develop smarter management strategies—for example, optimizing water supply networks or scheduling maintenance based on usage patterns.

Time series analysis is a powerful tool for analyzing random processes, which are collections of random variables indexed continuously over time. As illustrated in Fig. 4.50a, the daily WD data for the City of Guelph is plotted over several years, showing continuous fluctuations that represent a typical random process. Examples of such processes include drought severity, watering restrictions, fire demand, and other unforeseen events. Similar time-varying behavior is observed in many engineering and environmental variables, such as lake levels, precipitation, runoff, and wind velocity, which evolve over time and are typically measured at regular intervals (e.g., daily or monthly).

In contrast, discrete-event simulation is used to model systems where changes occur at specific time points. In Fig. 4.50b, we observe the annual population and corresponding average daily water demand, which are sampled at yearly intervals. This represents a discrete-time sequence. Examples of random sequences include

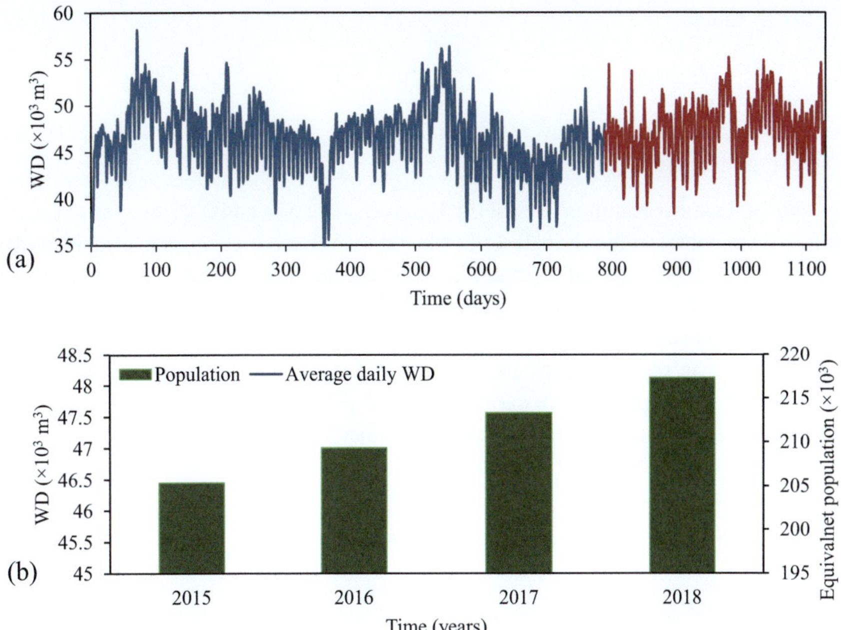

Fig. 4.50 **a** Daily water demand (WD) as a continuous time random process in the City of Guelph. **b** Discrete-time sequence showing annual average daily WD and population trends from 2015 to 2018 in the City of Guelph

failure events, flooding occurrences, or traffic congestion, where each event happens at a particular time and is counted discretely.

Together, the two figures illustrate the distinction between continuous-time random processes and discrete-time random sequences. Understanding both is essential for selecting the appropriate modeling approach—time series analysis for continuous phenomena and discrete-event simulation for event-driven systems.

4.6.1 Poisson Process

A Poisson process is a mathematical model that describes the occurrence of random events over a period of time. It is particularly useful in situations where we know the average rate at which events happen, but we cannot predict exactly when each event will occur.

Think of it as a model for counting how many times an event—like an earthquake, a power outage, or a website visit—happens in a fixed time frame (such as a year, a month, or an hour), assuming the events happen independently of each other and the average time between them remains constant.

In this model:

- The time between events is known on average (e.g., on average, there are 5 earthquakes per year),
- The exact timing of the events is random (e.g., you can't say when exactly the next earthquake will occur).

One of the most common uses of the Poisson process in engineering, geosciences, and environmental studies is to describe the number of occurrences of natural hazards. For example, the number of earthquakes, the number of hurricanes, or the number of tornadoes that happen in a given time period (say, one year) can be modeled using a Poisson distribution.

Let's say you're looking at earthquakes in western Canada. If, historically, there have been on average 3 severe earthquakes per year, and these earthquakes happen randomly throughout the year, then the number of earthquakes in one year is a Poisson-distributed random variable.

In such examples:

- The random variable is the number of events (e.g., earthquakes) in a given time period.
- This random variable is discrete, because you can count the number of occurrences (e.g., 0, 1, 2... earthquakes).
- The reference space, which is the time period over which you count (e.g., 1 year), is continuous—you can count events over any continuous duration of time.

Understanding the Poisson process enables engineers and scientists to predict the likelihood of a specific number of events occurring, plan infrastructure such as earthquake-resistant buildings based on probable risk, and design more reliable systems by accounting for the randomness of failure events. By mathematically modeling these occurrences, they can make better-informed decisions even in the face of uncertainty.

In many random processes—particularly those modeled by a Poisson process—the occurrence of an event at a given time does not affect the probability of an event occurring at the next time instance. This implies that the events are independent and the occurrence rate (e.g., number of events per year) remains constant over time. Such an assumption is fundamental when modeling natural phenomena like earthquakes or equipment failures, where the timing of past events does not influence future likelihoods, allowing for simplified and memoryless mathematical treatment.

The Poisson PMF models the likelihood of a given number of discrete events occurring within a fixed time or space interval, assuming events occur independently and at a constant average rate. Mathematically, it is expressed as:

$$\Pr(X = x; \lambda) = \frac{\lambda^x e^{-\lambda}}{x!} = \frac{(\nu t)^x e^{-\nu t}}{x!} \tag{4.44}$$

where λ represents the mean number of occurrences during the interval, x is the actual number of events, and e is Euler's constant. Since λ=νt, where ν is the mean occurrence rate and t is the time interval, the Poisson distribution uses a single parameter λ to define both its mean and variance. In upcoming examples, we'll explore how to apply this function to real-world scenarios.

The CDF for a Poisson distribution gives the probability that a random variable X is less than or equal to a specified value k. This is denoted as $\Pr(X \leq k)$ and is computed by summing the probabilities of all values from X = 0 up to X = k. Mathematically, this is expressed as a summation of the Poisson PMF:

$$\Pr(X \leq k) = \sum_{x=0}^{k} \frac{\lambda^x e^{-\lambda}}{x!} \tag{4.45}$$

where λ is the mean number of occurrences.

If, instead, we want to find the probability that X exceeds k, we can simply use the complement rule: $Pr(X > k) = 1 - Pr(X \leq k)$.

The Poisson process, or Poisson sequence, is a foundational model in stochastic processes defined by several key assumptions. First, events occur randomly and can happen at any point in time or space. Second, the occurrences of events in disjoint (non-overlapping) intervals are statistically independent, meaning the occurrence of one does not influence another. Third, the probability of a single event occurring within a small interval Δx is directly proportional to the size of that interval and is expressed using the constant mean rate λ, representing the expected number of events per unit time or space. Finally, the probability of more than one event occurring within the same small interval Δx is negligible, being of higher order relative to Δx.

To compute Poisson probabilities in Excel, you can use the built-in POISSON function. Use the following command to calculate the probability that a Poisson-distributed random variable is exactly equal to a specific value (as shown on the left side of the figure).

```
=POISSON(value, mean, FALSE)
```

Alternatively, use the following to compute the probability that the random variable is less than or equal to the given value (illustrated on the right side).

```
=POISSON(value, mean, TRUE)
```

This makes Excel a convenient tool for evaluating both the PMF and the CDF of the Poisson distribution.

4.6.1.1 Poisson Process for Estimating the Probability of Natural Hazard Events

Table 4.8 lists the number of natural hazards (e.g., floods, storms, wildfires) recorded each year in Ontario (Canada) over a 10 year period (2012–2021). Assume that the occurrence of these hazards follows a Poisson distribution.

Table 4.8 Number of natural hazards recorded yearly in Ontario (Canada)

Year	No. of natural hazards
2012	1
2013	0
2014	2
2015	1
2016	1
2017	3
2018	2
2019	3
2020	4
2021	4

(a) Based on the data provided, estimate the average annual rate (λ) of hazard events.
(b) Using this λ, calculate the probability that exactly 10 natural hazard events will occur in Ontario in 2022.
(c) Calculate the probability that exactly 3 events will occur in 2022.
(d) Plot the Poisson distribution for $\lambda = 2.1$. What can you say about the likelihood of observing more than 7 events in a year?
(e) If 5 events are observed in 2022, is the original $\lambda = 2.1$ still valid? If not, how would you update it using the new data?

Solution

(a) Based on the data provided, estimate the average annual rate (λ) of hazard events.

The average annual rate can be determined by first totaling the number of events over the 10-year period:

$$X = \sum_{i=1}^{10} x_i = 1 + 0 + 2 + 1 + 1 + 3 + 2 + 3 + 4 + 4 = 21$$

The average annual rate (λ) of hazard events is then determined simply by dividing the total number of events by the time period:

$$\lambda = \frac{21}{10} = 2.1$$

(b) Using this λ, calculate the probability that **exactly 10 natural hazard events** will occur in Ontario in 2022.

We use the Poisson probability mass function:

$$Pr(X = x; \lambda) = \frac{\lambda^x e^{-\lambda}}{x!}$$

Substitute:

- x=10
- λ=2.1

$$Pr(X = 10; \lambda = 2.1) = \frac{2.1^{10} \cdot e^{-2.1}}{10!} \approx 0.0000563$$

The probability of observing 10 events is extremely low, around 0.0056%, which implies this outcome is highly unlikely.

(c) Calculate the probability that **exactly 3 events** will occur in 2022.

Use the same Poisson formula, but now:

- x=3

$$Pr(X = 3; \lambda = 2.1) = \frac{2.1^{3} \cdot e^{-2.1}}{3!} \approx 0.189$$

There is an 18.9% chance of observing 3 natural hazard events in 2022.

(d) Plot the Poisson distribution for λ = 2.1. What can you say about the likelihood of observing more than 7 events in a year?

Using Eq. 4.45 with $\lambda = 2.1$, the discrete time sequence for Ontario's natural hazard data is plotted in Fig. 4.51. Analysis of the distribution indicates that the most likely number of hazard events is 1 to 4 events per year, with the tail of the distribution suggesting that observing 7 or more events is rare, but not impossible. To plot this figure, Code 4.24 was used.

Code 4.24 Poisson distribution for Ontario's natural hazard events

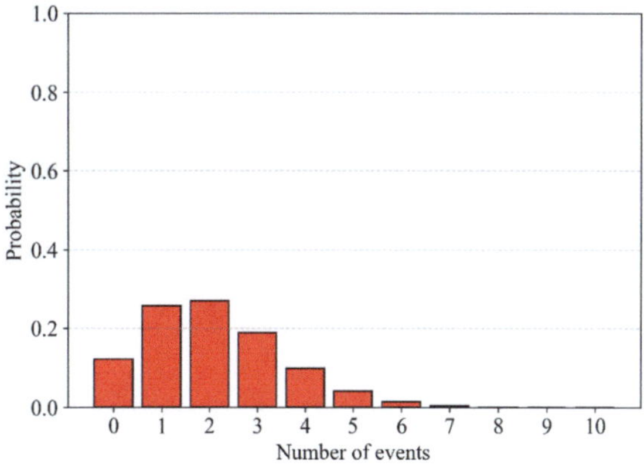

Fig. 4.51 Poisson distribution for Ontario's natural hazard events

```
import numpy as np
import matplotlib.pyplot as plt
from scipy.stats import poisson

# Set font family and size globally
plt.rcParams['font.family'] = 'Times New Roman'
plt.rcParams['font.size'] = 16

# Data
hazards = np.array([1, 0, 2, 1, 1, 3, 2, 3, 4, 4])
lambda_est = np.mean(hazards)
x = np.arange(0, 11)
pmf = poisson.pmf(x, mu=lambda_est)

# Plot
plt.figure(figsize=(7, 5))
plt.bar(x, pmf, color='red', edgecolor='black')
plt.xlabel('Number of events')
plt.ylabel('Probability')
plt.xticks(x)
plt.ylim(0, 1)
plt.grid(axis='y', linestyle='--', alpha=0.6)
plt.tight_layout()

# Save as TIFF and JPEG
plt.savefig('poisson_plot.tif', format='tiff', dpi=300)
plt.savefig('poisson_plot.jpeg', format='jpeg', dpi=300)

plt.show()
```

(e) If 5 events are observed in 2022, is the original $\lambda = 2.1$ still valid? If not, how would you update it using the new data?

We observed 5 events in 2021, and if we also observed 5 in 2022, that's more than double the average ($\lambda = 2.1$). This raises the question: Is $\lambda = 2.1$ still valid?

In this situation, it is very convenient to simply update the annual rate of hazard occurrence. To evaluate this, consider updating λ with the new data:

- Total hazards from 2012 to 2022 $= 21 + 5 = 26$
- New number of years $= 11$

$$\lambda_{\text{updated}} = \frac{26}{11} \approx 2.36$$

We can now recompute the probability of 10 or 3 events using this updated λ to reflect more recent conditions.

4.6.1.2 Poisson Process for Estimating the Probability of Rare Rainstorm Events

Consider a case where the City of Toronto (Canada) experiences severe rainstorms at an average rate of 0.1 rainstorms per year.

(a) What is the probability that the city will experience 2 severe rainstorms in the next year?
(b) What is the probability that the city will experience 2 severe rainstorms in the next 5 years?

Solution:

According to historical data, the average annual occurrence rate of severe rainstorms is:

$\lambda = 0.1$ rainstorms/year

We are interested in finding the probability of exactly 2 rainstorms, first over a 1 year period and then over a 5 year period. In both cases, we treat the number of events, X, as a Poisson-distributed random variable.

(a) **Probability of 2 rainstorms in the next 1 year**

Here, the number of events we are interested in is:

$$x = 2$$

And the time period is 1 year, so:

$$\lambda = 0.1$$

We apply the Poisson probability mass function:

$$P(X = x) = \frac{\lambda^x e^{-\lambda}}{x!}$$

Substituting the values:

$$P(X = 2) = \frac{0.1^2 e^{-0.1}}{2!}$$

$$P(X = 2) = \frac{0.01 \cdot e^{-0.1}}{2} \approx \frac{0.01 \cdot 0.9048}{2} \approx 0.00452$$

Thus, the probability of observing 2 severe rainstorms in one year is:

$$0.00452 \text{ or } 0.452\%$$

(b) **Probability of 2 rainstorms in the next 5 years**

Now, we are considering a 5-year period. The average number of events over this period becomes:

$$\lambda = 0.1 \times 5 = 0.5$$

We are still interested in:

$$\text{x} = 2$$

Again, applying the Poisson PMF:

$$P(X = 2) = \frac{0.5^2 e^{-0.5}}{2!}$$
$$P(X = 2) = \frac{0.25 \cdot e^{-0.5}}{2} \approx \frac{0.25 \cdot 0.6065}{2} \approx 0.07582$$

So, the probability of experiencing 2 severe rainstorms in the next 5 years is:

$$0.0758 \quad \text{or} \quad 7.58\%$$

4.6.1.3 Poisson Process for Dump Truck Arrivals at a Construction Site

At a dam construction site, it is observed that an average of three dump trucks arrive every 10 min. What is the probability that four or fewer dump trucks will arrive at the site within a 20 min interval?

Solution

Step 1: Identify Given Information

Average arrival rate per 10 min: 3 trucks
We are interested in a 20 min interval, so:

$$\lambda = 2 \times 3 = 6$$

(Since 20 min is double the base interval of 10 min.)
We want the probability of 4 or fewer dump trucks:

$$P(X \leq 4)$$

Step 2: Use the Poisson Probability Mass Function

The Poisson probability mass function is:

$$P(X = x) = \frac{\lambda^x e^{-\lambda}}{x!}$$

To calculate the cumulative probability for 0 through 4 trucks:

$$P(X \leq 4) = P(0) + P(1) + P(2) + P(3) + P(4)$$
$$P(X \leq 4) = \sum_{x=0}^{4} \frac{6^x e^{-6}}{x!}$$

Breaking down each term:

$$P(0) = \frac{6^0 e^{-6}}{0!} = \frac{1 \cdot e^{-6}}{1}$$
$$P(1) = \frac{6^1 e^{-6}}{1!} = \frac{6 \cdot e^{-6}}{1}$$
$$P(2) = \frac{6^2 e^{-6}}{2!} = \frac{36 \cdot e^{-6}}{2}$$
$$P(3) = \frac{6^3 e^{-6}}{3!} = \frac{216 \cdot e^{-6}}{6}$$
$$P(4) = \frac{6^4 e^{-6}}{4!} = \frac{1296 \cdot e^{-6}}{24}$$

Summing all these gives:

$$P(X \leq 4) \approx 0.285$$

Step 3: Verify Using Excel

Alternatively, Excel's built-in Poisson formula could be used to arrive at the same solution using the syntax noted below:

```
=POISSON(4, 6, TRUE)
```

This yields the same result:

$$P(X \leq 4) \approx 0.2850565$$

Thus, there is an approximately 28.5% chance that four or fewer dump trucks will arrive at the dam site within a 20 min period.

4.6.1.4 Poisson Process for Car Accident Claims at an Insurance Company

An insurance company has observed that, on average, 6.2 customers report a car accident claim per day. What is the probability that at least one customer will file a car accident claim on any given day?

Solution

Step 1: Identify the parameters

- Average number of claims on a per-day basis:

$$\lambda = 6.2$$

- If we let X be a Poisson-distributed random variable representing the number of daily car accident claims, then we are interested in the probability that at least one claim occurs, i.e.,

$$P(X \geq 1)$$

Step 2: Use the complement rule

It is easier to calculate the complement of this event:

$$P(X \geq 1) = 1 - P(X = 0)$$

Step 3: Apply the Poisson PMF formula

The Poisson probability mass function is:

$$P(X = x) = \frac{\lambda^x e^{-\lambda}}{x!}$$

Substituting for X=0 and λ=6.2:

$$P(X = 0) = \frac{6.2^0 \cdot e^{-6.2}}{0!} = \frac{1 \cdot e^{-6.2}}{1} \approx 0.00202$$

Step 4: Subtract from 1

$$P(X \geq 1) = 1 - P(X = 0) = 1 - 0.00202 = 0.99798$$

Final Answer

$$P(X \geq 1) \approx 99.8\%$$

There is a 99.8% probability that at least one customer will file a car accident claim on a given day.

4.6.1.5 Poisson Process for Left-hand Turn Lane Storage Design

In designing a left-hand turn lane at an intersection, traffic engineers need to ensure that the lane can accommodate a sufficient number of vehicles during a traffic light cycle. If the traffic light cycle time is 1 min and the average arrival rate of vehicles making left turns is 100 cars per hour, what is the minimum number of car lengths the storage lane must be designed for to accommodate left-turning vehicles 96% of the time?

Solution

Define the average arrival rate:

$$\nu = \frac{100}{60} \quad \text{cars per minute}$$

Determine the mean number of cars during a 1-minute light cycle:

$$\lambda = \nu \cdot t = \frac{100}{60} \cdot 1 = \frac{100}{60}$$

Set up the Poisson CDF condition:

We want to find the minimum number of car lengths, k, such that the probability of having at most k cars arriving during the 1-minute light cycle is at least 96%:

$$P(X \leq k) = \sum_{x=0}^{k} \frac{\lambda^x e^{-\lambda}}{x!} = 0.96$$

Trial-and-error to find k:

- Try $k = 3$:

$$P(X \leq 3) = \sum_{x=0}^{3} \frac{1}{x!}\left(\frac{100}{60} \cdot 1\right)^{x} e^{-100/60} = 0.93$$

  This is below 0.96.

- Try k = 4:

$$P(X \leq 4) = 0.968$$

  This is above 0.96 → meets the design criterion.

Therefore, to provide enough storage in the left-hand turn lane at this intersection 96% of the time, the lane must be designed to be a minimum of 4 car lengths.

4.6.1.6 Poisson Process for Evaluating Earthquake and Infrastructure Risks

In a metropolitan region, strong earthquakes follow a Poisson process, occurring on average once every 50 years. There are three bridges in this region, and each has a 30% chance of collapsing during a strong earthquake. Assume the bridge collapses are statistically independent of each other and of subsequent earthquakes. You are asked to answer the following:

(a) What is the probability that at most one strong earthquake occurs in the next 20 years?
(b) What is the probability that exactly one of the three bridges will collapse during a single earthquake?
(c) What is the probability that no bridge collapses due to earthquakes in the next 20 years?

Solution

Part (a)—Probability of At Most One Earthquake in 20 Years

Let X be the number of strong earthquakes in 20 years.

The Poisson rate λ, adjusted for a 20-year basis, is:

$$\lambda = \left(\frac{1}{50}\right) \times 20 = 0.4$$

We want:

$$P(X \leq 1) = P(X = 0) + P(X = 1)$$

Using the Poisson formula:

$$P(X = x) = \frac{\lambda^x e^{-\lambda}}{x!}$$

So,

$$\begin{aligned} P(X = 0) &= \frac{0.4^0 e^{-0.4}}{0!} = e^{-0.4} \approx 0.6703 \\ P(X = 1) &= \frac{0.4^1 e^{-0.4}}{1!} = 0.4e^{-0.4} \approx 0.2681 \\ P(X \leq 1) &= 0.6703 + 0.2681 = 0.938 \end{aligned}$$

There is a 93.8% chance that at most one strong earthquake will hit the region in the next 20 years.

Part (b)—Probability That Exactly One Bridge Collapses

Let Y be the number of bridges that collapse in a single strong earthquake. Each bridge has a collapse probability p = 0.3 (and thus a survival probability of 0.7).

We want:

$$P(Y = 1) = \binom{3}{1}(0.3)^1(0.7)^2 = 3 \cdot 0.3 \cdot 0.49 = 0.441$$

During such an earthquake, the chance that exactly one bridge will collapse is 44.1%.

This is the probability that a binomial random variable Y equals 1, meaning there is exactly 1 success in 3 independent trials, with the probability of success in each trial being $p = 0.3$.

$$P(Y = k) = \frac{n!}{k!(n-k)!} \cdot p^k \cdot (1-p)^{n-k} \tag{4.46}$$

where n is the number of trials, k represents the number of successes, p denotes the probability of success on a single trial, and 1-p shows the probability of failure.

Part (c)—Probability of no bridge collapse over 20 years

We must apply the Law of Total Probability.

Let X be the number of earthquakes in 20 years.

Let P_c=0.3 be the chance a bridge collapses during one earthquake.
Then, P(no collapse in one quake) = $(1-0.3)^3 = 0.343$
Let A_x be the event "x earthquakes occur."
Then,

$$\begin{aligned} P(\text{no bridge collapse}) &= \sum_{x=0}^{\infty} P(\text{no collapse}|x \text{ quakes}) P(X = x) \\ &= \sum_{x=0}^{\infty} (0.343)^x . \frac{(0.4)^x e^{-0.4}}{x!} \\ &= e^{-0.4} \cdot \sum_{x=0}^{\infty} \frac{(0.343 \cdot 0.4)^x}{x!} \\ &= e^{-0.4} \cdot e^{0.1372} = e^{-0.2628} \approx 0.769 \end{aligned}$$

The combined chance of no bridge collapses due to earthquakes over 20 years is 76.9%.

4.6.2 *Extreme Values Distribution*

In the field of civil engineering, one of the most critical challenges in design is balancing safety and cost. On one hand, we aim to ensure that structures—such as bridges, dams, buildings, and roads—can endure rare but extreme events like hurricanes, floods, or traffic surges. On the other hand, we must keep construction economically feasible. Designing a structure to withstand every imaginable extreme event would lead to excessive costs and impractical overdesign. Therefore, extreme value analysis is used to estimate the probability of rare but impactful occurrences, helping engineers make informed decisions that balance reliability with affordability.

Extreme value theory (EVT) deals with the statistical behavior of the maximum or minimum values in a dataset. Instead of focusing on average conditions, which are more common in daily operations, EVT focuses on the rare outliers—the extremes—which often control the structural limits and safety criteria.

Let us explore some applications of extreme value distributions across various branches of engineering:

1. **Ocean engineering: the highest waves**

In ocean and coastal engineering, structures such as offshore oil platforms, sea walls, dikes, and breakwaters must be designed to withstand the highest waves that might occur during severe storms or hurricanes. Engineers use statistical models to estimate the distribution of wave heights, particularly the maximum wave height expected within a given return period (e.g., once every 50 years). These models guide the sizing and stability analysis of marine structures, ensuring they remain safe during extreme ocean conditions.

2. **Structural engineering: wind loads, earthquakes, and design codes**

Building codes are often informed by extreme events that could pose a threat to structural integrity. For instance:

- Wind speeds are represented as extreme values with specific recurrence intervals (e.g., the wind speed expected once every 100 years).
- Design loads such as snow, earthquake ground motion, or ice accumulation are defined as extreme scenarios based on historical data.
- Seismic zones are mapped by assigning areas with different levels of equal seismic risk, helping determine how earthquake-resistant structures should be in different locations.

All of these scenarios depend on EVT to quantify the low-probability, high-consequence events that structures must survive.

3. **Highway and traffic engineering: kth largest traffic intensity**

In traffic and transportation engineering, we often deal with recurring loads, such as the volume of vehicles on a bridge. Rather than using the single largest traffic volume (which may be an outlier), designs often use the kth largest value (e.g., the 5th largest hourly traffic flow in a year). This approach accounts for variability without being overly conservative. Analyzing these traffic intensities relies on order statistics, a branch of probability closely related to EVT.

4. Failures in materials and systems: load and strength extremes

Failures in engineering systems—whether due to excessive loading, material weakness, or human error—are frequently associated with extreme values. For example:

- Maximum loads applied to a structure can exceed design thresholds, leading to failure.
- Minimum material strengths (e.g., a weak concrete batch) can fall short of required performance, increasing risk.

5. Meteorological Hazards: Floods, Droughts, and Storms

EVT is also indispensable for modeling natural hazards:

- Flood levels in river basins
- Drought intensities affecting the water supply
- Storm surge heights in coastal areas
- Tornado wind speeds and hailstorm intensities

Designing drainage systems, flood control channels, or emergency shelters depends on estimating the probability of rare meteorological events. EVT helps predict how rare such events are and guides risk mitigation strategies.

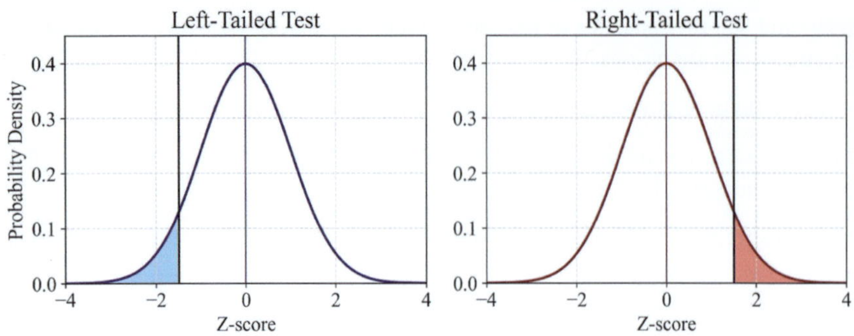

Fig. 4.52 Left-tailed and right-tailed hypothesis tests based on the standard normal distribution

In many real-world engineering applications, we are not always interested in the average or typical behavior of a system. Instead, we are often concerned about the most severe, rarest, or most extreme occurrences that could have significant impacts. These events—such as the strongest storm, largest earthquake, highest flood level, or maximum wind speed—are known as extreme values.

But what exactly do we mean by "extreme value" in a statistical or probabilistic sense?

In probability theory, an extreme value refers to either a very small value (far below average) or a very large value (far above average) that occurs with low probability. These values are found in the **tails** of a probability distribution, specifically, the left and right ends of the curve. These tails represent the unlikely but potentially critical outcomes that can challenge the safety and functionality of civil infrastructure.

For example, Fig. 4.52 shows a two-tailed normal distribution curve highlighting critical regions. On the right tail, you might find the highest flood level recorded in a century. While on the left tail, you might encounter extremely low temperatures that could affect material strength or freeze a pipeline.

Civil engineers must design structures that are not only functional and efficient under normal conditions, but also safe and reliable under rare but dangerous conditions. For instance:

- A bridge must withstand not just daily traffic but also the heaviest traffic jam expected in its lifetime.
- A dam should remain stable during the most intense rainfall recorded in the region.
- A skyscraper must be designed to resist extreme wind gusts and potential earthquake tremors.

This means engineers must account for low-probability, high-impact events—and that's exactly where extreme value analysis becomes essential.

It's important to note that the term "*extreme value*" may carry slightly different meanings depending on the context:

- In elementary statistics, it may refer to the minimum or maximum value in a dataset.

- In engineering statistics, it typically refers to rare observations in the tails of a probability distribution.
- In **calculus**, the term extrema refers to local minima and maxima of a function.

Despite these variations, the common idea is that extreme values represent unusual, rare, or critical outcomes that deviate significantly from the average.

To predict such values, engineers use Extreme Value Distributions (EVDs), which help estimate the maximum or minimum values likely to occur in a dataset. These distributions come from EVT (introduced earlier), which is focused specifically on analyzing and modeling rare events.

An EVD is a probability model that describes the likely behavior of the largest (or smallest) values in a data sample. Rather than modeling the average or typical case, these distributions are used to predict the limit—how high or low something can get. For example:

- How tall can the highest wave be?
- What is the strongest wind speed expected once every 50 years?
- What is the maximum flood level a river might reach?

These questions matter deeply in civil engineering design, where failure due to underestimating these extremes can lead to catastrophic consequences.

Depending on the characteristics of the data you're analyzing, one of three EVD types will typically apply. These are known as Gumbel (Type I), Fréchet (Type II), and Weibull (Type III) distributions.

The Gumbel distribution, also known as Extreme Value Type I, is used when the extremes—either maximums or minimums—are unbounded, meaning there is no known upper or lower limit. It comes in two forms: one for modeling maximum values, and the other for minimum values. The Gumbel distribution is widely applied in hydrology and meteorology for analyzing extreme events such as maximum rainfall, peak river discharges, or the strongest wind speeds, where the assumption is that no physical upper limit is known. A typical use case is estimating the height of the 100-year flood, which helps in floodplain mapping and infrastructure planning in flood-prone areas.

The Fréchet distribution, or Extreme Value Type II, is suited for cases where maximum values are possible, and large outliers are likely. This distribution has a heavy right tail, indicating that while extremely large values are rare, they are not impossible and must be accounted for. The Fréchet distribution is appropriate for phenomena where extreme deviations from the mean are expected, such as financial market crashes, large earthquakes, or material fatigue leading to failure. For example, engineers and seismologists might use the Fréchet distribution to model the largest recorded earthquakes, which can significantly exceed previously observed magnitudes.

The Weibull distribution, representing Extreme Value Type III, is used when the variable of interest is bounded above, meaning there is a known physical or natural upper limit. It is typically applied in scenarios where extremes cannot exceed a threshold, making it suitable for both minimum and maximum value modeling. This distribution is often used in material strength analysis, where components like steel rods have a maximum load capacity before failure. A common use case is predicting the minimum strength at which a batch of manufactured materials might fail under stress, critical for quality assurance and safety testing.

In engineering and environmental science, the application of EVDs follows a structured process. Practitioners begin by collecting historical data relevant to their domain, such as river flows, wind speeds, or wave heights. This data is then segmented into blocks, often by year or season, and the extreme values (e.g., annual maximum flood heights) are extracted. These extracted extremes are then fit to one of the EVD types, depending on the data characteristics. Once the model is calibrated, it is used to estimate the magnitude of rare events, such as the 100-year or 500-year flood level, helping guide decisions in infrastructure design, insurance, and risk mitigation.

The existence of three EVD types is rooted in statistical theory, particularly the Fisher-Tippett-Gnedenko theorem. This theorem states that the distribution of maxima (or minima) from sufficiently large samples will converge to one of three possible forms, regardless of the original distribution of the data. The type that best fits the extremes depends on the tail behavior and bounds of the dataset: if there's no limit to extremes, the Gumbel distribution (Type I) is appropriate; if there is a potential for extremely large values, the Fréchet distribution (Type II) is used; and if the extremes are bounded, the Weibull distribution (Type III) fits best.

4.6.2.1 Extreme Value Distributions Type I—Gumbel Distribution

The EVD Type I, commonly known as the Gumbel distribution, is widely used in practice to model extreme values of unbounded random variables. It plays a vital role in fields ranging from hydrology and structural safety to earthquake engineering and reliability analysis.

The Gumbel distribution comes in two distinct forms:

- Gumbel Max: Used to model the maximum values from a dataset (e.g., maximum annual flood level, peak wind speed).
- Gumbel Min: Used to model the minimum values from a dataset (e.g., lowest annual temperature, minimum structural strength).

These two forms are mathematically similar, but they describe different tails of the distribution. Let's take a closer look at both.

Gumbel Maximum Distribution (Type I for Maxima)

This form models the distribution of the largest values observed in a dataset. It is suitable when we are interested in events such as:

- The highest recorded rainfall in a year
- The peak river discharge in a season

- The maximum wind load experienced by a structure

The PDF, CDF, and moments (Mean and Standard Deviation) of this distribution are defined as follows:

$$f_X(x) = \alpha \exp(-\alpha(x-u) - \exp(-\alpha(x-u))) \tag{4.47}$$

$$F_X(x) = \exp(-\exp(-\alpha(x-u))) \tag{4.48}$$

$$\mu = u + \frac{0.577216}{\alpha} \tag{4.49}$$

$$\sigma = \frac{\pi}{\alpha\sqrt{6}} \tag{4.50}$$

where u is the location parameter (shifts the distribution), and $\alpha>0$ is the scale parameter (controls the spread).

This version of the distribution is particularly useful when designing protective infrastructure like levees, bridges, and retaining walls that must withstand worst-case scenarios.

Gumbel Minimum Distribution (Type I for Minima)

This form models the smallest values in a dataset. Some examples of typical applications of this model include:

- The weakest material strength in quality control
- The lowest daily temperature over several decades
- The minimum load-carrying capacity before failure

The PDF, CDF, and moments (Mean and Standard Deviation) of this distribution are defined as follows:

$$f_X(x) = \alpha \exp(\alpha(x-u) - \exp(\alpha(x-u))) \tag{4.51}$$

$$F_X(x) = 1 - \exp(-\exp(\alpha(x-u))) \tag{4.52}$$

$$\mu = u - \frac{0.577216}{\alpha} \tag{4.53}$$

$$\sigma = \frac{\pi}{\alpha\sqrt{6}} \tag{4.54}$$

where u is the location parameter, and $\alpha>0$ is the scale parameter

The Gumbel min distribution is ideal in applications where the minimum observed performance or capacity dictates the failure threshold—such as in fatigue analysis or early-life failures in products.

Application of the Gumbel Distribution in Extreme Wave Height Exceedance

The extreme wave height that is used in the design of an offshore oil storage facility for a design life of 25 years is a random variable with a mean of 30 ft and a standard deviation of 10 ft. Determine the probability that the extreme wave height will exceed 35 ft in 25 years.

Solution

Step 1: Derive the Gumbel distribution parameters

To use the Gumbel CDF, we first need to convert the given statistical parameters into Gumbel-specific parameters:

1. Alpha (α) is the scale parameter and can be computed using the standard deviation:

$$\alpha = \frac{\pi}{\sigma\sqrt{6}} = \frac{\pi}{10\sqrt{6}} \approx 0.1282$$

   The inverse of this value is used in the exponent of the Gumbel CDF:

$$\frac{1}{\alpha} \approx 0.1282$$

2. Location parameter (u) is derived by rearranging the formula for the mean of a Gumbel distribution:

$$\mu = u + \frac{0.5772}{\alpha} \quad \Rightarrow \quad u = \mu - \frac{0.5772}{\alpha} = 30 - \frac{0.5772}{0.1282} \approx 25.497$$

Step 2: Compute the probability using the Gumbel CDF

The cumulative distribution function for the Gumbel distribution is:

$$F_X(x) = \exp(-\exp(-\alpha(x - u)))$$

To find the probability that the wave height exceeds 35 ft:

- First, calculate $F_X(35)$, the probability that the wave height is less than or equal to 35 ft:v

$$F_X(35) = \exp(-\exp(-0.1282 \cdot (35 - 25.497))) \approx 0.744$$

- Then subtract this value from 1 to find the **exceedance** probability:

$$P(X \geq 35) = 1 - F_X(35) = 1 - 0.744 = 0.256$$

The result tells us that there is a 25.6% chance that the extreme wave height will exceed 35 ft over the design life of the facility. This type of analysis is crucial in civil and offshore engineering applications, where the safety, reliability, and resilience of structures must be designed against rare but potentially catastrophic events.

4.6.2.2 Extreme Value Distributions Type II—Fréchet Distribution

The Fréchet distribution is a type of EVD that is particularly suited to modeling maximum values of phenomena that are unbounded above but have a hard lower limit (e.g., zero or a physical minimum). Unlike the Gumbel distribution, which models extremes with exponential decay in both tails, the Fréchet distribution decays polynomially and is defined only for values greater than a positive threshold. It is especially useful when the maximum values grow significantly, and is widely used in fields where extreme values dominate system behavior.

Fréchet distributions are commonly used in:

- Hydrology: Modeling maximum daily or annual rainfalls, flood magnitudes, and river discharges.
- Lifetimes and Durability: Estimating the longest lifespan of materials.
- Sports Statistics: For instance, in horse racing, the distribution of fastest race times.
- Insurance and Risk Analysis: To model rare, catastrophic events (e.g., mega-disasters).

It is especially useful when the data contains heavy-tailed distributions, meaning the probability of observing extremely large values is significantly higher than that predicted by a normal distribution.

Suppose you're evaluating the potential for extreme flooding in a river basin. A Fréchet distribution can help you estimate the probability that a flood discharge will exceed a given threshold, even if you've never observed such a flood before. This is especially valuable when planning the design height of levees or spillways that must withstand rare but severe events. Similarly, if you're estimating maximum loads in a structure or highest wind speeds, the Fréchet distribution helps engineers make informed decisions about design safety margins, thereby reducing the likelihood of structural failure under extreme conditions.

Mathematical Formulation of the Fréchet Distribution

Let X_n be a random variable that follows the Fréchet distribution. Then the PDF is given by:

$$f_{X_n}(x) = \frac{k}{v_n}\left(\frac{v_n}{x}\right)^{k+1} \exp\left[-\left(\frac{v_n}{x}\right)^k\right] \tag{4.55}$$

where x is the extreme value of interest, k is the shape parameter (controls the tail heaviness), and v_n is the scale or location parameter (shifts the distribution).

The CDF of the Fréchet distribution, which can be used to determine the probability that the extreme value is less than some value x, is defined as follows:

$$F_{X_n}(x) = \exp\left[-\left(\frac{v_n}{x}\right)^k\right] \tag{4.56}$$

The moments (mean and variance) of this distribution are given by Eqs. 4.57 and 4.58 below:

$$\mu_{X_n} = v_n \Gamma\left(1 - \frac{1}{k}\right) \tag{4.57}$$

$$\sigma_{X_n}^2 = v_n^2\left[\Gamma\left(1 - \frac{2}{k}\right) - \Gamma^2\left(1 - \frac{1}{k}\right)\right], \quad \text{for } k \geq 2 \tag{4.58}$$

where Γ(t) is the Gamma function, defined as:

$$\Gamma(t) = \int_0^\infty r^{t-1} e^{-r} dr \tag{4.59}$$

where t is the input (shape) argument of the Gamma function and is typically a positive real number in this context (e.g., $t = 1 - \frac{1}{k} or\, t = 1 - \frac{2}{k}$). The variable r is a dummy variable of integration, meaning it has no physical meaning on its own and is used only to evaluate the integral from 0 to ∞.

It generalizes the factorial function to real and complex numbers.

The Fréchet distribution is defined only for positive values, meaning its domain is limited to x > 0; it assigns zero probability to any value less than or equal to zero. This makes it particularly suitable for modeling quantities that naturally cannot be negative, such as rainfall, wind speed, or structural loads. One of its most important properties is its heavy tail, which means that the probability of extremely large values decreases slowly rather than rapidly. Additionally, the distribution is highly flexible due to its shape parameter k, which controls the tail behavior. By adjusting this parameter, the Fréchet distribution can be tailored to fit a wide range of datasets where the occurrence of extreme maximum values plays a critical role in design and risk assessment.

Application of Fréchet Distribution in Estimating Wind Speed Exceedance

In a specific geographic location, engineers want to determine a threshold wind velocity that is exceeded with only a 2% chance annually. Based on long-term measurements, the maximum annual wind velocity is modeled using a Fréchet (Type II Extreme Value) distribution. The available data reports the mean wind speed as 55 mph and the standard deviation as 12.8 mph.

Solution

You are given:

- Mean of annual maximum wind speed: $\mu_{x_n} = 55$ mph
- Standard deviation: $\sigma_{x_n} = 12.8$ mph
- Desired exceedance probability: $P(Y \geq y) = 0.02$

We assume the data follow a Fréchet distribution with shape parameter k and scale parameter $\mathrm{v_n}$. We are asked to compute the value of y such that there is a 2% chance the wind speed exceeds it.

Step 1: Use the Theoretical Mean and Variance of Fréchet

For a Fréchet distribution:

- The mean is:

$$\mu_{X_n} = v_n \Gamma\left(1 - \frac{1}{k}\right)$$

- The variance is:

$$\sigma_{X_n}^2 = v_n^2\left[\Gamma\left(1 - \frac{2}{k}\right) - \Gamma^2\left(1 - \frac{1}{k}\right)\right], \quad \text{for } k \geq 2$$

Given:

- $\mu_{x_n} = 55$
- $\sigma_{x_n} = 12.8 \Rightarrow \sigma_{xn}^2 = 163.84$

Now, substitute into the equations:

1. Mean:

$$55 = v_n \cdot \Gamma\left(1 - \frac{1}{k}\right)$$

2. Variance

$$163.84 = v_n^2 \cdot \left[\Gamma\left(1 - \frac{2}{k}\right) - \Gamma^2\left(1 - \frac{1}{k}\right)\right]$$

Step 2: Eliminate v_n to Find k

Divide equation (2) by the square of equation (4.1):

$$\frac{163.84}{55^2} = \frac{\Gamma\left(1-\frac{2}{k}\right) - \Gamma^2\left(1-\frac{1}{k}\right)}{\Gamma^2\left(1-\frac{1}{k}\right)}$$

$$\Rightarrow \quad 0.0542 = \frac{\Gamma\left(1-\frac{2}{k}\right)}{\Gamma^2\left(1-\frac{1}{k}\right)} - 1$$

Step 3: Solve for v_n

Using Eq. (4.1) again:

$$55 = v_n \cdot \Gamma\left(1 - \frac{1}{6.5}\right)$$

Evaluate $\Gamma\left(1-\frac{1}{6.5}\right) \approx \Gamma(0.8462) \approx 1.112$

$$v_n = \frac{55}{1.112} \approx 49.4 \text{ mph}$$

Step 4: Solve for the threshold wind velocity

We use the CDF of the Fréchet distribution

$$F_Y(y) = 1 - e^{-(v_n/y)^k}$$

Given that:

$$P(Y \geq y) = 0.02 \Rightarrow F_Y(y) = 1 - 0.02 = 0.98$$

So:

$$0.98 = 1 - e^{-(49.4/y)^{6.5}} \Rightarrow e^{-(49.4/y)^{6.5}} = 0.02$$

Take natural log:

$$-(49.4/y)^{6.5} = \ln(0.02) \Rightarrow (49.4/y)^{6.5} = -\ln(0.02) \approx 3.912$$

Take 6.5th root:

$$\frac{49.4}{y} = 3.912^{1/6.5} \approx 1.085 \Rightarrow y = \frac{49.4}{1.085} \approx 91 \text{ mph}$$

The wind velocity that will be exceeded with probability 0.02 is approximately 91 mph.

4.6.2.3 Extreme Value Distributions Type III—Weibull Distribution

The next member of the EVD family is the Weibull distribution, categorized as Type III. This distribution is particularly valuable in reliability engineering, survival analysis, and hazard modeling because it provides the flexibility to capture both minimum and maximum extremes, depending on the parameterization. The Weibull distribution is defined by three continuous parameters: the shape parameter $\alpha > 0$, the scale parameter $\beta > 0$, and the location parameter γ. If the location parameter γ is set to zero, the distribution simplifies to the two-parameter Weibull model, which is widely used due to its simpler form and sufficient accuracy in many applications.

Mathematical Formulation of the Weibull Distribution

The probability density function (PDF) for the Weibull distribution is:

$$f(x) = \frac{\alpha}{\beta}\left(\frac{x-\gamma}{\beta}\right)^{\alpha-1} \exp\left[-\left(\frac{x-\gamma}{\beta}\right)^{\alpha}\right] \tag{4.60}$$

and the cumulative distribution function (CDF) is:

$$F(x) = 1 - \exp\left[-\left(\frac{x-\gamma}{\beta}\right)^{\alpha}\right], \quad \text{for } x \geq \gamma \tag{4.61}$$

The moments of the distribution—mean μ and variance σ^2—are determined using the gamma function $\Gamma(\cdot)$, and are given as:

$$\mu(x) = \gamma + \beta\left[\Gamma\left(1 + \frac{1}{\alpha}\right)\right] \tag{4.62}$$

$$\sigma^2(x) = \beta^2\left[\Gamma\left(1 + \frac{2}{\alpha}\right) - \Gamma^2\left(1 + \frac{1}{\alpha}\right)\right] \tag{4.63}$$

The use of the gamma function in these expressions accounts for the variability and skewness of the distribution. These characteristics make the Weibull distribution exceptionally versatile for modeling phenomena such as the time to failure of mechanical components, the life expectancy of structural elements, and environmental extremes like minimum daily temperatures or drought durations.

When the location parameter γ is equal to zero, a specific case of the Weibull distribution, known as the Two-Parameter Weibull Distribution, arises. In practice, setting $\gamma = 0$ simplifies the distribution without significantly limiting its usefulness for many engineering applications.

The mathematical relationships that define the two-parameter Weibull distribution are characterized by only two parameters:

- α: the shape parameter, which controls the form or steepness of the distribution.
- β: the scale parameter, which stretches or compresses the distribution along the x-axis.

The PDF of the two-parameter Weibull distribution is given by:

$$f(x) = \frac{\alpha}{\beta}\left(\frac{x}{\beta}\right)^{\alpha-1} \exp\left[-\left(\frac{x}{\beta}\right)^{\alpha}\right] \tag{4.64}$$

This function describes the likelihood of observing a specific value of the random variable X. For $x \geq 0$, this formula is valid and captures how the distribution's peak and spread are influenced by the values of α and β. A larger α typically results in a sharper peak, while a larger β broadens the distribution.

The CDF for the two-parameter Weibull distribution is expressed as:

$$F(x) = 1 - \exp\left[-\left(\frac{x}{\beta}\right)^{\alpha}\right], \quad \text{for } x \geq 0 \tag{4.65}$$

This function gives the probability that the variable X is less than or equal to a specific value x. The mean $\mu(x)$ and variance $\sigma^2(x)$ of the two-parameter Weibull distribution are defined as:

$$\mu(x) = \beta\left[\Gamma\left(1 + \frac{1}{\alpha}\right)\right] \tag{4.66}$$

$$\sigma^2(x) = \beta^2\left[\Gamma\left(1 + \frac{2}{\alpha}\right) - (\mu(x))^2\right] \tag{4.67}$$

Here, $\Gamma(\cdot)$ is the gamma function, which generalizes the factorial function for continuous values. These moment expressions are important because they allow engineers to quantify the central tendency and spread of failure times or minimum extreme observations. Such values are fundamental in designing components and systems with a high degree of safety and reliability.

Application of Weibull Distribution in Reliability Analysis of a Flexible Membrane

The time to failure for a flexible membrane follows the Weibull distribution with $\beta = 300$ months and $\alpha = 2$. What is the probability that the membrane will last more than 200 months?

Solution

The CDF of the Weibull distribution is defined as:

$$F(x) = 1 - \exp\left[-\left(\frac{x}{\beta}\right)^{\alpha}\right]$$

where x is the time (in months), $\beta = 300$ is the scale parameter, and $\alpha = 2$ is the shape parameter. We want to calculate:

$$P(X > 200) = 1 - F(200)$$

Substitute the values directly into the CDF equation:

$$F(200) = 1 - \exp\left(-\left(\frac{200}{300}\right)^{2}\right) = 1 - \exp(-0.4444) \approx 1 - 0.6412 = 0.3588$$

Thus, the probability that the membrane fails before 200 months is 35.88%.

Using the complement rule, the probability that the membrane lasts longer than 200 months is:

$$P(X > 200) = 1 - F(200) = 1 - 0.3588 - 0.6412 \rightarrow 64.12\%$$

This problem can also be conveniently addressed by using Excel's built-in function `WEIBULL.DIST` as follows:

```
=WEIBULL.DIST(200, 2, 300, TRUE)
```

This returns `0.35882`, which corresponds to F(200).

To find the survival probability P(X > 200), subtract this value from 1:

$$P(X > 200) = 1 - 0.35882 = 0.64118 \approx 64.1\%$$

Application of Weibull Distribution in Estimating Survival Probability

The lifetime (time to failure) of a flexible membrane is modeled as a Weibull-distributed random variable with shape parameter = 2 and scale parameter = 300 months.

What is the probability that the membrane survives beyond 200 months?

Solution

The CDF is given by:

$$F(x) = 1 - \exp\left[-\left(\frac{x}{\beta}\right)^{\alpha}\right]$$

The probability of surviving beyond time is:

$$P(X > x) = 1 - F(x)$$

To estimate α and β, we use the moment equations.

$$\mu = \beta \cdot \Gamma\left(1 + \frac{1}{\alpha}\right) \Rightarrow 1000 = \beta \cdot \Gamma\left(1 + \frac{1}{\alpha}\right)$$
$$\sigma^2 = \beta^2\left[\Gamma\left(1 + \frac{2}{\alpha}\right) - \Gamma^2\left(1 + \frac{1}{\alpha}\right)\right]$$

Given:

$$\mu = 1000,$$
$$\sigma = 400 \Rightarrow \sigma^2 = 160000$$
$$160000 = \beta^2\left[\Gamma\left(1 + \frac{2}{\alpha}\right) - \Gamma^2\left(1 + \frac{1}{\alpha}\right)\right]$$

To solve for α, we eliminate β from the equations. This can be done by taking logarithms:

From the first equation:

$$\ln(1000) = \ln(\beta) + \ln\left(\Gamma\left(1 + \frac{1}{\alpha}\right)\right)$$

From the second equation:

$$\ln(160000) = 2\ln(\beta) + \ln\left[\Gamma\left(1 + \frac{2}{\alpha}\right) - \Gamma^2\left(1 + \frac{1}{\alpha}\right)\right]$$

Subtracting twice the first equation from the second, we eliminate β and simplify

$$h(\alpha) = \ln\Gamma\left(1 + \frac{2}{\alpha}\right) - 2\ln\Gamma\left(1 + \frac{1}{\alpha}\right) - \ln(1.16) = 0$$

This is a nonlinear equation in α and must be solved numerically

We can solve $h(\alpha) = 0$ using Goal Seek in Excel (*Data > Forecast|What If Analysis > Goal Seek*):

1. Enter a formula for using GAMMALN.

```
=GAMMALN(1+2/B3)-2*GAMMALN(1+1/B3)-LN(1.16)
```

2. Set the formula cell to 0 by changing α.

$$\alpha \approx 2.688$$
$$\beta = \frac{1000}{\Gamma(1 + 1/\alpha)} \approx 1124.670$$

Now we compute the CDF:

$$F(2000) = 1 - \exp\left[-\left(\frac{2000}{1124.67}\right)^{2.688}\right] \approx 0.0091$$

So the probability that the engine survives beyond 2,000 hours is

$$P(X > 2000) = 1 - F(2000) = 1 - 0.0091 = 0.9909 \quad \text{or} \quad 99.09\%$$

The details of using Excel are provided in the following Fig. 4.53.

4.7 Selection of Distribution Functions

In civil engineering, the accurate representation of uncertainty is essential for safe and efficient design. One of the foundational steps in probabilistic modeling is the selection of an appropriate probability distribution function to describe a random variable. This distribution encapsulates the uncertainty and variability inherent in natural processes or engineered systems.

4.7.1 The Importance of Distribution Selection

The PDF of a random variable provides a mathematical description of the likelihood of occurrence of different outcomes. Typically, this is expressed using a standard distribution type—such as the Normal, Lognormal, Beta, or Gumbel distributions—defined by one or more parameters (e.g., mean, standard deviation, shape factors).

However, in most practical scenarios, the true underlying distribution of a phenomenon is not known in advance. While historical or observed data can offer some guidance, it is often insufficient for a clear-cut decision. Moreover, the results

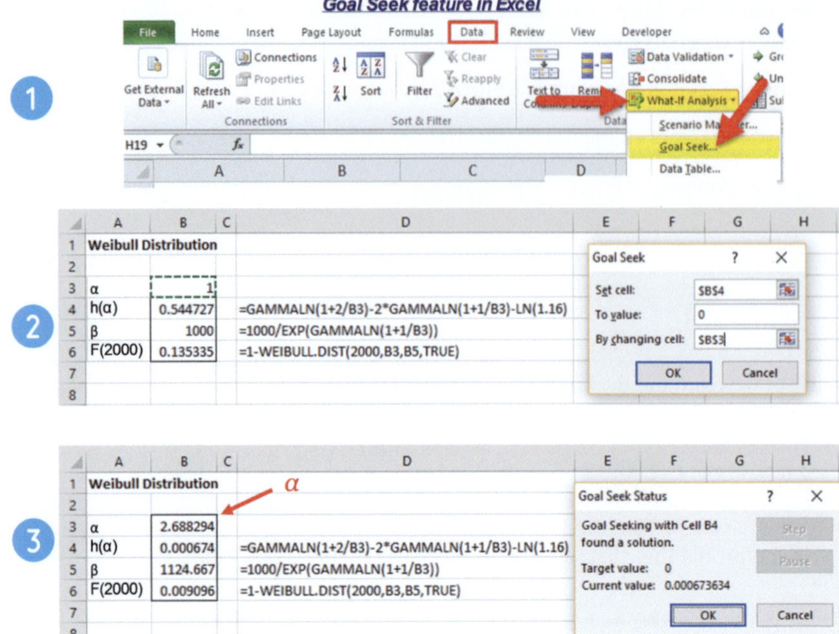

Fig. 4.53 Demonstration of using Excel's Goal Seek feature to estimate the shape parameter αand scale parameter β, followed by the evaluation of the survival probability for a component lasting beyond 2,000 hours using the Weibull distribution

of reliability and risk assessments are frequently sensitive to the tails of the distribution—the parts that represent rare but potentially catastrophic events. Therefore, selecting an appropriate distribution type is a decision of high consequence, and it must be made with care and justification.

4.7.2 Step-by-Step Procedure for Distribution Selection

The selection process generally follows three key steps:

1. Initial Selection Based on Prior Knowledge,

Begin by identifying a shortlist of possible distribution types. This selection is informed by:

- Previous studies or experience with similar types of problems,
- Physical understanding of the phenomenon,
- Analytical derivations or empirical laws.

At this stage, engineers may consult scientific literature, guidelines, or recognized databases to align with standard practices.

2. **Parameter Estimation from Observed Data**

After proposing potential distribution types, the next step is to fit them to available data. This involves estimating the parameters (e.g., mean, variance, shape, scale) using one or more of the following techniques:

- Method of Moments: Matching sample moments (e.g., mean and variance) to theoretical moments.
- Least-Squares Estimation: Minimizing the difference between observed and expected values.
- Maximum Likelihood Estimation (MLE): Finding parameters that maximize the probability (likelihood) of observing the given data.
- Visual Inspection: Plotting the data on probability paper or histograms to visually assess fit.

3. **Verification Through Goodness-of-Fit Tests**

Once candidate distributions are fitted to the data, they must be evaluated for adequacy. This is achieved using statistical goodness-of-fit tests such as:

- Chi-square (χ^2) test,
- Kolmogorov–Smirnov (K–S) test,
- Anderson–Darling (A–D) test.

These tests measure how well the observed data conform to the expected probabilities under a given distribution model. Each test has its strengths; for example, the Anderson–Darling test gives more weight to the tails of the distribution, making it especially useful for engineering applications where extreme events are of interest. The details of these tests, along with the mathematical formulations and corresponding coding, are provided in Sect. 4.5.1.2, titled 'Evaluating Goodness-of-Fit Using Statistical Tests'.

4.7.3 Best Practices for Final Selection

- Tail Sensitivity Consideration

If two or more distributions provide statistically comparable fits, the choice should favor the one that better models the behavior of the tail data. This is particularly critical in hazard analysis, where the rarest and most extreme events (e.g., 100-year floods, maximum wind speeds, earthquake magnitudes) can be the most impactful.

- Visual Support

Complement statistical tests with graphical tools such as probability plots, Q-Q plots, or cumulative distribution overlays. These visualizations help intuitively assess how well the model captures the behavior of the dataset.

- Documentation and Transparency

When a distribution is selected, it is essential to document:

- The rationale behind the choice,
- The methods used for parameter estimation,
- The outcome of fit tests,
- Any assumptions or approximations made.

This transparency ensures reproducibility and builds confidence in risk assessment.

The concepts presented in this chapter rely on established foundations in descriptive statistics, uncertainty quantification, and distribution analysis widely applied in engineering data interpretation and model validation (Anderson & Darling, 1952; Ang & Tang, 2007; Der Kiureghian & Ditlevsen, 2009; Ebtehaj & Bonakdari, 2022, 2024a, 2024b; Ebtehaj et al., 2020a, 2020b, 2020c, 2021; Fisher, 1922; Helton et al., 2004; Hintze & Nelson, 1998; International Organization for Standardization, 2006, 2023; Joint Committee for Guides in Metrology (JCGM), 2008; Massey, 1951; McGill et al., 1978; Montgomery & Runger, 2018; Pearson, 1900; Ross, 2019; SciPy community, 2026; Silverman, 1986; Tukey, 1977).

4.8 Summary

This chapter presented a comprehensive foundation for understanding and applying descriptive statistics, probability theory, and uncertainty analysis in civil and environmental engineering. The chapter began by emphasizing the importance of data exploration as the first step in any risk-based or probabilistic assessment. Descriptive statistics were introduced as essential tools for organizing, summarizing, and interpreting raw data, enabling engineers to extract meaningful information from observations before advancing to probabilistic modeling.

Measures of central tendency, dispersion, and percentiles were discussed alongside graphical techniques such as histograms, box plots, and violin plots. These methods were shown to provide complementary numerical and visual insights into data structure, variability, skewness, and the presence of outliers. Together, they form the basis for identifying data quality issues and understanding underlying patterns that influence engineering decisions.

The chapter then addressed uncertainty as an inherent feature of engineering systems. Key sources of uncertainty, including natural variability, model limitations, and data-related uncertainty, were examined within the context of civil engineering design and hazard assessment. The distinction between aleatory and epistemic uncertainty was highlighted to clarify how different uncertainties should be interpreted and managed. Random variables were introduced as the mathematical framework for representing uncertainty, leading to a discussion of probability density functions, cumulative distribution functions, and statistical moments.

Building on these fundamentals, the chapter explored probability distributions commonly used in engineering practice. Both discrete and continuous distributions were examined, with emphasis on their interpretation, parameterization, and practical relevance. Random processes were introduced to model time-dependent phenomena such as traffic arrivals, accident occurrences, and natural hazards, while extreme value concepts were presented to address low-probability, high-impact events critical to safe and resilient design.

The final sections focused on the selection and evaluation of probability distribution functions. Parameter estimation methods, including the method of moments, least-squares estimation, and maximum likelihood estimation, were presented alongside goodness-of-fit tests such as Chi-square, Kolmogorov–Smirnov, and Anderson–Darling. These tools provide a systematic framework for choosing appropriate probabilistic models and validating their suitability for engineering applications.

Overall, this chapter established the statistical and probabilistic foundations required for hazard analysis, reliability assessment, and risk-informed decision-making. By integrating descriptive statistics, uncertainty characterization, random processes, extreme value analysis, and distribution selection, engineers are equipped with a rigorous framework for moving from observed data to defensible, uncertainty-aware engineering judgments. These concepts form a critical bridge between raw data analysis and the advanced risk and reliability methods introduced in subsequent chapters.

References

Anderson, T. W., & Darling, D. A. (1952). Asymptotic theory of certain "goodness-of-fit" criteria based on stochastic processes. *The Annals of Mathematical Statistics, 23*(2), 193–212. https://doi.org/10.1214/aoms/1177729437

Ang, A. H.-S., & Tang, W. H. (2007). *Probability concepts in engineering: Emphasis on applications to civil and environmental engineering* (2nd ed.). Wiley. https://books.google.com/books?id=G4IoAQAAMAAJ

Der Kiureghian, A., & Ditlevsen, O. (2009). Aleatory or epistemic? Does it matter? *Structural Safety, 31*(2), 105–112. https://doi.org/10.1016/j.strusafe.2008.06.020

Ebtehaj, I., & Bonakdari, H. (2022). A reliable hybrid outlier robust non-tuned rapid machine learning model for multi-step ahead flood forecasting in Quebec, Canada. *Journal of Hydrology, 614*, 128592. https://www.sciencedirect.com/science/article/pii/S0022169422011623

Ebtehaj, I., & Bonakdari, H. (2024a). Generalized structure of group method of data handling: Novel technique for flash flood forecasting. *Water Resources Management, 38*(9), 3235–3253. https://doi.org/10.1007/s11269-024-03811-1

Ebtehaj, I., & Bonakdari, H. (2024b). CNN versus LSTM: A comparative study for hourly precipitation intensity prediction as a key factor for flood forecasting frameworks. *Atmosphere, 15*(9), 1082. https://www.mdpi.com/2073-4433/15/9/1082

Ebtehaj, I., Bonakdari, H., Zaji, A. H., & Gharabaghi, B. (2020a). Evolutionary optimization of neural network to predict sediment transport without sedimentation. *Complex and Intelligent Systems*, 1–16. https://doi.org/10.1007/s40747-020-00213-9

Ebtehaj, I., Bonakdari, H., Safari, M. J. S., Gharabaghi, B., Zaji, A. H., Madavar, H. R., & Mehr, A. D. (2020b). Combination of sensitivity and uncertainty analyses for sediment transport modeling in sewer pipes. *International Journal of Sediment Research, 35*(2), 157–170. https://www.sciencedirect.com/science/article/pii/S1001627918303810

Ebtehaj, I., Bonakdari, H., Zeynoddin, M., Gharabaghi, B., & Azari, A. (2020c). Evaluation of preprocessing techniques for improving the accuracy of stochastic rainfall forecast models. *International Journal of Environmental Science and Technology, 17*(1), 505–524. https://doi.org/10.1007/s13762-019-02361-z

Ebtehaj, I., Sammen, S. S., Sidek, L. M., Malik, A., Sihag, P., Al-Janabi, A. M. S., Chau, K. W., & Bonakdari, H. (2021). Daily water level prediction by using new hybrid machine learning models. *Engineering Applications of Computational Fluid Mechanics, 15*(1), 1343–1361. https://doi.org/10.1080/19942060.2021.1966837

Fisher, R. A. (1922). On the mathematical foundations of theoretical statistics. *Philosophical Transactions of the Royal Society of London. Series A, 222*(594–604), 309–368. https://doi.org/10.1098/rsta.1922.0009

Helton, J. C., Johnson, J. D., & Oberkampf, W. L. (2004). An exploration of alternative approaches to the representation of uncertainty in model predictions. *Reliability Engineering and System Safety, 85*(1–3), 39–71. https://doi.org/10.1016/j.ress.2004.03.025

Hintze, J. L., & Nelson, R. D. (1998). Violin plots: A box plot-density trace synergism. *The American Statistician, 52*(2), 181–184. https://doi.org/10.1080/00031305.1998.10480559

International Organization for Standardization. (2006). *Statistics—Vocabulary and symbols—Part 1: General statistical terms and terms used in probability (ISO 3534-1:2006).* https://www.iso.org/standard/40145.html

International Organization for Standardization. (2023). *Accuracy (trueness and precision) of measurement methods and results—Part 1: General principles and definitions (ISO 5725-1:2023).* https://www.iso.org/standard/69418.html

Joint Committee for Guides in Metrology (JCGM). (2008). *Evaluation of measurement data—Guide to the expression of uncertainty in measurement (JCGM 100:2008).* Bureau International des Poids et Mesures. https://www.bipm.org/documents/20126/2071204/JCGM_100_2008_E.pdf

Massey, F. J., Jr. (1951). The Kolmogorov-Smirnov test for goodness of fit. *Journal of the American Statistical Association, 46*(253), 68–78. https://doi.org/10.2307/2280095

McGill, R., Tukey, J. W., & Larsen, W. A. (1978). Variations of box plots. *The American Statistician, 32*(1), 12–16. https://doi.org/10.1080/00031305.1978.10479236

Montgomery, D. C., & Runger, G. C. (2018). *Applied statistics and probability for engineers* (7th ed.). Wiley. https://www.wiley.com/en-ca/Applied%2BStatistics%2Band%2BProbability%2Bfor%2BEngineers%2C%2B7th%2BEdition-p-9781119400363

Pearson, K. (1900). X. On the criterion that a given system of deviations from the probable in the case of a correlated system of variables is such that it can be reasonably supposed to have arisen from random sampling. *The London, Edinburgh, and Dublin Philosophical Magazine and Journal of Science, 50*(302), 157–175. https://doi.org/10.1080/14786440009463897

Ross, S. M. (2019). *A first course in probability* (10th ed.). Pearson. https://books.google.com/books?id=vQe-tAEACAAJ

SciPy Community. (2026). *Statistical functions (scipy.stats)—SciPy v1.17.0 Manual.* https://docs.scipy.org/doc/scipy/reference/stats.html

Silverman, B. W. (1986). *Density estimation for statistics and data analysis.* Chapman & Hall/CRC. https://doi.org/10.1201/9781315140919

Tukey, J. W. (1977). *Exploratory data analysis.* Addison-Wesley. https://books.google.com/books?id=UT9dAAAAIAAJ

Chapter 5
Decision Trees, Utility Theory, and Risk-Based Decisions in Civil Engineering

5.1 Introduction

In the domain of civil engineering, decision-making often involves complex trade-offs between environmental constraints, financial feasibility, regulatory compliance, and societal needs. Amid this intricate landscape, decision tree analysis has emerged as a powerful and intuitive method to navigate uncertainty and assess alternative solutions using a systematic approach. This chapter introduces the conceptual foundation, methodological structure, and real-world applicability of decision trees as a decision-support tool tailored for civil engineering problems.

Decision trees are rooted in the principles of structured decision-making and probabilistic modeling. They offer a graphical and hierarchical representation of sequential decisions, where each branch represents a possible course of action or event outcome. This format allows engineers and project managers to visualize decision paths clearly, calculate expected values based on uncertainties, and select optimal strategies based on predefined objectives such as cost minimization, risk reduction, or sustainability enhancement.

In the context of civil infrastructure planning, construction management, and environmental design, decision trees have proven to be particularly effective in evaluating options under uncertainty. For instance, when designing flood mitigation structures, selecting appropriate materials, or phasing construction projects, engineers are often faced with multiple outcomes influenced by stochastic variables such as climate patterns, market prices, or regulatory shifts. Decision trees help to quantify these variables and integrate them into the planning process by incorporating probability distributions, outcome values, and risk preferences.

This chapter not only introduces the mathematical and logical structure of decision trees but also emphasizes their educational utility. The examples provided are tailored to undergraduate students in civil engineering, using real-world analogs such as material selection for building construction, infrastructure investment planning, as well as maintenance scheduling. These scenarios are designed to illustrate how

H. Bonakdari et al., *Hazard and Risk Assessment in Civil Engineering Design*,
https://doi.org/10.1007/978-3-032-23600-5_5

complex decisions can be broken down into manageable steps, each accompanied by a quantitative evaluation.

This chapter also highlights how decision trees complement other modeling approaches such as simulation, sensitivity analysis, and multi-criteria decision-making (MCDM) techniques. By integrating these frameworks, students and practitioners can enhance their analytical capabilities and apply robust, evidence-based reasoning in the face of uncertainty.

In summary, the chapter aims to equip readers with the conceptual tools and applied knowledge necessary to formulate, analyze, and interpret decision problems using decision trees. Through structured examples, graphical illustrations, and practice problems, students will gain hands-on experience with this method, preparing them for more advanced decision analytics in their professional careers.

5.2 Decision Analysis Under Uncertainty: Tools and Techniques

Civil engineering projects encompass several critical phases: design, construction, operation, maintenance, and decommissioning (Fig. 5.1). Each phase involves strategic and technical decision-making, where engineers must evaluate numerous interrelated factors, such as environmental loads, geotechnical variability, hydrological extremes, material longevity, labor availability, and budget constraints, for example. In an ideal scenario, these decisions would be made with perfect foresight. However, civil engineers almost always operate in environments filled with incomplete, imprecise, and evolving information.

Uncertainty is a defining characteristic of engineering practice. Whether estimating flood return periods in hydrologic modeling, forecasting soil bearing capacity in geotechnical design, or anticipating the service life of aging infrastructure, engineers must routinely make high-stakes decisions without a complete knowledge of future conditions. These uncertainties arise from multiple sources (Fig. 5.2) including natural variability (e.g., weather, river flow, seismic activity); data limitations (e.g., sparse sampling, old topographic maps, sensor errors, high cost of data collection); model imperfections (e.g., empirical assumptions, simplified boundary conditions), and human factors (e.g., labor shortages, policy changes, or project delays). This inherent uncertainty makes decision-making challenging, as engineers lack complete data on future environmental conditions, material performance over time, or unforeseen construction issues. Decision Trees provide a structured, visual framework for navigating these uncertainties, enabling engineers to systematically evaluate each decision path in the face of incomplete or probabilistic information. Figure 5.2 shows some ways that Decision Trees can support each project phase.

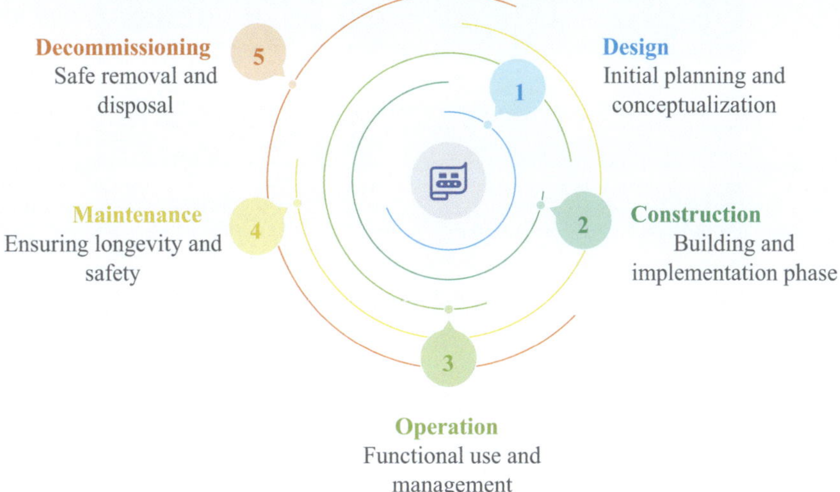

Fig. 5.1 Phases of civil engineering projects

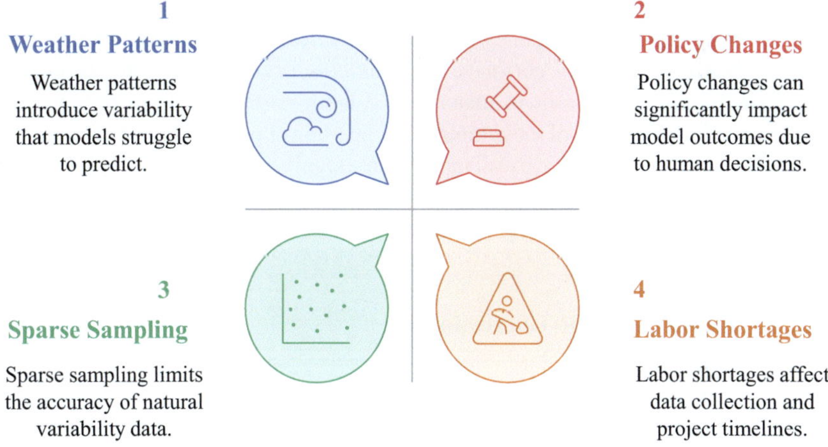

Fig. 5.2 Sources of uncertainties in engineering practice

5.2.1 Design Phase

In the design phase, engineers must make crucial decisions that define a project's core functionality, structural safety, serviceability, and long-term sustainability. These decisions influence everything from material selection and geometry to load-bearing strategies, drainage planning, and environmental resilience. Civil engineers must address multiple, often competing, objectives—such as minimizing costs while ensuring long-term safety and meeting evolving regulatory requirements.

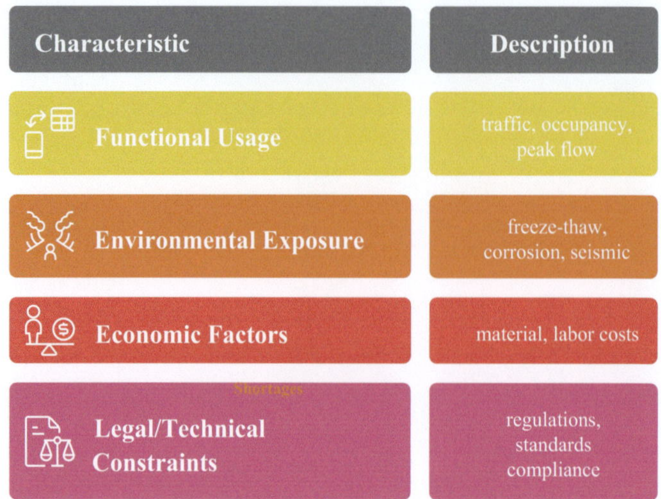

Fig. 5.3 Common factors necessitating assumptions during the design phase

Design decisions are shaped by assumptions (Fig. 5.3) about functional usage (such as traffic volumes, occupancy levels, or peak flow rates), environmental exposure (like freeze-thaw cycles, corrosive environments, or seismic activity), economic factors (including fluctuations in material and labor costs), and legal or technical constraints. However, most of these inputs are uncertain or probabilistic. For example, the actual capacity of a retaining wall may deviate significantly from the initial design assumptions due to unforeseen sub-strata in the surrounding soil structure or due to varying groundwater levels.

5.2.1.1 Managing Design Uncertainties

Decision Trees provide a structured, systematic way to explore different design scenarios and the consequences of these decisions that are brought about by uncertainty. By visualizing design options as branching pathways, each associated with conditional outcomes and their respective probabilities, engineers can assess the trade-offs among cost, performance, risk, and adaptability.

For instance, in the design of a coastal protection system, engineers might evaluate three alternatives: a reinforced concrete seawall, a steel sheet pile wall, or a composite-based hybrid system. Each alternative would be examined under different future sea level rise projections, such as 0.5, 1.0, and 1.5 m by the end of the century. For each scenario, associated factors such as expected service life, failure probability, long-term maintenance needs, and environmental compliance costs are estimated. The decision tree supports the evaluation of each option's long-term viability and cost-efficiency, making it easier to justify design decisions to stakeholders and regulators.

5.2.1.2 Anticipating Future Requirements

Infrastructure systems are expected to perform under future conditions that are rarely fully known at the time of design. A road designed today might eventually serve an expanded residential zone, a higher volume of freight vehicles, or an altered drainage network. Decision Trees help engineers simulate these future use scenarios and evaluate the impact on design choices.

Consider the example of designing a stormwater detention basin in a suburban development. Future urbanization and climate-induced rainfall intensification introduce uncertainty in runoff volumes. Engineers can model different combinations of urban growth rates and rainfall scenarios (e.g., the Coupled Model Intercomparison Project Phase 5 representative concentration pathway (RCP) 4.5 and 8.5 projections) as separate branches in a decision tree. Each branch incorporates the probability of occurrence, construction and retrofitting costs, overflow risks, and regulatory penalties. This allows the selection of a basin design that balances initial cost with adaptability, ensuring it can accommodate a reasonable range of future conditions without excessive over-design or vulnerability.

5.2.2 Construction Phase

The construction phase involves translating the design into a physical reality through procurement, site preparation, material assembly, and project execution. While detailed plans are developed during design, actual site conditions and implementation dynamics often introduce significant uncertainty. These can arise from unexpected subsurface conditions, supply chain delays, labor shortages, weather disruptions, and discrepancies between the as-designed and as-built environments.

Managing such uncertainties is critical, as they can lead to cost overruns, schedule delays, quality issues, or safety incidents. Decision Trees serve as a practical tool for structuring contingency planning and making informed, real-time decisions when facing construction-related risks.

5.2.2.1 Managing Construction Uncertainties

Decision Trees can help construction managers and engineers evaluate multiple execution strategies or mitigation measures when facing uncertainty in field conditions. For example, during excavation for a deep foundation, encountering unanticipated groundwater inflow may necessitate quick decisions between dewatering, redesigning the foundation type, or delaying construction until further geotechnical investigation is completed.

Each response path can be represented as a branch in a decision tree, with associated costs, time delays, risks to adjacent infrastructure, and environmental compliance implications. Probabilities are assigned based on historical data, expert judgment, or probabilistic geotechnical models. The tree allows for comparison of expected costs and consequences, supporting transparent, justifiable decision-making under pressure. In large-scale projects such as dam or tunnel construction, decision trees may be integrated into construction risk management plans. Various excavation methods (such as drill-and-blast versus tunnel boring machine) may be considered, each with different cost structures, timelines, and risk profiles under uncertain rock mass behavior. A significant advantage of using a decision tree is that it can be updated dynamically as construction progresses and new information becomes available.

5.2.2.2 Responding to Resource and Supply Variability

Construction activities are also susceptible to uncertainties in the availability and cost of resources such as steel, cement, and fuel. Global market fluctuations, transportation bottlenecks, or local political issues can delay delivery or inflate prices. Decision Trees can support procurement planning by modeling scenarios such as early bulk purchase, just-in-time delivery, or the use of alternative materials. For each path, engineers assess cost risk, scheduling flexibility, storage requirements, and contract penalties. This helps contractors and project managers identify the most resilient supply strategy within the broader context of project delivery. In rapidly evolving conditions, such as post-disaster infrastructure recovery or pandemic-related restrictions, real-time decision trees supported by digital construction management systems can be used to navigate changing labor availability, access restrictions, and health and safety requirements.

5.2.3 Operation Phase

The operation phase marks the period during which the infrastructure is in service: bridges carry traffic, water and wastewater treatment plants purify water, levees protect communities, and buildings are occupied and maintained. Although the structure is complete, engineers must still make numerous decisions to ensure long-term performance, user safety, environmental compliance, and economic efficiency. Uncertainty remains a major concern during operation, particularly regarding degradation over time, variable usage patterns, unexpected loading, and evolving regulatory or environmental conditions. These uncertainties can impact structural performance, service reliability, and maintenance needs. Decision Trees offer a structured way to evaluate alternative responses to emerging operational challenges. They support lifecycle management by helping engineers and operators systematically assess potential outcomes under different conditions and response strategies.

5.2.3.1 Managing Performance Degradation and Maintenance

Civil infrastructure deteriorates with time due to a variety of factors, including environmental exposure, material fatigue, chemical attack, or operational stress. One example of the deterioration of heavy civil infrastructure is the corrosion of reinforced concrete in bridges in the presence of chlorides, especially in cold regions where de-icing salts are used. In this case, it may be useful for an engineer to use a decision tree to determine the appropriate time and method of intervention. For a highway overpass whose barrier walls are showing early signs of spalling, three maintenance options might be considered:

- Immediate patch repair,
- Cathodic protection installation, or
- Monitoring and deferred intervention.

Each branch of the decision tree is associated with probabilities reflecting future deterioration rates, intervention costs, safety risks, and the likelihood of regulatory non-compliance with regional standards. To provide up-to-date risk assessment, these decision trees can incorporate data from visual inspections, structural health monitoring systems, and non-destructive testing to optimize maintenance timing.

5.2.3.2 Adapting to Changing Usage Patterns

Over the course of their useful life, infrastructure will often experience changes in use that diverge from their original design assumptions. A commuter bridge may face heavier truck traffic due to nearby industrial development, or a stormwater network might become overloaded due to urban expansion and climate change-driven increases in rainfall intensity. Decision trees help operational planners consider different usage scenarios and their consequences. For a wastewater treatment plant, for example, increasing influent loads might trigger decisions about flow equalization, retrofitting with higher rate processes, or the need for plant expansion. Each branch of the decision tree can be evaluated for capital investment needs, energy use, permit compliance, and reliability under peak conditions. By simulating future operational states, engineers can prioritize upgrades or policy changes that keep infrastructure functioning reliably and safely under uncertainty.

5.2.3.3 Emergency Response Planning

The operation phase also includes preparing for rare but high-consequence events, such as floods, earthquakes, pipeline ruptures, or system failures. Decision trees can be used to plan emergency response actions by mapping out possible failure paths, response times, and mitigation outcomes. For example, in a levee system protecting a flood-prone community, the branches of a decision tree may include information regarding the possibility and timing of:

- Pre-emptive pumping of reservoirs
- Temporary sandbag reinforcement
- Controlled overflow to designated zones
- Evacuation decisions.

Probabilities are derived from hydrological forecasts, historical data, and real-time sensor networks. The goal is to identify the most effective and timely combination of responses based on current and projected conditions. This form of decision analysis supports proactive infrastructure management and can lead to reduced response times during emergencies.

5.2.4 Maintenance and End-of-Life Considerations

Infrastructure assets are typically designed with a planned service life, often ranging from 30 years to over 100 years, depending on the structure and its intended use. However, the longevity of any given structure is influenced by environmental exposure, material behavior, usage intensity, and the effectiveness of maintenance interventions. As civil infrastructure ages, engineers must navigate decisions related to long-term upkeep, retrofitting, and eventually, decommissioning or adaptive reuse.

Uncertainties in this phase may involve future climate conditions, the political or economic context of asset ownership, unknown subsurface conditions, or unanticipated environmental consequences of dismantling. These uncertainties can greatly influence both the cost and environmental footprint of end-of-life decisions.

Decision trees offer a transparent method to evaluate alternatives under these uncertainties, enabling engineers to consider a range of future conditions and identify the most responsible and efficient path forward.

5.2.4.1 Evaluating End-of-Life Options

For infrastructure that has reached or exceeded its design life, civil engineers must assess whether to continue maintenance, perform major rehabilitation, repurpose the asset, or decommission it entirely. Each option comes with a set of uncertain outcomes, including safety risks, community impacts, regulatory obligations, and long-term environmental effects.

Consider the example of decommissioning an aging concrete dam. A decision tree can model the following options:

- Maintain the dam with structural retrofits
- Partially breach the structure to allow controlled river restoration
- Fully demolish and remove the dam
- Convert the site into a wetland or hydro-ecological park.

Each branch must reflect not only the technical feasibility and cost but also potential outcomes such as sediment flushing, downstream habitat disruption (or restoration), stakeholder acceptance, and regulatory compliance. The probabilities of each branch can be informed by hydrological modeling, sediment transport simulations, ecological risk assessments, and stakeholder consultation.

Decision trees help weigh the short-term economic and logistical costs against long-term societal and environmental impacts. They also allow incorporation of adaptive strategies, such as delaying final decommissioning until more data is available or environmental conditions stabilize.

5.2.4.2 Planning for Adaptive Reuse

In some cases, aging infrastructure can be repurposed to extend its utility while reducing demolition waste. For example, old industrial buildings may be converted into public spaces, rail corridors may be turned into greenways and pedestrian active-use trails, or decommissioned wastewater plants may be reused for stormwater management.

A decision tree can map out these reuse scenarios, comparing investment needs, regulatory hurdles, environmental risks, and potential return on investment. For each scenario, engineers consider how changing land use patterns, community needs, or sustainability goals affect the viability of conversion as compared to demolition.

For instance, an elevated highway slated for removal might instead be evaluated for conversion into a public linear park, with branches representing cost, structural retrofitting feasibility, and projected social benefit. Each alternative is explored with respect to uncertainty in urban development trends, funding availability, and stakeholder support.

5.3 How Decision Trees Work

Similar to graphical networks (GNs), Decision trees offer a comprehensive visual framework for analyzing complex decision-making problems. They allow engineers to visualize every possible course of action, including subsequent events and outcomes, in a structured and intuitive format. This is particularly valuable in civil engineering, where decisions often involve many variables, cascading uncertainties, and sequential stages—from site investigations to post-construction performance evaluations. By mapping out alternative pathways and their corresponding outcomes, engineers can anticipate the implications of each design, construction, or management decision. This global perspective is essential when dealing with infrastructure systems, where errors or misjudgments can result in high economic costs, environmental damage, or safety failures.

But what exactly is a decision tree? According to the *Oxford Dictionary of Statistics*, a decision tree is "a graphical representation of the alternatives in a decision-making problem." It organizes choices and possible consequences in a branching structure, capturing both decisions made under certainty and chance events that involve probabilistic outcomes. As further elaborated by Von Winterfeldt and Edwards (1986), "In decision analysis, a decision tree and the closely related influence diagram are used as a visual and analytical decision support tool, where the expected values (or expected utility) of competing alternatives are calculated."

In essence, a decision tree does more than display choices—it allows for quantitative reasoning under uncertainty. Engineers can use it to calculate the expected monetary value, risk exposure, lifecycle cost, or utility of each option based on likely scenarios. This transforms the decision-making process from intuitive, or ad hoc reasoning, into a structured, evidence-based analysis. In civil engineering, this type of tool is indispensable. Projects such as flood control infrastructure, high-rise foundations, transportation corridors, and water treatment facilities often confront significant uncertainties. These range from environmental variability (e.g., future flood intensity, seismic activity, rainfall patterns) to design risks (e.g., assumptions regarding peak use/demand) and construction-related risks (e.g., soil unpredictability, contractor delays, material shortages). A poorly informed decision at any stage, be it site selection, material choice, or maintenance strategy, can lead to severe consequences, including service interruptions, structural failures, inflated costs, or public safety hazards.

Given the high stakes, civil engineers need tools that help systematically evaluate risks and rewards over the entire project lifecycle. Decision trees fill this role by enabling (Fig. 5.4):

- Detailed scenario modeling for alternative actions
- Visualization of risk propagation through sequential events
- Assignment of probabilities to uncertain outcomes
- Computation of the expected value for comparison among options.

In high-impact projects, such as designing a dam with the potential for overtopping or determining the depth of bridge piles in an area with uncertain bedrock elevation, using a decision tree allows engineers to simulate possible failure modes and weigh different mitigation options before costly decisions are implemented.

Figure 5.5 shows a sample decision tree structured as a series of nodes and branches. Each decision node represents a point at which an engineer must choose among alternatives (e.g., selecting a design type). The following decision nodes are chance nodes, which illustrate possible future events, such as whether material strength will degrade prematurely or whether demand for the infrastructure will increase. Each outcome carries associated costs or risks that can be compared to inform the best course of action.

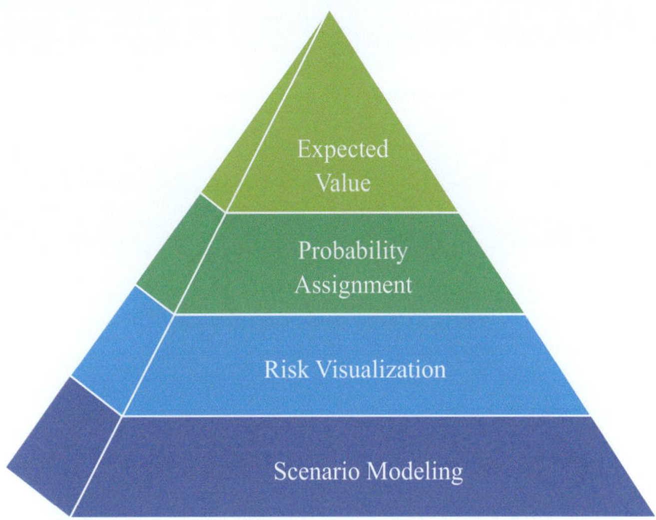

Fig. 5.4 Decision tree process for risk management assessment

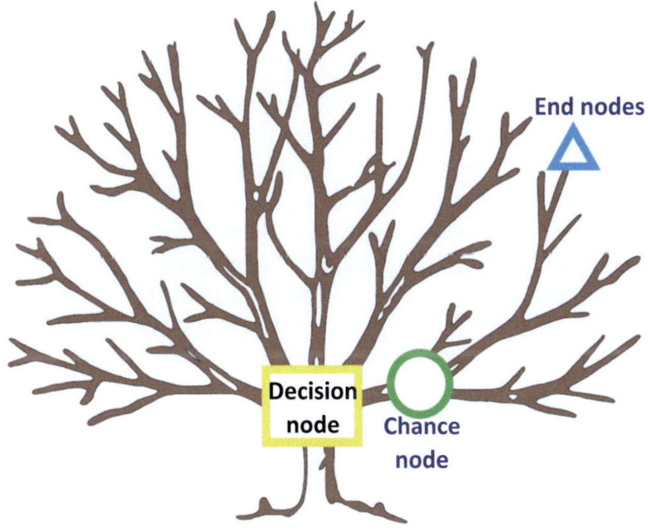

Fig. 5.5 Understanding the elements of a decision tree

5.3.1 Key Elements of the Decision Tree

In practical terms, a decision tree is composed of interconnected elements (Fig. 5.6) that map the full scope of the decision-making process:

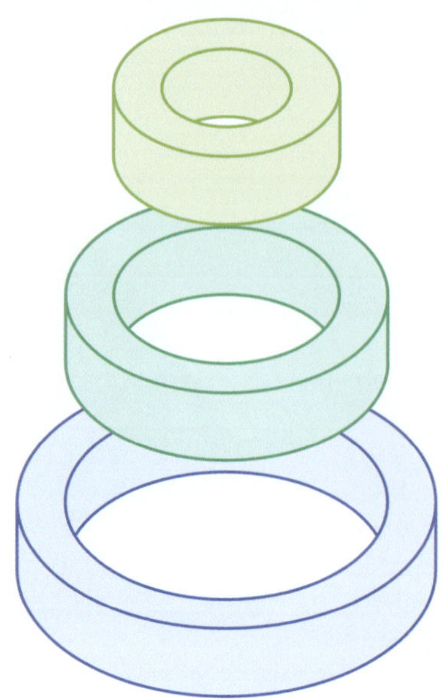

Fig. 5.6 Fundamental elements of a decision tree

- Decision nodes are where engineers choose among competing actions, such as selecting the type of foundation system (e.g., mat foundation, pile foundation, caisson, etc.) in an area with uncertain soil stratigraphy.
- Chance nodes model the external, or uncertain, events that may follow a decision, such as fluctuations in construction material prices, rainfall intensity during construction, or unexpected subgrade settlement.
- Outcome nodes represent the final state of each decision pathway, quantifying parameters like lifecycle cost, structural reliability, or compliance with sustainability goals, enabling an objective comparison between the various alternatives.

By structuring the problem in this way, engineers can navigate complexity in an objective manner with greater clarity and confidence, even in situations where outcomes are not fully predictable.

5.3.1.1 Decision Node

A decision node is the starting point of the decision-making process in a decision tree and is typically represented by a yellow box (e.g., Fig. 5.5). It marks a critical juncture where the engineer or project manager must choose among multiple mutually exclusive alternatives, each leading to a distinct branch in the tree. These alternatives represent controllable actions that influence the project's direction and are central to structuring complex engineering decisions into discrete, analyzable components.

In civil engineering applications, decision nodes capture pivotal choices such as determining the appropriate elevation of a levee, selecting between pile or shallow foundations, or deciding on the optimal alignment for a highway corridor. In structural engineering, for instance, a decision node may reflect selecting among different retaining wall systems—reinforced concrete, segmental block walls, or gabion baskets—each with distinct implications for cost, construction time, drainage, aesthetics, and sustainability. In transportation engineering, it might involve choosing between flexible and rigid pavement designs, while in hydraulic engineering, decisions could concern culvert types or channel lining materials. For environmental engineering applications, the decision node may represent the choice between alternative treatment technologies (such as dissolved air flotation, sedimentation, and ballasted sedimentation) at a water treatment plant.

From a visual and analytical perspective, the decision node serves as a gateway in the tree's hierarchical structure. Branches extending out of each decision node represent the alternative actions or engineering strategies. These may lead to further decision points or chance nodes, indicating subsequent uncertainties or dependent outcomes. For example, in a geotechnical assessment for a new structure, one branch might represent selecting a 10-m-deep foundation, while another explores a 15-m-deep option. Each decision will have differing effects on the stability, settlement behavior, and cost of the structure, which are reflected in the chance nodes.

By organizing decisions within a decision tree framework, engineers can adopt a step-by-step, hierarchical approach to navigate the project timeline. This method enables transparent evaluation of trade-offs based on performance criteria, regulatory compliance, lifecycle costs, constructability, and risk. It also facilitates logical sequencing aligned with design phases and construction planning, allowing early decisions to be traced through their downstream impacts on cost and performance.

5.3.1.2 Chance Nodes

Chance nodes, depicted as green circles (e.g., Fig. 5.5), represent uncertain events or external conditions that influence the outcome of a decision, but remain beyond the engineer's control. These uncertainties may arise from natural variability, limited site data, or unpredictable external influences such as regulatory changes or evolving environmental conditions. Incorporating chance nodes into a decision tree allows engineers to explicitly model uncertainty and assess the range of possible outcomes associated with a given decision path.

In civil and environmental engineering, chance nodes account for a broad spectrum of uncertainty types:

- **Geotechnical uncertainty**: variations in soil bearing capacity, groundwater inflow, or the potential for seismic-induced soil liquefaction.
- **Hydrologic uncertainty**: differences in flood return periods, sediment transport behavior, or long-term precipitation patterns under climate change scenarios.

- **Regulatory uncertainty**: future changes to discharge standards, zoning restrictions, or safety regulations that could affect design and construction compliance.

For example, after selecting a construction technique for a building foundation, a chance node may capture the probability of encountering differing subsurface conditions—such as stable, marginal, or weak soils—with associated likelihoods derived from field investigations, such as borehole logs and cone penetration test (CPT) data. In hydraulic infrastructure design, it might reflect the probability of experiencing various flood magnitudes, based on rainfall frequency analysis, or sediment transport scenarios informed by historical data and modeling.

To represent these uncertainties, each possible outcome at a chance node is assigned a numeric probability derived from statistical models, expert judgment, or risk assessments. This probabilistic approach supports robust and resilient decision-making. For instance, in stormwater system design, engineers might evaluate system performance under both 1-in-25 and 1-in-100 year rainfall events, ensuring acceptable functionality across a wide range of plausible futures.

By integrating chance nodes into the decision tree framework, engineers are equipped to develop solutions that are not only optimized for expected conditions, but also resilient against rare or adverse scenarios, enhancing both safety and long-term performance.

5.3.1.3 Outcome Nodes

Outcome nodes, also referred to as terminal nodes or payoff nodes, appear at the end of each path in a decision tree and are typically represented by a triangle (e.g., Fig. 5.5). These nodes signify the final consequences of a sequence of decisions and uncertain events, summarizing the overall performance of each possible scenario. Outcome nodes are critical for evaluating and comparing alternatives in terms of cost, utility, environmental impact, reliability, and social acceptability.

In engineering practice, these nodes allow for quantitative assessment of trade-offs. For instance:

- In flood control projects, one outcome node might weigh the savings from constructing a shorter floodwall against the potential financial losses caused by overtopping during a rare flood event.
- In urban drainage, different nodes could reflect varying degrees of flood mitigation success, from moderate costs and high performance to budget overruns with limited benefit.
- In pavement design, outcome nodes can express total lifecycle costs, incorporating construction, maintenance schedules, user delay costs, and rehabilitation intervals.

Each outcome node typically integrates a combination of factors, such as:

- Initial capital expenditures
- Operation and maintenance costs
- Consequences of failure or downtime penalties

- Regulatory compliance and reputational risks
- Resilience and adaptability under changing future conditions.

By comparing these quantified results across all paths, engineers can identify the most balanced, efficient, and feasible solutions. Outcome nodes can also be used to conduct sensitivity analysis, helping pinpoint which input assumptions (e.g., probability estimates or cost figures) affect the final decision most significantly. This enhances the robustness of engineering evaluations and aids in refining models. Finally, outcome nodes serve as an effective communication tool. They distill complex decision logic into clear visual summaries, facilitating dialogue with non-technical stakeholders, policymakers, and community members. This improves the transparency, accountability, and trustworthiness of the planning and decision-making process.

5.3.2 The Importance of the Decision Node

In civil engineering decision trees, the decision node serves as the critical point where multiple strategic options must be evaluated in light of uncertainty and potential outcomes. Highlighted in yellow in the figure, it draws attention to the moment where engineers must carefully balance competing objectives, such as minimizing cost, maximizing safety, and ensuring compliance with regulatory and environmental standards. This node is not merely a visual marker; it represents the gateway to the entire decision-making framework. The choice made at this point directly affects all subsequent paths, influencing everything from construction feasibility and material behavior to long-term structural integrity and maintenance demands. Whether selecting a flood protection strategy, determining pavement thickness for a roadway, or choosing between steel and concrete for a bridge superstructure, decisions at this node define the trajectory of project success. By explicitly visualizing the decision node, engineers are prompted to consider the broader implications of each available alternative. This promotes a systematic approach to assessing trade-offs, ensuring that decisions are not only technically sound but also aligned with project timelines, budgets, and stakeholder expectations. The decision node thus embodies more than a simple choice—it is a moment of strategic analysis, where risk, performance, and feasibility converge.

5.3.3 Engineering Decision Making Using Decision Trees

To form a decision tree, engineers can begin by analyzing alternatives under uncertainty, systematically laying out all feasible actions alongside the various external conditions that may influence outcomes, such as environmental hazards or load variations. Next, they can quantify the consequences of each decision-scenario pairing

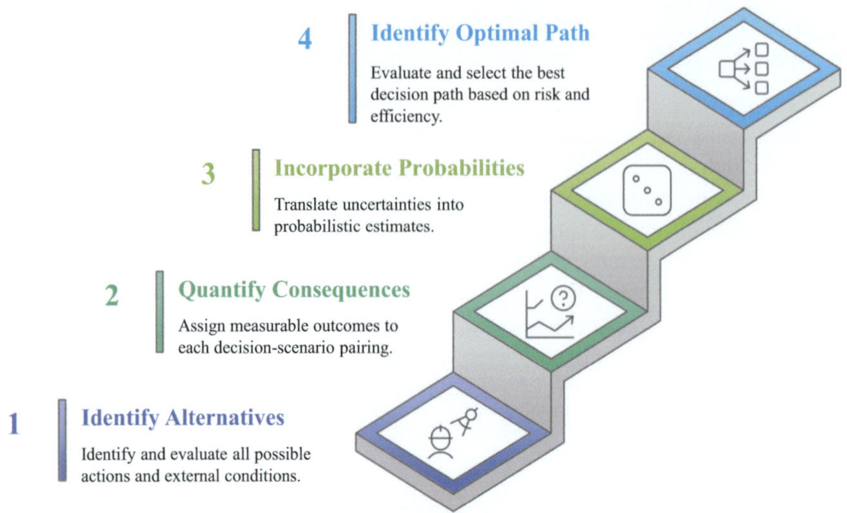

Fig. 5.7 Key steps in the engineering decision-making process using a decision tree

by assigning measurable outcomes—like cost, safety level, or construction time—to each combination of action and event. This is followed by incorporating probabilities into engineering judgments, where uncertainties are translated into probabilistic estimates using historical data, expert opinion, or simulation outputs, allowing for more informed comparisons. Finally, engineers can identify the optimal decision path by evaluating the expected utility or cost of each branch in the tree, selecting the alternative that offers the best balance between risk, performance, and resource efficiency (Fig. 5.7).

5.3.3.1 Analyzing Alternatives Under Uncertainty

Decision trees provide a structured platform for engineers to explore every feasible option under conditions of uncertainty. Each branch models a distinct decision pathway, with subsequent branches representing possible events or states of nature beyond the engineer's control. For example, in the planning of a river embankment system, an engineer might consider three design alternatives for levee height: 3, 4, and 5 m. From each of these decision branches, further sub-branches represent different flood scenarios, such as a 50-year, 100-year, or 500-year flood. It might be worthwhile to note that the possibilities for levee height are the decision node, while the 50-,100-, and 500-year return periods are part of the uncertainty that is modeled for each alternative. This visual and analytical framework ensures that no critical scenario is overlooked, helping engineers anticipate the full spectrum of consequences before construction begins. It also supports the design of adaptive infrastructure—such as modular walls or phased embankment extensions—based on evolving risk conditions.

5.3.3.2 Quantify the Consequences of Each Decision-Scenario Pairing

Every engineering decision carries measurable outcomes—be it structural performance, cost impact, environmental disruption, or safety implications. A decision tree enables the quantification of these outcomes using a payoff table, which assigns numerical values (e.g., cost in dollars, delay in days, failure probability) to each decision-scenario combination.

For instance, in tunnel construction through variable geology, one decision branch might represent the use of a tunnel boring machine, while another represents drill-and-blast. Each method has different costs, construction durations, and risks based on geological conditions. A payoff table can be constructed to estimate these outcomes under different subsurface scenarios (e.g., soft clay, fractured rock, or groundwater intrusion). This process allows for comparative analysis based on defined project criteria such as lowest cost, shortest schedule, or highest safety level.

5.3.3.3 Incorporating Probabilities into Engineering Judgments

Uncertain future events can often be described using probabilities, whether based on historical data, simulation models, or expert judgment. Decision trees allow these probabilities to be incorporated into the analysis, making it possible to compute the expected value for each path. The expected value is calculated by multiplying the utility (or cost) of each outcome by its probability and summing these products across all branches of a decision. This transforms subjective judgment into objective analysis.

For example, in urban drainage design, suppose there's a 20% chance of experiencing extreme rainfall due to climate change and an 80% chance of moderate rainfall. Two infrastructure options—standard culverts or oversized retention basins—can be compared by computing the expected cost of flooding and mitigation under each scenario. This allows the selection of the alternative that offers the best trade-off between upfront investment and long-term flood risk.

5.3.3.4 Identifying the Optimal Decision Path

Once all paths and their expected utilities or losses have been calculated, the optimal decision is the one that either maximizes benefit or minimizes cost/risk. This approach ensures that decisions are not made based on isolated factors, but rather on a complete assessment of trade-offs across all realistic future outcomes. In structural retrofitting for seismic resistance, for example, engineers might weigh three options: no action, partial reinforcement, or full base isolation. By analyzing projected costs, potential downtime, and failure probabilities for each earthquake intensity scenario, the decision tree helps identify the strategy that minimizes total expected loss over the structure's remaining lifespan. This optimization process is especially valuable in projects with tight budgets, public accountability, or long-term operational responsibilities, such as highway networks, water treatment plants, and large dams. It supports transparent, justifiable decisions that align with stakeholder expectations, risk tolerance, and sustainability objectives.

5.3.4 Applications and Advantages of Decision Trees in Civil Engineering

5.3.4.1 Why Decision Trees Matter in Civil Engineering

Decision Trees allow engineers to deconstruct complex decision-making problems into a series of logical, sequential steps. Each branch in the tree represents a distinct action or event, followed by possible outcomes—some of which are deterministic (decisions) and others probabilistic (chance events). This structure provides a comprehensive and systematic framework for analyzing a wide range of possible futures, which is especially critical in civil engineering, where projects are influenced by uncertainty at every stage of the lifecycle.

5.3.4.2 When Are Decision Trees Most Valuable?

In civil engineering, decision trees are particularly valuable for addressing problems characterized by:

- **Environmental uncertainty**, such as variable rainfall intensity, fluctuating river discharge, freeze-thaw cycles, or seismic hazards, which can influence structural loading, soil stability, and material degradation over time.
- **High-impact financial consequences** where suboptimal decisions may lead either to catastrophic failures (e.g., dam overtopping, bridge collapse) or to excessive overdesign that results in wasted resources and inflated project costs.
- **Interdependent, multi-stage decisions** such as site selection followed by foundation design and material procurement, where each step constrains or informs the next, and errors compound through the sequence (Fig. 5.8).

5.3.4.3 Decision Trees as Tools to Enhance Engineering Judgment

Beyond simply organizing choices, the structured, branching logic of a decision tree serves as a powerful analytical tool in civil engineering, offering distinct advantages

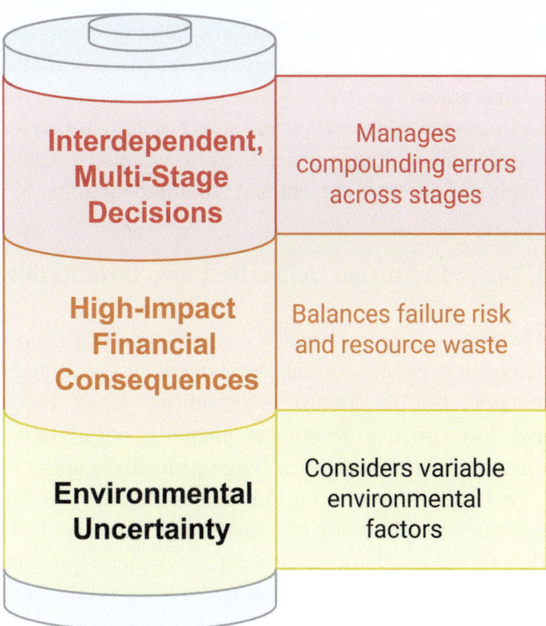

Fig. 5.8 Common problem types in civil engineering where decision trees are particularly effective

that enhance both the technical rigor and communicative clarity of the decision-making process. As shown in Fig. 5.9, decision trees:

- Support the integration of multidisciplinary data, combining hydrological models, geotechnical assessments, cost estimates, and material performance data into a unified decision framework.

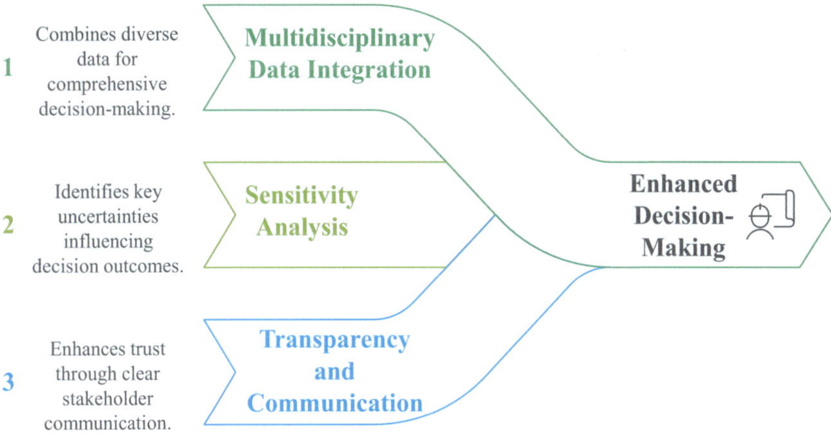

Fig. 5.9 Benefits of decision trees in civil engineering

- Facilitates sensitivity analysis, helping engineers identify which uncertainties (e.g., flood probability, concrete durability) have the greatest influence on decision outcomes.
- Promotes transparency and stakeholder communication, as the tree's visual structure can be easily interpreted by clients, regulators, and community members, enhancing trust in technical recommendations.

5.3.4.4 Improving Design for Long-Term Resilience

The visual clarity of decision trees helps engineers anticipate not just immediate consequences (e.g., initial costs) but also cascading effects, such as increased maintenance, user disruptions, or retrofitting needs. This enables more robust, resilient, and cost-effective design and planning, particularly in infrastructure systems that must perform reliably over long service lives under uncertain future conditions.

By explicitly modeling long-term outcomes, decision trees allow engineers to integrate lifecycle thinking into early-stage planning. For instance, when selecting materials for a coastal bridge subject to saltwater exposure and high winds, a decision tree can compare options like standard reinforced concrete, epoxy-coated steel, or fiber-reinforced polymers. While the initial cost of advanced materials may be higher, the tree structure enables a comprehensive evaluation of long-term performance, maintenance intervals, corrosion risks, and replacement timelines under varying environmental conditions. Probabilistic branches can account for scenarios such as accelerated sea-level rise or increased storm frequency due to climate change. By assigning costs and probabilities to these outcomes, the decision tree highlights which material choice offers the best balance between upfront investment and long-term durability, minimizing life-cycle costs while ensuring continued structural reliability.

5.3.5 Engineering Applications of Decision Trees in Complex Scenarios

Civil engineering projects are often characterized by uncertainty, complexity, and high-stakes decision-making. Decision trees provide a powerful analytical framework for addressing these challenges by modeling different options, outcomes, and their associated probabilities. This section explores how decision trees support engineering decisions across three major domains: managing environmental uncertainty, evaluating financial trade-offs, and navigating sequential decision-making processes. Through practical examples, we illustrate how engineers can apply decision trees to enhance clarity, optimize outcomes, and improve project resilience at every stage of the infrastructure lifecycle.

5.3.5.1 Environmental Uncertainty

Civil engineering projects are frequently exposed to environmental uncertainties that can drastically affect infrastructure performance. Factors such as rainfall intensity, river flow rates, seismic activity, and temperature fluctuations all influence design and operational choices. These uncertainties require engineers to consider a range of possible environmental conditions and their impacts on infrastructure to ensure durability, safety, and functionality. Here are some examples.

Flood Protection Systems

When designing a flood protection system, an engineer must account for potential variations in rainfall intensity and river flow rates, which may fluctuate due to seasonal patterns or climate change. A decision tree can be used to model different flood levels, such as a "normal" flood level, a "moderate" flood level, and an "extreme" flood level, each associated with a probability based on historical data (i.e., return period) or climate projections. Each flood level may require a different height and structural reinforcement for the protective barrier. By visualizing these options in a decision tree, the engineer can analyze the expected costs, benefits, and risks associated with each height level, ultimately identifying the design that minimizes both flood risk and unnecessary expense.

Seismic-Resistant Building Design

In earthquake-prone areas, civil engineers must design buildings and infrastructure that can withstand seismic forces. However, earthquake intensity is highly uncertain and varies depending on fault activity, soil conditions, and other factors. A decision tree can be used to evaluate different design options, such as building with flexible materials, implementing base isolators, or designing for higher resilience standards. Each branch of the tree represents a different level of earthquake intensity, from mild tremors to severe earthquakes. By assigning probabilities to these seismic events and estimating the costs of damage for each scenario, the decision tree helps engineers determine the optimal design approach to enhance building resilience without incurring excessive construction costs.

Stormwater Management Systems

Urban areas face increased risk of flooding due to rapid urbanization and the subsequent reduction in natural drainage. In designing stormwater management systems, engineers must account for a wide range of possible rainfall intensities, which could vary significantly due to climate change. A decision tree can model different rainfall scenarios, such as low, medium, and high-intensity storms. Each scenario can be evaluated for a specific stormwater management technique, such as permeable pavements, retention ponds, or green roofs. The decision tree allows engineers to evaluate the expected effectiveness and cost of each technique under different rainfall intensities, enabling them to design a system that can handle both frequent, low-intensity storms and rare, high-intensity storms.

Coastal Protection Infrastructure

Coastal areas are increasingly vulnerable to sea-level rise and storm surges, which are becoming more severe with climate change. Engineers designing coastal protection measures—such as sea walls, levees, or dunes—must anticipate varying storm surge heights and erosion rates. A decision tree can help by modeling different scenarios of sea-level rise and storm intensity over the project's lifespan. For instance, one branch could represent a modest increase in sea level with moderate storm intensity, while another could represent a higher sea level and more frequent extreme storms. By estimating the probability and impact of each scenario, engineers can use the decision tree to select coastal protection solutions that balance the cost of construction with the long-term benefits of reduced erosion and flood risk.

Bridge Foundation Stability

In regions with fluctuating river flows, bridge foundations must be designed to withstand potential scour effects, where high water flow can erode the riverbed around bridge piers. Engineers need to anticipate scenarios such as "normal," "high," and "extreme" river flow conditions, each with different probabilities. A decision tree can represent these flow conditions as branches, each associated with potential costs from damage or failure. Engineers can then compare the outcomes of different foundation depths or protection measures, selecting the option that provides adequate stability under extreme flow conditions without incurring unnecessary construction costs under normal conditions.

Design of Water Treatment Facilities

Engineers can greatly benefit from the rational discretization of possible outcomes provided by a decision tree when designing potable water purification facilities. During the design process, there are a number of critical decisions that must be made regarding the type of contaminants that must be removed, the extent of their removal, and the required capacity of the facility. To further complicate matters, these facilities are most commonly designed with long service lives, often reaching up to 100 years. This necessitates that engineers account for unknown future conditions in their preferred design alternative, including forecasting multiple scenarios for population growth estimates (and their corresponding peak demand factors) and more stringent regulatory environments. The decision tree can also include multiple scenarios for future drought conditions, which may necessitate alternative source water considerations, and for the impact of climate change on the source water quality. Climate change is known to be causing an increase in the amount of dissolved organic contaminants in source waters globally, as well as an increase in the color of natural waters (termed brownification), which may affect the optimal choice of treatment technology.

By capturing these variables and their likelihood, decision trees aid in developing infrastructure that is resilient against extreme environmental events, ensuring that the design accommodates both current and future environmental challenges. This approach allows civil engineers to systematically address uncertainty, resulting in designs that are robust, economically efficient, and adaptable to changing environmental conditions.

5.3.5.2 Financial Consequences and Cost–Benefit Trade-Offs

Financial considerations are critical in any civil engineering project, but especially so where decisions involve substantial investments. The accuracy of design choices has significant financial implications—under-design can lead to infrastructure failure, costly repairs, or large value charge orders during construction, while over-design results in unnecessary expenditure on materials and labor. Decision trees enable engineers to visualize and quantify these trade-offs, allowing them to evaluate the cost-effectiveness of each decision path. For instance, in constructing a bridge foundation, the engineer might consider different pile depths. A decision tree can capture scenarios where a shorter pile could fail under unexpected soil shifts (resulting in failure costs) versus scenarios where a longer pile could increase safety but add to the budget. This approach allows for a comprehensive analysis of financial risks and rewards, leading to decisions that optimize both performance and cost. Some additional examples of the applications of decision trees for financial decision making in the engineering design process are illustrated below.

Bridge Foundation Depth

When designing a bridge foundation, an engineer must decide on the depth of the foundation piles. Shallower piles have a lower material cost and are less costly to install, but they may not provide adequate stability if the soil shifts or experiences high loads, potentially resulting in failure and expensive repairs. Deeper piles increase stability but significantly add to construction costs. A decision tree can capture these trade-offs by modeling scenarios for "normal soil stability" and "unexpected soil shifts." Each path of the decision tree represents either the lower or higher pile depth, with associated costs of failure or safety. By incorporating probabilities of soil conditions and calculating expected costs for each depth, the decision tree allows the engineer to select a depth that minimizes financial risk while ensuring safety.

Pavement Type Selection

In road construction, engineers must choose between different pavement materials, each with distinct costs, durability, and maintenance requirements. For instance, a higher-quality asphalt may have a higher initial cost but require less frequent repairs, while a cheaper asphalt (such as tar-and-chip) may lead to more maintenance expenses in the long run. A decision tree can represent different types of pavements as branches, capturing scenarios like "low traffic load" and "high traffic load." Each scenario affects the pavement's lifespan and the frequency of required repairs. By assigning probabilities to these traffic conditions, engineers can calculate the total expected costs of each pavement type over the road's lifetime, helping them make a choice that balances initial costs with long-term savings.

Floodwall Height Decision

In flood-prone areas, engineers face the challenge of selecting the optimal height for a floodwall. A shorter wall is less expensive but may fail during extreme flood events, resulting in flood damage costs. A taller wall provides better protection but significantly increases construction costs. A decision tree can model different flood scenarios (e.g., "normal floods," "moderate floods," and "extreme floods"), each associated with probabilities based on historical data. The tree helps quantify the expected costs of each wall height by accounting for potential flood damage and construction expenses. This analysis allows engineers to choose a floodwall height that minimizes the combined cost of construction and potential flood damage, optimizing both safety and budget.

Retrofitting Earthquake-Resistant Structures

In earthquake-prone regions, engineers may need to retrofit buildings to improve resilience. Various retrofitting options, such as adding dampers, installing base isolators, or reinforcing structural beams, come with different costs and levels of protection. A decision tree can represent these retrofit options as branches, with subsequent branches representing potential earthquake intensities (e.g., "low intensity," "moderate intensity," and "high intensity"). Each earthquake scenario involves different damage costs if the structure fails, along with retrofitting expenses. By calculating the expected costs for each option, the decision tree helps engineers select the retrofit that balances upfront retrofitting costs with the potential savings from avoided damage during an earthquake.

Water Supply Pipeline Design

In water supply systems, engineers must decide on the pipe diameter to balance construction costs with operational efficiency. Smaller pipes are cheaper but may lead to higher friction losses, increased pumping costs, and insufficient capacity during peak demand periods. Larger pipes, while more expensive, reduce friction losses and are better suited for high-demand scenarios. A decision tree can model these scenarios by representing different pipe diameters as branches and subsequent branches capturing "low demand," "medium demand," and "high demand" scenarios. By quantifying the long-term operational savings and comparing them with initial construction costs, engineers can choose a pipe diameter that minimizes overall expenses over the system's lifecycle.

Selecting a Dam Height for Hydropower

Engineers designing a dam for hydropower production must determine the optimal height, balancing construction costs against the potential for energy generation. A taller dam increases energy output by creating a larger reservoir but incurs higher material and labor costs. A decision tree can capture scenarios for "average rainfall," "low rainfall," and "high rainfall" years, each affecting reservoir levels and energy production. Each branch of the decision tree represents a dam height, and subsequent branches represent the rainfall scenarios. By assigning probabilities to each rainfall scenario and calculating the expected revenue from energy production, engineers can identify the dam height that maximizes net benefits, considering both construction costs and revenue potential.

By using decision trees to evaluate these trade-offs, civil engineers can make well-informed choices that optimize performance while controlling costs. This approach allows for a comprehensive analysis of financial risks and rewards, leading to decisions that maximize cost-effectiveness without compromising safety or functionality.

5.3.5.3 Sequential Decision-Making

Many civil engineering projects involve a series of interdependent decisions, where each choice influences the next. This is especially common in large-scale projects, such as dam construction or urban water management, where the initial design phase impacts subsequent phases like construction methods and maintenance protocols. Decision trees excel at representing these sequential dependencies, mapping out the consequences of each choice and helping engineers anticipate potential challenges or required adjustments in later stages. For example, in a phased water management project, an initial decision on reservoir size affects future decisions on spillway design, which in turn impacts downstream flood risk. By using decision trees, engineers can visualize how early choices shape later outcomes, enabling them to optimize the entire decision sequence and mitigate potential downstream risks. Some examples of how decision trees can be applied to each phase of sequential projects are given in the following sections.

Phased Water Management Project

In a multi-phase water management project, an initial decision on the size of a reservoir determines future design considerations, such as spillway capacity and flood control measures. If the engineer opts for a larger reservoir, they must subsequently invest in a larger spillway to accommodate potential overflow, increasing construction costs but improving flood protection downstream. A decision tree can map out these choices sequentially, with each branch representing reservoir size and subsequent branches capturing spillway design options and their impacts on downstream flood risk. By visualizing how the choice of reservoir size cascades through the project, the decision tree enables engineers to evaluate each phase holistically, ensuring that the selected design minimizes flood risk while staying within budget.

Dam Construction Sequence

Building a dam involves several sequential decisions, starting from site selection and progressing through material selection, spillway design, and eventually, power generation capacity. For instance, an engineer might initially choose between two potential dam sites—one with favorable geological conditions but limited power potential due to low streamflow, and another with higher power potential but challenging soil conditions. Each site selection affects material requirements and construction techniques, which in turn influence the power generation capacity. A decision tree can help visualize each phase, from site selection to material choice and power output, capturing the interplay between choices. By systematically mapping each stage and its dependencies, engineers can ensure that initial decisions align with the project's overall goal of maximizing power output without exceeding cost and safety constraints.

Urban Wastewater Treatment Plant Design

In designing an urban wastewater treatment plant, engineers face sequential decisions that determine treatment efficiency, environmental impact, and scalability. For example, the initial choice of treatment technology (e.g., conventional activated sludge vs. membrane bioreactor) affects subsequent decisions, such as land area requirements, treatment capacity, which contaminants can be removed, as well as maintenance protocols. A decision tree can represent these sequential decisions, starting with the treatment technology choice, followed by plant layout, discharge protocols, and expansion options. By mapping out these stages, the decision tree helps engineers anticipate the long-term impact of each choice, ensuring that the selected technology and design layout can meet current and future wastewater demands.

Highway Expansion Project

In highway expansion projects, initial decisions regarding road width and the number of lanes influence later stages, such as traffic management, material choice, and maintenance strategies. If engineers opt to build a wider highway initially, it may reduce the need for future expansions but increase initial construction costs. Subsequent decisions, such as lane surfacing materials and traffic control measures, depend on the initial road width. A decision tree can represent each phase, starting with the choice of road width, followed by material selection, drainage design, and ongoing maintenance planning.

Seismic Retrofit of Urban Infrastructure

Retrofitting buildings for seismic resilience in urban areas often involves sequential decisions, from choosing retrofit techniques to selecting materials and planning maintenance schedules. For instance, an initial decision to reinforce the building structure with steel bracing instead of base isolation affects subsequent material and design choices, as well as long-term maintenance needs. A decision tree can map these stages, showing how the choice of reinforcement method affects costs, structural stability, and ongoing maintenance. This enables engineers to assess the trade-offs at each stage, ensuring the chosen retrofit strategy provides both immediate protection and long-term resilience within budget constraints.

Tunnel Construction in Variable Soil Conditions

When constructing a tunnel through different soil layers, engineers make sequential decisions regarding excavation methods, reinforcement, and waterproofing. The choice of excavation method for the first section, such as using a tunnel boring machine (TBM) versus a conventional drill-and-blast method, influences subsequent decisions about reinforcement types and waterproofing requirements. A decision tree can represent these stages, with branches capturing each excavation method, reinforcement option, and associated costs or risks. By visualizing how choices in the initial excavation phase impact later reinforcement and waterproofing strategies, engineers can optimize the decision sequence to balance safety, cost, and construction speed.

5.3.5.4 The Strategic Value of Decision Trees in Civil Engineering

In each of these examples, decision trees offer a structured framework for mapping out complex, sequential decision-making processes in civil engineering. They allow engineers to anticipate how each decision influences later stages, enabling them to identify an optimal path that balances immediate requirements with future demands. This approach is essential for achieving project objectives cohesively and efficiently, particularly in large-scale projects where interdependencies among decisions play a critical role in overall success.

As we have seen up until this point, decision trees serve as a critical decision-support tool in civil engineering, providing a structured methodology for managing uncertainty, evaluating financial trade-offs, and optimizing decisions across project phases. By visually representing each decision path and its corresponding outcomes, decision trees help engineers analyze complex scenarios involving variable environmental conditions, engineering constraints, and long-term operational considerations. This approach enables engineers to assess both short-term and long-term implications of their choices, ensuring alignment with project objectives such as safety, functionality, regulatory compliance, and cost-efficiency. Whether planning a flood protection system, selecting a foundation method under geotechnical uncertainty, or evaluating retrofit options for aging infrastructure, decision trees offer a transparent framework to systematically weigh alternatives. Even under incomplete or evolving information, decision trees promote a holistic perspective that balances risk, performance, and budget. By supporting data-driven, scenario-based planning, they ultimately contribute to the development of infrastructure that is not only technically sound and economically viable but also resilient and adaptable to future conditions.

5.4 Understanding Decision Tree Analysis: A Schematic Tool for Complex Decision-Making

Decision trees are particularly powerful in situations with uncertainty, as they incorporate both decisions (choices made by the decision-maker) and chance events (random events with probabilities) to represent a full range of possible scenarios. By using this approach, engineers and decision-makers in civil engineering can evaluate all potential outcomes, calculate expected values, and make data-driven, risk-aware choices.

A decision tree comprises decision nodes (points where a choice must be made), chance nodes (points of uncertainty), and outcome nodes (final results or payoffs). Each pathway represents a sequence of decisions and events, allowing engineers to understand how each choice impacts future stages of the project. This structure is especially useful when multiple interdependent variables are involved, such as material selection, environmental risk, construction sequencing, or policy compliance.

This method is especially valuable when:

- Multiple decisions must be made in sequence, where each stage influences the next.
- Probabilistic outcomes need to be modeled due to uncertainty in weather patterns, ground conditions, material performance, or usage demands.
- There is a need to transparently communicate complex trade-offs to stakeholders, policymakers, or funding agencies.

By making the decision-making process explicit, transparent, and data-driven, decision tree analysis helps civil engineers develop infrastructure systems that are technically feasible, economically efficient, and resilient to future risks.

Let's now explore how this method can be applied through detailed, real-world examples from civil engineering practice.

5.4.1 Example 1: Bridge Foundation Design Under Uncertain Soil Conditions

In bridge construction, foundation design is critical for ensuring structural stability and load transfer to the underlying soil. This decision becomes even more complex when subsurface conditions are uncertain or vary significantly across the site. Engineers often need to choose between a shallow foundation (such as spread footings or mats) and a deep foundation (such as piles or drilled shafts), each with distinct performance characteristics, construction requirements, and cost implications. A decision tree provides a systematic way to model these choices and assess their outcomes under uncertainty.

1. **Decision Node (Choice of Foundation Type)**:

   The decision node represents the primary engineering choice: whether to proceed with a shallow or deep foundation.

   - A **shallow foundation** transfers loads to soil strata near the surface, with examples being spread footing, and mat or raft foundations. These foundation types are generally less costly and easier to construct, relying on the upper soil layers to support loads. However, they may be vulnerable to settlement, bearing failure, or tilting if the soil has low strength or is subject to moisture variation or consolidation.
   - A **deep foundation** transfers loads to deeper, more stable soil or rock layers, making it suitable for sites with compressible or variable soils near the surface. Some examples of deep foundations include pile foundations, caissons, and pier footings. Although more expensive, it provides enhanced stability and reduced sensitivity to near-surface soil behavior.

2. **Chance Nodes (Soil Stability Scenarios)**:

   Following the foundation selection, the next factor is the uncertainty in soil stability, represented by chance nodes. If a geotechnical investigation reveals variability in soil properties across the site or insufficient borehole data, engineers may classify the site with probabilistic estimates of encountering either:

   - **Stable soil conditions** (e.g., dense sand, well-graded gravel, or stiff clay) with high allowable bearing capacity, typically above 150 kPa, and low compressibility.
   - **Unstable soil conditions** (e.g., soft clay, loose fill, or organic silt) with low bearing capacity (e.g., <75 kPa) and high potential for settlement or lateral spreading.

These probabilities may be based on standard penetration test (SPT) results, cone penetration test (CPT) profiles, or statistical models from regional soil data. For example, there may be a 60% chance of stable soil and a 40% chance of encountering weak or problematic layers.

3. **Outcome Nodes (Payoffs and Costs)**:

   The final branches of the tree lead to outcome nodes, which summarize the consequences, often in terms of financial cost, performance, or risk:

   - A **shallow foundation in stable soil** would result in minimal cost and excellent performance (e.g., \$800,000 total construction cost).
   - A **shallow foundation in unstable soil** could require costly post-construction mitigation (e.g., underpinning, grouting), leading to foundation failure risk or increased maintenance and liability (e.g., \$1,500,000 total cost).
   - A **deep foundation in stable soil** may be technically unnecessary, but it guarantees performance (e.g., \$1,300,000 cost), resulting in potential overdesign.
   - A **deep foundation in unstable soil** provides optimal performance and reliability under challenging ground conditions (e.g., \$1,350,000 cost), avoiding the risk of structural distress.

Engineers then calculate the expected cost of each foundation choice by weighting the cost of each scenario by the probability of the corresponding soil condition:

$$\textbf{Shallow foundation expected cost} = (0.6 \times \$800{,}000) + (0.4 \times \$1{,}500{,}000) = \$1{,}120{,}000$$

$$\textbf{Deep foundation expected cost} = (0.6 \times \$1{,}300{,}000) + (0.4 \times \$1{,}350{,}000) = \$1{,}320{,}000$$

While the shallow foundation shows a lower expected cost, its higher variance and failure risk may be unacceptable depending on the structure's importance (e.g., critical transportation link, lifeline infrastructure) and the engineer's or client's risk tolerance. The structure of this decision-making process under soil uncertainty is illustrated in Fig. 5.10.

This decision tree allows engineers to integrate technical, geotechnical, and economic data into a unified model that supports transparent, rational decision-making. It also facilitates communication with stakeholders and regulators by clearly showing the implications of each design choice under soil uncertainty. Beyond cost optimization, the approach ensures structural safety and serviceability are preserved, even when subsurface conditions are only partially understood.

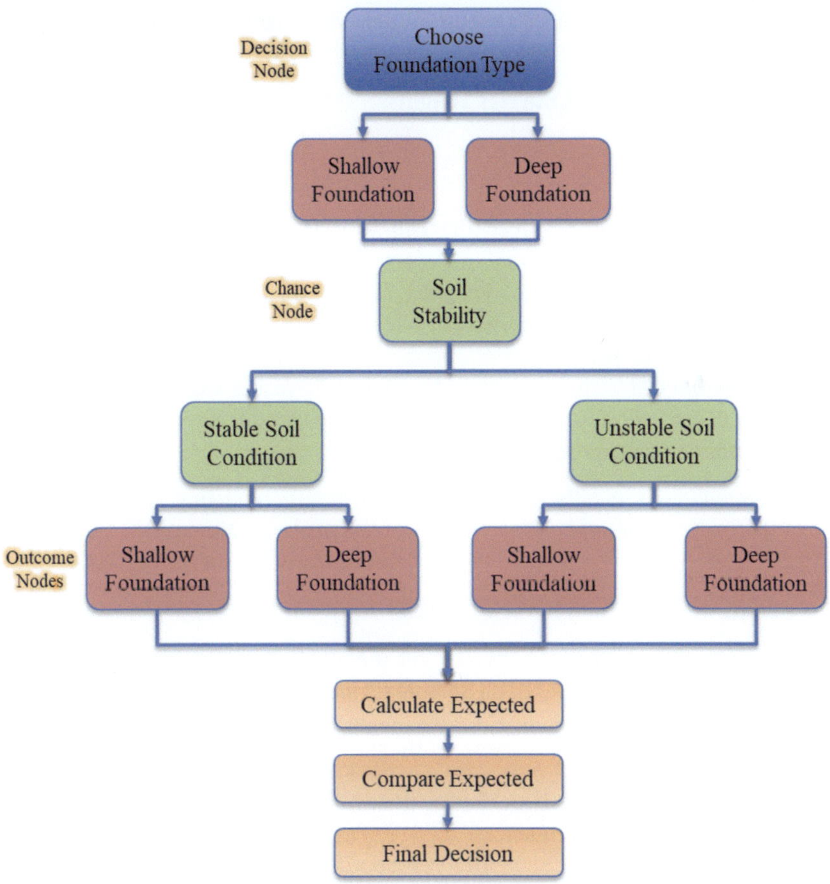

Fig. 5.10 Decision tree representation for bridge foundation design under uncertain soil conditions

5.4.2 Example 2: Flood Protection System Design with Variable Flood Levels

In flood-prone regions, designing an effective flood protection system is a key civil engineering challenge. One critical decision involves determining the optimal height of flood barriers—such as levees, floodwalls, or embankments—to protect infrastructure, communities, and natural resources. Since future flood severity is uncertain and influenced by climate variability, land use changes, and upstream hydrologic conditions, engineers must evaluate multiple flood scenarios when making design choices. Decision tree analysis provides a robust framework for comparing alternatives under uncertainty.

1. **Decision Node (Flood Barrier Height)**

   The decision node represents the selection of flood barrier height, with three design alternatives:

   - **Low barrier** (e.g., designed for a 1-in-50-year flood): Minimizes construction cost but may fail under rare or extreme flood events.
   - **Medium barrier** (e.g., designed for a 1-in-100-year flood): Balances construction cost and protection but still poses moderate flood risk under more extreme scenarios.
   - **High barrier** (e.g., designed for a 1-in-500-year flood): Offers maximum flood protection but incurs high capital costs and may face hydraulic or permitting constraints.

These options are typically evaluated using hydraulic modeling software such as HEC-RAS, incorporating flood elevation profiles, freeboard allowances, and climate-adjusted design flows.

2. **Chance Nodes (Flood Severity Levels)**

   Following the selection of barrier height, the next factor is the probabilistic nature of future flood events. Based on hydrologic modeling, statistical flood frequency analysis, and historical gauge records, engineers can define three flood severity scenarios:

   - **Normal flood** (e.g., 1-in-20-year flood): Occurs with high probability, say 60%, with low consequences.
   - **Moderate flood** (e.g., 1-in-100-year flood): Probability of occurrence is 30%.
   - **Extreme flood** (e.g., 1-in-500-year flood or higher): Probability is 10%, but consequences are severe.

These probabilities are often derived from annual peak flow data fitted to statistical distributions such as Gumbel, Log-Pearson Type III, or Generalized Extreme Value (GEV), and adjusted for projected climate change using RCP scenarios.

3. **Outcome Nodes (Resulting Costs and Benefits)**

   Each path in the decision tree results in an outcome representing the total expected cost, which combines:

   - **Construction cost** of the barrier, based on the volume of materials, foundation conditions, and labor.
   - **Flood damage cost**, which varies by severity and protection level (e.g., property loss, displacement costs, infrastructure repair).
   - **Residual risk**, which represents the likelihood of overtopping or failure, and long-term maintenance obligations.

Sample outcomes might be:

- **Low barrier + normal flood**: Total cost = $3 million (low construction, minimal damage).
- **Low barrier + moderate flood**: Total cost = $12 million (low construction cost + moderate flood damage).
- **Low barrier + extreme flood**: Total cost = $25 million (low construction cost + major flood damage).
- **Medium barrier + normal flood**: Total cost = $4 million (moderate construction + negligible damage).
- **Medium barrier + moderate flood**: Total cost = $6 million (moderate construction + moderate damage).
- **Medium barrier + extreme flood**: Total cost = $15 million (moderate construction + major damage).
- **High barrier + normal flood**: Total cost = $10 million (high construction, negligible damage).
- **High barrier + moderate flood**: Total cost = $7 million (high construction, negligible damage).
- **High barrier + extreme flood**: Total cost = $10 million (high construction, negligible damage).

Expected values are then calculated using probability-weighted averaging. For example:

- **Low barrier expected cost** = (0.6 × $3M) + (0.3 × $12M) + (0.1 × $25M) = $7.9M.
- **Medium barrier expected cost** = (0.6 × $4M) + (0.3 × $6M) + (0.1 × $15M) = $5.7M.
- **High barrier expected cost** = (0.6 × $10M) + (0.3 × $7M) + (0.1 × $10M) = $9.1M.

In this case, the medium barrier provides the lowest expected total cost, offering a reasonable compromise between upfront investment and flood damage prevention. The decision-making process for selecting optimal flood barrier height under uncertain flood severity is illustrated in Fig. 5.11.

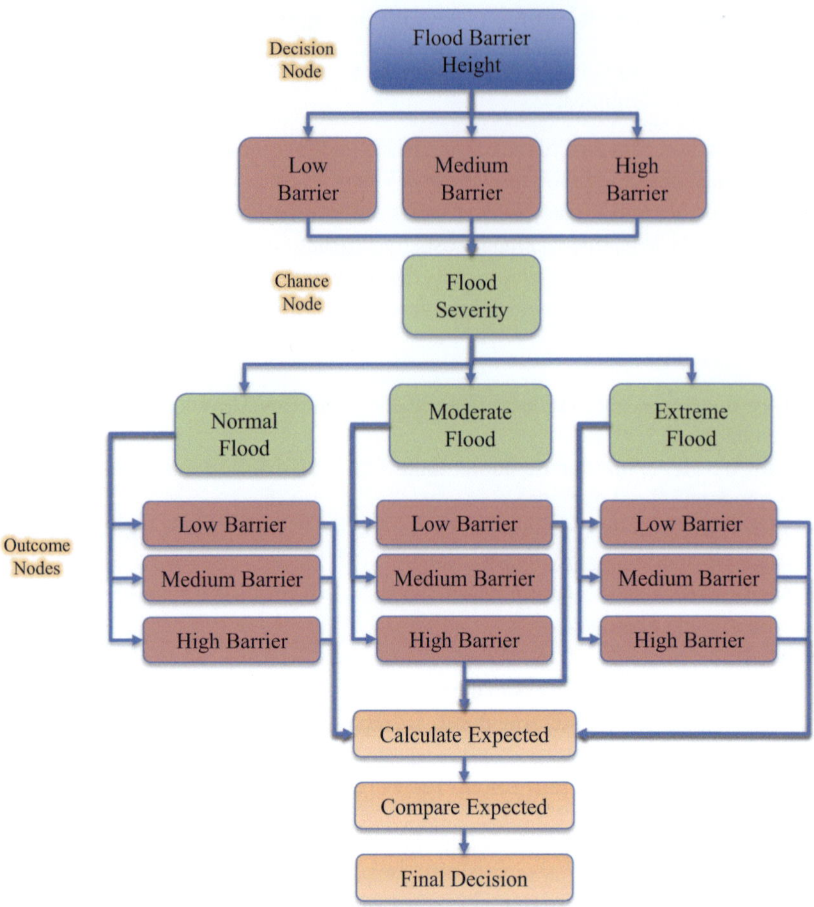

Fig. 5.11 Decision tree analysis for flood protection system design under variable flood levels

5.4.3 *Example 3: Pavement Material Selection in Road Construction*

In road construction, selecting an appropriate pavement material is essential for ensuring performance, longevity, and cost-effectiveness over the infrastructure's service life. Engineers must consider factors such as initial construction cost, expected traffic loading, climate conditions, and long-term maintenance requirements. Given the uncertainties in future vehicle loads and environmental exposure—especially in regions with freeze–thaw cycles or extreme heat—decision tree analysis provides a robust method to evaluate material options and associated risks. Figure 5.12 provides a conceptual overview of how real-world factors—such as traffic

Fig. 5.12 Conceptual representation of pavement material selection under varying real-world conditions, including traffic load and environmental exposure. Asphalt, o Portland cement concrete (PCC), and composite pavements are illustrated with corresponding climate and usage scenarios

intensity and environmental conditions—affect the selection of pavement materials in road construction.

1. **Decision Node (Pavement Material Choice)**

   The decision node represents the selection among three commonly used pavement types:

   - **Asphalt pavement**: Lower initial cost and easier to install, but more susceptible to rutting and fatigue cracking under heavy traffic. Performance is also sensitive to temperature fluctuations.
   - **Portland cement concrete (PCC)** pavement: Higher initial cost, but greater structural rigidity, resistance to deformation, and longer design life under heavy axle loads.
   - **Composite pavement**: A hybrid system combining asphalt over concrete or vice versa, aiming to balance structural performance and surface smoothness. Offers moderate cost with potentially extended service life if properly designed.

These choices are informed by mechanistic-empirical design standards (e.g., AASHTO Pavement ME Design), considering subgrade support, expected equivalent single axle loads (ESALs), and surface distress mechanisms.

2. **Chance Nodes (Traffic Load and Environmental Conditions)**

   After material selection, chance nodes capture uncertainties in service conditions:

   - **Traffic loading**:
     - Low (e.g., rural road, <5 million ESALs over 20 years)
     - High (e.g., urban arterial, >20 million ESALs)
     - Probabilities based on traffic projections from regional transportation models.
   - **Environmental exposure**:
     - Mild climate (e.g., stable temperature, low precipitation)
     - Harsh climate (e.g., frequent freeze–thaw, high summer temperatures, wet subgrade)
     - Probabilities estimated from historical climate data and local weather trends.

Each combination affects the pavement's deterioration rate, influencing cracking, rutting, thermal stress, and base/subbase moisture sensitivity.

3. **Outcome Nodes (Lifecycle Costs and Maintenance Requirements)**

   Each terminal node of the decision tree corresponds to a unique outcome, expressed in terms of:

   - **Initial construction cost** (e.g., per lane-kilometer),
   - **Annual maintenance cost** (e.g., crack sealing, resurfacing, joint repair),
   - **Rehabilitation interval** (e.g., mill and overlay, slab replacement), and
   - **Total lifecycle cost** over a 30-year design period.

Example outcome values:

- **Asphalt + high traffic + harsh climate**: Initial = \$400,000/km; Maintenance = \$30,000/year; Life = 12 years; Total lifecycle cost ≈ \$1.25M
- **Concrete + low traffic + mild climate**: Initial = \$700,000/km; Maintenance = \$10,000/year; Life = 35 years; Total lifecycle cost ≈ \$950,000
- **Composite + high traffic + harsh climate**: Initial = \$600,000/km; Maintenance = \$20,000/year; Life = 25 years; Total lifecycle cost ≈ \$1.1M.

Engineers can then compute expected lifecycle costs by assigning probabilities to each traffic-environment combination and averaging the weighted outcomes:

- Asphalt expected cost = 0.5 × \$850,000 + 0.5 × \$1.25M = \$1.05M
- Concrete expected cost = 0.5 × \$950,000 + 0.5 × \$1.15M = \$1.05M
- Composite expected cost = 0.5 × \$1.0M + 0.5 × \$1.1M = \$1.05M.

Though the expected costs may be similar, risk preferences, maintenance access, and agency priorities (e.g., sustainability, carbon footprint) could influence the final choice.

Decision tree analysis allows engineers to incorporate uncertainty in traffic patterns and environmental exposure into pavement material selection. This approach ensures a comprehensive comparison of short-term and long-term costs while accounting for the performance of the roadway under varied conditions. As a result, the final decision aligns with the agency's priorities for durability, budget stability, and minimal user disruption over the pavement's lifecycle.

5.4.4 *Example 4: Stormwater Detention Basin Design Under Uncertain Urban Growth and Rainfall Intensification*

Urban stormwater systems must manage increasing runoff volumes due to both the increased amount of impervious surfaces and intensifying rainfall patterns associated with climate change. Engineers designing detention basins face uncertainty regarding future land use and precipitation intensity. Decision tree analysis helps evaluate basin sizing options under multiple scenarios involving runoff generation and regulatory compliance. Figure 5.13 conceptually illustrates how engineers must anticipate future urban expansion and rainfall intensification when selecting detention basin capacities in stormwater system design.

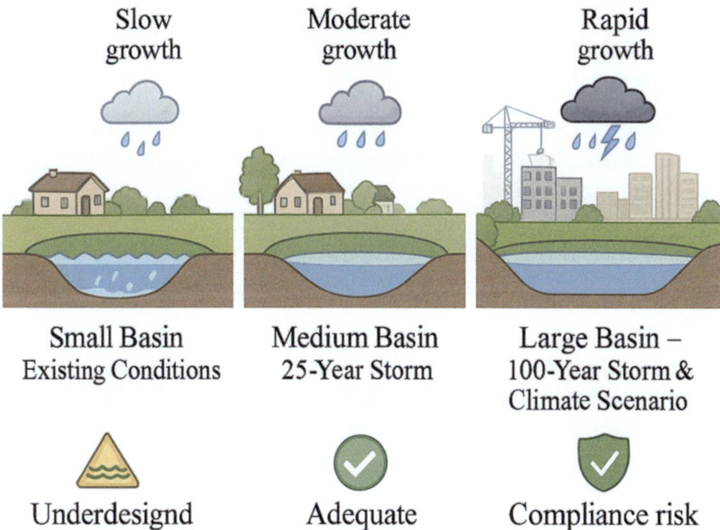

Fig. 5.13 Conceptual illustration of detention basin sizing strategies under uncertainty in urban growth and rainfall intensity

1. **Decision Node (Detention Basin Capacity Selection)**

   The engineer must select among three design capacities for the basin:

   - **Small basin**: Designed for existing conditions, with low upfront construction costs but limited flexibility for future capacity increases.
   - **Medium basin**: Sized for moderate urban growth and a 1-in-25-year storm, balancing cost and resilience.
   - **Large basin**: Sized for full build-out and a 1-in-100-year storm with climate-adjusted rainfall intensity, offering long-term adaptability but higher construction costs and land use requirements.

Each option is evaluated using hydrologic modeling tools such as SWMM or PCSWMM, with input from land-use projections and IDF (intensity-duration-frequency) curves.

2. **Chance Nodes (Urban Growth and Rainfall Scenarios)**

   The primary uncertainties are:

   - **Urban growth**:

     – Slow (e.g., minimal expansion over 30 years)—40% probability
     – Rapid (e.g., full development within 20 years)—60% probability

   - **Rainfall intensification**:

     – Baseline (historical trends)—50% probability
     – Climate-adjusted (20–30% increase in peak intensity)—50% probability.

The probabilities associated with the different urban growth and rainfall scenarios are based on land use planning documents and regional climate models (e.g., RCP4.5 or RCP8.5 scenarios).

3. **Outcome Nodes (Hydraulic Performance and Cost Implications)**

   Each combination of detention basin sizing, urban growth, and rainfall intensity results in:

   - **Construction cost** (excavation, outlet control structures, land acquisition),
   - **Regulatory compliance risk** (e.g., overflow frequency, CSO activation),
   - **Flood damage potential** (if the basin overflows during design storms).

Sample outcomes:

- Small basin + rapid growth + climate rainfall: frequent overflows, $600,000 damage over 20 years
- Medium basin + baseline conditions: adequate performance, $1.5M total lifecycle cost
- Large basin + slow growth: underutilization, high initial cost ($2.5M), minimal long-term benefits.

Expected cost analysis can be performed:

- Small basin expected cost = \$800,000 + (risk-weighted damage and upgrade costs)
- Medium basin expected cost = \$1.5M total (balanced case)
- Large basin expected cost = \$2.5M (high investment, low residual risk).

Decision trees help engineers compare design options by integrating uncertainty in land development and climate impacts. This approach supports resilient infrastructure planning that aligns with municipal risk tolerance, regulatory thresholds, and sustainability goals.

5.4.5 Example 5: Bridge Seismic Retrofit Decision Making Under Probabilistic Earthquake Risk

For aging bridges in seismically active regions, retrofitting is often required to meet modern performance standards. However, uncertainties about future earthquake occurrence, ground motion intensity, and structural response make it challenging to decide on the level of intervention. A decision tree can help to structure and quantify the choices, outcomes, and associated costs of the seismic retrofit.

1. **Decision Node (Retrofit Strategy Selection)**

   Engineers must choose one of three retrofit strategies:

   - No retrofit: This option involves no capital cost but a high degree of vulnerability.
   - Moderate retrofit: Strengthening key components (e.g., bearings, pier caps), reducing the risk of collapse during moderate shaking.
   - Full retrofit: Seismic isolation, column jacketing, and abutment enhancement for full compliance with current seismic codes.

Each strategy is analyzed using structural modeling tools (e.g., OpenSees, SAP2000) and fragility curves developed for similar bridge types.

2. **Chance Nodes (Earthquake Intensity Scenarios)**

   Future seismic hazard is modeled using ground motion probability distributions:

   - **Low-intensity earthquake (PGA < 0.15 g)**—60% probability
   - **Moderate-intensity earthquake (PGA ≈ 0.25 g)**—30% probability
   - **High-intensity earthquake (PGA > 0.35 g)**—10% probability

These probabilities can be developed using probabilistic seismic hazard maps and local geotechnical site classes.

3. **Outcome Nodes (Damage, Downtime, and Cost Consequences)**

   Each retrofit-decision–earthquake-intensity pairing leads to outcomes that include considerations regarding the:

   - **Initial retrofit cost**,
   - **Expected damage repair cost** (from nonlinear dynamic analysis),
   - **Downtime duration** and user delay costs,
   - **Life-safety implications** (collapse probability, rescue access).

Example values:

- No retrofit + high-intensity quake: \$500,000 retrofit savings, but \$25M in damage and full collapse risk
- Moderate retrofit + moderate-intensity quake: \$2M initial cost, \$1M damage, minor downtime
- Full retrofit + high-intensity quake: \$5M cost, negligible damage, full functionality retained.

Engineers compute expected lifecycle costs:

- No retrofit expected cost = \$6M
- Moderate retrofit expected cost = \$3.8M
- Full retrofit expected cost = \$5.5M.

By modeling seismic scenarios and response outcomes, decision trees guide engineers toward a retrofit strategy that minimizes both economic loss and safety risk under uncertainty. This framework is especially valuable for infrastructure owners balancing limited budgets against public safety obligations.

5.4.6 Example 6: Wastewater Treatment Technology Selection Under Regulatory Uncertainty

In the design and upgrade of municipal or industrial wastewater treatment plants, technology selection is a critical decision. It must not only address current contaminant loads and discharge limits but also anticipate future regulatory shifts, especially regarding nutrient removal, microplastics, and emerging contaminants. Decision tree analysis enables engineers to compare treatment alternatives under evolving environmental standards and operational uncertainties, supporting long-term sustainability and cost-efficiency. Figure 5.14 visualizes how regulatory uncertainty influences the

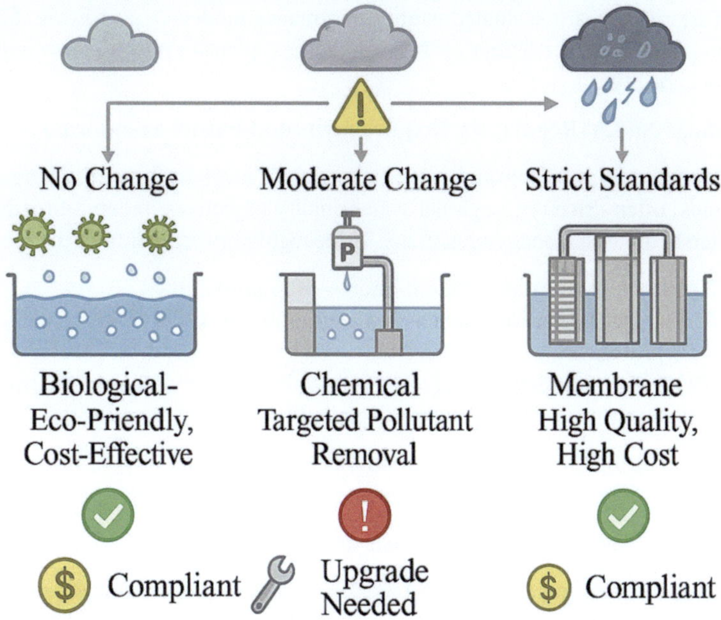

Fig. 5.14 Conceptual diagram illustrating wastewater treatment technology options (biological, chemical, and membrane based) under different future regulatory scenarios, highlighting potential risks, upgrade needs, and compliance strategies

choice of wastewater treatment technologies by linking each option to potential future standards and operational implications.

1. **Decision Node (Technology Selection)**

   Engineers must choose among three primary treatment technology options:

   - **Biological treatment** (e.g., conventional activated sludge, sequencing batch reactors): Eco-friendly and cost-effective for organic and nutrient removal, but performance can degrade under shock loading or stricter discharge limits (e.g., for phosphorus or ammonia).
   - **Chemical treatment** (e.g., coagulation–flocculation, chemical precipitation): Effective for specific pollutants (e.g., heavy metals, phosphorus), but involves higher operational costs and chemical use, with potential sludge handling implications.
   - **Membrane filtration** (e.g., microfiltration, ultrafiltration, membrane bioreactors): Offers high-quality effluent and robustness against changing influent quality, but requires high capital investment, energy consumption, and membrane maintenance.

Design alternatives are evaluated using performance models (e.g., Biowin, GPS-X) and compared against existing effluent guidelines, plant footprint constraints, and operator capacity.

2. **Chance Nodes (Regulatory Requirements and Future Standards)**

   After technology selection, the next uncertainty involves future discharge regulations, often driven by regional water quality objectives, federal directives, or watershed-based management plans. Three regulatory scenarios can be modeled:

   - **No change** to existing effluent limits—40% probability
   - **Moderate tightening**, such as reduced limits for BOD, TSS, or nutrients—40% probability
   - **Strict new standards**, including removal of micropollutants, pharmaceuticals, or nitrogen <3 mg/L (for example, PFAS)—20% probability.

Probabilities are based on regulatory trend analysis, consultations with environmental agencies, and review of adopted or proposed policies in similar jurisdictions.

3. **Outcome Nodes (Compliance Costs and Upgrade Needs)**

   Each technology-regulation combination results in specific outcomes in terms of:

   - Initial capital cost,
   - Operational and maintenance cost (O&M),
   - Need for retrofitting or advanced treatment (polishing) units (e.g., ozonation, GAC),
   - Non-compliance penalties or environmental risks.

Example outcomes:

- **Biological treatment + strict standards**: Initial = $10 M; upgrade cost = $5 M for membrane polishing; risk of non-compliance = high
- **Chemical treatment + moderate standards**: Initial = $12 M; O&M = $500 k/year; moderate compliance assurance
- **Membrane filtration + strict standards**: Initial = $18 M; O&M = $800 k/year; no upgrade required.

Engineers can compute expected lifecycle costs (20-year period), accounting for inflation, compliance penalties, and residual environmental risk:

- **Biological treatment expected cost** = $10M + (0.4 × $0) + (0.4 × $2M) + (0.2 × $5M) = $11.8M.
- **Chemical treatment expected cost** = $12M + (0.4 × $0) + (0.4 × $1M) + (0.2 × $3M) = $13M.
- **Membrane filtration expected cost** = $18M + minimal variation = $18M.

Although membrane filtration has the highest initial cost, its robustness against regulatory change may be justified for high-risk regions or plants discharging into sensitive water bodies. Conversely, biological or chemical options might be preferable in areas with more regulatory certainty or limited capital availability.

By integrating technology performance, regulatory forecasts, and economic risk into a unified structure, the decision tree supports proactive decision making that incorporates future uncertainty into capital project technology selection. It ensures that engineers consider both short-term investment and long-term adaptability, avoiding costly retrofits and regulatory non-compliance. This approach aligns with principles of sustainable water infrastructure planning and supports informed decision-making in the face of regulatory and environmental uncertainty.

5.4.7 Example 7: Pipeline Rehabilitation Strategy Selection Under Uncertain Deterioration Rates

Aging water and sewer pipeline networks often face structural deterioration due to corrosion, ground movement, root intrusion, and hydraulic overloading. Choosing the most appropriate rehabilitation strategy is complex, particularly when the rate and severity of deterioration are uncertain due to incomplete inspection data or variable environmental conditions. Decision tree analysis allows engineers to evaluate different rehabilitation options by modeling potential future deterioration scenarios and estimating the long-term performance and cost-effectiveness of each intervention. Figure 5.15 depicts common rehabilitation strategies and how their effectiveness varies under uncertain deterioration scenarios, providing a real-world context for infrastructure renewal decision-making.

1. **Decision Node (Rehabilitation Strategy Selection)**

   Engineers must select among three common rehabilitation strategies:

   - **Spot repair**: Involves targeted repair at known defect locations. It is low-cost but may not address undetected or future failures. Suitable for localized structural issues.
   - **Cured-in-place pipe (CIPP) lining**: A Trenchless rehabilitation method that creates a new pipe within the old pipe. Moderate cost, good structural reinforcement, and minimal excavation.
   - **Full pipe replacement**: Complete removal and replacement of the pipeline. Highest cost but ensures full capacity restoration and long service life.

Each strategy is analyzed using condition rating data (e.g., PACP grading), hydraulic modeling (e.g., EPANET or InfoWorks), and pipe material aging curves.

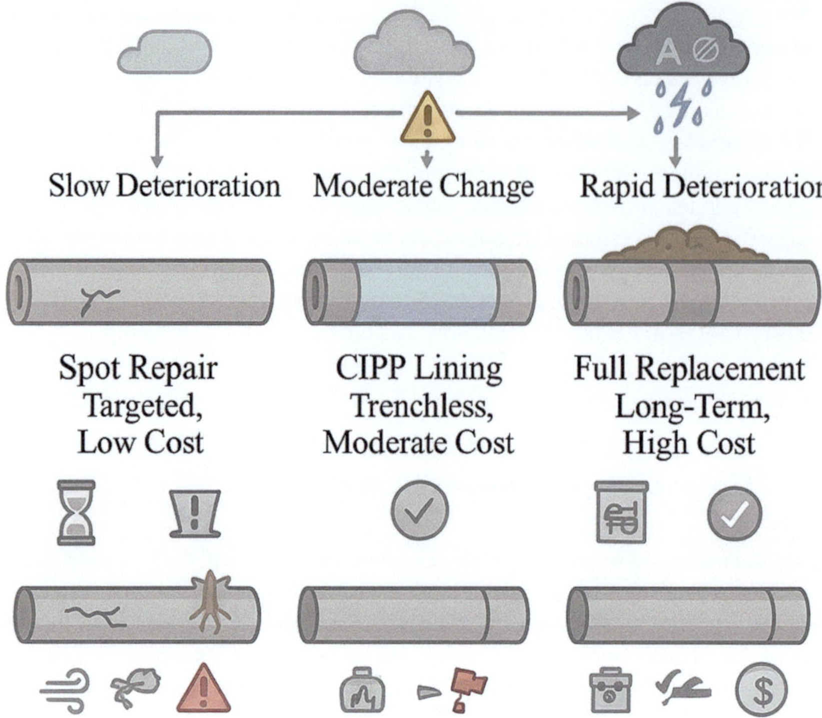

Fig. 5.15 Conceptual illustration of pipeline rehabilitation strategies—spot repair, cured-in-place pipe (CIPP) lining, and full replacement—shown under different deterioration conditions. The diagram emphasizes cost-risk trade-offs and expected performance under uncertainty

2. **Chance Nodes (Future Deterioration Scenarios)**

   The primary uncertainties involve:

   - **Deterioration rate**:
     - Slow deterioration (e.g., <1 PACP grade/year)—40% probability
     - Moderate deterioration (e.g., 1–2 PACP grades/year)—40% probability
     - Rapid deterioration (e.g., ≥2 PACP grades/year or sudden collapse risk)—20% probability.

These probabilities are based on historical failure records, pipe age, material (e.g., cast iron, vitrified clay), surrounding soil conditions, and CCTV inspection coverage. Additional factors include groundwater infiltration, corrosive soils, or stray current-induced corrosion in metallic pipelines.

3. **Outcome Nodes (Costs, Service Life, and Risk of Failure)**

   Each rehabilitation-degradation scenario leads to outcomes that consider the proponent's obligations with respect to:

- Capital cost of the rehabilitation action,
- Future failure cost (e.g., emergency repairs, service disruption, flooding, road damage),
- Service life extension (e.g., 10 years for spot repair vs. 50 years for full replacement),
- Residual risk of unaddressed defects or failure propagation.

Example outcomes:

- Spot repair + rapid deterioration: Low initial cost ($200/m), high failure risk → long-term cost = $1,500/m (due to emergency repairs)
- CIPP + moderate deterioration: Moderate cost ($500/m), 30-year life → lifecycle cost = $700/ m
- Full replacement + slow deterioration: High cost ($1,200/m), long-term performance → lifecycle cost = $1,300/m.

Expected value calculations:

- **Spot repair expected cost** = $0.4 \times \$300 + 0.4 \times \$800 + 0.2 \times \$1,500 = \$740/\text{m}$.
- **CIPP expected cost** = $0.4 \times \$600 + 0.4 \times \$700 + 0.2 \times \$900 = \$700/\text{m}$.
- **Full replacement expected cost** = consistent at $\$1,200 - 1,300/\text{m}$.

Although CIPP has a slightly higher upfront cost than spot repair, it offers better performance under uncertainty and avoids reactive maintenance cycles. Full replacement may be justified in high-risk locations or for critical pipelines where failure is unacceptable.

Decision tree analysis enables engineers to incorporate uncertainty in deterioration rates and failure consequences into pipeline asset management strategies. By comparing expected lifecycle costs, engineers can identify the most cost-effective rehabilitation approach that minimizes service disruption, maintenance costs, and infrastructure vulnerability. This structured method supports data-driven renewal planning in water utilities and municipal infrastructure programs.

5.5 Key Benefits of Decision Trees in Civil Engineering Decision-Making

The detailed examples presented earlier illustrate how decision trees serve as a practical and rigorous framework for guiding civil engineering decisions under uncertainty. This approach is especially effective in environments where engineers must balance cost, safety, performance, and long-term risk. The key advantages of using decision trees in civil engineering are provided in Fig. 5.16 and are elaborated upon in the subsections that follow.

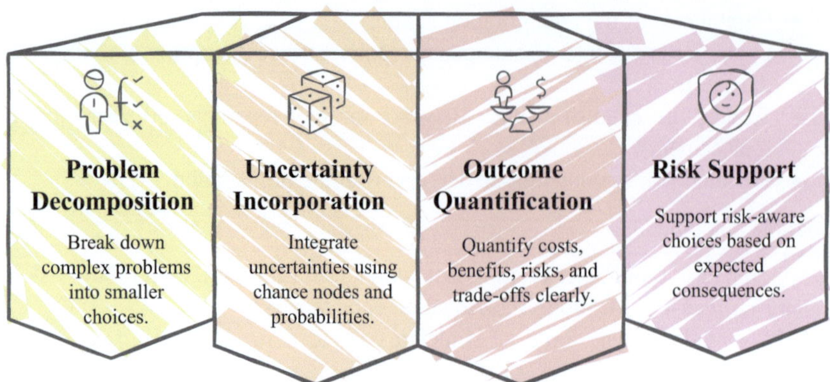

Fig. 5.16 Key advantages of using decision trees in civil engineering

5.5.1 *Breaking Down Complex Problems into Manageable Decisions*

Civil engineering projects often involve multiple interdependent decisions across design, construction, operation, and maintenance phases. Decision trees allow engineers to decompose these large, complex problems into a sequence of smaller, logically structured choices and probabilistic events. Each branch of the tree represents a manageable segment of the problem, making it easier to isolate critical decision points, evaluate alternatives, and track the consequences of each choice through the entire project lifecycle.

5.5.2 *Incorporating Uncertainties Through Probabilistic Modeling*

Many engineering decisions involve inherent uncertainties, such as fluctuating material costs, variable soil properties, unpredictable weather conditions, or evolving regulatory standards. Decision trees integrate these uncertainties using **chance nodes**, which assign probabilities to future events based on statistical models, expert judgment, or historical data. This allows engineers to analyze not only the deterministic outcomes of a decision, but also the probabilistic risks associated with less certain variables, ensuring decisions are robust against a wide range of future scenarios.

5.5.3 Quantifying Potential Outcomes for Transparent Evaluation

Decision trees provide a clear method for quantifying the costs, benefits, risks, and trade-offs associated with each possible course of action. Outcome nodes in the tree summarize expected values that include considerations such as capital expenditure, maintenance costs, lifespan, environmental impact, or safety performance. This quantitative structure supports data-driven comparison of options and facilitates clear communication of technical recommendations to stakeholders, including clients, regulatory agencies, and the public.

5.5.4 Supporting Risk-Aware and Rational Choices

Engineering design and planning must account for both financial and safety risks. By explicitly modeling risk and uncertainty, decision trees enable engineers to compare decisions based not just on cost minimization but also on **expected consequences**, such as the likelihood of failure, downtime, or non-compliance. This risk-aware approach aligns with infrastructure resilience goals and supports the development of solutions that maintain performance under adverse conditions, such as extreme loading, flooding, seismic events, or accelerated deterioration.

5.6 Utility and Risk in Decision Trees

Building on the structured approach and ability to handle uncertainties, decision trees in civil engineering extend beyond simply mapping out choices—they serve as a powerful analytical framework for evaluating both utility and risk across multiple decision paths. This dual consideration enables engineers to make decisions that are not only technically sound but also economically and strategically optimal. The concept of *utility* is central to decision tree analysis, as it captures the relative value or benefit associated with each possible outcome. In civil engineering, utility encompasses a wide range of project-specific criteria, both tangible and intangible. Tangible factors include construction costs, operation and maintenance expenses, lifecycle performance, and compliance with design standards. Intangible factors, though harder to quantify, are equally important—these may include user safety, system reliability, social impact, and resilience to extreme events. By assigning a utility value to each outcome node in a decision tree, engineers can objectively compare alternatives across varied scenarios. This is especially valuable in high-stakes contexts where minimizing risk (e.g., flood damage, foundation failure, structural collapse) must be balanced with maximizing return on investment and maintaining cost control. In such cases, the decision is not simply about choosing the least expensive option, rather, engineering must select the one that delivers the greatest overall value under uncertainty. Expected utility becomes the cornerstone of this analysis. For each decision path, the utility of possible outcomes is weighted by their probability of occurrence. This yields an *expected utility value* that integrates both the desirability of an outcome and the likelihood of its realization. Engineers can then compare expected utility values across all decision branches to identify the path that offers the highest expected benefit, or, conversely, the lowest expected cost or risk. This process transforms decision-making from intuition-driven judgment into a structured, evidence-based approach. It aligns decisions with core civil engineering objectives: safety, durability, sustainability, and cost-efficiency. Moreover, by transparently incorporating uncertainty and trade-offs, decision trees foster stakeholder trust and facilitate communication with regulators, funding agencies, and the public. In essence, decision trees empower civil engineers to navigate uncertainty with clarity and precision. By enabling the systematic evaluation of utility and risk, they guide the selection of solutions that are robust, resilient, and aligned with long-term infrastructure goals—even when conditions are unpredictable or data is incomplete.

5.6.1 Defining Utility in Civil Engineering Contexts

The concept of utility is foundational to decision tree analysis in engineering, particularly within civil engineering, where each decision can significantly influence both short-term project execution and long-term infrastructure performance. Utility represents the perceived or calculated value of a given outcome and can be framed in terms of benefits, satisfaction, or effectiveness derived from a particular decision (Fig. 5.17). It provides a way to capture and compare diverse project goals—ranging from financial savings to safety enhancements—within a unified framework.

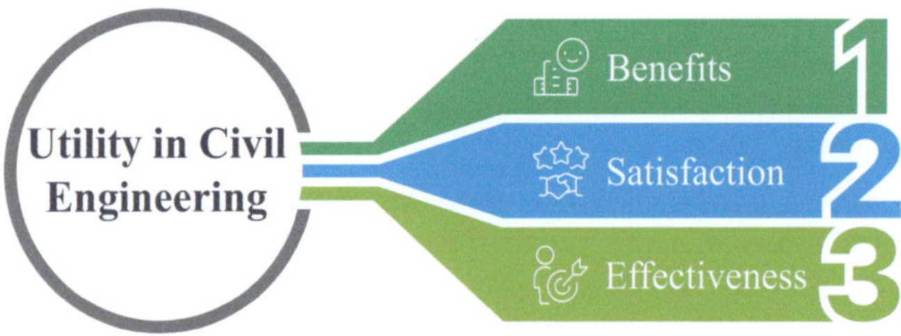

Fig. 5.17 Core components of utility in civil engineering decision-making

5.6.2 Tangible and Intangible Components of Utility

In civil engineering, utility incorporates both quantifiable factors and qualitative considerations, forming a comprehensive foundation for evaluating decision alternatives:

- **Tangible factors** may include initial construction costs, operation and maintenance expenses, expected service life, and return on investment. These are typically numerical and can be directly estimated using financial models, technical specifications, or historical data. For instance, initial costs reflect material and labor expenditures, while service life provides insight into the duration over which an infrastructure component will remain functional before replacement or rehabilitation is needed.
- **Intangible factors**, although not always easily quantifiable, are equally critical. These include risk reduction, such as minimizing the probability of failure under extreme conditions; environmental impacts, such as reducing greenhouse gas emissions through the choice of materials or construction practices; and social impacts, like minimizing disruption to surrounding communities during repair or retrofitting. While intangible factors may not appear in a cost spreadsheet, they significantly influence public perception, regulatory compliance, and long-term project sustainability.

Figure 5.18 organizes these utility elements across seven key characteristics—initial costs, expenses, service life, return, risk, environment, and social impact—categorizing each as either tangible, intangible, or both. This structured view helps engineers and stakeholders assess project alternatives holistically, ensuring that decisions align not only with technical and economic objectives but also with broader societal and environmental values.

For instance, when designing a flood protection system, engineers may compare an embankment with a lower overall height that is less expensive and quicker to construct with a higher embankment that offers increased protection. While the former has a lower upfront cost, its utility may be lower if it fails to prevent flooding in high-risk scenarios. Conversely, the higher embankment incurs more upfront investment but provides greater resilience, lower long-term risk, and potentially higher utility, especially when considering potential damages, insurance implications, and community impact.

5.6.3 Expected Utility: A Probabilistic Metric for Decision Evaluation

Expected utility is a key metric within decision trees, enabling engineers to evaluate each decision path by accounting for both the benefits and costs associated

Characteristic	Tangible Factors	Intangible Factors
Initial Costs	Includes construction costs	Not applicable
Expenses	Operation and maintenance costs	Not applicable
Service Life	Expected duration of service	Not applicable
Return	Return on investment	Not applicable
Risk	Not applicable	Reducing probability of failure
Environment	Not applicable	Lowering greenhouse gas emissions
Social	Not applicable	Community disruption during repairs

Fig. 5.18 Classification of utility characteristics in civil engineering decision-making. The tangible and intangible columns should note that they are merely "examples"

with a range of possible future scenarios, each characterized by an associated probability. This approach allows for a comprehensive comparison of alternatives that goes beyond simple deterministic outcomes, incorporating the variability and uncertainty inherent in real-world engineering projects. By calculating expected utility, engineers can systematically quantify the trade-offs between upfront investment, operational performance, long-term maintenance demands, and risk exposure. This is particularly valuable when making infrastructure decisions in contexts marked by uncertain external factors, such as variable groundwater levels in foundation design, fluctuating traffic volumes in pavement planning, or changing flood frequencies due to climate change. In civil engineering applications, high uncertainty often stems from environmental variability, limited site data, evolving regulations, or emerging societal expectations (e.g., sustainability, resilience). Expected utility helps engineers navigate these challenges by providing a structured method to balance potential gains—such as enhanced safety, longer service life, or regulatory compliance—with potential losses, including financial overruns, system underperformance, or environmental impact. For example, in seismic retrofit projects, engineers may evaluate whether moderate reinforcement is sufficient for probable earthquakes or whether full retrofitting is justified to guard against rare but catastrophic events. By selecting the option with the highest expected utility—or conversely, the option that combines the lowest expected lifecycle cost and risk-weighted impact—engineers ensure their choices are both technically defensible and economically optimized. This process aligns infrastructure decisions with core project objectives such as safety, reliability, and long-term cost-efficiency, while also aligning with stakeholder risk tolerance, public safety standards, and environmental commitments. In this way, expected utility analysis transforms uncertainty from a source of indecision into a quantifiable, manageable component of engineering strategy.

5.6.4 Practical Application of Utility and Risk Assessment in Civil Engineering Decision Trees

When engineers structure decision trees around utility and risk, they gain a systematic and transparent approach to evaluating complex decisions under conditions of uncertainty. This methodology enables civil engineers to weigh economic costs, performance outcomes, and project risks in a structured format, aligning choices with technical specifications, stakeholder expectations, and long-term infrastructure goals. The utility-risk framework is particularly valuable across diverse sub-disciplines in civil engineering, where uncertainty is inherent in every phase, from preliminary design to asset management.

In geotechnical engineering, utility and risk assessments are essential in foundation design, especially in the presence of incomplete or variable subsurface information. Engineers often must choose between shallow foundations (e.g., spread footings) and deep foundations (e.g., piles, caissons, or drilled shafts), each with having different cost implications and performance reliability depending on soil conditions. A well-structured decision tree begins with a decision node representing the foundation type, followed by chance nodes modeling probable soil conditions, such as dense granular soil, soft cohesive soil, or mixed fill layers. Each branch leads to outcome nodes reflecting total expected costs, including initial construction, the potential for differential settlement, soil remediation, and serviceability loss. By calculating the expected utility of each option, engineers can compare not just direct costs but also long-term risk exposure and reliability, critical for structures such as bridges, retaining walls, and towers.

In flood management, decision trees assist engineers in selecting protective infrastructure dimensions, most commonly floodwall or levee heights, by integrating hydrologic uncertainties with economic and societal impacts. Engineers must assess trade-offs between capital investment and risk reduction. A shorter floodwall might meet a 10-year flood standard at a lower cost but leave assets vulnerable to more severe events. A higher wall may withstand a 100-year flood event, but require greater land acquisition, visual impact mitigation, and more significant investment. Chance nodes represent flood return periods—10, 50, or 100-year events—each assigned probabilities using local gauge data or synthetic rainfall-runoff modeling. Outcome nodes quantify costs from flood damage, wall overtopping, or maintenance, enabling engineers to calculate expected utility for each wall height. This approach ensures the final design offers optimal resilience and aligns with climate adaptation and public safety goals.

In transportation engineering, decision trees play a crucial role in pavement type selection under variable usage and environmental conditions. Engineers may choose between asphalt, which has lower initial costs but higher susceptibility to rutting and oxidation, and concrete, which offers better performance under heavy traffic but requires significantly greater initial investment. A decision tree captures this by starting with a decision node representing material choice, followed by chance nodes such as traffic loading (low, medium, high) and climate conditions (freeze–thaw frequency, high temperatures, precipitation). These factors influence cracking rates, surface degradation, and the frequency of rehabilitation. Outcome nodes incorporate total lifecycle costs—including maintenance frequency, user delay costs, and rehabilitation intervals—and compare them against durability metrics and service reliability. By assessing the expected utility of each pavement system, engineers can identify the material that delivers the lowest long-term cost while maintaining structural and functional performance.

Across all these domains, decision trees serve as a decision-support framework that incorporates utility and risk into infrastructure planning, design, and operation. Rather than relying solely on deterministic calculations or conservative assumptions, engineers can visualize and evaluate how different decisions perform under a range of future conditions. This enables more adaptive, cost-efficient, and resilient solutions, especially when project decisions involve long service lives, limited budgets, and changing environmental or regulatory landscapes. By quantifying both the probabilistic nature of risks and the value of outcomes, decision trees help engineers make informed, rational decisions that reflect the real-world complexity of civil engineering practice. Whether designing for uncertain soil conditions, variable hydrologic inputs, or evolving usage demands, this approach promotes the creation of infrastructure systems that are not only functional and safe, but also sustainable and adaptable to changing needs over time.

5.7 Decision Tree Implementation Steps

Implementing a decision tree involves a systematic approach to break down complex decision-making processes into manageable, sequential steps. This method provides engineers with a clear framework for identifying alternatives, modeling uncertainties, and evaluating potential outcomes. By structuring decisions in this way, engineers can visualize the full spectrum of possible scenarios and assess each option's expected utility in a transparent and data-driven manner. This is especially valuable in civil engineering projects, where decisions often involve large capital cost expenditures, incomplete information, and long-term implications. Figure 5.19 illustrates the key steps involved in implementing a decision tree in civil engineering applications. These steps are followed by detailed examples that demonstrate how the methodology can be applied to real-world problems, ranging from flood protection design to material selection and infrastructure rehabilitation.

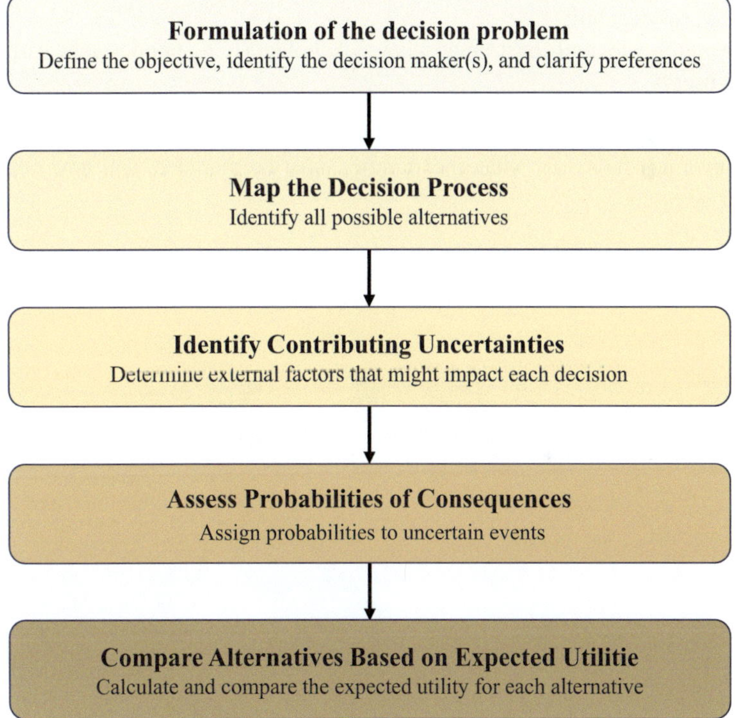

Fig. 5.19 Steps that impact each outcome node for Implementing decision tree's in engineering projects

5.7.1 Formulation of the Decision Problem

The first, and most critical, step in implementing a decision tree is the precise formulation of the decision problem. This involves defining the decision objective, identifying the decision maker(s), and establishing their preferences, constraints, and priorities. In civil engineering, this step ensures that the analysis is aligned with project-specific goals, such as minimizing lifecycle costs, ensuring structural reliability, meeting regulatory requirements, or enhancing environmental resilience.

Formulating the decision problem requires a thorough understanding of (Fig. 5.20):

- The scope of the engineering problem (e.g., design, maintenance, or retrofit)
- The suite of available alternatives that the engineer can consider
- The uncertainties involved (e.g., natural hazards, material performance and cost, or future demand)
- The objectives and criteria to be used for evaluating outcomes (e.g., cost, utility, safety, serviceability)
- The preferences and constraints of key stakeholders (e.g., public agencies, clients, end-users).

This step lays the groundwork for building the decision tree by determining what decisions must be made, what uncertainties must be modeled, and how outcomes will be assessed.

Fig. 5.20 Key elements in formulating an engineering decision problem for decision tree analysis

In a flood protection project, for instance, an engineer must determine the optimal height for a levee to protect a riverfront community. The decision problem can be formulated as:

"What levee height minimizes the expected flood damage while staying within the municipality's budget and meeting floodplain management regulations?"

The decision maker is the project engineer, working in collaboration with local authorities and environmental planners. Preferences include maximizing public safety, minimizing expected flood loss, and staying within an allocated capital expenditure (e.g., \$10 million). Additional considerations may include environmental permitting, long-term maintenance costs, and public perception. Formulating the problem in this structured way ensures that the subsequent steps in the decision tree reflect all key objectives and constraints.

5.7.2 *Mapping of the Decision Process*

After formulating the decision problem, the next step is to map out the decision process by identifying all feasible alternatives or courses of action available to the decision maker. These alternatives form the primary branches of the decision tree, originating from decision nodes. This step ensures that the full range of technically and economically viable solutions is considered before evaluating uncertainties and outcomes.

In civil engineering, mapping the decision process requires (Fig. 5.21):

- Reviewing design standards, codes, and constraints to ensure each alternative is permissible

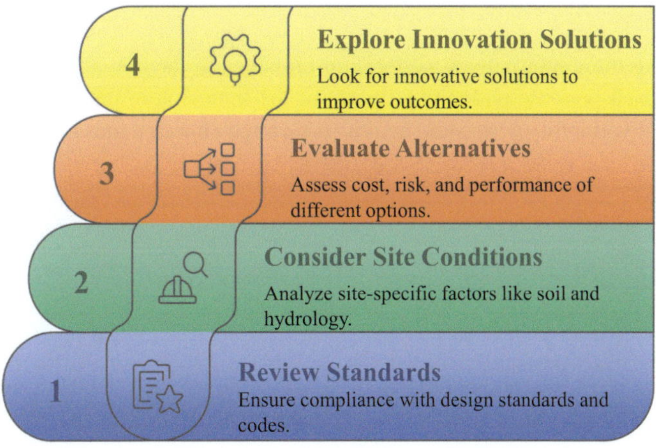

Fig. 5.21 Mapping the decision-making process in civil engineering

- Considering site-specific conditions (e.g., soil type, hydrology, structural loading, water quality data)
- Ensuring that alternatives vary meaningfully in cost, risk, constructability, and long-term performance
- Accounting for innovative or non-traditional solutions that may offer improved outcomes.

This stage outlines the immediate choices that the engineer will face while also preparing the foundation for integrating chance events and performance scenarios in the next steps.

In a bridge construction project, for instance, engineers must select the most appropriate foundation type based on geotechnical, hydraulic, and structural considerations while also conforming to applicable regulations. The identified alternatives may include:

- Shallow foundation (e.g., spread footings), suitable for strong, near-surface soils and low-to-moderate loads
- Piled foundation, ideal for transferring loads to deeper, more competent strata in soft or variable soils
- Raft foundation, a viable option when differential settlement must be minimized or where load distribution is critical across weak soils.

Each of these alternatives carries distinct implications:

- Cost: Shallow foundations typically have the lowest construction cost, while pile foundations are more expensive due to deeper excavation and specialized equipment.
- Performance: Pile foundations may offer higher reliability in weak or liquefiable soils.
- Feasibility: Raft foundations may be constrained by site geometry or excavation limitations.

Mapping these alternatives as decision branches in the tree allows engineers to visualize each possible path and prepare for the inclusion of uncertainty factors—such as soil variability or seismic activity—through chance nodes in the subsequent steps.

5.7.3 *Identification of Contributing Uncertainties*

Once the available decision alternatives are mapped, the next step is to identify the key uncertainties that could influence the outcome of each decision. In decision tree analysis, these uncertainties are represented as chance nodes, which follow decision nodes and account for variables outside the engineer's control. Accurately identifying and modeling these uncertainties is essential for creating a realistic and robust decision structure.

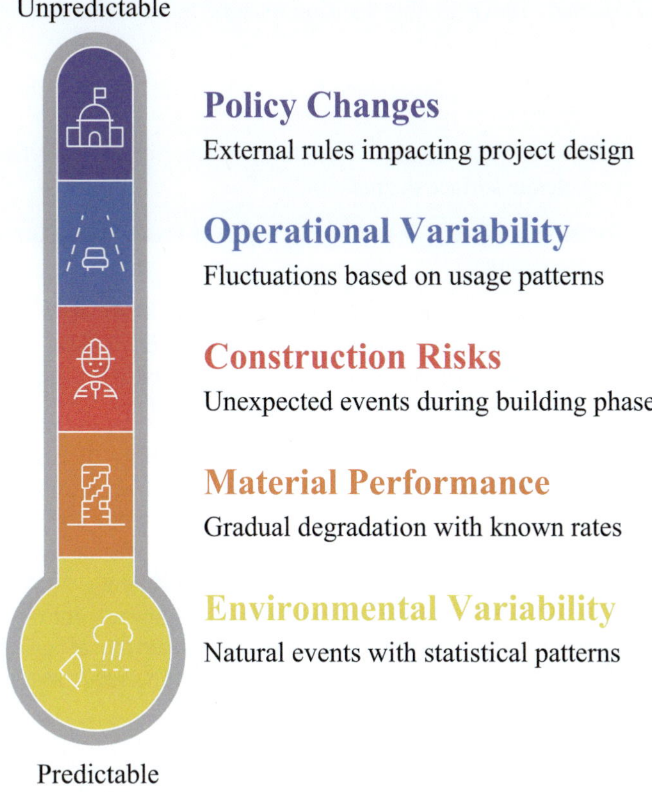

Fig. 5.22 Civil engineering uncertainties range from predictable to unpredictable

In civil engineering, contributing uncertainties may stem from (Fig. 5.22):

- **Environmental variability** (e.g., rainfall intensity, freeze–thaw cycles, seismic ground motion)
- **Material performance deviations** (e.g., concrete shrinkage, steel corrosion rates)
- **Construction risks** (e.g., groundwater inflow, unexpected subsoil conditions)
- **Operational variability** (e.g., future traffic volumes, urban expansion)
- **Policy or regulatory changes** (e.g., new environmental discharge standards, updated safety codes).

These uncertainties are not hypothetical—they are often supported by empirical data, simulation models, probabilistic analyses, or expert judgment. Each chance node should include branches that represent possible outcomes, with probabilities assigned to reflect the likelihood of each scenario.

Consider an example related to pavement design for a municipal roadway. In a pavement material selection project, engineers may need to choose between asphalt and concrete. However, this decision must be made in the face of two key uncertainties:

- Traffic load: Will the road experience high truck traffic or only light commuter traffic?
- Climate conditions: Will the region see mild weather or frequent freeze–thaw cycles that accelerate surface degradation?

After the decision node (e.g., asphalt selected), the decision tree branches into chance nodes representing:

- Traffic scenarios: "High traffic load" and "Low traffic load", with probabilities based on projected vehicle counts and local transportation models.
- Climate exposure: "Severe climate" and "Mild climate", based on regional climate data, elevation, and anticipated weather extremes.

These uncertainties directly influence the expected utility of each option by affecting maintenance frequency, pavement lifespan, surface performance, and life-cycle costs. For example, asphalt may have lower performance under high traffic and severe climate conditions, requiring more frequent resurfacing, while concrete might have higher resilience but with greater upfront investment.

By clearly identifying and incorporating these chance events into the decision tree, engineers can anticipate variability, build resilience into the design, and ensure that selected alternatives remain viable under different future scenarios.

5.7.4 Identification of Potential Consequences and Their Utility (Cost/Benefit)

After mapping decisions and identifying key uncertainties, the next step is to determine the potential consequences associated with each decision path. This involves tracing each pathway through the decision tree—from the initial decision nodes, through possible chance events, to its final outcome—and assigning a utility value to represent the net cost or benefit associated with each outcome. In civil engineering, utility often reflects a combination of economic, technical, and performance-based criteria, providing a basis for comparing alternatives objectively (Fig. 5.23).

Consequences should account for:

- Initial construction or implementation costs
- Long-term operation and maintenance costs
- Failure or repair costs under adverse conditions
- Performance benefits (e.g., increased reliability, longer service life)
- Social or environmental impact (e.g., displacement, emissions, visual intrusion).

Utility values can be expressed in monetary terms (e.g., net present value), composite indices (e.g., lifecycle performance score), or qualitative ratings converted

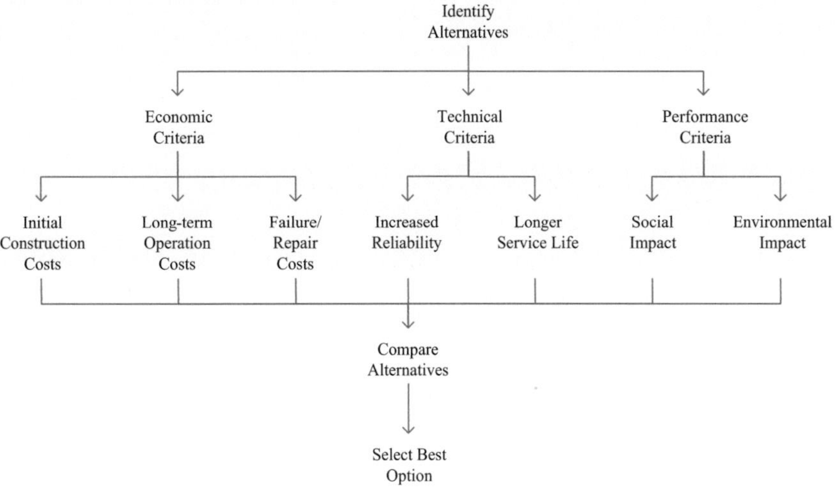

Fig. 5.23 Utility Assessment in engineering applications

into numeric scales for analysis. Assigning these values allows engineers to quantify the trade-offs involved in each decision and ensure alignment with project goals such as cost-efficiency, safety, resilience, and sustainability.

In a dam design project, for instance, engineers may be evaluating two main options:

- A dam with a low overall height that has a construction cost of $50 million, which provides partial flood control but may incur $15 million in potential flood damage over its lifespan if overtopped during rare events.
- A higher dam, with a construction cost of $75 million, which significantly reduces downstream flood risk, lowering expected damage to $2 million.

Each decision is evaluated under different flood scenarios, modeled using hydrologic data and return period analysis (e.g., 50 vs. 200-year flood events). For each scenario, the consequence includes both:

- Upfront capital cost (based on design specifications and unit costs)
- Expected damage cost (calculated using risk = probability × consequence from flood event simulations).

The utility assigned to each outcome could be the inverse of the total cost—e.g., higher utility for lower total cost—or a performance index that combines cost, flood protection level, and environmental criteria. For example, if the lower dam results in a total expected cost of $65 million and the higher dam in $77 million, utility scores might be adjusted to reflect non-financial benefits such as risk mitigation, regulatory compliance, or public safety. This structured evaluation allows engineers to:

- Compare alternatives on a common scale
- Prioritize options that deliver the best value for cost
- Communicate the rationale for choices to stakeholders with clarity and confidence.

5.7.5 Probability Assessment of Uncertain Events

For each identified uncertainty in the decision tree, engineers must assign probabilities to the possible outcomes at each chance node. These probabilities quantify the likelihood of external events that could impact project performance, cost, or safety. Accurate probability assessment is essential, as it directly influences the calculation of expected utility, helping prioritize solutions that best balance risk and reward.

Probabilities can be estimated using a variety of engineering and data-driven techniques, including:

- Historical data analysis, such as streamflow records, climate patterns, or failure rates from similar projects
- Numerical modeling or simulations, including hydrologic modeling, finite element analysis, or geotechnical Monte Carlo simulations
- Expert judgment, especially when empirical data is limited or when forecasting rare or unprecedented conditions (e.g., regulatory changes or emerging climate patterns).

Probabilities assigned to each branch of a chance node should be mutually exclusive and collectively exhaustive, ensuring that all possible outcomes are considered and the total probability at each node equals 1. These values may later be refined through sensitivity analysis, allowing engineers to test how changes in assumptions affect decision outcomes.

For example, in a flood management project, an engineer is tasked with determining the appropriate height for a new floodwall. The uncertainties lie in future flood severity, which is influenced by catchment hydrology, climate change, and land use dynamics. Based on historical data and flood frequency analysis, the engineer assigns probabilities to three flood scenarios:

- 10-year flood event: 10% annual probability
- 50-year flood event: 2% annual probability
- 100-year flood event: 1% annual probability.

Each floodwall height—low, medium, or high—is represented as a decision branch. For each wall height, the decision tree includes chance nodes with the flood event probabilities and the corresponding consequences (e.g., overtopping, damage cost, safety risk). For example:

- A low floodwall may prevent a 10-year flood but not a 50- or 100-year flood, resulting in high expected flood damage in those less likely but more severe scenarios.

- A medium floodwall may block the 10- and 50-year events but still allow overtopping under a 100-year flood.
- A high floodwall may fully protect against all modeled flood events, but it comes at a significantly higher construction and environmental cost.

By integrating these probabilities into the tree, engineers can calculate expected utility values for each option, enabling them to select the wall height that offers the most cost-effective protection given the statistical likelihood of future floods. This probability-based structure ensures that decision-making is not based on a single worst-case or average scenario, but rather reflects a realistic distribution of potential risks, improving both project reliability and stakeholder confidence.

5.7.6 Comparison of the Different Decision Alternatives Based on Their Expected Utilities

After all decision pathways have been defined, utilities assigned, and probabilities estimated for each outcome, engineers can evaluate the expected utility of each decision alternative. This is done by calculating the expected utility for each path, which involves multiplying the utility (or cost) of each potential consequence by its corresponding probability, then summing these products across all branches. The result is a single metric for each alternative that reflects its performance under uncertainty.

This comparison enables engineers to identify the alternative that best meets the project's objectives, whether that means minimizing total lifecycle cost, maximizing safety, or enhancing long-term resilience. Rather than relying on worst-case or average-case assumptions, the expected utility approach incorporates the full range of possible outcomes, each weighted by its likelihood of occurrence. This provides a more comprehensive, risk-informed basis for decision-making.

In civil engineering, this comparison is particularly valuable when decisions involve trade-offs between higher upfront costs and improved long-term performance. It also helps justify decisions to stakeholders by demonstrating that all major risks and consequences have been systematically considered.

In a flood risk mitigation project, engineers must choose between constructing a levee system or restoring upstream wetlands. The levee provides immediate and predictable protection with lower short-term costs, but it requires ongoing maintenance and offers limited adaptability under future climate variability. Wetland restoration involves higher initial investment and longer implementation time but provides flexible flood attenuation, ecosystem services, and reduced long-term maintenance. By assigning utilities to economic costs, flood damage reduction, environmental benefits, and system resilience, and evaluating outcomes under moderate and extreme flood scenarios, engineers calculate the expected utility of each alternative. The analysis shows that wetland restoration yields a higher expected utility over a multi-decade horizon, supporting a decision that emphasizes long-term risk reduction, adaptability, and co-benefits beyond direct flood control.

This final comparison step brings together all previous components of the decision tree—decision alternatives, uncertainties, outcomes, and probabilities—into a single, coherent analysis that helps civil engineers choose the optimal solution for complex, risk-sensitive infrastructure challenges.

5.8 Utility and Expected Utility in Decision Trees for Civil Engineering

In civil engineering, decision-making frequently involves high-stakes choices under uncertain conditions—whether selecting a foundation type for a bridge, determining the optimal flood protection level, or choosing a pavement material for long-term traffic performance. Each alternative carries unique risks, costs, and benefits. To make informed and rational decisions, engineers apply the concepts of utility and expected utility, which allow for a systematic evaluation of competing options across varying scenarios.

5.8.1 *Defining Utility in Civil Engineering*

Utility represents the value or preference associated with a particular outcome. It is not limited to financial cost alone but can include a wide range of performance metrics and stakeholder objectives (Fig. 5.24):

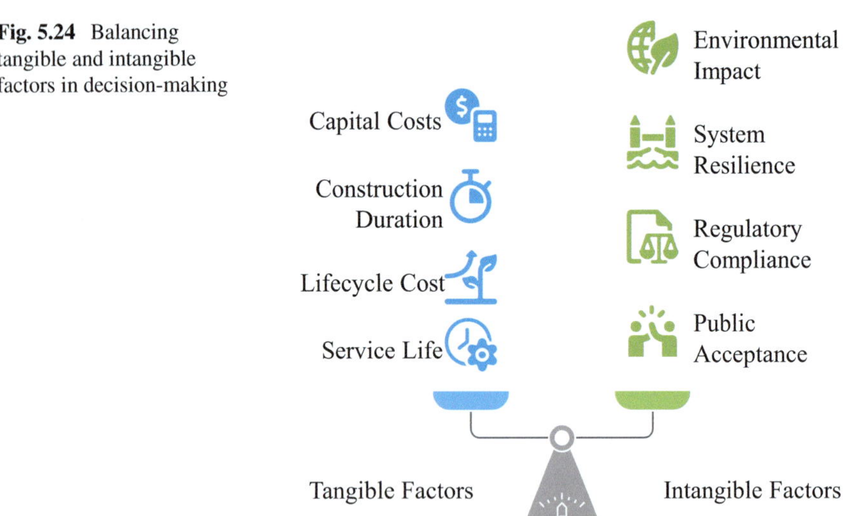

Fig. 5.24 Balancing tangible and intangible factors in decision-making

- **Tangible factors**: capital costs, operation and maintenance costs, construction duration, service life, lifecycle cost
- **Intangible factors**: safety, environmental impact, system resilience, regulatory compliance, public acceptance.

Utilities are typically normalized onto a common scale (e.g., 0–1) or represented directly in monetary terms if the analysis focuses on cost minimization.

Expected utility extends the concept of utility by incorporating uncertainty. Rather than evaluating decisions based only on single outcomes, expected utility evaluates all possible outcomes, each weighted by the probability of its occurrence. This transforms the decision process from a deterministic evaluation into a probabilistic, risk-aware optimization.

5.8.2 Defining and Calculating Expected Utility

Let d_j represent a specific decision option, such as choosing a shallow foundation, constructing a medium floodwall, or using asphalt pavement, h_i denote a possible state of nature or scenario, for example, stable soil conditions, the occurrence of a 100-year flood, or experiencing high traffic load. The utility or value associated with selecting decision d_j under scenario h_l is denoted by $U(d_j, h_l)$, which reflects the benefit or cost outcome of that decision in the given context. Additionally, each scenario h_i has an associated probability of occurrence, represented by $P(h_i)$, which quantifies the likelihood that a particular state of nature will materialize. Then, the expected utility $EU(d_j)$ of decision option d_j is given by the weighted sum:

$$EU(d_j) = \sum_{i=1}^{n} P(h_i) \times U(d_j, h_i) \tag{5.1}$$

where n is the number of possible uncertain outcomes for decision d_j. Each utility is weighted by the likelihood of its associated event. The result is a single expected value used for comparing decision alternatives.

Interpreting the expected utility depends on how utility is defined (Fig. 5.25). When utility is defined as a benefit, the decision option with the highest expected utility is considered the most desirable. Conversely, if utility is defined as a cost, the optimal decision corresponds to the one with the lowest expected cost, or in other words, the least negative utility. Expected utility provides engineers with a systematic method to directly compare various options, even when the associated risks, costs, and benefits differ among alternatives. This approach supports multi-criteria decision-making by integrating diverse types of impacts into a unified valuation framework.

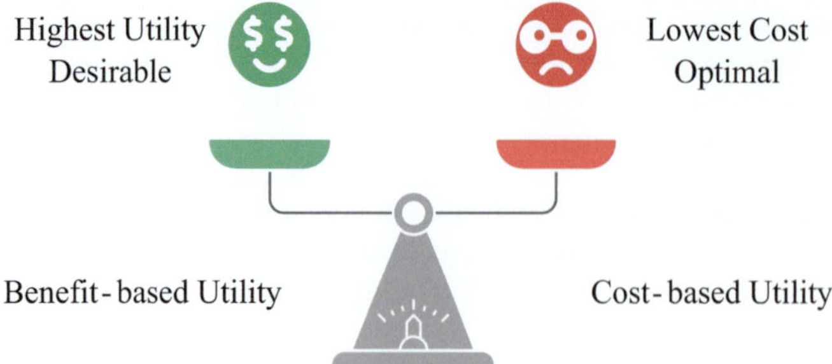

Fig. 5.25 Visualizing expected utility

5.9 Practical Application of Expected Utility in Civil Engineering

5.9.1 Example 1: Pile Foundation Depth Decision in Geotechnical Engineering

Problem Statement:

A construction engineer needs to select the appropriate depth for a steel pile foundation in an area with weak soil overlying a rock layer. The choices are to use either a 40-foot pile or a 50-foot pile (Fig. 5.26).

If the pile depth is insufficient, there is a risk of structural settlement or failure due to the soil's weak load-bearing capacity. However, deeper piles require more material and installation costs. The engineer needs to determine which pile depth will minimize total project costs while ensuring structural stability. The relevant decision-making parameters are outlined below (Fig. 5.27).

- **Decision Options**:
  - 40-foot pile
  - 50-foot pile

Fig. 5.26 Schematic of pile foundation depth decision

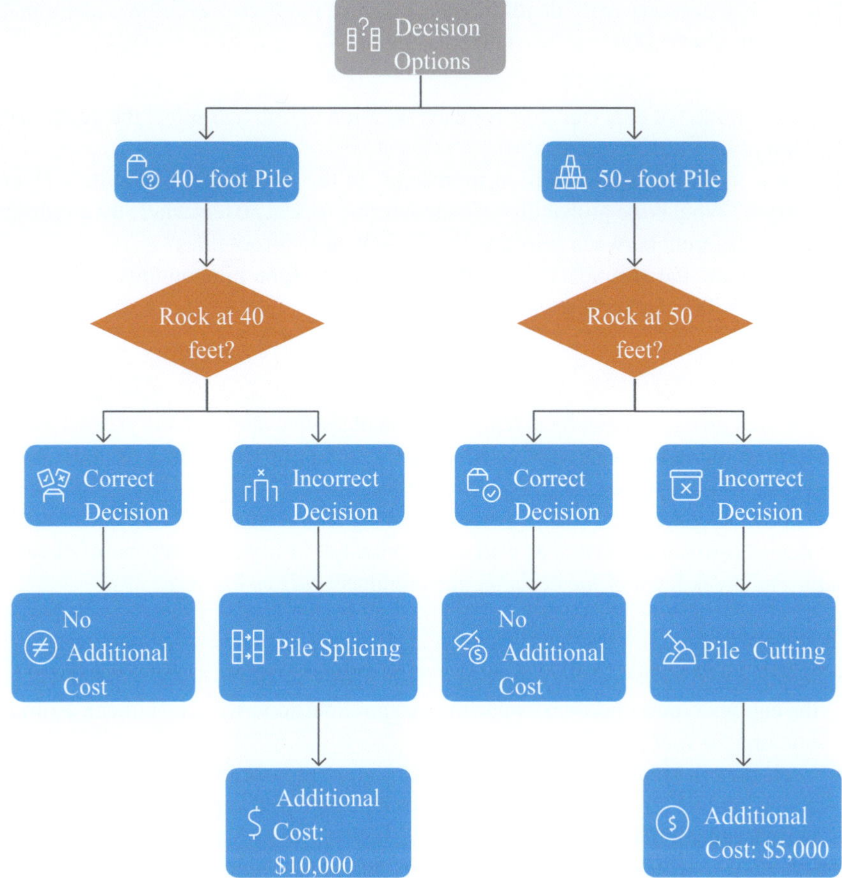

Fig. 5.27 Pile decision analysis

- **Chance Outcomes**:
  - **Rock layer is at 40 feet** (probability = 0.6).
  - **Rock layer is at 50 feet** (probability = 0.4).
- **Utility/Cost Values**:
  - **40-foot pile**:

    Correct decision (rock at 40 feet): no additional cost.
    Incorrect decision (rock at 50 feet): requires pile splicing, additional cost = $10,000.

  - **50-foot pile**:

    Correct decision (rock at 50 feet): no additional cost.

Unnecessary extra depth (rock at 40 feet): requires pile cutting, additional cost = $5,000.

(a) Determine the expected cost for each decision option and select the depth that minimizes total expected cost.
(b) How sensitive is the decision to changes in the probability of the rock layer depth? What is the probability of encountering rock at 50 feet where the engineer be indifferent between selecting the 40-foot and 50-foot piles?
(c) Calculate the utilities for the 40-foot and 50-foot pile options, given the probabilities and costs associated with splicing.

Solution

Figure 5.28 provides a graphical representation of the decision-making process using a decision tree in the context of selecting an optimal pile depth for a foundation in uncertain soil conditions. The layout of the decision tree highlights the sequence of decisions, outcomes, and consequences, allowing engineers to visualize each step in the decision-making process, as well as the potential utility (or cost/benefit) of each path. The breakdown of the decision tree components is as follows:

Action Alternatives (Decision Node):

- Represented by the yellow square, this node marks the initial decision point where the engineer must choose between the two possible actions: using either a 40-foot pile or a 50-foot pile.
- Each branch stemming from the decision node represents a distinct course of action, with each choice leading to different potential outcomes based on the underlying conditions.

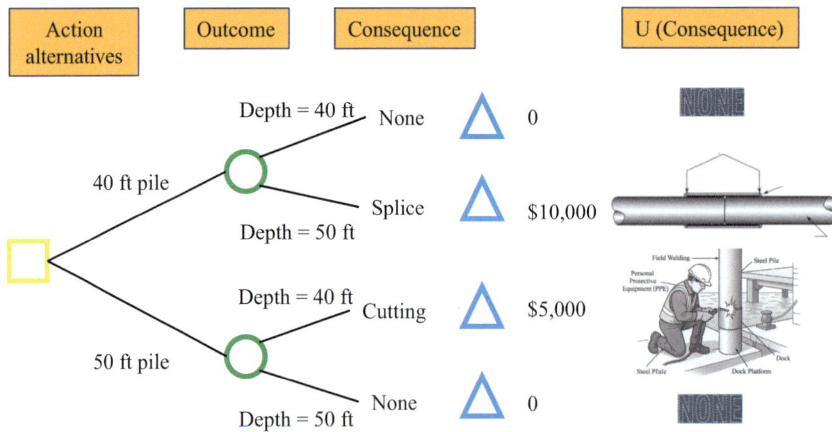

Fig. 5.28 Breakdown of the pile foundation depth decision components

Outcome (Chance Nodes):

- The green circles represent chance nodes, where uncertainty comes into play. Each chance node corresponds to a possible state of nature, such as the depth at which the rock layer lies.
- For each pile option, there are two possible outcomes:
  - The rock layer is at 40 feet
  - The rock layer is at 50 feet
- These outcomes are associated with different probabilities, reflecting the engineer's uncertainty about subsurface conditions based on the available borehole log data.

Consequence (Outcome Nodes):

- The blue triangles represent consequence nodes or outcome nodes. These nodes depict the end result of each decision and chance combination.
- For each combination of pile depth and rock layer depth, there is a corresponding consequence:
- If the 40-foot pile is chosen but the rock is at 50 feet, splicing the pile becomes necessary (illustrated by the splicing diagram)
- If the 50-foot pile is chosen and the rock is at 40 feet, there is a surplus of material that must be cut off (shown as a consequence of over-design)

Utility (U):

- Each outcome node is assigned a utility value (U), reflecting the cost or benefit associated with each consequence. The utility values help quantify the economic impact or value of each possible path. For example:
  - When the 40-foot pile is adequate (rock at 40 feet), there are no additional costs.
  - When splicing is required for the 40-foot pile (rock at 50 feet), additional labor and material costs are incurred yielding a utility of 10,000.

These utilities guide the engineer in selecting the path with the lowest expected cost, optimizing both safety and economic efficiency.

Figure 5.29 provides the mathematical format of the decision tree by incorporating probabilities for each outcome at the chance nodes. These probabilities, labeled $P(h_0) = 0.6$ and $P(h_1) = 0.4$, represent the likelihood of encountering different conditions, such as the rock layer being at a particular depth, after a decision on pile depth has been made. This additional layer of information allows for a more refined analysis of the decision problem, helping the engineer evaluate each option based on expected utility (or expected cost).

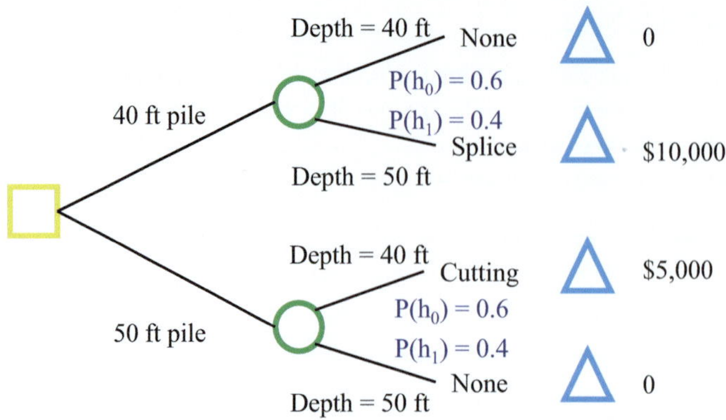

Fig. 5.29 Mathematical formulation of pile foundation depth decision tree

5.9.1.1 Solve Using Expected Utility

(a) **Expected Cost**

Identify Possible Outcomes and Costs:

- For a 40-foot pile:
  - Rock at 40 feet: No additional cost.
  - Rock at 50 feet: Additional cost of $10,000 (splicing needed).
- For a 50-foot pile:
  - Rock at 40 feet: Additional cost of $5,000 (unnecessary depth).
  - Rock at 50 feet: No additional cost.

Assign Probabilities:

Rock at 40 feet: Probability = 0.6.
Rock at 50 feet: Probability = 0.4.

Calculate Expected Cost for Each Option:

- Expected cost for 40-foot pile:

$$\text{Expected Cost} = (0.6 \times 0) + (0.4 \times 10{,}000) = 0 + 4{,}000 = 4{,}000$$

- Expected cost for 50-foot pile:

$$\text{Expected Cost} = (0.6 \times 5{,}000) + (0.4 \times 0) = 3{,}000 + 0 = 3{,}000$$

The expected cost associated with the 50-foot pile is $3,000, compared to $4,000 for the 40-foot option. Thus, the 50-foot pile emerges as the more cost-effective and risk-robust choice, minimizing expected costs under uncertainty regarding the depth of the rock layer. Based on this analysis, the engineer should opt for the 50-foot pile, as it strikes an effective balance between cost-efficiency and risk mitigation by lowering the chance of incurring additional expenses from pile splicing.

(b) **Sensitivity Analysis**:

Set up the equality for expected costs to find the threshold probability p at which the engineer would be indifferent:

$$\text{Expected Cost}(40\,ft) = \text{Expected Cost}\,(50\,ft)$$

$$0 \times (1 - p) + 10{,}000 \times p = 5{,}000 \times (1 - p) + 0 \times p$$

$$10{,}000 \times p = 5{,}000 - 5{,}000 \times p$$

$$p = 0.33$$

If the probability of encountering the rock layer at 50 feet is 0.33 or higher, the engineer should choose the 50-foot pile. If it is lower than 0.33, the 40-foot pile becomes the more cost-effective choice.

(c) **Utilities calculation**

The goal is to calculate the expected cost (utility) for each pile option based on the given probabilities and costs. Here, lower costs represent higher utility, meaning the option with the lowest expected cost will be the optimal choice. To calculate the utilities (in this case, expected costs) for both the 40-foot and 50-foot pile options, we'll follow a standard decision analysis approach using the expected value formula (Eq. 5.1):

Expected Utility (Cost) for the 40-foot Pile:

The expected cost EU_{40} for the 40-foot pile is calculated as follows:

1. If rock is at 40 feet (h_0):
   - Probability $P(h_0) = 0.60$
   - Cost = \$0 (no additional cost if the rock is at 40 feet)
   - Contribution to expected cost $= 0.60 \times 0 = 0$
2. If rock is at 50 feet (h_1):
   - Probability $P(h_1) = 0.40$
   - Cost = \$10,000 (for splicing the pile)
   - Contribution to expected cost $= 0.40 \times 10{,}000 = 4{,}000$.

Total expected cost for 40-foot pile:

$$EU_{40} = 0 + 4{,}000 = 4{,}000$$

Expected Utility (Cost) for the 50-foot Pile:

The expected cost EU_{50} for the 50-foot pile is calculated as follows:

1. If rock is at 40 feet (h_0):
   - Probability $P(h_0) = 0.60$
   - Cost = \$5,000 (for the unnecessary extra depth)
   - Contribution to expected cost $= 0.60 \times 5{,}000 = 3{,}000$
2. If rock is at 50 feet (h_1):
   - Probability $P(h_1) = 0.40$
   - Cost = \$0 (no additional cost if the rock is at 50 feet)
   - Contribution to expected cost $= 0.40 \times 0 = 0$.

   Total expected cost for 50-foot pile:

$$EU_{50} = 3{,}000 + 0 = 3{,}000$$

The 50-foot pile has a lower expected cost of \$3,000 compared to the 40-foot pile's expected cost of \$4,000. Therefore, the optimal choice is the 50-foot pile. The calculation shows that, although the 50-foot pile has a higher initial cost, it offers a safer investment by reducing the likelihood of incurring costly splicing.

5.9.1.2 Solve Using GeNIe's Influence Diagram

To solve the pile foundation depth decision problem using influence diagram of the GeNie software, the model should incorporate decision nodes representing the choice of pile depth, chance nodes reflecting the uncertainty in rock depth, and utility nodes capturing the associated cost outcomes. This structure allows for a systematic evaluation of alternatives under uncertainty, enabling informed decision-making based on expected utility. The detailed of the problem has been defined in Fig. 5.29.

We need to choose between:

- A 40-foot pile (lower initial cost, risk of being too short),
- A 50-foot pile (higher initial cost, safer if the rock is deeper).

Chance outcomes:

- Rock at 40 ft → probability 0.6
- Rock at 50 ft → probability 0.4.

Utilities:

- 40-ft pile + rock at 40 ft → \$0 extra cost
- 40-ft pile + rock at 50 ft → \$10,000 (splicing)
- 50-ft pile + rock at 40 ft → \$5,000 (cutting)

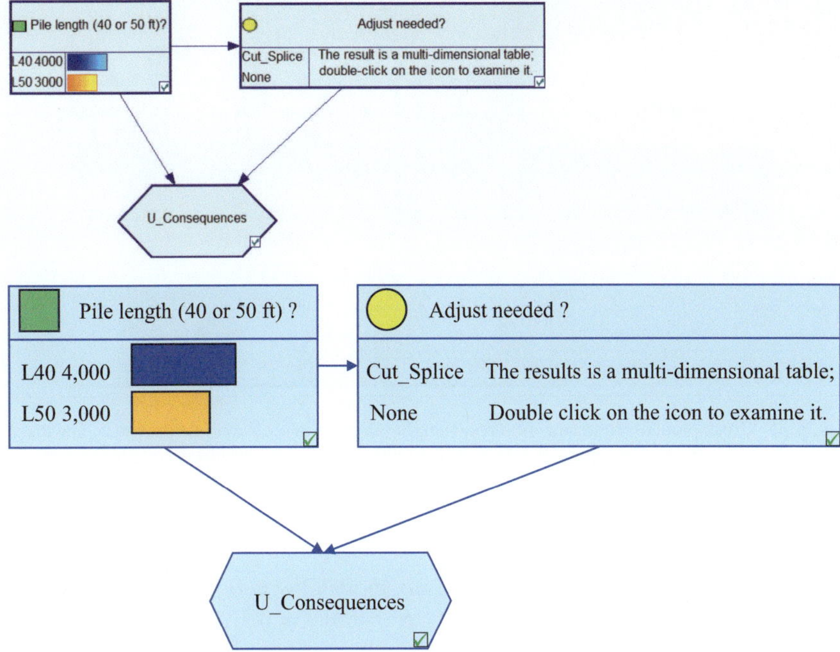

Fig. 5.30 Influence diagram in GeNIe to determine the expected cost for each decision option

- 50-ft pile + rock at 50 ft → $0 extra cost (Fig. 5.30).

5.9.2 *Example 2: Flood Risk Management—Determining the Height of a Floodwall*

A civil engineer is tasked with selecting the appropriate height for a floodwall to protect a residential area located in a floodplain. The choices are a 3-m floodwall or a 5-m floodwall. A taller floodwall provides more protection but comes with a significantly greater construction cost. However, future flood severity is uncertain. Figure 5.31 visually represents the real-world decision an engineer faces when choosing a floodwall height, highlighting the trade-offs between cost, protection, and uncertainty in future flood severity.

The engineer needs to determine which floodwall height minimizes the total expected cost, including both the construction cost and the cost of flood damage if the wall is overtopped.

Decision Options:

- 3-m wall
- 5-m wall.

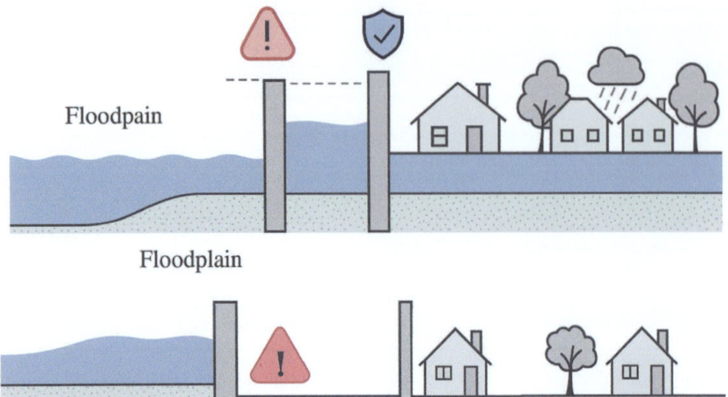

Fig. 5.31 Conceptual illustration of floodwall height selection in a residential floodplain. The image contrasts the protective outcomes of different floodwalls under different flood severity scenarios

Chance Outcomes:

- Moderate flood (peak water level = 3 m), probability = 0.7
- Severe flood (peak water level = 5 m), probability = 0.3.

Utility/Cost Values:

- 3-m wall:
  - Moderate flood: No damage → \$0
  - Severe flood: Overtopping causes damage → \$80,000
- 5-m wall:
  - Moderate flood: Overdesigned → \$20,000 (excess construction cost)
  - Severe flood: Fully protective → \$0.

Determine the expected cost for each decision option and select the height that minimizes total expected cost. How sensitive is the decision to changes in the probability of the rock layer depth?

5.9.2.1 Solve Using Expected Utility

Step 1: Identify Outcomes and Costs

- 3-m wall:
  - Moderate flood (0.7): No damage = \$0
  - Severe flood (0.3): Damage = \$80,000

- 5-m wall:
  - Moderate flood (0.7): Overdesign cost = \$20,000
  - Severe flood (0.3): No damage = \$0.

Step 2: Calculate Expected Cost for Each Option

- 3-m wall:

$$EU_{3m} = (0.7 \times 0) + (0.3 \times 80{,}000) = 0 + 24{,}000 = 24{,}000$$

- 5-m wall:

$$EU_{5m} = (0.7 \times 20{,}000) + (0.3 \times 0) = 14{,}000 + 0 = 14{,}000$$

Step 3: Decision

- Expected cost for the 3-m wall = \$24,000
- Expected cost for the 5-m wall = \$14,000.

Figure 5.32. illustrates a structured approach to evaluating two decision options. This decision tree represents the evaluation of flood protection strategies under uncertain flood severity conditions. The two decision options available to the engineer are to construct either a 3-m or a 5-m flood protection wall. The yellow square represents the decision node where this choice is made. Each branch from the decision node leads to a chance node (green circle), which models the uncertainty in flood magnitude: a moderate flood with a peak water level of 3 m (probability = 0.7), and a severe flood with a peak water level of 5 m (probability = 0.3). The blue triangles at the end of each path represent outcome nodes, where the associated cost (used here as a negative utility) is incurred. For the 3-m wall, the expected utility is calculated as $u(a_0) = 0.7 \times 0 + 0.3 \times 80{,}000 = 24{,}000$, since it protects fully during moderate floods but fails during severe ones, resulting in significant damage. In contrast, the 5-m wall costs more to construct, but prevents all damage in either flood scenario. It has an expected utility of $u(a_1) = 0.7 \times 20{,}000 + 0.3 \times 0 = 14{,}000$. Thus, despite the higher initial construction cost, the 5-m wall offers a more cost-effective and resilient solution under uncertain flood risk.

The engineer should choose the 5-m wall, which minimizes the expected cost and provides full protection under both flood scenarios. If we were interested in finding the probability, p, of a severe flood at which the engineer would be indifferent between the two options then we can set the expected costs equal to one another.

Let p = probability of severe flood, so (1 − p) = probability of moderate flood.

$$(1 - p) \times 0 + p \times 80{,}000 = (1 - p) \times 20{,}000 + p \times 0$$

Simplify:

$$80{,}000 \times p = 20{,}000 - 20{,}000 \times p$$

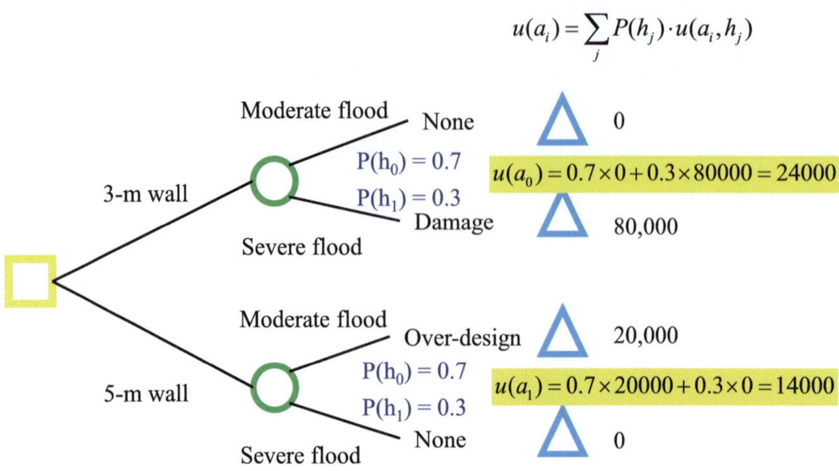

Fig. 5.32 Decision tree for evaluating flood wall height under uncertain flood severity

Then,

$$p = 0.2$$

Thus, if the probability of a severe flood exceeds 20%, the 5-m wall is the better choice. If it is below 20%, the 3-m wall becomes more cost-effective.

5.9.2.2 Solve Using GeNIe's Influence Diagram

This example can be also be solved using GeNIe, a graphical decision modeling tool that supports influence diagrams, Bayesian networks, and decision trees (Fig. 5.33). In this case, the decision problem involves selecting between two floodwall heights—3 and 5 m—under uncertain flood severity conditions. The workflow in GeNIe begins by creating a new influence diagram and inserting a decision node labeled "*Floodwall Height (3 m or 5 m)?*", where the user defines the two available design options. Next, a chance node is created titled *"Damage or Overdesigned?"* to represent the uncertain flood conditions, with two possible outcomes: a moderate flood (with a probability of 0.7) and a severe flood (with a probability of 0.3). These nodes are connected to reflect the influence of the floodwall choice on the resulting consequence. A utility node called "*U_Consequences*" is then created and linked to both the decision and chance nodes. This node captures the economic outcome of each scenario, quantified as cost values—$0 if the wall fully protects, $20,000 if it is overdesigned, and $80,000 if overtopped during a severe flood. After entering these values into the utility function table, GeNIe automatically computes the expected utility for each decision alternative and displays them in the *Floodwall Height* node. Here it can be seen that the 3-m

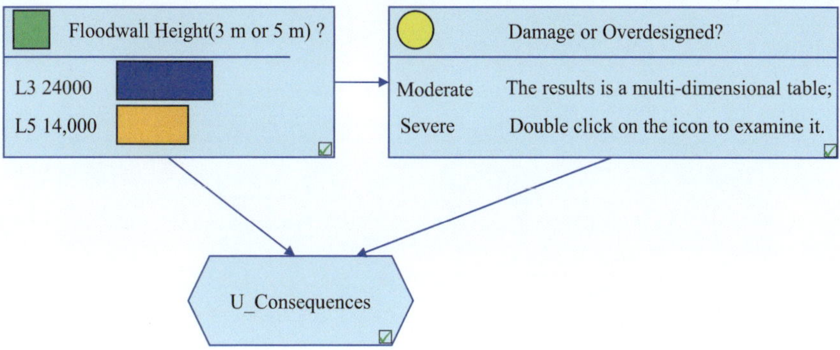

Fig. 5.33 Influence diagram in GeNIe for evaluating optimal floodwall height under flood risk uncertainty

wall has an expected cost of $24,000, while the 5-m wall has a lower expected cost of $14,000. This clear, visual workflow enables engineers to systematically evaluate trade-offs and make cost-effective, risk-informed infrastructure decisions.

5.9.3 Example 3: Pavement Material Decision in Road Design

An engineer must select the pavement material for a new urban arterial road. The two options are asphalt and concrete. Asphalt has a lower initial cost but may require more frequent maintenance under high traffic loads. Concrete has a higher upfront cost but offers greater durability, especially under heavier use. However, the future traffic volume is uncertain.

The objective is to minimize total expected lifecycle cost by evaluating each option under two possible traffic conditions: low traffic and high traffic. Figure 5.34 illustrates the practical challenge faced by engineers in selecting pavement material for an urban arterial road under uncertain future traffic loads.

(a) Decision Options:

- Asphalt
- Concrete

(b) Chance Outcomes (Traffic Scenarios):

- Low traffic (20,000 vehicles/day): 60% probability
- High traffic (40,000 vehicles/day): 40% probability

Fig. 5.34 Illustrates the practical challenge faced by engineers in selecting pavement material for an urban arterial road under uncertain future traffic loads

(c) Utility/Cost Values (30-year lifecycle total cost) (Table 5.1):

Below a step-by-step solution is provided using the cost in terms of the expected utility.

Step 1: Identify outcomes and their costs

- Asphalt + Low traffic: $500,000
- Asphalt + High traffic: $800,000
- Concrete + Low traffic: $700,000
- Concrete + High traffic: $750,000.

Step 2: Calculate expected utility (cost) for each decision

Asphalt:

$$EU_{Asphalt} = (0.6 \times 500{,}000) + (0.4 \times 800{,}000) = 300{,}000 + 320{,}000 = 620{,}000$$

Table 5.1 Lifecycle cost estimates for asphalt and concrete pavement types under varying traffic load conditions over a 30-year period

Pavement type	Traffic load	Probability	Lifecycle cost (USD)
Asphalt	Low traffic	0.6	$500,000
Asphalt	High traffic	0.4	$800,000
Concrete	Low traffic	0.6	$700,000
Concrete	High traffic	0.4	$750,000

Concrete:

$$EU_{Concrete} = (0.6 \times 700{,}000) + (0.4 \times 750{,}000) = 420{,}000 + 300{,}000 = 720{,}000$$

Step 3: Decision

- Expected cost for Asphalt = $620,000
- Expected cost for Concrete = $720,000.

Asphalt is the more cost-effective option in terms of expected lifecycle cost. Even though it performs worse under high traffic loadings requiring more maintenance and repair, the higher probability of low traffic makes it the preferred choice in this scenario.

(b) Sensitivity Analysis

Let's calculate the threshold probability of high traffic at which both options would have equal expected cost.

Let p = probability of high traffic, so (1 − p) = probability of low traffic.

Set expected costs equal:

$$(1 - p) \times (500{,}000) + p \times (800{,}000) - (1 - p) \times (700{,}000) + p \times (750{,}000)$$

Simplify both sides and set equal:

$$500{,}000 + 300{,}000 \times p = 700{,}000 + 50{,}000 \times p$$

$$p = 0.8$$

If the probability of high traffic exceeds 80%, concrete becomes the better choice for the pavement material. If the likelihood of high traffic remains below 80%, asphalt remains the optimal decision, yielding a lower total lifecycle cost.

5.9.4 Example 4: Stormwater Detention Basin Sizing Under Rainfall Uncertainty

A civil engineer must select the optimal size for a stormwater detention basin in a new residential development. The basin must reduce peak discharge during rainfall events and prevent downstream flooding. The options are:

- Small basin
- Medium basin

- Large basin.

Each basin has a different construction cost and level of flood protection. However, future rainfall intensities are uncertain due to climate change, and the engineer must account for three possible scenarios. Figure 5.35 introduces the challenge of basin sizing in the context of climate-driven rainfall variability, emphasizing the trade-offs between construction effort and long-term flood protection.

(a) Decision Options:

- Small basin
- Medium basin
- Large basin.

(b) Chance Outcomes (Rainfall Intensity Scenarios):

- Normal rainfall (10-year storm)—probability = 0.5
- Moderate rainfall (25-year storm)—probability = 0.3
- Intense rainfall (50-year storm)—probability = 0.2.

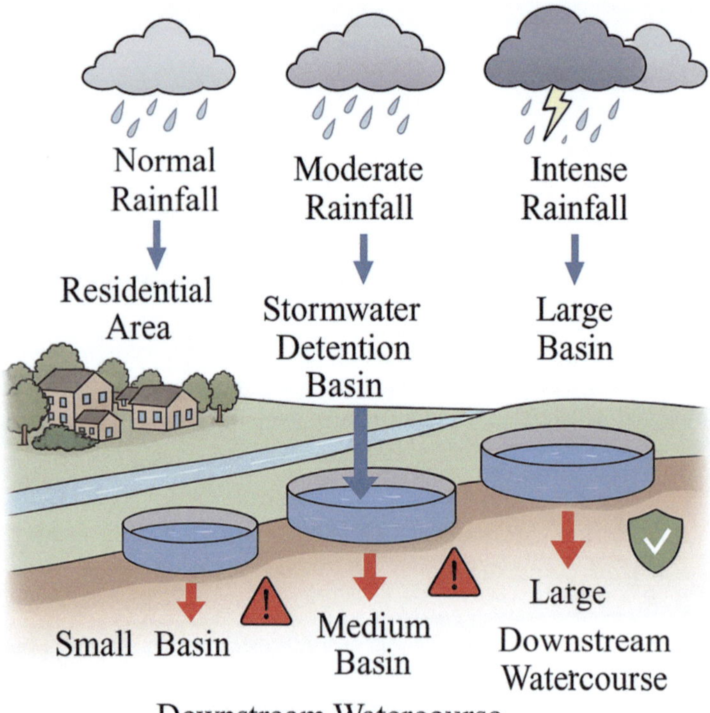

Fig. 5.35 Conceptual visualization of stormwater detention basin sizing under uncertain future rainfall intensities

(c) Utility/Cost Values (Construction + Damage Costs):

Table 5.2 presents cost estimate data for each possible scenario (i.e., combination of basin size and rainfall intensity) based on historical data.

Step 1: Calculate Expected Cost for Each Basin Size

Small Basin:

$$\begin{aligned} EU_{Small} &= (0.5 \times 300{,}000) + (0.3 \times 500{,}000) + (0.2 \times 800{,}000) \\ &= 150{,}000 + 150{,}000 + 160{,}000 = 460{,}000 \end{aligned}$$

Medium Basin:

$$\begin{aligned} EU_{Medium} &= (0.5 \times 500{,}000) + (0.3 \times 550{,}000) + (0.2 \times 700{,}000) \\ &= 250{,}000 + 165{,}000 + 140{,}000 = 555{,}000 \end{aligned}$$

Large Basin:

$$\begin{aligned} EU_{Large} &= (0.5 \times 800{,}000) + (0.3 \times 800{,}000) + (0.2 \times 800{,}000) \\ &= 400{,}000 + 240{,}000 + 160{,}000 = 800{,}000 \end{aligned}$$

Step 2: Decision

Based on the calculations for expected utility above, we can summarize the costs associated with each design choice in Table 5.3.

Even after accounting for the uncertainty in the rainfall intensity, the cost for the flood basin increases proportional to the size of the basin with the small basin having the lowest expected cost and the large basin having the greatest cost. In terms of cost,

Table 5.2 Utility and cost analysis based on basin size and rainfall scenarios

Basin size	Rainfall scenario	Probability	Construction cost	Flood damage cost	Total cost
Small basin	Normal (10-yr)	0.5	$300,000	$0	$300,000
Small basin	Moderate (25-yr)	0.3	$300,000	$200,000	$500,000
Small basin	Intense (50-yr)	0.2	$300,000	$500,000	$800,000
Medium basin	Normal	0.5	$500,000	$0	$500,000
Medium basin	Moderate	0.3	$500,000	$50,000	$550,000
Medium basin	Intense	0.2	$500,000	$200,000	$700,000
Large basin	Normal	0.5	$800,000	$0	$800,000
Large basin	Moderate	0.3	$800,000	$0	$800,000
Large basin	Intense	0.2	$800,000	$0	$800,000

Table 5.3 Expected total cost by flood basin size

Basin size	Expected cost
Small	$460,000
Medium	$555,000
Large	$800,000

the small basin is optimal, however, it offers no protection under more severe storms, which may be unacceptable depending on risk tolerance or regulatory constraints. The significantly greater expense of the large sized flood basin may only be justified in very high-risk or high-value areas.

(b) Sensitivity Analysis

Let's determine the probability threshold at which the medium basin becomes more cost-effective than the small basin in the case of a severe rainfall event. Let p be the probability of the 50-year storm.

Set expected costs equal:

$$(0.5 \times 300{,}000) + (0.3 \times 500{,}000) + (\mathrm{p} \times 800{,}000)$$
$$= (0.5 \times 500{,}000) + (0.3 \times 550{,}000) + (\mathrm{p} \times 700{,}000)$$

Simplify both sides and solve:

$$100{,}000\mathrm{p} = 165{,}000$$

$$p = 165$$

The outcome of this analysis indicates that, under the current probability structure, the small basin is the optimal economic choice as the probability of the extreme rainfall event cannot increase beyond p = 1. However, if public safety or flood control regulations prioritize protection from rare events, the medium or large basin may still be recommended despite higher expected cost.

5.9.5 *Example 5: Slope Stabilization Method Selection Under Soil Uncertainty*

An engineer must choose the optimal slope stabilization method for a hillside near a transportation corridor. The slope shows signs of instability, and soil investigations suggest variable subsurface conditions. The three stabilization options are being considered including (1) soil nailing, (2) retaining wall, and (3) re-grading (reshaping the slope).

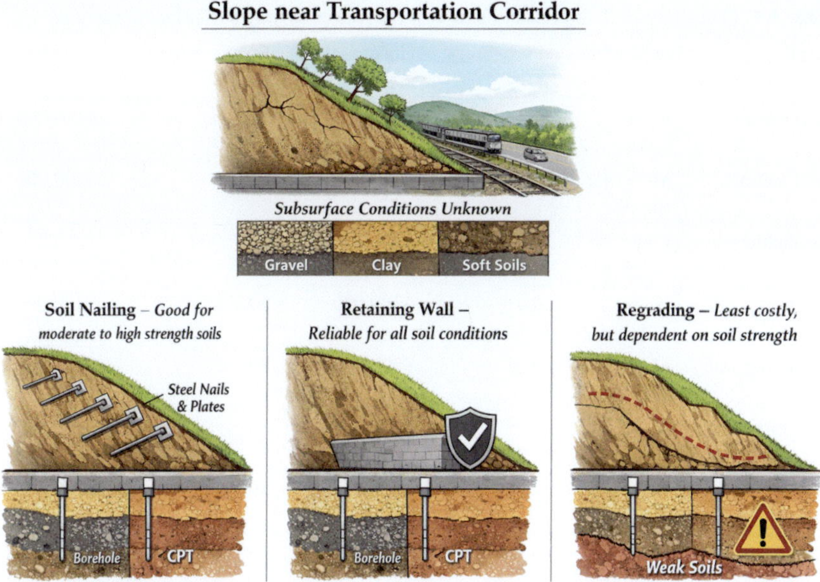

Fig. 5.36 Conceptual illustration of slope stabilization options for a hillside near a transportation corridor. Soil nailing, retaining walls, and re-grading are considered under uncertain subsurface strength conditions

Each option has a different cost and level of reliability depending on the actual subsurface conditions. Since the subsurface strength is uncertain, the engineer must account for three possible scenarios based on borehole and cone penetration test (CPT) data. Figure 5.36 provides a visual overview of the stabilization options available for a slope near a transportation corridor, highlighting how soil conditions influence the choice of method.

(a) Decision Options:

- Soil nailing
- Retaining wall
- Re-grading the slope

(b) Chance Outcomes (Subsurface Strength Conditions):

- High stability soil (e.g., dense sandy gravel)—40% probability
- Moderate stability soil (e.g., stiff clay)—35% probability
- Low stability soil (e.g., loose sand or soft clay)—25% probability

(c) Utility/Cost Values (Design + Failure Risk Costs) (Table 5.4):

Below, a step-by-step solution for the calculation of the expected utility of each design option is presented.

Table 5.4 Cost analysis of slope stabilization options under varying subsurface conditions

Option	Subsurface condition	Probability	Construction cost	Risk/failure cost	Total cost
Soil nailing	High stability	0.4	\$150,000	\$0	\$150,000
Soil nailing	Moderate stability	0.35	\$150,000	\$50,000	\$200,000
Soil nailing	Low stability	0.25	\$150,000	\$120,000	\$270,000
Retaining wall	High stability	0.4	\$300,000	\$0	\$300,000
Retaining wall	Moderate stability	0.35	\$300,000	\$0	\$300,000
Retaining wall	Low stability	0.25	\$300,000	\$0	\$300,000
Re-grading	High stability	0.4	\$100,000	\$0	\$100,000
Re-grading	Moderate stability	0.35	\$100,000	\$90,000	\$190,000
Re-grading	Low stability	0.25	\$100,000	\$180,000	\$280,000

Step 1: Compute Expected Costs for Each Option

Soil Nailing:

$$\begin{aligned} EU_{SoilNailing} &= (0.4 \times 150{,}000) + (0.35 \times 200{,}000) + (0.25 \times 270{,}000) \\ &= 60,000 + 70,000 + 67,500 = 197{,}500 \end{aligned}$$

Retaining Wall:

$$\begin{aligned} EU_{RetainingWall} &= (0.4 \times 300{,}000) + (0.35 \times 300{,}000) + (0.25 \times 300{,}000) \\ &= 120{,}000 + 105{,}000 + 75{,}000 = 300{,}000 \end{aligned}$$

Regrading:

$$\begin{aligned} EU_{Regrading} &= (0.4 \times 100{,}000) + (0.35 \times 190{,}000) + (0.25 \times 280{,}000) \\ &= 40{,}000 + 66{,}500 + 70{,}000 = 176{,}500 \end{aligned}$$

Step 2: Decision (Table 5.5)

Based purely on expected cost, reshaping the slope is the most economical choice. However, it also carries the highest risk in low stability conditions. If failure in low

Table 5.5 Summary of expected total cost by stabilization method

Stabilization method	Expected cost
Soil nailing	\$197,500
Retaining wall	\$300,000
Re-grading	\$176,500

stability soil is considered unacceptable by the local regulatory officials, soil nailing might be preferred despite the slightly higher expected cost.

(b) Sensitivity Analysis

To test how sensitive the decision is to the probability of having a low stability soil, let's find the threshold probability, p, at which soil nailing becomes equal in expected cost to re-grading assuming that the probability of finding a moderate stability soil remains constant.

Let p = probability of low stability. Then:

- Moderate stability = 0.35
- High stability = 1 − 0.35 − p.

Set expected costs equal:

$$(1 - 0.35 - p)(150{,}000) + (0.35)(200{,}000) + p(270{,}000)$$
$$= (1 - 0.35 - p)(100{,}000) + (0.35)(190{,}000) + p(280{,}000)$$

Solving this equation yields a threshold probability value of $p = 0.6$. This threshold highlights the sensitivity of the cost comparison to the assumed likelihood of low soil stability conditions. It suggests that if the probability of low stability is greater than 60%, soil nailing becomes more cost-effective than re-grading. The choice between re-grading and soil nailing depends on the client's risk tolerance and site conditions. Re-grading is a suitable option if the client is willing to accept a certain level of risk in areas where soil stability may be weaker. In contrast, soil nailing is recommended when enhanced stability is required—particularly in zones with moderate to low subsurface stability—or when there is a need to protect critical downstream infrastructure. Further geotechnical investigations can help refine the probability estimates and support a more informed decision.

5.9.6 *Example 6: Tunnel Support System Selection Under Geotechnical Uncertainty*

During the planning phase of a tunnel through a mountainous area, engineers must select an appropriate tunnel support system. The geological profile indicates varying ground conditions with different levels of stability. The objective is to select a support

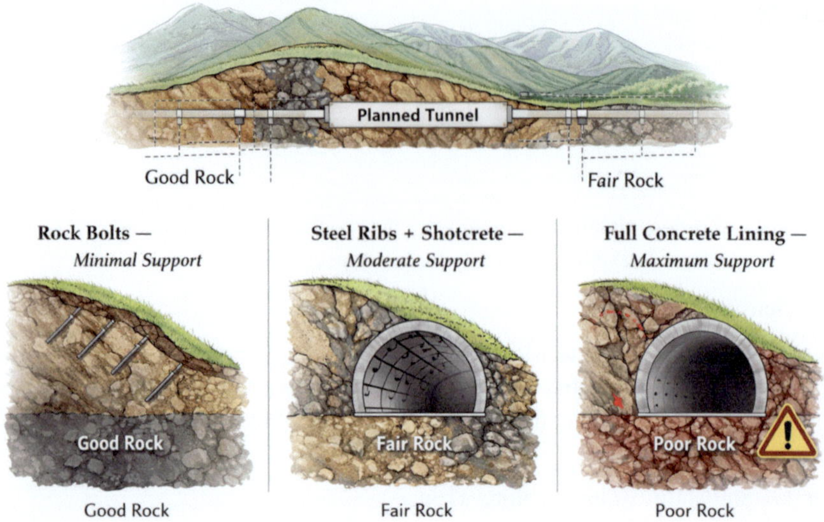

Fig. 5.37 Conceptual illustration of tunnel support options—rock bolts, steel ribs with shotcrete, and full concrete lining—applied under varying subsurface stability conditions encountered in mountainous tunneling projects

system that ensures structural safety while minimizing total expected cost, including both installation and potential failure or deformation costs. The three feasible options are (1) rock bolts only (minimal support), (2) steel ribs with shotcrete (moderate support), and (3) full concrete lining (maximum support).

Subsurface stability is uncertain and based on preliminary geological surveys. Figure 5.37 illustrates the relationship between tunnel support systems and geotechnical variability, showing how different rock conditions influence support design decisions.

(a) Decision Options:

- Rock bolts
- Steel ribs + shotcrete
- Full concrete lining

(b) Chance Outcomes (Ground Stability Conditions):

- Good rock (Class I–II)—30% probability
- Fair rock (Class III)—50% probability
- Poor rock (Class IV–V)—20% probability

(c) Utility/Cost Values (Support + Failure/Repair Costs) (Table 5.6).

A step-by-step solution to the problem is presented below by calculating the expected utility.

Table 5.6 Installation and risk-related cost estimates for different tunnel support options across varying ground conditions

Support option	Ground class	Probability	Installation cost	Risk/repair cost	Total cost
Rock bolts	Good	0.3	\$100,000	\$0	\$100,000
Rock bolts	Fair	0.5	\$100,000	\$100,000	\$200,000
Rock bolts	Poor	0.2	\$100,000	\$300,000	\$400,000
Steel ribs + shotcrete	Good	0.3	\$200,000	\$0	\$200,000
Steel ribs + shotcrete	Fair	0.5	\$200,000	\$30,000	\$230,000
Steel ribs + shotcrete	Poor	0.2	\$200,000	\$100,000	\$300,000
Full concrete lining	Good	0.3	\$400,000	\$0	\$400,000
Full concrete lining	Fair	0.5	\$400,000	\$0	\$400,000
Full concrete lining	Poor	0.2	\$400,000	\$0	\$400,000

Step 1: Calculate Expected Costs for Each Option

Rock Bolts:

$$\begin{aligned} EU_{RockBolts} &= (0.3 \times 100{,}000) + (0.5 \times 200{,}000) + (0.2 \times 400{,}000) \\ &= 30{,}000 + 100{,}000 + 80{,}000 = 210{,}000 \end{aligned}$$

Steel Ribs + Shotcrete:

$$\begin{aligned} EU_{Steel+Shotcrete} &= (0.3 \times 200{,}000) + (0.5 \times 230{,}000) + (0.2 \times 300{,}000) \\ &= 60{,}000 + 115{,}000 + 60{,}000 = 235{,}000 \end{aligned}$$

Full Concrete Lining:

$$\begin{aligned} EU_{Full\ Lining} &= (0.3 \times 400{,}000) + (0.5 \times 400{,}000) + (0.2 \times 400{,}000) \\ &= 120{,}000 + 200{,}000 + 80{,}000 = 400{,}000 \end{aligned}$$

Step 2: Decision

Table 5.7 summarizes the expected costs for each design alternative after accounting for the uncertainty in the rock stability.

Table 5.7 Expected total costs of tunnel support options

Support option	Expected cost (USD)
Rock bolts	\$210,000
Steel ribs + shotcrete	\$235,000
Full concrete lining	\$400,000

The rock bolts provide the lowest expected cost, but the risk of failure in poor ground is high. It is likely that the project will have a low risk tolerance for deformation and in such a case, it is likely required to consider steel ribs + shotcrete or a full concrete lining.

(b) Sensitivity Analysis

Let's calculate the threshold probability of encountering poor rock at which the steel ribs + shotcrete become more cost-effective than rock bolts.

Let p be the probability of encountering poor rock. Consider that the probability of encountering fair rock remains 0.5, then the probability of good rock is given by 0.5 − p.

Set expected costs equal:

$$\begin{aligned}&(0.5 - p)(100{,}000) + 0.5(200{,}000) + p(400{,}000)\\&= (0.5 - p)(200{,}000) + 0.5(230{,}000) + p(300{,}000)\end{aligned}$$

Solving this yields:

$$p = 0.325$$

If the probability of poor rock exceeds 32.5%, steel ribs + shotcrete become the preferred choice over rock bolts. This emphasizes the importance of accurate geological risk assessment. The final recommendation depends on the likelihood of encountering poor ground and the project's risk tolerance. Rock bolts are recommended if poor ground conditions are unlikely (i.e., when $p < 0.325$) and the associated risk is deemed acceptable. For projects expecting moderate-to-poor ground or when a more cautious approach is preferred, steel ribs combined with shotcrete offer a balanced solution. Full lining should be reserved for situations where maximum security is essential, such as beneath riverbeds, in urban areas, or for protecting critical infrastructure.

5.9.7 Example 7: Water Treatment Method Selection Under Regulatory Uncertainty

A municipality plans to upgrade its water treatment plant and must choose one of three treatment technologies to meet both current and anticipated future drinking water standards. However, there is uncertainty regarding how stringent future regulations will become, especially concerning emerging contaminants (e.g., pharmaceuticals, PFAS, and microplastics).

The goal is to select a treatment method that minimizes long-term cost while ensuring regulatory compliance and operational flexibility. Figure 5.38 illustrates the challenge of selecting a water treatment method that remains compliant under possible future regulations targeting emerging contaminants such as PFAS and microplastics.

(a) Decision Options:

- Conventional treatment (coagulation, flocculation, sedimentation, chlorination)
- Advanced oxidation process (AOP) (ozone + UV)
- High-pressure membrane filtration system (nanofiltration/reverse osmosis)

(b) Chance Outcomes (Regulatory Scenarios):

- No regulatory change (current standards maintained)—50% probability
- Moderate tightening (limits on microcontaminants)—30% probability

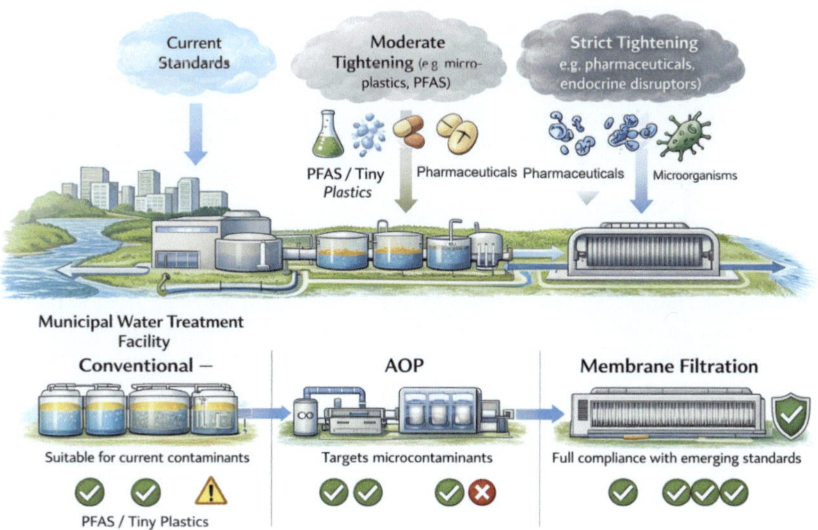

Fig. 5.38 Conceptual illustration of three water treatment technology options in the face of future regulatory uncertainty

- Strict tightening (emerging contaminant removal mandated)—20% probability

(c) Utility/Cost Values (Initial + Upgrade Costs, Lifecycle over 20 years) (Table 5.8).

Step-by-Step Solution

Step 1: Calculate Expected Cost for Each Treatment Option

Conventional Treatment:

$$\begin{aligned} EU_{Conventional} &= (0.5 \times 10) + (0.3 \times 15) + (0.2 \times 22) \\ &= 5 + 4.5 + 4.4 = 13.9 million USD \end{aligned}$$

AOP (Advanced Oxidation Process):

$$\begin{aligned} EU_{AOP} &= (0.5 \times 13) + (0.3 \times 15) + (0.2 \times 18) \\ &= 6.5 + 4.5 + 3.6 = 14.6 million USD \end{aligned}$$

Membrane Filtration:

$$EU_{Membrane} = 1.0 \times 18 = 18 million USD$$

Table 5.8 Cost breakdown of water treatment options under different regulatory tightening scenarios

Treatment option	Scenario	Probability	Capital & O&M	Upgrade/ compliance cost	Total cost
Conventional	No change	0.5	$10M	$0	$10M
Conventional	Moderate tightening	0.3	$10M	$5M	$15M
Conventional	Strict tightening	0.2	$10M	$12M	$22M
AOP	No change	0.5	$13M	$0	$13M
AOP	Moderate tightening	0.3	$13M	$2M	$15M
AOP	Strict tightening	0.2	$13M	$5M	$18M
Membrane filtration	All scenarios	1	$18M	$0	$18M

Note Membrane filtration fully complies with all potential future standards, so no upgrades are needed

Table 5.9 Expected total costs of water treatment options

Treatment option	Expected cost (Million USD)
Conventional	13.9
AOP	14.6
Membrane filtration	18

Step 2: Decision (Table 5.9)

The conventional treatment system has the lowest expected cost ($13.9M) but poses significant upgrade risks under stricter future regulations. Thc AOP system is slightly more expensive but provides better resilience against moderate future tightening. The membrane system is future-proof but least cost-effective in expectation.

(b) Sensitivity Analysis

Let's determine the threshold probability of strict regulation at which AOP becomes more cost-effective than conventional treatment.

Let p = probability of strict tightening. Then:

- Moderate = 0.3
- No change = 0.7 − p.

Set expected costs equal:

$$(0.7 - p)(10) + (0.3 \times 15) + (p \times 22) = (0.7 - p)(13) + (0.3 \times 15) + (p \times 18)$$

Simplifying:

$$p = 0.3$$

If the probability of strict regulatory tightening exceeds 30%, the AOP system becomes more cost-effective than the conventional system. This highlights the importance of anticipating policy changes in infrastructure planning. The final recommendation should be guided by anticipated regulatory changes and the project's risk and adaptability requirements. Conventional treatment is appropriate if strict regulatory tightening is unlikely and the municipality has a high tolerance for risk or if the anticipated timeline for regulatory tightening is far in the future. The AOP are recommended when moderate-to-strict regulatory changes are reasonably expected and long-term adaptability is a priority for the municipality. Membrane filtration should be selected when the project must be fully future-proof—such as for critical drinking water sources or to meet international compliance—and when cost is not the primary constraint.

5.9.8 Example 8: Building Envelope System Selection Under Energy Code Uncertainty

A construction engineer must select an exterior wall system for a mid-rise commercial building. Four facade options are available, each differing in initial cost, thermal performance, and adaptability to future energy efficiency regulations. Since energy codes are tightening over time, the engineer must consider potential future requirements for insulation, air leakage control, and climate performance.

The objective is to minimize long-term cost, which includes the initial investment and the potential cost of future retrofits for compliance. Figure 5.39 illustrates the key exterior wall options available for mid-rise buildings and their readiness for future energy efficiency codes, emphasizing the importance of long-term adaptability in facade system selection.

(a) Decision Options:

- Brick veneer with conventional insulation
- EIFS (Exterior Insulation and Finish System)
- Precast concrete panels with continuous insulation
- Double-skin facade system

(b) Chance Outcomes (Energy code scenarios in next 10 years):

- No change in code (baseline retained)—40% probability
- Moderate tightening (higher R-value required)—35% probability

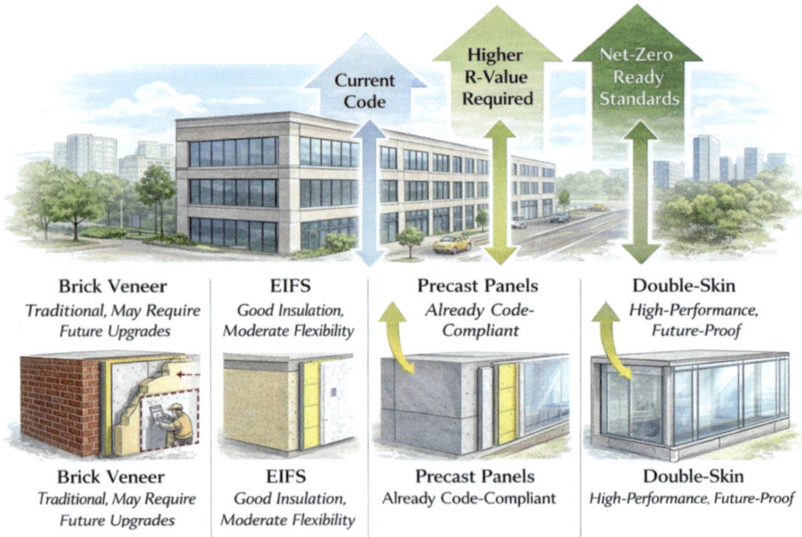

Fig. 5.39 Conceptual comparison of four facade system options for a commercial building under increasing energy code scenarios

- Aggressive tightening (net-zero ready performance)—25% probability

(c) Utility/Cost Table (Lifecycle Cost = Construction + Upgrade Costs) (Table 5.10).

Step-by-Step Solution

Step 1: Calculate Expected Cost per Square Meter

Brick veneer:

$$EU_{Brick} = (0.4 \times 180) + (0.35 \times 230) + (0.25 \times 300) = 72 + 80.5 + 75 = 227.5$$

EIFS:

$$EU_{EIFS} = (0.4 \times 200) + (0.35 \times 220) + (0.25 \times 260) = 80 + 77 + 65 = 222$$

Precast concrete panels:

$$EU_{Precast} = 1.0 \times 240 = 240$$

Double-skin facade:

$$EU_{Double-skin} = 1.0 \times 320 = 320$$

Table 5.10 Cost components of facade system options under varying policy tightening scenarios

System	Scenario	Probability	Initial Cost ($/m^2)	Retrofit cost ($/m^2)	Total cost ($)
Brick veneer	No change	0.4	180	0	180
Brick veneer	Moderate tightening	0.35	180	50	230
Brick veneer	Aggressive tightening	0.25	180	120/	300
EIFS	No change	0.4	200	0	200
EIFS	Moderate tightening	0.35	200	20	220
EIFS	Aggressive tightening	0.25	200	60	260
Precast concrete panels	All scenarios	1	240	$0	240
Double-skin facade	All scenarios	1	320	$0	320

Note Precast panels and double-skin facade already meet all future code levels and require no upgrade

Step 2: Decision (Table 5.11)

EIFS provides the lowest expected lifecycle cost, balancing energy performance and cost-effectiveness under uncertainty. Brick veneer is slightly cheaper upfront but has higher expected retrofit costs. Precast and double-skin facades are future-proof but more expensive.

(b) Sensitivity Analysis

Let's find the probability threshold of aggressive code tightening at which precast concrete panels become more cost-effective than EIFS.

Let p = probability of aggressive tightening. Then:

- Moderate = 0.35
- No change = 0.65 − p.

Set expected cost of EIFS equal to precast (240):

$$(0.65 - p)(200) + 0.35(220) + p(260) = 240$$

$$p = 0.55$$

If the probability of aggressive energy code tightening exceeds 55%, precast panels become more cost-effective than EIFS. This shows the importance of projecting regulatory trajectories in high-performance building design. The final recommendation depends on the project's risk level, performance goals, and budget priorities. EIFS is the preferred option for low-to-moderate risk projects offering a strong balance between cost and performance. Brick veneer may be chosen when immediate cost savings are a priority and flexibility for future upgrades is acceptable. Precast concrete panels are suitable when future-proofing is important and a moderate cost premium can be justified.

Table 5.11 Expected unit lifecycle costs for different building envelope systems

Envelope system	Expected lifecycle cost ($\$/m^2$)
Brick veneer	$227.50
EIFS	$222.00
Precast concrete panels	$240.00
Double-skin facade	$320.00

5.9.9 Example 9: Stormwater Infrastructure Planning Under Climate Change Uncertainty

A city is upgrading its stormwater drainage system to cope with increasing rainfall intensity due to climate change. Engineers must select one of five design options for the main drainage corridor. Each design offers a different level of capacity, resilience, and cost. Climate projections indicate uncertainty about future rainfall intensity and frequency over the next 50 years.

The goal is to select the option that minimizes total lifecycle cost, which includes initial construction and potential future damage or retrofit costs under different climate scenarios. Figure 5.40 visualizes the different drainage design strategies a city may consider in response to climate uncertainty, illustrating trade-offs in resilience and system adaptability over a 50-year planning horizon.

(a) Decision Options (Design Capacity):

1. Baseline (current standard, 10-year storm)
2. Moderate upgrade to baseline (+20% capacity)
3. High resilience upgrade of baseline (+40% capacity)
4. Nature-based system (green infrastructure +20%)
5. Smart drainage system (real-time controls +30%)

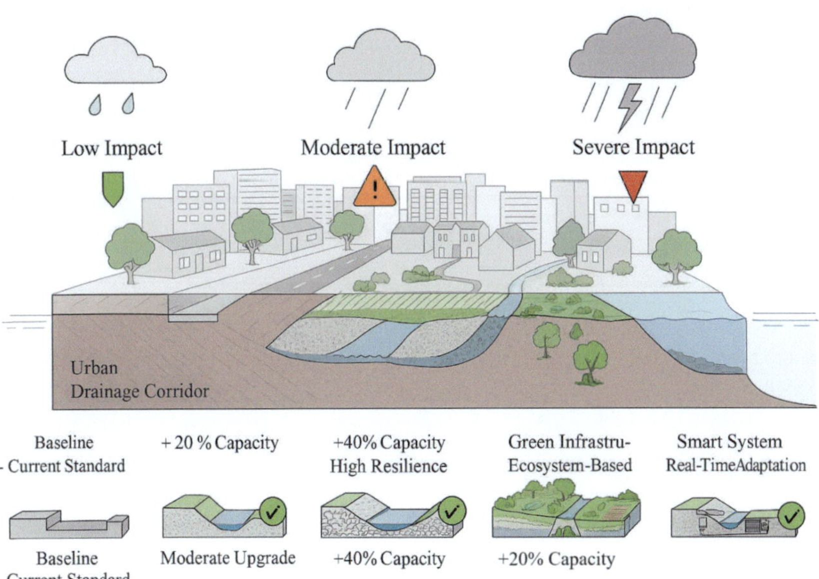

Fig. 5.40 Conceptual illustration of five stormwater drainage system designs evaluated under increasing rainfall scenarios due to climate change

(b) Chance Outcomes (Future Climate Scenarios):

- Low climate impact (No significant change)—30% probability
- Moderate impact (25% increase in peak rainfall)—40% probability
- Severe impact (50% increase in peak rainfall)—30% probability

(c) Utility/Cost Table (Total Lifecycle Cost = Construction + Damage/Retrofit Cost) (Table 5.12).

Step-by-Step Solution

Step 1: Calculate Expected Costs for Each Option

Baseline:

$$EU_{Baseline} = (0.3 \times 10) + (0.4 \times 15) + (0.3 \times 25) = 3 + 6 + 7.5 = 16.5\text{M}$$

+20% Capacity:

$$EU_{+20\%} = (0.3 \times 12) + (0.4 \times 14) + (0.3 \times 22) = 3.6 + 5.6 + 6.6 = 15.8\text{M}$$

Table 5.12 Cost breakdown of various stormwater design options under different severity scenarios

Design option	Scenario	Probability	Construction cost (M $)	Damage/upgrade cost (M $)	Total cost (M $)
Baseline	Low	0.3	10	0	10
Baseline	Moderate	0.4	10	5	15
Baseline	Severe	0.3	10	15	25
+20% capacity	Low	0.3	12	0	12
+20% capacity	Moderate	0.4	12	2	14
+20% capacity	Severe	0.3	12	10	22
+40% capacity	All	1	16	0	16
Nature-based system	Low	0.3	14	0	14
Nature-based system	Moderate	0.4	14	2	16
Nature-based system	Severe	0.3	14	6	20
Smart drainage system	Low	0.3	15	0	15
Smart drainage system	Moderate	0.4	15	0	15
Smart drainage system	Severe	0.3	15	3	18

Table 5.13 Expected lifecycle costs of stormwater management design options

Option	Expected lifecycle cost (M$)
Baseline	16.5
+20% capacity	15.8
+40% capacity	16.0
Nature-based system	16.6
Smart drainage system	15.9

+40% Capacity:

$$EU + 40\% = 16\text{M}$$

Nature-based system:

$$EU_{Green} = (0.3 \times 14) + (0.4 \times 16) + (0.3 \times 20) = 4.2 + 6.4 + 6 = 16.6\text{M}$$

Smart drainage system:

$$EU_{Smart} = (0.3 \times 15) + (0.4 \times 15) + (0.3 \times 18) = 4.5 + 6 + 5.4 = 15.9\text{M}$$

Step 2: Decision (Table 5.13)

The +20% capacity upgrade provides the lowest expected lifecycle cost under current climate uncertainty assumptions. It represents a moderate investment with reasonable protection against both moderate and severe scenarios. The smart drainage system performs nearly as well and may offer added flexibility. The baseline design has high risk and cost under severe conditions.

(b) **Sensitivity Analysis**

Let's find at what probability of severe climate impact the +40% capacity option becomes more cost-effective than +20% capacity.

Let p = probability of severe impact. Then:

- moderate = 0.4
- low = 0.6 − p.

Set expected costs equal:

$$(0.6 - p)(12) + 0.4(14) + p(22) = 16$$

$$p = 0.32$$

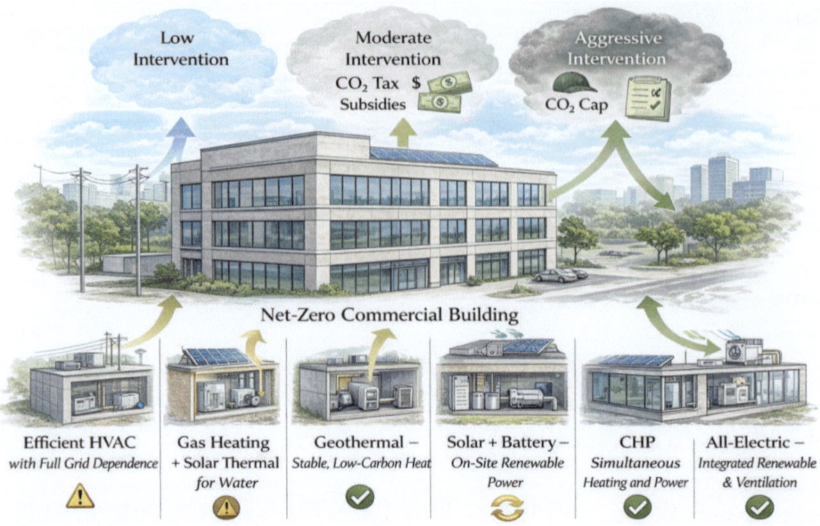

Fig. 5.41 Conceptual illustration of six energy system configurations for a net-zero commercial building

If the probability of severe climate impact exceeds 32%, the +40% capacity design becomes more cost-effective than the 20% upgrade.

The final recommendation should align with the project's resilience goals, operational flexibility, and broader environmental values. A +20% capacity increase offers a cost-effective solution with moderate resilience and is suitable for most conventional scenarios. Smart drainage systems are ideal for dynamic urban environments where real-time performance monitoring and adaptability are valuable. If severe climate scenarios are considered likely (with $p > 32\%$), opting for a +40% capacity provides greater long-term protection. Nature-based systems should be considered when co-benefits such as improved aesthetics, biodiversity, and community stakeholder engagement justify the additional investment.

5.9.9.1 Example 10: Energy System Selection for Net-Zero Building Under Future Policy Uncertainty

An engineer is designing the energy system for a new net-zero commercial building. Six different energy system configurations are available, each with varying levels of upfront cost, operational efficiency, carbon emissions, and reliance on grid electricity. The challenge is to choose the system that minimizes total expected cost while maintaining alignment with potential future carbon pricing and energy market regulations. Figure 5.41 illustrates the range of energy system design options for a net-zero commercial building, highlighting how carbon pricing and regulatory uncertainty may influence long-term performance and cost-effectiveness.

(a) Decision Options:

1. Grid electricity + high-efficiency HVAC
2. Natural gas + solar thermal for water heating
3. Geothermal heat pump system
4. Photovoltaic cell (PV) + battery storage
5. Combined heat and power (CHP)
6. All-electric with PV and heat recovery ventilation (HRV)

(b) Chance Outcomes (Future Energy/Carbon Policy Scenarios):

- Low intervention (current market continues)—30%
- Moderate intervention (carbon pricing + incentives)—40%
- Aggressive intervention (mandatory carbon caps)—30%

(c) Utility/Cost Table (Total Lifecycle Cost = CapEx + OpEx + Carbon Compliance Over 30 Years) (Table 5.14).

Step-by-Step Solution

Step 1: Calculate Expected Cost for Each Option

Grid + HVAC:

$$EU = (0.3 \times 4.0) + (0.4 \times 4.5) + (0.3 \times 6.5) = 1.2 + 1.8 + 1.95 = 4.95\text{M}$$

Table 5.14 Scenario-based cost analysis of energy system design options, including capital expenditures (CapEx), carbon and operational costs, and total projected costs under varying decarbonization scenarios

Option	Scenario	Probability	CapEx (M $)	Carbon/Op cost (M $)	Total cost (M $)
Grid + HVAC	Low	0.3	1.5	2.5	4
	Moderate	0.4	1.5	3	4.5
	Aggressive	0.3	1.5	5	6.5
Gas + Solar Thermal	Low	0.3	2	1.5	3.5
	Moderate	0.4	2	2	4
	Aggressive	0.3	2	4	6
Geothermal	All scenarios	1	3.5	0.8	4.3
PV + Battery	All scenarios	1	3.8	0.5	4.3
CHP	Low	0.3	2.5	1.5	4
	Moderate	0.4	2.5	2.8	5.3
	Aggressive	0.3	2.5	4.5	7
All-Electric (PV + HRV)	Low	0.3	4	0.4	4.4
	Moderate	0.4	4	0.3	4.3
	Aggressive	0.3	4	0.2	4.2

Gas + Solar Thermal:

$$EU = (0.3 \times 3.5) + (0.4 \times 4.0) + (0.3 \times 6.0) = 1.05 + 1.6 + 1.8 = 4.45\text{M}$$

Geothermal:

$$EU = 1.0 \times 4.3 = 4.3\text{M}$$

PV + Battery:

$$EU = 1.0 \times 4.3 = 4.3\text{M}$$

CHP:

$$EU = (0.3 \times 4.0) + (0.4 \times 5.3) + (0.3 \times 7.0) = 1.2 + 2.12 + 2.1 = 5.42\text{M}$$

All-Electric (PV + HRV):

$$EU = (0.3 \times 4.4) + (0.4 \times 4.3) + (0.3 \times 4.2) = 1.32 + 1.72 + 1.26 = 4.3\text{M}$$

Step 2: Decision (Table 5.15)

The Geothermal, PV + Battery, and All-Electric (PV + HRV) systems all provide the lowest expected lifecycle cost of $4.3M. Among them, PV + HRV has the most favorable outcome under aggressive policies. CHP and Grid-reliant systems become very costly under strict carbon scenarios. Even a modest increase in carbon pricing significantly affects grid-dependent and fossil fuel-based systems. If carbon compliance costs rise more than projected, PV + HRV becomes the safest investment.

The final recommendation depends on the balance between upfront investment, long-term resilience, and sustainability goals. Geothermal or a combination of PV and HRV systems offer an optimal balance between initial cost and long-term resilience, making them strong candidates for sustainable building performance. Gas combined

Table 5.15 Expected total costs of various energy system options

Energy system option	Expected cost (M $)
Grid + HVAC	4.95
Gas + Solar Thermal	4.45
Geothermal	4.3
PV + Battery	4.3
CHP	5.42
All-Electric (PV + HRV)	4.3

with solar thermal may be appropriate when capital constraints are a primary concern and future system retrofits are acceptable. However, CHP or grid-only HVAC systems should generally be avoided unless short-term economic factors take clear precedence over long-term environmental and operational sustainability.

5.9.10 Example 11: Structural Framing System Selection for a Critical Facility Under Multi-hazard Uncertainty

A structural engineer must select a framing system for a critical healthcare facility located in a region subject to seismic, windstorm, and flood hazards. The decision involves balancing initial construction cost, performance under multiple hazards, and lifecycle upgrade or repair costs. Six structural system options are available, each with different resilience profiles and capital investments. Minimize the total expected lifecycle cost of the facility by accounting for potential damage, code upgrades, and downtime costs under multi-hazard uncertainty. Figure 5.42 illustrates how different structural framing systems respond to varying levels of multi-hazard exposure, guiding engineers in balancing resilience and cost when designing critical facilities.

(a) Decision Options (Framing Systems):

1. Reinforced Concrete Moment Frame (RCMF)
2. Steel Moment Frame (SMF)
3. Steel Braced Frame (SBF)
4. Composite Steel–Concrete Frame (CSC)
5. Base-Isolated RC Frame (BI-RC)
6. Mass Timber Frame with Steel Core (MT-SC)

(b) Chance Outcomes (Hazard Scenarios—Multi-hazard Dependent):

Each hazard scenario has associated probabilities and can trigger structural/ nonstructural damage and regulatory upgrades (Table 5.16).

(c) Cost Matrix (Million USD over 50 years) (Table 5.17):

Step-by-Step Solution

Step 1: Compute Expected Lifecycle Cost (ELC)

RCMF:

$$ELC = 0.25 \times 9.5 + 0.40 \times 10.8 + 0.35 \times 14.7 = 2.375 + 4.32 + 5.145 = 11.84\text{M}$$

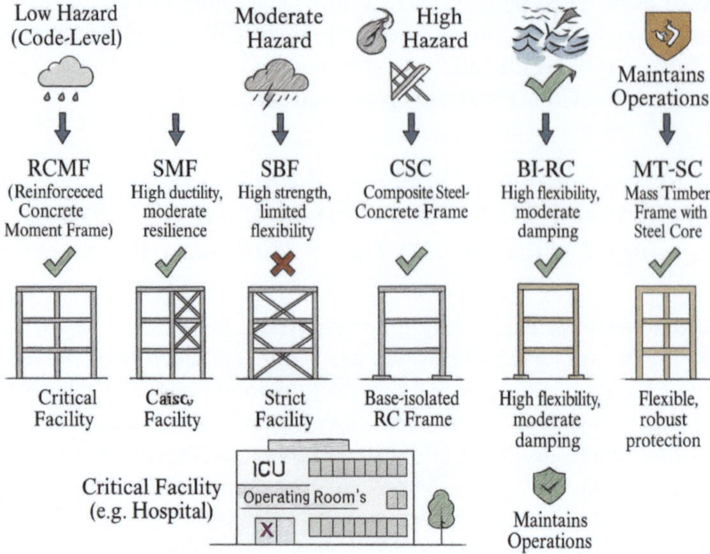

Fig. 5.42 Conceptual visualization of structural system options for a critical healthcare facility exposed to seismic, windstorm, and flood hazards. The illustration highlights differences in structural performance, damage potential, and continuity of critical operations under increasing hazard severity

Table 5.16 Hazard level classification with associated probabilities and expected performance outcomes, ranging from standard functionality to major damage and extended downtime

Low hazard (code-level events only)	25%	No major loss, standard performance
Moderate hazard (1-in-250 events)	40%	Some repair needed, limited downtime
High hazard (1-in-1000 or compound)	35%	Significant damage, code upgrade, long downtime

Downtime costs are included only for hospital critical services (e.g., ICU disruption = $100 k/month)

SMF:

$$ELC = 0.25 \times 9.1 + 0.40 \times 10.9 + 0.35 \times 15.0 = 2.275 + 4.36 + 5.25 = 11.89\text{M}$$

SBF:

$$ELC = 0.25 \times 8.4 + 0.40 \times 9.9 + 0.35 \times 15.3 = 2.1 + 3.96 + 5.355 = 11.42\text{M}$$

Table 5.17 Scenario-based cost assessment of structural system options, detailing capital expenditures (CapEx), damage and upgrade costs, downtime-related impacts, and total projected costs across low, moderate, and high hazard scenarios

System	Scenario	CapEx	Damage/upgrade	Downtime	Total cost
RCMF	Low	9	0.5	0	9.5
	Moderate	9	1.5	0.3	10.8
	High	9	4.5	1.2	14.7
SMF	Low	8.5	0.6	0	9.1
	Moderate	8.5	2	0.4	10.9
	High	8.5	5	1.5	15
SBF	Low	8	0.4	0	8.4
	Moderate	8	1.6	0.3	9.9
	High	8	5.5	1.8	15.3
CSC	Low	10	0.3	0	10.3
	Moderate	10	1.2	0.2	11.4
	High	10	3	0.8	13.8
BI-RC	Low	11.5	0.2	0	11.7
	Moderate	11.5	0.8	0	12.3
	High	11.5	1.5	0.3	13.3
MT-SC	Low	9.8	0.4	0	10.2
	Moderate	9.8	1	0.3	11.1
	High	9.8	2.5	0.9	13.2

CSC:

$$ELC = 0.25 \times 10.3 + 0.40 \times 11.4 + 0.35 \times 13.8 = 2.575 + 4.56 + 4.83 = 11.97\text{M}$$

BI-RC:

$$ELC = 0.25 \times 11.7 + 0.40 \times 12.3 + 0.35 \times 13.3 = 2.925 + 4.92 + 4.655 = 12.50\text{M}$$

MT-SC:

$$ELC = 0.25 \times 10.2 + 0.40 \times 11.1 + 0.35 \times 13.2 = 2.55 + 4.44 + 4.62 = 11.61\text{M}$$

Step 2: Decision (Table 5.18)

The Steel Braced Frame (SBF) has the lowest expected lifecycle cost, making it the optimal choice under the given uncertainty structure. The MT-SC hybrid is a close second and may be preferred if sustainability or embodied carbon are project

Table 5.18 Expected lifecycle costs of various structural systems, reflecting total projected expenditures over time based on construction, damage, and downtime factors across hazard scenarios

Structural system	Expected lifecycle cost (M USD)
Reinforced concrete moment frame (RCMF)	11.84
Steel moment frame (SMF)	11.89
Steel braced frame (SBF)	11.42
Composite steel–concrete (CSC)	11.97
Base-isolated RC (BI-RC)	12.5
Mass timber + steel core (MT-SC)	11.61

priorities. The BI-RC system, while very resilient, is significantly more expensive and may only be justifiable for essential post-disaster function facilities. If the probability of High Hazard increases beyond 50%, systems like BI-RC and CSC may become cost-optimal due to superior performance and minimal downtime.

5.10 Sensitivity Analysis in Decision Trees

5.10.1 Definition and Purpose

Sensitivity analysis in decision trees is a systematic method to assess how changes in input parameters such as probabilities of events or the utilities (i.e., costs, benefits, or other payoffs) associated with outcomes affect the optimal decision. By varying these inputs within plausible ranges, sensitivity analysis reveals the robustness of a decision and identifies critical variables that significantly influence the outcome. This process is crucial in decision-making under uncertainty, as it highlights whether small changes in assumptions could lead to alteration of the preferred solution.

In civil engineering, sensitivity analysis is particularly valuable due to the inherent uncertainties in projects, including (Fig. 5.43):

- Environmental conditions: Variable weather patterns, seismic risks, or flood probabilities.
- Material behavior: Uncertainty in material strength, durability, or fatigue.
- Regulatory changes: Evolving building codes, carbon pricing, or environmental regulations.
- Economic factors: Fluctuations in construction costs, maintenance expenses, or funding availability.
- Future usage: Uncertain demand for infrastructure, such as traffic loads or population growth.

The purpose of sensitivity analysis is to assess the robustness of decisions under uncertainty by identifying the input variables that most strongly influence

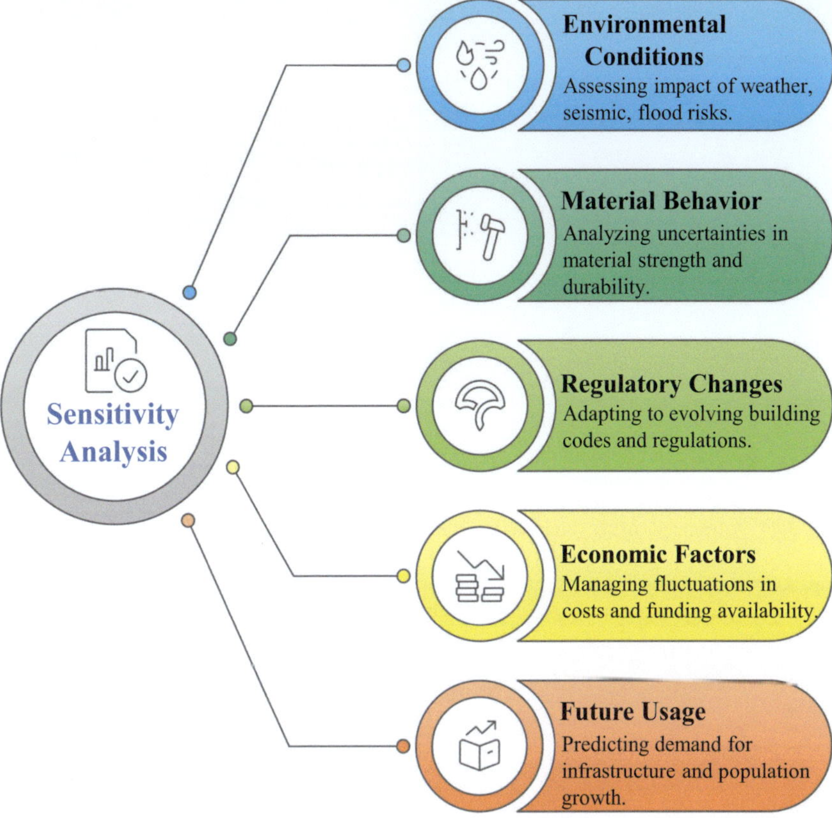

Fig. 5.43 Common uncertainties involved in civil engineering projects

the outcome. This information guides efforts to prioritize data collection, model refinement, or uncertainty reduction for the most influential variables. Additionally, sensitivity analysis enhances transparency in project decision making by quantifying the risks and sensitivities associated with various recommendations, permitting stakeholders to gain a clearer understanding of potential outcomes. Ultimately, sensitivity analysis supports risk-aware decision-making by revealing the conditions under which alternative strategies may become more favorable (Fig. 5.44).

By conducting sensitivity analysis, engineers can avoid over-reliance on single-point estimates and make informed decisions that account for variability and risk. Sensitivity analysis in decision trees serves several critical objectives (Fig. 5.45):

- **Testing the Robustness of the Preferred Decision**: Determines whether the optimal decision remains the best choice if input parameters (e.g., probabilities or costs) vary within a reasonable range. For example, if a bridge design is chosen based on a 1% annual flood probability, will it remain optimal if the probability increases to 2%?

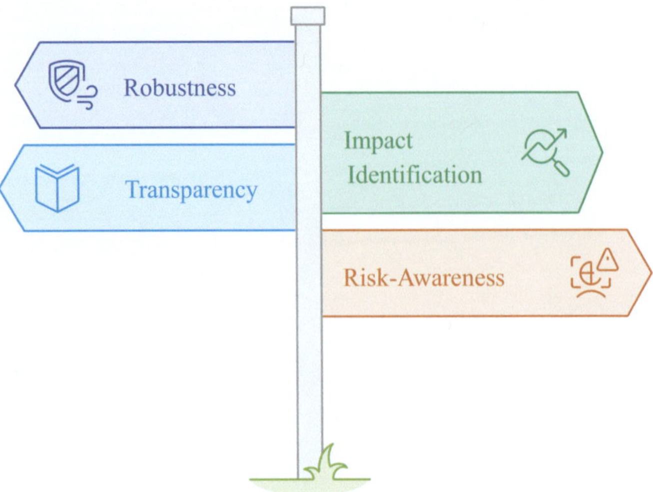

Fig. 5.44 Advantages of sensitivity analysis in civil engineering projects

Fig. 5.45 Critical objectives of the sensitivity analysis in decision trees

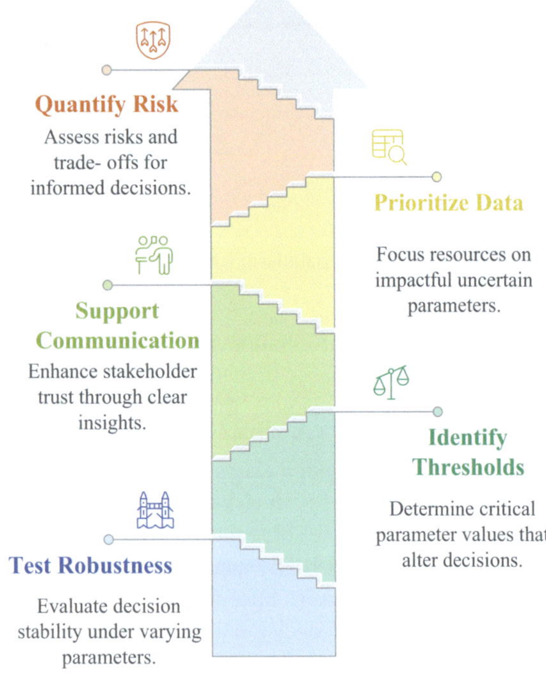

- **Identify Threshold Values**: Finds specific values of a parameter (e.g., a probability or cost) at which the preferred decision changes to another alternative. This helps define the boundaries of a decision's validity.
- **Support Stakeholder Communication**: Provides clear, data-driven insights to explain how sensitive a recommendation is to underlying assumptions, fostering trust and facilitating communication with key stakeholder groups including clients, regulators, or funding bodies.
- **Prioritize Data Collection**: Identifies which uncertain parameters have the most significant impact on the decision, guiding engineers to focus resources on collecting more accurate data for those variables (e.g., conducting additional soil tests if soil stability is a critical factor).
- **Quantify Risk and Uncertainty**: Enables engineers to quantify the risks associated with uncertain inputs and evaluate trade-offs between cost, safety, and performance.

5.10.2 Key Parameters for Sensitivity Analysis

Sensitivity analysis in decision trees is a vital technique for evaluating the robustness of decisions under uncertainty, particularly in civil engineering, where projects face complex risks from environmental, economic, and regulatory factors. By systematically varying key input parameters, sensitivity analysis identifies which uncertainties exert the greatest influence on the optimal decision, highlights conditions under which preferences may shift, and supports informed, risk-aware choices. In decision trees, this process involves testing parameters such as probabilities, costs, and external scenarios to ensure decisions remain viable across plausible ranges of uncertainty. The following sections detail five key parameters for sensitivity analysis (Fig. 5.46), namely, the probability of chance events, utility or cost values, regulatory or environmental scenarios, time horizons, and discount rates. Each of these parameters are explored in detail in terms of their definition, sources of uncertainty, civil engineering examples, implications for decision trees, and practical considerations. These parameters are critical for designing resilient infrastructure, prioritizing data collection, and communicating trade-offs to stakeholders, ensuring decisions are both robust and adaptable to future changes.

5.10.2.1 Probability of Chance Events

The probability of chance events in decision trees refers to the likelihood (expressed as a value between 0 and 1) of uncertain outcomes occurring at chance nodes, such as the occurrence of natural disasters, material failures, or regulatory approvals, which directly influence the expected utility or cost of each decision path. Uncertainty arises from multiple sources: limited historical data (e.g., incomplete flood records), future variability due to climate change or urbanization (e.g., increasing storm frequency),

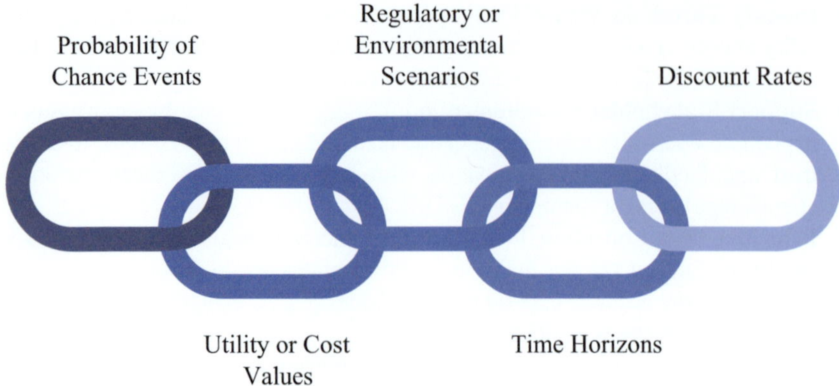

Fig. 5.46 Key parameters for sensitivity analysis

subjective expert judgment when data is scarce, and assumptions in statistical models (e.g., assuming a Poisson distribution for earthquakes).

Examples in civil engineering include the probability of a 1% annual chance flood impacting a levee, a 0.2% likelihood of a seismic event damaging a bridge, the chance of concrete fatigue failure under heavy traffic loads, or the probability of securing environmental permits for a new highway.

The implications for decision trees are significant, as small changes in high-impact probabilities can shift the optimal decision; for instance, a slight increase in flood probability might justify a more expensive floodwall over natural drainage improvements. Sensitivity analysis, such as one-way analysis (varying flood probability from 0.01 to 0.1) or threshold analysis (finding the probability where two options are equal), reveals which probabilities are critical, guiding risk mitigation or data refinement.

Practical considerations include prioritizing data collection for sensitive probabilities (e.g., conducting hydrological studies to refine flood risk estimates), adopting conservative estimates for safety–critical infrastructure (e.g., designing for worst-case seismic scenarios), and using sensitivity results to communicate risks to stakeholders, such as explaining why a low-probability, high-impact event justifies costly resilience measures. For example, in a decision tree for a coastal protection system, sensitivity analysis might show that the decision to build a seawall relies on the storm probability, prompting investment in advanced climate modeling to reduce uncertainty.

5.10.2.2 Utility or Cost Values

The definition of utility or cost values in decision trees encompasses the numerical metrics assigned to outcomes, representing monetary costs (e.g., construction expenses), benefits (e.g., economic gains), or non-monetary factors (e.g., environmental impacts) quantified through methods like multi-criteria decision analysis, which are used to compute the expected utility of each decision option. Uncertainty stems from market fluctuations (e.g., volatile material prices due to supply chain issues), estimation errors in early project phases (e.g., unforeseen site conditions), economic factors like inflation or funding constraints, and subjective quantification of non-monetary utilities (e.g., assigning a dollar value to reduced travel time).

Examples in civil engineering include construction costs for a new bridge ($50M), repair costs after a flood ($10M), annual maintenance expenses for a highway ($2M/year), societal benefits like reduced congestion from a transit system ($5M/year in economic value), and environmental costs such as carbon emissions penalties or habitat restoration ($1M).

The implications for decision trees are critical, as utility values drive trade-offs between options; high sensitivity to construction costs might favor a cheaper design, while sensitivity to repair costs could justify a resilient one. Sensitivity analysis techniques, such as tornado diagrams (ranking cost impacts) or two-way sensitivity (exploring construction vs. repair costs), identify which values most influence the decision, ensuring robustness or highlighting risks.

Practical considerations involve improving cost estimation through detailed quantity surveys or risk-based modeling, exploring value engineering to reduce sensitive costs (e.g., using alternative materials), and testing different weightings for non-monetary utilities (e.g., prioritizing safety over cost) to align with stakeholder priorities. For instance, in a decision tree for a dam, sensitivity analysis might reveal that high repair costs after failure make a robust design preferable, prompting engineers to negotiate bulk material discounts or incorporate sustainable materials to balance costs and environmental impacts.

5.10.2.3 Regulatory or Environmental Scenarios

The definition of regulatory or environmental scenarios in decision trees refers to external conditions—such as government policies, building codes, or environmental changes—that affect the probabilities, costs, or structure of the decision tree, introducing uncertainty in long-term infrastructure planning. Uncertainty arises from evolving policies (e.g., unpredictable changes in environmental regulations), climate change impacts (e.g., uncertain sea level rise projections), economic incentives like subsidies or carbon taxes, and technological advancements that alter compliance requirements (e.g., new energy-efficient materials).

Examples in civil engineering include stricter seismic building codes increasing construction costs, carbon pricing policies adding $0.5M to a concrete-heavy project, hydrological shifts due to climate change requiring larger flood defenses, and environmental regulations restricting construction in wetlands, necessitating costly mitigation measures.

The implications for decision trees are profound, as scenarios can reshape the tree by adding new branches (e.g., compliance failure) or altering costs and probabilities; sensitivity to a carbon tax might shift preference to a low-carbon design. Scenario analysis (comparing "high-regulation" vs. "low-regulation" futures) or one-way sensitivity analysis (varying compliance costs), ensures decisions are robust across plausible futures and identifies the points where different alternatives become optimal.

Practical considerations include designing adaptable infrastructure (e.g., modular floodwalls to accommodate future regulations), engaging stakeholders to anticipate policy shifts (e.g., consulting regulators), and using sensitivity results to justify proactive measures, such as exceeding current codes to future-proof a building. For example, in a decision tree for a skyscraper, sensitivity analysis might show that a new energy code significantly increases costs, prompting engineers to adopt prefabricated components that meet both current and anticipated standards, ensuring long-term viability.

5.10.2.4 Time Horizons

The definition of time horizons in decision trees refers to the duration over which outcomes are evaluated, such as the lifespan of a structure, the planning period for a project, or the time until a major event (e.g., a 100-year flood), which affects the discounting of costs and benefits in present-value calculations. Uncertainty arises from variable project lifespans (e.g., a bridge's durability depends on maintenance), statistical uncertainty in event timing (e.g., when a rare flood occurs), obsolescence due to technological or regulatory changes, and differing economic horizons based on funding constraints.

Examples in civil engineering include a 50-year lifespan for a highway versus a 30-year design, planning for a 100-year flood versus a 500-year event, lifecycle analysis of a building's energy performance over 20 versus 40 years, and designing coastal defenses for sea level rise over 50 versus 100 years.

The implications for decision trees are significant, as time horizons influence the weight of future costs and benefits; a longer horizon might justify a durable design due to reduced maintenance, while a shorter one favors lower upfront costs. Sensitivity analysis, such as varying the lifespan from 40 to 60 years or comparing short-term versus long-term scenarios, reveals whether a decision is robust or overly dependent on optimistic horizon assumptions.

Practical considerations include using lifecycle costing to account for lifespan uncertainty, designing for rarer events (e.g., 1,000-year floods) if sensitivity is high, and aligning horizons with stakeholder goals (e.g., budget limits or sustainability targets). For instance, in a decision tree for a highway, sensitivity analysis might show that a 20-year horizon favors concrete pavement material, but a 10-year horizon supports cheaper alternatives such as conventional asphalt pavement design.

5.10.2.5 Discount Rates

The definition of discount rates in decision trees is the rate (e.g., 3%) used to convert future costs and benefits into their present value, reflecting the time value of money, opportunity costs, or societal preferences for immediate versus future outcomes, critical for long-term infrastructure decisions. Uncertainty stems from economic variability (e.g., fluctuating interest rates or inflation), policy-driven rates mandated by agencies (e.g., U.S. OMB's 3–7% range), and subjective risk premiums for high-risk projects.

Examples in civil engineering include discounting the maintenance costs of a dam over 50 years, evaluating the present value of flood defenses preventing damages in 2075, comparing upfront costs of energy-efficient building features to future savings, and discounting revenue from a toll road in a public–private partnership.

The implications for decision trees are substantial, as discount rates exponentially affect long-term outcomes; a small increase from 2 to 5% can drastically reduce the present value of future benefits, altering the optimal decision. Sensitivity analysis, such as varying the rate from 2 to 5% or testing policy-mandated rates, identifies whether a decision is robust or sensitive to rate assumptions, highlighting trade-offs between upfront costs and long-term gains.

Practical considerations include testing a range of rates to account for economic uncertainty, aligning with agency guidelines for funding approval, and using low rates for sustainability-focused projects to prioritize future benefits, such as climate resilient infrastructure. For example, in a decision tree for a solar farm, sensitivity analysis might show that a 2% rate justifies high initial costs due to long-term energy savings, but a 5% rate favors a cheaper fossil fuel option, prompting engineers to advocate for green financing with lower rates to support sustainability goals.

5.10.3 Techniques in Sensitivity Analysis

Sensitivity analysis in decision trees is a critical methodology for evaluating the robustness and reliability of decisions under uncertainty, particularly in civil engineering, where projects face multifaceted risks from environmental variability, economic fluctuations, regulatory changes, and long-term performance uncertainties. By systematically altering input parameters—such as probabilities of events, costs, benefits, or external scenarios—sensitivity analysis reveals how sensitive a decision is to these changes, identifying which variables have the most significant impact on the outcome and under what conditions the preferred decision might shift. This process is essential for ensuring that infrastructure decisions, such as designing bridges, flood defenses, or transportation networks, are not only optimal under baseline assumptions but also resilient to plausible variations in uncertain inputs. The techniques used in sensitivity analysis provide structured approaches to quantify uncertainty, prioritize data collection, and communicate risks to stakeholders, enabling engineers to make informed, risk-aware choices that balance cost, safety, and performance.

For instance, consider a decision tree evaluating whether to build a floodwall (Option A) or rely on natural drainage improvements (Option B) to protect a coastal city. The decision hinges on uncertain parameters like the probability of a flood (e.g., 2% annually), the cost of flood damage (e.g., $10M), and future regulatory requirements (e.g., stricter environmental standards). Sensitivity analysis explores how changes in these parameters affect the expected cost of each option. If increasing the flood probability to 5% shifts the preference to the floodwall, this indicates high sensitivity to that parameter, prompting engineers to refine flood risk estimates or design a more robust solution. Similarly, imagine a decision tree for choosing between a concrete bridge (cheaper but less durable) and a steel bridge (costlier but more resilient). Sensitivity analysis might test the impact of varying construction costs or seismic event probabilities, revealing whether the concrete bridge remains optimal if costs rise by 20% or if seismic risks are underestimated. These conceptual examples underscore the role of sensitivity analysis in uncovering critical uncertainties, ensuring decisions are robust, and guiding resource allocation toward reducing the most impactful unknowns.

The techniques of sensitivity analysis—ranging from simple one-way variations to complex probabilistic simulations—offer diverse tools to explore these uncertainties. Each technique serves a specific purpose, from isolating the effect of a single variable to capturing interactions among multiple parameters or assessing decisions under discrete future scenarios. In civil engineering, these techniques are particularly valuable for addressing the long-term implications of infrastructure projects, where uncertainties like climate change, material performance, or regulatory evolution can significantly alter outcomes. By applying these methods, engineers can quantify the range of possible outcomes, identify thresholds where decisions change, and provide stakeholders with clear, data-driven insights into the risks and trade-offs involved. The following sections will detail these techniques, exploring their definitions, applications, advantages, limitations, and practical considerations, with a focus on their relevance to civil engineering decision-making.

5.10.3.1 One-Way Sensitivity Analysis

One-way sensitivity analysis is a foundational tool in decision analysis that systematically varies a single uncertain parameter—such as probability, cost, benefit, discount rate, or time horizon—while holding all other parameters constant. For each variation, the expected value of a decision outcome (e.g., expected utility, expected net benefit, or expected cost) is recalculated to assess how sensitive the optimal choice is to that particular parameter. This allows decision-makers to isolate and understand the influence of individual factors on the final decision and assess the robustness of the preferred alternative under uncertainty. In civil engineering, this method is widely used in project planning, risk mitigation, and design optimization where parameters are uncertain or based on estimates. For example, uncertainties in hydrological parameters, seismic risks, material prices, and construction timelines often necessitate one-way sensitivity analysis to inform resilient and cost-effective decisions. One-way sensitivity analysis is most appropriate in the early stages of decision-making, especially when a particular parameter is suspected to significantly influence the outcome. In real-world decision-making, precise estimates for probabilities, costs, or benefits are often difficult to obtain due to uncertainty in future events, expert judgment variability, or limited data. One-way sensitivity analysis allows decision-makers to:

- Evaluate robustness of a decision to uncertainty in a key parameter.
- Identify threshold values—critical points where the preferred option may switch.
- Understand model behavior under varying conditions.
- Prioritize data collection by focusing on the identified influential variables and their ranges of interest (Fig. 5.47).

The one-way sensitivity analysis procedure involves:

Step 1-Construct the Decision Tree

- Develop a decision tree that outlines the available decision alternatives, associated costs or benefits, and uncertain outcomes represented by probabilities.

Step 2-Select the Parameter for Analysis

- Choose one uncertain input variable (e.g., probability of failure, cost of a repair, expected demand) for sensitivity testing.

Step 3-Define a Plausible Range

- Identify a reasonable range of values for the chosen parameter based on expert knowledge, historical data, or modeling scenarios.

Step 4-Recalculate Expected Values

- For each value within the specified range, recalculate the expected value (cost or utility) for each decision alternative.

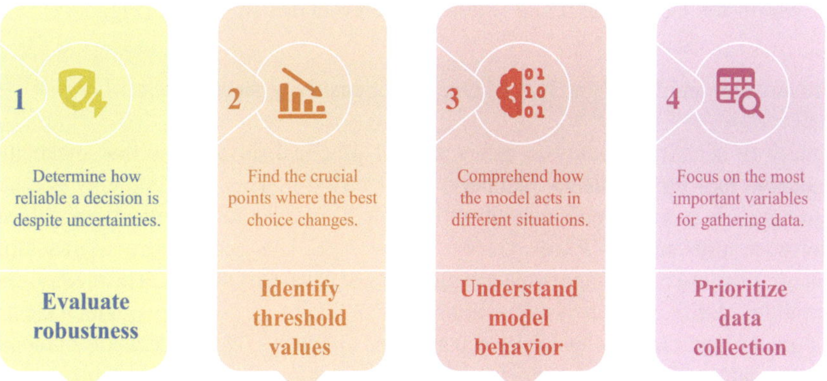

Fig. 5.47 Benefits of a one-way sensitivity analysis

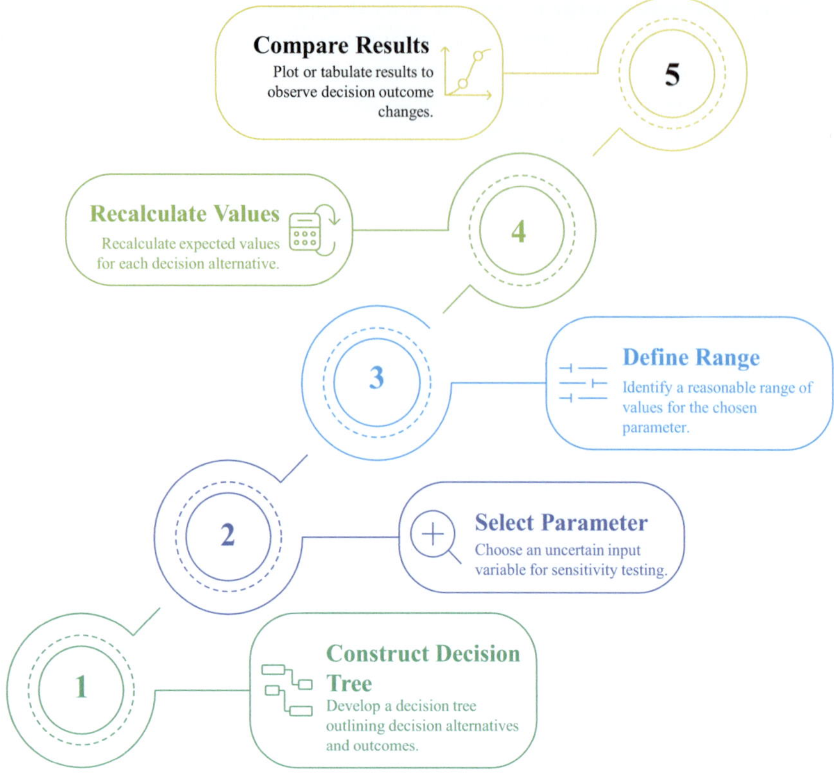

Fig. 5.48 The procedure for one-way sensitivity analysis

Step 5-Compare and Interpret Results

- Plot or tabulate the results to observe how the decision outcome changes with the parameter. Determine if and where the preferred decision shifts (Fig. 5.48).

Advantages and Disadvantages of One-Way Sensitivity Analysis

One-way sensitivity analysis offers several practical advantages that make it a valuable tool for engineers, planners, and decision-makers. Its simplicity is a key strength—it can be implemented with minimal computational effort using basic software tools such as Excel, GeNIe, or TreeAge. This accessibility allows rapid integration into early-stage project evaluations or feasibility studies. The method also provides clarity through visualizations such as line or tornado plots, which clearly illustrate how changes in a single parameter affect the overall decision outcome. These outputs are easy to interpret and communicate, making them suitable for both technical experts and non-technical stakeholders. Another major benefit is the ability to obtain early insight into which parameters most strongly influence the evaluation

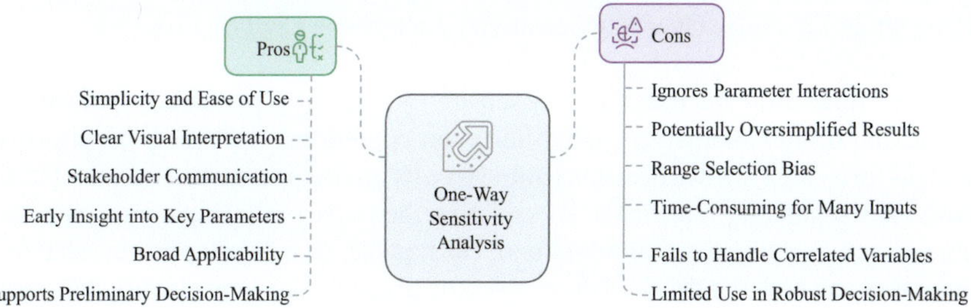

Fig. 5.49 Advantages and disadvantages of one-wave sensitivity analysis

of decision alternatives. In civil engineering projects, where data collection is often constrained by limited budgets, time, and site accessibility, this capability allows resources to be strategically focused on reducing uncertainty in the most influential parameters, thereby improving decision quality without unnecessary data expenditures. The technique's broad applicability makes it useful across multiple areas of civil engineering, from evaluating different structural designs and materials to assessing environmental impacts or managing construction budgets. Furthermore, it supports effective communication with policymakers, funding bodies, and the public by translating technical uncertainties into digestible 'what-if' narratives that inform planning and investment decisions.

Despite its utility, one-way sensitivity analysis has notable limitations that must be considered. Critically, this approach only examines one variable at a time and therefore ignores potential interactions between parameters—interactions that are common in real-world civil engineering problems, such as the joint effect of fluctuating material prices and project timelines. This can lead to oversimplified conclusions, particularly in complex systems where outcomes depend on threshold effects or nonlinear relationships (e.g., dam failure beyond a critical flood level). Additionally, the analysis is sensitive to range selection bias; results may vary significantly depending on the chosen parameter bounds, which are often based on expert judgment or limited data. A comprehensive understanding typically requires multiple analyses, one for each key parameter, which can be time-consuming and may still miss synergistic effects. Moreover, in cases where input variables are correlated—such as labor and equipment costs or rainfall and soil saturation—evaluating them independently may lead to misleading interpretations. Therefore, while one-way sensitivity analysis is an excellent starting point, it should ideally be complemented by more advanced techniques (e.g., multi-way sensitivity analysis, Monte Carlo simulation, or Bayesian networks) for robust decision support in civil engineering. Figure 5.49 shows the main advantages and disadvantages of a one-way sensitivity analysis.

Conceptual Examples One-Way Sensitivity Analysis in Civil Engineering

In civil engineering practice, one-way sensitivity analysis is particularly valuable for evaluating design alternatives under uncertain conditions. For example, infrastructure projects often involve decisions influenced by environmental hazards, fluctuating costs, and regulatory thresholds. By isolating the impact of a single uncertain variable—such as the probability of a flood, earthquake, or material price escalation—engineers can better understand the risk profile of each option and identify the tipping points at which one design becomes more favorable than another. This approach enhances transparency in decision-making and supports more resilient infrastructure planning by revealing which variables warrant further study or monitoring. Whether applied to transportation infrastructure, water management systems, or construction materials, one-way sensitivity analysis serves as a straightforward but powerful tool for prioritizing design choices and guiding investment decisions under uncertainty. Let's now explore a detailed example.

One-Way Sensitivity Analysis of Flood Protection Strategies

A coastal city is deciding whether to invest in a reinforced seawall (Option A) or implement enhanced natural drainage systems (Option B) to protect against flooding. The decision is driven by the uncertainty in the annual probability of a severe flood, which could cause significant damage to infrastructure and property. The city wants to choose the option with the lowest expected cost, considering both initial investment and potential flood-related repair costs. The key uncertain parameter is the flood probability, and the city seeks to understand how sensitive the decision is to changes in this probability. Figure 5.50 illustrates the city's challenge in selecting a flood protection strategy given uncertain flood probabilities, setting the stage for evaluating expected costs and sensitivity to climate projections.

Decision Tree Structure: Representative costs for the two options are provided below and are typical of a planning-level cost estimate developed during the project planning stage

- Chance Node: A severe flood occurs with probability $(1 - p)$.
- Option A (Seawall):
  - Construction cost: \$25 million.
  - Repair cost if flood occurs: \$2 million (seawall mitigates most damage).
- Option B (Drainage):
  - Construction cost: \$8 million.
  - Repair cost if flood occurs: \$18 million (drainage is less effective at preventing damage).
- Parameter for Sensitivity Analysis: Annual flood probability p, estimated at 0.03 (3%) but potentially ranging from 0.01 (1%) to 0.1 (10%) based on hydrological models and climate projections.

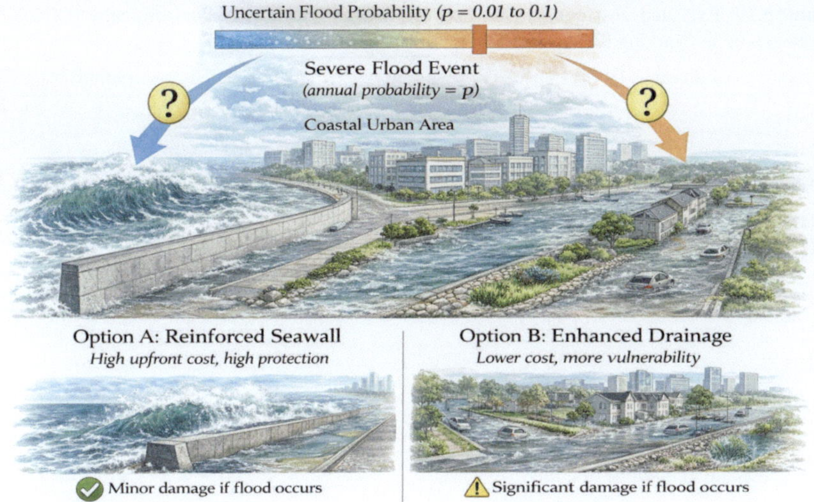

Fig. 5.50 Conceptual illustration of two flood mitigation strategies—reinforced seawall and enhanced natural drainage—under an uncertain severe flood probability

Questions to Address:

1. How does the expected cost of each option vary as the flood probability, p, changes from 0.01 to 0.1?
2. At what flood probability, if any, does the preferred option change (i.e., the threshold probability)?
3. How sensitive is the decision to the flood probability, and what does this imply for data collection or design choices?
4. What are the practical implications of the analysis for the city's decision-making process?

Step-by-Step Solution Using One-Way Sensitivity Analysis

Step 1: Define the Decision Tree Structure

The decision problem involves a deterministic decision node (choosing Option A or B), followed by a chance node representing the occurrence of a severe flood.

Let assume:

- Annual probability of a flood occurring = p
- Probability of no flood = 1 – p.

Step 2: Formulate Expected Cost Equations

The expected cost combines the construction cost and the probabilistic repair cost. The city will calculate the expected cost for both options as functions of p:

Table 5.19 Expected cost comparison between Seawall (Option A) and drainage (Option B) strategies across varying flood probabilities

Flood probability p	Expected cost option A*	Expected cost option B*
0.01	25 + 2(0.01) = 25.02	8 + 18(0.01) = 8.18
0.02	25.04	8.36
0.03	25.06	8.54
0.04	25.08	8.72
0.05	25.1	8.9
0.06	25.12	9.08
0.07	25.14	9.26
0.08	25.16	9.44
0.09	25.18	9.62
0.1	25.2	9.8

* All costs are in millions of dollars

Expected Cost for Option A:

$$EC_A(p) = Construction\ Cost_A + p \times Repair\ Cost_A$$

$$EC_A(p) = 25 + 2p$$

Expected Cost for Option B:

$$EC_B(p) = Construction\ Cost_B + p \times Repair\ Cost_B$$

$$EC_B(p) = 8 + 18p$$

Step 3: Perform the Sensitivity Analysis

To evaluate how changes in p affect the decision, the city varies p from **0.01 to 0.10** (1–10%) a range derived from hydrological models and climate projections.

The table below shows the expected cost for each option at different values of p (Table 5.19).

Step 4: Identify the Threshold Probability

To determine the flood probability at which the preferred option changes (i.e., where the expected costs are equal), we solve for p in:

$$25 + 2p = 8 + 18p$$

$$17 = 16p$$

$$p = 17/16$$

$$P = 1.0625$$

This threshold is above 1, which is not a realistic probability. Therefore, within the range $p \in [0.01, 0.10]$, Option B (Drainage) always has a lower expected cost than Option A.

Step 5: Interpret the Sensitivity and Results

The preferred decision does not change within the entire plausible range of p. Even at the highest assumed flood probability (10%), Option B remains significantly more cost-effective than Option A.

Cost Gap Between Options

- At $p = 0.01$: Option A is \$16.84M more expensive.
- At $p = 0.10$: Option A is \$15.40M more expensive.

This confirms that the decision is not sensitive to the value of p within the assumed range.

Step 6: Practical Implications for the City

The preferred strategy for the city is to proceed with Option B—enhancing natural drainage—since it consistently offers a lower expected cost across all realistic flood probability values. Given the robustness of this decision, there is no compelling need to invest in more precise estimates of the annual flood probability (p), as additional hydrological data collection would not significantly alter the outcome. This choice provides financial flexibility, enabling the city to redirect the \$15–17 million in savings toward other infrastructure needs, or to reduce the debt servicing obligations by eliminating the need to borrow this capital for the sea wall construction. From a policy perspective, while more reinforced infrastructure such as seawalls may offer higher levels of protection, their substantial cost is only justified under extreme flood risk scenarios.

This example demonstrates the power of one-way sensitivity analysis in guiding decision-making under uncertainty. By analyzing the impact of a single parameter—flood probability—on expected costs, the city confidently identifies a robust, cost-effective strategy without needing exhaustive probabilistic modeling. Such analyses are especially useful for public sector planning, where budget constraints and climate uncertainty coexist. This mathematical illustration of one-way sensitivity analysis demonstrates its power in evaluating decision robustness. By varying the flood probability from 0.01 to 0.1, we confirmed that the drainage system (Option B) is optimal, with no threshold within the entire range of realistic probabilities.

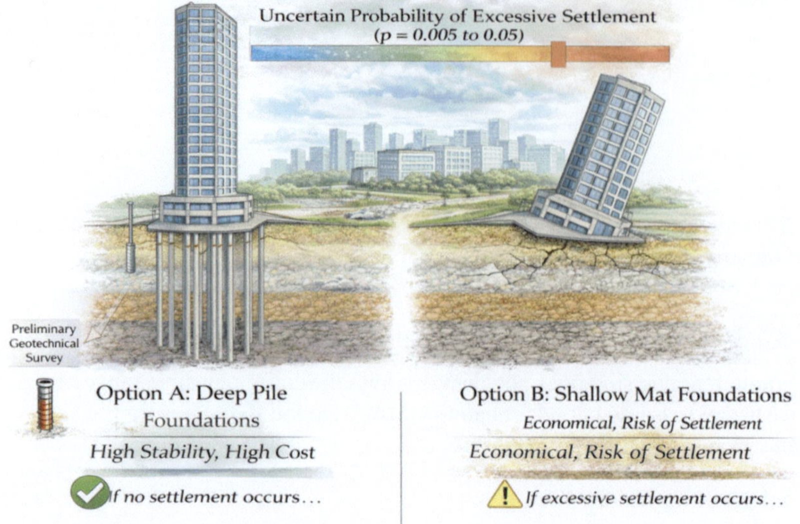

Fig. 5.51 Evaluating deep pile and shallow mat foundation systems for a high-rise building on variable soils. The image illustrates the different settlement risks and structural consequences under uncertain geotechnical conditions

One-Way Sensitivity Analysis of Foundation Design Decisions

A civil engineering firm is designing the foundation for a high-rise building in a region with variable soil conditions. The firm must decide between two foundation options: deep pile foundations (Option A) or shallow mat foundations (Option B). The decision hinges on the probability of excessive soil settlement, which could lead to structural damage requiring costly repairs. Deep pile foundations are more expensive but have greater resiliency to settlement, while shallow mat foundations are cheaper but risk higher repair costs if settlement occurs. The key uncertain parameter is the probability of excessive soil settlement, and the firm seeks to understand how sensitive the decision is to changes in this probability. Figure 5.51 introduces the foundation design decision.

Decision Tree Structure

- Chance Node: Excessive soil settlement occurs with probability p, or no excessive settlement occurs with probability (1 − p).
- Option A (Deep Pile Foundations):
  - Construction cost: $12M.
  - Repair cost if excessive settlement occurs: $1 million (piles mitigate most damage).

- Option B (Shallow Mat Foundations):
  - Construction cost: \$6 million.
  - Repair cost if excessive settlement occurs: \$8 million (mat foundations are less effective against settlement).
- Parameter for Sensitivity Analysis: Annual probability of excessive soil settlement, p, estimated at 0.02 (2%) based on preliminary geotechnical surveys but potentially ranging from 0.005 (0.5%) to 0.05 (5%) due to variability in soil data and potential seismic influences.

Questions to Address:

1. How does the expected cost of each foundation option vary as the settlement probability p changes from 0.005 to 0.05?
2. At what settlement probability, if any, does the preferred option change (i.e., the threshold probability)?
3. How sensitive is the decision to the settlement probability, and what does this imply for geotechnical investigations or design choices?
4. What are the practical implications of the analysis for the firm's decision-making process?

Step 1: Define the Expected Cost Functions

Let:

- p = probability of excessive soil settlement
- 1 − p = probability of no excessive settlement.

Option A: Deep Pile Foundations

$$EC_A(p) = 12 + p \times 1 = 12 + p$$

Option B: Shallow Mat Foundations

$$EC_B(p) = 6 + p \times 8 = 6 + 8p$$

Step 2: Vary p from 0.005 to 0.05

We compute the expected costs for each option as p varies within the plausible range (Table 5.20):

$$p \in \{0.005, 0.01, 0.015, 0.02, 0.025, 0.03, 0.035, 0.04, 0.045, 0.05\}$$

Table 5.20 Expected cost calculations for two foundation design options under varying settlement probabilities

Settlement probability, p	Option A: Cost 12 + p	Option B: Cost 6 + 8p
0.005	12.005	6.04
0.01	12.01	6.08
0.015	12.015	6.12
0.02	12.02	6.16
0.025	12.025	6.2
0.03	12.03	6.24
0.035	12.035	6.28
0.04	12.04	6.32
0.045	12.045	6.36
0.05	12.05	6.4

All values are in millions of dollars

Step 3: Find the Threshold Probability

To find when the preferred option changes, we solve for the value of p where:

$$EC_A(p) = EC_B(p)$$

$$12 + p = 6 + 8p$$

$$6 = 7p$$

$$p = 6/7$$

$$p \approx 0.857$$

The analysis reveals a threshold probability of $p = 0.857(85.7\%)$, at which the preferred decision would switch. However, this value lies far outside the plausible range of settlement probabilities, which is typically between 0.5 and 5%. As a result, Option B remains the preferred strategy under all realistic scenarios, confirming the robustness of the decision and eliminating the need for further refinement of the probability estimate.

Answers to the Questions

1. How does the expected cost vary as p increases?
   - Option A increases linearly: from \$12.005M to \$12.050M.
   - Option B also increases linearly but at a faster rate: from \$6.04M to \$6.40M.

Despite both increasing, Option B remains significantly cheaper throughout.

2. At what settlement probability does the preferred option change?

The threshold is p = 0.857. Since this value exceeds the range of realistic probabilities, Option B (mat foundation) is preferred for all practical values of p.

3. How sensitive is the decision to p?

The decision is not sensitive to changes in p between 0.005 and 0.05. Even at the upper bound p = 0.05, Option B is $5.65 million cheaper than Option A.

Further efforts to refine the estimate of p would not change the decision, so extensive geotechnical studies for the purpose of a more refined estimate of p are not be cost-effective.

4. What are the practical implications for the firm's decision-making process?

The firm should select Option B, the shallow mat foundation, based on expected cost minimization. Although deep piles provide greater bearing capacity than a mat foundation, their high initial cost is not justified under any plausible soil risk level. The firm may consider reallocating funds saved by choosing Option B into: (1) improved soil compaction or ground improvement measures, (2) monitoring systems to detect early signs of settlement, and (3) structural insurance to hedge against rare settlement events.

This mathematical illustration of one-way sensitivity analysis in a geotechnical context demonstrates its utility in evaluating decision robustness for foundation design. By varying the settlement probability from 0.005 to 0.05, we confirmed that shallow mat foundations (Option B) are optimal, with no threshold within realistic probabilities. The moderate sensitivity, shown by the narrowing cost gap, highlights the importance of precise geotechnical data and flexible design strategies. This step-by-step process—defining equations, calculating costs, finding thresholds, and interpreting results—provides a clear, actionable framework for civil engineers to assess uncertainties in soil-related risks, prioritize geotechnical investigations, and communicate findings to stakeholders.

One-Way Sensitivity Analysis of Roofing Material Selection

A construction firm is designing the roof for a large industrial warehouse and must decide between two roofing materials: steel roofing (Option A) or asphalt shingle roofing (Option B). The decision depends on the probability of extreme wind events (e.g., hurricanes or severe storms) that could damage the roof, requiring costly repairs. Steel roofing is more expensive upfront but more durable against wind damage, while asphalt shingles are cheaper but more vulnerable to wind-related damage. The key uncertain parameter is the annual probability of an extreme wind event, and the firm seeks to understand how sensitive the decision is to changes in this probability. Figure 5.52 illustrates the engineering decision between two roofing systems under uncertain wind hazard probability, emphasizing trade-offs between cost, durability, and exposure to climate-driven risk.

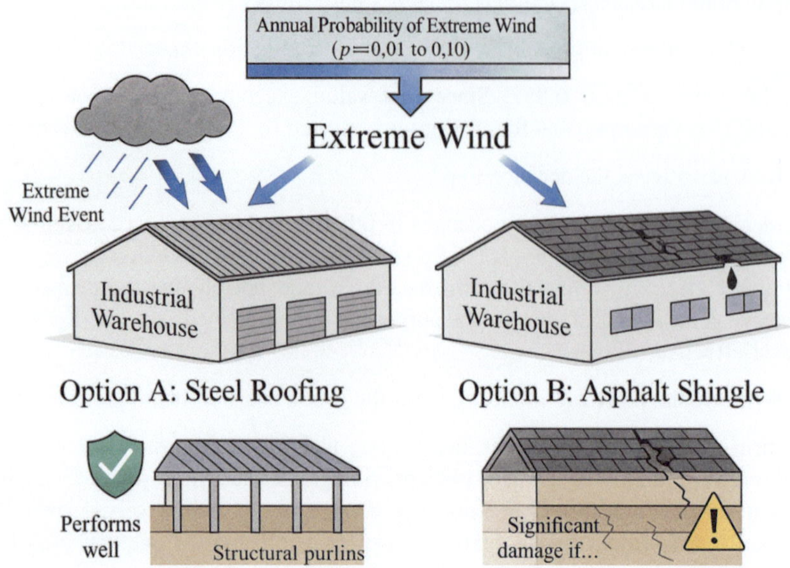

Fig. 5.52 Conceptual illustration of steel and asphalt shingle roofing systems for an industrial warehouse under extreme wind event uncertainty

Decision Tree Structure

- Chance Node: An extreme wind event occurs with probability p, or no extreme wind event occurs with probability (1 − p).
- Option A (Steel Roofing):
  - Installation cost: $5 million.
  - Repair cost if wind event occurs: $0.5 million (steel is highly resistant to wind damage).
- Option B (Asphalt Shingle Roofing):
  - Installation cost: $2 million.
  - Repair cost if wind event occurs: $3 million (asphalt shingles are less resistant, requiring significant repairs).
- Parameter for Sensitivity Analysis: Annual probability of an extreme wind event p, estimated at 0.04 (4%) based on regional meteorological data but potentially ranging from 0.01 (1%) to 0.1 (10%) due to climate variability and potential increases in storm intensity.

Questions to Address

1. How does the expected cost of each roofing option vary as the wind event probability p changes from 0.01 to 0.1?

Table 5.21 Expected cost comparison of steel roofing and asphalt shingles under varying wind event probabilities, highlighting differential sensitivity to wind-related risk

Wind event probability p	Steel roofing: 5 + 0.5p	Asphalt shingles: 2 + 3p
0.01	5.005	2.03
0.02	5.01	2.06
0.03	5.015	2.09
0.04	5.02	2.12
0.05	5.025	2.15
0.06	5.03	2.18
0.07	5.035	2.21
0.08	5.04	2.24
0.09	5.045	2.27
0.1	5.05	2.3

All values are in millions of dollars

2. At what wind event probability, if any, does the preferred option change (i.e., the threshold probability)?
3. How sensitive is the decision to the wind event probability, and what does this imply for material selection or further analysis?
4. What are the practical implications of the analysis for the firm's decision-making process?

Step 1: Define the Expected Cost Functions

Let:

- p = annual probability of an extreme wind event
- 1 − p = probability of no extreme wind event.

Option A: Steel Roofing

$$EC_A(p) = 5 + 0.5p$$

Option B: Asphalt Shingle Roofing

$$EC_B(p) = 2 + 3p$$

Step 2: Vary p from 0.01 to 0.10

We compute the expected cost of both options across a range of wind event probabilities (Table 5.21).

Step 3: Find the Threshold Probability

To determine when the preferred option changes, we solve for the value of p where:

$$EC_A(p) = EC_B(p)$$

$$5 + 0.5p = 2 + 3p$$

$$3 = 2.5p$$

$$p = 3/2.5$$

$$p = 1.2$$

The calculated threshold wind event probability is $p = 1.2$ (120%), which is not feasible since probabilities cannot exceed 1. Given this, Option B is always preferred within the realistic probability range of $p \in [0.01, 0.10]$. This confirms that the decision is robust and not sensitive to the variability in wind event probabilities within plausible bounds.

Answers to the Questions

1. How does the expected cost vary as p increases?
   - Steel Roofing (Option A) increases very slowly, from \$5.005M to \$5.050M.
   - Asphalt Shingles (Option B) increases more sharply, from \$2.03M to \$2.30M.

Despite the increasing repair costs with p, Option B remains significantly cheaper across the full range.

2. At what probability does the preferred option change?

The threshold occurs at $p = 1.2$, which is beyond any realistic scenario. Thus, Option B is always the preferred choice based on expected cost.

3. How sensitive is the decision to the wind event probability?

The decision is not sensitive within the tested range (1–10%). Even at the highest probability of 10%, Option B is \$2.75 million cheaper than Option A. This implies the decision is robust, and improved precision in estimating p would not impact the choice.

4. What are the practical implications for the firm's decision-making process?

The firm should confidently select asphalt shingle roofing (Option B) based on cost-effectiveness. While steel is more durable, its high installation cost outweighs its limited savings in repair costs even in high-risk scenarios. Funds saved from choosing asphalt shingles can be reallocated toward (1) Reinforcement techniques (e.g., clips, underlayment), (2) Insurance to cover rare extreme events, and (3) Scheduled maintenance to extend the life of the roof.

This mathematical illustration of one-way sensitivity analysis in a construction context demonstrates its utility in evaluating decision robustness for roofing material selection. By varying the wind event probability from 0.01 to 0.1, we confirmed that asphalt shingle roofing (Option B) is optimal, with no threshold within realistic probabilities. The moderate sensitivity, shown by the narrowing cost gap, highlights the importance of precise meteorological data and durable material specifications. This step-by-step process—defining equations, calculating costs, finding thresholds, and interpreting results—provides a clear, actionable framework for civil engineers to assess uncertainties in construction-related risks, prioritize data collection, and communicate findings to stakeholders.

One-Way Sensitivity Analysis of Stormwater Management Options

A city is preparing to upgrade its stormwater drainage system in response to projected increases in rainfall intensity due to climate change. The engineering team is evaluating two alternatives. Option A involves constructing a large underground retention tank, which demands a substantial upfront investment but provides high long-term resilience against extreme storm events. In contrast, Option B proposes deploying permeable pavements and bioswales citywide—a form of distributed green infrastructure that is more cost-effective initially, yet less reliable during intense rainfall, potentially resulting in flooding and associated damage costs. The choice between these two strategies hinges on the probability of extreme rainfall events exceeding historical levels. To guide decision making, the city aims to conduct a sensitivity analysis to assess how the relative cost-effectiveness of each option shifts in response to changes in this probability. This approach will help determine whether the decision is robust or if further climate modeling and risk analysis are needed to reduce uncertainty. Figure 5.53 presents the contrasting stormwater strategies available to the city, framing the resilience and cost trade-offs between centralized and decentralized drainage systems under future climate uncertainty.

Decision Tree Structure

- Chance Node: An extreme rainfall event occurs with probability p, or does not occur with probability 1 − p.
- Option A:
  - Construction cost: \$20 million
  - Damage/repair cost if flood occurs: \$1 million
- Option B:
  - Construction cost: \$10 million
  - Damage/repair cost if flood occurs: \$7 million
- Uncertain Parameter:
  - Probability of extreme rainfall event p estimated at 0.03 (3%), but may range from 0.01 to 0.1 due to climate model uncertainty.

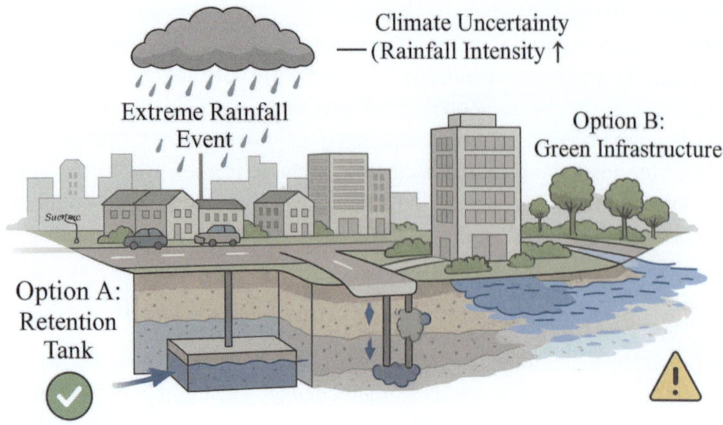

Fig. 5.53 Conceptual illustration of two stormwater infrastructure strategies—a large underground retention tank and a distributed green infrastructure system—evaluated under increasing rainfall intensity due to climate change

Questions to Address

1. How does the expected cost of each stormwater option vary as the flood probability p changes from 0.01 to 0.10?
2. At what flood probability (if any) does the preferred option change?
3. How sensitive is the decision to changes in flood probability?
4. What are the implications for design selection and climate resilience planning?

Step-by-Step Solution

Step 1: Define Expected Cost Equations

Let p be the probability of an extreme rainfall event.

Option A (Retention Tank)

$$EC_A(p) = 20 + 1 \cdot p$$

$$EC_A(p) = 20 + p$$

Option B (Green Infrastructure)

$$EC_B(p) = 10 + 7 \cdot p$$

$$EC_B(p) = 10 + 7p$$

Table 5.22 Expected cost comparison of flood protection Options A and B as flood probability p varies from 0.01 to 0.10, illustrating differences in sensitivity to flood-related risk

Flood probability p	Option A: 20 + p	Option B: 10 + 7p
0.01	20.01	10.07
0.02	20.02	10.14
0.03	20.03	10.21
0.04	20.04	10.28
0.05	20.05	10.35
0.06	20.06	10.42
0.07	20.07	10.49
0.08	20.08	10.56
0.09	20.09	10.63
0.1	20.1	10.7

All values in millions of dollars

Step 2: Vary p from 0.01 to 0.10 (Table 5.22).

Step 3: Calculate the Threshold Probability

Find the value of p where the expected costs are equal:

$$20 + p = 10 + 7p$$

$$10 = 6p$$

$$p = 10/6$$

$$p = 1.667$$

This threshold is well beyond 100%, meaning the preferred option will not change within the considered range.

Step 4: Interpret Results

Q1: Cost Variation

- Option A increases slowly with p, from $20.01M to $20.10M.
- Option B increases faster, from $10.07M to $10.70M.

Despite the growth, Option B is consistently cheaper in all tested cases.

Q2: Threshold Probability

The threshold value of p = 1.667 implies Option A never becomes cheaper in the given range of (p ∈ [0.01, 0.10].

Q3: Sensitivity

The decision is not sensitive to flood probability in this range. Even at p = 0.10, Option A is still almost $9.4M more expensive than Option B.

Q4: Practical Implications

The city can confidently choose Option B (green infrastructure), saving substantial costs without sacrificing robustness. Because the decision is not sensitive to variations in p, detailed probabilistic modeling may not be needed unless new risks (e.g., major climate shifts) emerge. The city could invest savings into monitoring, public education, or supplementary mitigation, enhancing overall system resilience.

One-Way Sensitivity Analysis for Concrete Mix Selection in Cold Climate

An engineer is tasked with designing a reinforced concrete office tower in a cold Canadian climate, where the structure will be subject to frequent freeze–thaw cycles and chloride exposure from de-icing salts. To ensure long-term performance, the engineer must choose from four concrete mix designs, each differing in initial construction cost, durability, and reliability under uncertain field conditions. Given real-world variability in factors such as batch quality, curing practices, and supplier consistency, each mix may result in one of three performance outcomes: high quality (requiring no repairs), moderate quality (involving minor surface cracking and repairs), or poor quality (failing to meet durability specifications, leading to early major repairs). The engineer must evaluate these options using a risk-informed approach that accounts for both cost and performance under uncertainty. The objective is to select the mix with the lowest expected lifecycle cost, accounting for the probability of each performance outcome and its financial implications. Figure 5.54 illustrates how cold-climate exposure and variability in concrete mix application influence durability outcomes, guiding engineers in selecting robust materials for long-term structural performance (Tables 5.23, 5.24 and 5.25).

Questions to Address

1. What is the expected lifecycle cost of each concrete mix?
2. Which mix is the most cost-effective under current assumptions?
3. How does the decision respond to increased risk of poor performance?
4. At what point would a more durable (but costlier) mix become preferable?
5. What are the implications for material selection and construction risk management?

Step-by-Step Solution

Step 1: Define the Expected Cost Formula

$$\text{Expected Cost (EC)} = \sum (\text{Probability} \times \text{Total Cost}) \tag{5.2}$$

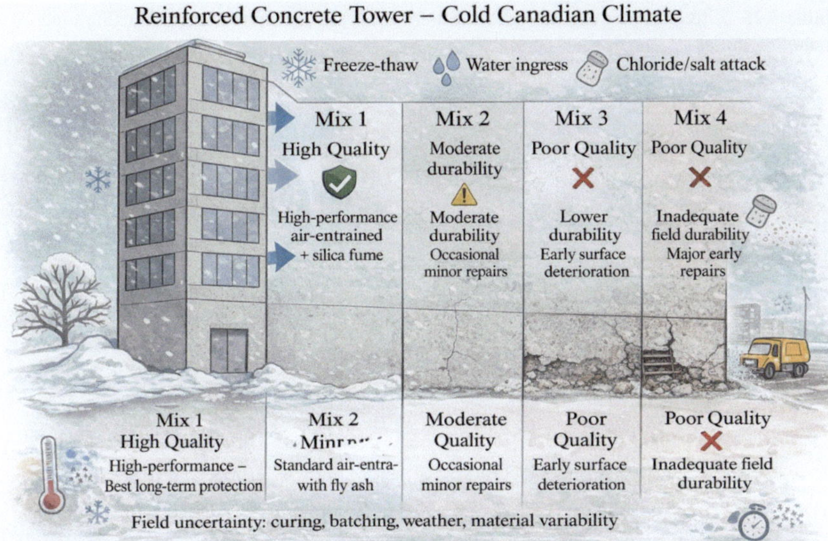

Fig. 5.54 Conceptual comparison of four concrete mix designs for a reinforced concrete tower in a cold climate

Table 5.23 Descriptions of concrete mix design options

Option	Description
A	Conventional mix (30 MPa, local source)
B	Air-entrained mix (35 MPa, improved durability)
C	High-performance mix (40 MPa, silica fume)
D	Self-consolidating concrete (SCC, factory-batched)

Table 5.24 Material performance scenarios for different concrete mix options

Quality outcome	Description	A (Conventional) (%)	B (Air-entrained) (%)	C (HPC) (%)	D (SCC) (%)
High	Meets/exceeds spec (no repairs)	60	70	90	95
Moderate	Minor cracks; repair in 10 yrs	30	20	10	5
Poor	Below spec; major repairs	10	10	0	0

Table 5.25 Lifecycle cost estimates for different concrete mix options across quality outcomes, including initial and repair costs per 1,000 m^3, expressed in thousands of Canadian dollars

Option	Outcome	Initial cost	Repair cost	Total cost
A	High	100	0	100
A	Moderate	100	20	120
A	Poor	100	80	180
B	High	110	0	110
B	Moderate	110	10	120
B	Poor	110	60	170
C	High	130	0	130
C	Moderate	130	10	140
C	Poor	–	–	–
D	High	140	0	140
D	Moderate	140	10	150
D	Poor	–	–	–

Step 2: Calculate Expected Cost for Each Mix

Option A (Conventional Mix)

$$EC_A = 0.60 \times 100 + 0.30 \times 120 + 0.10 \times 180 = 60 + 36 + 18 = 114$$

Option B (Air-Entrained Mix)

$$EC_B = 0.70 \times 110 + 0.20 \times 120 + 0.10 \times 170 = 77 + 24 + 17 = 118$$

Option C (High-Performance Concrete)

$$EC_C = 0.90 \times 130 + 0.10 \times 140 = 117 + 14 = 131$$

Option D (Self-Consolidating Concrete)

$$EC_D = 0.95 \times 140 + 0.05 \times 150 = 133 + 7.5 = 140.5$$

Step 3: Summarize the Results

Preferred Mix: Option A (Conventional mix) has the lowest expected lifecycle cost, making it the most economical option under current performance assumptions.

Step 4: Sensitivity Insight

Option A carries 10% risk of early major repair. What happens if this risk increases?

Table 5.26 Expected lifecycle costs of concrete mix options

Option	Description	Expected cost (CAD $1,000 m^3)
A	Conventional mix	114
B	Air-entrained mix	118
C	High-performance concrete	131
D	Self-consolidating concrete	140.5

Let's compute the expected cost for Option A at different poor outcome probabilities. To do so, the poor and moderate probabilities outlined in Table 5.27 are used to calculate EC using the same approach as in Step 2 shown above. The probability of high-quality concrete is simply defined as 1-p(poor)-p(medium) because these conditions are mutually exclusive and exhaustive.

Threshold Insight: Using linear interpretation, it can be seen from the results of Table 5.27 that if probability of poor performance exceeds approximately17%, Option B (air-entrained mix) becomes more cost-effective.

Step 5: Engineering Implications

- Option A is the most economical under current assumptions and is suitable for non critical or controlled-exposure elements.
- If durability or quality control is uncertain, Option B provides better long-term value.
- Options C and D offer outstanding reliability but are not cost-justified unless:
  - The structure has critical exposure zones.
  - There is very high consequence of failure (e.g., parking garage decks, podium slabs).
  - The project values low maintenance over time more than upfront savings.

Recommendation: Perform material trials, field QA programs, and consider supplier track record to estimate realistic risks for early deterioration. Adjust the mix choice accordingly.

The outcome of the sensitivity analysis for the cold weather concrete mix design scenario can be summarized as follows:

Question 1: What is the expected lifecycle cost of each concrete mix?

The calculation of the lifecycle cost for each alternative was demonstrated in Step 2 above, ranging from 114$ for mix design A to 140.5$ for mix design D. The complete cost breakdown for the different alternatives is provided in Table 5.26.

Question 2: Which mix is the most cost-effective under current assumptions?

Option A (Conventional mix) is the most cost-effective, with an expected lifecycle cost of $114/m^3. Based on the available data, this mix design has a 60% chance of no repair and, although it carries a 10% chance of major repair, its low initial cost offsets the risk under current assumptions.

Table 5.27 Sensitivity analysis of expected cost for concrete mix Option A under varying probabilities of poor and moderate performance outcomes

Poor %	Moderate %	Expected cost A
5%	35%	$111
10%	30% (base)	$114
15%	25%	$117
20%	20%	$120
25%	15%	$123

Table 5.28 Effect of increasing poor outcome probability on the expected cost of concrete mix Option A, highlighting cost sensitivity to performance risk

Poor outcome probability	Expected cost of Option A
10% (Base)	$114
15%	$117
20%	$120
25%	$123

Question 3: How does the decision respond to increased risk of poor performance?

A sensitivity analysis showed that, for a fixed probability of high performance concrete (i.e., 60%), the expected costs of Option A increased as the probability of poor performance increases (Table 5.27).

When the poor outcome probability exceeds approximately 17%, Option B (air-entrained mix) becomes more cost-effective due to its greater durability and lower repair costs in poor scenarios (Table 5.28).

Question 4: At what point would a more durable (but costlier) mix become preferable?

The break-even threshold is reached when the probability of poor performance for Option A exceeds ~17%. At that point, Option B, which costs more initially but has better durability (lower risk of poor performance), becomes the optimal choice.

This means that if site conditions, material variability, or supplier quality suggest a greater than 1 in 6 chance of early deterioration, the engineer should consider switching to Option B.

Question 5: What are the implications for material selection and construction risk management?

- Option A is appropriate for projects with tight budgets and reasonable confidence in material quality.
- If construction occurs under variable environmental or curing conditions, or if supplier QA/QC is uncertain, it is safer to choose Option B, even at a slightly higher cost.
- Options C and D offer very high reliability but are not economically optimal unless:
  - The project involves critical infrastructure, or

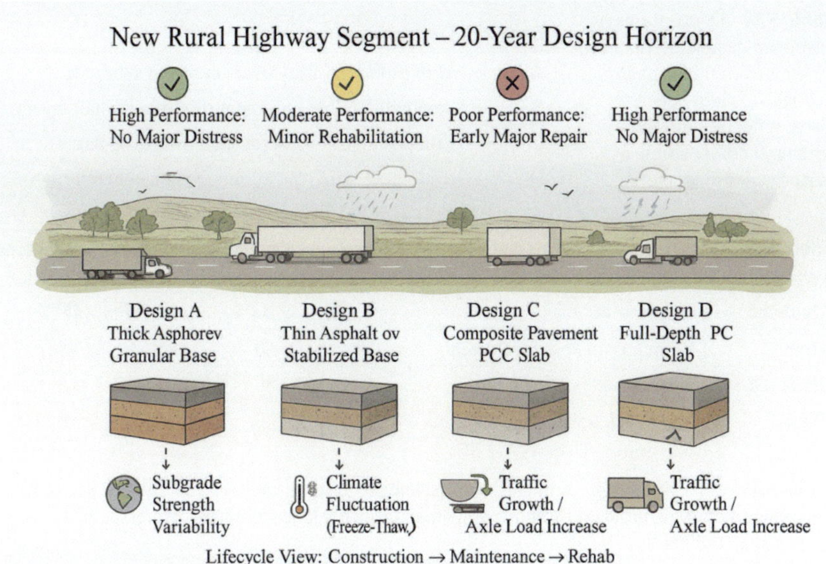

Fig. 5.55 Conceptual comparison of four pavement design alternatives for a rural highway exposed to uncertain traffic, climate, and subgrade conditions

– The owner prioritizes minimal long-term maintenance and is willing to invest more upfront.

The selection of concrete mix should not rely on cost alone but also on the risk tolerance of the project, the anticipated exposure conditions, and the quality assurance framework in place during construction.

One-Way Sensitivity Analysis for Pavement Design in Highway Construction

A transportation engineer must select one of four pavement design alternatives for a new rural highway section expected to carry moderate to high truck traffic over the next 20 years. Each design option has different initial construction costs, maintenance schedules, and durability under environmental and loading uncertainties. Due to variability in soil subgrade strength, climate conditions, and traffic growth, the actual long-term performance of the pavement is uncertain. Outcomes are categorized into high performance (no major distress), moderate performance (minor rehabilitation required), and poor performance (major rehabilitation needed early). The objective is to select the design with the lowest expected lifecycle cost, accounting for both initial and future repair expenditures under different probabilistic performance scenarios. Figure 5.55 presents the pavement design options under consideration for a rural highway project, illustrating how variability in field conditions may lead to different long-term performance outcomes (Tables 5.29, 5.30, 5.31 and 5.32).

Table 5.29 Descriptions of pavement design alternatives, ranging from thin asphalt overlays to rigid concrete slabs, reflecting differences in structural composition and performance characteristics

Option	Description
A	Thin asphalt overlay over stabilized subgrade
B	Standard flexible pavement (hot-mix asphalt layers)
C	Composite pavement (asphalt over PCC base)
D	Rigid pavement (thick jointed plain concrete slab)

Table 5.30 Performance scenarios and associated probabilities for each pavement design alternative, reflecting long-term durability and rehabilitation requirements over a 20-year period

Outcome	Description	A (%)	B (%)	C (%)	D (%)
High	No major distress for 20 years	50	60	85	95
Moderate	Surface rehab in year 10	40	30	15	5
Poor	Major rehab within first 5 years	10	10	0	0

Table 5.31 Lifecycle cost estimates for pavement design alternatives across performance outcomes, including initial construction and repair costs per lane-kilometer, expressed in thousands of Canadian dollars

Option	Outcome	Initial cost	Repair cost	Total cost
A	High	600	0	600
A	Moderate	600	150	750
A	Poor	600	500	1100
B	High	700	0	700
B	Moderate	700	100	800
B	Poor	700	300	1000
C	High	900	0	900
C	Moderate	900	80	980
D	High	1100	0	1100
D	Moderate	1100	100	1200

Table 5.32 Expected lifecycle costs of pavement design alternatives, incorporating weighted performance outcomes and associated repair costs, expressed in thousands of Canadian dollars per lane-kilometer

Option	Description	Expected cost (CAD $1,000 s)
A	Thin asphalt overlay	710
B	Standard flexible pavement	760
C	Composite pavement	912
D	Rigid pavement	1105

Table 5.33 Sensitivity analysis of expected lifecycle cost for Pavement Option A under varying probabilities of poor and moderate performance outcomes

Poor %	Moderate %	EC_A (Expected cost)
10%	40% (base)	$710
15%	35%	$735
20%	30%	$760
25%	25%	$785

Questions to Address

1. What is the expected lifecycle cost of each pavement alternative?
2. Which pavement design is the most cost-effective under current assumptions?
3. How does the decision respond to uncertainty in early failure probability?
4. At what point would another option become preferable?
5. What are the implications for pavement selection and long-term risk?

Step 1: Define the Expected Cost Formula

The expected lifecycle cost is calculated using Eq. (5.2)

Step 2: Calculate Expected Cost for Each Design

Option A (Thin Overlay)

$$EC_A = 0.50 \times 600 + 0.40 \times 750 + 0.10 \times 1100 = 300 + 300 + 110 = 710$$

Option B (Standard Flexible Pavement)

$$EC_B = 0.60 \times 700 + 0.30 \times 800 + 0.10 \times 1000 = 420 + 240 + 100 = 760$$

Option C (Composite Pavement)

$$EC_C = 0.85 \times 900 + 0.15 \times 980 = 765 + 147 = 912$$

Option D (Rigid Pavement)

$$EC_D = 0.95 \times 1100 + 0.05 \times 1200 = 1045 + 60 = 1105$$

Step 3: Summarize and Compare Results

Preferred Option: Option A has the lowest expected lifecycle cost.

Step 4: Sensitivity Insight

Option A is cost-effective, but it has 10% risk of early major rehabilitation.

Let's assess what happens if the probability of poor performance increases (Table 5.33).

At 20% poor performance, Option A's expected cost equals Option B's.

Step 5: Engineering Implications

- Option A is the most economical under current conditions.
- If there's greater risk of poor subgrade or traffic overloading, the probability of early failure could increase. At >20% poor performance, Option B becomes a better choice.
- Option C and D are robust but too costly unless:
  - There are extremely high truck volumes,
  - Minimal maintenance windows,
  - Or the project prioritizes service life over cost.

For rural roads with moderate risk and budget constraints, Option A is preferred. For roads in uncertain soil conditions or high load zones, consider Option B or hybrid designs.

Answers to the Five Key Questions

1. What is the expected lifecycle cost of each pavement alternative?

   Table 5.32 shows the answered of this question.
2. Which pavement design is the most cost-effective under current assumptions?

   Option A (thin overlay) is the most cost-effective, offering the lowest expected cost with acceptable performance in a moderate-risk context.
3. How does the decision respond to uncertainty in early failure probability?
   - As the probability of early failure increases, the expected cost of Option A also increases.
   - At 20% poor performance, Option A = Option B.
   - If risk of failure is underestimated or field conditions worsen, Option A may become less favorable.
4. At what point would another option become preferable?
   - If probability of poor performance for Option A exceeds 20%, then Option B becomes more economical.
   - This provides a clear risk threshold for switching designs.
5. What are the implications for pavement selection and long-term risk?
   - Choose Option A when subgrade is well-prepared and traffic growth is predictable.
   - Use Option B in uncertain conditions or where longer performance certainty is needed.
   - Options C and D offer longer life but should only be used when cost is less of a constraint or in strategic corridors.

5.10.3.2 Two-Way Sensitivity Analysis

Two-way sensitivity analysis is an advanced technique in decision tree modeling used to evaluate the combined influence of two uncertain input parameters on a decision outcome. Unlike one-way sensitivity analysis, which varies a single parameter while holding all others constant, two-way sensitivity analysis explores how different combinations of two parameters—such as probabilities, costs, benefits, or discount rates—impact the expected utility or cost of each decision alternative. This allows decision-makers to understand not only the individual influence of each variable but also their interaction effects, which may significantly alter the preferred course of action.

The analysis is typically visualized using 2D decision maps, heatmaps, or contour plots, where regions are shaded according to which decision alternative yields the lowest expected cost or highest expected utility for a given parameter pair. This enables the identification of threshold boundaries where decision preferences change, as well as robust regions where the same option remains optimal across a wide range of values. In civil engineering, where uncertainties often stem from environmental risks, economic volatility, or evolving regulatory standards, two-way sensitivity analysis provides a more comprehensive understanding of risk trade-offs than one-dimensional methods.

Two-way sensitivity analysis becomes especially valuable in situations where two input parameters are highly uncertain and difficult to estimate precisely, are interdependent or correlated—such as the relationship between natural hazard probability and associated damages—and jointly influence the project's economic outcomes or safety performance. These conditions highlight the importance of evaluating the combined effects of uncertain variables to better support robust decision-making. Figure 5.56 illustrates the key conditions under which two-way sensitivity analysis matters.

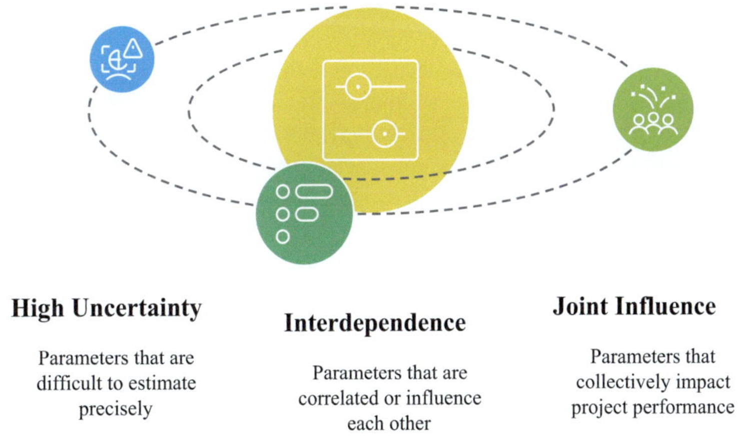

Fig. 5.56 Conditions under which two-way sensitivity analysis matters

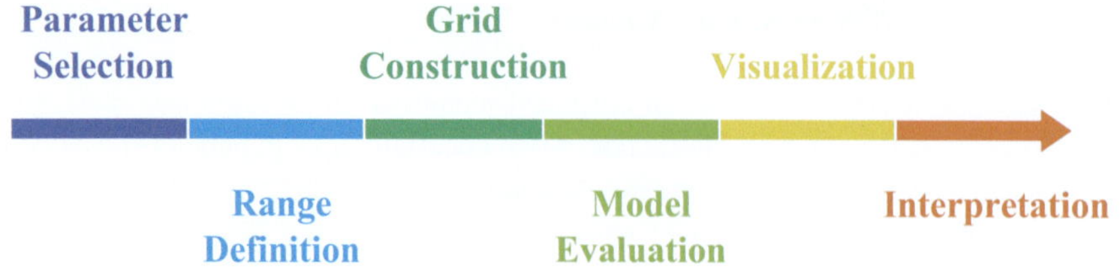

Fig. 5.57 Two-way sensitivity analysis process

This method is widely applied in risk-based infrastructure planning, cost–benefit evaluations, and resilience assessments, where decision-makers must consider the compound effects of natural variability and economic uncertainty.

The two-way sensitivity analysis process consists of a structured sequence of steps designed to evaluate how variations in two key parameters influence decision outcomes. As illustrated in Fig. 5.57, the first step is parameter selection, where two critical and potentially influential variables—such as flood frequency and associated damage costs—are identified based on the decision context. This is followed by range definition, in which realistic value ranges for each parameter are established using historical observations, expert judgments, or future scenario projections. Next, a parameter grid is constructed—commonly in the form of a 10×10 matrix—to represent a spectrum of possible combinations, resulting in multiple simulation runs.

At each grid point, the model undergoes evaluation to recalculate the expected outcomes (e.g., cost, utility, or risk level) for all competing alternatives. The results are then synthesized through visualization techniques, such as heatmaps or contour plots, which make it easier to identify optimal choices across the entire range of input combinations. Finally, the interpretation step focuses on analyzing the structure and boundaries of decision regions, highlighting zones of preference shifts and thresholds beyond which one alternative becomes more favorable than another.

This systematic approach not only enhances understanding of the decision landscape but also helps civil engineers identify break-even regions, where the final choice is sensitive to simultaneous changes in both variables. Compared to one-way analysis, this approach captures the combined effect of uncertainty in two variables at the same time. Furthermore, it enables the detection of high-risk or high-opportunity zones, providing insights into which parameter combinations are likely to yield the most critical outcomes in terms of safety, cost, or performance.

In civil engineering, two-way sensitivity analysis is especially valuable for projects where decision outcomes are shaped by the combined uncertainty of two dominant factors, such as environmental risks and economic costs. Infrastructure systems are often exposed to natural hazards—floods, earthquakes, extreme weather—whose probabilities and impacts are not easily predicted in isolation. By simultaneously varying two uncertain parameters, engineers can assess not only the robustness of a preferred design but also the conditions under which alternative strategies may become more effective or economical. This dual-parameter approach enables a deeper understanding of trade-offs and interactions that govern project

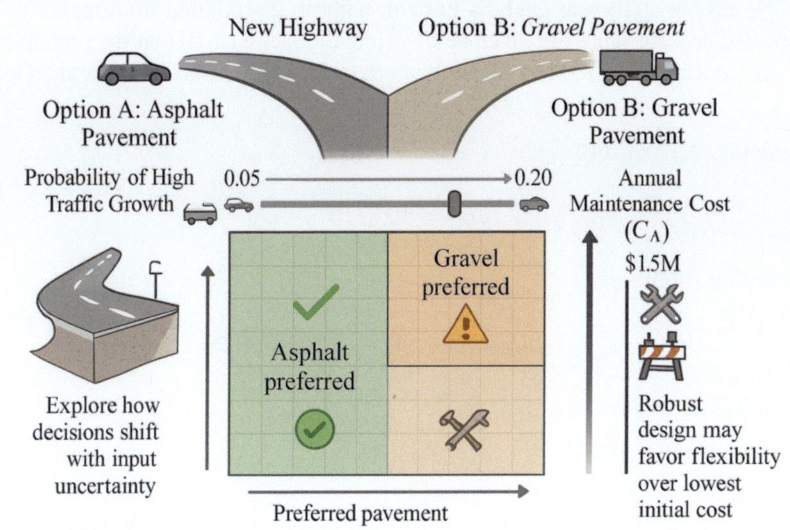

Fig. 5.58 Conceptual framework for two-way sensitivity analysis in pavement selection

viability, allowing for better-informed planning and more resilient infrastructure investments. Whether selecting structural reinforcements, evaluating flood protection measures, or deciding on long-term maintenance plans, two-way sensitivity analysis serves as a practical decision-support tool that connects engineering design to real-world variability and risk.

Two-Way Sensitivity Analysis in Pavement Selection for a Highway Project

A transportation agency must decide between two pavement types for a new highway section (Fig. 5.58):

- Option A: Asphalt Pavement
  - Initial construction cost: $10 million
  - Annual maintenance cost if high traffic growth occurs: $C_A \in [0.5, 1.5]$ in millions of dollars
- Option B: Gravel Pavement
  - Initial construction cost: $6 million
  - Annual maintenance cost if high traffic growth occurs: fixed at $2.5 million.

Uncertain Parameters

- p: Probability of high traffic growth, ranging from 0.05 to 0.2
- C_A: Annual maintenance cost for asphalt, ranging from $0.5M to $1.5M.

We assume a 10-year analysis horizon with no discounting for simplicity. The expected cost formulas are given below. Note the factor of 10 that the probabilistic term is multiplied by to account for the horizon period in both Option A and Option B.

- Option A (Asphalt):

$$EU_A = 10 + 10 \cdot p \cdot C_A$$

- Option B (Gravel):

$$EU_B = 6 + 10 \cdot p \cdot 2.5 = 6 + 25p$$

Step-by-Step Analysis.

1. Grid Construction

   Vary:

   - $p \in \{0.05, 0.1, 0.15, 0.2\}$
   - $C_A \in \{0.5, 1.0, 1.5\}$.

This results in 12 parameter combinations (4 × 3).

2. Cost Calculation Examples.

   - $p = 0.05$, $C_A = 0.5$:

$$EU_A = 10 + 10 \cdot 0.05 \cdot 0.5 = 10.25$$

$$EUB = 6 + 25 \cdot 0.05 = 7.25 \rightarrow \text{Choose Option B}$$

   - $p = 0.2$, $C_A = 0.5$:

$$EU_A = 10 + 10 \cdot 0.2 \cdot 0.5 = 11$$

$$EU_B = 6 + 25 \cdot 0.2 = 11 \rightarrow \text{Tie}$$

   - $p = 0.2$, $C_A = 1.5$:

$$EU_A = 10 + 10 \cdot 0.2 \cdot 1.5 = 13$$

$$EU_B = 11 \rightarrow \text{Choose Option B}$$

To compute all 12 combinations of p and C_A, one can conveniently use either Excel or Python. In Excel, simply create two columns listing all combinations of p and C_A and then define equations for the expected cost of each option using the following syntax:

- EUA = 10 + 10 * p * CA
- EUB = 6 + 25 * p.

These formulas can be applied using Excel's cell referencing to populate EUA and EUB for each parameter combination. However, using Python can be more efficient, especially for larger grids or more complex extensions. With a basic comprehension of simple loops or list algorithms, you can iterate through the parameter ranges and compute the results programmatically.

After computing EUA and EUB across all combinations, the output can be conveniently visualized using heatmaps as a means to ease interpretation of the results. Moreover, calculating the ratio EU_A/EU_B provides an easy metric to interpret the relative cost-effectiveness:

- Values greater than 1 indicate that Asphalt (Option A) is more expensive.
- Values less than 1 imply that Gravel (Option B) incurs higher expected costs.

This visualization supports better-informed decision-making under uncertainty. The syntax for producing a heatmap displaying the results of the comparison is provided in Code 5.1. Lines 1–4 import the essential libraries used in the anal ysis: `pandas` is used for managing and storing tabular data, `seaborn` and `matplotlib.pyplot` are used for creating advanced plots like heatmaps, and `itertools` is used to generate all possible combinations of the input parameters in a compact and efficient way.

Lines 7–8 define the two key variables for the sensitivity analysis: the probability `p` of high traffic growth, and the possible values of the asphalt maintenance cost `CA`. These are specified as discrete sets of values—`p` ranges from 0.05 to 0.2, while `CA` ranges from \$0.5M to \$1.5M, representing the potential variability in pavement deterioration under heavy traffic. It should be noted that the user can simply alter the numerical value of these variables based on the relevant values for similar problems.

Line 11 uses `itertools.product` to generate a full grid of (p, CA) combinations—this Cartesian product yields 12 unique pairs, covering all possible scenarios within the defined ranges.

Lines 14–19 define a loop to compute the expected cost of each pavement option across the parameter grid. For each (p, CA) pair, the expected cost for asphalt pavement (`EUA`) is calculated using the formula: `EUA = 10 + 10 * p * CA`, reflecting the \$10M base cost plus the expected value of maintenance costs over 10 years. For gravel pavement, the expected cost (`EUB`) is computed as `EUB = 6 + 25 * p`, derived from \$6M base cost plus 10 years of fixed \$2.5M maintenance under high traffic. Additionally, the ratio `UCA/UCB = EUA/EUB` is computed to support comparative evaluation.

Line 22 converts the computed results into a `pandas.DataFrame`, labeling the columns appropriately for later analysis and plotting. Special formatting using LaTeX-like syntax (e.g., `C$_A$`) ensures that these column names appear with the correct style in the generated plots with, in this case, the underscore key being used to indicate subscript text.

Lines 25–26 create a simplified version of the DataFrame for exporting to Excel, using plain-text column names to avoid formatting issues. The `.to_excel()` function saves the table to an external `.xlsx` file named using the first input argument, allowing stakeholders to inspect or use the results in other tools like Excel.

Lines 29–31 transform the DataFrame into three separate pivot tables for heatmap visualization. Each table reshapes the data so that `p` values become the rows and `CA` values the columns, while the cell values contain `EUA`, `EUB`, and the ratio `EUA/EUB` respectively. This layout is ideal for two-dimensional visual comparisons.

Line 34 uses the function `plt.rcParams.update` to control the default aesthetic properties of all plots in Python. Here, the first input argument identifies the font family to be Times New Roman and the second argument sets the font size to 16, ensuring visual consistency and readability in figures intended for publication or presentations.

Line 37 initializes a `matplotlib` figure with three side-by-side subplots using `plt.subplots()`. The figure is set to a width of 20 and height of 6 inches to comfortably accommodate all three heatmaps with separate colorbars.

Line 40 is used to generate the heatmap for Option A's expected cost (`EUA`) using the `sns.heatmap()` function. Annotations are enabled to show numeric values inside each cell, with the Oranges-Red (`'OrRd'`) colormap enhancing visual gradients. In Lines 41–43 the axes are labeled and titled with math-formatted strings. Lines 46–49 produce a similar heatmap for Option B (`EUB`), maintaining consistent styling and structure.

Lines 52–55 create the third heatmap, this time for the ratio of expected utilities (`EUA/EUB`). This ratio helps interpret relative cost-efficiency: a value less than 1 indicates that asphalt is cheaper, greater than 1 suggests gravel is more economical, and exactly 1 represents cost parity.

Line 57 applies tight layout adjustments to prevent overlapping of labels and elements across the three plots.

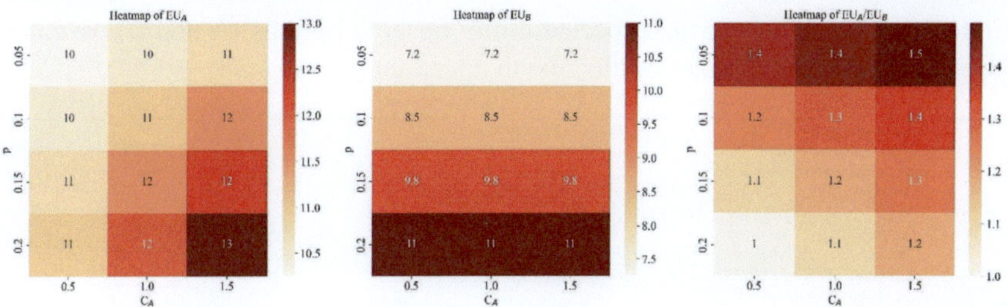

Fig. 5.59 Heatmaps of expected costs and cost ratio for pavement options under two-way sensitivity analysis

	A	B	C	D	E
1	p	CA	EUA	EUB	EUA/EUB
2	0.05	0.5	10.25	7.25	1.413793
3	0.05	1	10.5	7.25	1.448276
4	0.05	1.5	10.75	7.25	1.482759
5	0.1	0.5	10.5	8.5	1.235294
6	0.1	1	11	8.5	1.294118
7	0.1	1.5	11.5	8.5	1.352941
8	0.15	0.5	10.75	9.75	1.102564
9	0.15	1	11.5	9.75	1.179487
10	0.15	1.5	12.25	9.75	1.25641
11	0.2	0.5	11	11	1
12	0.2	1	12	11	1.090909
13	0.2	1.5	13	11	1.181818
14					

Fig. 5.60 Excel output table of all parameter combinations and calculated metrics

Line 58 displays the figure, rendering the heatmaps in the output window or notebook interface.

Lines 61–62 save the final figure in both JPEG and TIFF formats at 300 dpi. These high-resolution formats are suitable for inclusion in reports, slides, or publications.

The output of this code includes two primary components: (1) a figure consisting of three heatmap subplots (Fig. 5.59) which visually display the expected costs of both pavement options and their ratio across all parameter combinations; and (2) an Excel file containing the full dataset of computed values, systematically organized and exported for further analysis (Fig. 5.60) provides a representation of this file.

Code 5.1 Calculation of the expected utility for both options (A and B) pavement type decision-making under traffic growth uncertainty

```
import pandas as pd
import seaborn as sns
import matplotlib.pyplot as plt
import itertools

# Define the sets
p_values = [0.05, 0.1, 0.15, 0.2]
CA_values = [0.5, 1.0, 1.5]

# Generate all combinations
combinations = list(itertools.product(p_values, CA_values))

# Calculate EUA, EUB, and UCA/UCB
results = []
for p, CA in combinations:
    EUA = 10 + 10 * p * CA
    EUB = 6 + 25 * p
    UCA_UCB = EUA / EUB
    results.append([p, CA, EUA, EUB, UCA_UCB])

# Create DataFrame
df = pd.DataFrame(results, columns=['p', 'C$_A$', 'EU$_A$', 'EU$_B$', 'EU$_A$/EU$_B$'])

# Save to Excel with simplified column names for Excel export
excel_df = pd.DataFrame(results, columns=['p', 'CA', 'EUA', 'EUB', 'EUA/EUB'])
excel_df.to_excel('EUA_EUB_results.xlsx', index=False)

# Pivot for heatmap (p as rows, CA as columns)
pivot_eua = df.pivot(index='p', columns='C$_A$', values='EU$_A$')
pivot_eub = df.pivot(index='p', columns='C$_A$', values='EU$_B$')
pivot_uca_ucb = df.pivot(index='p', columns='C$_A$', values='EU$_A$/EU$_B$')

# Set font globally
plt.rcParams.update({'font.family': 'Times New Roman', 'font.size': 16})

# Plot all heatmaps in one figure with separate color bars
fig, axs = plt.subplots(1, 3, figsize=(20, 6))

# EU_A heatmap
sns.heatmap(pivot_eua, annot=True, cmap='OrRd', ax=axs[0], cbar=True, annot_kws={"size": 16})
axs[0].set_title('Heatmap of EU$_A$', fontsize=16)
axs[0].set_xlabel('C$_A$', fontsize=16)
axs[0].set_ylabel('p', fontsize=16)

# EU_B heatmap
sns.heatmap(pivot_eub, annot=True, cmap='OrRd', ax=axs[1], cbar=True, annot_kws={"size": 16})
axs[1].set_title('Heatmap of EU$_B$', fontsize=16)
axs[1].set_xlabel('C$_A$', fontsize=16)
axs[1].set_ylabel('p', fontsize=16)

# UCA/UCB heatmap
sns.heatmap(pivot_uca_ucb, annot=True, cmap='OrRd', ax=axs[2], cbar=True, annot_kws={"size": 16})
axs[2].set_title('Heatmap of EU$_A$/EU$_B$', fontsize=16)
axs[2].set_xlabel('C$_A$', fontsize=16)
axs[2].set_ylabel('p', fontsize=16)

plt.tight_layout()
plt.show()

# Save the figure in JPEG and TIF formats only
fig.savefig('combined_heatmaps_individual_colorbars.jpg', format='jpeg', dpi=300)
fig.savefig('combined_heatmaps_individual_colorbars.tif', format='tiff', dpi=300)
```

Boundary Equation

To find the boundary line where both options have equal expected cost:

$$EU_A = EU_B$$

$$10 + 10pC_A = 6 + 25p$$

$$C_A = \frac{25p - 4}{10p}$$

$$C_A = 2.5 - \frac{0.4}{p}$$

This defines a critical boundary curve:

- At $p = 0.2$: $C_A = 2.5 - 2 = 0.5$
- At $p = 0.1$: $C_A = 2.5 - 4 = -1.5 \rightarrow$ Infeasible.

Only at high p and low C_A does asphalt become competitive.

Result Interpretation

- Dominance Region: Gravel (Option B) dominates most of the parameter space.
- Indifference Point: At $p = 0.2$, $C_A = 0.5$, both options have equal cost.
- Sensitivity: Given the current situation, the decision was not sensitive to the probability of high traffic growth over a 10 year period as the cost was equal, or lower, for gravel construction in all cases. It is expected that should the asphalt maintenance costs change, the decision will become sensitive to the traffic growth.
- Practical Insight: To favor asphalt, the agency must ensure low maintenance costs and face high traffic growth.

Implications

- Data Prioritization: Improve forecasts of traffic growth probability and maintenance cost estimates.
- Design Strategy: Unless confident in low C_A, gravel pavement offers a robust and cost-effective solution.
- Visualization: A heatmap with p on the x-axis and C_A on the y-axis can illustrate decision regions, with a curve $C_A = 2.5 - 0.4/p$ marking the boundary.

This example highlights the power of two-way sensitivity analysis in civil engineering decisions where uncertainty exists in both demand projections and long-term maintenance economics. By examining the interaction between traffic growth and maintenance costs, engineers can better design resilient infrastructure and optimize lifecycle investments. The method provides a structured framework not only for selecting the best alternative under uncertainty but also for identifying where improved data can most influence the decision. This strategic clarity is essential in modern infrastructure planning.

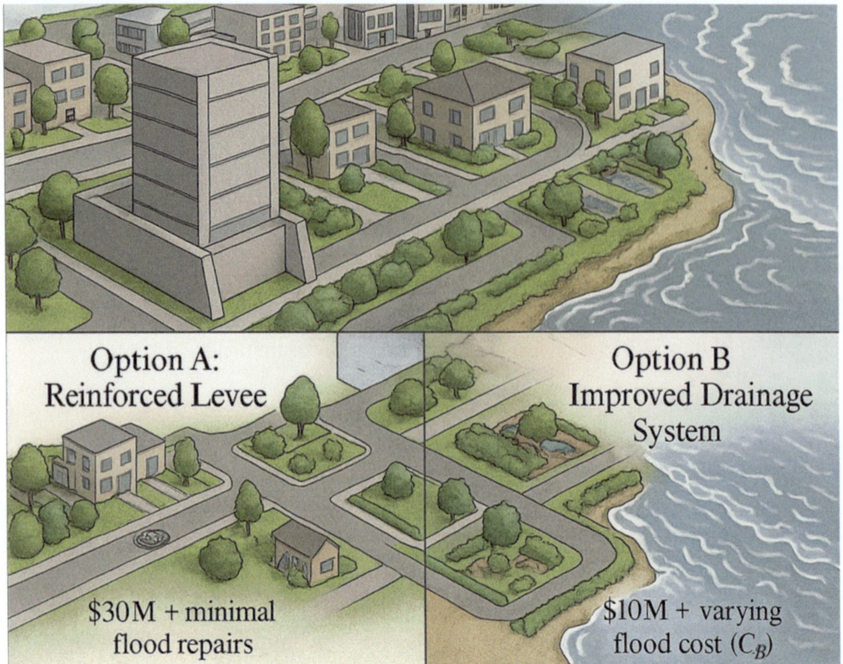

Fig. 5.61 Conceptual setup of two-way sensitivity analysis for flood protection decisions

Two-Way Sensitivity Analysis of Flood Protection Strategies Under Uncertainty in Flood Probability and Damage Cost

A coastal city is deciding whether to build a reinforced levee (Option A) or implement improved natural drainage systems (Option B) to protect against flooding. The decision depends on two uncertain parameters: the annual probability of a severe flood and the cost of flood damage if the drainage system fails to prevent damage. The levee is more expensive to construct but significantly reduces flood damage, while the drainage system is cheaper but less effective, leading to higher damage costs if a flood occurs. The city aims to choose the option with the lowest expected cost using two-way sensitivity analysis to assess how the decision is affected by simultaneous variations in flood probability and damage cost. As illustrated in Fig. 5.61, a two-way sensitivity framework helps to determine which flood protection strategy is more cost-effective under different combinations of flood risk probability and flood damage severity.

Decision Tree Structure

- Chance Node: A severe flood occurs with probability p, or no flood occurs with probability (1 − p).
- Option A (Levee):

  - Construction cost: $30 million.
  - Repair cost if flood occurs: $1 million (levee mitigates most damage).
- Option B (Drainage):
  - Construction cost: $10 million.
  - Repair cost if flood occurs: C_B million (variable, as drainage is less effective).
- Parameters for Sensitivity Analysis:
  - Annual flood probability p, estimated at 0.05 (5%) but ranging from 0.01 (1%) to 0.1 (10%) based on hydrological models and climate projections.
  - Flood damage cost for Option B (C_B), estimated at $15M but ranging from $10M to $20M due to uncertainty in property values, infrastructure damage, and economic impacts.

Questions to Address

1. How does the expected cost of each option vary as flood probability p (0.01 to 0.1) and damage cost C_B ($10M to $20M) change simultaneously?
2. What combinations of p and C_B result in Option A (Levee) or Option B (Drainage) being optimal, and where is the boundary between these decision regions?
3. How sensitive is the decision to the combined effects of flood probability and damage cost, and what does this imply for data collection or design choices?
4. What are the practical implications of the analysis for the city's decision-making process?

Step 1: Define Expected Cost Equations

Option A (Levee)

- No flood (1 − p): Cost = $30M (construction only)
- Flood (p): Cost = $30M + $1M = $31M
- Expected Cost (EU_A):

$$EU_A = (1 - p) \cdot 30 + p \cdot 31 = 30 + p$$

Option B (Drainage)

- No flood (1 − p): Cost = $10M (construction only)
- Flood (p): Cost = $10M + C_B
- Expected Cost (EU_B):

$$EU_B = (1 - p) \cdot 10 + p \cdot (10 + C_B) = 10 + p \cdot C_B$$

Step 2: Define the Parameter Grid

Use a 5 × 5 grid of key values:

- Flood probability (p): [0.01, 0.03, 0.05, 0.07, 0.10]
- Flood damage cost C_B: [$10M, $12.5M, $15M, $17.5M, $20M].

Step 3: Calculate Expected Costs and Decisions

We compute expected costs and compare both options for each parameter combination. Two sample calculations are provided below for illustrative purposes:

Sample Calculations:

Example 1: $p = 0.01$, $C_B = 10$

$$EU_A = 30 + 0.01 = 30.01M,\ EU_B = 10 + 0.01 \cdot 10 = 10.10M \Rightarrow \text{Option B is better}$$

Example 2: $p = 0.10$, $C_B = 20$

$$EU_A = 30 + 0.10 = 30.10M,\ EU_B = 10 + 0.10 \cdot 20 = 12.00M \Rightarrow \text{Option B is better}$$

This calculation is repeated for all 25 combinations of p and C_B with the full data table confirming that Option B (Drainage) is preferred in all cases (Table 5.34). In practice, a finer grid (e.g., 50 × 50) would provide smoother boundaries.

Step 4: Derive the Decision Boundary

To find the cutoff line where both options have the same cost:

$$EU_A = EU_B$$

$$30 + p = 10 + p \cdot C_B$$

$$p \cdot C_B = 20 + p$$

$$C_B = 20/p + 1$$

Using this relationship, the threshold value of C_B at which the two options are equal for various key values of p are calculated and summarized in Table 5.35.

The breakeven cost C_B is far outside the \$10M–\$20M range, meaning Option B is always more cost-effective within realistic conditions.

Step 5: Interpretation and Recommendations

For all combinations of $p \in [0.01, 0.10]$ and $C_B \in [10M, 20M]$, Option B (Drainage) yields the lowest expected cost.

The city should not invest in a levee unless:

- Flood probability increases drastically, or
- Damage costs exceed \$200M per event.

Table 5.34 Comparison of expected costs for flood protection Options A and B under varying flood probabilities and damage cost assumptions

Flood probability, p	Damage cost C_B (M $)	Expected cost A (30 + p)	Expected cost B ($10 + p \cdot C_B$)	Preferred option
0.01	10	30.01	10.1	B (Drainage)
0.01	12.5	30.01	10.13	B (Drainage)
0.01	15	30.01	10.15	B (Drainage)
0.01	17.5	30.01	10.18	B (Drainage)
0.01	20	30.01	10.2	B (Drainage)
0.03	10	30.03	10.3	B (Drainage)
0.03	12.5	30.03	10.38	B (Drainage)
0.03	15	30.03	10.45	B (Drainage)
0.03	17.5	30.03	10.53	B (Drainage)
0.03	20	30.03	10.6	B (Drainage)
0.05	10	30.05	10.5	B (Drainage)
0.05	12.5	30.05	10.63	B (Drainage)
0.05	15	30.05	10.75	B (Drainage)
0.05	17.5	30.05	10.88	B (Drainage)
0.05	20	30.05	11	B (Drainage)
0.07	10	30.07	10.7	B (Drainage)
0.07	12.5	30.07	10.88	B (Drainage)
0.07	15	30.07	11.05	B (Drainage)
0.07	17.5	30.07	11.23	B (Drainage)
0.07	20	30.07	11.4	B (Drainage)
0.1	10	30.1	11	B (Drainage)
0.1	12.5	30.1	11.25	B (Drainage)
0.1	15	30.1	11.5	B (Drainage)
0.1	17.5	30.1	11.75	B (Drainage)
0.1	20	30.1	12	B (Drainage)

Table 5.35 Threshold damage cost values (C_B) at which the expected costs of flood protection Options A and B are equal

p	Threshold C_B
0.01	2001
0.03	668.33
0.05	401
0.07	286.57
0.1	201

Based on this analysis, the four questions at the beginning of the problem may now be answered.

Question 1: How does the expected cost of each option vary as flood probability p and damage cost C_B change simultaneously?

Option A (Levee): The expected cost increases linearly with flood probability p, with the formula: $EU_A = 30 + p$.

Since the levee significantly reduces flood damage, the impact of increasing p is minimal (only + \$1M per 100% increase in p).

For Option B (Drainage), the expected cost increases as a function of both the flood probability, p, and the damage costs, C_B according to the relationship given by:

$$EU_B = 10 + p \cdot C_B$$

Thus, as either variable increases, the cost rises faster than in Option A, especially if both are high. However, for the tested ranges ($p \leq 0.10$, $C_B \leq \$20M$), the increase in EU_B remains modest, and Option B remains more economical throughout.

Question 2: What combinations of p and C_B result in Option A (Levee) or Option B (Drainage) being optimal, and where is the boundary between these decision regions?

Option B (Drainage) is always optimal within the entire grid tested:

- $p \in \{0.01, 0.03, 0.05, 0.07, 0.10\}$
- $C_B \in \{10, 12.5, 15, 17.5, 20\}$.

The decision boundary occurs where:

$$EU_A = EU_B \Rightarrow C_B = 20/p + 1$$

For instance:

- $p = 0.10 \Rightarrow CB = 201$
- $p = 0.05 \Rightarrow CB = 401$.

These threshold values are far outside the realistic range of C_B values (\$10M–\$20M) predicted for region of interest in this coastal city, so the boundary lies well beyond the domain of practical scenarios.

Question 3: How sensitive is the decision to the combined effects of flood probability and damage cost, and what does this imply for data collection or design choices?

Sensitivity is low in the examined range:

- Even with the maximum tested values ($p = 0.10$, $C_B = \$20M$), Option B is significantly cheaper.
- The decision only changes in extreme scenarios, like flood damages exceeding \$200M.

The implication of this low sensitivity is that there is no immediate need for highly precise estimates of p or C_B in the current planning window. However, if climate change projections worsen or urban expansion increases exposure, then:

- Accurate modeling of C_B and extreme flood probabilities becomes more important.
- Re-evaluating the decision may become necessary in the future.

Question 4: What are the practical implications of the analysis for the city's decision-making process?

The city should confidently proceed with Option B (Improved Drainage) based on cost-effectiveness under all tested conditions. Levee construction is not economically justified unless flood risk or damage estimates increase by an order of magnitude. The city can:

- Invest saved capital into complementary flood resilience strategies (e.g., early warning systems, zoning) or, alternatively, reduce debt servicing costs by borrowing less capital.
- Monitor changes in flood patterns or property exposure that might warrant future re-evaluation.
- Avoid overdesign and promote sustainable, nature-based infrastructure.

Two-Way Sensitivity Analysis in Bridge Retrofit Planning

A provincial transportation agency must decide between two risk management strategies for an aging highway bridge in a seismically active region. Option A involves retrofitting the bridge to current seismic standards at an upfront cost of $20 million, reducing potential earthquake damage to $5 million if such an event occurs, and incurring no cost if no earthquake happens. Option B adopts a monitoring and delay approach, requiring a $2 million investment in sensors and regular inspections, while accepting the risk that a damaging earthquake would result in the full cost of failure—denoted as C_B, which may range from $50 million to $150 million, depending on traffic volume, detour implications, and potential casualties. No cost is incurred under Option B if no earthquake occurs. The key uncertainties are the annual probability of a damaging earthquake (p), estimated between 1 and 10%, and the bridge failure cost C_B. The objective is to determine which option minimizes the expected lifecycle cost over a 1-year risk window, accounting for both economic efficiency and seismic risk exposure. Figure 5.62 illustrates the conceptual setup of the seismic decision problem, where the choice between proactive retrofitting and monitoring depends on earthquake risk (p) and the potential societal cost of bridge failure (C_B).

Questions to Address

1. How do the expected costs of each option change as earthquake probability p and failure cost C_B change simultaneously?

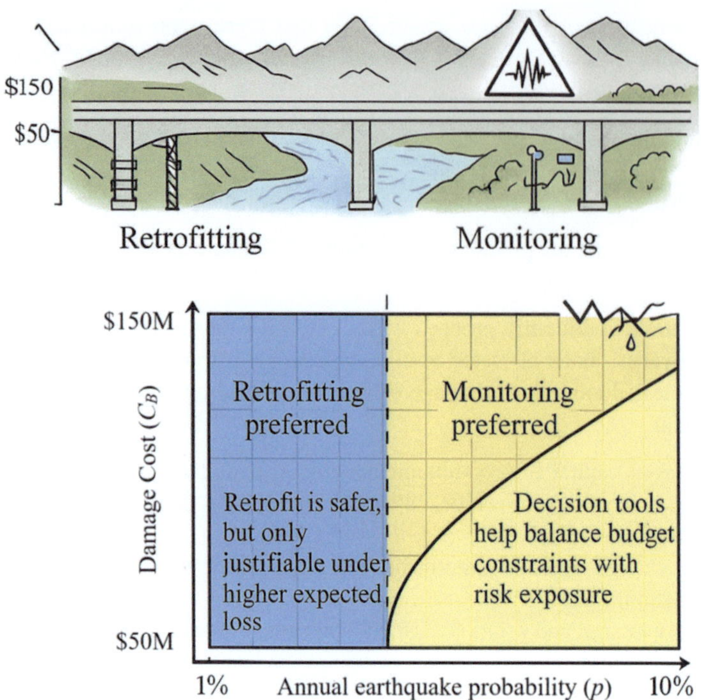

Fig. 5.62 Visual framework of two-way sensitivity analysis comparing seismic retrofit versus monitoring strategies for a critical highway bridge. The decision depends on both the annual probability of a damaging earthquake and the societal cost of potential failure

2. For which combinations of p and C_B is retrofitting (Option A) preferred over monitoring (Option B)?
3. How sensitive is the decision to these uncertainties, and what does this suggest about data priorities?
4. What are the practical implications for bridge management and infrastructure investment?

Step 1: Define Expected Cost Equations

Option A: Retrofit

- No earthquake (1 − p): Cost = \$20M
- Earthquake (p): Cost = \$20M + \$5M = \$25M
- Expected cost:

$$EU_A = (1 - p) \cdot 20 + p \cdot 25 = 20 + 5p$$

Option B: Monitor and Delay

- No earthquake (1 − p): Cost = \$2M

- Earthquake (p): Cost = \$2M + CBC_BCB
- Expected cost:

$$EUB = (1 - p) \cdot 2 + p \cdot (2 + C_B) = 2 + p \cdot CB$$

Step 2: Define the Parameter Grid

Use a 5 × 5 grid for illustrative coverage:

- Earthquake probability p: [0.01, 0.03, 0.05, 0.07, 0.10]
- Bridge failure cost C_B: [\$50M, \$100M, \$125M, \$150M, \$200M].

Step 3: Calculate Expected Costs and Optimal Decision (Table 5.36).

Result: Option B dominates the entire 5 × 5 grid, except when failure cost is \$200M at high p.

Step 4: Find the Decision Boundary (Table 5.37).

$$EU_A = EU_B$$

$$20 + 5p = 2 + p \cdot C_B$$

$$p \cdot C_B = 18 + 5p$$

$$C_B = (18 + 5p)/p$$

$$C_B = 18/p \mid 5$$

All break-even C_B values exceed \$150M for all $p \leq 0.10$, so Option B is always preferred in this scenario unless failure cost exceeds \$185M at p = 0.10.

Step 5: Interpretation and Practical Recommendations

Answer to Question 1:

- Option A (Retrofit) increases linearly with p.
- Option B (Monitor) is more sensitive, increasing with both p and C_B.

Still, B remains cheaper within tested values.

Answer to Question 2:

- Option A is only optimal if: CB > 18/p + 5

For example, if p = 0.10, then C_B must exceed \$185M.

- For CB ≤ 150M, Option B is always cheaper.

Table 5.36 Expected utility cost comparison between retrofitting (Option A) and monitoring (Option B) under varying flood probabilities and damage cost values (C_B), identifying the preferred option based on cost-effectiveness

p	C_B	$EU_A = 20 + 5p$	$EU_B = 2 + p{\cdot}CB$	Preferred option
0.01	50	20.05	2.5	B (Monitor)
0.01	100	20.05	3	B (Monitor)
0.01	125	20.05	3.25	B (Monitor)
0.01	150	20.05	3.5	B (Monitor)
0.01	200	20.05	4	B (Monitor)
0.03	50	20.15	3.5	B (Monitor)
0.03	100	20.15	5	B (Monitor)
0.03	125	20.15	5.75	B (Monitor)
0.03	150	20.15	6.5	B (Monitor)
0.03	200	20.15	8	B (Monitor)
0.05	50	20.25	4.5	B (Monitor)
0.05	100	20.25	7	B (Monitor)
0.05	125	20.25	8.25	B (Monitor)
0.05	150	20.25	9.5	B (Monitor)
0.05	200	20.25	12	B (Monitor)
0.07	50	20.35	5.5	B (Monitor)
0.07	100	20.35	9	B (Monitor)
0.07	125	20.35	10.75	B (Monitor)
0.07	150	20.35	12.5	B (Monitor)
0.07	200	20.35	16	B (Monitor)
0.1	50	20.5	7	B (Monitor)
0.1	100	20.5	12	B (Monitor)
0.1	125	20.5	14.5	B (Monitor)
0.1	150	20.5	17	B (Monitor)
0.1	200	20.5	22	A (Retrofit)

Table 5.37 Break-even damage cost values (C_B) at which the expected costs of retrofitting and monitoring become equal, across different flood probabilities pp

p	Break-even C_B
0.01	1805
0.03	605
0.05	365
0.07	263.6
0.1	185

Answer to Question 3:

The decision is robust unless seismic risk or consequences are severely underestimated. If future seismic modeling shows increases in either p or failure impact, data collection and reassessment become essential.

Answer to Question 4: The agency should choose Option B (Monitor & Delay) under current risk levels. Continue tracking urban growth and updated seismic forecasts to detect when a retrofit becomes necessary. A full retrofit may only be justified under catastrophic cost forecasts or updated risk zones.

Two-Way Sensitivity Analysis in Foundation Design Under Seismic Loading

In designing the foundation for a critical structure under seismic loading, a geotechnical engineer must choose between Option A, a pile foundation, and Option B, a shallow mat foundation with ground improvement. This decision is governed by two key uncertainties: the probability of liquefaction-induced failure (p) during the structure's design life, estimated between 1 and 10%, and the potential cost of failure (C_B) for the shallow foundation, which can range from \$5 million to \$25 million, depending on site accessibility, structural importance, and recovery delays. Option A involves a \$6 million construction cost and \$0.5 million in repairs if liquefaction occurs, with no cost if it doesn't. Option B requires only \$3 million upfront but bears the full failure cost C_B if liquefaction occurs, and no cost otherwise. A two-way sensitivity analysis is necessary to assess how changes in both p and C_B influence the optimal decision. The objective is to identify the foundation option that minimizes expected lifecycle cost under uncertain seismic conditions, ensuring both cost-efficiency and structural resilience. Figure 5.63 provides a visual overview of the foundation decision problem, highlighting the trade-off between higher upfront investment in pile foundations and the potential repair and societal costs associated with shallow mat foundations in liquefiable soils under seismic loading.

Questions to Address

1. How do expected costs vary with liquefaction probability p and failure cost C_B?
2. For which combinations of p and C_B is Option A (Pile) or Option B (Mat + Ground Improvement) preferred?
3. How sensitive is the decision to seismic risk and potential failure consequences?
4. What are the practical design implications for geotechnical risk management?

Step 1: Define Expected Cost Equations

Option A: Pile Foundation

- No liquefaction (1 − p): Cost = \$6M
- Liquefaction (p): Cost = \$6M + \$0.5M = \$6.5M
- Expected cost:

$$EU_A = (1 - p) \cdot 6 + p \cdot 6.5 = 6 + 0.5p$$

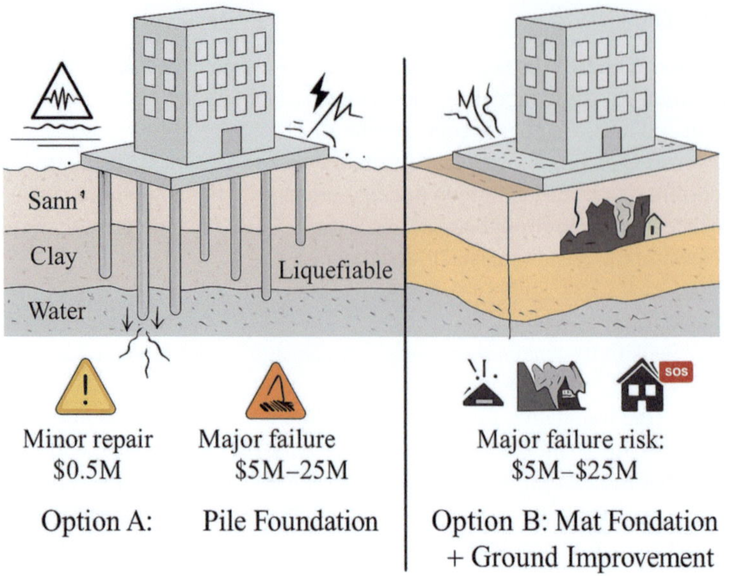

Fig. 5.63 Conceptual illustration of a two-way sensitivity analysis comparing pile foundations and shallow mat foundations with ground improvement under liquefaction risk

Option B: Mat Foundation + Ground Improvement

- No liquefaction (1 − p): Cost = $3M
- Liquefaction (p): Cost = $3M + CBC_BCB
- Expected cost:

$$EU_B = (1 - p) \cdot 3 + p \cdot (3 + C_B) = 3 + p \cdot C_B$$

Step 2: Define the Parameter Grid

Use a 5 × 10 grid for simplicity:

- Liquefaction probability p: [0.01, 0.03, 0.05, 0.07, 0.10]
- Failure cost C_B: [$5M, $10M, $15M, $20M, $25M, $25M, $30M, $35M, $40M, $45M].

Step 3: Calculate Expected Costs and Optimal Decision

Based on the provided results in Table 5.38, option B (Mat foundation) is consistently preferred across most of the parameter space, especially at lower values of traffic growth probability (p) and construction cost (C_B). However, Option A (Pile foundation) begins to emerge as the more favorable choice when both the probability of high traffic growth reaches 10% (p = 0.10) and the associated cost of Option B exceeds $35 million. Thus, Option A only becomes optimal in high-cost, high-risk scenarios. This threshold behavior highlights the importance of jointly considering both uncertainty and cost escalation in foundation design decisions (Table 5.39).

Step 4: Find the Decision Boundary

$$EU_A = EU_B$$

$$6 + 0.5p = 3 + p \cdot C_B$$

$$p \cdot C_B = 3 + 0.5p$$

$$C_B = (3 + 0.5p)/p$$

$$C_B = 3/p + 0.5$$

Thus, for Option A to be preferable at $p = 0.10$, C_B must exceed \$30.5M.

Step 5: Interpretation and Practical Recommendations

Answer to Question 1:

- Option A is insensitive to both variables (adds only \$0.5M per unit p).
- Option B is highly sensitive to both p and C_B, but still remains cheaper for all realistic ranges.

Answer to Question 2:

- Option B (Mat + Ground Improvement) dominates for $C_B \leq 25$M and $p \leq 0.10$.
- Option A only becomes optimal when both p and damage cost are extreme, e.g., $C_B \geq 35$M and $p \geq 0.10$.

Answer to Question 3:

- The decision is not very sensitive for typical geotechnical uncertainty levels.
- No need for high-precision liquefaction models unless seismicity or criticality of structure increases.

Answer to Question 4:

Engineers can confidently use mat foundations with moderate ground improvement. Only consider piles for critical structures, very poor soils, or high-occupancy risks. Routine liquefaction risk monitoring and post-earthquake inspection should suffice for cost-effective resilience.

Table 5.38 Comparison of expected utility costs for retrofitting (Option A) and material monitoring (Option B) under varying probabilities p and cost of failure scenarios (C_B), showing the preferred strategy based on minimum expected cost

p	CB	$EU_A = 6 + 0.5p$	$EU_B = 3 + p \cdot C_B$	Preferred option
0.01	5	6.005	3.05	B (Mat)
0.01	10	6.005	3.1	B (Mat)
0.01	15	6.005	3.15	B (Mat)
0.01	20	6.005	3.2	B (Mat)
0.01	25	6.005	3.25	B (Mat)
0.01	30	6.005	3.3	B (Mat)
0.01	35	6.005	3.35	B (Mat)
0.01	40	6.005	3.4	B (Mat)
0.01	45	6.005	3.45	B (Mat)
0.01	40	6.005	3.4	B (Mat)
0.03	5	6.015	3.15	B (Mat)
0.03	10	6.015	3.3	B (Mat)
0.03	15	6.015	3.45	B (Mat)
0.03	20	6.015	3.6	B (Mat)
0.03	25	6.015	3.75	B (Mat)
0.03	30	6.015	3.9	B (Mat)
0.03	35	6.015	4.05	B (Mat)
0.03	40	6.015	4.2	B (Mat)
0.03	45	6.015	4.35	B (Mat)
0.03	40	6.015	4.2	B (Mat)
0.05	5	6.025	3.25	B (Mat)
0.05	10	6.025	3.5	B (Mat)
0.05	15	6.025	3.75	B (Mat)
0.05	20	6.025	4	B (Mat)
0.05	25	6.025	4.25	B (Mat)
0.05	30	6.025	4.5	B (Mat)
0.05	35	6.025	4.75	B (Mat)
0.05	40	6.025	5	B (Mat)
0.05	45	6.025	5.25	B (Mat)
0.05	40	6.025	5	B (Mat)
0.07	5	6.035	3.35	B (Mat)
0.07	10	6.035	3.7	B (Mat)
0.07	15	6.035	4.05	B (Mat)
0.07	20	6.035	4.4	B (Mat)
0.07	25	6.035	4.75	B (Mat)

(continued)

Table 5.38 (continued)

p	CB	$EU_A = 6 + 0.5p$	$EU_B = 3 + p \cdot C_B$	Preferred option
0.07	30	6.035	5.1	B (Mat)
0.07	35	6.035	5.45	B (Mat)
0.07	40	6.035	5.8	B (Mat)
0.07	45	6.035	6.15	A (Pile)
0.07	40	6.035	5.8	B (Mat)
0.1	5	6.05	3.5	B (Mat)
0.1	10	6.05	4	B (Mat)
0.1	15	6.05	4.5	B (Mat)
0.1	20	6.05	5	B (Mat)
0.1	25	6.05	5.5	B (Mat)
0.1	30	6.05	6	B (Mat)
0.1	35	6.05	6.5	A (Pile)
0.1	40	6.05	7	A (Pile)
0.1	45	6.05	7.5	A (Pile)
0.1	40	6.05	7	A (Pile)

Table 5.39 *Threshold* values of failure cost (C_B) at which the expected utility of retrofitting equals that of material monitoring, calculated for various probabilities p of failure occurrence

p	Threshold C_B
0.01	300.5
0.03	100.17
0.05	60.5
0.07	43.36
0.1	30.5

Two-Way Sensitivity Analysis in Wastewater Treatment Plant Upgrade

A municipality must decide how to reduce nutrient pollution (nitrogen and phosphorus) discharged into a nearby lake to comply with tightening environmental regulations. Figure 5.64 introduces the real-world challenge of nutrient pollution control in municipal water management, where engineering and nature-based solutions are weighed against uncertain regulatory risks and environmental outcomes.

The two technology options being considered by the municipality for removal of the nutrients include:

- Option A: Advanced biological nutrient removal (BNR) system upgrade.
- Option B: Constructed wetlands as a nature-based solution for tertiary treatment.

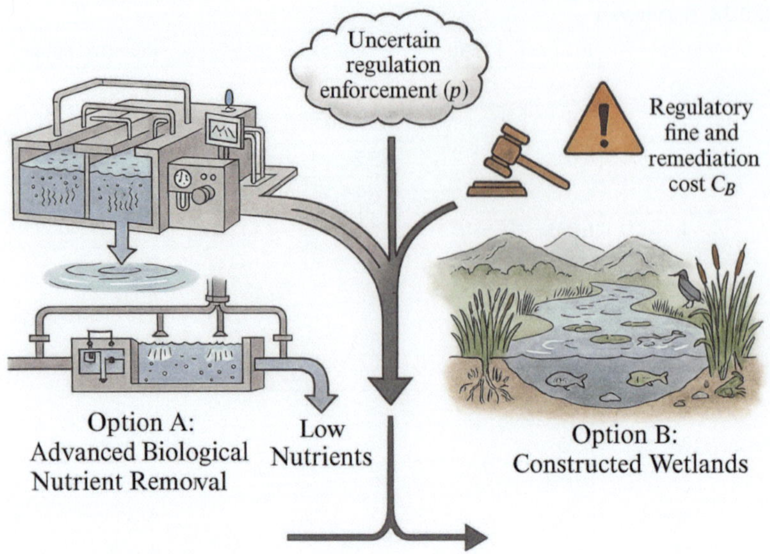

Fig. 5.64 Conceptual overview of two nutrient pollution control options considered by a municipality: a high-efficiency advanced BNR system and a nature-based constructed wetland

The decision depends on two uncertain factors:

- p: Probability that regulatory fines will be enforced for exceeding nutrient discharge limits—ranging from 0.01 (1%) to 0.10 (10%) based on changing environmental enforcement trends.
- C_B: Cost of fines and environmental damage if violations occur—ranging from \$1M to \$6M, depending on lake eutrophication levels and legal penalties.

Option A: BNR Upgrade

- Capital and O&M cost: \$12M
- If violation occurs (rare due to high efficiency): residual penalty of \$0.2M.

Option B: Constructed Wetlands

- Capital and O&M cost: \$6M
- If violation occurs: full fine and remediation cost C_B.

Objective: Select the most cost-effective long-term investment that ensures compliance and protects environmental quality.

Questions to Address

1. How do the expected lifecycle costs vary as fine probability, p, and cost C_B change?

2. For which combinations of p and C_B is Option A (BNR) or Option B (Wetlands) optimal?
3. What is the breakeven condition where both options have equal expected cost?
4. How sensitive is the decision to policy uncertainty and ecological response?
5. What implications does this analysis have for resilient and sustainable wastewater planning?
6. Should risk-averse environmental objectives (e.g., ecosystem protection) influence the final decision beyond cost alone?

Step 1: Define Expected Cost Equations

Option A: BNR System

- No violation (1 − p): Cost = \$12M
- Violation (p): Cost = \$12M + \$0.2M = \$12.2M
- Expected cost:

$$EU_A = (1 - p) \cdot 12 + p \cdot 12.2 = 12 + 0.2p$$

Option B: Constructed Wetlands

- No violation (1 − p): Cost = \$6M
- Violation (p): Cost − \$6M + C_B
- Expected cost:

$$EU_B = (1 - p) \cdot 6 + p \cdot (6 + CB) = 6 + p \cdot C_B$$

Step 2: Define the Parameter Grid

We select:

- Probability p: [0.01, 0.03, 0.05, 0.07, 0.10]
- Penalty cost C_B: [\$40M, \$50M, \$55M, \$60M, \$65M].

→In this case, for illustrative purposes, we will consider a coarse 5 × 5 grid (25 scenarios).

Step 3: Calculate Expected Costs and Optimal Decision (Table 5.40).

So, Option B is preferred for all reasonable ranges of p and C_B, and Option A becomes optimal only if C_B > \$60M and p = 0.10.

Step 4: Find the Decision Boundary

The boundary condition is determined by setting the expected costs of the two alternative treatment technologies equal to one another.

$$EU_A = EU_B$$

$$12 + 0.2p = 6 + p \cdot C_B$$

Table 5.40 Summary of expected costs and preferred treatment alternatives for various probabilities and costs of regulatory fines

p	C_B	$EU_A = 12 + 0.2p$	$EU_B = 6 + p \cdot C_B$	Preferred option
0.01	40	12.002	6.4	B (Wetland)
0.01	50	12.002	6.5	B (Wetland)
0.01	55	12.002	6.55	B (Wetland)
0.01	60	12.002	6.6	B (Wetland)
0.01	65	12.002	6.65	B (Wetland)
0.03	40	12.006	7.2	B (Wetland)
0.03	50	12.006	7.5	B (Wetland)
0.03	55	12.006	7.65	B (Wetland)
0.03	60	12.006	7.8	B (Wetland)
0.03	65	12.006	7.95	B (Wetland)
0.05	40	12.01	8	B (Wetland)
0.05	50	12.01	8.5	B (Wetland)
0.05	55	12.01	8.75	B (Wetland)
0.05	60	12.01	9	B (Wetland)
0.05	65	12.01	9.25	B (Wetland)
0.07	40	12.014	8.8	B (Wetland)
0.07	50	12.014	9.5	B (Wetland)
0.07	55	12.014	9.85	B (Wetland)
0.07	60	12.014	10.2	B (Wetland)
0.07	65	12.014	10.55	B (Wetland)
0.1	40	12.02	10	B (Wetland)
0.1	50	12.02	11	B (Wetland)
0.1	55	12.02	11.5	B (Wetland)
0.1	60	12.02	12	B (Wetland)
0.1	65	12.02	12.5	A (BNR)

$$p \cdot C_B = 6 + 0.2p$$

$$C_B = (6 + 0.2p)/p$$

$$C_B = 6/p + 0.2$$

Using this relationship, we are able to plot the regulatory fine amounts necessary for the two treatment technologies to have the same cost under a given probability of compliance failure (Table 5.41).

Table 5.41 Break-even values of damage cost (C_B) for BNR and Wetland across different event probabilities

p	Threshold C_B
0.01	600.2
0.03	200.2
0.05	120.2
0.07	86.06
0.1	60.2

So, Option A is only justified when $C_B > \$60.2$ M at $p = 0.10$, which is well beyond normal fines or estimated environmental damages.

Step 5: Interpretation and Practical Recommendations

Answer to Question 1:

- Option A changes little with p (just + \$0.2M over entire range).

  Option B is highly sensitive to both p and C_B, however, even at upper extremes ($p = 0.10$, $C_B = \$6$M), it is still the more economical treatment alternative in this scenario.

Answer to Question 2:

- Option B is optimal for all realistic combinations.
- Option A only becomes cost-effective at very high fines (>\$60M) and strong regulatory enforcement ($p = 0.10+$), which is unlikely based on the provided information.

Answer to Question 3:

The breakeven formula ($C_B = 6/p + 0.2$) shows the decision is robust. Break-even C_B values are far above real-world ranges, confirming strong preference for Option B.

Answer to Question 4:

- Low sensitivity in decision boundary means:
  - The city does not need to invest in costly uncertainty reduction efforts.
  - But it should continue monitoring policy and lake health to anticipate any change in risk perception.

Answer to Question 5:

- Constructed wetlands are a cost-effective, nature-based solution that perform well under uncertainty.
- They offer resilience, co-benefits for biodiversity, and are publicly favored.
- The city should pair wetlands with adaptive monitoring to ensure compliance and sustainability.

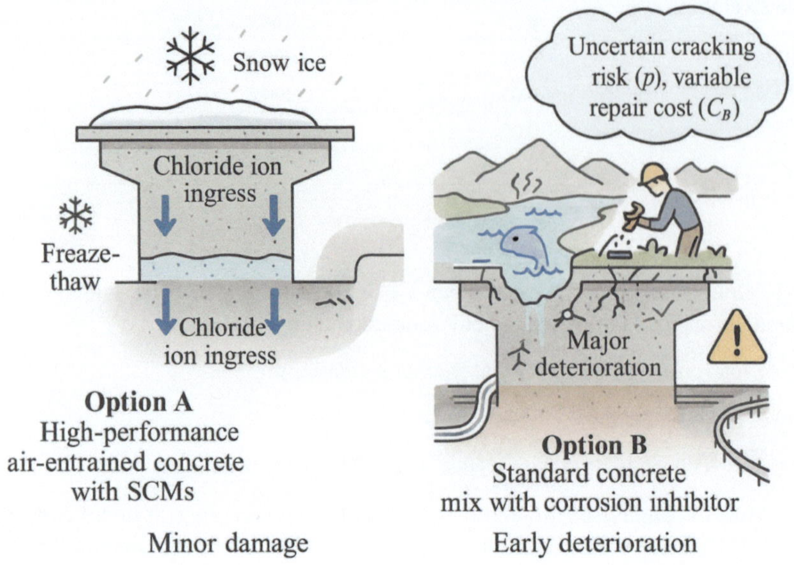

Fig. 5.65 Conceptual illustration of two concrete mix designs for a bridge deck in a cold climate

Answer to Question 6:

- If ecological conservation or reputational risk are valued more than just cost:
  - BNR may be worth considering as a zero-risk, high-performance option.
  - Decision-makers may use multi-criteria analysis that goes beyond expected monetary cost.

Two-Way Sensitivity Analysis for Concrete Mix Selection in Bridge Deck Design

A civil engineering firm must choose between two concrete mix designs for constructing a bridge deck exposed to freeze–thaw cycles, chloride ingress, and de-icing salts in a cold climate. The decision involves balancing upfront cost, service life performance, and the risk of premature deterioration. Figure 5.65 provides a visual comparison of how different concrete mixes perform under freeze–thaw and chloride exposure, highlighting the durability considerations and repair implications critical to bridge deck design in cold climates.

Two options are available:

- Option A: High-performance air-entrained concrete with supplementary cementitious materials (SCMs)
- Option B: Standard mix with chemical corrosion inhibitor.

Two uncertain factors affect the expected lifecycle cost:

- p: Probability of early cracking/spalling in Option B within 20 years (ranges from 0.01 to 0.10), due to uncertain quality control and chloride loads.
- C_B: Repair and maintenance cost if deterioration occurs (ranges from \$0.5M to \$3.0M depending on damage extent and traffic disruption).

Option A: High-Performance Mix

- Upfront cost: \$1.8M
- Maintenance cost if cracking occurs: \$0.1M (only surface treatment needed).

Option B: Standard Mix with Inhibitor

- Upfront cost: \$1.2M
- Full-depth patching and deck closure if cracking occurs: C_B.

Objective: Select the concrete mix with the lowest expected total lifecycle cost, considering long-term risk and repair implications.

Questions to Address

1. How do lifecycle costs vary as deterioration probability p and failure cost C_B change?
2. Which option is preferred across different combinations of these uncertainties?
3. What is the break-even line where both options yield equal expected cost?
4. How sensitive is the decision to variability in maintenance cost estimation and durability risk?
5. What should be prioritized in quality control, monitoring, or material testing to reduce uncertainty?
6. How would this decision change under more aggressive climate exposure or longer design life?

Step 1: Define Expected Cost Equations

Option A: High-Performance Concrete

- No deterioration: Cost = \$1.8M
- Deterioration occurs: Cost = \$1.8M + \$0.1M = \$1.9M
- Expected cost:

$$EU_A = (1 - p) \cdot 1.8 + p \cdot 1.9 = 1.8 + 0.1p$$

Option B: Standard Mix with Inhibitor

- No deterioration: Cost = \$1.2M
- Deterioration occurs: Cost = \$1.2M + C_B
- Expected cost:

$$EU_B = (1 - p) \cdot 1.2 + p \cdot (1.2 + C_B) = 1.2 + p \cdot C_B$$

Table 5.42 Expected cost comparison between high-performance (Option A) and standard (Option B) material strategies under varying failure probabilities p and damage costs C_B, with preferred option based on lower expected utility cost

p	CB	$EU_A = 1.8 + 0.1p$	$EU_B = 1.2 + p \cdot C_B$	Preferred option
0.01	2	1.801	1.22	B (Standard)
0.01	4	1.801	1.24	B (Standard)
0.01	6	1.801	1.26	B (Standard)
0.01	8	1.801	1.28	B (Standard)
0.01	10	1.801	1.3	B (Standard)
0.03	2	1.803	1.26	B (Standard)
0.03	4	1.803	1.32	B (Standard)
0.03	6	1.803	1.38	B (Standard)
0.03	8	1.803	1.44	B (Standard)
0.03	10	1.803	1.5	B (Standard)
0.05	2	1.805	1.3	B (Standard)
0.05	4	1.805	1.4	B (Standard)
0.05	6	1.805	1.5	B (Standard)
0.05	8	1.805	1.6	B (Standard)
0.05	10	1.805	1.7	B (Standard)
0.07	2	1.807	1.34	B (Standard)
0.07	4	1.807	1.48	B (Standard)
0.07	6	1.807	1.62	B (Standard)
0.07	8	1.807	1.76	B (Standard)
0.07	10	1.807	1.9	B (Standard)
0.1	2	1.81	1.4	B (Standard)
0.1	4	1.81	1.6	B (Standard)
0.1	6	1.81	1.8	B (Standard)
0.1	8	1.81	2	A (High-Performance)
0.1	10	1.81	2.2	A (High-Performance)

Step 2: Define the Parameter Grid

Use a 5 × 5 grid:

- Deterioration probability p: [0.01, 0.03, 0.05, 0.07, 0.10]
- Failure cost C_B: [$2M, $4M, $6M, $8M, $10M].

Step 3: Calculate Expected Costs and Preferred Option (Table 5.42).

Step 4: Find the Decision Boundary (Table 5.43).

$$EU_A = EU_B$$

Table 5.43 Threshold values of damage cost (C_B) at which the expected costs of high-performance and standard material strategies are equal, calculated across different probabilities pp of failure

p	Threshold C_B
0.01	60.1
0.03	20.1
0.05	12.1
0.07	8.67
0.1	6.1

$$1.8 + 0.1p = 1.2 + p \cdot C_B$$

$$p \cdot C_B = 0.6 + 0.1p$$

$$C_B = (0.6 + 0.1p)/p$$

$$C_B = 0.6/p + 0.1$$

Even at p = 0.10, the failure cost C_B must exceed \$6.1M for Option A to be justified.

Step 5: Interpretation and Practical Recommendations

Answer to Question 1:

- Option A is very stable, increasing only slightly with p.
- Option B grows linearly with both p and C_B, but remains cheaper for all combinations tested.

Answer to Question 2: Option B is consistently preferred unless deterioration costs exceed ~\$6M at the highest risk level (p = 0.10).

Answer to Question 3: Break-even formula: CB = 0.6/p + 0.1. For all practical ranges, the required CB is much higher than actual repair costs, confirming B is optimal.

Answer to Question 4: The decision is not sensitive to minor variations in repair cost. Focus should be placed on better durability risk forecasting if more extreme chloride exposure is expected (e.g., marine structures).

Answer to Question 5: Quality assurance testing (e.g., chloride diffusion tests, air void spacing) is more critical for Option B. If concrete durability can't be guaranteed through standard practice, Option A becomes more appealing.

Answer to Question 6: If design life is extended (e.g., from 20 to 75 years) or if future climate increases freeze–thaw cycles, the probability p would increase. Under more aggressive conditions, Option A may become justified or multi-criteria decision-making would be more appropriate.

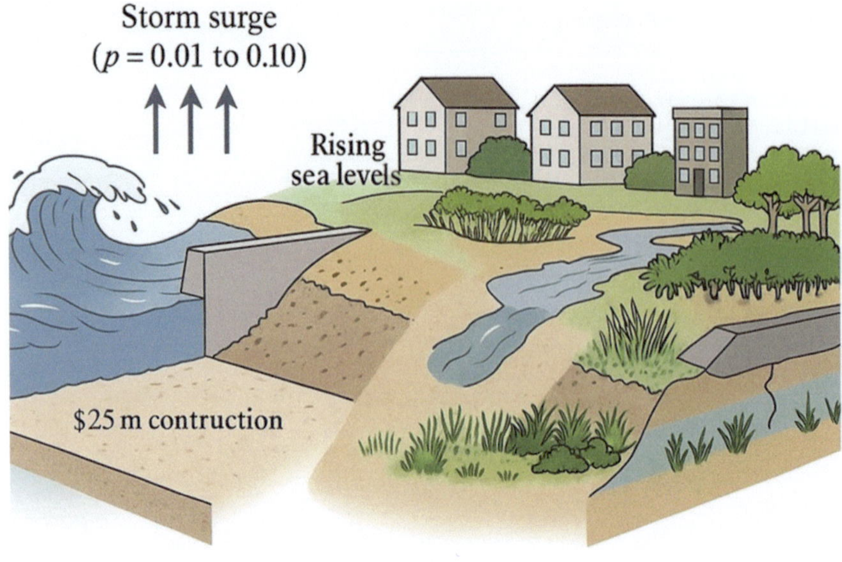

Fig. 5.66 Conceptual comparison of three coastal flood mitigation strategies: **a** a high-capital seawall offering strong defense, **b** low-cost nature-based systems with higher performance uncertainty, and **c** a hybrid solution combining engineered and ecological elements

Two-Way Sensitivity Analysis for Flood Risk Mitigation Strategy

A coastal municipality is evaluating three flood risk mitigation strategies to reduce the impacts of storm surges and sea level rise on residential and commercial zones.

The options under consideration are:

- Option A: Build a permanent concrete seawall
- Option B: Implement nature-based solutions such as dunes, wetlands, and mangrove restoration
- Option C: Use a hybrid strategy combining smaller engineered barriers with nature-based buffers.

The decision must account for uncertainties in future flood frequency and economic damage per event. Figure 5.66 illustrates the design context of the three flood risk mitigation options under consideration. These physical configurations help clarify how each strategy balances upfront investment with varying levels of protection under flood uncertainty.

Two key uncertain parameters:

- p: Annual probability of severe coastal flooding, projected between 0.01 (1%) and 0.10 (10%)

Table 5.44 Mitigation options summary

Option	Base cost (M$)	Damage if a flood occurs	Description
A—Seawall	$25M	$2M (residual overtopping/ maintenance)	High capital, strong defense
B—Nature-based	$12M	C_D (unpredictable performance)	Low cost, high variability
C—Hybrid	$18M	30% of C_D	Moderate cost and protection

- C_D: Damage cost per flood event in millions of dollars, ranging from $5M to $30M depending on population exposure and infrastructure vulnerability (Table 5.44).

Select the option with the lowest expected lifecycle cost, incorporating both fixed costs and expected flood-related damages. In your answer, address the following questions:

1. How do expected total costs change across different flood probabilities and damage levels?
2. Which strategy is optimal under different climate and exposure conditions?
3. What are the break-even conditions between each pair of options?
4. How does the variability in nature-based performance influence the overall decision?
5. What policy or planning recommendations emerge from the sensitivity analysis?
6. Should non-monetary co-benefits (e.g., biodiversity, carbon storage) shift the decision toward greener solutions?

Step 1: Define Expected Cost Equations

Option A: Seawall

- Expected cost:

$$EU_A = 25 + p \cdot 2 = 25 + 2p$$

Option B: Nature-based

- Expected cost:

$$EU_B = 12 + p \cdot C_D$$

Option C: Hybrid

- Expected cost:

$$EU_C = 18 + p \cdot (0.3 \cdot C_D) = 18 + 0.3p \cdot C_D$$

Table 5.45 Expected cost comparison among three flood protection options under varying flood probabilities (p) and damage cost values (C_D)

p	C_D	$EU_A = 25 + 2p$	$EU_B = 12 + p \cdot C_D$	$EU_C = 18 + 0.3p \cdot C_D$	Preferred option
0.01	50	25.02	12.5	18.15	B
0.01	100	25.02	13	18.3	B
0.01	150	25.02	13.5	18.45	B
0.01	200	25.02	14	18.6	B
0.01	250	25.02	14.5	18.75	B
0.03	50	25.06	13.5	18.45	B
0.03	100	25.06	15	18.9	B
0.03	150	25.06	16.5	19.35	B
0.03	200	25.06	18	19.8	B
0.03	250	25.06	19.5	20.25	B
0.05	50	25.1	14.5	18.75	B
0.05	100	25.1	17	19.5	B
0.05	150	25.1	19.5	20.25	B
0.05	200	25.1	22	21	C
0.05	250	25.1	24.5	21.75	C
0.07	50	25.14	15.5	19.05	B
0.07	100	25.14	19	20.1	B
0.07	150	25.14	22.5	21.15	C
0.07	200	25.14	26	22.2	C
0.07	250	25.14	29.5	23.25	C
0.1	50	25.2	17	19.5	B
0.1	100	25.2	22	21	C
0.1	150	25.2	27	22.5	C
0.1	200	25.2	32	24	C
0.1	250	25.2	37	25.5	A

Step 2: Define the Parameter Grid

- Flood probability p: [0.01, 0.03, 0.05, 0.07, 0.10]
- Damage per flood C_D: [\$5M, \$10M, \$15M, \$20M, \$30M]
  - →For illustrative purposes a coarse 5×5 grid yielding 25 combinations is considered.

Step 3: Calculate Expected Costs and Preferred Option (Table 5.45).

Step 4: Identify Break-Even Points

Table 5.46 Break-even damage cost values (C_D) at which the expected costs of mitigation Options A & B, A & C, and B & C are equal

p	Threshold C_D		
	A & B	A & C	B & C
0.01	1302.00	2340.00	857.14
0.03	435.33	784.44	285.71
0.05	262.00	473.33	171.43
0.07	187.71	340.00	122.45
0.1	132.00	240.00	85.71

Again, the break even points between the alternatives can be determiend by setting the expression for expected utility equal to one another.

Between A and B:

$$25 + 2p = 12 + p \cdot C_D \Rightarrow p(C_D - 2) = 13 \Rightarrow p = 13/(C_D - 2)$$

Between A and C:

$$25 + 2p = 18 + 0.3p \cdot C_D \Rightarrow p(0.3C_D - 2) = 7 \Rightarrow p = 7/(0.3C_D - 2)$$

Between B and C:

$$12 + p \cdot C_D = 18 + 0.3p \cdot C_D \Rightarrow 0.7p \cdot C_D = 6 \Rightarrow p = 6/0.7C_D$$

These formulas can be used to determine the threshold across any value of C_D (Table 5.46).

Step 5: Interpretation and Practical Recommendations

Answer to Question 1:

- Option B has the lowest expected cost in most scenarios with low to moderate risk.
- Option C becomes preferable in high-risk, high-damage contexts.
- Option A is most robust, but cost-prohibitive unless flood probability and damage are very high.

Answer to Question 2:

- B is best when $p < 0.07$ and $C_D < \$20M$
- C becomes optimal when $p > 0.07$ and $C_D > \$25M$
- A is rarely optimal unless damages exceed \$50M and $p > 0.10$.

Answer to Question 3: Use the break-even equations to set policy triggers (e.g., "If new projections raise annual flood probability to 0.08, switch to hybrid strategy.").

Answer to Question 4:

- Nature-based options have higher uncertainty in protective performance but excel under low enforcement or financial constraints.
- Including probabilistic performance ranges could further refine this model (e.g., 60–80% attenuation variability).

Answer to Question 5:

- Cities should pair early green infrastructure with adaptive retreat and zoning.
- Hybrid strategies offer a balance and support both resilience and biodiversity.

Answer to Question 6:

- Co-benefits of Option B (carbon sequestration, habitat creation, recreational spaces) may outweigh its expected cost for climate adaptation planning.
- Multi-objective frameworks should be used for final selection.

5.10.3.3 Threshold Analysis

Threshold analysis is a targeted form of sensitivity analysis used to determine the precise value of an uncertain parameter at which the preferred decision shifts from one alternative to another. Unlike one-way or two-way sensitivity analyses that explore how outcomes such as cost, performance, or risk vary over a range of input values, threshold analysis focuses on identifying tipping points—specific values of uncertain variables where the optimal decision reverses. This approach is particularly useful when dealing with binary or discrete decisions, such as whether to implement a high-cost, high-reliability infrastructure system or to opt for a more flexible but potentially risk-prone alternative. By isolating these critical values, decision-makers can test the robustness of their choices against uncertainty and better understand the minimum conditions under which a given alternative becomes superior.

The process of conducting threshold analysis involves several structured steps. First, the analyst defines the decision alternatives and establishes the performance metric of interest, typically cost, utility, environmental impact, or risk. Next, the uncertain parameter (or parameters) likely to influence the decision is identified—this may include climate variables, failure probabilities, interest rates, or construction material costs. Then, mathematical functions are formulated for each alternative, expressing the total cost or benefit as a function of the uncertain variable. The threshold is identified by setting these functions equal and solving for the uncertain parameter. This yields the exact point at which the two alternatives yield equal outcomes, beyond or below which one becomes preferable.

For example, consider a municipality choosing between building a permanent levee system or relying on temporary flood barriers to protect a neighborhood from river flooding. The cost of the permanent levee is fixed and known, while the cost of temporary barriers depends on the probability of flooding, which is uncertain due to climate variability. Threshold analysis allows the municipality to calculate the flood probability value at which the long-term costs of the temporary barrier approach match those of the levee system. If the actual flood probability exceeds this value, the permanent levee is the better option. This threshold becomes a policy-relevant boundary: if new climate projections indicate a higher likelihood of extreme floods, the decision can be confidently revisited.

One of the primary advantages of threshold analysis is its decision-oriented nature. Rather than analyzing broad outcome trends, it directly informs whether a decision should be reconsidered based on specific shifts in context or assumptions. This makes it highly valuable for decision-making under uncertainty, especially in infrastructure projects where stakes are high and misjudgments are costly. It is also computationally efficient. In many cases, the threshold can be found analytically, using simple algebraic equations, without requiring iterative simulations or extensive modeling. Furthermore, threshold analysis highlights which variables are most influential in altering the decision—helping engineers and planners focus monitoring efforts or further refine estimates for those critical parameters.

However, threshold analysis also has several limitations. Firstly, this approach typically handles one parameter at a time and does not explore how multiple uncertainties may interact. In complex systems with interdependent risks (e.g., flood probability and damage severity), a single-variable threshold may oversimplify reality. Additionally, threshold analysis does not provide information about how gradual changes in input values affect outcomes—it only identifies the break-even point. For broader understanding, one- or two-way sensitivity analysis may still be necessary to map the full behavior of the system across different scenarios. The advantages and limitations of threshold analysis are reflected in Fig. 5.67, which visualizes the break-even point at which one alternative becomes more favorable than another under varying assumptions. This figure exemplifies the decision-oriented nature of threshold analysis and computational simplicity, while also illustrating its inherent constraint of only being able to analyze a single parameter at a time without capturing interactions or gradual shifts in outcomes.

As illustrated in Fig. 5.68, threshold analysis is particularly effective in situations where a single input parameter has a dominant influence on the decision outcome, is highly uncertain or expected to change over time—such as in response to climate variability or market trends—and represents a critical policy trigger, regulatory limit, or economic tipping point that could shift the preferred course of action.

Threshold analysis is widely applied in decision-making for infrastructure design, climate adaptation planning, and environmental compliance, where identifying the exact point at which an investment or action becomes justified is crucial. The procedure (Fig. 5.69) follows a clear logical path that begins with defining the competing decision options—such as permanent versus temporary flood protection—followed by developing cost, utility, or risk functions for each alternative in terms of the uncertain parameter. The threshold is then identified by equating these functions (e.g., Total Cost_A = Total Cost_B) and solving for the parameter value that balances the two. Once the threshold is established, the actual or projected value of the parameter is compared against it to determine the optimal course of action. This result is then translated into a decision rule (e.g., "If flood probability >2.5%, build levee; otherwise, deploy seasonal barriers") to guide practical implementation. Finally, scenario

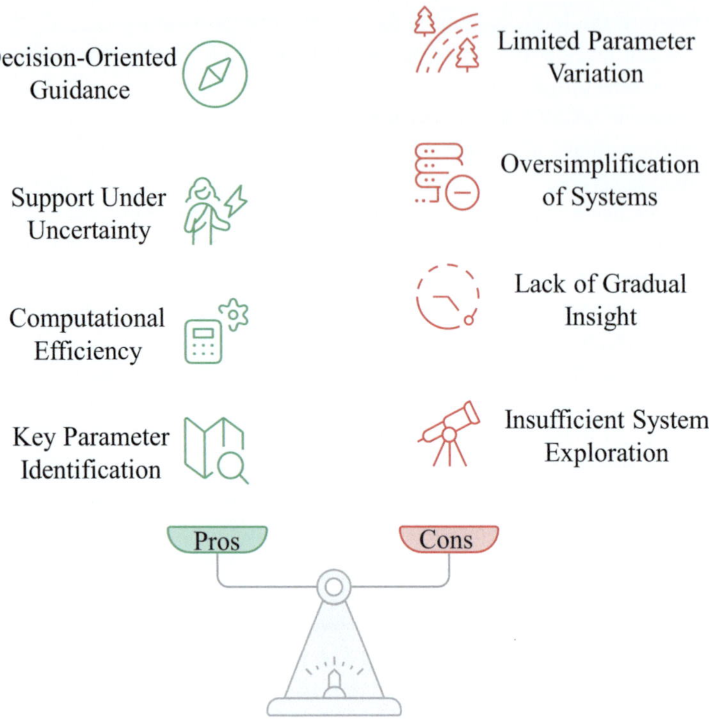

Fig. 5.67 The main advantages and limitations of the threshold analysis

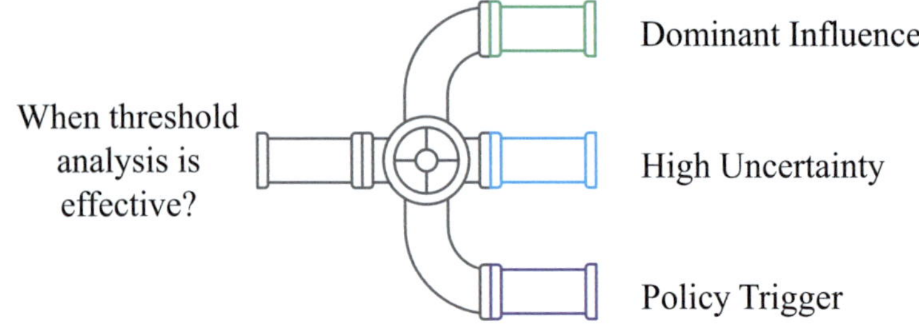

Fig. 5.68 Key conditions under which threshold analysis is most effective

testing may be conducted to evaluate the robustness of the threshold under varying assumptions and uncertainties.

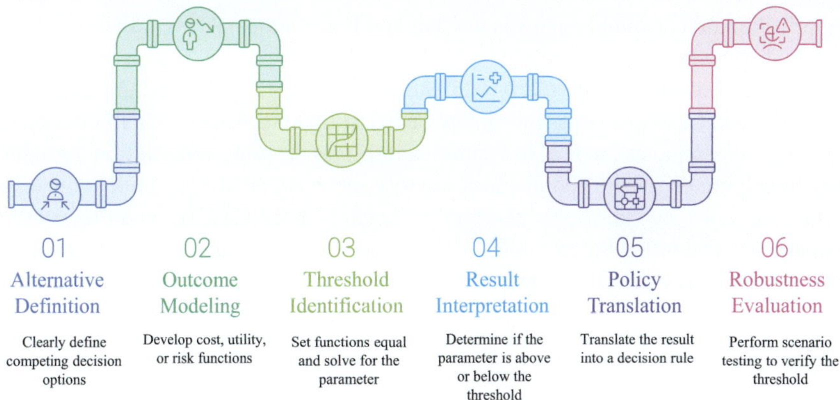

Fig. 5.69 Threshold analysis-based decision-making process

This technique helps civil engineers and planners establish clear decision boundaries and justify investments based on future conditions or risk exposure. It is especially useful when the outcome space is binary (e.g., invest or delay, upgrade or maintain) and when the cost of misjudging the uncertain parameter is high. Threshold analysis is also commonly used in regulatory frameworks (e.g., environmental concentration limits, resilience criteria) where action is required only when a variable exceeds a specified level.

Comparing Threshold Analysis with Traditional Sensitivity Methods

Contrasting threshold analysis with other forms of sensitivity analysis brings to light how each method addresses uncertainty and decision dynamics. One-way sensitivity analysis varies a single input over a plausible range and observes changes in the output, providing useful information about how sensitive the result is to individual parameters. It often produces tornado diagrams that show which variables have the greatest influence on outcome variability. Two-way sensitivity analysis takes this further by varying two parameters simultaneously, often visualized with heat maps or decision regions, revealing how combinations of uncertainties affect the preferred choice. In contrast, threshold analysis zooms in on the precise value at which decisions change. It does not show the shape of the outcome curve but instead answers the more focused question: "When does my decision stop being valid?".

By pinpointing when a decision should change, threshold analysis adds strategic value to traditional cost–benefit frameworks and complements broader sensitivity techniques like one-way and two-way analyses.

Application of Threshold Analysis in Civil and Environmental Engineering

Threshold analysis is particularly well-suited to strategic infrastructure planning, environmental management, and policy decision-making where a small change in one critical parameter can lead to a significant shift in the preferred solution. Its value lies in helping engineers and planners identify the exact conditions under which an action becomes economically justified, technically necessary, or environmentally imperative. The following examples illustrate how this method can be applied across various domains of civil and environmental engineering.

Flood Risk Management

In urban flood control projects, municipalities often choose between building permanent defenses (e.g., levees, floodwalls) and implementing temporary or adaptive solutions (e.g., deployable barriers, wetland buffers). The key uncertainty is typically the future probability of extreme rainfall or river discharge, influenced by climate change. Threshold analysis helps determine the critical flood return period (or exceedance probability) beyond which the permanent infrastructure becomes more cost-effective over its lifespan. For instance, if the probability of a 100-year flood occurring in the next 30 years exceeds a calculated threshold, the city might justify the upfront investment in permanent structures.

Bridge Design and Load Forecasting

Bridge design often involves selecting materials and structural systems based on anticipated loads, maintenance costs, and life-cycle performance. A typical threshold analysis might examine the critical increase in traffic volume (due to population growth or economic development) at which a prestressed concrete design becomes more economical than a steel alternative. This threshold informs decisions about whether to overdesign a structure now or plan for retrofitting in the future, balancing capital investment with long-term flexibility.

Water Supply and Treatment Capacity

In water treatment or supply planning, engineers often face decisions about whether to expand treatment plant capacity now or defer expansion until demand increases. Here, the threshold value could be the population size, per capita water demand, or peak daily flow rate that justifies the investment. By identifying this threshold, municipalities can align infrastructure expansion with actual growth, avoiding both overbuilding and under-capacity risks.

Environmental Impact Mitigation

When evaluating environmental remediation strategies, such as soil decontamination or wetland restoration, threshold analysis can be used to assess pollutant concentration thresholds or land-use intensities at which the environmental cost of inaction surpasses the cost of intervention. For instance, a threshold level of nitrate in groundwater may trigger the need for buffer zone implementation or changes in agricultural practices to comply with health regulations or sustainability goals.

Renewable Energy and Building Retrofits

In sustainable building design, one common application of threshold analysis is evaluating the point at which investments in high-efficiency insulation or renewable energy systems (e.g., solar panels, geothermal heat pumps) become economically favorable. This threshold is often linked to energy price escalation rates, carbon tax levels, or equipment lifespan. Planners use this analysis to decide whether to invest in net-zero upgrades now or delay based on current economic and regulatory assumptions.

Pavement Design Under Uncertain Climate Conditions

In pavement engineering, climate-related uncertainties such as increased freeze–thaw cycles or precipitation levels can degrade traditional asphalt performance. Threshold analysis can identify the critical number of freeze–thaw days per year beyond which using air-entrained concrete or polymer-modified asphalt becomes more economical than conventional designs. This is particularly important in regions experiencing rapid climate shifts or transitioning from dry to wet zones.

Construction Method Selection

Contractors and owners often choose between in-house construction and subcontracting based on cost, duration, and risk. A threshold analysis might calculate the labor cost inflation rate or project delay penalty at which subcontracting becomes preferable. This allows contractors to make forward-looking decisions based on economic scenarios rather than static cost comparisons.

In all these applications, threshold analysis provides a clear decision boundary, helping civil and environmental engineers translate uncertainty into actionable thresholds. It is especially useful when combined with other techniques, such as scenario analysis, cost–benefit analysis, or probabilistic risk assessment, to support comprehensive and resilient infrastructure decisions. By pinpointing when decisions should change, threshold analysis ensures that public and private sector investments remain both timely and justifiable under evolving conditions.

In civil engineering, threshold analysis is especially valuable for projects where a single uncertain factor has the potential to alter the optimal design or strategy. Unlike broader sensitivity methods that explore how output metrics vary, threshold analysis focuses on finding the exact value of a parameter at which one decision becomes more favorable than another. This makes it particularly suitable for scenarios involving climate uncertainty, cost escalation, or regulatory thresholds, where decision reversals are highly sensitive to one key variable.

Infrastructure systems, such as bridges, flood control structures, water treatment plants, and pavements, are often subject to discrete investment decisions (e.g., whether to upgrade, maintain, or defer). These decisions are heavily influenced by projected environmental conditions or market fluctuations. Threshold analysis helps engineers pinpoint critical conditions under which it becomes advantageous to shift strategies, thereby aligning investments with projected future states rather than current observations.

Detailed Case Studies in Threshold Decision-Making

These examples demonstrate how threshold analysis supports proactive and adaptive decision-making in civil engineering. By determining exactly when a decision flips from one alternative to another, engineers can prioritize monitoring, improve forecasting accuracy, and communicate decisions in a clear, quantitative framework. Whether the concern is climate risk, urban expansion, material degradation, or environmental compliance, threshold analysis offers a precise and powerful way to convert uncertainty into action thresholds. The general workflow for applying threshold analysis involves the following steps (Fig. 5.70):

1. Model Formulation:

   Define the decision problem clearly, specifying the competing alternatives (e.g., Option A vs. Option B) and identifying the key uncertain parameter that could influence the preferred choice. This parameter may relate to environmental conditions (e.g., flood probability, temperature extremes), economic factors (e.g., material costs, energy prices), or policy thresholds (e.g., regulatory limits, growth targets).
2. Cost or Utility Function Development:

   Formulate a mathematical expression for the total cost, benefit, or utility of each alternative as a function of the uncertain parameter. These expressions should reflect life-cycle costs, maintenance, operational risks, or penalties associated with non-compliance or system failure.
3. Equation Balancing:

   Set the outcome functions of the two alternatives equal to one another (e.g., Total_Cost_A = Total_Cost_B) and solve for the uncertain parameter. This yields the threshold value where the decision outcome is indifferent between the two options.
4. Feasibility Verification:

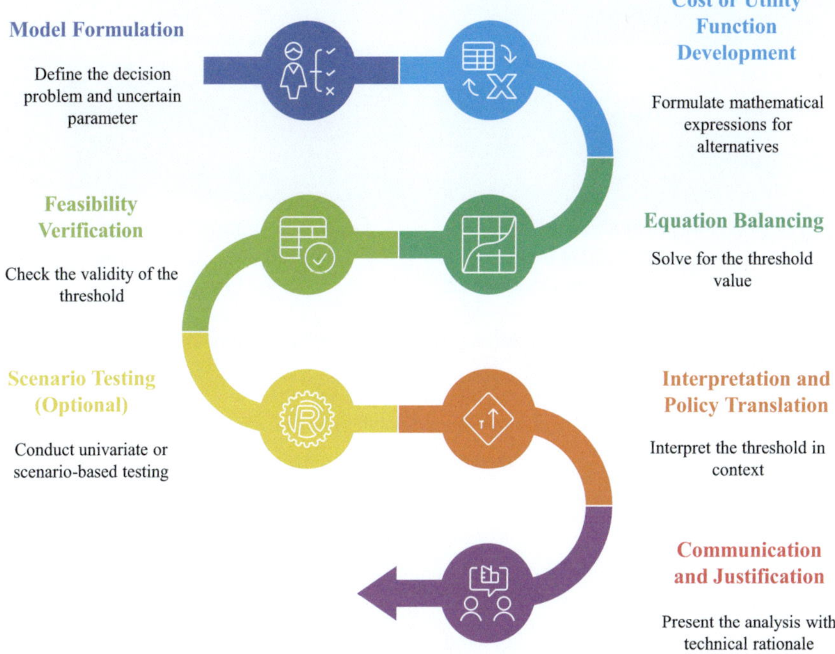

Fig. 5.70 Mastering threshold analysis: a step-by-step guide

Check whether the calculated threshold value is practically valid and meaningful within the defined range of the parameter. If the threshold falls outside the plausible or acceptable range, it may indicate that one option dominates across all scenarios.

5. Scenario Testing (Optional):

   If multiple variables influence the outcome, conduct univariate or scenario-based testing to examine how the threshold value behaves under different assumptions (e.g., changing discount rates, repair costs, or environmental regulations). This can add depth to the analysis and inform robustness checks.

6. Interpretation and Policy Translation:

   Interpret the threshold in context: If the actual or projected value of the parameter exceeds (or falls below) the threshold, select the corresponding optimal alternative. Express the result in clear terms for decision-makers (e.g., "If annual flood probability exceeds 2.5%, build levee").

7. Communication and Justification:

   Present the threshold analysis with appropriate technical rationale and visualization, if needed (e.g., a breakpoint chart). This ensures that stakeholders understand the logic behind the decision, enhancing transparency and support for future actions.

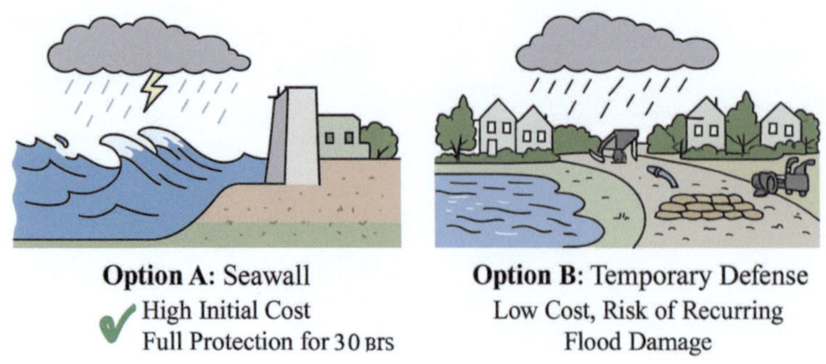

Fig. 5.71 Conceptual illustration of threshold analysis for selecting flood protection strategies

Table 5.47 Data for threshold analysis of coastal flood protection strategies

Parameter	Option A (Seawall)	Option B (Temporary protection)
Initial cost	\$18 million	\$3 million
Annual maintenance	\$0.2 million/year	\$0.8 million/year
Expected flood damage per event	\$0	\$12 million
Time horizon (year)	30	30
Discounting	Ignored	Ignored

Example 1: Threshold Analysis of Coastal Flood Protection Strategies

A coastal municipality is considering two flood protection strategies for a residential area vulnerable to storm surge events and sea-level rise:

- Option A: Build a permanent seawall. This option has a high up-front cost but offers full protection for the next 30 years.
- Option B: Continue with temporary flood protection using sandbags and mobile pumps. This option has lower initial costs but is subject to recurring flood risks and annual maintenance.

The uncertain parameter is the annual flood probability, denoted by p, due to climate change and evolving coastal dynamics.

The municipality wants to determine the threshold flood probability $p*$ at which Option A becomes more cost-effective than Option B over a 30-year analysis period (ignoring discounting for simplicity). Figure 5.71 introduces the threshold analysis framework, helping visualize how flood probability influences the preferred strategy between temporary flood defense and long-term seawall investment (Table 5.47).

Questions to Address

1. What is the threshold value of annual flood probability $p*$ where the total cost of both options is equal?
2. What happens if the actual flood probability is lower or higher than this threshold?

3. How does this analysis inform infrastructure investment and policy?
4. Is the decision sensitive to changes in flood damage or maintenance costs?

Step 1: Total Cost of Option A (Seawall)

Total $\mathrm{Cost_A}$ = Initial Cost + Annual Maintenance × Years

$$Total\ Cost_A = 18 + (0.2 \times 30) = 18 + 6 = \$24\,\mathrm{M}$$

Since the seawall offers full protection, no flood damage is expected.

Step 2: Total Cost of Option B (Temporary Protection)

- Initial Cost: \$3 million
- Annual Maintenance: \$0.8M/year × 30 years = \$24 million
- Expected Flood Damage per Year = p × 12 million.

$$Total\ Cost_B = 3 + 24 + 30 \times 12p = 27 + 360pM\,\$$$

Step 3: Set Costs Equal to Solve for Threshold p∗

$$Total\ Cost_A = Total\ Cost_B$$

$$24 = 27 + 360p \Rightarrow p = -3/360 = -0.0083$$

This result is invalid, because probability cannot be negative.

Interpretation: Even at 0% flood probability, Option A (seawall) is still cheaper than Option B over 30 years. This suggests the seawall is always more economical under these assumptions.

Step 4: Test Alternative Assumptions

Let's test if higher seawall costs or lower damage costs would yield a realistic threshold.

Case 1: Seawall More Expensive

Suppose Option A's cost increases to \$30 million:

$$TotalCost_A = 30 + 6 = 36 \text{ million}$$

Set equal:

$$36 = 27 + 360p \Rightarrow p* = 0.025$$

Valid threshold, if flood probability is higher than 2.5% per year, the seawall becomes more cost-effective.

Case 2: Lower Flood Damage (e.g., \$4M/event)

$$\textit{Expected flood damage over } 30\,\textit{years} = 30 \times 4p = 120p$$

$$TotalCost_B = 3 + 24 + 120p = 27 + 120p$$

Set equal:

$$24 = 27 + 120p \Rightarrow p = -3/120 = -0.025$$

Again, no realistic threshold—Option A dominates.

Let's keep the most reasonable case:

- Option A total cost = \$36M (e.g., higher-cost seawall)
- Option B cost = 27 + 360.

Threshold calculation:

$$36 = 27 + 360p \Rightarrow p* = 0.025 or 2.5\%$$

1. What is the threshold value of annual flood probability p∗?

Under the adjusted cost scenario (more expensive seawall at \$30M initial cost), the threshold probability is:

$$p* = (36 - 27)/360 = 0.025(or 2.5\%)$$

This means that if flooding is expected more than once every 40 years, Option A (seawall) becomes more economical.

2. What happens if the actual flood probability is lower or higher than this threshold?
    - If p < 2.5% (e.g., floods occur once every 50+ years): Option B (temporary protection) is more economical.
    - If p > 2.5% (e.g., floods occur more frequently): Option A (seawall) is more economical.

   Hence, the optimal strategy depends on expected flood frequency.

3. How does this analysis inform infrastructure investment and policy?

This analysis provides a quantitative decision rule:

"If projected flood probability exceeds 2.5%, invest in a permanent seawall."

It enables:

- Forward-looking, evidence-based planning
- Justification for proactive investments before damage occurs
- Guidance for climate risk assessments and flood modeling efforts

4. Is the decision sensitive to changes in flood damage or maintenance costs?

Yes, the threshold is very sensitive:

- If expected flood damage is reduced (e.g., \$4M/event), the cost advantage of Option A increases, and no threshold exists—Option A always dominates.
- If Option A's cost increases (e.g., from \$24M to \$36M), a valid threshold emerges.
- If Option B's annual maintenance increases, the threshold probability shifts lower, again favoring the seawall.

Thus, the result depends critically on the balance of upfront cost versus risk exposure.

This threshold analysis identifies 2.5% as the critical flood probability at which building a seawall becomes more cost-effective than relying on temporary protection. This helps guide investment under uncertainty and translates risk projections into actionable decisions. When combined with flood forecasting and climate modeling, threshold analysis provides a powerful tool for optimizing civil infrastructure planning in the face of environmental change.

Example 2: Threshold Analysis of Seismic Retrofit Strategies for Public Infrastructure

A city in a seismically active region is evaluating whether to retrofit a critical public building—an old municipal hospital—to meet current earthquake resistance standards. The building currently meets only minimal seismic safety criteria, making it vulnerable during moderate or large earthquakes.

The city is considering two alternatives:

- Option A: Perform a comprehensive seismic retrofit that brings the structure up to current building code standards.
- Option B: Defer retrofitting and continue routine maintenance, accepting the risk of potential damage during an earthquake.

The key uncertainty is the annual probability of a damaging earthquake, denoted as p, which is estimated using regional seismic hazard models but remains imprecise due to fault complexity.

The municipality wants to find the threshold earthquake probability above which investing in the retrofit becomes the more cost-effective and responsible option. Figure 5.72 visually outlines the decision framework for seismic retrofitting, showing how increasing earthquake probability affects the cost-effectiveness of investing in

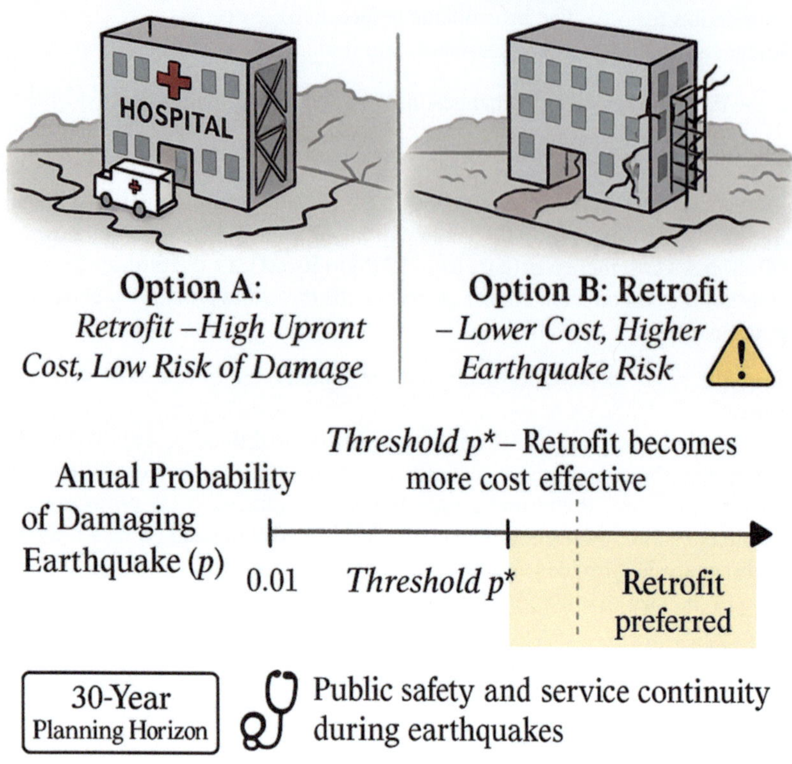

Fig. 5.72 Conceptual illustration of a threshold analysis comparing seismic retrofit versus deferred maintenance for a critical public hospital

Table 5.48 Data for threshold analysis of seismic retrofit strategies for public hospital

Parameter	Option A (Retrofit)	Option B (Status Quo)
Initial cost	\$12 million	\$1 million (routine maintenance)
Annual maintenance cost	\$0.1 million/year	\$0.2 million/year
Expected earthquake damage	\$0 (fully retrofitted)	\$25 million per damaging event
Time horizon	30 years	30 years
Discounting	Ignored	Ignored

structural upgrades versus maintaining the current condition of a vulnerable public facility (Table 5.48).

Questions to Address

1. What is the threshold annual earthquake probability where both options have equal cost?
2. What happens if the actual probability is lower or higher than this threshold?
3. What implications does the result have for risk-informed investment in infrastructure?
4. How sensitive is the decision to variation in damage cost or retrofit cost?

Step 1: Total Cost of Option A (Seismic Retrofit)

Total Cost_A = Initial Cost + Annual Maintenance × Years

$$TotalCost_A = 12 + (0.1 \times 30) = 12 + 3 = \$15\text{M}$$

No earthquake damage expected due to full compliance with seismic design codes.

Step 2: Total Cost of Option B (Status Quo)

- Initial Cost: $1 million
- Annual Maintenance: $0.2M/year × 30 = $6M
- Expected Earthquake Damage per Year = p × 25 M.

$$TotalCostB = 1 + 6 + 30 \times 25p = 7 + 750pM\$$$

Step 3: Set Costs Equal to Solve for Threshold

$$TotalCost_A = TotalCost_B \Rightarrow 15 = 7 + 750p \Rightarrow p* = 0.0107\ or\ 1.07\%$$

Step 4: Interpretation

- If the annual probability of a damaging earthquake exceeds 1.07%, then retrofitting the hospital is the more cost-effective option.
- If the probability of the earthquake is lower than 1.07%, continuing with regular maintenance is a less costly alternative (although, introduces more risk due to the significant cost per damaging event).

Step 5: Sensitivity Testing

Let's examine sensitivity to two key assumptions:

Case 1: Lower Damage Cost (e.g., $15M)

$$TotalCost_B = 1 + 6 + \ (30 \times 15p) = 7 + 450p$$

Set equal:

$$15 = 7 + 450p \Rightarrow p* = 8/450 = 0.0178 or 1.78\%$$

A lower damage cost raises the threshold, making retrofitting less urgent.

Case 2: Retrofit Cost Increases (e.g., \$18M)

$$TotalCost_A = 18 + 3 = M21$$

Set equal:

$$21 = 7 + 750p \Rightarrow p* = 0.0187 or 1.87\%$$

A higher retrofit cost raises the threshold, reducing cost-effectiveness of Option A. Based on the results of this analysis we can discuss the questions provided in the problem statement:

1. What is the threshold value of annual earthquake probability?

   1. $p* = (15 - 7)/750 = 8/750 = 0.0107 = 1.07\%$

   If the probability of a damaging earthquake is higher than 1.07% per year (or roughly once every 93 years), Option A becomes more economical.
2. What happens if the actual earthquake probability is lower or higher than this threshold?

   - If p < 1.07%: Keep current strategy (Option B); lower cost, acceptable risk.
   - If p > 1.07%: Retrofit the hospital as Option A is cheaper and safer.

3. What implications does the result have for infrastructure investment and policy?

   This analysis gives a clear decision rule:

   "If the annual risk of earthquake damage exceeds 1.07%, invest in seismic retrofitting"

It supports:

- Proactive mitigation planning.
- Justification for capital investment.
- Compliance with public safety mandates.
- Better allocation of limited resilience budgets.

4. How sensitive is the decision to changes in cost assumptions?

   - Lowering damage costs raises p∗, making retrofitting less urgent.
   - Increasing retrofit costs also raises p∗.
   - The decision is very sensitive to projected losses from earthquake events and cost of retrofitting.

   Thus, accurate seismic risk modeling and realistic cost estimates are crucial.

Threshold analysis reveals that if the annual probability of a damaging earthquake is greater than 1.07%, then seismic retrofitting of the municipal hospital is the more cost-effective choice. This empowers city planners to make risk-informed decisions and align public safety investments with seismic hazard projections. The technique provides a transparent, quantitative foundation for infrastructure resilience planning in earthquake-prone regions.

Example 3: Threshold Analysis of Phosphorus Control Strategies in Wastewater Treatment

A regional wastewater authority is considering whether to upgrade its treatment plant with advanced nutrient removal (ANR) technology to reduce phosphorus discharge into a nearby lake. The lake is experiencing signs of eutrophication, threatening aquatic ecosystems and drinking water supplies.

The decision is between two alternatives:

- Option A: Upgrade the plant with ANR to consistently meet anticipated future discharge limits.
- Option B: Continue with standard treatment, risking non-compliance and potential fines or retrofit in the future.

The uncertain parameter is the future phosphorus discharge limit imposed by environmental regulators. The limit is expected to tighten, but its exact value is uncertain due to ongoing scientific reviews and policy negotiations.

The wastewater authority wants to determine the threshold phosphorus limit C* (in mg/L) below which it becomes more economical to upgrade now rather than risk future penalties and emergency retrofits. Figure 5.73 illustrates the wastewater authority's decision framework under regulatory uncertainty, showing how future phosphorus limits drive the threshold at which investing in advanced treatment becomes the economically preferable option (Table 5.49).

Questions to Address

1. What is the threshold phosphorus limit C* (mg/L) at which Option A becomes more cost-effective than Option B?
2. What happens if the future phosphorus limit is stricter or more lenient than the threshold?
3. How much do fines versus retrofit costs contribute to the total cost of Option B?
4. How do changes in operation & maintenance (O&M) costs affect the decision?
5. How does this threshold analysis help support environmental compliance strategy and risk management?

Step 1: Define Total Cost of Option A (Upgrade Now)

$$TotalCost_A = InitialCost + O\&M \times Years$$

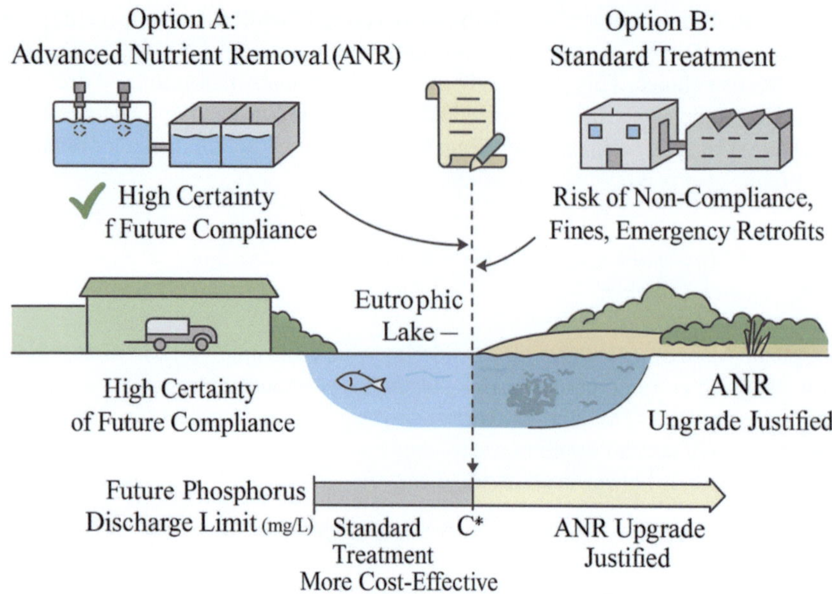

Fig. 5.73 Conceptual threshold analysis comparing advanced nutrient removal and standard wastewater treatment strategies under uncertain future phosphorus discharge limits

Table 5.49 Given data for threshold analysis of phosphorus control strategies in wastewater treatment

Parameter	Option A (Upgrade now)	Option B (Delay and retrofit)
Initial cost	\$10 million	\$1 million (minor upgrades)
Annual O&M cost	\$0.6 million	\$0.4 million
Retrofit cost (if forced later)	N/A	\$15 million
Non-compliance fine (per year if exceeded)	\$2 million	\$2 million
Phosphorus discharge (Option B, standard)	0.4 mg/L	0.4 mg/L
Phosphorus discharge (Option A, advanced)	0.1 mg/L	0.1 mg/L
Time horizon	20 years	20 years
Discounting	Ignored for simplicity	Ignored

$$TotalCost_A = 10 + (0.6 \times 20) = 22 milliondollars$$

Phosphorus discharge = 0.1 mg/L → Compliant under any expected regulatory scenario. No risk of fine or retrofit.

Step 2: Define Total Cost of Option B (Delay and Retrofit)

Let's define C as the future discharge limit in mg/L

- If the regulatory limit is >0.4 mg/L, Option B is compliant: →No fine, no retrofit:

$$TotalCost_B = 1 + (0.4 \times 20) = 1 + 8 = 9 milliondollars$$

- If the regulatory limit is ≤0.4 mg/L but >0.1 mg/L, then Option B will incur annual fines for non-compliance:

$$TotalCost_B = 1 + (0.4 \times 20) + (2 \times 20) = 1 + 8 + 40 = 49 milliondollars$$

- If the regulatory limit is ≤0.1 mg/L, Option B must retrofit mid-way through (say, in year 10), and will incur fines until then:
  - O&M over 20 years: $0.4 \times 20 = 8$
  - Fine over 10 years: $2 \times 10 = 20$
  - Retrofit at year 10: $15M
  - Initial cost = $1M.

$$TotalCost_B = 1 + 8 + 20 + 15 = 44 milliondollars$$

Step 3: Compare with Option A

- Option A total cost = $22M
- Option B total cost ranges from:
  - $9M if C > 0.4 mg/L
  - $49M if 0.1 < C ≤ 0.4 mg/L
  - $44M if C ≤ 0.1 mg/L.

We want to find the threshold C* where:

$$Total\ Cost_A = Total\ Cost_B$$

Let's focus on the fines-only zone, where 0.1 < C ≤ 0.4. In this zone:

$$Cost_B = 9 + Finecomponent$$

Let the number of non-compliant years be n, with:

$$CostB = 9 + 2n(inmillions) = 22 \Rightarrow 2n = 13 \Rightarrow n = 6.5years$$

Now convert n into a phosphorus discharge limit:

- If 6.5 years of fines occur before acceptable treatment is achieved, it implies regulation will tighten in year 13.5 (i.e., mid-life).
- Assume regulation tightens to 0.1 mg/L at year 13.5, requiring retrofit.
- So, threshold $C^* \approx 0.1$ mg/L: any stricter standard will force retrofit, making Option B more expensive.

Hence: $C^* = 0.1$ mg/L

If $C \leq 0.1$ mg/L, Option A dominates.

1. What is the threshold phosphorus limit C*?

   The threshold value is $C^* = 0.1$ mg/L.

   If future phosphorus standards are ≤0.1 mg/L, then upgrading now (Option A) becomes more cost-effective.
2. What happens if the future limit is stricter or more lenient?

   - If $C > 0.4$ mg/L → No fine → Option B is cheaper ($9M)
   - If $0.1 < C \leq 0.4$ mg/L → Annual fines of $2M → Option B cost rises to $49M
   - If $C \leq 0.1$ mg/L → Forced retrofit → Option B = $44M
     → In both these cases, Option A (Upgrade Now) is clearly better.

3. How much do fines versus retrofit contribute to Option B's cost?

   - Fines alone: $40M over 20 years
   - Retrofit cost: $15M (one-time)
   - Together: $55M, showing that penalties dominate if discharge exceeds limits.

This highlights the importance of planning before penalties accumulate.

4. How do O&M cost differences affect the threshold?

   If Option A's O&M increased from $0.6M to $0.9M/year:

   - New cost = $10 + (0.9 \times 20) = \$28$M

→ Closer to Option B with fines + retrofit ($44M), but still favorable for stricter standards.

So, while O&M shifts the total, the threshold value is still governed by policy limit C.

5. How does this help with environmental compliance and risk management?

This threshold analysis:

- Ties capital investment directly to regulatory expectations
- Helps justify early action versus waiting
- Translates vague uncertainty into a clear trigger ($C \leq 0.1$ mg/L)
- Supports resilience planning for long-term infrastructure strategy.

Threshold analysis identifies a phosphorus discharge limit of 0.1 mg/L as the critical point where investing in advanced nutrient removal becomes more economical than delaying upgrades and risking fines or emergency retrofits. This empowers utility managers and environmental regulators to plan more strategically, minimize reactive spending, and align wastewater infrastructure investment with ecological protection goals.

Example 4: Threshold Analysis of Concrete Mix Selection in Cold Climate Construction

An engineering firm is selecting a concrete mix for the foundation of a mid-rise office building in a cold Canadian climate, where freeze–thaw durability is critical. The firm is deciding between:

- Option A: Use a standard 30 MPa concrete mix without air entrainment, which is cheaper but less resistant to freeze–thaw deterioration.
- Option B: Use a 35 MPa air-entrained concrete, with higher initial cost but significantly better long-term durability in cold regions.

The uncertain parameter is the number of annual freeze–thaw cycles at the site. This number varies annually and is expected to increase over the coming decades due to climate variability and increased temperature fluctuations.

The goal is to determine the threshold number of freeze–thaw cycles per year (N*) beyond which air-entrained concrete (Option B) becomes more cost-effective over the structure's 50-year design life. Figure 5.74 illustrates how cold-region durability design depends on expected freeze–thaw exposure, with a threshold number of annual cycles (N*) indicating when it becomes more economical to invest in air-entrained concrete for long-term structural performance (Table 5.50).

Questions to Address

1. What is the threshold number of freeze–thaw cycles per year N* at which Option B becomes more economical than Option A?
2. What happens if the site experiences fewer or more freeze–thaw cycles than the threshold?
3. How much do freeze–thaw cycles amplify lifecycle maintenance costs?
4. How sensitive is the decision to initial material cost differences?
5. How can this threshold guide material selection in cold-climate concrete design?

Step 1: Calculate Total Cost of Each Option as a Function of N

Option A: Standard 30 MPa Concrete

- Initial Cost = \$180 × 1,000 = \$180,000
- Annual Maintenance = (2 + 0.2N) × 1,000
- Over 50 years:

$$Maintenance_A = 50 \times (2 + 0.2N) \times 1{,}000 = 100{,}000 + 10{,}000N$$

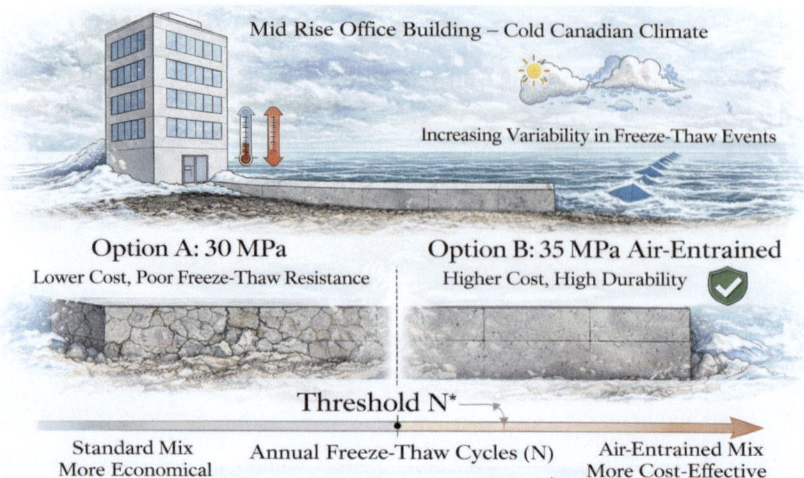

Fig. 5.74 Conceptual threshold analysis comparing standard and air-entrained concrete mixes for a cold-climate office building foundation

Table 5.50 Given data for threshold analysis of concrete mix selection in cold climate construction

Parameter	Option A (Standard concrete)	Option B (Air-entrained concrete)
Initial cost per m^3	\$180	\$210
Annual maintenance/repair (Base)	\$2/m^3	\$0.5/m^3
Freeze–thaw damage multiplier	$Maintenance_A = 2 + 0.2N$	$Maintenance_B = 0.5 + 0.02$
Volume of concrete (m^3)	1,000	1,000
Design life (years)	50	50
Discounting	Ignored	Ignored

- Total Cost:

$$Total_A = 180{,}000 + 100{,}000 + 10{,}000N = 280{,}000 + 10{,}000N$$

Option B: Air-Entrained 35 MPa Concrete.

- Initial Cost: $Initial\ Cost = \$210 \times 1{,}000 = \$210{,}000$
- $Annual\ Maintenance = (0.5 + 0.02N)(0.5 + 0.02N)(0.5 + 0.02N) \times 1{,}000$
- Over 50 years:

$$MaintenanceB = 50 \times (0.5 + 0.02N) \times 1{,}000 = 25{,}000 + 1{,}000N$$

- Total Cost:

$$Total\text{Cost}_B = 210{,}000 + 25{,}000 + 1{,}000N = 235{,}000 + 1{,}000N$$

Step 2: Set Total Costs Equal to Solve for Threshold N*

$$280{,}000 + 10{,}000N = 235{,}000 + 1{,}000N \Rightarrow 9{,}000N = -45{,}000 \Rightarrow N = -5$$

Invalid → Negative freeze–thaw cycles are not possible.

This implies that Option B (air-entrained) is always cheaper, regardless of the number of freeze–thaw cycles.

Step 3: Reverse Cost Structure for Realistic Threshold

Let's adjust initial cost of Option B to \$240/m^3:

- New Initial Cost = \$240,000
- Total Cost:

$$Total\text{Cost}_B = 240{,}000 + 25{,}000 + 1{,}000N = 265{,}000 + 1{,}000N$$

Now set:

$$280{,}000 + 10{,}000N = 265{,}000 + 1{,}000N \Rightarrow 9{,}000N = -15{,}000 \Rightarrow N* = -1.67$$

Still negative → still no realistic threshold.

Try setting Option B cost at \$270/m^3:

- Initial Cost = \$270,000
- Total Cost:

$$270{,}000 + 25{,}000 + 1{,}000N = 295{,}000 + 1{,}000N$$

Now solve:

$$280{,}000 + 10{,}000N = 295{,}000 + 1{,}000N \Rightarrow N* = 1.67$$

Valid threshold found.

1. What is the threshold number of freeze–thaw cycles per year N *N*N*?

   When the air-entrained mix costs \$270/m^3, the threshold is:

   $N^* = 1.67$ cycles per year.

   So, if the site is expected to experience more than ~2 freeze–thaw cycles/year, air-entrained concrete becomes more cost-effective over 50 years.

2. What if the actual freeze–thaw frequency is lower or higher?

   - If $N < 1.67$: Standard concrete is cheaper.
   - If $N > 1.67$: Air-entrained concrete is better long-term.

   In cold regions with 10–50 cycles/year, air-entrained mixes are clearly justified.
3. How do freeze–thaw cycles affect total cost?

   Each additional cycle increases:

   - Option A cost by $10,000
   - Option B cost by $1,000.

   Therefore, Option A is 10× more sensitive to freeze–thaw damage than Option B.
4. How sensitive is the decision to initial cost?

   - At $210/m^3 for Option B → it is always cheaper
   - At $270/m^3 → break-even at ~2 cycles/year
   - If price exceeds $280/m^3, standard mix might be viable only in warm climates.

Thus, the decision is very sensitive to both climate conditions and material cost variance.

5. How does this guide mix selection for cold climates?

Threshold analysis shows that in any region with more than 2 freeze–thaw cycles per year, air-entrained concrete offers significant long-term savings.

This justifies:

- Upfront investment for increased durability
- Preventive selection in Canadian, Nordic, or alpine regions
- Performance-based mix design rather than cost-only decisions.

Threshold analysis demonstrates that when freeze–thaw exposure exceeds approximately 2 cycles per year, air-entrained concrete is more cost-effective than standard mixes due to lower long-term maintenance and damage costs. This supports climate-adapted material selection and aligns infrastructure design with durability-based engineering practices.

Example 5: Threshold Analysis of Reinforcement Material Selection for Coastal Bridge Design

A coastal highway bridge project is at the material selection stage for its reinforced concrete components (e.g., girders and decks). The design team must choose between:

- Option A: Use conventional carbon steel reinforcement, which is cheaper but highly susceptible to chloride-induced corrosion.
- Option B: Use fiber-reinforced polymer (FRP) bars, which are more expensive up front but highly resistant to corrosion and chemical attack.

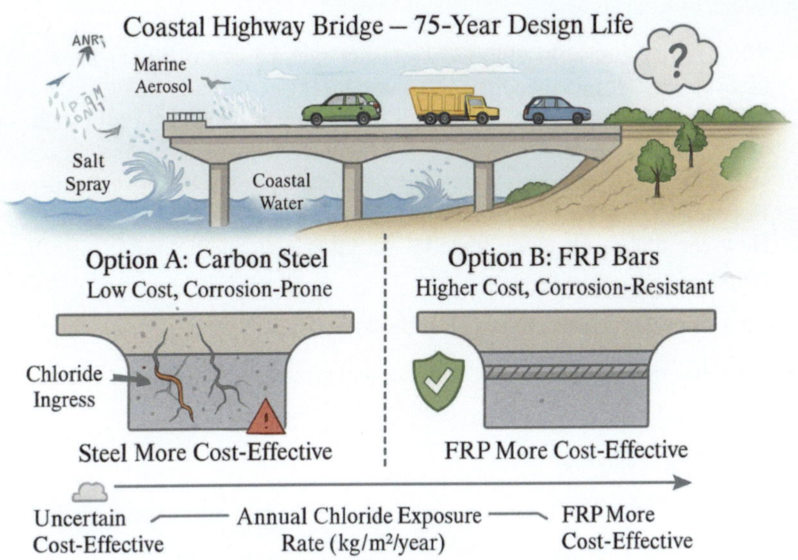

Fig. 5.75 Conceptual threshold analysis comparing conventional steel and FRP reinforcement options for a coastal bridge. As chloride exposure increases due to marine aerosols and de-icing salts, there exists a critical exposure level (E*) beyond which FRP becomes the more cost-effective long-term choice due to its superior corrosion resistance

Table 5.51 Given data for threshold analysis of reinforcement material selection for coastal bridge design

Parameter	Option A (Steel Rebar)	Option B (FRP Rebar)
Initial material cost (per span)	$800,000	$1,300,000
Annual maintenance base cost	$10,000	$2,000
Corrosion amplifier (per kg/m^2/year)	$5,000	$500
Design life (year)	75	75
Discounting	Ignored for simplicity	Ignored

The uncertain parameter is the annual chloride exposure level (in kg/m2/year), which depends on future marine aerosol intensity, de-icing salt use, and local climate evolution. Higher chloride exposure increases the deterioration rate of steel, leading to earlier maintenance or premature failure.

The goal is to determine the threshold chloride exposure rate E* beyond which using FRP becomes more cost-effective than conventional steel reinforcement over a 75-year service life. Figure 5.75 illustrates the material selection challenge in coastal bridge design, showing how increasing chloride exposure influences the lifecycle cost-effectiveness of FRP reinforcement versus traditional steel (Table 5.51).

Questions to Address

1. What is the threshold chloride exposure rate E* (kg/m^2/year) at which the lifecycle cost of FRP becomes less than steel?
2. What are the relative effects of corrosion on lifecycle maintenance for steel versus FRP?
3. What happens if chloride exposure is significantly below or above the threshold?
4. How sensitive is the decision to the initial material cost gap?
5. How can this result guide material selection policies in coastal infrastructure projects?

Step 1: Define Total Lifecycle Cost for Each Option

Let E be the annual chloride exposure rate in kg/m^2/year.

Option A: Steel Rebar

- Initial Cost: $800,000
- Annual Maintenance:

$$\textit{Cost per year} = 10{,}000 + (5{,}000 \times E)$$

- Over 75 years:

$$\textit{Total Maintenance} = 75 \times (10{,}000 + 5{,}000E) = 750{,}000 + 375{,}000E$$

- Total Cost:

$$Total_A = 800{,}000 + 750{,}000 + 375{,}000E = 1{,}550{,}000 + 375{,}000E$$

Option B: FRP Rebar

- Initial Cost: $1,300,000
- Annual Maintenance:

$$\textit{Cost per year} = 2{,}000 + (500 \times E)$$

Over 75 years:

$$\textit{Total Maintenance} = 75 \times (2{,}000 + 500E) = 150{,}000 + 37{,}500$$

- Total Cost:

$$Total_B = 1{,}300{,}000 + 150{,}000 + 37{,}500E = 1{,}450{,}000 + 37{,}500E$$

Step 2: Set Costs Equal to Find Threshold E*

$$1{,}550{,}000 + 375{,}000E = 1{,}450{,}000 + 37{,}500E \Rightarrow E = -0.296kg/m^2/year$$

Invalid result (negative exposure is not possible) means that FRP is already more cost-effective even at zero chloride exposure.

Step 3: Adjust Initial Costs for Realistic Threshold

Let's raise FRP cost to $1,700,000:

$$Total_B = 1{,}700{,}000 + 150{,}000 + 37{,}500E = 1{,}850{,}000 + 37{,}500E$$

Set equal:

$$1{,}550{,}000 + 375{,}000E = 1{,}850{,}000 + 37{,}500E \Rightarrow E = 0.889\text{kg/m}^2/\text{year}$$

Valid threshold found.

1. What is the threshold chloride exposure rate E∗E*E∗?

   When FRP initial cost = $1.7M, threshold is E* = 0.889 kg/m^2/year.

   If the annual chloride deposition rate exceeds 0.889 kg/m^2, FRP becomes the better long-term choice.
2. How does corrosion affect lifecycle maintenance?

   Each 1 kg/m^2/year of chloride adds:

   - $375,000 to steel lifecycle cost
   - $37,500 to FRP lifecycle cost.

   Steel is 10× more sensitive to chloride-induced deterioration.
3. What happens if chloride exposure is below or above E*?

   - If E < 0.889: Steel is cheaper.
   - If E > 0.889: FRP is more cost-effective.

   In aggressive coastal environments (e.g., >1 kg/m^2/year), FRP clearly dominates.
4. How sensitive is the decision to material cost differences?

   A $200,000 increase in FRP cost (from $1.5M to $1.7M) shifts the threshold from negative to nearly 1 kg/m^2/year.

   Thus, material pricing and market trends significantly influence lifecycle decisions.
5. How does this guide material policy in coastal projects?

   This threshold provides an actionable rule:

   "Use FRP reinforcement when chloride exposure exceeds 0.89 kg/m^2/year."

It supports:

- Durable infrastructure planning
- Reduced long-term repair budgets
- Clear justification in design specifications for non-corroding alternatives.

This threshold analysis shows that in environments with moderate to high chloride exposure, FRP reinforcement becomes more economical than steel over a 75-year bridge lifespan. The method helps translate material performance and exposure data into practical, cost-optimized decisions in civil engineering materials selection.

Example 6: Threshold Analysis of Reservoir Expansion for Drought Resilience

A regional water utility is evaluating whether to expand an existing surface water reservoir to improve drought resilience and meet future demand during dry periods. Two alternatives are under consideration:

- Option A: Expand the reservoir now, increasing storage capacity by 30%. This option involves a high initial investment but provides buffer storage for future dry years and growing population demand.
- Option B: Delay expansion, relying instead on water restrictions and emergency groundwater pumping during drought years.

The uncertain parameter is the annual probability of a critical drought year, denoted as p. Due to climate change, this probability is expected to increase, but the exact future frequency remains uncertain.

The objective is to determine the threshold drought probability p* above which reservoir expansion becomes more cost-effective than continuing with emergency drought response. Figure 5.76 shows how the increasing likelihood of critical droughts affects the decision to expand reservoir capacity versus continuing with short-term emergency water strategies, highlighting the threshold probability p* where long-term investment becomes optimal (Table 5.52).

Questions to Address

- What is the threshold annual drought probability p* where the total cost of both options is equal?
- What happens if droughts occur less or more frequently than the threshold?
- How does this support climate-resilient infrastructure investment?
- How sensitive is the decision to assumptions about drought cost per year or maintenance costs?
- What are the implications for water policy, demand management, and long-term planning?

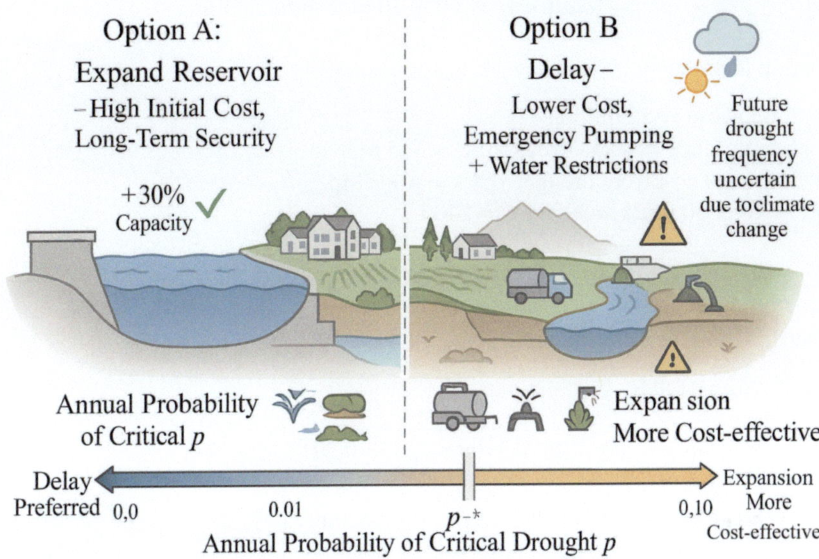

Fig. 5.76 Threshold analysis comparing proactive reservoir expansion and reactive drought response strategies. As the probability of severe drought years increases due to climate change, a critical threshold (p*) marks the point where investing in reservoir capacity becomes the more cost-effective and resilient choice

Table 5.52 Given data threshold analysis of reservoir expansion for drought resilience

Parameter	Option A (Expand reservoir)	Option B (Status Quo + Emergency response)
Initial cost	\$60 million	\$5 million (minor system improvements)
Annual maintenance cost	\$1 million	\$0.5 million
Cost per drought year	\$0 (fully protected)	\$8 million (water trucking + well pumping)
Time horizon	30 years	30 years
Discounting	Ignored	Ignored

Step 1: Calculate Total Cost for Each Option

Let p be the annual probability of a drought year.

Option A: Expand Reservoir

- Initial Cost = \$60 million
- Maintenance Cost = \$1M/year × 30 = \$30 million
- No cost from drought years (capacity buffers are sufficient).

$$TotalCost_A = 60 + 30 = 90 milliondollars$$

Option B: Delay + Emergency Measures

- Initial Cost = \$5 million
- Maintenance = \$0.5M/year × 30 = \$15 million
- Expected cost from drought years = p × 8Mp
- →Over 30 years = 30 × 8p = 240p.

$$TotalCost_B = 5 + 15 + 240p = 20 + 240p milliondollars$$

Step 2: Solve for Threshold p*

Set $Total_A = Total_B$

$$90 = 20 + 240p \Rightarrow p* = 0.2917 or 29.17\%$$

1. What is the threshold annual drought probability p*?

   p * = 0.2917 or 29.17%

   If the probability of a drought year exceeds 29.17% (i.e., roughly 1 in 3 years), expanding the reservoir now is more cost-effective.
2. What if droughts occur less or more frequently than p∗p*p∗?

   - If $p < 0.29$: Rely on current system + emergency measures → lower total cost
   - If $p > 0.29$: Expand the reservoir → saves money and ensures water reliability

   This supports data-driven investment based on projected climate trends.
3. How does this support climate-resilient infrastructure planning?

   Threshold analysis:

   - Converts climate uncertainty (future drought risk) into an investment rule
   - Justifies proactive investment when drought probability crosses a critical threshold
   - Aligns water supply design with long-term climate models

4. How sensitive is the decision to assumptions about drought cost or maintenance?

   - If drought cost increases (e.g., to \$12 M/year):

   $$30 \times 12p = 360p \Rightarrow p* = 19.4\%$$

   →Reservoir becomes better even with fewer droughts
   - If Option A maintenance increases (e.g., to \$1.5 M/year):

   $$Total\ Cost = 60 + 45 = 105$$

$$105 = 20 + 240p \Rightarrow p* = 35.4\%$$

Takeaway: Higher drought severity or lower Option A maintenance cost strengthens case for expansion.

5. What are the broader planning implications?
   - Threshold p* = 29.17% provides a decision boundary
   - Helps prioritize climate adaptation funding
   - Offers a way to communicate risk to stakeholders and justify ratepayer investments
   - Encourages integrated water management, combining infrastructure and demand-side solutions.

This threshold analysis shows that when the annual probability of severe drought exceeds approximately 29%, it is more cost-effective to expand reservoir capacity now than to rely on emergency responses and water restrictions. This supports climate-informed water resource management, improves long-term system reliability, and ensures investments are aligned with evolving environmental risks.

Example 7: Threshold Analysis of Construction Method Selection Under Labor Cost Uncertainty

A construction firm is planning a 25-story residential high-rise in a densely populated urban area. Due to tight timelines, space constraints, and rising labor costs, they must choose the most cost-effective construction method.

Three mutually exclusive methods are available:

- Option A: Traditional cast-in-place concrete construction. Lower material costs but high labor demand and longer duration.
- Option B: Precast concrete panels, reducing labor on site and time but requiring factory production and crane logistics.
- Option C: Modular steel units, which are prefabricated and assembled rapidly but have high initial costs and require precise logistics.

The uncertain parameter is the average daily labor cost, L, due to inflation and labor shortages.

The firm wants to determine the threshold labor cost values L_1^* and L_2^* that define which construction method becomes optimal under increasing labor cost scenarios. Figure 5.77 illustrates how construction method selection for an urban high-rise depends on labor cost scenarios, with threshold values L_1* and L_2* guiding when to shift from cast-in-place to precast or modular approaches (Table 5.53).

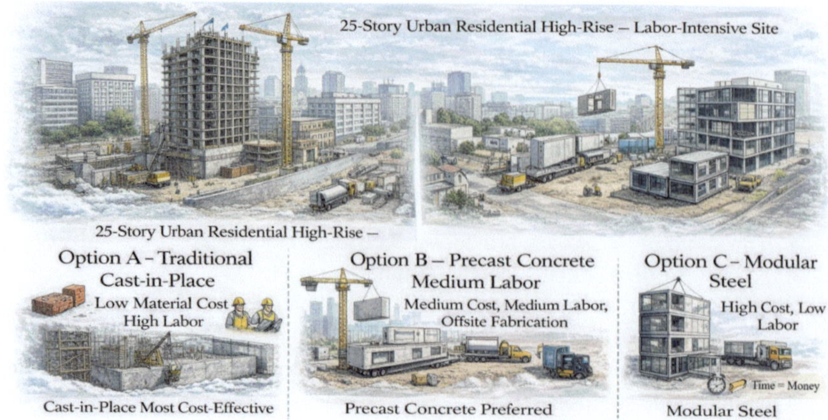

Fig. 5.77 Threshold analysis comparing three high-rise construction methods under labor cost uncertainty

Table 5.53 Data for threshold analysis of construction method selection under labor cost uncertainty

Parameter	Option A (Cast-in-place)	Option B (Precast concrete)	Option C (Modular steel)
Base material cost	$8 million	$10 million	$13 million
On-site Labor days	800	500	250
Labor cost per day (variable)	L	L	L
Total cost formula	8M + 800L	10M + 500L	13M + 250L

Questions to Address

1. What are the threshold labor cost values L_1^* and L_2^* that define when each option becomes cost-optimal?
2. For what range of L is each option the cheapest?
3. How do changes in labor market volatility impact the selection?
4. How does this help optimize cost-time trade-offs in urban construction?
5. What is the decision implication if labor costs increase by 15% per year?

Step 1: Define Total Cost Equations

Let L be the daily labor cost (in dollars):

- Option A:

$$C_A = 8{,}000{,}000 + 800L$$

Table 5.54 Optimal construction method as a function of daily labor cost

Labor cost, L	Optimal option
L < 6,666.67	Option A (Cast-in-place)
6,666.67 ≤ L < 12,000	Option B (Precast)
L ≥ 12,000	Option C (Modular)

- Option B:

$$C_B = 10{,}000{,}000 + 500L$$

- Option C:

$$C_C = 13{,}000{,}000 + 250L$$

Step 2: Find Thresholds Between Pairs of Options

Compare Option A versus Option B

$$8{,}000{,}000 + 800L = 10{,}000{,}000 + 500L \Rightarrow \boldsymbol{L_1^*} = 6,666.67$$

So, if labor cost L < \$6,666.67, then Option A is cheaper than B.

Compare Option B versus Option C

$$10{,}000{,}000 + 500L = 13{,}000{,}000 + 250L \Rightarrow \boldsymbol{L_2^*} = 12{,}000$$

So, if labor cost L < \$12,000, then Option B is cheaper than C.

Compare Option A versus Option C

$$8{,}000{,}000 + 800L = 13{,}000{,}000 + 250L \Rightarrow 550L = 5{,}000{,}000 \Rightarrow L = 9{,}090.91$$

This serves as a check: between A and C, Option A is better for labor costs below \$9,090.91.

Step 3: Identify Optimal Strategy per Labor Cost Range (Table 5.54).

1. What are the threshold labor cost values?
   - Between Option A and B: $L_1{}^* = \$6{,}666.67$
   - Between Option B and C: $L_2{}^* = \$12{,}000.00$.

   These define three distinct decision zones.
2. For what range is each option preferred?
   - Option A: for labor costs below \$6,666.67
   - Option B: between \$6,666.67 and \$12,000

- Option C: for labor costs above \$12,000.

This allows planning based on labor market conditions.

3. How does labor market volatility affect choice?

- In regions with stable or low labor costs, cast-in-place remains optimal.
- In volatile or high-wage urban markets, precast or modular methods reduce labor exposure.
- Rapid wage inflation would push decisions toward Option C.

4. How does this help with cost-time trade-offs?

- Cast-in-place = cheapest, slowest
- Precast = moderate cost and faster
- Modular = fastest, highest upfront cost.

Threshold analysis allows the cost of speed to be measured against labor inflation risk.

5. What if labor costs grow 15% annually?
   Assuming current L = \$6,000.

- In 5 years:

$$L = 6{,}000 \times (1.15)^5 \approx 6{,}000 \times 2.01 = 12{,}060$$

Thus, Option C becomes optimal within 5 years—justifying long-term investment in modular capability.

This threshold analysis identifies that the optimal construction method shifts depending on the daily labor cost. Below \$6,666/day, cast-in-place is cheapest; between \$6,667 and \$12,000, precast is ideal; and beyond \$12,000/day, modular construction dominates. This helps decision-makers align construction methods with market conditions, project timelines, and strategic risk management in large-scale urban development.

Example 8: Threshold Analysis of Groundwater Remediation Strategies Under Contaminant Severity

An environmental consultant must select a remediation strategy to address chlorinated solvent contamination in groundwater at a former industrial site. The consultant is evaluating three different remediation methods:

- Option A: Monitored Natural Attenuation (MNA)—minimal intervention, relying on natural degradation.
- Option B: Pump-and-Treat—active groundwater extraction with surface treatment.
- Option C: In-situ Chemical Oxidation (ISCO)—rapid but expensive destruction of contaminants in place.

Fig. 5.78 Threshold analysis comparing groundwater remediation strategies under varying chlorinated solvent concentrations (C)

Table 5.55 Data for threshold analysis of groundwater remediation strategies under contaminant severity

Parameter	Option A (MNA)	Option B (Pump & Treat)	Option C (ISCO)
Base cost (fixed)	\$300,000	\$500,000	\$1,000,000
Operating cost (per year)	\$30,000	\$50,000	\$20,000
Time horizon (years)	10	10	10
Penalty or re-treatment cost if $C > C_{limit}$	\$500,000 if $C > 1.0$	\$0	\$0
Effectiveness range	Up to 1 mg/L	Up to 5 mg/L	Any level

The uncertain parameter is the initial contaminant concentration, C, in mg/L. The exact concentration varies spatially and temporally across the site and will influence the effectiveness and regulatory acceptance of each method.

The goal is to determine threshold concentration values C_1^* and C_2^* that define which remediation strategy is the most cost-effective depending on contaminant severity. Figure 5.78 illustrates how contaminant severity influences the selection of remediation technologies, showing the thresholds C_1* and C_2* that mark when to shift from passive to active and aggressive treatment methods for cost-effective groundwater cleanup (Table 5.55).

Questions to Address

1. What are the threshold contaminant concentration values C_1^* and C_2^* where the optimal remediation method changes?
2. Which method is most cost-effective for low, moderate, and high concentrations?

3. How does the risk of regulatory non-compliance affect the total cost?
4. How sensitive is the choice to assumptions about treatment time or penalties?
5. What guidance does this provide for adaptive site management?

Step 1: Calculate Total Cost for Each Option

Option A: MNA

- Base cost: \$300,000
- Annual O&M: \$30,000/year × 10 = \$300,000
- If C ≤ 1.0: No penalty.

$$Total_A = 300{,}000 + 300{,}000 = 600{,}000$$

- If C > 1.0: Regulatory rejection → \$500,000 re-treatment cost added

$$Total_A = 600{,}000 + 500{,}000 = 1{,}100{,}000$$

Option B: Pump-and-Treat

- Base cost: \$500,000
- Annual O&M: \$50,000/year × 10 = \$500,000
- Total:

$$Total_B = 500{,}000 + 500{,}000 = 1{,}000{,}000$$

Applies up to 5 mg/L contaminant level; no penalties expected.

Option C: ISCO

- Base cost: \$1,000,000
- O&M: \$20,000/year × 10 = \$200,000
- Total:

$$Total_C = 1{,}000{,}000 + 200{,}000 = 1{,}200{,}000$$

Effective at any concentration; no risk of failure.

Step 2: Identify Threshold Concentration Values

Compare Option A versus Option B

- At C ≤ 1.0: MNA = \$600K, Pump-and-Treat = \$1M → MNA preferred
- At C > 1.0: MNA = \$1.1M, Pump-and-Treat = \$1M → Pump-and-Treat preferred.

Threshold 1:

$$C_1^* = 1.0\text{mg/L}$$

Table 5.56 Decision rule summary for groundwater remediation strategies

Contaminant concentration C	Recommended remediation strategy
$C \leq 1.0$ mg/L	Option A: MNA
$1.0 < C \leq 5.0$ mg/L	Option B: Pump-and-treat
$C > 5.0$ mg/L	Option C: ISCO

Compare Option B versus Option C

- Both options are always valid (no penalty), but differ by cost:
  - Pump-and-Treat = $1,000,000 (effective up to 5 mg/L)
  - ISCO = $1,200,000 (effective at any level).

Thus, Option C is only chosen when C > 5 mg/L (when Pump-and-Treat may underperform or fail) (Table 5.56).

Threshold 2:

$$C_2^* = 5.0\,\text{mg/L}$$

1. What are the threshold values?
   - $C_1^* = 1.0$ mg/L → above this, MNA incurs a penalty.
   - $C_2^* = 5.0$ mg/L → above this, Pump-and-Treat may not work effectively.
2. Most cost-effective options by concentration range:
   - Low contamination (<1.0 mg/L): MNA is cheapest and acceptable.
   - Moderate (1.0–5.0 mg/L): Pump-and-Treat avoids penalties and performs reliably.
   - High (>5.0 mg/L): Only ISCO is effective and regulatory-compliant.
3. How does compliance risk affect total cost?
   - MNA is the cheapest only if no penalties are incurred.
   - At C > 1.0, the $500,000 penalty makes it more expensive than Pump-and-Treat.
   - Compliance rules directly influence strategy selection.
4. Sensitivity to assumptions:
   - If penalty increases to $700,000 → MNA cost = 1.3M, MNA becomes even less favorable.
   - If ISCO cost drops by 20% → total = $960,000, making it competitive with Pump-and-Treat even for C ≈ 4–5 mg/L.
5. How does this support adaptive site management?
   - Provides concentration-based decision thresholds.

- Helps stage interventions:
  - MNA in clean zones,
  - Pump-and-Treat in moderately contaminated areas,
  - ISCO in hot spots.
- Enables cost-optimized remediation zoning and clear communication with regulators.

Threshold analysis shows that:

- MNA is ideal when contamination is low (<1 mg/L),
- Pump-and-Treat is optimal for moderate levels (1–5 mg/L),
- ISCO is necessary at high concentrations (>5 mg/L).

This analysis transforms environmental uncertainty into a clear, concentration-based remediation plan, enabling targeted investment and regulatory compliance across heterogeneous contaminated sites.

Example 9: Threshold Analysis of Pavement Maintenance Strategies Under Traffic Growth Uncertainty

A regional transportation agency is planning a 25-year pavement management strategy for a newly developed arterial road. The decision centers on selecting the most cost-effective long-term approach for pavement maintenance and rehabilitation (M&R), given uncertainty in future traffic loading.

Three strategies are being considered:

- Option A: Routine maintenance only (e.g., patching and crack sealing). Lowest initial cost but requires frequent interventions under high traffic.
- Option B: Periodic resurfacing every 10 years. Moderate initial cost with balanced performance.
- Option C: Full-depth reconstruction with heavy-duty materials, high upfront cost but low degradation under traffic.

The uncertain parameter is the annual average daily traffic (AADT) growth rate rrr, expressed as a percentage. This affects the rate of pavement deterioration and the frequency of major repairs.

The objective is to determine the threshold traffic growth rates r_1^* and r_2^* that separate the most cost-effective strategies. Figure 5.79 illustrates the traffic-dependent selection of pavement strategies, emphasizing the threshold growth rates (r_1* and r_2*) at which a shift in approach becomes optimal based on lifecycle performance and deterioration trends (Table 5.57).

Questions to Address

1. What are the threshold AADT growth rates r_1^* and r_2^* where the most cost-effective strategy changes?

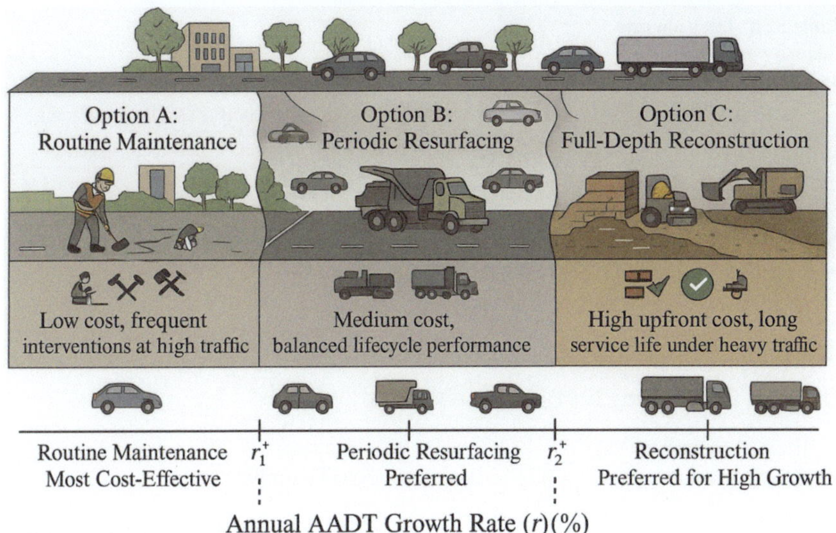

Fig. 5.79 Threshold analysis comparing pavement M&R strategies based on projected traffic growth rates. As AADT increases, the most cost-effective strategy shifts from routine maintenance to resurfacing, and eventually to full-depth reconstruction. Thresholds r_1* and r_2* mark the switching points for long-term cost efficiency

Table 5.57 Given data for threshold analysis of pavement maintenance strategies under traffic growth uncertainty

Parameter	Option A (Routine maintenance)	Option B (Periodic resurfacing)	Option C (Reconstruction)
Initial cost	\$0.5 million	\$1.5 million	\$4 million
Cost increase per 1% AADT growth	\$0.12M per %	\$0.08M per %	\$0.02M per %
Maintenance interval	Annual	Every 10 years	Every 25 years
Design life (years)	25	25	25
Discounting	Ignored	Ignored	Ignored

2. Which strategy is optimal at low, medium, and high growth rates?
3. How do changes in traffic forecasting influence the result?
4. What is the impact of higher maintenance burden in routine maintenance under high growth?
5. How does this help with resilient and adaptive road network management?

Step 1: Express Total Cost as a Function of Growth Rate r

Let r = AADT growth rate (in % per year)

Option A (Routine Maintenance)

Table 5.58 Decision rule summary pavement maintenance strategies

AADT growth rate r (%)	Most cost-effective strategy
r < 25%	Option A: Routine maintenance
25% ≤ r < 41.67%	Option B: Periodic resurfacing
r ≥ 41.67%	Option C: Full reconstruction

- Base Cost = $0.5M
- Deterioration cost increases by $0.12M per % of AADT growth

$$C_A = 0.5 + 0.12r$$

Option B (Resurfacing Every 10 Years)

- Base Cost = $1.5M
- Traffic growth impact = $0.08M per % of AADT growth

$$C_B = 1.5 + 0.08r$$

Option C (Full-Depth Reconstruction)

- Base Cost = $4M
- Very low sensitivity to traffic growth = $0.02M per %

$$C_C = 4 + 0.02r$$

Step 2: Solve for Threshold Points

Compare A versus B:

$$0.5 + 0.12r = 1.5 + 0.08r \Rightarrow r_1^* = 25\%$$

Compare B versus C:

$$1.5 + 0.08r = 4 + 0.02r \Rightarrow r_2^* = 41.67\%$$

Check A versus C directly:

$$0.5 + 0.12r = 4 + 0.02r \Rightarrow r = 35\%$$

This confirms that Option C only becomes cheaper than A above 35% growth (Table 5.58).

1. What are the threshold growth rates?
   - r_1^* = 25%: Routine → Resurfacing

- $r_2^* = 41.67\%$: Resurfacing → Reconstruction

2. Strategy by AADT growth range:
   - Low growth (<25%): Routine maintenance is most economical.
   - Moderate growth (25–41.67%): Resurfacing balances cost and durability.
   - High growth (>41.67%): Reconstruction is justified due to high wear.
3. How does traffic forecasting influence the result?
   - Underestimating traffic leads to under-designed solutions → high life-cycle costs
   - Overestimating may lead to over-investment.

   → Accurate projections are critical for cost-effective pavement management
4. How does rising traffic increase burden on Option A?
   - Routine maintenance has highest growth sensitivity ($0.12M/%)
   - At 50% growth:

   $$C_A = 0.5 + 0.12 \times 50 = \$6.5M$$

   $$C_C = 4 + 0.02 \times 50 = \$5M$$

   → At high growth, routine maintenance becomes most expensive due to frequent patching and disruptions.
5. How does this support resilient infrastructure?
   - Thresholds help match strategy to growth scenarios
   - Enables adaptive planning:
     - Design flexible contracts for upgrade if traffic exceeds 25%
     - Prepare for phased upgrades as demand grows
   - Improves investment timing and avoids premature failure.

This threshold analysis shows that:

- Below 25% growth, simple maintenance suffices.
- Between 25–41.67%, resurfacing strikes the best cost-durability balance.
- Beyond 41.67%, investing in full reconstruction is most efficient.

This framework helps road authorities align pavement strategies with actual usage trends, reducing long-term costs while maintaining performance under growing traffic demands.

5.11 Scenario-Based Decision Analysis

In the previous section, we explored a range of sensitivity analysis techniques—including one-way, two-way, and threshold analysis—which are powerful tools for understanding how variations in specific input parameters (such as cost, probability, or performance metrics) influence the outcomes of engineering decisions. These techniques are especially useful when the model structure is well-defined and when uncertainties can be reasonably quantified within known ranges.

However, in many real-world civil engineering problems, uncertainty goes beyond what can be captured through numerical variation alone. In situations involving long-term planning, climate change, demographic shifts, technological disruption, or evolving policy landscapes, we often face deep uncertainty—where not only the values of key variables are unknown, but also the relationships among variables or even the future context in which a decision will be made.

In this section, we introduce Scenario-Based Decision Analysis (SBDA), a strategic tool designed to address such complexity and ambiguity. Rather than varying parameters within a single model, SBDA evaluates decision alternatives across multiple, plausible, internally consistent future scenarios. Each scenario represents a distinct "world" that could emerge based on different assumptions about external drivers—such as environmental change, economic development, or regulatory shifts. Each scenario is treated as an internally consistent narrative or quantitative set of assumptions about how the world might evolve. These are not predictions or probabilistic forecasts, but exploratory constructs that help decision-makers test the performance of strategies under contrasting conditions. The aim is to identify options that are robust—performing reasonably well across all scenarios—or adaptive, allowing future adjustment as the real world unfolds.

SBDA is especially useful in civil engineering where long-term infrastructure investments must contend with uncertainties such as future rainfall intensity under climate change, sea level rise projections, demand growth or urban expansion, and regulatory or budgetary changes. By exploring how choices play out across multiple scenarios, engineers can better prepare for a wide range of possible futures, rather than optimizing for a single expected outcome that may never materialize.

5.11.1 Why Is Scenario-Based Decision Analysis Important in Civil Engineering?

Civil engineering projects often have long time horizons, high capital investments, and significant consequences for public safety, environmental integrity, and societal wellbeing. Decisions about infrastructure, whether concerning flood defenses, wastewater systems, urban development, or transportation networks, must account

Fig. 5.80 Key benefits of scenario-based decision analysis in engineering and planning

for future conditions that cannot be precisely predicted. Traditional sensitivity analysis, while valuable, assumes a degree of predictability and control that is not always present in complex, uncertain systems.

SBDA offers a complementary approach by enabling engineers and planners to explore how decisions perform across a range of possible future environments, rather than focusing solely on variability in individual input parameters. The main advantages of SBDA are illustrated in Fig. 5.80. This method helps identify robust solutions that, while not necessarily optimal in any single scenario, perform reasonably well across multiple futures. It also enhances understanding of the vulnerabilities and trade-offs associated with each decision alternative, supporting the development of adaptive or flexible strategies that can evolve over time as future conditions become clearer.

5.11.2 *What Is Scenario-Based Decision Analysis?*

SBDA is a decision-support methodology that examines how different courses of action (e.g., design options, policy interventions, investment strategies) perform under a set of well-defined, contrasting scenarios. These scenarios may be constructed using expert judgment, stakeholder input, model projections, or historical analogs, and are typically qualitative or semi-quantitative in nature.

SBDA is built on several core features that distinguish it from traditional decision-making approaches. The foundational characteristics of SBDA are summarized in Fig. 5.81, highlighting its scenario evaluation, performance comparison, robustness criteria, and focus on long-term adaptability and resilience. The process of SBDA begins with the development of scenarios, each representing a possible future shaped by specific assumptions about key drivers such as climate variables, demographic trends, or regulatory shifts. Performance comparison is then carried out by estimating the outcomes of each decision alternative under every scenario, capturing a wide range of potential conditions. Rather than relying solely on expected values, SBDA uses decision criteria based on robustness metrics—such as minimax, minimax regret, or satisficing—to evaluate options. It should be noted that minimax selects the option that minimizes the worst-case outcome across all scenarios, minimax regret minimizes the maximum loss relative to the best scenario-specific outcome, and satisficing identifies alternatives that meet predefined acceptable performance thresholds rather than optimizing outcomes. Finally, SBDA emphasizes adaptability and resilience, promoting strategies that can either withstand adverse conditions or be adjusted over time as uncertainties evolve.

SBDA is particularly useful in risk-prone or climate-sensitive projects, such as, coastal and riverine flood management, urban drainage and stormwater design, large-scale infrastructure development, and environmental and resource sustainability planning.

While sensitivity analysis explores how variations in input parameters affect the outcomes within a single model framework, SBDA goes a step further by testing the stability of decisions across a range of possible futures or model structures. Rather than focusing solely on parameter uncertainty, SBDA considers structural, environmental, or contextual uncertainties—such as climate shifts, policy changes, or socioeconomic developments—that may fundamentally alter the decision landscape. In this way,

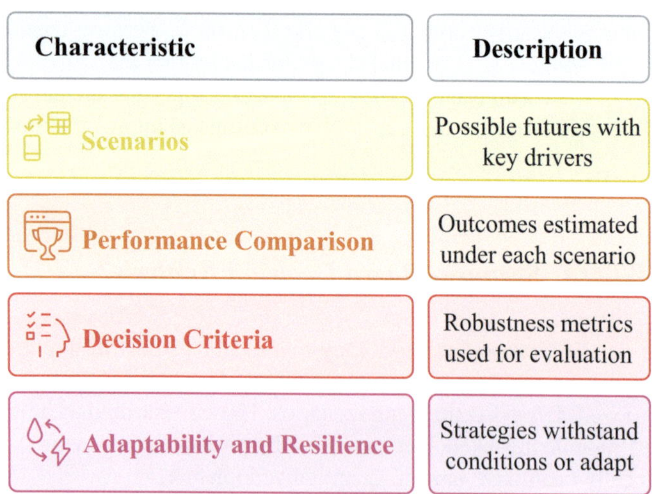

Fig. 5.81 Key characteristics of scenario-based decision analysis

- Sensitivity analysis answers:
  - *"What happens if this variable changes?"*
- Scenario-based analysis answers:
  - *"What happens if the world changes in this way?"*

In this way, SBDA complements sensitivity analysis by addressing deeper uncertainties and helping identify strategies that remain effective or adaptable across diverse long-term scenarios.

Both tools are valuable and often complementary in decision-making under uncertainty. Sensitivity analysis is best suited for situations where uncertainties are quantifiable, allowing decision-makers to explore how variations in specific parameters influence outcomes within a single model. In contrast, SBDA is more appropriate when uncertainties are structural or strategic, involving potential shifts in system behavior, policy environments, or long-term drivers. By combining both methods, analysts can capture a broader spectrum of uncertainty—from fine-tuned parameter changes to fundamentally different future conditions.

5.11.3 The Application of SBDA in Real-World Civil Engineering Problems

5.11.3.1 Flood Protection Strategy in a Coastal City

To ensure long-term flood resilience, a coastal city is evaluating three infrastructure investment options for protection against sea-level rise and increasing storm surge risks over a 30-year project horizon. The options—detailed in Table 5.59—differ in their design approach, upfront capital investment, and ongoing maintenance requirements.

To account for future uncertainty, three climate scenarios are defined (Table 5.60). These range from a conservative sea-level rise projection to a high-end scenario with more frequent and severe storm surges, based on mid-century (2055) forecasts. Figure 5.82 illustrates the strategic shift in flood protection planning based on projected sea-level rise and storm surge severity. As climate risks intensify from S1 to S3, the hybrid infrastructure approach becomes increasingly justified in terms of long-term lifecycle costs.

Table 5.59 The municipality's infrastructure options for flood protection

Option	Description	Initial cost ($M)	Maintenance ($M/year)
A	Reinforced levee system	100	2
B	Coastal wetland restoration	70	1.5
C	Hybrid system (levee + wetlands)	130	2.5

Table 5.60 Projected sea-level rise and storm surge frequencies by 2055 under different climate scenario

Scenario	Description	Sea-level rise by 2055 (m)	Storm surge frequency
S1	Conservative climate outlook	+0.3	1-in-100-year
S2	Moderate warming	+0.6	1-in-50-year
S3	High-end sea-level rise	+1.0	1-in-20-year

Hydrodynamic modeling was used to estimate total lifecycle costs for each infrastructure option under each climate scenario. These costs include initial construction, long-term maintenance, and projected flood damages, as summarized in Table 5.61.

To identify the most suitable option under uncertainty, we apply two widely used decision-making criteria:

- Minimax Rule (Robustness Criterion):
  - This rule selects the option with the *lowest worst-case cost*. Looking at the total lifecycle costs across the three climate scenarios, the maximum cost for each option is:

    Option A: $230M
    Option B: $310M
    Option C: $220M

    Since Option C has the *lowest maximum cost*, it is selected as the most robust choice under the Minimax rule.
- Minimax Regret Rule (Opportunity Loss Criterion):
  - A regret matrix is created by calculating the difference between each option's cost and the *lowest cost* in each scenario. The maximum regret for each option is then calculated as Table 5.62. According to the Minimax Regret rule, the best choice is the option with the *lowest maximum regret*, which is Option A.

By comparing both the robustness and regret-based decision criteria, planners can assess trade-offs between performance under worst-case conditions and potential for missed opportunities, supporting a more resilient and informed flood protection investment strategy.

Interpretation:

- Option C is most robust to high-risk climate futures.
- Option A minimizes regret, especially under conservative scenarios.
- A staged or adaptive plan using Option B initially, then upgrading to C, may balance cost and resilience.

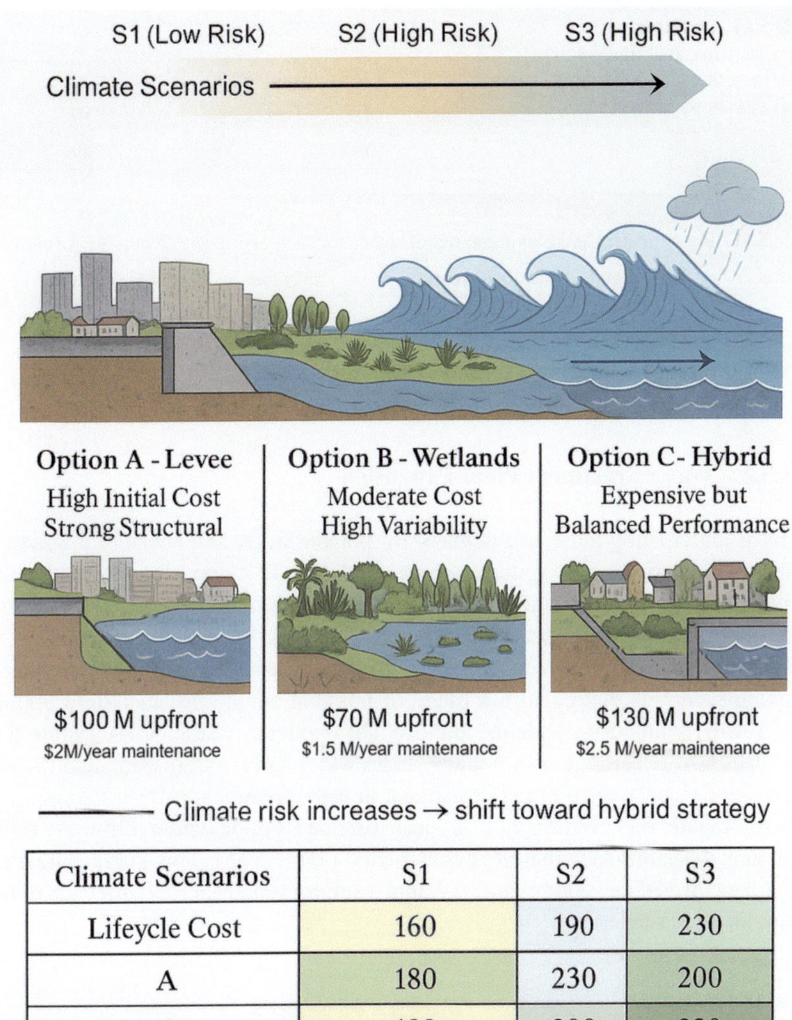

Climate Scenarios	S1	S2	S3
Lifeycle Cost	160	190	230
A	180	230	200
C	130	200	220

Climate risk increases → shift toward hybrid strategy

Fig. 5.82 Threshold analysis of flood protection strategies under three projected climate scenarios (S1 to S3)

Table 5.61 Performance matrix showing total lifecycle costs (in millions of dollars) across climate scenarios

Option	S1	S2	S3
A	160	190	230
B	180	230	310
C	190	200	220

Table 5.62 Regret matrix and maximum regret values for each option across climate scenarios

Option	S1 regret	S2 regret	S3 regret	Max regret
A	0	0	10	10
B	20	40	90	90
C	30	10	0	30

5.11.3.2 Water Treatment Plant Expansion

A city is anticipating increased demand for potable water and needs to upgrade its treatment plant to ensure long-term supply reliability. To support long-term planning for potable water treatment infrastructure, the city is evaluating three upgrade options over a 20-year horizon. These options vary in terms of cost, capacity, and technological flexibility (Table 5.63). Future uncertainty is addressed by defining three distinct planning scenarios that capture a range of possible conditions, including population growth, regulatory standards, and drought frequency (Table 5.64). Figure 5.83 visualizes how uncertainties in population growth, regulatory pressure, and drought frequency influence the optimal investment in water treatment infrastructure.

To compare the performance of each upgrade option under different future scenarios, a performance matrix is constructed (Table 5.65). This matrix integrates both financial cost and compliance risk into a single composite score for each option across all three scenarios:

Table 5.63 Comparison of water teeatment plant upgrade options

Option	Description	Initial cost ($M)	Capacity (MLD)	Tech flexibility
A	Expand existing facility	40	25	Low
B	Build second plant	60	30	Medium
C	Modular membrane tech units	80	40	High

Table 5.64 Future planning scenarios for water teeatment needs

Scenario	Population growth	Contaminant regulations	Drought frequency
S1	Low	Current standards	Rare
S2	Medium	Stricter DBP limits	Occasional
S3	High	Advanced standards	Frequent

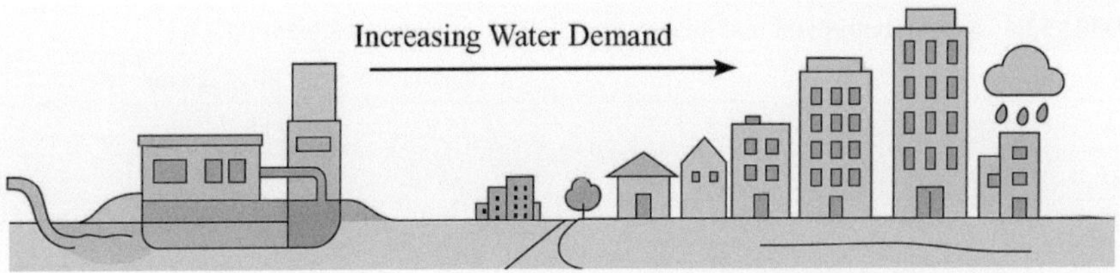

Fig. 5.83 Comparative threshold analysis of three water treatment upgrade strategies across future planning scenarios

Table 5.65 Performance matrix showing combined cost and compliance risk scores across scenarios (costs in millions)

Option	S1	S2	S3
A	45	70	90
B	55	65	75
C	80	85	90

- Scenario S1 represents a low-growth, low-risk future.
- Scenario S2 reflects moderate challenges, including tighter DBP standards.
- Scenario S3 assumes significant population growth and the strictest regulatory and environmental pressures.

In general, lower performance scores indicate better overall outcomes, balancing cost and risk. This table enables the city to compare each option's strengths and vulnerabilities across potential futures. To select the most appropriate investment strategy, two decision-making criteria are typically applied:

1. Minimax Rule (Robustness Approach):

   This criterion selects the option with the lowest worst-case (maximum) score, prioritizing protection against the most adverse conditions.
2. Minimax Regret Rule (Opportunity Loss Approach):

   A regret matrix is derived by comparing each option's score in a scenario to the best (lowest) score for that scenario. The option with the lowest maximum regret across scenarios is then chosen, minimizing the potential for future regret in decision-making.

These structured approaches help ensure that the chosen upgrade is not only cost-effective but also resilient to future uncertainties in demand, regulation, and climate conditions. Performance of all options under different scenarios have been provided in Table 5.65.

To guide decision-making under uncertainty, we use the Minimax Regret Rule, which focuses on minimizing potential opportunity loss. The regret matrix in Table 5.66 quantifies the difference between each option's performance and the best possible performance within each scenario. For each option, the maximum regret across all scenarios is calculated in this table. According to the Minimax Regret

Table 5.66 Regret matrix and maximum regret scores for water treatment options

Option	S1	S2	S3	Max regret
A	0	5	15	15
B	10	0	0	10
C	35	20	15	35

criterion, the option with the lowest maximum regret is considered the most robust under uncertain future conditions. In this case, Option B has the smallest maximum regret of 10, making it the preferred choice using this decision rule.

Interpretation

- Option B balances risk and performance across all futures.
- While Option C is more future-proof, its high cost makes it less efficient unless S3 becomes reality.
- The city could start with Option B and monitor water quality and growth to decide on future modular upgrades.

5.11.3.3 Pavement Design for Highway Expansion

A transportation agency is planning the construction of a 25-km highway and must select the most cost-effective pavement type over a 40-year lifecycle. Three pavement options are under consideration, varying in their initial costs and maintenance requirements (Table 5.67). Figure 5.84 illustrates how the preferred pavement strategy changes with scenario severity, highlighting a shift from asphalt to concrete as traffic, climate, and material price pressures increase.

To account for uncertainties in future conditions, three planning scenarios are defined in Table 5.68. These capture variability in traffic demand, climate stressors (freeze–thaw cycles), and material cost escalation over time:

Using engineering-economic modeling, the total lifecycle costs of each pavement option under all scenarios are presented in Table 5.69. The agency then applies multiple decision criteria to guide selection:

Decision Criteria:

1. Minimax Rule (Robustness):

    - This rule focuses on minimizing the worst-case cost:

Table 5.67 Comparison of pavement design options for highway construction

Option	Description	Initial cost ($M)	Annual rehab cost ($M/year)
A	Asphalt pavement	25	1.2
B	Concrete pavement	40	0.5
C	Composite (AC + PCC)	35	0.8

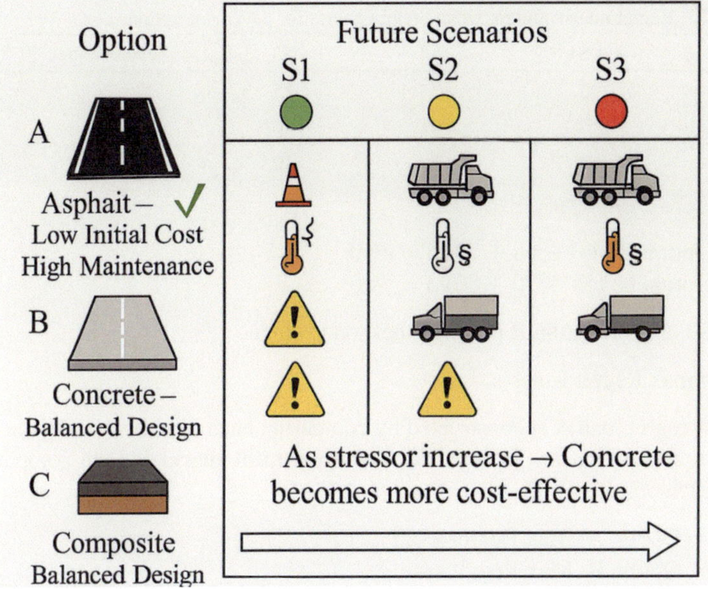

Fig. 5.84 Cenario-based threshold analysis of three pavement design strategies over a 40-year lifecycle

Table 5.68 Future scenarios for highway design based on traffic, climate, and economic variables

Scenario	Traffic growth	Freeze–thaw cycles	Material price inflation
S1	Low	Mild	Stable
S2	Medium	Moderate	Moderate
S3	High	Severe	High

Table 5.69 Performance matrix showing total lifecycle costs (in millions of dollars) for each pavement option under varying future scenarios

Option	S1	S2	S3
A	65	75	95
B	60	68	70
C	63	70	80

- A = 95
- B = 70
- C = 80

So, Option B is selected as the most robust, with the lowest maximum cost.

2. Expected Value Rule (assuming equal scenario probabilities):

- Option A: (65 + 75 + 95)/3 = 78.3

Table 5.70 Regret matrix and maximum regret values for pavement options

Option	S1	S2	S3	Max regret
A	5	7	25	25
B	0	0	0	0
C	3	2	10	10

- Option B: (60 + 68 + 70)/3 = 66.0
- Option C: (63 + 70 + 80)/3 = 71.0

  So, again, Option B performs best on average.

3. Minimax Regret Rule:

   - A regret matrix is constructed by comparing each option's cost to the lowest in each scenario (Table 5.70). The **maximum regret** is then computed for each option:

     – Option A: Max Regret = 25
     – Option B: Max Regret = 0
     – Option C: Max Regret = 10.

     So, Option B is the best choice under this rule as well.

Final Recommendation: Across all three criteria—worst-case robustness, expected cost, and opportunity loss—Option B (Concrete pavement) consistently emerges as the most cost-effective and resilient investment for the highway project.

Interpretation

- Concrete pavement (Option B) is the most robust and lowest regret option under all scenarios.
- While more expensive upfront, it provides cost stability under uncertainty in traffic and climate.
- SBDA justifies higher initial investment in exchange for long-term resilience and predictability.

5.11.4 Metrics for Robustness and Regret Analysis

In SBDA, the goal is not always to optimize for a single most probable future, but rather to identify decisions that are robust across a wide range of plausible scenarios. Two prominent evaluation approaches used in this context are robustness analysis and regret analysis. Robustness analysis emphasizes resilience by focusing on the worst-case outcomes, ensuring that a chosen decision performs acceptably even under adverse conditions. On the other hand, regret analysis evaluates the potential cost of not having selected the best possible strategy in hindsight, aiming to minimize the gap between the chosen decision and the optimal one under each scenario. These tools allow civil engineers and planners to make decisions that are more resilient to uncertainty, especially in long-term infrastructure design, climate-sensitive projects, and resource allocation under ambiguity.

5.11.4.1 Robustness Metric (Minimax Rule)

The robustness of an alternative is measured by the worst outcome it produces across all scenarios. A more robust option is one that performs relatively well even in the most adverse conditions.

Let C_{ij} be the outcome (e.g., cost) of Option i under Scenario j. Then:

$$\text{Robustness Score}_i = \max_j (C_{ij}) \tag{5.3}$$

Select the option i^* such that:

$$i^* = \arg\min_i \max_j (C_{ij}) \tag{5.4}$$

In this context, lower scores indicate better performance, reflecting outcomes that are more favorable or less risky. The underlying assumption is that the decision-maker is risk-averse, meaning they prioritize strategies that minimize the likelihood of catastrophic failures. This approach is particularly effective in high-stakes scenarios where the consequences of failure are severe, such as dam collapses, uncontrolled flooding, or failures in critical infrastructure like water treatment plants. By focusing on avoiding worst-case outcomes, this interpretation aligns with a cautious and protective decision-making mindset.

5.11.4.2 Regret Metric (Minimax Regret Rule)

Regret is the difference between the outcome of a chosen alternative and the best possible outcome in each scenario. The maximum regret reflects the worst "what-if" loss had we known the true scenario in advance.

For each scenario j, identify the best outcome:

$$C_j^* = \min_i (C_{ij}) \tag{5.5}$$

Compute regret for each option:

$$R_{ij} = C_{ij} - C_j^* \tag{5.6}$$

Then, for each option:

$$\text{Max Regret}_i = \max_j (R_{ij}) \tag{5.7}$$

Choose the option i^* that minimizes regret:

$$i^* = \arg\min_i \max_j (R_{ij}) \tag{5.8}$$

This approach reflects the concept of opportunity loss incurred by not selecting the optimal strategy in hindsight. It is particularly suited for decision-makers with a moderate level of risk aversion who aim to reduce feelings of regret after the outcomes are realized. This method is especially valuable in decision-making contexts that involve political accountability, multiple stakeholders, or where public transparency is essential, as it promotes choices that are justifiable and balanced across various possible futures.

5.11.4.3 Additional Metrics (Optional in SBDA)

While robustness and regret are the most common metrics, SBDA may incorporate other analytical tools for specific contexts:

Satisficing

This approach ensures that a decision satisfies minimum acceptable criteria across the majority or all considered scenarios. It involves selecting options where the performance metric $C_{ij} \leq \text{threshold} \quad \forall j$, meaning that for each scenario j, the outcome remains within an acceptable limit. This method is particularly useful in multi-objective decision-making problems, such as those requiring a balance between economic cost and environmental performance, where no single optimal solution exists, but satisfactory performance across various dimensions is essential.

Trade-Off Plots

Trade-off plots are visual tools, typically presented as scatter plots (e.g., cost versus emissions), that illustrate how different decision options perform across multiple scenarios. They are particularly valuable in Multi-Criteria Decision Analysis (MCDA), as they help decision-makers identify and evaluate the compromises between conflicting objectives. By visualizing the distribution of outcomes, trade-off plots facilitate a clearer understanding of the relationships among competing criteria, supporting more informed and balanced decision-making.

Expected Value (with Probabilities)

When the probabilities of different scenarios (P_j) are known, the expected value for each decision option can be calculated using the formula $EV_i = \sum_j P_j \cdot C_{ij}$. This approach provides a weighted average of outcomes across all scenarios, reflecting both the performance and the likelihood of each scenario. The preferred decision is the one with the lowest expected value, indicating the most favorable average outcome over time. This method is particularly useful when reliable probabilistic information is available, allowing for a more quantitative and risk-informed decision-making process.

5.11.4.4 Metric Selection Guidance

Different decision metrics are best suited for different decision-making contexts (Fig. 5.85). Robustness is ideal when extreme or irreversible outcomes must be avoided, emphasizing safety and resilience. Regret is appropriate when stakeholders are particularly sensitive to missed opportunities and aim to minimize hindsight dissatisfaction. Satisficing is most useful when a minimum level of performance must be guaranteed across all future scenarios, ensuring essential thresholds are met. Lastly, Expected Value is the preferred metric when scenario probabilities are clearly defined and credible, allowing for probabilistic averaging to guide optimal choices.

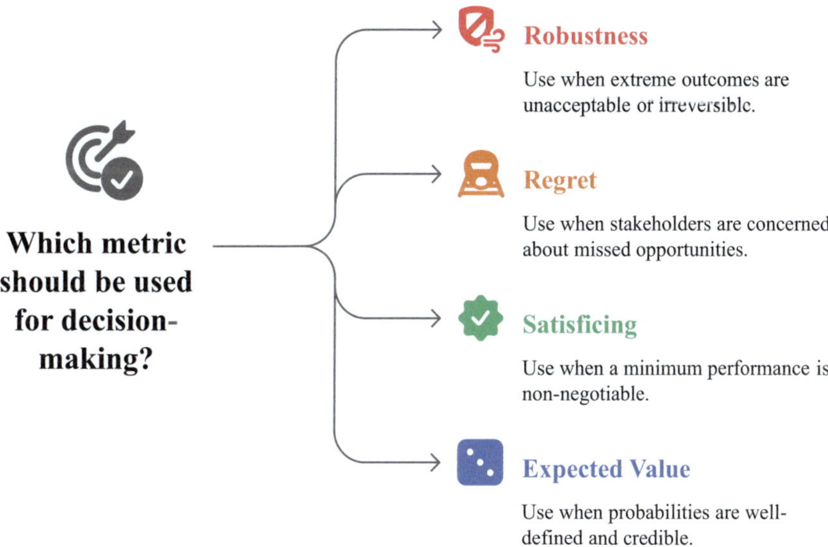

Fig. 5.85 Selection of decision-making metrics under uncertainty

5.11.5 Integration of SBDA with Adaptive Planning and Stakeholder Engagement

In complex civil engineering projects, the future is rarely static. Infrastructure systems must operate under dynamic conditions, including shifting climate patterns, urban expansion, evolving regulations, and changing community needs. While SBDA offers a structured method to evaluate decision alternatives under deep uncertainty, its full potential is realized when combined with adaptive planning frameworks and stakeholder engagement processes.

This integration supports iterative, flexible, and participatory decision-making, improving both the resilience of infrastructure systems and the legitimacy of decisions made in uncertain environments.

5.11.5.1 Adaptive Planning

Adaptive planning is a dynamic decision-making approach that emphasizes flexibility and the capacity to adjust actions over time in response to evolving information, unexpected events, or shifting conditions. Unlike static, one-time decisions, adaptive strategies are intentionally structured as sequences of steps that include predefined opportunities for revision, scaling, or redirection. In the context of civil engineering, adaptive planning can take several practical forms. For instance, infrastructure investments may be phased—such as constructing floodwalls in stages as risk levels change. Similarly, systems can be designed with modularity in mind, allowing components like water treatment plants to be expanded as demand grows. Additionally, contingency triggers may be embedded within the plan, such as setting specific thresholds that, once crossed, automatically prompt upgrades or further interventions. This approach enables more resilient, cost-effective, and responsive infrastructure development (Fig. 5.86).

5.11.5.2 How SBDA Supports Adaptive Planning

The SBDA naturally complements adaptive planning by providing critical insights that enhance flexibility and responsiveness. First, it reveals decision robustness across time horizons by allowing planners to evaluate how different strategies perform under both near-term and long-term scenarios. This helps identify which actions are reliably beneficial and safe to implement immediately, and which ones are better deferred or made conditional based on how the future unfolds. Second, SBDA identifies tipping points and triggers by highlighting scenario-specific thresholds—such as sea-level rise exceeding 0.6 m—where a current strategy might fail or require augmentation. These thresholds can then be used as adaptive triggers, guiding when to pivot to alternative strategies or increase system capacity, thereby strengthening long-term resilience and efficiency.

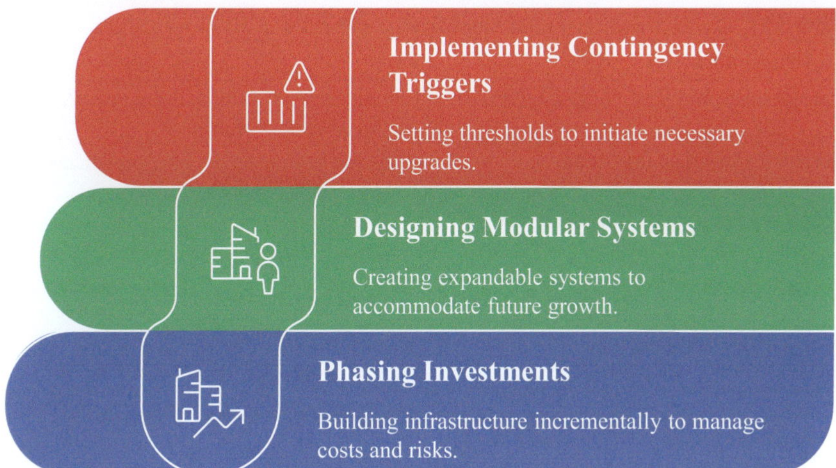

Fig. 5.86 Practical forms of the adaptive planning in civil engineering

5.11.5.3 Engaging Stakeholders in Scenario-Based Decisions

Engineering decisions often affect a wide range of stakeholders, including local governments, utilities, residents, regulatory agencies, environmental groups, and funding bodies. SBDA facilitates inclusive, transparent, and credible decision-making in several key ways: (1) scenario co-development, (2) trade-off visualization, (3) clarifies uncertainty, and (4) supports consensus building.

SBDA enhances stakeholder engagement by integrating their diverse perspectives into the decision-making process. First, through scenario co-development, stakeholders actively participate in constructing plausible future scenarios based on their own knowledge, concerns, and priorities. This participatory approach not only makes the analysis more relevant but also increases acceptance and support for the outcomes. Second, SBDA enables trade-off visualization by illustrating how different strategies perform across a range of futures. This helps stakeholders understand the implications of conflicting values, such as balancing cost with environmental benefits or choosing between immediate savings and long-term resilience. Third, SBDA clarifies uncertainty by openly presenting a range of possible futures without claiming precision in forecasting. This transparency acknowledges the inherent limitations of modeling and fosters greater trust in the decision process. Finally, SBDA supports consensus building by offering a framework where, even in the absence of agreement regarding the most likely future scenario, stakeholders can support robust or adaptive strategies that are acceptable under multiple scenarios. This collaborative approach ultimately strengthens decision legitimacy and effectiveness.

Table 5.71 Key tools and frameworks supporting the integration of SBDA with adaptive planning

Tool/framework	Purpose
Dynamic adaptive policy pathways (DAPP)	Maps sequences of decision options over time based on scenario evolution
Robust decision making (RDM)	Emphasizes performance across many futures, using simulation ensembles
Multi-stakeholder scenario workshops	Engages community and expert input in scenario development and evaluation
Decision scaling	Identifies climate or system thresholds that affect strategy success
Monitoring and feedback loops	Enables real-time adjustment of strategies during implementation

5.11.5.4 Tools and Methods for Integration

To effectively integrate SBDA with adaptive planning and stakeholder engagement, a range of tools and frameworks can be employed to enhance decision-making under uncertainty (Table 5.71). Dynamic Adaptive Policy Pathways (DAPP) is used to map sequences of decision options over time, allowing planners to adapt as scenarios evolve. Robust Decision Making (RDM) focuses on evaluating how strategies perform across a wide array of possible futures by leveraging simulation ensembles, supporting decisions that remain effective under diverse conditions. Multi-Stakeholder Scenario Workshops facilitate collaborative scenario development and evaluation, incorporating both expert and community input to ensure relevance and legitimacy. Decision Scaling helps identify critical climate or system thresholds that can influence the success or failure of strategies, guiding more targeted and responsive planning. Finally, Monitoring and Feedback Loops provide mechanisms for real-time adjustments during implementation, ensuring that strategies remain aligned with emerging conditions and stakeholder priorities. Together, these tools support a flexible, inclusive, and resilient planning approach.

5.11.5.5 Step-by-Step Framework for Applying Scenario-Based Decision Analysis

Before generating specific examples, it is essential to understand the general workflow for applying SBDA, particularly when integrating adaptive planning and stakeholder engagement. This structured approach ensures that uncertainty is comprehensively addressed, decisions are flexible, and the process is inclusive and transparent. Below is a recommended nine-step procedure to apply SBDA effectively in civil engineering contexts (Fig. 5.87).

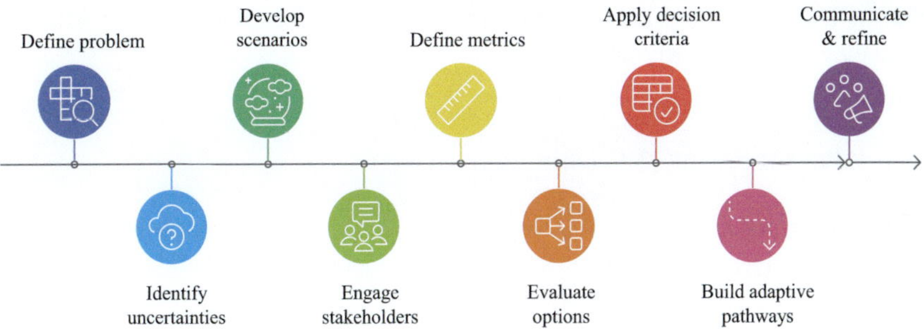

Fig. 5.87 Step-by-step framework for applying SBDA in civil engineering

Step 1: Define the Decision Problem

Begin by clearly articulating the decision problem, including its primary objectives—such as minimizing costs, reducing flood risks, or ensuring system reliability. This step involves identifying the set of viable decision alternatives, for example, traditional levees, green infrastructure, or hybrid solutions. It is essential to establish the relevant planning horizon, whether short-, medium-, or long-term, and to outline any operational or contextual constraints, including budgetary limits, environmental regulations, or technical feasibility. Clearly framing the problem sets the foundation for a structured and meaningful analysis.

Step 2: Identify Key Uncertainties

In this step, identify the external factors that may significantly affect decision outcomes. These uncertainties often fall into categories such as climate (e.g., rainfall variability, temperature changes, sea level rise), demographics (e.g., urbanization trends, population growth), economics (e.g., fluctuations in material costs, infrastructure demand), and policy or regulatory shifts (e.g., changes in zoning laws or emissions regulations). The focus should be on high-impact, high-uncertainty variables that introduce substantial unpredictability to the system's future state.

Step 3: Develop Plausible Scenarios

Once key uncertainties are identified, construct a set of three to five internally consistent scenarios that capture different plausible future conditions. These scenarios should combine various uncertainties to reflect complex realities—such as a future with both high rainfall and rapid urban development. To ensure validity and relevance, incorporate stakeholder input, expert judgment, and formal foresight tools like the Delphi method. These scenarios form the basis for evaluating how different strategies perform under varying future conditions.

Step 4: Engage Stakeholders Early

Stakeholder engagement should begin early in the process to ensure the analysis reflects diverse perspectives and concerns. Identify key stakeholders—such as municipal planners, utility managers, regulators, and community representatives—and actively involve them in scenario development, priority setting, and defining evaluation criteria. Early engagement fosters transparency, trust, and a sense of shared ownership in the decision-making process, ultimately leading to more widely supported outcomes.

Step 5: Define Performance Metrics

Establish a set of metrics to evaluate how each decision alternative performs across scenarios. These may include lifecycle costs, system reliability, environmental impact, social acceptability, and adaptability. The chosen metrics should be quantifiable or at least ordinally ranked and must be applicable to all alternatives being considered. Clear and consistent metrics allow for meaningful comparisons and support evidence-based decision-making.

Step 6: Evaluate Decision Alternatives Under Each Scenario

Using appropriate tools—such as engineering models, simulations, or expert estimations—evaluate each alternative's performance within each scenario. The results should be organized into a performance matrix, where each cell represents the outcome of a given alternative under a specific scenario. This structured representation enables easy comparison and forms the basis for robustness and regret analysis in subsequent steps.

Step 7: Apply Robustness and Regret Criteria

Assess the performance of alternatives using robust decision-making metrics. This may involve applying the minimax criterion to identify the option with the best worst-case performance, or the minimax regret approach to find the option with the lowest maximum regret across scenarios. The satisficing approach can also be used to identify options that meet minimum acceptable thresholds in all future scenarios. If probabilities are available, calculate expected values to support probabilistic decision-making. These evaluations help identify strategies that remain acceptable across a wide range of uncertain futures.

Step 8: Design Adaptive Pathways (if Applicable)

Based on insights from the scenario analysis, develop adaptive strategies that unfold over time. This may include phased plans or conditional actions, such as "implement Option A now, and switch to Option C if demand exceeds a certain threshold by year Y." Identify specific triggers or tipping points within the scenarios that would prompt a reassessment or redirection of the strategy. When long-term flexibility is required, tools like the Dynamic Adaptive Policy Pathways (DAPP) method can be employed to structure and communicate these pathways.

Step 9: Communicate Results and Refine

Finally, present the analysis results in clear, accessible formats such as performance matrices, regret tables, scenario graphs, and adaptive pathway diagrams. These visual and tabular summaries help convey complex information effectively to stakeholders. Facilitate stakeholder review sessions to gather feedback and assess the need for refinement—this could include updating scenarios, introducing new alternatives, or adjusting the evaluation metrics. Iteration ensures the analysis remains relevant, credible, and aligned with stakeholder expectations.

5.11.6 Applications of SBDA in Civil Engineering Practice

5.11.6.1 Example 1—Flood Protection Strategy in a Coastal City

A coastal municipality is evaluating strategies to protect the downtown area from future coastal flooding. The city has experienced several storm surge events in the past two decades, and scientific reports suggest that sea-level rise (SLR) and storm surge intensity are expected to worsen under climate change. With a 30-year planning horizon and a limited infrastructure budget, the city must choose among three flood protection options provided in Table 5.72. Figure 5.88 provides a conceptual representation of the decision context. The city's objectives center on achieving a cost-effective, resilient, and adaptable infrastructure strategy. Specifically, the city aims to minimize the total lifecycle cost by reducing long-term maintenance expenses and potential damage, ensuring dependable protection against future extreme weather events while maintaining the flexibility necessary to adapt its infrastructure if future environmental conditions become more severe than currently anticipated. The key questions to be addressed are as follow:

1. What are the most relevant uncertainties affecting flood protection decisions?
2. What plausible future scenarios should be considered?
3. How does each flood protection option perform under each scenario?
4. Which option is most robust across all futures?

Table 5.72 Summary of coastal flood protection strategies evaluated by a municipality under climate change Uncertainty

Option	Description	Initial cost ($M)	Annual maintenance ($M/year)
A	Reinforced concrete levee	100	2
B	Coastal Wetland Buffer restoration	70	1.5
C	Hybrid system (Levee + Wetlands)	130	2.5

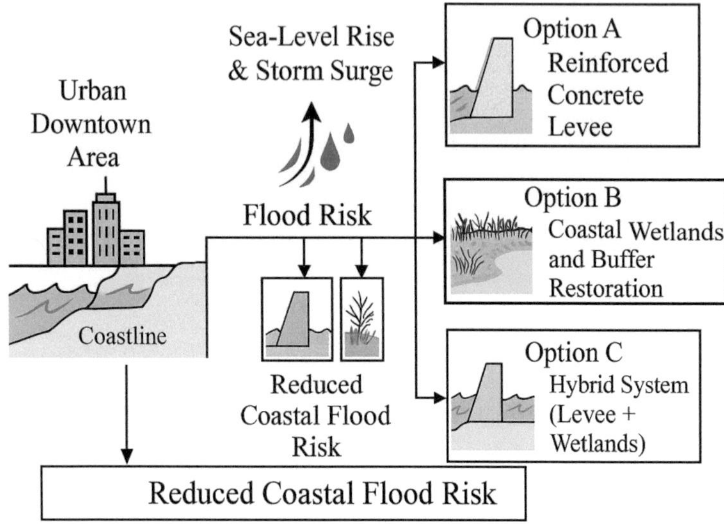

Fig. 5.88 Comparison of three coastal flood protection strategies over a 30-year planning horizon

5. Which option has the least regret if future conditions turn out to be worse than expected?
6. Can an adaptive pathway be developed to allow flexible investment over time?

Step-by-Step Solution Using SBDA

Step 1: Define the Decision Problem

- Objective: Select a flood protection strategy that minimizes total lifecycle cost while providing reliable protection over a 30-year horizon.
- Alternatives:
  - A: Levee (high upfront, strong defense)
  - B: Wetlands (low-cost, eco-friendly, less reliable)
  - C: Hybrid (moderate flexibility, higher cost)

- Constraints:
  - Budget is limited to $150M.
  - Must comply with federal environmental impact guidelines.
- Timeframe: 30 years

Step 2: Identify Key Uncertainties

- Sea-Level Rise (SLR): Unknown trajectory over the next 30 years.
- Storm Surge Frequency: Influenced by ocean warming and atmospheric variability.

These uncertainties cannot be assigned reliable probabilities due to model spread and climate variability.

Step 3: Develop Plausible Scenarios

Table 5.73 outlines three climate scenarios used to evaluate coastal flood protection strategies. Each scenario reflects varying assumptions about future sea-level rise and the frequency of extreme storm surge events, ranging from conservative to high-end projections aligned with climate change models.

Step 4: Engage Stakeholders

The city's objectives are focused on developing a resilient and forward-thinking infrastructure plan that balances financial prudence with robust performance. One of the primary goals is to minimize the total lifecycle cost, which includes not only upfront investment but also long-term maintenance and potential damage costs. Another critical objective is to ensure reliable protection against increasingly frequent and intense extreme weather events. Recognizing the uncertainty of future climate conditions, the city also aims to preserve adaptability in its infrastructure choices, allowing for modifications if future risks prove more severe than anticipated.

To support these goals, a diverse group of stakeholders participated in a scenario workshop. The participants included municipal engineers, who prioritized cost-effectiveness and system reliability; environmental NGOs, who advocated for nature-based and ecologically sustainable solutions; residents and business owners, who expressed concerns about flood insurance rates and potential disruptions to their daily lives; and city council members, who emphasized the importance of public accountability and transparent decision-making. The workshop sought to validate

Table 5.73 Description of sea-level rise and storm surge scenarios for flood risk assessment

Scenario	Description	Sea-level rise (m)	Storm surge recurrence
S1	Conservative climate outlook	+0.3	1-in-100 year
S2	Moderate warming	+0.6	1-in-50 year
S3	high-end sea-level rise	+1.0	1-in-20 year

* Derived from regional IPCC projections and local coastal modeling

three proposed planning scenarios, establish consensus on key decision metrics—namely cost and reliability—and stress the importance of adaptability as a guiding principle in long-term planning.

Step 5: Define Performance Metrics.

- Total Lifecycle Cost = Initial + Maintenance + Flood Damage
- Maximum Acceptable Damage per Scenario = $100M.

  All costs are in present values (no discounting).

Step 6: Evaluate Options Under Each Scenario

Table 5.74 summarizes the initial investment, maintenance costs, and projected flood damage costs for three flood protection options (A, B, and C) under three climate scenarios (S1–S3). The damage estimates reflect the varying severity of sea-level rise and storm surge events, supporting a comparative evaluation of total expected costs over the planning horizon. In addition, Table 5.75 presents the total lifecycle costs (in million dollars) for each flood protection option (A, B, and C) under three climate scenarios (S1–S3). These totals include initial construction, maintenance, and projected flood damage costs, enabling a holistic comparison of long-term financial implications.

Step 7: Apply Robustness and Regret Criteria

A. Robustness (Minimax Rule)

Table 5.76 shows the worst-case total lifecycle cost for each flood protection option under the most adverse climate scenario. The minimax rule is applied to identify the most robust option by minimizing potential losses in the worst-case scenario. Option A is most robust (lowest worst-case cost).

Table 5.74 Estimated lifecycle costs and flood damages for each flood protection option under different climate scenarios

Option	Initial ($M)	Maintenance ($M)	Damage S1		
			S1	S2	S3
A	100	60	0	20	70
B	70	45	20	70	120
C	130	75	0	10	50

Table 5.75 Total estimated lifecycle costs of flood protection options across climate scenarios

Option	S1	S2	S3
A	160	180	230
B	135	185	235
C	205	215	255

Table 5.76 Robustness analysis using the minimax (worst-case) criterion

Option	Worst-case cost
A	230
B	235
C	255

B. Regret Matrix

Table 5.77 displays the regret values for each option under three climate scenarios (S1–S3), along with the maximum regret for each option. Regret is defined as the difference between an option's cost and the cost of the optimal option under the same scenario. The max regret column identifies the option with the least potential for hindsight-based decision regret. Option A also has lowest maximum regret under worst-case (S3) scenario.

Step 8: Design Adaptive Pathways

The city has adopted an adaptive pathways approach to infrastructure design, structured as a two-stage plan to balance cost, environmental impact, and future uncertainty. The initial investment focuses on wetland restoration (Option B), with an estimated cost of $70 million. This nature-based solution serves as a foundational measure, offering immediate environmental benefits while providing a moderate level of protection against flooding and sea level rise.

To support long-term adaptability, a robust monitoring plan will be implemented, tracking indicators such as the annual mean sea level and the frequency of storm surges exceeding 0.75 m. A critical trigger has been defined: if sea level rise surpasses 0.5 m or a 1-in-50-year storm event occurs within the next 15 years, the plan calls for an upgrade to a more robust hybrid infrastructure system (Option C), requiring an additional $60 million investment. This staged strategy spreads out financial commitments over time, helps avoid overdesign based on uncertain projections, and enhances the city's environmental credibility by prioritizing sustainable and responsive planning.

Step 9: Communicate Results and Refine

In the final step of the planning process, the focus shifts to synthesizing the findings and clearly conveying them to stakeholders for feedback and refinement. This stage involves the creation of key deliverables that illustrate both the technical and strategic

Table 5.77 Regret analysis of flood protection options across climate scenarios

Option	Regret			
	S1	S2	S3	Max
A	25	0	0	25
B	0	5	5	5
C	70	30	20	70

aspects of the proposed adaptive plan. These include a performance matrix, presented in the form of a bar chart, which compares the outcomes of different options based on predefined decision metrics, and an adaptive pathway timeline that visually maps out the sequence and timing of potential infrastructure upgrades.

Additionally, a stakeholder summary is prepared to explain the rationale behind each option and support informed decision-making. This summary highlights why Option A is considered robust in terms of cost and reliability under moderate conditions, and why Option B offers greater flexibility in adapting to uncertain future changes. It also outlines the specific trigger conditions—such as thresholds for sea level rise or storm events—that would prompt an upgrade to a more advanced system. This communication step ensures transparency, promotes stakeholder engagement, and provides a foundation for refining the plan based on collaborative input.

Answers to the key questions

1. What are the most relevant uncertainties affecting flood protection decisions?

The two **primary uncertainties** that strongly influence the performance and reliability of the flood protection strategies are:

- Sea-Level Rise (SLR):
  - The magnitude of sea-level rise over the 30-year planning horizon is highly uncertain due to varying greenhouse gas emission trajectories, ocean thermal expansion, and polar ice melt rates. SLR directly affects whether a levee or natural buffer will be overtopped during future storm surge events.
- Storm Surge Frequency and Intensity:
  - Storm surge events are driven by a combination of atmospheric pressure, wind speed, and tidal conditions. With climate change, both the frequency and severity of extreme coastal flooding events are expected to increase, but exact projections remain uncertain.

These uncertainties are difficult to quantify probabilistically and are therefore represented using three plausible scenarios reflecting different climate pathways.

2. What plausible future scenarios should be considered?

Three internally consistent climate scenarios (Table 5.78) were developed through a combination of expert judgment, stakeholder engagement, and regional climate model outputs. These scenarios were chosen to span a wide range of possible futures without assigning exact probabilities.

3. How does each flood protection option perform under each scenario?

To assess the effectiveness of each flood protection option, a comprehensive evaluation was conducted across multiple future scenarios. The core performance metric used for comparison was the total lifecycle cost, a holistic measure that captures both short- and long-term financial implications. This includes the initial capital cost

Table 5.78 Climate change scenarios defined by SLR and storm surge frequency

Scenario	Sea-level rise (SLR) (m)	Storm surge return period	Description
S1	+0.3	1-in-100 year	Conservative: minor climate change impacts
S2	+0.6	1-in-50 year	Moderate: middle-of-the-road projection
S3	+1.0	1-in-20 year	High-end: aggressive sea-level rise and more frequent flooding

required to implement each option, projected maintenance expenses over a 30-year period, and the estimated costs associated with flood damages that could occur under each scenario.

By analyzing these three cost components together, decision-makers can better understand the trade-offs between upfront investment and long-term resilience. This approach allows for a realistic comparison of how each option performs not only in average conditions but also under extreme or uncertain future scenarios. It ensures that the selected strategy offers not just cost-effectiveness, but also durability and adaptability in the face of evolving climate risks.

Table 5.79 compares the total estimated lifecycle costs—including construction, maintenance, and flood damages—for three flood protection strategies (Levee, Wetlands, and Hybrid) under three climate scenarios (S1–S3). The values illustrate how each option performs financially as climate risks intensify.

The evaluation of the three flood protection options across different future scenarios revealed distinct strengths and trade-offs for each. Option A (Levee) delivers full protection under Scenario 1 (S1), incurs moderate losses in Scenario 2 (S2), and experiences significant but still manageable losses in the more extreme Scenario 3 (S3). Its performance highlights a balance between cost and protection in less severe futures, though it may require upgrades if conditions worsen.

Option B, which involves wetland restoration, stands out for having the lowest overall cost in Scenario 1. However, its performance deteriorates sharply in more severe conditions. In Scenario 2, it suffers large damages, and in Scenario 3, it incurs unacceptable flood losses totaling around $120 million—exceeding the city's defined risk tolerance. This makes Option B a flexible but potentially risky choice without planned upgrades.

Table 5.79 Total lifecycle costs of flood protection options under different climate scenarios

Option	Total cost (M $)		
	S1	S2	S3
A (Levee)	160	180	230
B (Wetlands)	135	185	235
C (Hybrid)	205	$215	255

On the other hand, Option C, a hybrid system combining structural and nature-based solutions, offers the highest level of protection across all scenarios. Yet, its substantial initial and maintenance costs render it less appealing under conservative or moderately adverse futures. This makes it best suited for high-risk scenarios, or as an upgrade path if trigger conditions are met.

4. Which option is most robust across all futures?

Robustness is assessed using the Minimax rule, which selects the option with the lowest worst-case cost (Table 5.76). Option A (Levee) is the most robust option—it minimizes the worst-case outcome across all scenarios. Even under the most extreme climate future (S3), it limits total cost to $230M, which is below the other options.

5. Which option has the least regret if future conditions turn out to be worse than expected?

Regret is calculated as the difference between each option's cost and the best-performing option for each scenario. Although Option A is most robust, Option B has the lowest maximum regret (Table 5.77). However, this is misleading because Option B fails to meet minimum performance thresholds in S3, incurring flood damage of $120M, which is beyond the city's tolerance of $100M.

While Option B presents the lowest theoretical regret based on cost-performance calculations, Option A emerges as the most practical low-regret choice when accounting for real-world constraints and risk tolerance. Crucially, Option A never exceeds the flood damage thresholds established by the city, ensuring consistent protection across all scenarios. Its regret in Scenario 1 is only moderate, where Option B is marginally less expensive, but this trade-off is outweighed by the reliability and robustness it offers in more severe futures. Moreover, Option A sustains acceptable performance even under extreme conditions, making it a resilient and dependable choice in the face of future uncertainty. Given its balance between cost, risk mitigation, and adaptability, Option A stands out as the least regretful viable solution, aligning both with practical implementation needs and long-term resilience goals.

6. Can an adaptive pathway be developed to allow flexible investment over time?

Yes. Based on the SBDA results and stakeholder priorities, the city can adopt a phased, adaptive investment pathway as follow:

- Initial Investment (Year 0–10):
  - Begin with Option B (Wetlands) for $70M.
  - It provides cost-effective protection under Scenario 1 and builds environmental value.
- Monitoring System:
  - Track sea-level rise annually.
  - Use surge monitoring to detect increasing event frequency.
  - Install trigger points:

  - If SLR exceeds 0.5 m or 1-in-50-year events occur twice within 10 years, begin upgrading.

- Adaptive Upgrade (Year 10–15):
  - If triggers are met, transition to Option C (Hybrid).
  - Cost of upgrading from wetlands to hybrid is estimated at $60M.
  - This preserves earlier investment and ensures readiness for severe future conditions.

An adaptive pathway combining Option B and C is feasible and strategically sound. It minimizes initial cost, satisfies environmental goals, and retains the flexibility to respond to worsening climate conditions. This pathway reflects robust, flexible, and stakeholder-supported planning.

5.11.6.2 Example 2—Wastewater Treatment Plant Upgrade Under Environmental and Regulatory Uncertainty

A mid-sized city is currently operating a conventional activated sludge (CAS) wastewater treatment plant that is reaching the limits of both its hydraulic capacity and its ability to effectively remove nutrients. With population growth and increased development contributing to rising wastewater inflows, the existing infrastructure is under mounting pressure. Compounding the issue, new environmental regulations mandate more stringent effluent quality standards, particularly regarding the discharge of nitrogen and phosphorus—key nutrients that, in excess, contribute to eutrophication in receiving water bodies.

In response, the city's environmental engineering department must identify and implement an upgrade strategy that not only expands the plant's capacity to handle higher volumes of wastewater but also ensures compliance with enhanced treatment requirements. The planning horizon is 25 years. The chosen solution must strike a balance between technical effectiveness, operational feasibility, cost-efficiency, and long-term adaptability to future regulatory or environmental changes. The decision will be central to safeguarding public health, protecting local ecosystems, and supporting sustainable urban development. A conceptual overview of the strategic upgrade alternatives under uncertain regulatory and capacity conditions is illustrated in Fig. 5.89. This figure outlines the diverging pathways associated with each treatment option, highlighting their relative adaptability, resilience, and long-term suitability in response to increasing environmental pressure and wastewater inflow. Different options for wastewater treatment based on cost and upgrade flexibility have been presented in Table 5.80. The key questions that should be addressed are as follows:

1. What environmental and regulatory uncertainties affect wastewater treatment planning?
2. What future scenarios are plausible and should be considered?
3. How does each treatment option perform across those scenarios?

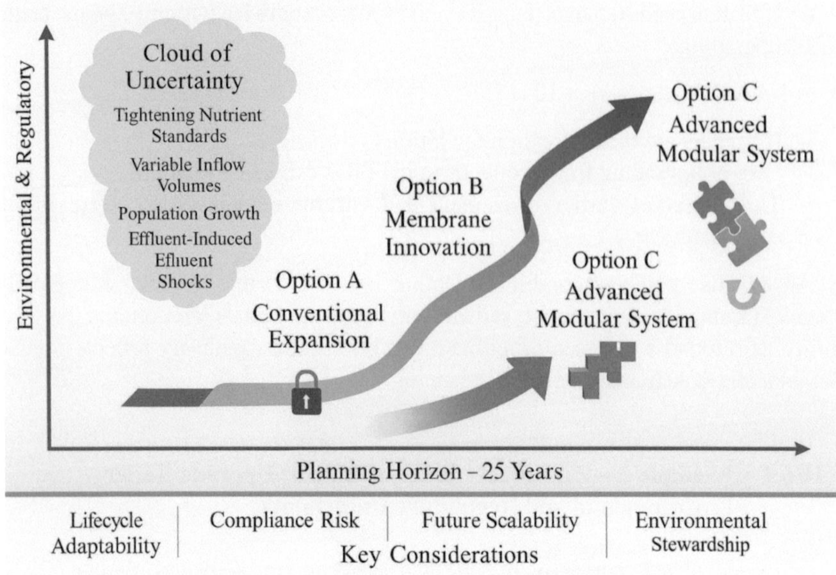

Fig. 5.89 Conceptual representation of strategic upgrade options for a municipal wastewater treatment facility under evolving environmental regulations and inflow demand

Table 5.80 Comparison of wastewater treatment options based on cost and upgrade flexibility

Option	Description	Initial cost ($M)	Annual O&M cost ($M)	Flexibility for future upgrades
A	Expand existing activated sludge (AS)	45	5	Low
B	Membrane bioreactor (MBR)	65	6.5	Medium
C	Modular advanced treatment units (ATUs)	85	7	High (scalable and upgradable)

4. Which option is most robust to uncertainty?
5. Which option results in the least regret if regulations become more stringent?
6. Can an adaptive pathway be proposed to manage both environmental risk and financial constraints?

Step-by-Step Solution Using SBDA

Step 1: Define the Decision Problem

The city's environmental engineering department begins by clearly defining the decision problem to guide the selection of an optimal upgrade strategy for the wastewater treatment plant. The primary objective is to ensure long-term regulatory compliance with both current and anticipated effluent standards, particularly for nitrogen and phosphorus, while simultaneously minimizing total lifecycle costs under conditions of future uncertainty, including population growth, climate variability, and evolving regulations.

Three upgrade alternatives have been identified:

- Option A: Expand Activated Sludge (AS)—This is the lowest-cost option but offers limited flexibility to accommodate future changes or tightening standards.
- Option B: Adopt Membrane Bioreactors (MBR)—This represents a moderate-cost solution that offers moderate flexibility, improving both treatment performance and plant compactness.
- Option C: Deploy Advanced Treatment Units (ATUs)—Though the most expensive, this option provides high flexibility, making it better suited for long-term adaptability and more aggressive nutrient removal goals.

Key constraints shaping the decision include a budget cap of $100 million and the non-negotiable requirement to comply with both present and foreseeable effluent discharge regulations. The planning horizon for evaluating the performance and sustainability of each option is set at 25 years, ensuring that the selected strategy supports both immediate needs and future resilience.

Step 2: Identify Key Uncertainties

To make a robust and future ready decision, the city must identify the critical uncertainties that could significantly influence the performance and suitability of each wastewater treatment upgrade option over the 25-year planning horizon.

- **Regulatory Stringency**

  One of the most pressing uncertainties is the future stringency of effluent regulations, particularly concerning nitrogen and phosphorus discharge limits. While current standards are known, environmental agencies may tighten these limits over time in response to evolving ecological concerns, watershed nutrient loading, or federal mandates. Stricter regulations would demand higher nutrient removal efficiency, potentially rendering low-flexibility options inadequate without costly retrofits.
- **Population Growth and Inflow Volume**

  Another key uncertainty involves future population growth and the resulting increase in wastewater inflow. Rising populations and urban development will increase both the hydraulic load and nutrient concentration entering the treatment plant. These changes can put pressure on the ability of the plant to meet effluent guidelines, especially if not anticipated in the design. The rate and scale of these changes are difficult to predict with precision, yet they have direct implications for treatment reliability and system resilience.

Addressing these uncertainties is essential in selecting a treatment option that not only meets today's needs but remains, viable and cost-effective under a range of possible future conditions.

Step 3: Develop Plausible Scenarios

Table 5.81 defines three potential future scenarios combining different levels of population growth and nutrient discharge standards for nitrogen (N) and phosphorus (P). Each scenario illustrates the resulting implications for treatment system performance, guiding infrastructure design and regulatory compliance strategies.

Step 4: Engage Stakeholders

To ensure that the decision-making process reflects diverse perspectives and practical constraints, the city organized a series of stakeholder engagement workshops. These sessions brought together a broad coalition of participants, including utility engineers and plant operators, who contributed technical insights on treatment feasibility and operational challenges; environmental regulators, who clarified current and anticipated compliance expectations; public health officers, who emphasized the importance of protecting community health and water quality; and representatives from the city council and finance department, who focused on fiscal responsibility and public accountability.

Through these collaborative discussions, stakeholders collectively agreed on two primary performance metrics to guide the evaluation of upgrade options:

- Total Lifecycle Cost—encompassing capital investment, long-term operation and maintenance expenses, and any future retrofitting needs.
- Compliance Risk—representing the probability of failing to meet effluent standards over the 25-year planning horizon, particularly under scenarios of increasing regulatory stringency and rising inflows.

By establishing a shared framework for assessing alternatives, the stakeholder engagement process helped align technical, financial, regulatory, and public health priorities, laying the foundation for a transparent and balanced infrastructure decision.

Table 5.81 Plausible future scenarios for wastewater treatment planning

Scenario	Description	Population growth	Nutrient limits (N, P mg/L)	Implication
S1	Conservative growth + current standards	Low	10 (N), 1.0 (P)	Minimal change
S2	Moderate growth + stricter standards	Medium	5 (N), 0.5 (P)	Requires better removal
S3	Rapid growth + ultra-stringent regulation	High	3 (N), 0.1 (P)	Requires advanced tech

Step 5: Define Performance Metrics

Based on the stakeholder engagement workshops, two two key performance metrics that integrate financial and regulatory considerations over the 25-year planning horizon were used to evaluate the feasibility of the different upgrade options.

1. Total Lifecycle Cost

   This metric captures the comprehensive financial impact of each alternative by summing three cost components:

   - Initial Capital Investment
   - Operation and Maintenance (O&M) Costs over 25 years
   - Penalties and Retrofit Costs incurred if the system fails to meet future regulatory standards

   This approach ensures that decision-makers consider not only upfront expenditures but also the long-term economic consequences of each choice.
2. Compliance Risk Score

   This qualitative metric translates regulatory reliability into a simple scale to reflect the risk of non-compliance:

   - 0 = Full compliance with current and projected regulations (no action needed)
   - 1 = Minor retrofit required to maintain compliance (manageable cost and effort)
   - 2 = Major retrofit and/or fines due to significant non-compliance (high risk and cost).

These performance metrics enable a clear and structured comparison of options, facilitating the identification of strategies that balance cost-effectiveness with regulatory resilience.

Step 6: Evaluate Options Under Each Scenario

Table 5.82 presents lifecycle cost estimates (in millions of dollars) for three wastewater treatment options, including initial capital costs, 25-year operation and maintenance (O&M) expenses, and scenario-dependent retrofit costs. Total costs are calculated for each future scenario (S1–S3), reflecting the financial impact of regulatory tightening and population growth. Moreover, Table 5.83 summarizes the compliance risk scores for each treatment option under three future scenarios (S1–S3). Scores indicate the likelihood of failing to meet nutrient discharge standards, with higher values reflecting greater regulatory risk. Option C demonstrates the highest robustness with zero compliance risk across all scenarios.

Step 7: Apply Robustness and Regret Criteria

Table 5.84 presents the worst-case total lifecycle costs for each treatment option across all future scenarios. Applying the minimax rule, Option A emerges as the most robust choice in cost terms, offering the lowest maximum expenditure under uncertainty.

Table 5.82 Lifecycle cost estimates for treatment options under evolving scenario assumptions

Option	Initial	O&M (25 yrs)	Retrofit	Total		
				S1	S2	S3
A	45	125	+20 (S2), +40 (S3)	170	190	210
B	65	162.5	+10 (S3)	227.5	227.5	237.5
C	85	175	0	260	260	260

Table 5.83 Compliance risk scores for treatment options across future scenarios

Option	S1	S2	S3
A	0	1	2
B	0	0	1
C	0	0	0

Table 5.84 Robustness evaluation of treatment options using the minimax cost criterion

Option	Worst-case cost
A	210
B	237.5
C	260

Table 5.85 shows the regret values for each treatment option under three future scenarios (S1–S3), with regret defined as the additional cost incurred compared to the lowest-cost option in each scenario. The maximum regret highlights the worst-case deviation from optimal cost. Option A demonstrates zero regret across all scenarios, indicating superior cost performance—though this analysis excludes compliance risks.

Table 5.86 presents adjusted lifecycle costs for each treatment option under Scenario S3 by incorporating compliance risk penalties. A cost of $15 million is added per compliance risk point. The resulting composite scores balance financial cost with regulatory risk. While Option A remains the most cost-effective, Option C offers guaranteed compliance and may be favored by risk-averse stakeholders.

Table 5.85 Regret analysis of treatment options based on lifecycle costs

Option	Regret			
	S1	S2	S3	Max
A	0	0	0	0
B	57.5	37.5	27.5	57.5
C	90	70	50	90

Table 5.86 Risk-weighted composite cost scores incorporating compliance penalties

Option	S3 cost	Risk penalty	Composite
A	210	$2 \times 15 = 30$	240
B	237.5	$1 \times 15 = 15$	252.5
C	260	0	260

Step 8: Design adaptive pathway

To address uncertainty in future regulatory requirements and wastewater inflow, the city has developed an adaptive pathway, a flexible, phased approach that aligns investments with actual conditions over time rather than relying on a single, fixed infrastructure choice.

- Initial Phase (Years 0–10)

  The plan begins with constructing Option A (an expanded activated sludge system) at a cost of $45 million. This option meets immediate capacity and compliance needs without overcommitting resources. To prepare for potential changes, the city will install real-time nitrogen sensors and inflow meters to actively monitor treatment performance and plant loading. This data will serve as a foundation for informed, evidence-based upgrades.
- Trigger Condition for Upgrade

  A predefined trigger will initiate the next phase of investment if one of the following conditions is met:

  – Inflow exceeds 90% of hydraulic **capacity**, indicating rapid population growth or increased demand.
  – Stricter effluent regulations are enacted, requiring enhanced nitrogen or phosphorus removal.

- Upgrade Phase (Years 10–15)

  If triggered, the city will implement one of two upgrades depending on updated needs and financial flexibility:

- Retrofit to Membrane Bioreactors (MBR) with an additional $25 million, providing a balance between cost and performance.
- Add modular Advanced Treatment Units (ATUs) at a higher cost of $40 million, offering maximum flexibility and nutrient removal capability.

This phased strategy provides multiple advantages:

- It avoids overdesign, reducing upfront financial burden while still enabling future scalability.
- It meets near-term regulatory and capacity requirements with a cost-effective initial investment.
- It retains the ability to adapt to evolving policies, technologies, and population growth, ensuring the system remains resilient and compliant well into the future.

Step 9: Communicate results and refine

In the final step of the decision-making process, the city's environmental engineering department focuses on clearly communicating the results of the analysis and gathering feedback for possible refinements. This step ensures transparency, builds stakeholder confidence, and supports informed decision-making across departments and the community.

Key Deliverables:

1. Performance Matrix

   A detailed matrix presents each upgrade option's total lifecycle cost alongside its compliance risk score, allowing stakeholders to compare trade-offs and identify the most resilient and cost-effective strategies.
2. Adaptive Pathway Timeline
   A visual timeline illustrates the phased approach over the 25-year horizon, showing when key investments (initial build, monitoring, upgrades) would occur based on observed conditions. This helps convey the flexibility and scalability of the proposed plan.
3. Stakeholder Briefing Document
   A concise summary is prepared for all involved parties—including city council, finance officials, and environmental regulators—explaining:

   - Why Option A is the most cost-effective choice in the short term, satisfying current needs within the budget.
   - Why Option B (MBR) or Option C (ATUs) may be required later to meet stricter effluent standards or rising inflows.
   - The trigger conditions that will prompt an upgrade decision, such as exceeding 90% plant capacity or the introduction of more stringent nutrient discharge regulations.

By presenting these deliverables, the city ensures that stakeholders are well-informed and that the upgrade strategy remains flexible, transparent, and responsive to both technical needs and public expectations.

Final Answers to the Questions

1. What environmental and regulatory uncertainties affect wastewater treatment planning?

   - Future nutrient standards (especially N, P)
   - Population and industrial growth affecting flow
   - Affects capacity and compliance

2. What future scenarios are plausible?

   - S1: Conservative growth, current limits (10 mg/L N, 1 mg/L P)
   - S2: Moderate growth, stricter (5 mg/L N, 0.5 mg/L P)
   - S3: Rapid growth, ultra-stringent (3 mg/L N, 0.1 mg/L P)

Table 5.87 Summary performance comparison of treatment options

Option	Cost range ($M)	Compliance risk	Adaptability
A	170–210	Moderate–high	Low
B	227.5–237.5	Low–moderate	Medium
C	260	Zero	High

3. How do options perform?

   Table 5.87 provides a high-level comparison of the three treatment options based on total lifecycle cost range, compliance risk, and adaptability. Option A is the most cost-efficient but offers limited flexibility and higher regulatory risk. Option B represents a middle ground in both cost and performance. Option C, while the most expensive, stands out for its zero compliance risk and high adaptability, making it a strong candidate for future-proof infrastructure.
4. Most robust to uncertainty?

   Option A (lowest worst-case cost), not robust to compliance risk under stricter futures.
5. Least regret if regulations become stricter?

   Option A demonstrates the lowest worst-case cost, making it the most attractive choice from a purely financial standpoint under uncertain future conditions. Its low initial investment and moderate operating costs ensure that even in adverse scenarios, the city avoids excessive spending—offering a degree of economic resilience.

   However, Option A is not fully robust when considering compliance risk in stricter regulatory futures. If nutrient discharge standards become significantly tighter or if inflow volumes increase beyond expectations, Option A may fail to meet performance requirements without substantial retrofitting. This exposes the city to the risk of non-compliance, potential fines, and emergency capital expenditures.

   Therefore, while Option A offers strong cost predictability under uncertainty, its limited flexibility and regulatory risk mean it should be pursued only within an adaptive framework—with monitoring in place and clear triggers for timely upgrades to more advanced options like MBR or ATUs.
6. Can an adaptive pathway be proposed?

   Yes, an adaptive pathway is not only possible but highly recommended given the uncertainty surrounding future regulatory requirements and wastewater inflow volumes. Indeed, an adaptive pathway starting with Option A and integrating conditional upgrades to Options B or C ensures the city remains both fiscally responsible and environmentally compliant, even in the face of future uncertainty.

5.11.6.3 Example 3—Foundation Strategy for a High-Rise in a Seismic Zone

A construction firm is preparing to build a 25-storey commercial high-rise in a moderate-to-high seismic risk region. Due to site variability and uncertain seismic conditions over the building's 100-year design life, selecting an appropriate foundation system is critical to ensure both safety and cost-efficiency. Table 5.88 summarizes three foundation design options for a seismically sensitive structure, evaluating their initial construction cost, inherent seismic resistance, and potential for future upgrades. Given that seismic safety standards evolve with updates to national hazard maps, the selection must balance current performance with adaptability to future code requirements. The selection of an appropriate foundation system for high-rise construction in seismic zones is a multifaceted decision involving geotechnical, structural, and regulatory considerations. As illustrated in Fig. 5.90, the process integrates seismic risk assessment, soil variability, future design code evolution, and practical construction constraints to determine the most resilient and cost-effective foundation strategy. The key questions to be addressed are as follow:

1. What are the main uncertainties in choosing a foundation strategy in seismic regions?
2. What plausible future seismic or regulatory scenarios should be considered?
3. How do the foundation options perform under each scenario?
4. Which option is most robust to long-term seismic risk and regulation changes?
5. Which option minimizes regret if the worst-case scenario occurs?
6. Can a staged/adaptive foundation approach be implemented to manage risk and budget constraints?

Step-by-Step Solution Using SBDA

Step 1: Define the Decision Problem

The decision problem centers on selecting the most suitable foundation design for a critical infrastructure project, one that ensures long-term seismic safety, remains cost-effective over its lifecycle, and can adapt to future changes in regulatory standards or seismic hazard assessments. Given the structure's importance and its intended

Table 5.88 Foundation alternative options based on cost, seismic performance, and upgrade potential

Option	Description	Initial cost ($M)	Seismic resistance	Upgrade potential
A	Shallow mat foundation	20	Low	None
B	Deep piles + grade beams	30	Medium	Moderate (retrofit possible)
C	Base-isolated raft foundation	45	High	High (scalable damping)

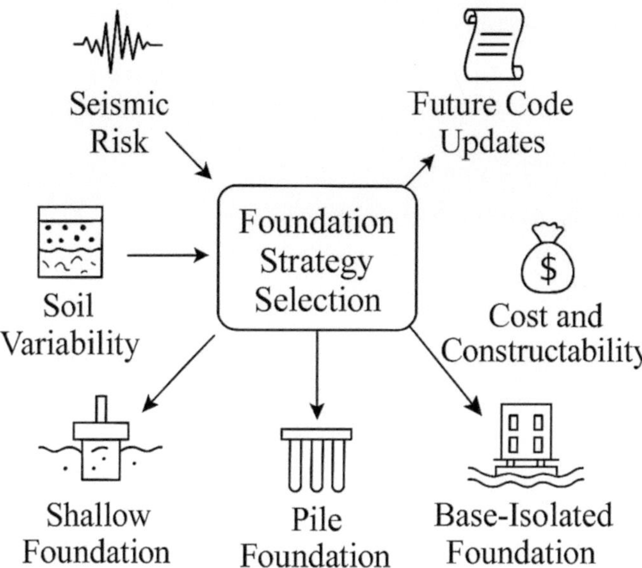

Fig. 5.90 Conceptual diagram of foundation system selection for a high-rise building under seismic uncertainty

design life of 100 years, the foundation must be resilient not only to current seismic risks but also to potential shifts in hazard profiles due to climate change, tectonic re-evaluation, or evolving engineering standards.

The primary objective of this decision problem is to select a foundation design that ensures both immediate and long-term structural reliability in the face of seismic hazards. Specifically, the selected design must:

- Meet long-term seismic performance and safety requirements, ensuring structural stability and occupant protection throughout the structure's 100-year design life.
- Minimize total lifecycle cost, accounting not only for initial construction expenses but also for long-term maintenance and the potential need for retrofitting due to evolving seismic standards or unforeseen ground behavior.
- Accommodate future changes in regulations or hazard conditions, allowing for adaptability to shifts in the National Building Code of Canada (NBCC), updated seismic risk assessments, or site-specific hazard changes, thereby reducing the risk of obsolescence and unplanned interventions.

To address the project's long-term seismic and economic objectives, three foundation design alternatives have been identified, each with distinct characteristics in terms of performance, cost, and adaptability:

- Option A: Mat Foundation

  A low-cost, shallow-depth foundation ideal for low seismic zones or stiff soils. Offers limited performance in high seismic areas but is attractive from a budgetary standpoint.
- Option B: Deep Pile System

  A standard solution for moderate seismic risk zones, offering greater load-bearing and lateral resistance by anchoring into deeper, more stable strata. It balances performance and cost.
- Option C: Base-Isolated Raft

  A high-cost but high-performance solution using seismic base isolation to minimize earthquake-induced forces. Ideal for critical facilities requiring uninterrupted operation and maximum safety.

The foundation design decision must adhere to two critical constraints that shape the feasibility and acceptability of each alternative:

- Budget Constraint: The preferred foundation solution should cost under $50 million to align with financial limitations.
- Regulatory Compliance: All options must meet or exceed the seismic provisions of the National Building Code of Canada (NBCC).

Given the 100-year design horizon and the evolving nature of seismic science and regulation, the selected option must provide both initial reliability and future adaptability, ensuring structural integrity and public safety over the long term.

Step 2: Identify Key Uncertainties

In the context of a 100-year design life, long-term planning must account for uncertainties that could significantly influence both the performance and regulatory acceptability of the foundation system over time. Two primary sources of uncertainty have been identified:

1. Seismic Hazard Level: Future ground motion intensities may increase due to tectonic activity or revised models.
2. Regulatory Standards: Future updates to NBCC may impose stricter foundation or damping system requirements for tall buildings.

Step 3: Develop Plausible Scenarios

Table 5.89 outlines three plausible future seismic design scenarios, incorporating projected peak ground acceleration levels and corresponding building code stringency. These scenarios reflect expected updates to seismic hazard classifications and regulatory standards aligned with upcoming National Building Code of Canada (NBCC) cycles, guiding resilient structural design planning.

Table 5.89 Seismic design scenarios based on ground acceleration and code evolution

Scenario	Description	Ground acceleration (g)	Code stringency
S1	Conservative forecast	0.25	Current code
S2	Moderate hazard increase	0.35	Revised code (2025)
S3	Severe earthquake zone Reclassification	0.50	Strict code (2030+)

Step 4: Engage Stakeholders

A comprehensive stakeholder engagement process was conducted to ensure that the foundation design decision reflects diverse priorities. The workshops and consultations involved key groups with vested interests in both the short-term construction and long-term performance of the structure:

- Structural engineers and geotechnical consultants, who provided insights into site-specific soil behavior, seismic performance, and technical feasibility of each foundation alternative.
- Regulatory authorities and insurers, who emphasized the importance of code compliance, liability mitigation, and risk-based assessment of design choices.
- Developers and investors, whose focus was on financial viability, lifecycle cost, and alignment with future market expectations.
- Public safety advocates, who prioritized resilience, occupant safety, and system reliability during rare but high-impact seismic events.

Through these engagements, a clear consensus emerged: while initial construction cost remains an important consideration, it should not outweigh more critical priorities. Stakeholders agreed that safety, regulatory compliance, and the flexibility to retrofit in response to future seismic or code changes are essential criteria in the decision-making process. This shared understanding shaped the evaluation framework and highlighted the need for a robust yet adaptable foundation strategy.

Step 5: Define Performance Metrics

To objectively assess the foundation design alternatives, two primary performance metrics were established. These metrics provide a structured way to compare the trade-offs between cost, safety, and long-term resilience across varying future scenarios.

1. Total Lifecycle Cost

This metric captures the comprehensive financial burden of each foundation option over the 100-year design life. It includes:

- Initial construction cost
- Expected retrofit cost if future upgrades are needed to meet revised seismic codes or hazard levels
- Insurance premiums, which are adjusted based on the seismic performance and perceived risk of each foundation type

Table 5.90 Lifecycle cost estimates of foundation options under seismic scenarios

Option	Initial	Retrofit Cost (S2/S3)	Insurance	Total		
				S1	S2	S3
A	20	+10 (S2), +25 (S3)	5	25	35	50
B	30	0 (S2), +5 (S3) only	3	33	33	38
C	45	0	1	46	46	46

This approach ensures that the evaluation reflects not just upfront spending, but also the long-term economic implications of choosing a less adaptable or risk-prone design.

2. Compliance Risk Score

This qualitative metric assesses each option's ability to meet current and future NBCC seismic standards, under various uncertainty scenarios:

- 0 = Full compliance under all modeled hazard and regulatory scenarios (no retrofitting required)
- 1 = Partial compliance; may require moderate retrofitting to maintain compliance under stricter codes
- 2 = Non-compliant or at high risk of failure, necessitating major intervention or posing unacceptable life safety risks

Together, these metrics provide a robust decision framework that balances cost-effectiveness with seismic safety, adaptability, and regulatory foresight.

Step 6: Evaluate Options Under Each Scenario

A. Lifecycle Cost Estimates (in $M)

Table 5.90 presents total lifecycle cost estimates (in millions of dollars) for three foundation options across three seismic hazard scenarios (S1–S3). Costs include initial construction, projected retrofit expenses triggered by evolving code requirements, and long-term insurance premiums. Option C shows the highest upfront cost but avoids future retrofits and maintains the lowest insurance burden.

B. Compliance Risk Scores

Table 5.91 presents the compliance risk scores for each foundation option under three seismic design scenarios (S1–S3). Scores reflect the likelihood that each option will fail to meet future seismic code requirements, with higher values indicating increased regulatory risk. Option C demonstrates full compliance across all scenarios, offering the greatest code resilience.

Table 5.91 Compliance risk scores for foundation options across seismic scenarios

Option	S1	S2	S3
A	0	1	2
B	0	0	1
C	0	0	0

Table 5.92 Robustness assessment of foundation options using the minimax cost criterion

Option	Worst-case cost
A	50
B	38
C	46

Table 5.93 Regret analysis of foundation options based on lifecycle costs

Option	Regret			
	S1	S2	S3	Max
A	0	2	12	12
B	8	0	0	8
C	21	13	8	21

Step 7: Apply Robustness and Regret Criteria

A. Robustness (Minimax Cost Rule)

Table 5.92 displays the worst-case total lifecycle cost for each foundation option under the most demanding seismic scenario. Applying the minimax rule, Option B emerges as the most robust choice in cost terms, offering the lowest maximum expenditure across all considered futures.

B. Regret Matrix (Based on Cost)

Tabe 5.93 presents the regret values for each foundation option under three seismic scenarios (S1–S3), where regret represents the additional cost incurred compared to the least-cost option for each scenario. The maximum regret column highlights the worst-case deviation. Option B shows the lowest maximum regret, making it the most regret-averse choice under cost uncertainty.

Step 8: Design Adaptive Pathway

To manage seismic uncertainty and future-proof the foundation system, a phased adaptive strategy has been proposed. This approach allows the city to minimize initial investment while preserving the ability to upgrade efficiently if seismic risks or regulations increase over time.

Phased Foundation Strategy

Phase 1 (Years 0–5): Immediate Implementation

- Construct Option B: Implement the deep pile system, a well-established and cost-effective solution suitable for current seismic risk levels and within the budget threshold.
- Install Monitoring Sensors: Integrate accelerometers and ground motion sensors to track real-time seismic activity and structural response.
- Design for Future Retrofit: Engineer pile caps and tie beams to be compatible with future base isolator installations, enabling smoother and more economical retrofitting if required.

Next phase will be initiated if either of the following conditions are met:

- The National Building Code of Canada (NBCC) releases updated seismic hazard maps indicating peak ground acceleration (PGA) $\geq$ 0.4g in the region.
- A regional seismic event occurs with a measured PGA > 0.3g, suggesting a shift in local hazard behavior or site amplification effects.

Phase 2 (Years 5–20): Conditional Retrofit

- Retrofit with Base Isolators: For an estimated cost of $10 million, base isolators are added to reduce lateral displacements during seismic events.
- This upgrade enhances structural resilience, reduces damage potential, and effectively extends the service life of the facility.

The adaptive foundation strategy offers several key advantages that make it a practical and future-ready solution for long-term infrastructure planning:

- Minimizes upfront capital expenditure by avoiding immediate investment in high-cost isolation systems.
- Maintains compliance and safety under current conditions while remaining ready for future updates.
- Preserves flexibility to adapt infrastructure based on real-world seismic data and evolving codes.

This adaptive pathway strikes a prudent balance between fiscal responsibility, seismic safety, and long-term resilience, ensuring that the foundation system remains robust, upgrade-ready, and regulation-compliant across its 100-year design horizon.

Step 9: Communicate Results and Refine

In the final step of the decision-making process, the project team focuses on effectively communicating the analysis results to key decision-makers and stakeholders, ensuring transparency, understanding, and the opportunity for refinement based on feedback. The goal is to facilitate an informed choice that aligns with technical, financial, and public safety priorities. To support this decision, the project team compiled and presented a set of targeted materials designed to clearly convey the trade-offs, adaptive strategies, and associated risks of each option as follow:

1. Cost + Risk Performance Matrix

   A comparative matrix outlining each foundation option's total lifecycle cost and compliance risk score, providing a clear visualization of the trade-offs between financial investment and regulatory/safety performance.
2. Adaptive Retrofit Flowchart

   A visual roadmap that illustrates the phased foundation strategy, including initial implementation, trigger conditions, and potential upgrade pathways. This helps stakeholders understand how flexibility is embedded into the design process.
3. Risk Communication Brief

   A concise summary explaining the strategic implications of each option:

   - Option A: Lowest initial cost but carries significant compliance and seismic performance risks in the long term.
   - Option B: Strikes a balanced approach, offering moderate cost, retrofit-ready design, and flexibility for future upgrades.
   - Option C: Provides the highest safety, but at a high upfront cost that may be unjustified unless extreme seismic conditions are confirmed.

Based on the presented evidence and discussions, stakeholders support Option B with built-in upgrade triggers, recognizing it as the solution that best balances long-term resilience, regulatory adaptability, and cost-effectiveness. This approach provides robust protection now while allowing strategic enhancement if seismic risks increase—ensuring the foundation system remains reliable and future-ready for its full 100-year lifespan.

Answers to the questions

1. What are the main uncertainties in choosing a foundation strategy?

   - Changes in PGA (Peak Ground Acceleration)
   - Updates to building code requirements

2. What plausible seismic scenarios should be considered?

   - S1: Conservative (0.25g), current NBCC
   - S2: Moderate increase (0.35g), 2025 code
   - S3: High risk reclassification (0.50g), 2030+ code

3. How do options perform under each scenario?

   - A: Non-compliant in S3, high retrofit costs
   - B: Reliable and upgradable
   - C: Always compliant, most expensive

4. Which is most robust?

   Option B—lowest worst-case cost ($38M)
5. Which minimizes regret?

   Option B—lowest max regret ($8M), with compliance and retrofit flexibility

6. Can an adaptive approach be used?

   Yes—start with Option B, retrofit to C if risk increases.

5.11.6.4 Example 4—Urban Drainage System Upgrade Under Climate Change Uncertainty

A mid-sized city is increasingly experiencing more frequent and intense rainfall events, causing urban flooding and stressing existing stormwater infrastructure. The city needs to upgrade its drainage system, but there is deep uncertainty about how climate change will affect rainfall intensity, duration, and frequency over the next 30 years. Table 5.94 outlines three design options for upgrading a mid-sized city's stormwater drainage system in response to increasing flood risks. Each option is evaluated based on initial capital cost, annual maintenance cost, and climate adaptability. The alternatives range from conventional infrastructure improvements to advanced adaptive systems integrating green infrastructure and smart technologies, supporting resilience planning over a 30-year horizon. A conceptual overview of the alternative drainage strategies considered in this study is illustrated in Fig. 5.91. This figure emphasizes the differences in system design logic and climate adaptability—ranging from conventional upgrades to smart hybrid solutions—without committing to specific quantitative estimates. It serves to visually communicate the core characteristics of each approach within the broader context of deep climate uncertainty affecting urban flood resilience planning. The key questions that should be addressed are as follows:

1. What climate-related uncertainties impact the drainage design decision?
2. What future climate scenarios should be considered?
3. How does each design option perform across these scenarios?
4. Which option is most robust to climate variability?
5. Which option minimizes regret if the most extreme rainfall scenario occurs?
6. Is it possible to adopt a phased, adaptive drainage strategy?

Table 5.94 Urban drainage upgrade alternatives under climate uncertainty

Option	Description	Initial cost ($M)	Annual maintenance ($M)	Climate adaptability
A	Conventional pipe network upgrade	40	2	Low
B	Distributed green infrastructure (GI)	55	2.5	Medium
C	Smart adaptive hybrid system	70	3	High (IoT + storage)

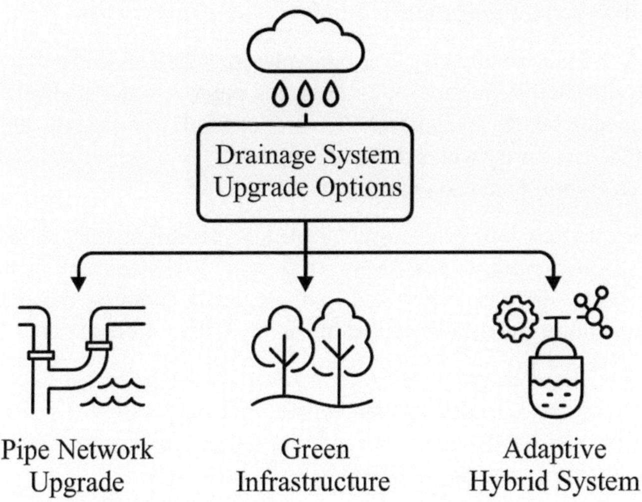

Fig. 5.91 Conceptual representation of urban drainage system upgrade alternatives under climate uncertainty

Step-by-Step Solution Using SBDA

Step 1: Define the Decision Problem

The decision problem focuses on identifying the most suitable urban drainage solution to manage increasing flood risks associated with climate uncertainty, aging infrastructure, and stricter environmental standards. The solution must provide long-term flood resilience, maintain cost-effectiveness, and comply with water quality discharge regulations over a 30-year planning horizon. The main objective is to select a drainage solution that:

- Minimizes flood risk across a range of future climate scenarios (e.g., increased rainfall intensity, storm surges)
- Remains cost-effective, accounting for both capital expenditure (capEx) and ongoing operational and maintenance costs (opEx)
- Adapts to climate variability and evolving hydrological patterns over time

To reach the aforementioned objectives, three alternative drainage solutions have been identified:

- Option A: Conventional Pipe Resizing

  Involves upsizing stormwater pipes to handle greater flow volumes. This traditional approach is straightforward and relatively low-tech but can be costly and inflexible, especially under high-uncertainty futures.

- Option B: Green Infrastructure

  Uses nature-based solutions such as bioswales, rain gardens, and permeable pavements to slow runoff, promote infiltration, and improve water quality. This option is ecologically beneficial and cost-effective for small to mid-scale areas, though performance may vary with extreme events
- Option C: Hybrid Smart System

  Combines physical infrastructure with real-time monitoring and control technologies (e.g., smart gates, sensors, adaptive storage) to dynamically manage flows. While initially more expensive and technologically complex, this option offers high adaptability and data-driven optimization under uncertain climate futures.

It should be noted that there are several key constraints that must be considered when evaluating and selecting the appropriate drainage solution. These constraints ensure that the chosen alternative not only meets the technical objectives but also aligns with financial, regulatory, and operational boundaries:

- **Budget**: The combined capital and operational cost must not exceed $100 million over the 30-year period.
- **Regulatory Compliance**: The selected system must meet water quality discharge guidelines, ensuring pollutants like sediment, nutrients, and heavy metals are adequately managed.
- **Timeframe**: The solution must provide effective service for a minimum of 30 years, supporting both short-term flood protection and long-term climate resilience.

Step 2: Identify Key Uncertainties

When planning a long-term urban drainage solution, it is critical to account for uncertainties that could significantly affect system performance and cost-effectiveness over the 30-year lifespan. Two major sources of uncertainty have been identified:

1. Future Rainfall Extremes

The most prominent uncertainty relates to changes in rainfall intensity, duration, and frequency, commonly represented by IDF (Intensity–Duration–Frequency) curves. Climate change is expected to alter precipitation patterns, potentially increasing the occurrence of short-duration, high-intensity storms or prolonged rainfall events. However, the magnitude and timing of these changes are difficult to predict with precision, making it challenging to size or design drainage infrastructure with confidence.

2. Urban Development Density

As the city continues to grow, population increases and urban densification will likely result in a higher percentage of impervious surfaces (e.g., roads, rooftops, parking lots). This change directly increases surface runoff and peak flow rates during storm events. The rate and distribution of urban development are uncertain, which adds complexity to forecasting future drainage loads and designing a system that remains effective under changing land use conditions.

These uncertainties reinforce the importance of selecting a drainage strategy that offers adaptability, allowing for performance optimization as more data becomes available and conditions evolve over time.

Step 3: Develop Plausible Climate Scenarios

Table 5.95 defines three future scenarios combining projected climate change impacts and urban development trends. Each scenario includes estimates of peak 24-h rainfall intensity and increases in impervious surface area, which influence runoff volumes and drainage system performance. These scenarios guide robust design and risk assessment for urban flood mitigation over the long term.

Step 4: Engage Stakeholders

To ensure that the drainage solution reflects a broad range of perspectives and meets both technical and community needs, a comprehensive stakeholder engagement process was undertaken. This step was essential in identifying shared priorities, surfacing concerns, and building consensus around key performance criteria under future uncertainty.

The following stakeholder groups were actively engaged to ensure the proposed solution reflected a broad range of expertise, values, and community needs:

- City Planners and Engineers

  Provided input on infrastructure feasibility, integration with existing systems, and alignment with urban development goals. Their expertise helped frame realistic implementation pathways and maintenance considerations.
- Environmental NGOs

  Advocated for sustainable and nature-based solutions that support ecological health, improve water quality, and contribute to urban climate resilience. They emphasized minimizing environmental degradation and promoting green infrastructure.

Table 5.95 Plausible climate and urbanization scenarios for stormwater planning

Scenario	Description	Peak 24-h rainfall (mm)	Impervious area increase (%)
S1	Conservative warming	100	5
S2	Moderate warming & urban sprawl	125	15
S3	High-end warming & urbanization	150	30

- Local Residents and Businesses

  Raised concerns about flood risk, property protection, and service disruption during and after extreme weather events. Their input highlighted the importance of visible benefits, community co-benefits, and minimal long-term maintenance burdens.
- Climate Scientists and Modelers

  Contributed insights on future rainfall projections, urban climate trends, and hydrologic modeling. They helped quantify uncertainty in rainfall extremes and identify adaptive thresholds relevant for system design.

Through collaborative discussions, stakeholders supported a performance-driven approach that considers lifecycle cost, flood mitigation effectiveness, regulatory compliance, and adaptive capacity. Their engagement played a vital role in shaping a balanced and forward-looking drainage strategy.

During the engagement process, stakeholders collectively emphasized several key priorities that should guide the selection and design of the urban drainage solution. These priorities reflect a balance between long-term resilience, equity, and fiscal responsibility:

- Adaptability

  Stakeholders agreed that the chosen system must be flexible and responsive to future uncertainties, including climate change, regulatory updates, and urban expansion. Solutions that allow for modular upgrades, real-time monitoring, or scalable components were favored over rigid, one-time designs.
- Protection of Vulnerable Areas

  Ensuring the safety of flood-prone and socially vulnerable neighborhoods was identified as a top priority. These areas often face the greatest risk from stormwater flooding and have limited capacity to recover from disruptions. As such, drainage strategies must prioritize equity and risk reduction for these communities.
- Avoid Overspending on Overdesigned Systems

  Stakeholders cautioned against investing heavily in large-scale infrastructure that may prove unnecessary if extreme conditions do not materialize. Instead, they preferred cost-effective solutions that provide reliable performance now but retain the option for enhancement if and when future conditions demand it.

These priorities reinforce the need for an adaptive, balanced approach, one that delivers effective protection without locking the city into inflexible or financially burdensome infrastructure.

Step 5: Define Performance Metrics

To evaluate the drainage alternatives in a structured and transparent manner, two key performance metrics have been established. These metrics capture both economic and functional aspects of system performance over the 30-year planning horizon, allowing for informed comparisons under uncertainty.

1. Total Lifecycle Cost

This metric represents the full financial burden of each drainage option over its operational lifespan. It includes:

- Initial capital investment (construction and implementation costs)
- Ongoing maintenance and operational expenses
- Expected flood damage costs, which account for projected losses (e.g., property damage, service disruptions) from storm events that exceed the system's capacity.

By integrating both upfront and long-term consequences, this metric ensures that decision-makers consider not only the cheapest option today, but the most economically sustainable solution over time.

2. Flood Failure Rate

This metric quantifies system reliability by calculating the percentage of modeled storm events (based on future climate scenarios) in which the drainage system is overwhelmed—resulting in surface flooding or system bypass. A lower failure rate indicates higher resilience and more consistent protection, particularly for vulnerable areas.

Together, these performance metrics provide a balanced lens through which to assess cost-effectiveness and flood protection reliability, both of which are essential for making a robust, future-oriented infrastructure decision.

Step 6: Evaluate Options Under Each Scenario

Table 5.96 presents the 30-year lifecycle cost estimates for three stormwater infrastructure upgrade options, including initial capital costs, maintenance, and scenario-dependent flood damage. Total costs are calculated for each climate scenario (S1–S3), reflecting how future rainfall intensification and urbanization impact financial performance. Option C, though highest in upfront cost, results in the lowest total cost under the most severe scenario (S3).

Table 5.97 shows the projected flood failure rates—defined as the probability of system underperformance resulting in flooding—for each stormwater management option across three climate scenarios (S1–S3). Option C demonstrates the highest reliability, maintaining low failure rates even under severe rainfall and urbanization conditions.

Table 5.96 Lifecycle cost estimates of urban drainage options across climate scenarios

Option	Initial	Maint. (30 yrs)	Flood Damage (S1–S3)	Total		
				S1	S2	S3
A	40	60	20, 60, 100	120	160	200
B	55	75	10, 30, 60	140	160	190
C	70	90	5, 10, 15	165	170	175

Table 5.97 Estimated flood failure rates of drainage options under climate scenarios

Option	S1 (%)	S2 (%)	S3 (%)
A	10	25	40
B	5	15	25
C	2	4	7

Table 5.98 Robustness evaluation of drainage options using the minimax total cost criterion

Option	Worst-case cost
A	200
B	190
C	175

Table 5.99 Regret analysis of drainage options based on lifecycle costs

Option	Regret			
	S1	S2	S3	Max
A	0	0	25	25
B	20	0	15	20
C	45	10	0	45

Step 7: Apply Robustness and Regret Criteria

Table 5.98 summarizes the worst-case (maximum) total lifecycle costs for each drainage upgrade option across all climate scenarios. According to the minimax rule, Option C is the most robust choice, offering the lowest maximum cost and thus the greatest protection against future uncertainty in rainfall intensity and urban growth.

Table 5.99 presents the regret values for each stormwater infrastructure option under three climate scenarios (S1–S3). Regret is calculated as the additional cost incurred relative to the lowest-cost option in each scenario. The maximum regret indicates the worst-case decision penalty. Option B demonstrates the lowest maximum regret, making it the most regret-averse choice under cost uncertainty.

Step 8: Design Adaptive Pathway

To manage uncertainty in future rainfall extremes and urban development, a two-phase adaptive drainage strategy has been developed. This approach allows the city to begin with a low-impact, cost-effective solution and scale up only if real-world conditions warrant additional investment—ensuring both resilience and fiscal efficiency.

Two-Phase Adaptive Strategy

Phase 1 (Years 0–10): Initial Implementation

- Begin with Option B: Green Infrastructure, focusing on distributed solutions such as bioswales, permeable pavements, and rain gardens.
- Employ a modular build approach, allowing for gradual expansion as needed.
- Install rainfall monitoring systems and initiate a tracking program for flood-related complaints to assess system performance over time.

Transition to Phase 2 if either of the following trigger conditions is met:

- Peak rainfall events exceeding 130 mm occur three or more times within a 10-year period
- Annual flood-related complaints exceed 10 incidents, indicating persistent performance gaps or growing public concern.

These conditions signal a significant increase in system stress or community impact, justifying the need for a technological upgrade.

Phase 2 (Years 10–20): Conditional Retrofit

- Upgrade to Option C: Hybrid Smart System, incorporating:
  - Smart sensors for real-time monitoring of flow, depth, and rainfall
  - Detention tanks to temporarily store excess runoff
  - Control valves for dynamic flow regulation and system optimization
- Estimated cost for this upgrade phase is approximately $30 million, a fraction of the total budget cap, leaving room for further adaptation if necessary.

The adaptive drainage strategy offers several compelling benefits that align with the city's goals of resilience, fiscal responsibility, and regulatory compliance:

- Avoids overdesign by deferring high-cost upgrades until justified by observed trends
- Provides scalable protection that evolves with climate and urbanization
- Supports data-informed decision-making, enhancing both public confidence and regulatory alignment.

This phased, flexible pathway ensures the drainage system remains cost-effective, environmentally sustainable, and resilient to future challenges over the 30-year design horizon.

Step 9: Communicate Results and Refine

The final step in the planning process involved presenting a comprehensive set of results to stakeholders and decision-makers to ensure transparency, support informed evaluation, and allow for refinements based on feedback. To facilitate this, the team prepared a series of deliverables as:

- Cost versus Risk Matrices

  Comparative visualizations that show how each drainage option performs in terms of total lifecycle cost and flood failure rate across different climate and urbanization scenarios. These matrices help stakeholders quickly understand trade-offs between investment and resilience.
- Floodplain Maps by Scenario

  Scenario-based geospatial maps illustrating projected flood extents under each option. These maps provide a visual reference for assessing the protection of vulnerable areas and the impact of different rainfall intensities or land-use changes.

During the stakeholder briefing, the results were summarized to clarify the implications of each alternative:

- Option A (Conventional Pipe Resizing) is the least resilient, offering limited adaptability and higher long-term risk under intensifying storm conditions.
- Option B (Green Infrastructure) is scalable and flexible, allowing for phased expansion and real-time monitoring, and performs well under moderate future scenarios.
- Option C (Hybrid Smart System) is the most robust, but its high initial cost may not be justified without clear evidence of worsening climate impacts.

Based on this analysis and stakeholder consensus, the city has decided to proceed with Option B, supported by an adaptive trigger plan that calls for upgrading to a smart hybrid system (Option C) if specific climate thresholds are exceeded in the future. This outcome strikes a strong balance between affordability, system flexibility, and long-term climate resilience.

Answers to the Key Questions

1. What climate-related uncertainties impact the drainage design decision?
   - Rainfall extremes: intensity, duration, and frequency
   - Urbanization pressures increasing impervious surface area
2. What future climate scenarios should be considered?
   - S1: Minor warming, light rainfall increase
   - S2: Moderate emissions + urban sprawl
   - S3: Severe climate with 150 mm rainfall and dense urbanization
3. How does each design option perform?

Table 5.100 provides a qualitative assessment of each stormwater design option across three key criteria: robustness to future uncertainty, cost efficiency, and flood risk mitigation. Option A offers short-term cost advantages but lacks resilience under future climate scenarios. Option B provides a balanced middle-ground performance. Option C, while costlier upfront, excels in both robustness and long-term flood protection, making it the most resilient and future-ready option.

Table 5.100 Overall performance summary of drainage design options

Option	Robustness	Cost efficiency	Flood risk mitigation
A	Low	High (short-term)	Poor (fails in S2/S3)
B	Medium	Moderate	Moderate to good
C	High	Costly upfront	Excellent

4. Which option is most robust to climate variability?

   Option C—robust even under worst-case (S3)
5. Which option minimizes regret?

   Option B—lowest maximum regret (20M)
6. Is it possible to adopt a phased, adaptive drainage strategy?

   Yes—Start with Option B, upgrade to C only if climate extremes emerge.

5.11.6.5 Example 5—Municipal Solid Waste Management Under Future Uncertainty

A regional municipality is tasked with selecting a long-term waste treatment strategy that addresses growing environmental, social, and regulatory pressures. The core objectives are to reduce dependence on landfilling, ensure compliance with environmental mandates, and advance broader sustainability goals such as resource recovery and emissions reduction.

This decision has become urgent due to several converging factors. First, the current landfill site is approaching its capacity limits, raising concerns about long-term waste disposal viability. Second, there is growing public opposition to continued landfill expansion, driven by concerns over environmental justice, odor, traffic, and land use. Lastly, the decision is complicated by key uncertainties, including future waste generation rates, which are influenced by population growth and consumption patterns; levels of participation in recycling programs, which fluctuate based on education, incentives, and convenience; and the enforcement and impact of future zero-waste policies, which remain difficult to predict but could drastically reshape waste streams.

Given these pressures and uncertainties, the municipality must identify a strategy that not only performs well under current conditions but also remains robust and adaptable over the coming decades.

Table 5.101 presents three municipal solid waste management options under consideration. Each alternative is assessed based on its initial investment, annual operation and maintenance (O&M) costs, potential for greenhouse gas emission reduction, and level of public acceptance. The comparison highlights trade-offs between economic feasibility, environmental performance, and societal support, aiding in the selection of a balanced and sustainable waste strategy.

Table 5.101 Evaluation of solid waste management alternatives based on cost, emissions, and social acceptance

Option	Description	Initial cost ($M)	O&M ($M)	Emission reduction	Public acceptance
A	Landfill expansion + basic recycling	30	4	Low	Low
B	Mechanical–biological treatment (MBT)	60	5.5	Medium	Moderate
C	Waste-to-energy (WTE, incineration)	90	6.5	High	Low–moderate

The Key questions that should be addressed are as follows:

1. What uncertainties affect waste management decisions?
2. What scenarios should be considered?
3. How do options perform across scenarios?
4. Which is most robust to environmental/regulatory risk?
5. Which minimizes regret if zero-waste policies are enforced?
6. Can the system be made flexible and adaptive?

The selection of an appropriate municipal solid waste treatment strategy is complicated by uncertainties in future waste generation, citizen participation in recycling programs, and the degree of enforcement of zero-waste policies. To conceptually represent the decision context, Fig. 5.92 presents a schematic of the strategic alternatives under consideration. This visual framework emphasizes the influence of systemic uncertainty on the relative performance and societal acceptability of landfill expansion, MBT, and WTE solutions.

Step-by-Step Solution Using SBDA

Step 1: Define the Decision Problem

The municipality faces a complex decision regarding the future of its regional waste management system. The primary objective is to select a strategy that minimizes environmental impact, manages long-term costs, and reduces the potential for social backlash, all while ensuring reliable waste treatment over a 25-year planning horizon. To achieve this, three alternative options are under consideration:

- Option A: Extend Landfill + Basic Recycling

  A continuation of the current model, involving modest landfill expansion and maintenance of existing recycling services. This is the most familiar and initially least expensive option, but it offers limited long-term resilience and is likely to face intense public opposition and increasing environmental scrutiny.
- Option B: Mechanical–Biological Treatment (MBT)

  Involves sorting mixed waste and processing organic fractions through biological stabilization (e.g., composting or anaerobic digestion). This option offers

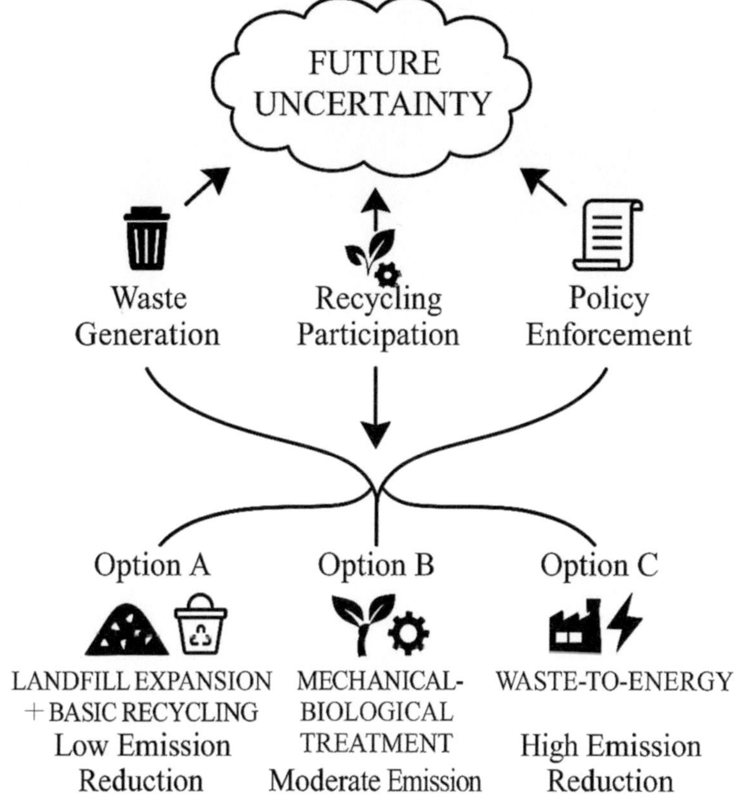

Fig. 5.92 Conceptual schematic of municipal solid waste management strategy selection under future uncertainty

moderate diversion rates, is more environmentally friendly, and may reduce the volume of residual waste sent to landfills.

- Option C: Waste-to-Energy (WTE) Incineration with Power Generation

  Converts non-recyclable waste into usable energy through combustion, reducing landfill volume significantly. This option offers the highest diversion potential but may provoke public resistance due to emissions concerns and typically involves higher capital costs.

The decision is constrained by two key factors: the existing landfill has only 10 years of remaining capacity, and there is a possibility that a 75% waste diversion mandate will be enforced in the near future. These constraints necessitate the selection of a system that can either substantially reduce landfill dependence or enable rapid adaptation to meet stringent diversion targets.

Over the 25-year timeframe, the chosen system must perform reliably despite uncertainties in waste generation, public behavior, and future policy enforcement—making adaptability and long-term sustainability critical to the decision-making process.

Step 2: Identify Key Uncertainties

To make a resilient and forward-looking decision, the municipality must consider critical uncertainties that could significantly affect the performance, feasibility, and public acceptance of each waste treatment option over the 25-year horizon.

1. Waste Generation Volume

One major uncertainty lies in the future volume of municipal solid waste generated. This is closely tied to population growth, urban density, and consumer behavior, including packaging trends, product lifespans, and material choices. If waste generation increases faster than projected, the system may become overwhelmed; conversely, successful waste reduction initiatives could reduce the throughput required. Planning for a system that performs under both scenarios is essential.

2. Regulatory Pressure

Another key uncertainty involves the evolution of environmental regulations. The region may soon face stricter mandates, such as landfill bans on organics or recyclables, or the enforcement of a 75% diversion target. There is also the potential for carbon pricing, extended producer responsibility (EPR) programs, or zero-waste legislation, all of which could redefine operational and financial requirements for waste systems. The inability to meet such requirements could lead to fines, retrofits, or the loss of public and regulatory trust.

These uncertainties highlight the need for a solution that is not only cost-effective and low-impact today, but also flexible enough to adapt to shifting waste volumes and tightening regulations in the decades ahead.

Step 3: Develop Plausible Scenarios

Table 5.102 outlines three potential future scenarios combining changes in waste generation, diversion regulations, and policy frameworks. The scenarios range from stable growth with voluntary measures to a transformative zero-waste strategy involving strict regulatory enforcement and landfill bans. These projections support long-term planning and stress-testing of waste management alternatives under evolving environmental and policy conditions.

Step 4: Engage Stakeholders

An inclusive stakeholder engagement process was conducted to ensure that the selection of a waste treatment strategy reflects both technical expertise and public priorities. Participants included waste engineers, who contributed operational and design perspectives; health and environmental regulators, who focused on compliance and long-term sustainability; local residents, who voiced concerns about quality of life

Table 5.102 Plausible future scenarios for solid waste management planning

Scenario	Description	Waste volume (t)	Diversion regulation	Policy type
S1	Stable Growth, Moderate Policy	50,000	50% voluntary	Incentives-based
S2	Increased Waste, Aggressive Policy	70,000	75% mandatory	Regulatory
S3	Urban Boom, Zero-Waste Transition	90,000	90% mandatory	Landfill Ban

and environmental justice; and recyclers, who emphasized material recovery and market viability.

Throughout the consultations, several core concerns emerged. Cost transparency was a consistent theme, with participants seeking clear, long-term financial implications of each option. Air quality concerns—especially related to emissions from Waste-to-Energy systems—were raised, particularly by residents and public health advocates. Legal compliance was flagged as a non-negotiable requirement, especially in light of potential diversion mandates and environmental enforcement. Finally, stakeholders stressed the importance of minimizing social backlash, noting that any solution perceived as unjust, polluting, or imposed without community input could face strong public resistance.

These discussions reinforced the need for a strategy that is technically sound, legally compliant, publicly acceptable, and financially clear, with room for future adaptation as regulatory and societal expectations evolve.

Step 5: Define Performance Metrics

To evaluate the three waste treatment options in a structured and transparent manner, a set of performance metrics has been established. These metrics reflect the priorities identified by stakeholders—cost-efficiency, regulatory compliance, and social acceptance—and are designed to support robust comparison under future uncertainties.

1. Lifecycle Cost

This metric captures the total financial burden over the system's 25-year lifespan, calculated as:

Lifecycle Cost = Initial Capital Investment + Operations & Maintenance (O&M) Costs + Potential Penalties (e.g., for non-compliance with diversion mandates or emissions limits).

This comprehensive cost view ensures that both short-term and long-term financial implications are considered in the decision.

2. Compliance Risk Score

This qualitative score reflects each option's ability to meet current and potential future regulations. The scale is defined as:

- 0 = Fully Compliant under all projected regulatory scenarios

- 1 = Partially Compliant, requiring moderate upgrades or adjustments
- 2 = Non-Compliant under likely future mandates, posing legal and operational risksThis metric emphasizes regulatory resilience and helps flag options that may require costly or disruptive retrofits.

3. Public Acceptance

This metric assesses the anticipated level of community support or resistance, based on stakeholder feedback and public sentiment trends. Rated qualitatively as:

- Low: High potential for public opposition or legal challenge
- Medium: Some concerns or controversy, but generally manageable
- High: Broad support or strong alignment with public values

This dimension is particularly important for identifying options likely to face social backlash, project delays, or reputational risks.

Together, these metrics provide a well-rounded framework for comparing the environmental, financial, regulatory, and social dimensions of each waste treatment strategy.

Step 6: Evaluate Options Under Each Scenario

Table 5.103 presents 25-year lifecycle cost estimates (in millions of dollars) for three solid waste management options, including initial capital expenditures, operation and maintenance (O&M) costs, and scenario-specific penalty costs associated with non-compliance to diversion or landfill regulations. Scenario S3 reflects the most stringent policy environment. Option C, while the most expensive upfront, incurs no penalties, offering full compliance across all scenarios.

Table 5.104 outlines the compliance risk associated with each waste management option under three future policy scenarios (S1–S3). Risk scores indicate the likelihood of failing to meet regulatory requirements, with higher values reflecting greater non-compliance risk. Option C maintains full compliance across all scenarios, demonstrating the highest regulatory resilience.

Step 7: Apply Robustness and Regret Criteria

A. Robustness (Minimax Cost Rule)

Table 5.105 presents the worst-case total lifecycle costs for each waste management alternative across all policy scenarios (S1–S3). Based on the minimax rule, Option

Table 5.103 Lifecycle cost estimates of waste management alternatives under future policy scenarios

Option	Initial	O&M (25 yrs)	Penalty			Total		
			S1	S2	S3	S1	S2	S3
A	30	100	0	20	50	130	150	180
B	60	137.5	0	0	15	197.5	197.5	212.5
C	90	162.5	0	0	0	252.5	252.5	252.5

Table 5.104 Compliance risk scores for waste management options across policy scenarios

Option	S1	S2	S3
A	0	1	2
B	0	0	1
C	0	0	0

A is the most robust in cost terms, offering the lowest maximum expenditure, though it carries higher compliance risks under stricter future regulations.

Table 5.106 shows the regret values for each waste management option under three policy scenarios (S1–S3), where regret represents the additional cost incurred relative to the lowest-cost option in each scenario. The maximum regret indicates the worst-case deviation from optimality. Option A has the lowest maximum regret—especially favorable if future regulations remain lenient—but may face compliance challenges under stricter scenarios.

Step 8: Design Adaptive Waste Management Pathway

To accommodate regulatory uncertainty and evolving waste generation trends, an adaptive two-phase waste management strategy has been proposed. This approach balances near-term feasibility with long-term flexibility, allowing the municipality to meet current needs while staying prepared for future mandates.

Phase 1 (Years 0–10): Initial Implementation

The pathway begins with Option B: Mechanical–Biological Treatment (MBT) at an estimated cost of $60 million. This system enables moderate diversion by mechanically sorting recyclables and biologically stabilizing organic waste. During this phase, the municipality will also expand recycling programs to increase diversion rates and reduce residual waste sent to landfill. MBT offers immediate environmental benefits and aligns with public preference for lower-impact solutions.

Table 5.105 Robustness evaluation of waste management options using the minimax cost criterion

Option	Worst-case cost
A	180
B	212.5
C	252.5

Table 5.106 Regret analysis of waste management options across policy scenarios

Option	S1	S2	S3	Max regret
A	0	0	30	30
B	67.5	0	15	67.5
C	122.5	55	0	122.5

Trigger for Phase 2

A shift to the next phase is initiated if either of the following conditions occur:

- A diversion target greater than 75% is formally mandated
- A landfill ban on organics or recyclables is implemented

These conditions reflect tightening regulatory pressure that MBT alone may not be able to meet.

Phase 2 (Years 10–20): Conditional Upgrade

If the trigger is activated, the strategy transitions to a hybrid system by retrofitting the MBT facility with a Waste-to-Energy (WTE) combustion line, including advanced emissions control technology. This retrofit, estimated at $40 million, significantly enhances diversion by converting non-recyclable waste into energy, while reducing landfill dependency and satisfying stricter environmental regulations.

This adaptive pathway allows the city to avoid overinvesting upfront, while ensuring it can scale up strategically if regulatory demands intensify. It also balances cost, compliance, and community acceptance, maintaining a flexible framework for long-term waste management success.

Step 9: Communicate Results and Refine

The final stage focused on presenting the analysis to stakeholders and decision-makers through clear, accessible materials that support informed evaluation and iterative refinement. The team delivered a performance matrix comparing the three options across key metrics: lifecycle cost, compliance risk, and public acceptance. This allowed for side-by-side evaluation of trade-offs under current and projected future conditions.

In addition, visual comparisons of compliance risks were provided, illustrating each option's vulnerability to regulatory shifts—particularly under stricter diversion mandates or landfill bans. To support flexible long-term planning, the team also shared an adaptive pathway diagram that outlined how the city could begin with Option B and transition to Option C if policy thresholds are crossed.

Following review and discussion, stakeholders reached consensus on proceeding with Option B (MBT) as the initial system, supported by a conditional upgrade plan to Option C (Waste-to-Energy) if future regulations or performance indicators warrant the shift. This approach was recognized for its balance of cost-efficiency, environmental responsibility, and adaptability to evolving policy and public expectations.

Answers to Key Questions

1. What uncertainties affect waste management decisions?

   The decision is shaped by two key uncertainties:

   - Waste volume growth, which depends on population, consumption trends, and behavioral changes in waste generation.

Table 5.107 Comparative summary of waste management options by flexibility, compliance risk, and cost

Option	Flexibility	Compliance risk	Cost
A	Low	High (fails in S2/S3)	Low
B	Medium	Marginal in S3	Medium
C	High	Fully compliant	High

- Regulatory tightening, including the potential introduction of stricter diversion mandates or landfill bans on recyclables and organics.

2. What scenarios should be considered?

   Three future scenarios were defined to assess system performance under varying conditions:

   - S1: Low growth with voluntary diversion programs (minimal policy pressure).
   - S2: Moderate growth paired with a 75% diversion mandate (likely near-future regulation).
   - S3: Urban boom accompanied by landfill bans and rapid regulatory tightening.

3. How do options perform across scenarios?
   A comparison of performance by flexibility, compliance risk, and cost is summarized in Table 5.107:
4. Which is most robust to environmental/regulatory risk?

   Option B is considered the most robust, offering a balanced combination of flexibility and cost-effectiveness while maintaining reasonable compliance potential across all scenarios.
5. Which minimizes regret if zero-waste policies are enforced?

   Option B again minimizes regret. It avoids non-compliance in Scenario 3 while preventing overspending in Scenario 1, where heavy infrastructure may not be necessary.
6. Can the system be made flexible and adaptive?

   Yes. The proposed strategy starts with Option B, integrates trend monitoring, and includes a conditional upgrade to Option C if triggers such as landfill bans or diversion mandates are enacted. This ensures both near-term feasibility and long-term adaptability under uncertain futures.

5.11.6.6 Example 6—Sustainable Material Selection for a Public Building Project

The municipal government's plan to construct a new multi-purpose public building, serving as a library, community center, and office space, presents an opportunity to align infrastructure development with long-term sustainability goals, specifically the target of carbon neutrality by 2050. The city must carefully select a material strategy that balances environmental impact, budgetary limits, and structural durability. The selected strategy should:

- Minimize embodied carbon and total lifecycle environmental impact,
- Remain within the allocated budget, ensuring cost-effectiveness for taxpayers,
- Maintain acceptable levels of durability, especially given the building's expected lifespan and functional demands.

The decision is influenced by several uncertainties:

- Future carbon pricing policies, which could significantly affect the cost and desirability of high-carbon materials.
- Technological advancements in low-carbon material production (e.g., carbon-sequestering concrete or more efficient steel recycling).
- Public pressure and policy incentives related to green building certification (e.g., LEED, Net-Zero).
- The long-term performance and weathering of newer materials like mass timber under the local climate.

The conceptual pathways explored in this project are illustrated in Fig. 5.93, highlighting material choices under uncertainty and their alignment with long-term sustainability goals.

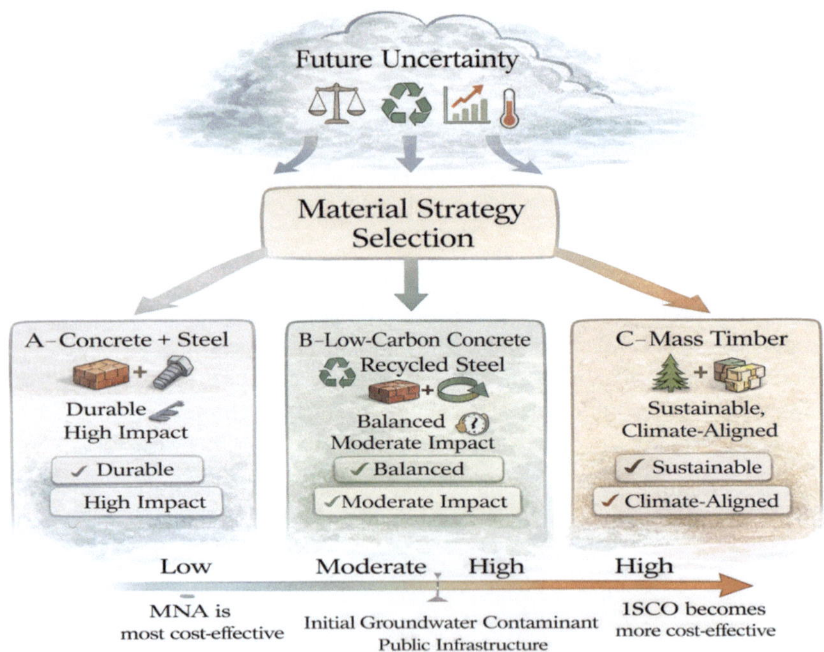

Fig. 5.93 Conceptual schematic of sustainable material strategy pathways for a public building

Table 5.108 compares three structural material strategies for a new construction project, assessing their initial costs, embodied carbon footprints, durability, and alignment with LEED and Net-Zero certification goals. While Option A is cost-effective and highly durable, it has the highest environmental impact. Option B offers a balanced compromise between emissions, cost, and certification potential. Option C, using mass timber, excels in sustainability performance and certification alignment, making it the most future-ready option despite higher initial cost and lower durability in certain conditions.

Key questions that should be addressed are as follows:

1. What sustainability-related uncertainties impact material selection?
2. What plausible future environmental and policy scenarios should be considered?
3. How do the material strategies perform across these scenarios?
4. Which option is most robust under climate policy risk?
5. Which option minimizes regret if carbon taxes or certifications become strict?
6. Can a phased or adaptive strategy be developed to allow flexibility in sustainable design decisions?

Step-by-Step Solution Using SBDA

Step 1: Define the Decision Problem

The municipal government must select a material strategy for constructing a new multi-purpose public building that will serve as a library, community center, and office space. The core objective is to identify a material approach that aligns with both the city's environmental commitments, notably carbon neutrality by 2050, and financial constraints, while remaining resilient to evolving policies and expectations over a 50-year design life. The alternatives under considerations are as:

- Option A: Conventional Concrete + Steel

Table 5.108 Comparative evaluation of structural material options based on sustainability, cost, and performance

Option	Material strategy	Initial cost ($M)	Embodied carbon ($kgCO_2/m^2$)	Durability	LEED/net-zero potential
A	Conventional concrete + steel	35	500	High	Low
B	Low-carbon concrete + recycled steel	40	320	Medium	Medium
C	Mass timber (CLT, Glulam)	45	150	Medium–low	High

The most traditional and affordable option, offering high structural durability but with the highest embodied carbon and limited alignment with sustainability targets.

- Option B: Low-Carbon Concrete + Recycled Steel

  A moderate step toward sustainability, this option reduces embodied carbon and improves LEED potential without major cost increases, while maintaining acceptable durability.

- Option C: Mass Timber (CLT, Glulam)

  The most environmentally progressive choice, with significantly reduced embodied carbon and high compatibility with LEED/Net-Zero certification, though at a higher upfront cost and moderate durability under local climate stressors.

The constraints re also as follows:

- The total capital budget must remain below $50 million, covering materials, labor, and supporting infrastructure.
- The selected material must be capable of supporting future compliance with LEED or Net-Zero certification standards, which may become mandatory or highly incentivized within the planning horizon.

The challenge lies in selecting a strategy that meets current project needs while remaining adaptable and justifiable under future uncertainties in policy, technology, and material performance.

Step 2: Identify Key Uncertainties

Selecting a long-term material strategy for a public building requires careful consideration of uncertainties that could significantly influence both cost and sustainability performance over the 50-year design life. Three key uncertainties have been identified:

1. Carbon Tax Policy Evolution

   The trajectory of carbon pricing is uncertain. If carbon taxes or emissions penalties increase over time, materials with high embodied carbon—like conventional concrete and steel—may become substantially more expensive. Conversely, materials with low-carbon profiles may gain cost advantages or receive incentives. The financial viability of each option will be affected by how aggressively these policies evolve.

2. Green Certification Requirements

   It is unclear whether LEED, Net-Zero, or similar green certification standards will remain voluntary or become mandatory, either through regulation or as a prerequisite for public funding or community approval. If certification becomes a requirement, options that align poorly with these standards (such as conventional material mixes) may require costly retrofits or jeopardize project approval.

3. Material Performance Uncertainty

   Emerging materials like mass timber offer strong sustainability advantages, but their long-term performance under local climatic conditions (e.g., humidity, freeze–thaw cycles, fire resistance) may still carry unknowns. There is also uncertainty about maintenance requirements, material availability, and public perception over time. These factors could affect lifecycle cost, structural integrity, and user confidence.

These uncertainties underscore the need for a strategy that is flexible, future-proof, and adaptable, balancing today's constraints with tomorrow's expectations.

Step 3: Develop Plausible Sustainability Scenarios

Table 5.109 outlines three future scenarios reflecting varying levels of policy ambition related to carbon pricing, green building certification requirements, and adoption of sustainable construction innovations. These scenarios range from a continuation of current policies to a transformative pathway aligned with 2050 net-zero goals, supporting resilient material and design decision-making under deep sustainability uncertainty.

Step 4: Engage Stakeholders

To ensure the material selection process reflects a balance of technical expertise, fiscal accountability, and environmental responsibility, a diverse group of stakeholders was engaged in the decision-making process. Participants included architects, who emphasized design feasibility and material integration; contractors, who provided insights on sourcing, constructability, and schedule implications; finance department representatives, who focused on capital planning and cost oversight; and environmental advocacy groups, who championed sustainability and policy alignment.

Through workshops and consultations, stakeholders identified a set of shared priorities that should guide material strategy decisions:

- Embodied Carbon: Reducing the building's upfront environmental impact was seen as essential for aligning with long-term climate goals and minimizing future carbon-related liabilities.

Table 5.109 Plausible future scenarios for sustainability-driven construction policy

Scenario	Description	Carbon tax ($/ton CO_2)	Certification	Innovation adoption
S1	Current policies hold	$50	LEED optional	Slow
S2	Moderate policy shift	$100	LEED required	Medium
S3	Aggressive climate policy (2050)	$200	Net-zero mandatory	Fast

- Transparency: Stakeholders called for clear, evidence-based comparisons of each material option's carbon footprint, cost, and long-term performance.
- Cost Responsibility: Ensuring the chosen material remains within the project's capital budget while accounting for long-term maintenance and potential regulatory costs was considered critical.
- Policy Risk Mitigation: With increasing uncertainty around carbon pricing and building performance standards, stakeholders supported a strategy that would minimize regulatory exposure and avoid costly retrofits in the future.

This input reinforced the importance of selecting a material strategy that balances environmental leadership with financial prudence, while remaining adaptable to evolving construction practices and policy landscapes.

Step 5: Define Performance Metrics

To evaluate the material strategy options in a structured and forward-looking manner, a set of performance metrics was defined. These metrics capture not only upfront costs and carbon impacts but also risks related to certification viability and material maturity, providing a balanced framework for comparison under future uncertainty.

1. Environmental Cost

This metric estimates the financial impact of embodied carbon, accounting for both direct material expenses and potential exposure to future carbon pricing. It is calculated as:

$$\begin{aligned}\text{Environmental Cost} &= (\text{Embodied Carbon} \times \text{Projected Carbon Tax Rate}) \\ &\quad + \text{Initial Material Cost}\end{aligned}$$

This approach reflects how choices made today could result in higher future liabilities if carbon taxes increase over the building's life.

2. Certification Score

This qualitative metric assesses the alignment of each material strategy with green building certification systems such as LEED or Net-Zero. Rated as:

- High: Strong alignment; likely to support certification with minimal additional investment
- Medium: Partial alignment; may require design enhancements or material substitutions
- Low: Poor alignment; significant changes needed to meet certification thresholds

This score helps evaluate each option's readiness for potential policy shifts or funding requirements tied to environmental certification.

1. Construction Risk

This metric reflects the uncertainty associated with construction quality, material performance, and contractor familiarity. Scored on a scale of:

- 0 = Proven: Well-established material with widespread construction experience and minimal risk
- 1 = Moderate: Some innovation or site-specific challenges expected
- 2 = Emerging: Limited precedent; higher risk of delays, learning curve, or technical setbacks

Together, these metrics allow the project team to assess trade-offs between cost, policy resilience, and construction reliability, and to identify a material strategy that is both sustainable and practically achievable.

Step 6: Evaluate Options Under Each Scenario

Table 5.110 presents the environmental lifecycle costs of three structural material options, accounting for initial costs and carbon-related expenses under varying carbon tax scenarios (S1–S3). Carbon costs are calculated based on embodied emissions ($kgCO_2/m^2$) and scenario-specific tax rates. Option A remains the least expensive under current and moderate policy environments, but under aggressive climate policy (S3), its cost increases significantly, making Option B more competitive and Option C the most aligned with long-term sustainability goals despite its higher upfront investment.

Table 5.111 assesses each material option's alignment with building certification requirements under three future policy scenarios (S1–S3), alongside a qualitative risk score indicating long-term regulatory and reputational exposure. Option A lacks compliance under stricter future policies, while Option B shows moderate adaptability. Option C consistently meets certification standards across all scenarios and carries the highest sustainability assurance, reflected in its top risk score.

Step 7: Apply Robustness and Regret Criteria

A. Robustness (Minimax Environmental Cost)

Table 5.112 evaluates the worst-case (maximum) environmental lifecycle cost for each structural material option across all carbon pricing scenarios (S1–S3). Based

Table 5.110 Environmental lifecycle cost evaluation under carbon pricing scenario

Option	Initial	Carbon	S1 ($50)	S2 ($100)	S3 ($200)
A	35	500	42.5	50	65
B	40	320	46	52	64
C	45	150	52.5	60	75

Table 5.111 Certification alignment and risk scores across sustainability scenarios

Option	S1 Cert	S2 Cert	S3 Cert	Risk score
A	Low	Fail	Fail	0
B	Medium	Pass	Partial	1
C	High	Pass	Pass	2

solely on cost, Option A is the most robust, offering the lowest maximum expenditure. However, this does not account for regulatory or certification compliance risks, which may offset its cost advantage under aggressive climate policies.

B. Regret Matrix (Cost-Based)

Table 5.113 presents the regret values for each option under three carbon pricing scenarios (S1–S3), where regret is defined as the additional cost relative to the lowest-cost alternative in each scenario. The maximum regret highlights the worst-case cost penalty. Option A shows zero regret across all scenarios, making it the most regret-averse option in purely cost-based terms—though it carries significant sustainability compliance risks under future policy shifts.

C. Sustainability-Weighted Decision

Table 5.114 presents adjusted total costs under Scenario S3, incorporating penalties for non-compliance with LEED or Net-Zero certification—assumed at $5M per project. Option A remains the lowest-cost option but fails to meet sustainability criteria. Option C achieves full compliance but at the highest cost. Option B strikes the best balance, achieving partial compliance with moderate penalties, making it the most cost-effective and sustainability-aligned compromise.

Step 8: Design Adaptive Strategy

To manage policy uncertainty and maintain alignment with long-term sustainability goals, an adaptive material and energy strategy has been developed. This phased approach enables the city to move forward confidently within budget, while preserving the flexibility to respond to regulatory or funding changes without compromising project goals.

Phase 1: Design & Procurement

The project begins with *Option B: Low-Carbon Concrete + Recycled Steel* as the baseline material strategy. This option strikes a balance between moderate embodied

Table 5.112 Robustness assessment using the minimax environmental lifecycle cost criterion

Option	Cost
A	65
B	64
C	75

Table 5.113 Regret analysis of structural material options based on environmental lifecycle costs

Option	S1	S2	S3	Max regret
A	0	0	0	0
B	3.5	2	1	3.5
C	10	10	10	10

Table 5.114 Sustainability-weighted lifecycle cost evaluation with certification penalties

Option	S3 Total + penalty	Result
A	$65 + 5 = 70$	✗
B	$64 + 2.5 = 66.5$	☑ Best compromise
C	$75 + 0 = 75$	☑ Compliant but costly

carbon, durability, and cost control. The structural design will be optimized to accommodate partial wood substitution, ensuring compatibility with future upgrades.

A trigger condition has been defined:

- If LEED or Net-Zero certification becomes mandatory before the construction phase begins, the project will upgrade to Option C (Mass Timber).
- This would involve an estimated additional cost of $5 million, but would enable the city to qualify for green building grants, expedited permitting, or other policy-linked incentives that can offset cost increases and accelerate delivery.

Phase 2: Post-Occupancy Enhancements

To further advance carbon neutrality targets beyond material selection, the strategy includes a planned post-occupancy phase that incorporates renewable energy systems, such as:

- Photovoltaic (PV) panels for onsite electricity generation
- Battery storage systems to manage energy demand and grid interaction.

These additions enhance the building's operational sustainability and help move the project toward Net-Zero energy performance, in alignment with the city's 2050 climate objectives.

This adaptive strategy allows the municipality to act decisively today while future-proofing the design against policy shifts and community expectations, ensuring both compliance and cost-effectiveness over the building's lifecycle.

Step 9: Communicate Results and Refine

In the final stage of the decision process, key findings were presented to stakeholders through a series of clear, comparative tools to support understanding, transparency, and informed feedback. The project team provided a suite of deliverables, including lifecycle carbon versus cost charts, which visualized the trade-offs between environmental impact and financial investment for each material option over the building's lifespan. These charts made it easier to contextualize future carbon cost exposure under varying policy scenarios.

In parallel, certification readiness heatmaps were shared to highlight how well each strategy aligned with evolving green building standards, such as LEED and Net-Zero. These visual tools identified Option C (mass timber) as the strongest performer in certification scenarios, with Option B offering moderate but improvable alignment. Finally, a phased decision tree was introduced to explain the adaptive pathway, showing how the project could transition from Option B to Option C if regulatory triggers—like a mandatory green certification—came into effect before construction.

As a result of this evidence-based communication, stakeholders approved Option B as the base strategy, with a phased upgrade to mass timber incorporated as a conditional pathway. This decision reflects a balanced response to environmental, financial, and policy considerations, and ensures the building remains flexible, cost-responsible, and aligned with long-term sustainability objectives.

Final Answers to the Key Questions

1. What are the most relevant uncertainties affecting flood protection decisions?

   - Carbon pricing, which will influence lifecycle cost of carbon-intensive materials.
   - Green certification regulations that may become mandatory.
   - Performance reliability of mass timber under climate and fire exposure.

2. What plausible future scenarios should be considered?

   - S1: Minimal change in carbon tax and voluntary certification.
   - S2: Mandatory LEED for public buildings and $100/ton CO_2.
   - S3: Net-zero building code and aggressive carbon pricing ($200/ton).

3. How does each flood protection option perform under each scenario?

   - Option A is cost-effective but fails under S2 and S3 (non-compliance).
   - Option B balances cost and readiness.
   - Option C performs best in environmental and certification metrics but has the highest cost and material risk.

4. Which option is most robust across all futures?

   Option B is most robust when considering both lifecycle cost and policy compliance under future scenarios.
5. Which option has the least regret if future conditions turn out to be worse than expected?

   Option B has moderate regret in cost but avoids penalties under new certification standards.
6. Can an adaptive pathway be developed to allow flexible investment over time?

   Yes. Begin with Option B, and add mass timber components and energy upgrades only if regulations change.

This adaptive pathway balances immediate feasibility with future sustainability compliance.

The decision-analysis framework presented in this chapter builds upon established theories of structured decision-making, utility assessment, and uncertainty quantification widely applied in engineering planning and risk evaluation (Amiri et al., 2024, 2025; BayesFusion, n.d.; Clemen & Reilly, 2013; Der Kiureghian & Ditlevsen, 2009; Ebtehaj & Bonakdari, 2022; Ebtehaj et al., 2024; FEMA, n.d.; FHWA, 2023; Howard, 1988; IEC, 2019; IPCC, 2022; ISO, 2017, 2018; JCGM, 2008; Keeney & Raiffa, 1976; NIST & SEMATECH, 2003; Pratt, 1964; Qureshi et al., 2025; Raiffa, 1968; Saltelli et al., 2008; Soltani et al., 2021; Soltani et al., 2024; USGS, 2018; von Winterfeldt & Edwards, 1986).

5.12 Summary

This chapter has demonstrated that decision trees constitute a rigorous yet intuitive decision-support framework for managing uncertainty in civil engineering projects across their entire lifecycle. It has emphasized that civil engineering decisions—ranging from design and construction to operation, maintenance, and end-of-life planning—are inherently uncertain due to environmental variability, data limitations, evolving regulations, and economic constraints. Within this context, decision trees have been shown to provide a structured and transparent approach for explicitly modeling uncertainty and evaluating competing alternatives using quantitative and probabilistic reasoning.

The chapter established the conceptual foundations of decision trees by describing their graphical structure and core components, including decision nodes, chance nodes, and outcome nodes. This structure enables engineers to decompose complex, multi-stage problems into manageable sequences of decisions and probabilistic events. By assigning probabilities to uncertain outcomes and utilities or costs to final consequences, decision trees were shown to transform subjective judgment into evidence-based analysis, thereby supporting rational, consistent, and defensible decision-making.

A central contribution of this chapter has been its systematic examination of decision-making under uncertainty across the various phases of civil engineering projects. During the design phase, decision trees were shown to support material selection, infrastructure sizing, and adaptability to future demands, particularly in the presence of climate change and urban growth uncertainty. In the construction phase, they were demonstrated to facilitate risk-informed responses to unforeseen site conditions, supply chain disruptions, and execution strategies. During operation and maintenance, decision trees were shown to guide performance monitoring, maintenance timing, emergency response planning, and adaptation to evolving usage patterns. The chapter further highlighted their relevance in end-of-life and adaptive reuse decisions, where long-term environmental, social, and economic consequences must be carefully balanced.

The chapter placed strong emphasis on expected utility theory as the analytical foundation of decision tree analysis. Utility was defined broadly to encompass both tangible factors—such as capital cost, maintenance expenditure, and service life—and intangible considerations, including safety, environmental impact, resilience, and social acceptance. By evaluating alternatives using expected utility as a probability-weighted measure of outcomes, the chapter demonstrated how engineers can compare competing options holistically, rather than relying on deterministic or worst-case assumptions. This approach was shown to be particularly valuable for high-stakes infrastructure decisions where long-term performance and risk tolerance are critical.

Through an extensive set of real-world civil engineering examples—including foundation design under soil uncertainty, flood protection system sizing, pavement material selection, stormwater basin design, seismic retrofitting, tunnel support systems, water and wastewater treatment technology selection, and energy-efficient building envelope design—the chapter demonstrated the practical application of decision trees across diverse engineering contexts. These examples illustrated how decision trees integrate technical analysis, probabilistic reasoning, and economic evaluation to support resilient, cost-effective, and defensible engineering solutions.

Finally, the chapter outlined a step-by-step methodology for implementing decision trees in engineering practice, encompassing problem formulation, alternative identification, uncertainty modeling, utility assessment, probability assignment, and comparison of expected utilities. By doing so, it equipped students and practitioners with a repeatable and transparent framework for navigating complex engineering decisions. Overall, this chapter has established decision trees as a foundational tool for modern civil engineering practice, enabling risk-aware, lifecycle-based, and sustainability-oriented decision-making under uncertainty.

References

Amiri, A., Soltani, K., Ebtehaj, I., & Bonakdari, H. (2024). A novel machine learning tool for current and future flood susceptibility mapping by integrating remote sensing and geographic information systems. *Journal of Hydrology, 632*, 130936. https://www.sciencedirect.com/science/article/pii/S0022169424003305

Amiri, A., Ebtehaj, I., Soltani, K., Gumiere, S. J., & Bonakdari, H. (2025). Future soil erosion trends in Canadian agricultural lands from runoff and sustainability impacts. *Scientific Reports, 15*(1), 23184. https://www.nature.com/articles/s41598-025-05947-5

BayesFusion. (n.d.). *Building an influence diagram with GeNIe* (tutorial). Retrieved February 18, 2026, from https://support.bayesfusion.com/docs/GeNIe/id_tutorial.html

Clemen, R. T., & Reilly, T. (2013). *Making hard decisions with decisiontools* (3rd ed.). Cengage Learning. https://www.cengage.ca/c/making-hard-decisions-with-decisiontools-3e-clemen-reilly/9780538797573/

Der Kiureghian, A., & Ditlevsen, O. (2009). Aleatory or epistemic? Does it matter? *Structural Safety, 31*(2), 105–112. https://doi.org/10.1016/j.strusafe.2008.06.020

Ebtehaj, I., & Bonakdari, H. (2022). A reliable hybrid outlier robust non-tuned rapid machine learning model for multi-step ahead flood forecasting in Quebec, Canada. *Journal of Hydrology, 614*, 128592. https://www.sciencedirect.com/science/article/pii/S0022169422011623

Ebtehaj, I., Fortin, J., Bonakdari, H., & Grégoire, G. (2024). How climate change will shape pesticide application in Quebec's golf courses: insights with deep learning based on assessing CMIP5 and CMIP6. *Applied Sciences, 14*(20), 9209. https://www.mdpi.com/2076-3417/14/20/9209

Federal Emergency Management Agency. (n.d.). *Benefit-cost analysis (BCA).* Retrieved February 18, 2026, from https://www.fema.gov/grants/tools/benefit-cost-analysis

Federal Highway Administration. (2023). *Lifecycle cost analysis realcost user manual* (FHWA-HRT-23-021). U.S. Department of Transportation. https://highways.dot.gov/sites/fhwa.dot.gov/files/FHWA-HRT-23-021.pdf

Howard, R. A. (1988). Decision analysis: Practice and promise. *Management Science, 34*(6), 679–695. https://doi.org/10.1287/mnsc.34.6.679

Intergovernmental Panel on Climate Change. (2022). Summary for policymakers. In H.-O. Pörtner, D. C. Roberts, E. S. Poloczanska, K. Mintenbeck, M. Tignor, A. Alegría, M. Craig, S. Langsdorf, S. Löschke, V. Möller, & A. Okem (Eds.), *Climate change 2022: Impacts, adaptation and vulnerability* (AR6 WGII). Cambridge University Press. https://doi.org/10.1017/9781009325844.001

International Electrotechnical Commission. (2019). *Risk management—Risk assessment techniques* (IEC Standard No. 31010:2019). https://www.iso.org/standard/72140.html

International Organization for Standardization. (2017). *Buildings and constructed assets—Service life planning—Part 5: Life-cycle costing* (ISO Standard No. 15686-5:2017). https://www.iso.org/standard/61148.html

International Organization for Standardization. (2018). *Risk management—Guidelines* (ISO Standard No. 31000:2018). https://www.iso.org/standard/65694.html

Joint Committee for Guides in Metrology (JCGM). (2008). *Evaluation of measurement data—Guide to the expression of uncertainty in measurement (JCGM 100:2008).* Bureau International des Poids et Mesures. https://www.bipm.org/documents/20126/2071204/JCGM_100_2008_E.pdf

Keeney, R. L., & Raiffa, H. (1976). *Decisions with multiple objectives: Preferences and value tradeoffs.* Cambridge University Press. https://www.cambridge.org/core/books/decisions-with-multiple-objectives/DEF338459C327778C3F8C4C4A682032F

National Institute of Standards and Technology (NIST), & SEMATECH. (2003). *NIST/SEMATECH e-handbook of statistical methods* (NIST Handbook 151; last updated 2022). https://doi.org/10.18434/M32189

Pratt, J. W. (1964). Risk aversion in the small and in the large. *Econometrica, 32*(1–2), 122–136. https://doi.org/10.2307/1913738

Qureshi, M. U. A., Amiri, A., Ebtehaj, I., Guimere, S. J., Cunderlik, J., & Bonakdari, H. (2025). Coupling HEC-RAS and AI for River morphodynamics assessment under changing flow regimes: Enhancing disaster preparedness for the Ottawa River. *Hydrology, 12*(2), 25. https://www.mdpi.com/2306-5338/12/2/25

Raiffa, H. (1968). *Decision analysis: Introductory lectures on choices under uncertainty.* Addison-Wesley. https://books.google.com/books?id=mxlPzVW9g-cC

Saltelli, A., Ratto, M., Andres, T., Campolongo, F., Cariboni, J., Gatelli, D., Saisana, M., & Tarantola, S. (2008). *Global sensitivity analysis: The primer.* Wiley. https://www.wiley.com/en-us/Global%2BSensitivity%2BAnalysis%3A%2BThe%2BPrimer-p-9780470059975

Soltani, K., Ebtehaj, I., Amiri, A., Azari, A., Gharabaghi, B., & Bonakdari, H. (2021). Mapping the spatial and temporal variability of flood susceptibility using remotely sensed normalized difference vegetation index and the forecasted changes in the future. *Science of The Total Environment, 770*, 145288. https://www.sciencedirect.com/science/article/pii/S0048969721003545

Soltani, K., Amiri, A., Ebtehaj, I., Cheshmehghasabani, H., Fazeli, S., Gumiere, S. J., & Bonakdari, H. (2024). Advanced forecasting of drought zones in Canada using deep learning and CMIP6 projections. *Climate, 12*(8), 119. https://www.mdpi.com/2225-1154/12/8/119

U.S. Geological Survey. (2018). *Guidelines for determining flood flow frequency—Bulletin 17C.* https://www.usgs.gov/publications/guidelines-determining-flood-flow-frequency-bulletin-17c

von Winterfeldt, D., & Edwards, W. (1986). *Decision analysis and behavioral research.* Cambridge University Press. https://books.google.com/books?id=eUo7AAAAIAAJ

Chapter 6
Sensitivity and Scenario-Based Decision Analysis in Civil Engineering

6.1 Introduction

Civil engineering decisions are rarely made under conditions of certainty. Whether selecting a foundation system, sizing a drainage network, or planning long-term flood protection, engineers must commit to designs while key inputs remain imperfectly known. Material costs fluctuate, regulatory frameworks evolve, hydrologic regimes shift, and societal expectations change. Traditional deterministic design, where single "best estimate" values are inserted into equations, often conceals the fragility of decisions. A solution that appears optimal under assumed parameters may perform poorly when those assumptions deviate from reality. Consequently, modern engineering practice increasingly emphasizes decision robustness rather than nominal optimality. This chapter introduces two complementary approaches for managing uncertainty in engineering decision-making: sensitivity analysis and scenario-based decision analysis (SBDA). Together, they provide a structured framework for understanding how uncertain parameters influence outcomes and for selecting alternatives that remain acceptable across a wide range of plausible futures.

Sensitivity analysis focuses on *parameter influence*. Instead of treating input variables as fixed, the engineer systematically perturbs probabilities, costs, discount rates, environmental conditions, and time horizons to observe how outcomes change. Within decision-tree formulations, this process reveals which assumptions truly control the decision and which are comparatively unimportant. One-way and two-way analyses quantify the stability of a preferred alternative, while threshold analysis identifies tipping points, the exact conditions under which a different decision becomes preferable. These techniques convert a static decision tree into a diagnostic tool that exposes vulnerability and guides data-collection priorities. However, sensitivity analysis alone cannot capture the complexity of real infrastructure planning. Many uncertainties are not independent parameters but coherent future states of the world. Climate change, population growth, regulatory tightening, and technological adoption unfold as interacting trajectories rather than isolated variables. For

H. Bonakdari et al., *Hazard and Risk Assessment in Civil Engineering Design*,
https://doi.org/10.1007/978-3-032-23600-5_6

such conditions, scenario-based decision analysis becomes necessary. SBDA evaluates engineering alternatives across internally consistent future scenarios instead of relying on single probabilistic forecasts. The objective shifts from maximizing expected value to ensuring acceptable performance under diverse plausible futures. Within this framework, engineers evaluate solutions using robustness and regret metrics. The robustness (minimax) criterion identifies alternatives whose worst-case performance is least harmful, suitable for high-consequence infrastructure. The minimax regret criterion seeks options that minimize the penalty of choosing incorrectly when future conditions differ from expectations. These perspectives better align with public infrastructure responsibilities, where failure costs may be societal rather than purely financial.

The chapter also emphasizes integration with adaptive planning. Modern infrastructure cannot always be fully optimized at the outset; instead, it should be designed to evolve. By linking SBDA with staged investments, monitoring triggers, and stakeholder engagement, engineers can implement flexible strategies that adjust as uncertainties resolve over time. This transforms decision-making from a one-time prediction problem into a managed learning process. To ground these concepts in practice, the chapter presents multiple civil engineering applications, including flood protection, water treatment expansion, pavement design, seismic foundation selection, urban drainage upgrades, and sustainable material choice. These examples demonstrate how analytical frameworks translate into defensible engineering decisions when facing environmental variability, regulatory uncertainty, and long service lives. Ultimately, this chapter reframes engineering judgment: the goal is not to predict the future perfectly, but to choose solutions that perform satisfactorily across many credible futures. Sensitivity analysis diagnoses uncertainty; scenario-based decision analysis manages it; adaptive planning operationalizes it. Together they form a decision methodology suited to contemporary infrastructure challenges, where resilience and accountability are as critical as efficiency.

6.1.1 Sensitivity Analysis in Decision Trees: Definition and Purpose

Sensitivity analysis in decision trees is a systematic method to assess how changes in input parameters such as probabilities of events or the utilities (i.e., costs, benefits, or other payoffs) associated with outcomes affect the optimal decision. By varying these inputs within plausible ranges, sensitivity analysis reveals the robustness of a decision and identifies critical variables that significantly influence the outcome. This process is crucial in decision-making under uncertainty, as it highlights whether small changes in assumptions could lead to alteration of the preferred solution.

In civil engineering, sensitivity analysis is particularly valuable due to the inherent uncertainties in projects, including (Fig. 6.1):

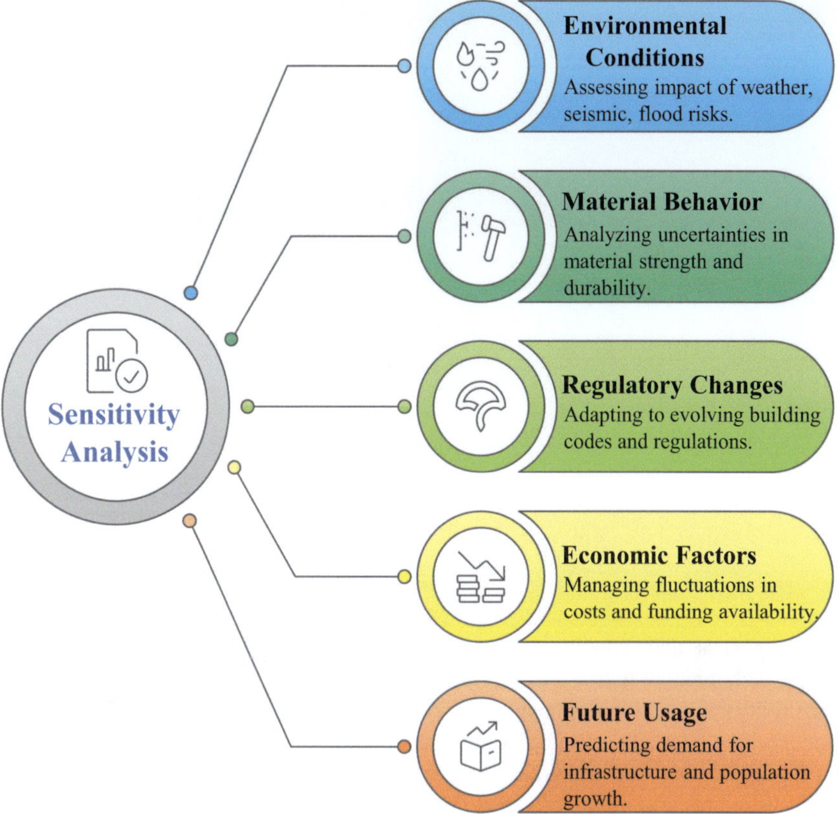

Fig. 6.1 Common uncertainties involved in civil engineering projects

- Environmental conditions: Variable weather patterns, seismic risks, or flood probabilities.
- Material behavior: Uncertainty in material strength, durability, or fatigue.
- Regulatory changes: Evolving building codes, carbon pricing, or environmental regulations.
- Economic factors: Fluctuations in construction costs, maintenance expenses, or funding availability.
- Future usage: Uncertain demand for infrastructure, such as traffic loads or population growth.

The purpose of sensitivity analysis is to assess the robustness of decisions under uncertainty by identifying the input variables that most strongly influence the outcome. This information guides efforts to prioritize data collection, model refinement, or uncertainty reduction for the most influential variables. Additionally,

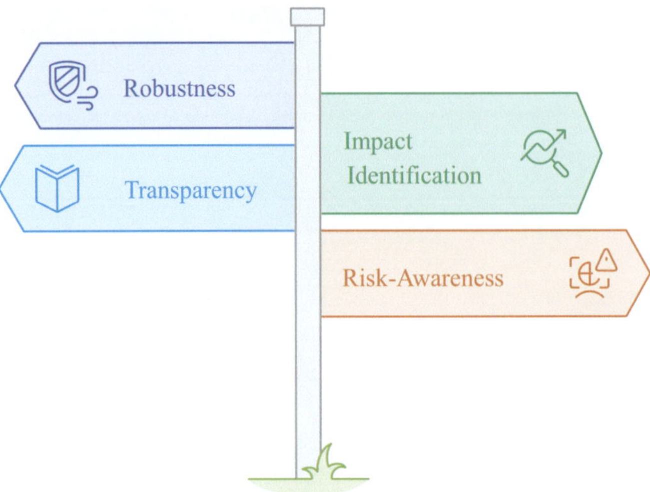

Fig. 6.2 Advantages of sensitivity analysis in civil engineering projects

sensitivity analysis enhances transparency in project decision making by quantifying the risks and sensitivities associated with various recommendations, permitting stakeholders to gain a clearer understanding of potential outcomes. Ultimately, sensitivity analysis supports risk-aware decision-making by revealing the conditions under which alternative strategies may become more favorable (Fig. 6.2).

By conducting sensitivity analysis, engineers can avoid over-reliance on single-point estimates and make informed decisions that account for variability and risk. Sensitivity analysis in decision trees serves several critical objectives (Fig. 6.3):

- **Testing the Robustness of the Preferred Decision**: Determines whether the optimal decision remains the best choice if input parameters (e.g., probabilities or costs) vary within a reasonable range. For example, if a bridge design is chosen based on a 1% annual flood probability, will it remain optimal if the probability increases to 2%?
- **Identify Threshold Values**: Finds specific values of a parameter (e.g., a probability or cost) at which the preferred decision changes to another alternative. This helps define the boundaries of a decision's validity.
- **Support Stakeholder Communication**: Provides clear, data-driven insights to explain how sensitive a recommendation is to underlying assumptions, fostering trust and facilitating communication with key stakeholder groups including clients, regulators, or funding bodies.
- **Prioritize Data Collection**: Identifies which uncertain parameters have the most significant impact on the decision, guiding engineers to focus resources on collecting more accurate data for those variables (e.g., conducting additional soil tests if soil stability is a critical factor).

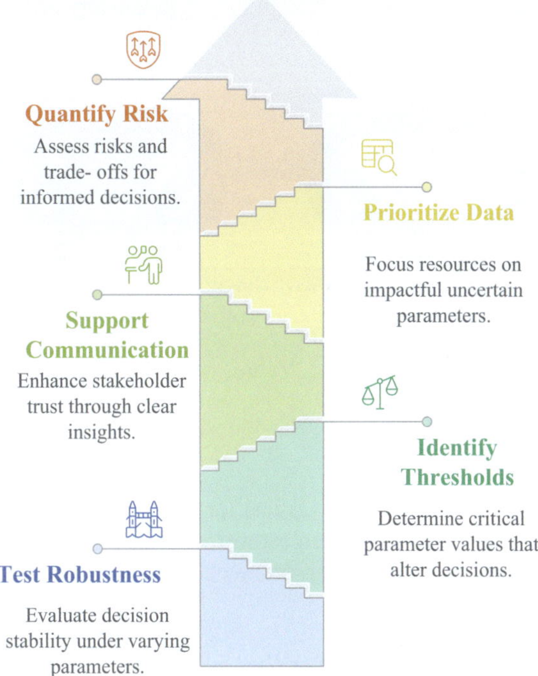

Fig. 6.3 Critical objectives of the sensitivity analysis in decision trees

- **Quantify Risk and Uncertainty**: Enables engineers to quantify the risks associated with uncertain inputs and evaluate trade-offs between cost, safety, and performance.

6.1.2 Key Parameters for Sensitivity Analysis

Sensitivity analysis in decision trees is a vital technique for evaluating the robustness of decisions under uncertainty, particularly in civil engineering, where projects face complex risks from environmental, economic, and regulatory factors. By systematically varying key input parameters, sensitivity analysis identifies which uncertainties exert the greatest influence on the optimal decision, highlights conditions under which preferences may shift, and supports informed, risk-aware choices. In decision trees, this process involves testing parameters such as probabilities, costs, and external scenarios to ensure decisions remain viable across plausible ranges of uncertainty. The following sections detail five key parameters for sensitivity analysis (Fig. 6.4), namely, the probability of chance events, utility or cost values, regulatory or environmental scenarios, time horizons, and discount rates. Each of these parameters are explored in detail in terms of their definition, sources of uncertainty, civil engineering examples, implications for decision trees, and practical considerations.

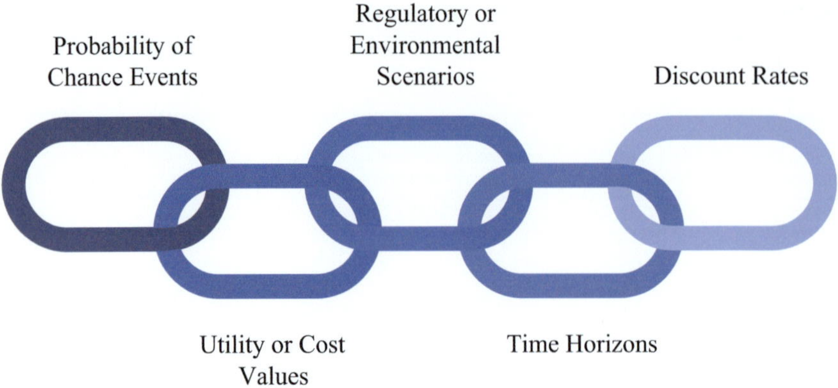

Fig. 6.4 Key parameters for sensitivity analysis

These parameters are critical for designing resilient infrastructure, prioritizing data collection, and communicating trade-offs to stakeholders, ensuring decisions are both robust and adaptable to future changes.

6.1.2.1 Probability of Chance Events

The probability of chance events in decision trees refers to the likelihood (expressed as a value between 0 and 1) of uncertain outcomes occurring at chance nodes, such as the occurrence of natural disasters, material failures, or regulatory approvals, which directly influence the expected utility or cost of each decision path. Uncertainty arises from multiple sources: limited historical data (e.g., incomplete flood records), future variability due to climate change or urbanization (e.g., increasing storm frequency), subjective expert judgment when data is scarce, and assumptions in statistical models (e.g., assuming a Poisson distribution for earthquakes).

Examples in civil engineering include the probability of a 1% annual chance flood impacting a levee, a 0.2% likelihood of a seismic event damaging a bridge, the chance of concrete fatigue failure under heavy traffic loads, or the probability of securing environmental permits for a new highway.

The implications for decision trees are significant, as small changes in high-impact probabilities can shift the optimal decision; for instance, a slight increase in flood probability might justify a more expensive floodwall over natural drainage improvements. Sensitivity analysis, such as one-way analysis (varying flood probability from 0.01 to 0.1) or threshold analysis (finding the probability where two options are equal), reveals which probabilities are critical, guiding risk mitigation or data refinement.

Practical considerations include prioritizing data collection for sensitive probabilities (e.g., conducting hydrological studies to refine flood risk estimates), adopting conservative estimates for safety-critical infrastructure (e.g., designing for worst-case seismic scenarios), and using sensitivity results to communicate risks to stakeholders, such as explaining why a low-probability, high-impact event justifies costly resilience measures. For example, in a decision tree for a coastal protection system, sensitivity analysis might show that the decision to build a seawall relies on the storm probability, prompting investment in advanced climate modeling to reduce uncertainty.

6.1.2.2 Utility or Cost Values

The definition of utility or cost values in decision trees encompasses the numerical metrics assigned to outcomes, representing monetary costs (e.g., construction expenses), benefits (e.g., economic gains), or non-monetary factors (e.g., environmental impacts) quantified through methods like multi-criteria decision analysis, which are used to compute the expected utility of each decision option. Uncertainty stems from market fluctuations (e.g., volatile material prices due to supply chain issues), estimation errors in early project phases (e.g., unforeseen site conditions), economic factors like inflation or funding constraints, and subjective quantification of non-monetary utilities (e.g., assigning a dollar value to reduced travel time).

Examples in civil engineering include construction costs for a new bridge ($50M), repair costs after a flood ($10M), annual maintenance expenses for a highway ($2M/year), societal benefits like reduced congestion from a transit system ($5M/year in economic value), and environmental costs such as carbon emissions penalties or habitat restoration ($1M).

The implications for decision trees are critical, as utility values drive trade-offs between options; high sensitivity to construction costs might favor a cheaper design, while sensitivity to repair costs could justify a resilient one. Sensitivity analysis techniques, such as tornado diagrams (ranking cost impacts) or two-way sensitivity (exploring construction vs. repair costs), identify which values most influence the decision, ensuring robustness or highlighting risks.

Practical considerations involve improving cost estimation through detailed quantity surveys or risk-based modeling, exploring value engineering to reduce sensitive costs (e.g., using alternative materials), and testing different weightings for non-monetary utilities (e.g., prioritizing safety over cost) to align with stakeholder priorities. For instance, in a decision tree for a dam, sensitivity analysis might reveal that high repair costs after failure make a robust design preferable, prompting engineers to negotiate bulk material discounts or incorporate sustainable materials to balance costs and environmental impacts.

6.1.2.3 Regulatory or Environmental Scenarios

The definition of regulatory or environmental scenarios in decision trees refers to external conditions—such as government policies, building codes, or environmental changes—that affect the probabilities, costs, or structure of the decision tree, introducing uncertainty in long-term infrastructure planning. Uncertainty arises from evolving policies (e.g., unpredictable changes in environmental regulations), climate change impacts (e.g., uncertain sea level rise projections), economic incentives like subsidies or carbon taxes, and technological advancements that alter compliance requirements (e.g., new energy-efficient materials).

Examples in civil engineering include stricter seismic building codes increasing construction costs, carbon pricing policies adding $0.5M to a concrete-heavy project, hydrological shifts due to climate change requiring larger flood defenses, and environmental regulations restricting construction in wetlands, necessitating costly mitigation measures.

The implications for decision trees are profound, as scenarios can reshape the tree by adding new branches (e.g., compliance failure) or altering costs and probabilities; sensitivity to a carbon tax might shift preference to a low-carbon design. Scenario analysis (comparing "high-regulation" vs. "low-regulation" futures) or one-way sensitivity analysis (varying compliance costs), ensures decisions are robust across plausible futures and identifies the points where different alternatives become optimal.

Practical considerations include designing adaptable infrastructure (e.g., modular floodwalls to accommodate future regulations), engaging stakeholders to anticipate policy shifts (e.g., consulting regulators), and using sensitivity results to justify proactive measures, such as exceeding current codes to future-proof a building. For example, in a decision tree for a skyscraper, sensitivity analysis might show that a new energy code significantly increases costs, prompting engineers to adopt prefabricated components that meet both current and anticipated standards, ensuring long-term viability.

6.1.2.4 Time Horizons

The definition of time horizons in decision trees refers to the duration over which outcomes are evaluated, such as the lifespan of a structure, the planning period for a project, or the time until a major event (e.g., a 100 year flood), which affects the discounting of costs and benefits in present-value calculations. Uncertainty arises from variable project lifespans (e.g., a bridge's durability depends on maintenance), statistical uncertainty in event timing (e.g., when a rare flood occurs), obsolescence due to technological or regulatory changes, and differing economic horizons based on funding constraints.

Examples in civil engineering include a 50 year lifespan for a highway versus a 30 year design, planning for a 100 year flood vs. a 500 year event, lifecycle analysis of a building's energy performance over 20 versus 40 years, and designing coastal defenses for sea level rise over 50 versus 100 years.

The implications for decision trees are significant, as time horizons influence the weight of future costs and benefits; a longer horizon might justify a durable design due to reduced maintenance, while a shorter one favors lower upfront costs. Sensitivity analysis, such as varying the lifespan from 40 to 60 years or comparing short-term vs. long-term scenarios, reveals whether a decision is robust or overly dependent on optimistic horizon assumptions.

Practical considerations include using lifecycle costing to account for lifespan uncertainty, designing for rarer events (e.g., 1,000 year floods) if sensitivity is high, and aligning horizons with stakeholder goals (e.g., budget limits or sustainability targets). For instance, in a decision tree for a highway, sensitivity analysis might show that a 20 year horizon favors concrete pavement material, but a 10 year horizon supports cheaper alternatives such as conventional asphalt pavement design.

6.1.2.5 Discount Rates

The definition of discount rates in decision trees is the rate (e.g., 3%) used to convert future costs and benefits into their present value, reflecting the time value of money, opportunity costs, or societal preferences for immediate vs. future outcomes, critical for long-term infrastructure decisions. Uncertainty stems from economic variability (e.g., fluctuating interest rates or inflation), policy-driven rates mandated by agencies (e.g., U.S. OMB's 3–7% range), and subjective risk premiums for high-risk projects.

Examples in civil engineering include discounting the maintenance costs of a dam over 50 years, evaluating the present value of flood defenses preventing damages in 2075, comparing upfront costs of energy-efficient building features to future savings, and discounting revenue from a toll road in a public-private partnership.

The implications for decision trees are substantial, as discount rates exponentially affect long-term outcomes; a small increase from 2 to 5% can drastically reduce the present value of future benefits, altering the optimal decision. Sensitivity analysis, such as varying the rate from 2 to 5% or testing policy-mandated rates, identifies whether a decision is robust or sensitive to rate assumptions, highlighting trade-offs between upfront costs and long-term gains.

Practical considerations include testing a range of rates to account for economic uncertainty, aligning with agency guidelines for funding approval, and using low rates for sustainability-focused projects to prioritize future benefits, such as climate-resilient infrastructure. For example, in a decision tree for a solar farm, sensitivity analysis might show that a 2% rate justifies high initial costs due to long-term energy savings, but a 5% rate favors a cheaper fossil fuel option, prompting engineers to advocate for green financing with lower rates to support sustainability goals.

6.1.3 Techniques in Sensitivity Analysis

Sensitivity analysis in decision trees is a critical methodology for evaluating the robustness and reliability of decisions under uncertainty, particularly in civil engineering, where projects face multifaceted risks from environmental variability, economic fluctuations, regulatory changes, and long-term performance uncertainties. By systematically altering input parameters—such as probabilities of events, costs, benefits, or external scenarios—sensitivity analysis reveals how sensitive a decision is to these changes, identifying which variables have the most significant impact on the outcome and under what conditions the preferred decision might shift. This process is essential for ensuring that infrastructure decisions, such as designing bridges, flood defenses, or transportation networks, are not only optimal under baseline assumptions but also resilient to plausible variations in uncertain inputs. The techniques used in sensitivity analysis provide structured approaches to quantify uncertainty, prioritize data collection, and communicate risks to stakeholders, enabling engineers to make informed, risk-aware choices that balance cost, safety, and performance.

For instance, consider a decision tree evaluating whether to build a floodwall (Option A) or rely on natural drainage improvements (Option B) to protect a coastal city. The decision hinges on uncertain parameters like the probability of a flood (e.g., 2% annually), the cost of flood damage (e.g., $10M), and future regulatory requirements (e.g., stricter environmental standards). Sensitivity analysis explores how changes in these parameters affect the expected cost of each option. If increasing the flood probability to 5% shifts the preference to the floodwall, this indicates high sensitivity to that parameter, prompting engineers to refine flood risk estimates or design a more robust solution. Similarly, imagine a decision tree for choosing between a concrete bridge (cheaper but less durable) and a steel bridge (costlier but more resilient). Sensitivity analysis might test the impact of varying construction costs or seismic event probabilities, revealing whether the concrete bridge remains optimal if costs rise by 20% or if seismic risks are underestimated. These conceptual examples underscore the role of sensitivity analysis in uncovering critical uncertainties, ensuring decisions are robust, and guiding resource allocation toward reducing the most impactful unknowns.

The techniques of sensitivity analysis—ranging from simple one-way variations to complex probabilistic simulations—offer diverse tools to explore these uncertainties. Each technique serves a specific purpose, from isolating the effect of a single variable to capturing interactions among multiple parameters or assessing decisions under discrete future scenarios. In civil engineering, these techniques are particularly valuable for addressing the long-term implications of infrastructure projects, where uncertainties like climate change, material performance, or regulatory evolution can significantly alter outcomes. By applying these methods, engineers can quantify the range of possible outcomes, identify thresholds where decisions change, and provide stakeholders with clear, data-driven insights into the risks and trade-offs involved.

The following sections will detail these techniques, exploring their definitions, applications, advantages, limitations, and practical considerations, with a focus on their relevance to civil engineering decision-making.

6.1.3.1 One-Way Sensitivity Analysis

One-way sensitivity analysis is a foundational tool in decision analysis that systematically varies a single uncertain parameter—such as probability, cost, benefit, discount rate, or time horizon—while holding all other parameters constant. For each variation, the expected value of a decision outcome (e.g., expected utility, expected net benefit, or expected cost) is recalculated to assess how sensitive the optimal choice is to that particular parameter. This allows decision-makers to isolate and understand the influence of individual factors on the final decision and assess the robustness of the preferred alternative under uncertainty. In civil engineering, this method is widely used in project planning, risk mitigation, and design optimization where parameters are uncertain or based on estimates. For example, uncertainties in hydrological parameters, seismic risks, material prices, and construction timelines often necessitate one-way sensitivity analysis to inform resilient and cost-effective decisions. One-way sensitivity analysis is most appropriate in the early stages of decision-making, especially when a particular parameter is suspected to significantly influence the outcome. In real-world decision-making, precise estimates for probabilities, costs, or benefits are often difficult to obtain due to uncertainty in future events, expert judgment variability, or limited data. One-way sensitivity analysis allows decision-makers to (Fig. 6.5):

- Evaluate robustness of a decision to uncertainty in a key parameter.
- Identify threshold values—critical points where the preferred option may switch.
- Understand model behavior under varying conditions.

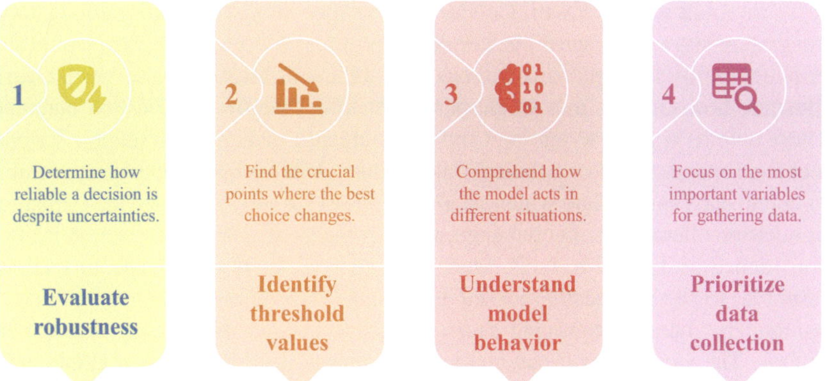

Fig. 6.5 Benefits of a one-way sensitivity analysis

- Prioritize data collection by focusing on the identified influential variables and their ranges of interest.

The one-way sensitivity analysis procedure involves:

Step 1-Construct the Decision Tree

- Develop a decision tree that outlines the available decision alternatives, associated costs or benefits, and uncertain outcomes represented by probabilities.

Step 2-Select the Parameter for Analysis

- Choose one uncertain input variable (e.g., probability of failure, cost of a repair, expected demand) for sensitivity testing.

Step 3-Define a Plausible Range

- Identify a reasonable range of values for the chosen parameter based on expert knowledge, historical data, or modeling scenarios.

Step 4-Recalculate Expected Values

- For each value within the specified range, recalculate the **expected value** (cost or utility) for each decision alternative.

Step 5-Compare and Interpret Results

- Plot or tabulate the results to observe how the decision outcome changes with the parameter. Determine if and where the preferred decision shifts (Fig. 6.6).

Advantages and Disadvantages of One-Way Sensitivity Analysis

One-way sensitivity analysis offers several practical advantages that make it a valuable tool for engineers, planners, and decision-makers. Its simplicity is a key strength—it can be implemented with minimal computational effort using basic software tools such as Excel, GeNIe, or TreeAge. This accessibility allows rapid integration into early-stage project evaluations or feasibility studies. The method also provides clarity through visualizations such as line or tornado plots, which clearly illustrate how changes in a single parameter affect the overall decision outcome. These outputs are easy to interpret and communicate, making them suitable for both technical experts and non-technical stakeholders. Another major benefit is the ability to obtain early insight into which parameters most strongly influence the evaluation of decision alternatives. In civil engineering projects, where data collection is often constrained by limited budgets, time, and site accessibility, this capability allows resources to be strategically focused on reducing uncertainty in the most influential parameters, thereby improving decision quality without unnecessary data expenditures. The technique's broad applicability makes it useful across multiple areas of civil engineering, from evaluating different structural designs and materials to assessing environmental impacts or managing construction budgets. Furthermore, it

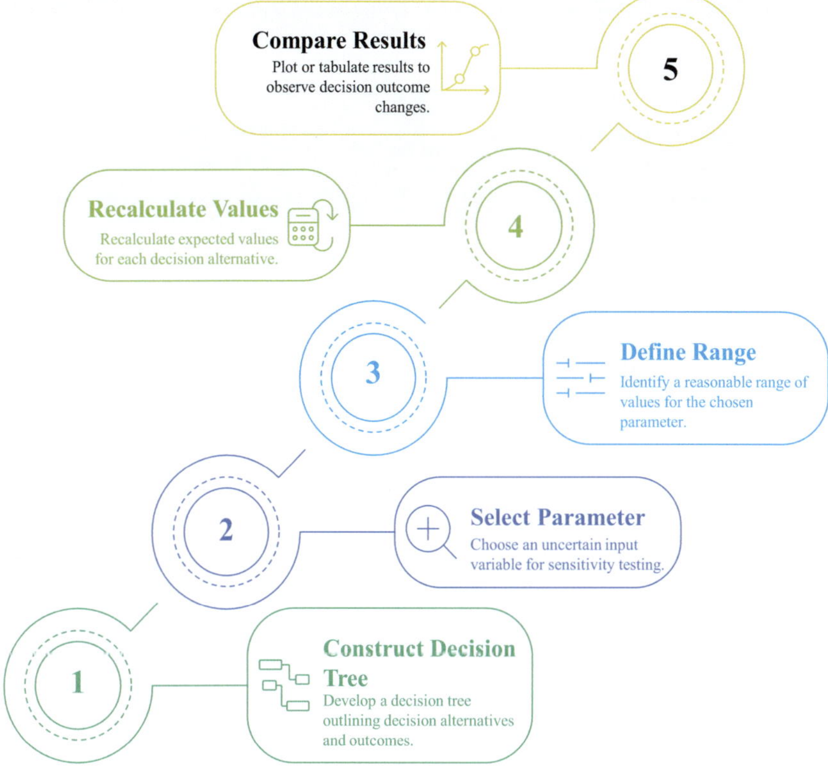

Fig. 6.6 The procedure for one-way sensitivity analysis

supports effective communication with policymakers, funding bodies, and the public by translating technical uncertainties into digestible 'what-if' narratives that inform planning and investment decisions.

Despite its utility, one-way sensitivity analysis has notable limitations that must be considered. Critically, this approach only examines one variable at a time and therefore ignores potential interactions between parameters—interactions that are common in real-world civil engineering problems, such as the joint effect of fluctuating material prices and project timelines. This can lead to oversimplified conclusions, particularly in complex systems where outcomes depend on threshold effects or nonlinear relationships (e.g., dam failure beyond a critical flood level). Additionally, the analysis is sensitive to range selection bias; results may vary significantly depending on the chosen parameter bounds, which are often based on expert judgment or limited data. A comprehensive understanding typically requires multiple analyses, one for each key parameter, which can be time-consuming and may still miss synergistic effects. Moreover, in cases where input variables are correlated—such as labor and equipment costs or rainfall and soil saturation—evaluating them independently may lead to misleading interpretations. Therefore, while one-way sensitivity analysis

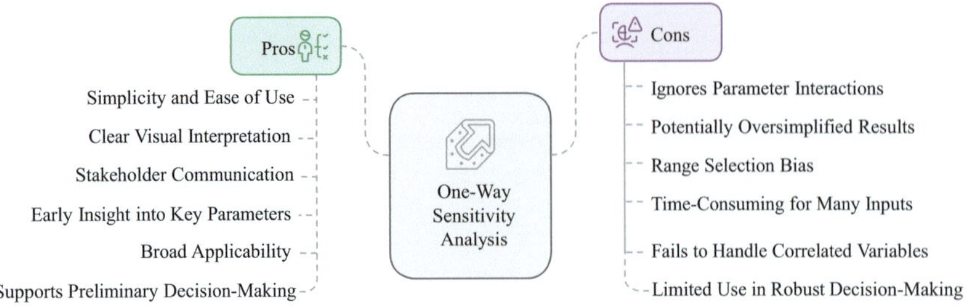

Fig. 6.7 Advantages and disadvantages of one-wave sensitivity analysis

is an excellent starting point, it should ideally be complemented by more advanced techniques (e.g., multi-way sensitivity analysis, Monte Carlo simulation, or Bayesian networks) for robust decision support in civil engineering. Fig. 6.7 shows the main advantages and disadvantages of a one-way sensitivity analysis.

Conceptual Examples One-Way Sensitivity Analysis in Civil Engineering

In civil engineering practice, one-way sensitivity analysis is particularly valuable for evaluating design alternatives under uncertain conditions. For example, infrastructure projects often involve decisions influenced by environmental hazards, fluctuating costs, and regulatory thresholds. By isolating the impact of a single uncertain variable—such as the probability of a flood, earthquake, or material price escalation—engineers can better understand the risk profile of each option and identify the tipping points at which one design becomes more favorable than another. This approach enhances transparency in decision-making and supports more resilient infrastructure planning by revealing which variables warrant further study or monitoring. Whether applied to transportation infrastructure, water management systems, or construction materials, one-way sensitivity analysis serves as a straightforward but powerful tool for prioritizing design choices and guiding investment decisions under uncertainty. Let's now explore a detailed example.

One-Way Sensitivity Analysis of Flood Protection Strategies

A coastal city is deciding whether to invest in a reinforced seawall (Option A) or implement enhanced natural drainage systems (Option B) to protect against flooding. The decision is driven by the uncertainty in the annual probability of a severe flood, which could cause significant damage to infrastructure and property. The city wants to choose the option with the lowest expected cost, considering both initial investment and potential flood-related repair costs. The key uncertain parameter is the flood probability, and the city seeks to understand how sensitive the decision is to changes in this probability. Figure 6.8 illustrates the city's challenge in selecting a flood

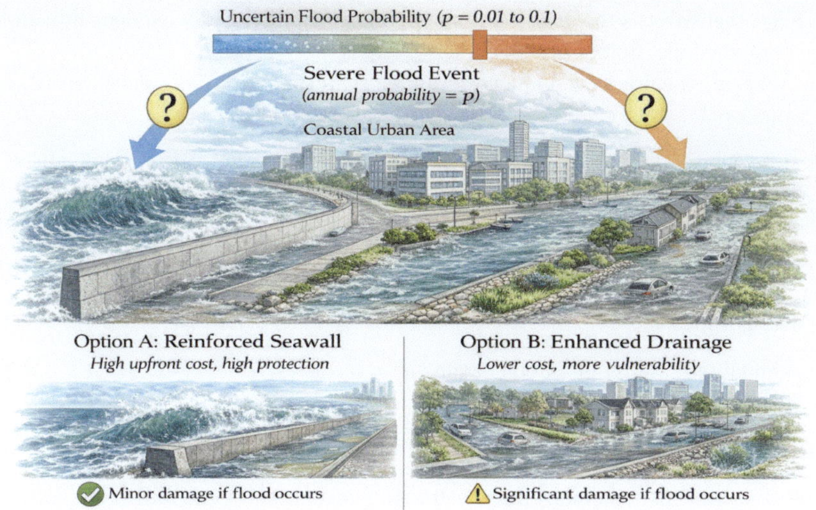

Fig. 6.8 Conceptual illustration of two flood mitigation strategies—reinforced seawall and enhanced natural drainage—under an uncertain severe flood probability

protection strategy given uncertain flood probabilities, setting the stage for evaluating expected costs and sensitivity to climate projections.

Decision Tree Structure: Representative costs for the two options are provided below and are typical of a planning-level cost estimate developed during the project planning stage

- Chance Node: A severe flood occurs with probability (1−p).
- Option A (Seawall):
  - Construction cost: $25 million.
  - Repair cost if flood occurs: $2 million (seawall mitigates most damage).
- Option B (Drainage):
  - Construction cost: $8 million.
  - Repair cost if flood occurs: $18 million (drainage is less effective at preventing damage).
- Parameter for Sensitivity Analysis: Annual flood probability p, estimated at 0.03 (3%) but potentially ranging from 0.01 (1%) to 0.1 (10%) based on hydrological models and climate projections.

  Questions to Address:

1. How does the expected cost of each option vary as the flood probability, p, changes from 0.01 to 0.1?
2. At what flood probability, if any, does the preferred option change (i.e., the threshold probability)?

3. How sensitive is the decision to the flood probability, and what does this imply for data collection or design choices?
4. What are the practical implications of the analysis for the city's decision-making process?

Step-by-Step Solution Using One-Way Sensitivity Analysis

Step 1: Define the Decision Tree Structure

The decision problem involves a deterministic decision node (choosing Option A or B), followed by a chance node representing the occurrence of a severe flood.

Let assume:

- Annual probability of a flood occurring = p
- Probability of no flood = 1-p

Step 2: Formulate Expected Cost Equations

The expected cost combines the construction cost and the probabilistic repair cost. The city will calculate the expected cost for both options as functions of p:

Expected Cost for Option A:

$$EC_A(p) = Construction\ Cost_A + p \times Repair\ Cost_A$$
$$EC_A(p) = 25 + 2p$$

Expected Cost for Option B:

$$EC_B(p) = Construction\ Cost_B + p \times Repair\ Cost_B$$
$$EC_B(p) = 8 + 18p$$

Step 3: Perform the Sensitivity Analysis

To evaluate how changes in p affect the decision, the city varies p from 0.01 to 0.10 (1% to 10%) a range derived from hydrological models and climate projections.

The table below shows the expected cost for each option at different values of p (Table 6.1):

Step 4: Identify the Threshold Probability

To determine the flood probability at which the preferred option changes (i.e., where the expected costs are equal), we solve for p in:

$$25 + 2p = 8 + 18p$$
$$17 = 16p$$
$$p = 17/16$$
$$P = 1.0625$$

Table 6.1 Expected cost comparison between Seawall (Option A) and Drainage (Option B) strategies across varying flood probabilities

Flood Probability p	Expected cost option A	Expected cost option B
0.01	$25 + 2(0.01) = 25.02$	$8 + 18(0.01) = 8.18$
0.02	25.04	8.36
0.03	25.06	8.54
0.04	25.08	8.72
0.05	25.1	8.9
0.06	25.12	9.08
0.07	25.14	9.26
0.08	25.16	9.44
0.09	25.18	9.62
0.1	25.2	9.8

*All costs are in millions of dollars

This threshold is above 1, which is not a realistic probability. Therefore, within the range $p \in [0.01, 0.10]$, Option B (Drainage) always has a lower expected cost than Option A.

Step 5: Interpret the Sensitivity and Results

The preferred decision does not change within the entire plausible range of p. Even at the highest assumed flood probability (10%), Option B remains significantly more cost-effective than Option A.

Cost Gap Between Options

- At $p = 0.01$: Option A is \$16.84M more expensive.
- At $p = 0.10$: Option A is \$15.40M more expensive.

This confirms that the decision is not sensitive to the value of p within the assumed range.

Step 6: Practical Implications for the City

The preferred strategy for the city is to proceed with Option B—enhancing natural drainage—since it consistently offers a lower expected cost across all realistic flood probability values. Given the robustness of this decision, there is no compelling need to invest in more precise estimates of the annual flood probability (p), as additional hydrological data collection would not significantly alter the outcome. This choice provides financial flexibility, enabling the city to redirect the \$15–17 million in savings toward other infrastructure needs, or to reduce the debt servicing obligations by eliminating the need to borrow this capital for the sea wall construction. From a policy perspective, while more reinforced infrastructure such as seawalls may offer higher levels of protection, their substantial cost is only justified under extreme flood risk scenarios.

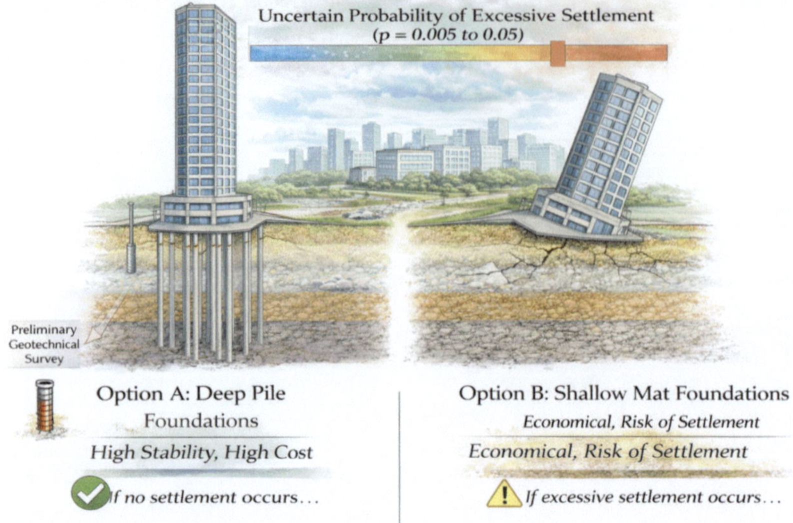

Fig. 6.9 Evaluating deep pile and shallow mat foundation systems for a high-rise building on variable soils. The image illustrates the different settlement risks and structural consequences under uncertain geotechnical conditions

This example demonstrates the power of one-way sensitivity analysis in guiding decision-making under uncertainty. By analyzing the impact of a single parameter—flood probability—on expected costs, the city confidently identifies a robust, cost-effective strategy without needing exhaustive probabilistic modeling. Such analyses are especially useful for public sector planning, where budget constraints and climate uncertainty coexist. This mathematical illustration of one-way sensitivity analysis demonstrates its power in evaluating decision robustness. By varying the flood probability from 0.01 to 0.1, we confirmed that the drainage system (Option B) is optimal, with no threshold within the entire range of realistic probabilities.

One-Way Sensitivity Analysis of Foundation Design Decisions

A civil engineering firm is designing the foundation for a high-rise building in a region with variable soil conditions. The firm must decide between two foundation options: deep pile foundations (Option A) or shallow mat foundations (Option B). The decision hinges on the probability of excessive soil settlement, which could lead to structural damage requiring costly repairs. Deep pile foundations are more expensive but have greater resiliency to settlement, while shallow mat foundations are cheaper but risk higher repair costs if settlement occurs. The key uncertain parameter is the probability of excessive soil settlement, and the firm seeks to understand how sensitive the decision is to changes in this probability. Figure 6.9 introduces the foundation design decision.

Decision Tree Structure

- Chance Node: Excessive soil settlement occurs with probability p, or no excessive settlement occurs with probability (1-p).
- Option A (Deep Pile Foundations):
  - Construction cost: $12 M.
  - Repair cost if excessive settlement occurs: $1 million (piles mitigate most damage).
- Option B (Shallow Mat Foundations):
  - Construction cost: $6 million.
  - Repair cost if excessive settlement occurs: $8 million (mat foundations are less effective against settlement).
- Parameter for Sensitivity Analysis: Annual probability of excessive soil settlement, p, estimated at 0.02 (2%) based on preliminary geotechnical surveys but potentially ranging from 0.005 (0.5%) to 0.05 (5%) due to variability in soil data and potential seismic influences.

Questions to Address:

1. How does the expected cost of each foundation option vary as the settlement probability p changes from 0.005 to 0.05?
2. At what settlement probability, if any, does the preferred option change (i.e., the threshold probability)?
3. How sensitive is the decision to the settlement probability, and what does this imply for geotechnical investigations or design choices?
4. What are the practical implications of the analysis for the firm's decision-making process?

Step 1: Define the Expected Cost Functions

Let:

- p = probability of excessive soil settlement
- 1−p = probability of no excessive settlement

Option A: Deep Pile Foundations

$$EC_A(p) = 12 + p \times 1 = 12 + p$$

Option B: Shallow Mat Foundations

$$EC_B(p) = 6 + p \times 8 = 6 + 8p$$

Table 6.2 Expected Cost Calculations for Two Foundation Design Options under Varying Settlement Probabilities

Settlement probability, p	Option A: Cost 12 + p	Option B: Cost 6 + 8 p
0.005	12.005	6.04
0.01	12.01	6.08
0.015	12.015	6.12
0.02	12.02	6.16
0.025	12.025	6.2
0.03	12.03	6.24
0.035	12.035	6.28
0.04	12.04	6.32
0.045	12.045	6.36
0.05	12.05	6.4

All values are in millions of dollars

Step 2: Vary p from 0.005 to 0.05

We compute the expected costs for each option as p varies within the plausible range (Table 6.2):

$$p \in \{0.005,\ 0.01,\ 0.015,\ 0.02,\ 0.025,\ 0.03,\ 0.035,\ 0.04,\ 0.045,\ 0.05\}$$

Step 3: Find the Threshold Probability

To find when the preferred option changes, we solve for the value of p where:

$$EC_A(p) = EC_B(p)$$
$$12 + p = 6 + 8p$$
$$6 = 7p$$
$$p = 6/7$$
$$p \approx 0.857$$

The analysis reveals a threshold probability of p = 0.857 (85.7%), at which the preferred decision would switch. However, this value lies far outside the plausible range of settlement probabilities, which is typically between 0.5 and 5%. As a result, Option B remains the preferred strategy under all realistic scenarios, confirming the robustness of the decision and eliminating the need for further refinement of the probability estimate.

Answers to the Questions

1. How does the expected cost vary as p increases?

   - Option A increases linearly: from \$12.005M to \$12.050M.
   - Option B also increases linearly but at a faster rate: from \$6.04M to \$6.40M.

   Despite both increasing, Option B remains significantly cheaper throughout.
2. At what settlement probability does the preferred option change?

   The threshold is $p = 0.857$. Since this value exceeds the range of realistic probabilities, Option B (mat foundation) is preferred for all practical values of p.
3. How sensitive is the decision to p?

   The decision is not sensitive to changes in p between 0.005 and 0.05. Even at the upper bound $p = 0.05$, Option B is \$5.65 million cheaper than Option A.

   Further efforts to refine the estimate of p would not change the decision, so extensive geotechnical studies for the purpose of a more refined estimate of p are not be cost-effective.
4. What are the practical implications for the firm's decision-making process?

The firm should select Option B, the shallow mat foundation, based on expected cost minimization. Although deep piles provide greater bearing capacity than a mat foundation, their high initial cost is not justified under any plausible soil risk level. The firm may consider reallocating funds saved by choosing Option B into: (1) improved soil compaction or ground improvement measures, (2) monitoring systems to detect early signs of settlement, and (3) structural insurance to hedge against rare settlement events.

This mathematical illustration of one-way sensitivity analysis in a geotechnical context demonstrates its utility in evaluating decision robustness for foundation design. By varying the settlement probability from 0.005 to 0.05, we confirmed that shallow mat foundations (Option B) are optimal, with no threshold within realistic probabilities. The moderate sensitivity, shown by the narrowing cost gap, highlights the importance of precise geotechnical data and flexible design strategies. This step-by-step process—defining equations, calculating costs, finding thresholds, and interpreting results—provides a clear, actionable framework for civil engineers to assess uncertainties in soil-related risks, prioritize geotechnical investigations, and communicate findings to stakeholders.

One-Way Sensitivity Analysis of Roofing Material Selection

A construction firm is designing the roof for a large industrial warehouse and must decide between two roofing materials: steel roofing (Option A) or asphalt shingle roofing (Option B). The decision depends on the probability of extreme wind events (e.g., hurricanes or severe storms) that could damage the roof, requiring costly repairs. Steel roofing is more expensive upfront but more durable against wind damage, while asphalt shingles are cheaper but more vulnerable to wind-related damage. The key uncertain parameter is the annual probability of an extreme wind event, and the

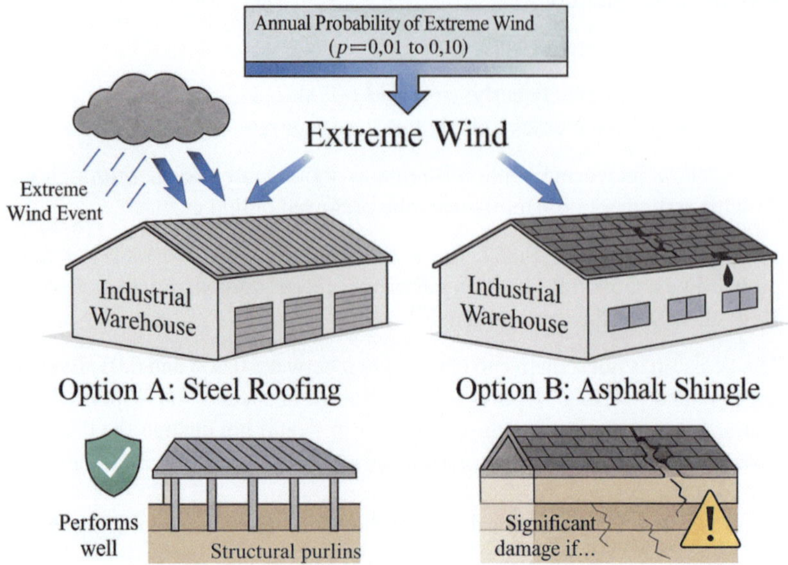

Fig. 6.10 Conceptual illustration of steel and asphalt shingle roofing systems for an industrial warehouse under extreme wind event uncertainty

firm seeks to understand how sensitive the decision is to changes in this probability. Figure 6.10 illustrates the engineering decision between two roofing systems under uncertain wind hazard probability, emphasizing trade-offs between cost, durability, and exposure to climate-driven risk.

Decision Tree Structure

- Chance Node: An extreme wind event occurs with probability p, or no extreme wind event occurs with probability (1−p).
- Option A (Steel Roofing):
  - Installation cost: $5 million.
  - Repair cost if wind event occurs: $0.5 million (steel is highly resistant to wind damage).
- Option B (Asphalt Shingle Roofing):
  - Installation cost: $2 million.
  - Repair cost if wind event occurs: $3 million (asphalt shingles are less resistant, requiring significant repairs).
- Parameter for Sensitivity Analysis: Annual probability of an extreme wind event p, estimated at 0.04 (4%) based on regional meteorological data but potentially ranging from 0.01 (1%) to 0.1 (10%) due to climate variability and potential increases in storm intensity.

Table 6.3 Expected cost comparison of steel roofing and asphalt shingles under varying wind event probabilities, highlighting differential sensitivity to wind-related risk

Wind event probability p	Steel Roofing: 5 + 0.5 p	Asphalt Shingles: 2 + 3 p
0.01	5.005	2.03
0.02	5.01	2.06
0.03	5.015	2.09
0.04	5.02	2.12
0.05	5.025	2.15
0.06	5.03	2.18
0.07	5.035	2.21
0.08	5.04	2.24
0.09	5.045	2.27
0.1	5.05	2.3

All values are in millions of dollars

Questions to Address

1. How does the expected cost of each roofing option vary as the wind event probability p changes from 0.01 to 0.1?
2. At what wind event probability, if any, does the preferred option change (i.e., the threshold probability)?
3. How sensitive is the decision to the wind event probability, and what does this imply for material selection or further analysis?
4. What are the practical implications of the analysis for the firm's decision-making process?

Step 1: Define the Expected Cost Functions

Let:

- p = annual probability of an extreme wind event
- 1−p = probability of no extreme wind event

Option A: Steel Roofing

$$EC_A(p) = 5 + 0.5p$$

Option B: Asphalt Shingle Roofing

$$EC_B(p) = 2 + 3p$$

Step 2: Vary p from 0.01 to 0.10

We compute the expected cost of both options across a range of wind event probabilities (Table 6.3):

Step 3: Find the Threshold Probability

To determine when the preferred option changes, we solve for the value of p where:

$$
\begin{aligned}
EC_A(p) &= EC_B(p) \\
5 + 0.5p &= 2 + 3p \\
3 &= 2.5p \\
p &= 3/2.5 \\
p &= 1.2
\end{aligned}
$$

The calculated threshold wind event probability is $p = 1.2$ (120%), which is not feasible since probabilities cannot exceed 1. Given this, Option B is always preferred within the realistic probability range of $p \in [0.01, 0.10]$. This confirms that the decision is robust and not sensitive to the variability in wind event probabilities within plausible bounds.

Answers to the Questions

1. How does the expected cost vary as p increases?
   - Steel Roofing (Option A) increases very slowly, from \$5.005 M to \$5.050 M.
   - Asphalt Shingles (Option B) increases more sharply, from \$2.03 M to \$2.30 M.

   Despite the increasing repair costs with p, Option B remains significantly cheaper across the full range.
2. At what probability does the preferred option change?

   The threshold occurs at $p = 1.2$, which is beyond any realistic scenario. Thus, Option B is always the preferred choice based on expected cost.
3. How sensitive is the decision to the wind event probability?

   The decision is not sensitive within the tested range (1–10%). Even at the highest probability of 10%, Option B is \$2.75 million cheaper than Option A. This implies the decision is robust, and improved precision in estimating p would not impact the choice.
4. What are the practical implications for the firm's decision-making process?

The firm should confidently select asphalt shingle roofing (Option B) based on cost-effectiveness. While steel is more durable, its high installation cost outweighs its limited savings in repair costs even in high-risk scenarios. Funds saved from choosing asphalt shingles can be reallocated toward (1) Reinforcement techniques (e.g., clips, underlayment), (2) Insurance to cover rare extreme events, and (3) Scheduled maintenance to extend the life of the roof.

This mathematical illustration of one-way sensitivity analysis in a construction context demonstrates its utility in evaluating decision robustness for roofing material selection. By varying the wind event probability from 0.01 to 0.1, we confirmed that asphalt shingle roofing (Option B) is optimal, with no threshold within realistic

Fig. 6.11 Conceptual illustration of two stormwater infrastructure strategies—a large underground retention tank and a distributed green infrastructure system—evaluated under increasing rainfall intensity due to climate change

probabilities. The moderate sensitivity, shown by the narrowing cost gap, highlights the importance of precise meteorological data and durable material specifications. This step-by-step process—defining equations, calculating costs, finding thresholds, and interpreting results—provides a clear, actionable framework for civil engineers to assess uncertainties in construction-related risks, prioritize data collection, and communicate findings to stakeholders.

One-Way Sensitivity Analysis of Stormwater Management Options

A city is preparing to upgrade its stormwater drainage system in response to projected increases in rainfall intensity due to climate change. The engineering team is evaluating two alternatives. Option A involves constructing a large underground retention tank, which demands a substantial upfront investment but provides high long-term resilience against extreme storm events. In contrast, Option B proposes deploying permeable pavements and bioswales citywide—a form of distributed green infrastructure that is more cost-effective initially, yet less reliable during intense rainfall, potentially resulting in flooding and associated damage costs. The choice between these two strategies hinges on the probability of extreme rainfall events exceeding historical levels. To guide decision-making, the city aims to conduct a sensitivity analysis to assess how the relative cost-effectiveness of each option shifts in response to changes in this probability. This approach will help determine whether the decision is robust or if further climate modeling and risk analysis are needed to reduce uncertainty. Figure 6.11 presents the contrasting stormwater strategies available to the city, framing the resilience and cost trade-offs between centralized and decentralized drainage systems under future climate uncertainty.

Decision Tree Structure

- Chance Node: An extreme rainfall event occurs with probability p, or does not occur with probability 1−p.
- Option A:
  - Construction cost: $20 million
  - Damage/repair cost if flood occurs: $1 million
- Option B:
  - Construction cost: $10 million
  - Damage/repair cost if flood occurs: $7 million
- Uncertain Parameter:
  - Probability of extreme rainfall event p estimated at 0.03 (3%), but may range from 0.01 to 0.1 due to climate model uncertainty.

Questions to Address

1. How does the expected cost of each stormwater option vary as the flood probability p changes from 0.01 to 0.10?
2. At what flood probability (if any) does the preferred option change?
3. How sensitive is the decision to changes in flood probability?
4. What are the implications for design selection and climate resilience planning?

Step-by-Step Solution

Step 1: Define Expected Cost Equations

Let p be the probability of an extreme rainfall event.
Option A (Retention Tank)

$$EC_A(p) = 20 + 1 \cdot p$$
$$EC_A(p) = 20 + p$$

Option B (Green Infrastructure)

$$EC_B(p) = 10 + 7 \cdot p$$
$$EC_B(p) = 10 + 7p$$

Step 2: Vary p from 0.01 to 0.10 (Table 6.4)

Step 3: Calculate the Threshold Probability

Find the value of p where the expected costs are equal:

$$20 + p = 10 + 7p$$
$$10 = 6p$$

Table 6.4 Expected cost comparison of flood protection Options A and B as flood probability p varies from 0.01 to 0.10, illustrating differences in sensitivity to flood-related risk

Flood probability p	Option A: 20 + p	Option B: 10 + 7 p
0.01	20.01	10.07
0.02	20.02	10.14
0.03	20.03	10.21
0.04	20.04	10.28
0.05	20.05	10.35
0.06	20.06	10.42
0.07	20.07	10.49
0.08	20.08	10.56
0.09	20.09	10.63
0.1	20.1	10.7

All values in millions of dollars

$$p = 10/6$$
$$p = 1.667$$

This threshold is well beyond 100%, meaning the preferred option will not change within the considered range.

Step 4: Interpret Results

Q1: Cost Variation

- Option A increases slowly with p, from \$20.01M to \$20.10M.
- Option B increases faster, from \$10.07M to \$10.70M.

Despite the growth, Option B is consistently cheaper in all tested cases.

Q2: Threshold Probability

The threshold value of $p = 1.667$ implies Option A never becomes cheaper in the given range of $(p \in [0.01, 0.10]$.
Q3: Sensitivity

The decision is not sensitive to flood probability in this range. Even at $p = 0.10$, Option A is still almost \$9.4M more expensive than Option B.

Q4: Practical Implications

The city can confidently choose Option B (green infrastructure), saving substantial costs without sacrificing robustness. Because the decision is not sensitive to variations in p, detailed probabilistic modeling may not be needed unless new risks (e.g., major climate shifts) emerge. The city could invest savings into monitoring, public education, or supplementary mitigation, enhancing overall system resilience.

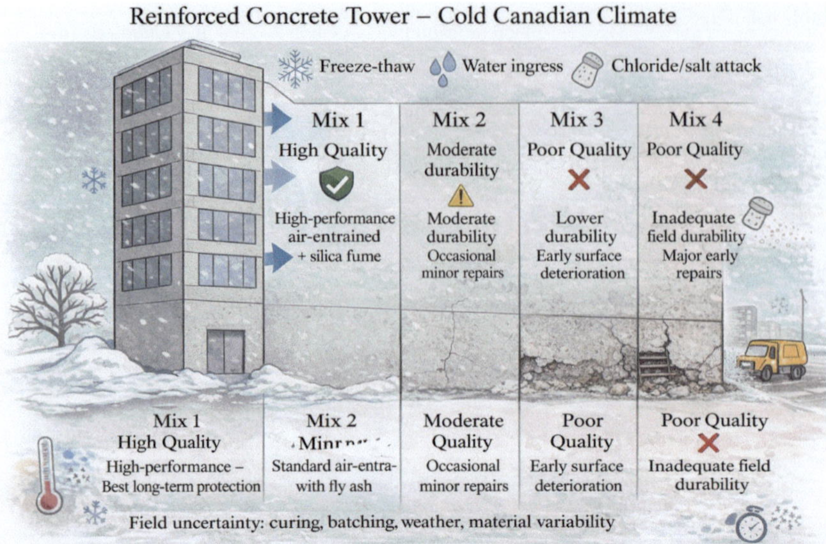

Fig. 6.12 Conceptual comparison of four concrete mix designs for a reinforced concrete tower in a cold climate

One-Way Sensitivity Analysis for Concrete Mix Selection in Cold Climate

An engineer is tasked with designing a reinforced concrete office tower in a cold Canadian climate, where the structure will be subject to frequent freeze-thaw cycles and chloride exposure from de-icing salts. To ensure long-term performance, the engineer must choose from four concrete mix designs, each differing in initial construction cost, durability, and reliability under uncertain field conditions. Given real-world variability in factors such as batch quality, curing practices, and supplier consistency, each mix may result in one of three performance outcomes: high quality (requiring no repairs), moderate quality (involving minor surface cracking and repairs), or poor quality (failing to meet durability specifications, leading to early major repairs). The engineer must evaluate these options using a risk-informed approach that accounts for both cost and performance under uncertainty. The objective is to select the mix with the lowest expected lifecycle cost, accounting for the probability of each performance outcome and its financial implications. Figure 6.12 illustrates how cold-climate exposure and variability in concrete mix application influence durability outcomes, guiding engineers in selecting robust materials for long-term structural performance (Tables 6.5, 6.6 and 6.7).

Questions to Address

1. What is the expected lifecycle cost of each concrete mix?
2. Which mix is the most cost-effective under current assumptions?
3. How does the decision respond to increased risk of poor performance?
4. At what point would a more durable (but costlier) mix become preferable?

Table 6.5 Descriptions of concrete mix design options

Option	Description
A	Conventional mix (30 MPa, local source)
B	Air-entrained mix (35 MPa, improved durability)
C	High-performance mix (40 MPa, silica fume)
D	Self-consolidating concrete (SCC, factory-batched)

Table 6.6 Material performance scenarios for different concrete mix options

Quality outcome	Description	A (Conventional) (%)	B (Air-entrained) (%)	C (HPC) (%)	D (SCC) (%)
High	Meets/exceeds spec (no repairs)	60	70	90	95
Moderate	Minor cracks; repair in 10 yrs	30	20	10	5
Poor	Below spec; major repairs	10	10	0	0

Table 6.7 Lifecycle cost estimates for different concrete mix options across quality outcomes, including initial and repair costs per 1,000 m^3, expressed in thousands of Canadian dollars

Option	Outcome	Initial cost	Repair cost	Total cost
A	High	100	0	100
A	Moderate	100	20	120
A	Poor	100	80	180
B	High	110	0	110
B	Moderate	110	10	120
B	Poor	110	60	170
C	High	130	0	130
C	Moderate	130	10	140
C	Poor	–	–	–
D	High	140	0	140
D	Moderate	140	10	150
D	Poor	–	–	–

5. What are the implications for material selection and construction risk management?

Step-by-Step Solution

Step 1: Define the Expected Cost Formula

$$\text{Expected Cost (EC)} = \sum (\text{Probability} \times \text{Total Cost}) \tag{6.1}$$

Step 2: Calculate Expected Cost for Each Mix

Option A (Conventional Mix)

$$EC_A = 0.60 \times 100 + 0.30 \times 120 + 0.10 \times 180 = 60 + 36 + 18 = 114$$

Option B (Air-Entrained Mix)

$$EC_B = 0.70 \times 110 + 0.20 \times 120 + 0.10 \times 170 = 77 + 24 + 17 = 118$$

Option C (High-Performance Concrete)

$$EC_C = 0.90 \times 130 + 0.10 \times 140 = 117 + 14 = 131$$

Option D (Self-Consolidating Concrete)

$$EC_D = 0.95 \times 140 + 0.05 \times 150 = 133 + 7.5 = 140.5$$

Step 3: Summarize the Results

Preferred Mix: Option A (Conventional mix) has the lowest expected lifecycle cost, making it the most economical option under current performance assumptions.

Step 4: Sensitivity Insight

Option A carries 10% risk of early major repair. What happens if this risk increases?

Let's compute the expected cost for Option A at different poor outcome probabilities. To do so, the poor and moderate probabilities outlined in Table 6.9 are used to calculate EC using the same approach as in Step 2 shown above. The probability of high-quality concrete is simply defined as 1-p(poor)-p(medium) because these conditions are mutually exclusive and exhaustive.

Threshold Insight: Using linear interpretation, it can be seen from the results of Table 6.9 that if probability of poor performance exceeds approximately 17%, Option B (air-entrained mix) becomes more cost-effective.

Step 5: Engineering Implications

- Option A is the most economical under current assumptions and is suitable for non-critical or controlled-exposure elements.
- If durability or quality control is uncertain, Option B provides better long-term value.
- Options C and D offer outstanding reliability but are not cost-justified unless:

Table 6.8 Expected lifecycle costs of concrete mix options

Option	Description	Expected cost (CAD $1,000 m^3)
A	Conventional mix	114
B	Air-entrained mix	118
C	High-performance concrete	131
D	Self-consolidating concrete	140.5

Table 6.9 Sensitivity analysis of expected cost for concrete mix option a under varying probabilities of poor and moderate performance outcomes

Poor (%)	Moderate (%)	Expected cost A
5	35	$111
10	30 (base)	$114
15	25	$117
20	20	$120
25	15	$123

- The structure has critical exposure zones.
- There is very high consequence of failure (e.g., parking garage decks, podium slabs).
- The project values low maintenance over time more than upfront savings.

Recommendation: Perform material trials, field QA programs, and consider supplier track record to estimate realistic risks for early deterioration. Adjust the mix choice accordingly.

The outcome of the sensitivity analysis for the cold weather concrete mix design scenario can be summarized as follows:

Question 1: What is the expected lifecycle cost of each concrete mix?

The calculation of the lifecycle cost for each alternative was demonstrated in Step 2 above, ranging from 114$ for mix design A to 140.5$ for mix design D. The complete cost breakdown for the different alternatives is provided in Table 6.8.

Question 2: Which mix is the most cost-effective under current assumptions?

Option A (Conventional mix) is the most cost-effective, with an expected lifecycle cost of $114/m^3. Based on the available data, this mix design has a 60% chance of no repair and, although it carries a 10% chance of major repair, its low initial cost offsets the risk under current assumptions.

Question 3: How does the decision respond to increased risk of poor performance?

A sensitivity analysis showed that, for a fixed probability of high performance concrete (i.e., 60%), the expected costs of Option A increased as the probability of poor performance increases (Table 6.9).

Table 6.10 Effect of increasing poor outcome probability on the expected cost of concrete mix option A, highlighting cost sensitivity to performance risk

Poor outcome probability (%)	Expected cost of option A
10 (Base)	$114
15	$117
20	$120
25	$123

When the poor outcome probability exceeds approximately 17%, Option B (air-entrained mix) becomes more cost-effective due to its greater durability and lower repair costs in poor scenarios (Table 6.10).

Question 4: At what point would a more durable (but costlier) mix become preferable?

The break-even threshold is reached when the probability of poor performance for Option A exceeds ~17%. At that point, Option B, which costs more initially but has better durability (lower risk of poor performance), becomes the optimal choice.

This means that if site conditions, material variability, or supplier quality suggest a greater than 1 in 6 chance of early deterioration, the engineer should consider switching to Option B.

Question 5: What are the implications for material selection and construction risk management?

- Option A is appropriate for projects with tight budgets and reasonable confidence in material quality.
- If construction occurs under variable environmental or curing conditions, or if supplier QA/QC is uncertain, it is safer to choose Option B, even at a slightly higher cost.
- Options C and D offer very high reliability but are not economically optimal unless:
  - The project involves critical infrastructure, or
  - The owner prioritizes minimal long-term maintenance and is willing to invest more upfront.

The selection of concrete mix should not rely on cost alone but also on the risk tolerance of the project, the anticipated exposure conditions, and the quality assurance framework in place during construction.

One-Way Sensitivity Analysis for Pavement Design in Highway Construction

A transportation engineer must select one of four pavement design alternatives for a new rural highway section expected to carry moderate to high truck traffic over the next 20 years. Each design option has different initial construction costs, maintenance schedules, and durability under environmental and loading uncertainties. Due to variability in soil subgrade strength, climate conditions, and traffic growth, the actual long-term performance of the pavement is uncertain. Outcomes are categorized into

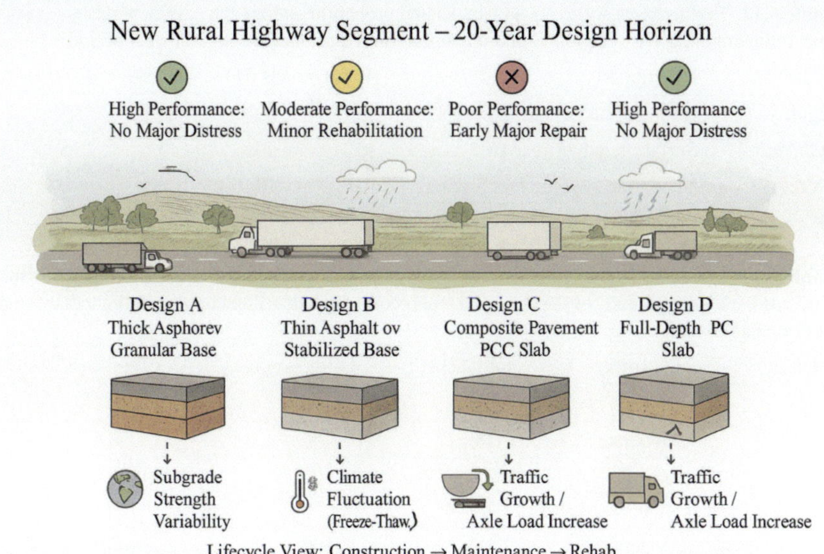

Fig. 6.13 Conceptual comparison of four pavement design alternatives for a rural highway exposed to uncertain traffic, climate, and subgrade conditions

Table 6.11 Descriptions of pavement design alternatives, ranging from thin asphalt overlays to rigid concrete slabs, reflecting differences in structural composition and performance characteristics

Option	Description
A	Thin asphalt overlay over stabilized subgrade
B	Standard flexible pavement (hot-mix asphalt layers)
C	Composite pavement (asphalt over PCC base)
D	Rigid pavement (thick jointed plain concrete slab)

high performance (no major distress), moderate performance (minor rehabilitation required), and poor performance (major rehabilitation needed early). The objective is to select the design with the lowest expected lifecycle cost, accounting for both initial and future repair expenditures under different probabilistic performance scenarios. Figure 6.13 presents the pavement design options under consideration for a rural highway project, illustrating how variability in field conditions may lead to different long-term performance outcomes (Tables 6.11, 6.12 and 6.13).

Questions to Address

1. What is the expected lifecycle cost of each pavement alternative?
2. Which pavement design is the most cost-effective under current assumptions?
3. How does the decision respond to uncertainty in early failure probability?
4. At what point would another option become preferable?
5. What are the implications for pavement selection and long-term risk?

Table 6.12 Performance scenarios and associated probabilities for each pavement design alternative, reflecting long-term durability and rehabilitation requirements over a 20 year period

Outcome	Description	A (%)	B (%)	C (%)	D (%)
High	No major distress for 20 years	50	60	85	95
Moderate	Surface rehab in year 10	40	30	15	5
Poor	Major rehab within first 5 years	10	10	0	0

Table 6.13 Lifecycle cost estimates for pavement design alternatives across performance outcomes, including initial construction and repair costs per lane-kilometer, expressed in thousands of Canadian dollars

Option	Outcome	Initial cost	Repair cost	Total cost
A	High	600	0	600
A	Moderate	600	150	750
A	Poor	600	500	1100
B	High	700	0	700
B	Moderate	700	100	800
B	Poor	700	300	1000
C	High	900	0	900
C	Moderate	900	80	980
D	High	1100	0	1100
D	Moderate	1100	100	1200

Step 1: Define the Expected Cost Formula

The expected lifecycle cost is calculated using Eq. (6.1)

Step 2: Calculate Expected Cost for Each Design

Option A (Thin Overlay)

$$EC_A = 0.50 \times 600 + 0.40 \times 750 + 0.10 \times 1100 = 300 + 300 + 110 = 710$$

Option B (Standard Flexible Pavement)

$$EC_B = 0.60 \times 700 + 0.30 \times 800 + 0.10 \times 1000 = 420 + 240 + 100 = 760$$

Option C (Composite Pavement)

$$EC_C = 0.85 \times 900 + 0.15 \times 980 = 765 + 147 = 912$$

Option D (Rigid Pavement)

$$EC_D = 0.95 \times 1100 + 0.05 \times 1200 = 1045 + 60 = 1105$$

Table 6.14 Expected lifecycle costs of pavement design alternatives, incorporating weighted performance outcomes and associated repair costs, expressed in thousands of Canadian dollars per lane-kilometer

Option	Description	Expected cost (CAD $1,000s)
A	Thin asphalt overlay	710
B	Standard flexible pavement	760
C	Composite pavement	912
D	Rigid pavement	1105

Step 3: Summarize and Compare Results

Preferred Option: Option A has the lowest expected lifecycle cost.

Step 4: Sensitivity Insight

Option A is cost-effective, but it has 10% risk of early major rehabilitation.
Let's assess what happens if the probability of poor performance increases:
At 20% poor performance, Option A's expected cost equals Option B's.

Step 5: Engineering Implications

- Option A is the most economical under current conditions.
- If there's greater risk of poor subgrade or traffic overloading, the probability of early failure could increase. At >20% poor performance, Option B becomes a better choice.
- Option C and D are robust but too costly unless:
  - There are extremely high truck volumes,
  - Minimal maintenance windows,
  - Or the project prioritizes service life over cost.

For rural roads with moderate risk and budget constraints, Option A is preferred. For roads in uncertain soil conditions or high load zones, consider Option B or hybrid designs.

Answers to the Five Key Questions

1. What is the expected lifecycle cost of each pavement alternative?

   Table 6.14 shows the answered of this question.
2. Which pavement design is the most cost-effective under current assumptions?

   Option A (thin overlay) is the most cost-effective, offering the lowest expected cost with acceptable performance in a moderate-risk context (Table 6.15).
3. How does the decision respond to uncertainty in early failure probability?

   - As the probability of early failure increases, the expected cost of Option A also increases.
   - At 20% poor performance, Option A = Option B.

Table 6.15 Sensitivity analysis of expected lifecycle cost for pavement option A under varying probabilities of poor and moderate performance outcomes

Poor (%)	Moderate (%)	EC_A (Expected cost)
10	40 (Base)	$710
15	35	$735
20	30	$760
25	25	$785

- If risk of failure is underestimated or field conditions worsen, Option A may become less favorable.

4. At what point would another option become preferable?
   - If probability of poor performance for Option A exceeds 20%, then Option B becomes more economical.
   - This provides a clear risk threshold for switching designs.
5. What are the implications for pavement selection and long-term risk?
   - Choose Option A when subgrade is well-prepared and traffic growth is predictable.
   - Use Option B in uncertain conditions or where longer performance certainty is needed.
   - Options C and D offer longer life but should only be used when cost is less of a constraint or in strategic corridors.

6.1.3.2 Two-Way Sensitivity Analysis

Two-way sensitivity analysis is an advanced technique in decision tree modeling used to evaluate the combined influence of two uncertain input parameters on a decision outcome. Unlike one-way sensitivity analysis, which varies a single parameter while holding all others constant, two-way sensitivity analysis explores how different combinations of two parameters—such as probabilities, costs, benefits, or discount rates—impact the expected utility or cost of each decision alternative. This allows decision-makers to understand not only the individual influence of each variable but also their interaction effects, which may significantly alter the preferred course of action.

The analysis is typically visualized using 2D decision maps, heatmaps, or contour plots, where regions are shaded according to which decision alternative yields the lowest expected cost or highest expected utility for a given parameter pair. This enables the identification of threshold boundaries where decision preferences change, as well as robust regions where the same option remains optimal across a wide

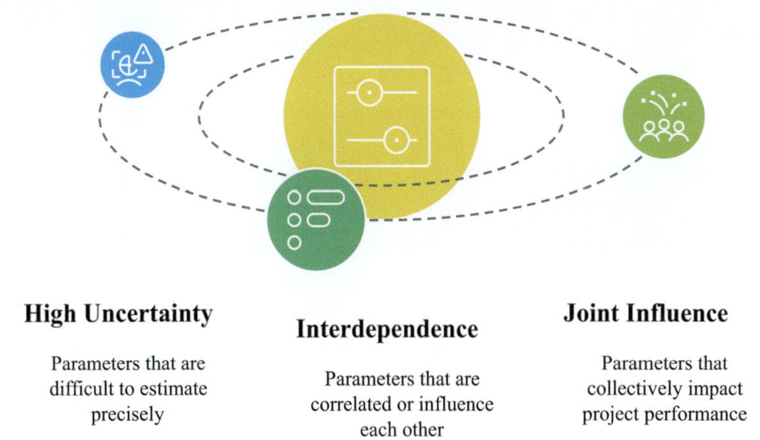

Fig. 6.14 Conditions under which two-way sensitivity analysis matters

range of values. In civil engineering, where uncertainties often stem from environmental risks, economic volatility, or evolving regulatory standards, two-way sensitivity analysis provides a more comprehensive understanding of risk trade-offs than one-dimensional methods.

Two-way sensitivity analysis becomes especially valuable in situations where two input parameters are highly uncertain and difficult to estimate precisely, are interdependent or correlated—such as the relationship between natural hazard probability and associated damages—and jointly influence the project's economic outcomes or safety performance. These conditions highlight the importance of evaluating the combined effects of uncertain variables to better support robust decision-making. Figure 6.14 illustrates the key conditions under which two-way sensitivity analysis matters.

This method is widely applied in risk-based infrastructure planning, cost-benefit evaluations, and resilience assessments, where decision-makers must consider the compound effects of natural variability and economic uncertainty.

The two-way sensitivity analysis process consists of a structured sequence of steps designed to evaluate how variations in two key parameters influence decision outcomes. As illustrated in Fig. 6.15, the first step is parameter selection, where two critical and potentially influential variables—such as flood frequency and associated damage costs—are identified based on the decision context. This is followed by range definition, in which realistic value ranges for each parameter are established using historical observations, expert judgments, or future scenario projections. Next, a parameter grid is constructed—commonly in the form of a 10 × 10 matrix—to represent a spectrum of possible combinations, resulting in multiple simulation runs.

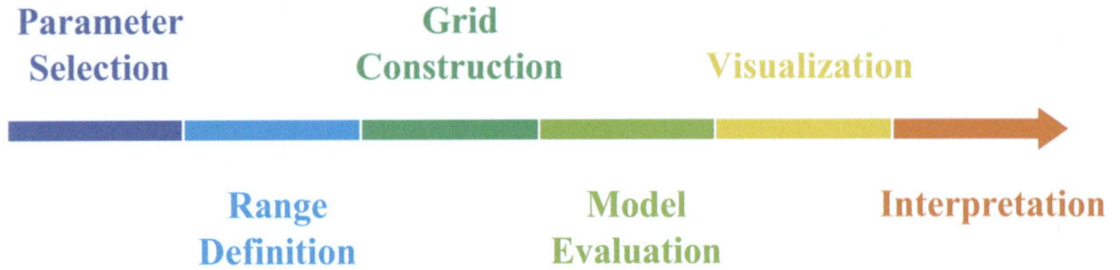

Fig. 6.15 Two-way sensitivity analysis process

At each grid point, the model undergoes evaluation to recalculate the expected outcomes (e.g., cost, utility, or risk level) for all competing alternatives. The results are then synthesized through visualization techniques, such as heatmaps or contour plots, which make it easier to identify optimal choices across the entire range of input combinations. Finally, the interpretation step focuses on analyzing the structure and boundaries of decision regions, highlighting zones of preference shifts and thresholds beyond which one alternative becomes more favorable than another.

This systematic approach not only enhances understanding of the decision landscape but also helps civil engineers identify break-even regions, where the final choice is sensitive to simultaneous changes in both variables. Compared to one-way analysis, this approach captures the combined effect of uncertainty in two variables at the same time. Furthermore, it enables the detection of high-risk or high-opportunity zones, providing insights into which parameter combinations are likely to yield the most critical outcomes in terms of safety, cost, or performance.

In civil engineering, two-way sensitivity analysis is especially valuable for projects where decision outcomes are shaped by the combined uncertainty of two dominant factors, such as environmental risks and economic costs. Infrastructure systems are often exposed to natural hazards—floods, earthquakes, extreme weather—whose probabilities and impacts are not easily predicted in isolation. By simultaneously varying two uncertain parameters, engineers can assess not only the robustness of a preferred design but also the conditions under which alternative strategies may become more effective or economical. This dual-parameter approach enables a deeper understanding of trade-offs and interactions that govern project viability, allowing for better-informed planning and more resilient infrastructure investments. Whether selecting structural reinforcements, evaluating flood protection measures, or deciding on long-term maintenance plans, two-way sensitivity analysis serves as a practical decision-support tool that connects engineering design to real-world variability and risk.

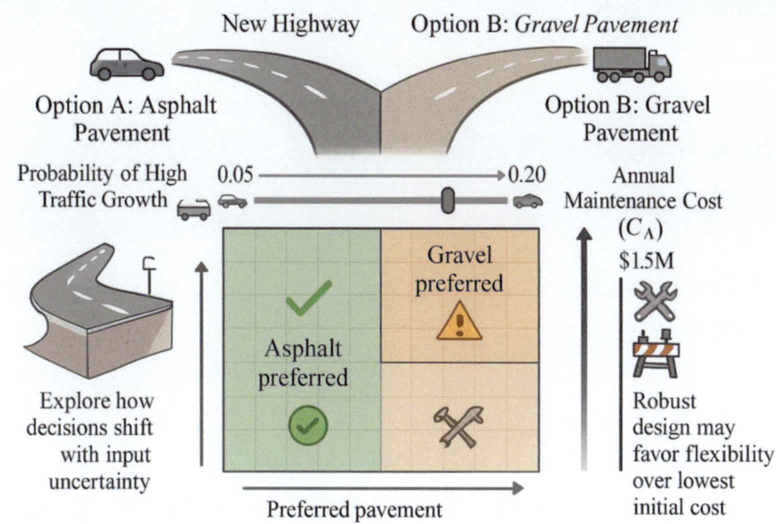

Fig. 6.16 Conceptual framework for two-way sensitivity analysis in pavement selection

Two-Way Sensitivity Analysis in Pavement Selection for a Highway Project

A transportation agency must decide between two pavement types for a new highway section (Fig. 6.16):

- Option A: Asphalt Pavement
  - Initial construction cost: $10 million
  - Annual maintenance cost if high traffic growth occurs: $C_A \in [0.5, 1.5]$ in millions of dollars
- Option B: Gravel Pavement
  - Initial construction cost: $6 million
  - Annual maintenance cost if high traffic growth occurs: fixed at $2.5 million.

Uncertain Parameters

- p: Probability of high traffic growth, ranging from 0.05 to 0.2
- C_A: Annual maintenance cost for asphalt, ranging from $0.5M to $1.5M.

We assume a 10 year analysis horizon with no discounting for simplicity. The expected cost formulas are given below. Note the factor of 10 that the probabilistic term is multiplied by to account for the horizon period in both Option A and Option B.

- Option A (Asphalt):

$$EU_A = 10 + 10 \cdot p \cdot C_A$$

- Option B (Gravel):

$$EU_B = 6 + 10 \cdot p \cdot 2.5 = 6 + 25p$$

Step-by-Step Analysis

1. Grid Construction

Vary:

- $p \in \{0.05, 0.1, 0.15, 0.2\}$
- $C_A \in \{0.5, 1.0, 1.5\}$

This results in 12 parameter combinations (4 × 3).

2. Cost Calculation Examples.

   - $p = 0.05$, $C_A = 0.5$:
     $EU_A = 10 + 10 \cdot 0.05 \cdot 0.5 = 10.25 EUB = 6 + 25 \cdot 0.05 = 7.25 \rightarrow$ Choose Option B
   - $p = 0.2$, $C_A = 0.5$
     $EU_A = 10 + 10 \cdot 0.2 \cdot 0.5 = 11{:}EU_B = 6 + 25 \cdot 0.2 = 11 \rightarrow$ Tie
   - $p = 0.2$, $C_A = 1.5$:
     $EU_A = 10 + 10 \cdot 0.2 \cdot 1.5 = 13 EU_B = 11 \rightarrow$ Choose Option B

To compute all 12 combinations of p and C_A, one can conveniently use either Excel or Python. In Excel, simply create two columns listing all combinations of p and C_A and then define equations for the expected cost of each option using the following syntax:

- EUA = 10 + 10 * p * CA
- EUB = 6 + 25 * p

These formulas can be applied using Excel's cell referencing to populate EUA and EUB for each parameter combination. However, using Python can be more efficient, especially for larger grids or more complex extensions. With a basic comprehension of simple loops or list algorithms, you can iterate through the parameter ranges and compute the results programmatically.

After computing EUA and EUB across all combinations, the output can be conveniently visualized using heatmaps as a means to ease interpretation of the results. Moreover, calculating the ratio EU_A/EU_B provides an easy metric to interpret the relative cost-effectiveness:

- Values greater than 1 indicate that Asphalt (Option A) is more expensive.
- Values less than 1 imply that Gravel (Option B) incurs higher expected costs.

This visualization supports better-informed decision-making under uncertainty. The syntax for producing a heatmap displaying the results of the comparison is provided in Code 5.1. Lines 1–4 import the essential libraries used in the analysis: `pandas` is used for managing and storing tabular data, `seaborn` and `matplotlib.pyplot` are used for creating advanced plots like heatmaps, and `itertools` is used to generate all possible combinations of the input parameters in a compact and efficient way.

Lines 7–8 define the two key variables for the sensitivity analysis: the probability `p` of high traffic growth, and the possible values of the asphalt maintenance cost `CA`. These are specified as discrete sets of values—`p` ranges from 0.05 to 0.2, while `CA` ranges from $0.5M to $1.5M, representing the potential variability in pavement deterioration under heavy traffic. It should be noted that the user can simply alter the numerical value of these variables based on the relevant values for similar problems.

Line 11 uses `itertools.product` to generate a full grid of (p, CA) combinations—this Cartesian product yields 12 unique pairs, covering all possible scenarios within the defined ranges.

Lines 14–19 define a loop to compute the expected cost of each pavement option across the parameter grid. For each (p, CA) pair, the expected cost for asphalt pavement (`EUA`) is calculated using the formula: `EUA = 10 + 10 * p * CA`, reflecting the $10M base cost plus the expected value of maintenance costs over 10 years. For gravel pavement, the expected cost (`EUB`) is computed as `EUB = 6 + 25 * p`, derived from $6M base cost plus 10 years of fixed $2.5M maintenance under high traffic. Additionally, the ratio `UCA/UCB = EUA / EUB` is computed to support comparative evaluation.

Line 22 converts the computed results into a `pandas.DataFrame`, labeling the columns appropriately for later analysis and plotting. Special formatting using LaTeX-like syntax (e.g., `C$_A$`) ensures that these column names appear with the correct style in the generated plots with, in this case, the underscore key being used to indicate subscript text.

Lines 25–26 create a simplified version of the DataFrame for exporting to Excel, using plain-text column names to avoid formatting issues. The `.to_excel()` function saves the table to an external `.xlsx` file named using the first input argument, allowing stakeholders to inspect or use the results in other tools like Excel.

Lines 29–31 transform the DataFrame into three separate pivot tables for heatmap visualization. Each table reshapes the data so that `p` values become the rows and `CA` values the columns, while the cell values contain `EUA`, `EUB`, and the ratio `EUA/EUB` respectively. This layout is ideal for two-dimensional visual comparisons.

Line 34 uses the function `plt.rcParams.update` to control the default aesthetic properties of all plots in Python. Here, the first input argument identifies the font family to be Times New Roman and the second argument sets the font size to 16, ensuring visual consistency and readability in figures intended for publication or presentations.

Line 37 initializes a `matplotlib` figure with three side-by-side subplots using `plt.subplots()`. The figure is set to a width of 20 and height of 6 inches to comfortably accommodate all three heatmaps with separate colorbars.

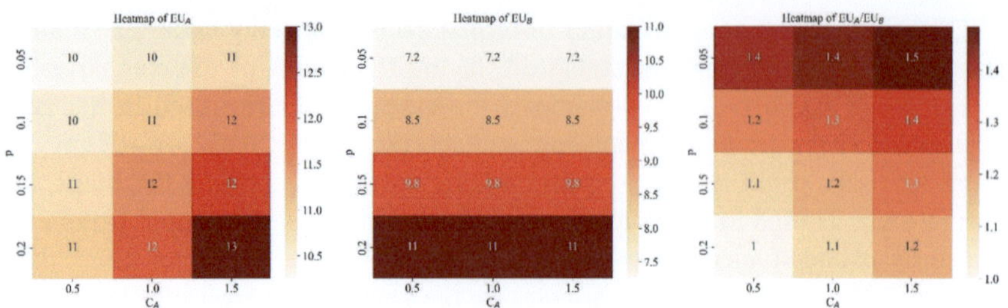

Fig. 6.17 Heatmaps of expected costs and cost ratio for pavement options under two-way sensitivity analysis

Line 40 is used to generate the heatmap for Option A's expected cost (`EUA`) using the `sns.heatmap()` function. Annotations are enabled to show numeric values inside each cell, with the Oranges-Red (`'OrRd'`) colormap enhancing visual gradients. In Lines 41–43 the axes are labeled and titled with math-formatted strings. Lines 46–49 produce a similar heatmap for Option B (`EUB`), maintaining consistent styling and structure.

Lines 52–55 create the third heatmap, this time for the ratio of expected utilities (`EUA/EUB`). This ratio helps interpret relative cost-efficiency: a value less than 1 indicates that asphalt is cheaper, greater than 1 suggests gravel is more economical, and exactly 1 represents cost parity.

Line 57 applies tight layout adjustments to prevent overlapping of labels and elements across the three plots.

Line 58 displays the figure, rendering the heatmaps in the output window or notebook interface.

Lines 61–62 save the final figure in both JPEG and TIFF formats at 300 dpi. These high-resolution formats are suitable for inclusion in reports, slides, or publications.

The output of this code includes two primary components: (1) a figure consisting of three heatmap subplots (Fig. 6.17) which visually display the expected costs of both pavement options and their ratio across all parameter combinations; and (2) an Excel file containing the full dataset of computed values, systematically organized and exported for further analysis (Fig. 6.18) provides a representation of this file.

Code 6.1 Calculation of the expected utility for both options (A and B) Pavement Type Decision-Making under Traffic Growth Uncertainty

p	CA	EUA	EUB	EUA/EUB
0.05	0.5	10.25	7.25	1.413793
0.05	1	10.5	7.25	1.448276
0.05	1.5	10.75	7.25	1.482759
0.1	0.5	10.5	8.5	1.235294
0.1	1	11	8.5	1.294118
0.1	1.5	11.5	8.5	1.352941
0.15	0.5	10.75	9.75	1.102564
0.15	1	11.5	9.75	1.179487
0.15	1.5	12.25	9.75	1.25641
0.2	0.5	11	11	1
0.2	1	12	11	1.090909
0.2	1.5	13	11	1.181818

Fig. 6.18 Excel output table of all parameter combinations and calculated metrics

```
import pandas as pd
import seaborn as sns
import matplotlib.pyplot as plt
import itertools

# Define the sets
p_values = [0.05, 0.1, 0.15, 0.2]
CA_values = [0.5, 1.0, 1.5]

# Generate all combinations
combinations = list(itertools.product(p_values, CA_values))

# Calculate EUA, EUB, and UCA/UCB
results = []
for p, CA in combinations:
    EUA = 10 + 10 * p * CA
    EUB = 6 + 25 * p
    UCA_UCB = EUA / EUB
    results.append([p, CA, EUA, EUB, UCA_UCB])

# Create DataFrame
df = pd.DataFrame(results, columns=['p', 'C$_A$', 'EU$_A$', 'EU$_B$', 'EU$_A$/EU$_B$'])

# Save to Excel with simplified column names for Excel export
excel_df = pd.DataFrame(results, columns=['p', 'CA', 'EUA', 'EUB', 'EUA/EUB'])
excel_df.to_excel('EUA_EUB_results.xlsx', index=False)

# Pivot for heatmap (p as rows, CA as columns)
pivot_eua = df.pivot(index='p', columns='C$_A$', values='EU$_A$')
pivot_eub = df.pivot(index='p', columns='C$_A$', values='EU$_B$')
pivot_uca_ucb = df.pivot(index='p', columns='C$_A$', values='EU$_A$/EU$_B$')

```

```
# Set font globally
plt.rcParams.update({'font.family': 'Times New Roman', 'font.size': 16})

# Plot all heatmaps in one figure with separate color bars
fig, axs = plt.subplots(1, 3, figsize=(20, 6))

# EU_A heatmap
sns.heatmap(pivot_eua, annot=True, cmap='OrRd', ax=axs[0], cbar=True, annot_kws={"size": 16})
axs[0].set_title('Heatmap of EU$_A$', fontsize=16)
axs[0].set_xlabel('C$_A$', fontsize=16)
axs[0].set_ylabel('p', fontsize=16)

# EU_B heatmap
sns.heatmap(pivot_eub, annot=True, cmap='OrRd', ax=axs[1], cbar=True, annot_kws={"size": 16})
axs[1].set_title('Heatmap of EU$_B$', fontsize=16)
axs[1].set_xlabel('C$_A$', fontsize=16)
axs[1].set_ylabel('p', fontsize=16)

# UCA/UCB heatmap
sns.heatmap(pivot_uca_ucb, annot=True, cmap='OrRd', ax=axs[2], cbar=True, annot_kws={"size": 16})
axs[2].set_title('Heatmap of EU$_A$/EU$_B$', fontsize=16)
axs[2].set_xlabel('C$_A$', fontsize=16)
axs[2].set_ylabel('p', fontsize=16)

plt.tight_layout()
plt.show()

# Save the figure in JPEG and TIF formats only
fig.savefig('combined_heatmaps_individual_colorbars.jpg', format='jpeg', dpi=300)
fig.savefig('combined_heatmaps_individual_colorbars.tif', format='tiff', dpi=300)
```

Boundary Equation

To find the boundary line where both options have equal expected cost:

$$EU_A = EU_B$$
$$10 + 10pC_A = 6 + 25p$$
$$C_A = \frac{25p - 4}{10p}$$
$$C_A = 2.5 - \frac{0.4}{p}$$

This defines a critical boundary curve:

- At p = 0.2: $C_A = 2.5 - 2 = 0.5$
- At p = 0.1: $C_A = 2.5 - 4 = -1.5 \rightarrow$ Infeasible

Only at high p and low C_A does asphalt become competitive.

Result Interpretation

- Dominance Region: Gravel (Option B) dominates most of the parameter space.
- Indifference Point: At $p = 0.2$, $C_A = 0.5$, both options have equal cost.
- Sensitivity: Given the current situation, the decision was not sensitive to the probability of high traffic growth over a 10 year period as the cost was equal, or lower, for gravel construction in all cases. It is expected that should the asphalt maintenance costs change, the decision will become sensitive to the traffic growth.
- Practical Insight: To favor asphalt, the agency must ensure low maintenance costs and face high traffic growth.

Implications

- Data Prioritization: Improve forecasts of traffic growth probability and maintenance cost estimates.
- Design Strategy: Unless confident in low C_A, gravel pavement offers a robust and cost-effective solution.
- Visualization: A heatmap with p on the x-axis and C_A on the y-axis can illustrate decision regions, with a curve $C_A = 2.5 - 0.4/p$ marking the boundary.

This example highlights the power of two-way sensitivity analysis in civil engineering decisions where uncertainty exists in both demand projections and long-term maintenance economics. By examining the interaction between traffic growth and maintenance costs, engineers can better design resilient infrastructure and optimize lifecycle investments. The method provides a structured framework not only for selecting the best alternative under uncertainty but also for identifying where improved data can most influence the decision. This strategic clarity is essential in modern infrastructure planning.

Two-Way Sensitivity Analysis of Flood Protection Strategies Under Uncertainty in Flood Probability and Damage Cost

A coastal city is deciding whether to build a reinforced levee (Option A) or implement improved natural drainage systems (Option B) to protect against flooding. The decision depends on two uncertain parameters: the annual probability of a severe flood and the cost of flood damage if the drainage system fails to prevent damage. The levee is more expensive to construct but significantly reduces flood damage, while the drainage system is cheaper but less effective, leading to higher damage costs if a flood occurs. The city aims to choose the option with the lowest expected cost using two-way sensitivity analysis to assess how the decision is affected by simultaneous variations in flood probability and damage cost. As illustrated in Fig. 6.19, a two-way sensitivity framework helps to determine which flood protection strategy is more cost-effective under different combinations of flood risk probability and flood damage severity.

Decision Tree Structure

- Chance Node: A severe flood occurs with probability p, or no flood occurs with probability $(1-p)$.
- Option A (Levee):
  - Construction cost: \$30 million.
  - Repair cost if flood occurs: \$1 million (levee mitigates most damage).
- Option B (Drainage):
  - Construction cost: \$10 million.
  - Repair cost if flood occurs: C_B million (variable, as drainage is less effective).

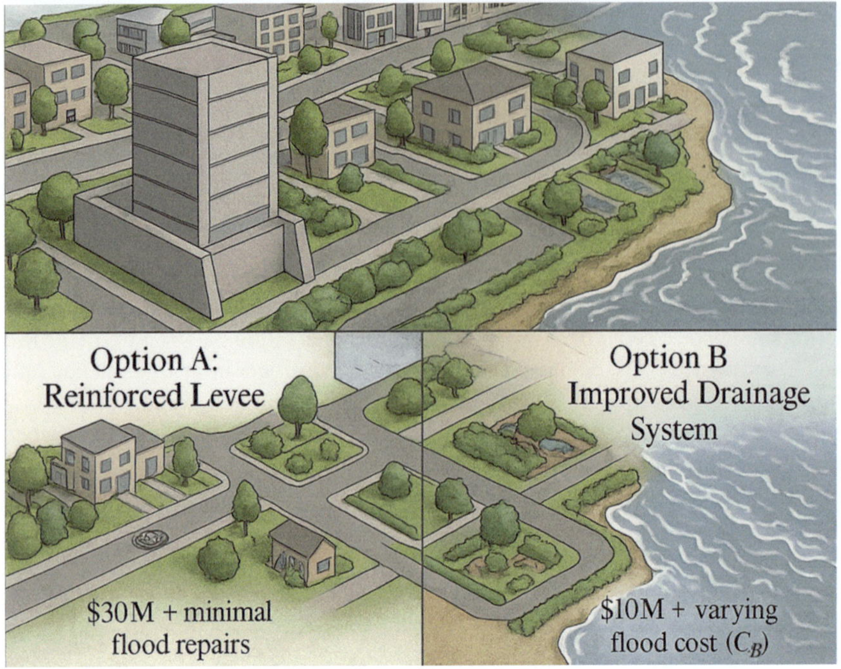

Fig. 6.19 Conceptual setup of two-way sensitivity analysis for flood protection decisions

- Parameters for Sensitivity Analysis:
  - Annual flood probability p, estimated at 0.05 (5%) but ranging from 0.01 (1%) to 0.1 (10%) based on hydrological models and climate projections.
  - Flood damage cost for Option B (C_B), estimated at $15M but ranging from $10M to $20M due to uncertainty in property values, infrastructure damage, and economic impacts.

Questions to Address

1. How does the expected cost of each option vary as flood probability p (0.01 to 0.1) and damage cost C_B ($10M to $20M) change simultaneously?
2. What combinations of p and C_B result in Option A (Levee) or Option B (Drainage) being optimal, and where is the boundary between these decision regions?
3. How sensitive is the decision to the combined effects of flood probability and damage cost, and what does this imply for data collection or design choices?
4. What are the practical implications of the analysis for the city's decision-making process?

Step 1: Define Expected Cost Equations

Option A (Levee)

- No flood (1−p): Cost = \$30M (construction only)
- Flood (p): Cost = \$30M + \$1M = \$31M
- Expected Cost (EU_A):

$$EU_A = (1 - p) \cdot 30 + p \cdot 31 = 30 + p$$

Option B (Drainage)

- No flood (1−p): Cost = \$10M (construction only)
- Flood (p): Cost = \$10M + C_B
- Expected Cost (EU_B):

$$EU_B = (1 - p) \cdot 10 + p \cdot (10 + C_B) = 10 + p \cdot C_B$$

Step 2: Define the Parameter Grid

Use a 5 × 5 grid of key values:

- Flood probability (p): [0.01,0.03, 0.05,0.07, 0.10]
- Flood damage cost C_B: [\$10M, \$12.5M, \$15M, \$17.5M, \$20M]

Step 3: Calculate Expected Costs and Decisions

We compute expected costs and compare both options for each parameter combination. Two sample calculations are provided below for illustrative purposes:

Sample Calculations:

Example 1 $p = 0.01,\ C_B = 10$

$$EU_A = 30 + 0.01 = 30.01M,\ EU_B = 10 + 0.01 \cdot 10 = 10.10M \Rightarrow \text{Option B is better}$$

Example 2 $p = 0.10,\ C_B = 20$

$$EU_A = 30 + 0.10 = 30.10M,\ EU_B = 10 + 0.10 \cdot 20 = 12.00M \Rightarrow \text{Option B is better}$$

This calculation is repeated for all 25 combinations of p and C_B with the full data table confirming that Option B (Drainage) is preferred in all cases (Table 6.34). In practice, a finer grid (e.g., 50×50) would provide smoother boundaries (Table 6.16).

Step 4: Derive the Decision Boundary

To find the cutoff line where both options have the same cost:

$$EU_A = EU_B$$
$$30 + p = 10 + p \cdot C_B$$

Table 6.16 Comparison of expected costs for flood protection options A and B under varying flood probabilities and damage cost assumptions

Flood probability, p	Damage cost C_B (M $)	Expected cost A (30 + p)	Expected cost B (10 + p·C_B)	Preferred option
0.01	10	30.01	10.1	B (Drainage)
0.01	12.5	30.01	10.13	B (Drainage)
0.01	15	30.01	10.15	B (Drainage)
0.01	17.5	30.01	10.18	B (Drainage)
0.01	20	30.01	10.2	B (Drainage)
0.03	10	30.03	10.3	B (Drainage)
0.03	12.5	30.03	10.38	B (Drainage)
0.03	15	30.03	10.45	B (Drainage)
0.03	17.5	30.03	10.53	B (Drainage)
0.03	20	30.03	10.6	B (Drainage)
0.05	10	30.05	10.5	B (Drainage)
0.05	12.5	30.05	10.63	B (Drainage)
0.05	15	30.05	10.75	B (Drainage)
0.05	17.5	30.05	10.88	B (Drainage)
0.05	20	30.05	11	B (Drainage)
0.07	10	30.07	10.7	B (Drainage)
0.07	12.5	30.07	10.88	B (Drainage)
0.07	15	30.07	11.05	B (Drainage)
0.07	17.5	30.07	11.23	B (Drainage)
0.07	20	30.07	11.4	B (Drainage)
0.1	10	30.1	11	B (Drainage)
0.1	12.5	30.1	11.25	B (Drainage)
0.1	15	30.1	11.5	B (Drainage)
0.1	17.5	30.1	11.75	B (Drainage)
0.1	20	30.1	12	B (Drainage)

$$p \cdot C_B = 20 + p$$
$$C_B = 20/p + 1$$

Using this relationship, the threshold value of C_B at which the two options are equal for various key values of p are calculated and summarized in Table 6.17.

The breakeven cost C_B is far outside the \$10M–\$20M range, meaning Option B is always more cost-effective within realistic conditions.

Step 5: Interpretation and Recommendations

For all combinations of $p \in [0.01, 0.10]$ and $C_B \in [10M, 20M]$, Option B (Drainage) yields the lowest expected cost.

Table 6.17 Threshold damage cost values (C_B) at which the expected costs of flood protection options A and B are equal

p	Threshold C_B
0.01	2001
0.03	668.33
0.05	401
0.07	286.57
0.1	201

The city should not invest in a levee unless:

- Flood probability increases drastically, or
- Damage costs exceed $200M per event.

Based on this analysis, the four questions at the beginning of the problem may now be answered.

Question 1: How does the expected cost of each option vary as flood probability p and damage cost C_B change simultaneously?

Option A (Levee): The expected cost increases linearly with flood probability p, with the formula:$EU_A = 30 + p$

Since the levee significantly reduces flood damage, the impact of increasing p is minimal (only +$1M per 100% increase in p).

For Option B (Drainage), the expected cost increases as a function of both the flood probability, p, and the damage costs, C_B according to the relationship given by:

$$EU_B = 10 + p \cdot C_B$$

Thus, as either variable increases, the cost rises faster than in Option A, especially if both are high. However, for the tested ranges ($p \leq 0.10$, $C_B \leq \$20M$), the increase in EU_B remains modest, and Option B remains more economical throughout.

Question 2: What combinations of p and C_B result in Option A (Levee) or Option B (Drainage) being optimal, and where is the boundary between these decision regions?

Option B (Drainage) is always optimal within the entire grid tested:

- $p \in \{0.01, 0.03, 0.05, 0.07, 0.10\}$
- $C_B \in \{10, 12.5, 15, 17.5, 20\}$

The decision boundary occurs where:

$$EU_A = EU_B \Rightarrow C_B = 20/p + 1$$

For instance:

- $p = 0.10 \Rightarrow CB = 201$
- $p = 0.05 \Rightarrow CB = 401$

These threshold values are far outside the realistic range of C_B values ($10M–$20M) predicted for region of interest in this coastal city, so the boundary lies well beyond the domain of practical scenarios.

Question 3: How sensitive is the decision to the combined effects of flood probability and damage cost, and what does this imply for data collection or design choices?

Sensitivity is low in the examined range:

- Even with the maximum tested values ($p = 0.10$, $C_B = \$20$ M), Option B is significantly cheaper.
- The decision only changes in extreme scenarios, like flood damages exceeding $200 M.

The implication of this low sensitivity is that there is no immediate need for highly precise estimates of p or C_B in the current planning window. However, if climate change projections worsen or urban expansion increases exposure, then:

- Accurate modeling of C_B and extreme flood probabilities becomes more important.
- Re-evaluating the decision may become necessary in the future.

Question 4: What are the practical implications of the analysis for the city's decision-making process?

The city should confidently proceed with Option B (Improved Drainage) based on cost-effectiveness under all tested conditions. Levee construction is not economically justified unless flood risk or damage estimates increase by an order of magnitude. The city can:

- Invest saved capital into complementary flood resilience strategies (e.g., early warning systems, zoning) or, alternatively, reduce debt servicing costs by borrowing less capital.
- Monitor changes in flood patterns or property exposure that might warrant future re-evaluation.
- Avoid overdesign and promote sustainable, nature-based infrastructure.

Two-Way Sensitivity Analysis in Bridge Retrofit Planning

A provincial transportation agency must decide between two risk management strategies for an aging highway bridge in a seismically active region. Option A involves retrofitting the bridge to current seismic standards at an upfront cost of $20 million, reducing potential earthquake damage to $5 million if such an event occurs, and incurring no cost if no earthquake happens. Option B adopts a monitoring and delay approach, requiring a $2 million investment in sensors and regular inspections, while accepting the risk that a damaging earthquake would result in the full cost of failure—denoted as C_B, which may range from $50 million to $150 million, depending on traffic volume, detour implications, and potential casualties. No cost is incurred under Option B if no earthquake occurs. The key uncertainties are the annual probability

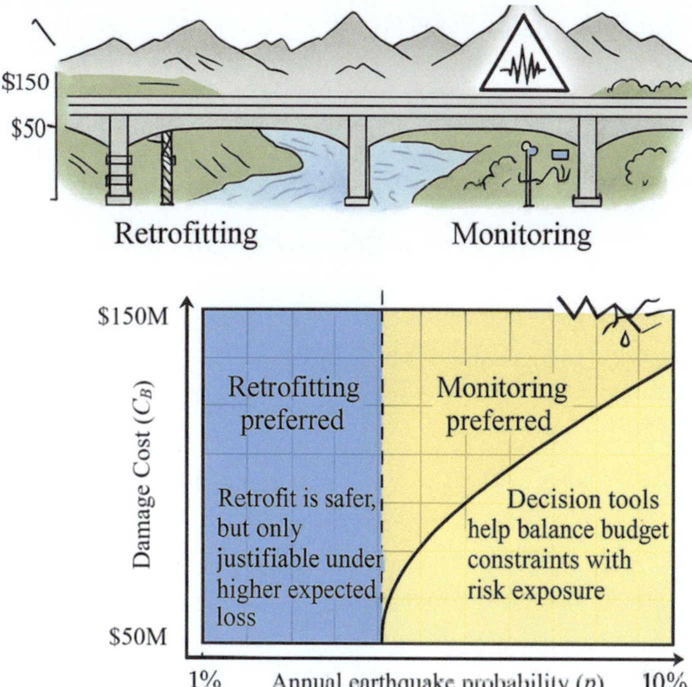

Fig. 6.20 Visual framework of two-way sensitivity analysis comparing seismic retrofit versus monitoring strategies for a critical highway bridge. The decision depends on both the annual probability of a damaging earthquake and the societal cost of potential failure

of a damaging earthquake (p), estimated between 1 and 10%, and the bridge failure cost C_B. The objective is to determine which option minimizes the expected lifecycle cost over a 1-year risk window, accounting for both economic efficiency and seismic risk exposure. Figure 6.20 illustrates the conceptual setup of the seismic decision problem, where the choice between proactive retrofitting and monitoring depends on earthquake risk (p) and the potential societal cost of bridge failure (C_B).

Questions to Address

1. How do the expected costs of each option change as earthquake probability p and failure cost C_B change simultaneously?
2. For which combinations of p and C_B is retrofitting (Option A) preferred over monitoring (Option B)?
3. How sensitive is the decision to these uncertainties, and what does this suggest about data priorities?
4. What are the practical implications for bridge management and infrastructure investment?

Step 1: Define Expected Cost Equations

Option A: Retrofit

- No earthquake (1−p): Cost = \$20M
- Earthquake (p): Cost = \$20M + \$5M = \$25M
- Expected cost:

$$EU_A = (1 - p) \cdot 20 + p \cdot 25 = 20 + 5p$$

Option B: Monitor and Delay

- No earthquake (1−p): Cost = \$2M
- Earthquake (p): Cost = \$2M + CBC_BCB
- Expected cost:

$$EUB = (1 - p) \cdot 2 + p \cdot (2 + C_B) = 2 + p \cdot CB$$

Step 2: Define the Parameter Grid

Use a 5×5 grid for illustrative coverage:

- Earthquake probability p: [0.01,0.03, 0.05,0.07, 0.10]
- Bridge failure cost C_B: [\$50M, \$100M, \$125M, \$150M, \$200M]

Step 3: Calculate Expected Costs and Optimal Decision (Table 6.18)

Result: Option B dominates the entire 5 × 5 grid, except when failure cost is \$200M at high p.

Step 4: Find the Decision Boundary (Table 6.19)

$$\begin{aligned} EU_A &= EU_B \\ 20 + 5p &= 2 + p \cdot C_B \\ p \cdot C_B &= 18 + 5p \\ C_B &= (18 + 5p)/p \\ C_B &= 18/p + 5 \end{aligned}$$

All break-even C_B values exceed \$150M for all $p \leq 0.10$, so Option B is always preferred in this scenario unless failure cost exceeds \$185M at $p = 0.10$.

Step 5: Interpretation and Practical Recommendations

Answer to Question 1:

- Option A (Retrofit) increases linearly with p.
- Option B (Monitor) is more sensitive, increasing with both p and C_B.

Still, B remains cheaper within tested values.

Table 6.18 Expected utility cost comparison between retrofitting (Option A) and monitoring (Option B) under varying flood probabilities and damage cost values (C_B), identifying the preferred option based on cost-effectiveness

p	C_B	$EU_A = 20 + 5p$	$EU_B = 2 + p \cdot CB$	Preferred option
0.01	50	20.05	2.5	B (Monitor)
0.01	100	20.05	3	B (Monitor)
0.01	125	20.05	3.25	B (Monitor)
0.01	150	20.05	3.5	B (Monitor)
0.01	200	20.05	4	B (Monitor)
0.03	50	20.15	3.5	B (Monitor)
0.03	100	20.15	5	B (Monitor)
0.03	125	20.15	5.75	B (Monitor)
0.03	150	20.15	6.5	B (Monitor)
0.03	200	20.15	8	B (Monitor)
0.05	50	20.25	4.5	B (Monitor)
0.05	100	20.25	7	B (Monitor)
0.05	125	20.25	8.25	B (Monitor)
0.05	150	20.25	9.5	B (Monitor)
0.05	200	20.25	12	B (Monitor)
0.07	50	20.35	5.5	B (Monitor)
0.07	100	20.35	9	B (Monitor)
0.07	125	20.35	10.75	B (Monitor)
0.07	150	20.35	12.5	B (Monitor)
0.07	200	20.35	16	B (Monitor)
0.1	50	20.5	7	B (Monitor)
0.1	100	20.5	12	B (Monitor)
0.1	125	20.5	14.5	B (Monitor)
0.1	150	20.5	17	B (Monitor)
0.1	200	20.5	22	A (Retrofit)

Table 6.19 Break-even damage cost values (C_B) at which the expected costs of retrofitting and monitoring become equal, across different flood probabilities pp

p	Break-even C_B
0.01	1805
0.03	605
0.05	365
0.07	263.6
0.1	185

Answer to Question 2:

- Option A is only optimal if:$CB > 18/p + 5$

  For example, if $p = 0.10$, then C_B must exceed \$185M.

- For $CB \leq 150$ M, Option B is always cheaper.

Answer to Question 3:

The decision is robust unless seismic risk or consequences are severely underestimated. If future seismic modeling shows increases in either p or failure impact, data collection and reassessment become essential.

Answer to Question 4:

The agency should choose Option B (Monitor & Delay) under current risk levels. Continue tracking urban growth and updated seismic forecasts to detect when a retrofit becomes necessary. A full retrofit may only be justified under catastrophic cost forecasts or updated risk zones.

Two-Way Sensitivity Analysis in Foundation Design Under Seismic Loading

In designing the foundation for a critical structure under seismic loading, a geotechnical engineer must choose between Option A, a pile foundation, and Option B, a shallow mat foundation with ground improvement. This decision is governed by two key uncertainties: the probability of liquefaction-induced failure (p) during the structure's design life, estimated between 1 and 10%, and the potential cost of failure (C_B) for the shallow foundation, which can range from \$5 million to \$25 million, depending on site accessibility, structural importance, and recovery delays. Option A involves a \$6 million construction cost and \$0.5 million in repairs if liquefaction occurs, with no cost if it doesn't. Option B requires only \$3 million upfront but bears the full failure cost C_B if liquefaction occurs, and no cost otherwise. A two-way sensitivity analysis is necessary to assess how changes in both p and C_B influence the optimal decision. The objective is to identify the foundation option that minimizes expected lifecycle cost under uncertain seismic conditions, ensuring both cost-efficiency and structural resilience. Figure 6.21 provides a visual overview of the foundation decision problem, highlighting the trade-off between higher upfront investment in pile foundations and the potential repair and societal costs associated with shallow mat foundations in liquefiable soils under seismic loading.

Questions to Address

1. How do expected costs vary with liquefaction probability p and failure cost C_B?
2. For which combinations of p and C_B is Option A (Pile) or Option B (Mat + Ground Improvement) preferred?
3. How sensitive is the decision to seismic risk and potential failure consequences?
4. What are the practical design implications for geotechnical risk management?

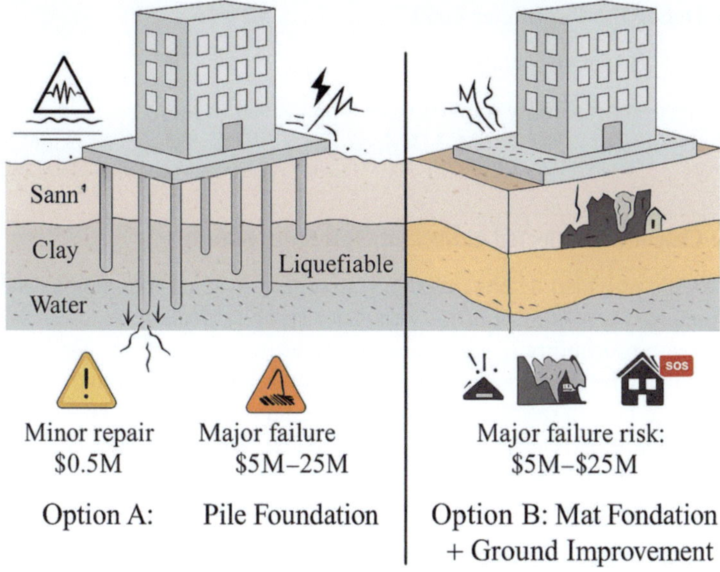

Fig. 6.21 Conceptual illustration of a two-way sensitivity analysis comparing pile foundations and shallow mat foundations with ground improvement under liquefaction risk

Step 1: Define Expected Cost Equations

Option A: Pile Foundation

- No liquefaction (1−p): Cost = \$6M
- Liquefaction (p): Cost = \$6M + \$0.5M = \$6.5M
- Expected cost:

$$EU_A = (1 - p) \cdot 6 + p \cdot 6.5 = 6 + 0.5p$$

Option B: Mat Foundation + Ground Improvement

- No liquefaction (1−p): Cost = \$3M
- Liquefaction (p): Cost = \$3M + CBC_BCB
- Expected cost:

$$EU_B = (1 - p) \cdot 3 + p \cdot (3 + C_B) = 3 + p \cdot C_B$$

Step 2: Define the Parameter Grid

Use a 5×10 grid for simplicity:

- Liquefaction probability p: [0.01,0.03, 0.05,0.07, 0.10]
- Failure cost C_B: [$5M, $10M, $15M, $20M, $25M, $25M, $30M, $35M, $40M, $45M]

Step 3: Calculate Expected Costs and Optimal Decision

Based on the provided results in Table 6.20, option B (Mat foundation) is consistently preferred across most of the parameter space, especially at lower values of traffic growth probability (p) and construction cost (C_B). However, Option A (Pile foundation) begins to emerge as the more favorable choice when both the probability of high traffic growth reaches 10% (p = 0.10) and the associated cost of Option B exceeds $35 million. Thus, Option A only becomes optimal in high-cost, high-risk scenarios. This threshold behavior highlights the importance of jointly considering both uncertainty and cost escalation in foundation design decisions.

Step 4: Find the Decision Boundary (Table 6.21)

$$
\begin{aligned}
EU_A &= EU_B \\
6 + 0.5p &= 3 + p \cdot C_B \\
p \cdot C_B &= 3 + 0.5p \\
C_B &= (3 + 0.5p)/p \\
C_B &= 3/p + 0.5
\end{aligned}
$$

Thus, for Option A to be preferable at p = 0.10, C_B must exceed $30.5M.

Step 5: Interpretation and Practical Recommendations

Answer to Question 1:

- Option A is insensitive to both variables (adds only $0.5M per unit p).
- Option B is highly sensitive to both p and C_B, but still remains cheaper for all realistic ranges.

Answer to Question 2:

- Option B (Mat + Ground Improvement) dominates for $C_B \leq 25M$ and $p \leq 0.10$.
- Option A only becomes optimal when both p and damage cost are extreme, e.g., $C_B \geq 35M$ and $p \geq 0.10$.

Answer to Question 3:

- The decision is not very sensitive for typical geotechnical uncertainty levels.
- No need for high-precision liquefaction models unless seismicity or criticality of structure increases.

Table 6.20 Comparison of expected utility costs for retrofitting (Option A) and material monitoring (Option B) under varying probabilities p and cost of failure scenarios (C_B), showing the preferred strategy based on minimum expected cost

p	CB	$EU_A = 6 + 0.5p$	$EU_B = 3 + p \cdot C_B$	Preferred option
0.01	5	6.005	3.05	B (Mat)
0.01	10	6.005	3.1	B (Mat)
0.01	15	6.005	3.15	B (Mat)
0.01	20	6.005	3.2	B (Mat)
0.01	25	6.005	3.25	B (Mat)
0.01	30	6.005	3.3	B (Mat)
0.01	35	6.005	3.35	B (Mat)
0.01	40	6.005	3.4	B (Mat)
0.01	45	6.005	3.45	B (Mat)
0.01	40	6.005	3.4	B (Mat)
0.03	5	6.015	3.15	B (Mat)
0.03	10	6.015	3.3	B (Mat)
0.03	15	6.015	3.45	B (Mat)
0.03	20	6.015	3.6	B (Mat)
0.03	25	6.015	3.75	B (Mat)
0.03	30	6.015	3.9	B (Mat)
0.03	35	6.015	4.05	B (Mat)
0.03	40	6.015	4.2	B (Mat)
0.03	45	6.015	4.35	B (Mat)
0.03	40	6.015	4.2	B (Mat)
0.05	5	6.025	3.25	B (Mat)
0.05	10	6.025	3.5	B (Mat)
0.05	15	6.025	3.75	B (Mat)
0.05	20	6.025	4	B (Mat)
0.05	25	6.025	4.25	B (Mat)
0.05	30	6.025	4.5	B (Mat)
0.05	35	6.025	4.75	B (Mat)
0.05	40	6.025	5	B (Mat)
0.05	45	6.025	5.25	B (Mat)
0.05	40	6.025	5	B (Mat)
0.07	5	6.035	3.35	B (Mat)
0.07	10	6.035	3.7	B (Mat)
0.07	15	6.035	4.05	B (Mat)
0.07	20	6.035	4.4	B (Mat)
0.07	25	6.035	4.75	B (Mat)

(continued)

Table 6.20 (continued)

p	CB	$EU_A = 6 + 0.5p$	$EU_B = 3 + p \cdot C_B$	Preferred option
0.07	30	6.035	5.1	B (Mat)
0.07	35	6.035	5.45	B (Mat)
0.07	40	6.035	5.8	B (Mat)
0.07	45	6.035	6.15	A (Pile)
0.07	40	6.035	5.8	B (Mat)
0.1	5	6.05	3.5	B (Mat)
0.1	10	6.05	4	B (Mat)
0.1	15	6.05	4.5	B (Mat)
0.1	20	6.05	5	B (Mat)
0.1	25	6.05	5.5	B (Mat)
0.1	30	6.05	6	B (Mat)
0.1	35	6.05	6.5	A (Pile)
0.1	40	6.05	7	A (Pile)
0.1	45	6.05	7.5	A (Pile)
0.1	40	6.05	7	A (Pile)

Table 6.21 Threshold values of failure cost (C_B) at which the expected utility of retrofitting equals that of material monitoring, calculated for various probabilities p of failure occurrence

p	Threshold C_B
0.01	300.5
0.03	100.17
0.05	60.5
0.07	43.36
0.1	30.5

Answer to Question 4:

Engineers can confidently use mat foundations with moderate ground improvement. Only consider piles for critical structures, very poor soils, or high-occupancy risks. Routine liquefaction risk monitoring and post-earthquake inspection should suffice for cost-effective resilience.

Two-Way Sensitivity Analysis in Wastewater Treatment Plant Upgrade

A municipality must decide how to reduce nutrient pollution (nitrogen and phosphorus) discharged into a nearby lake to comply with tightening environmental regulations. Figure 6.22 introduces the real-world challenge of nutrient pollution control in municipal water management, where engineering and nature-based solutions are weighed against uncertain regulatory risks and environmental outcomes.

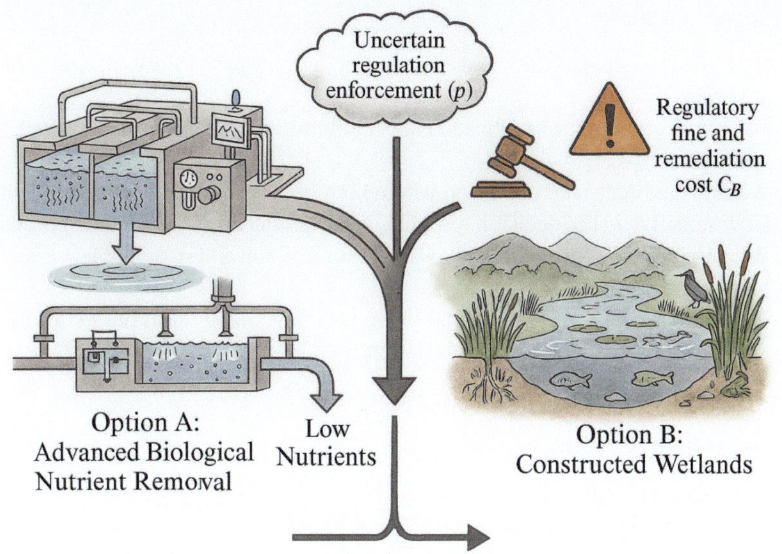

Fig. 6.22 Conceptual overview of two nutrient pollution control options considered by a municipality: a high-efficiency advanced BNR system and a nature-based constructed wetland

The two technology options being considered by the municipality for removal of the nutrients include:

- Option A: Advanced biological nutrient removal (BNR) system upgrade.
- Option B: Constructed wetlands as a nature-based solution for tertiary treatment.

The decision depends on two uncertain factors:

- p: Probability that regulatory fines will be enforced for exceeding nutrient discharge limits—ranging from 0.01 (1%) to 0.10 (10%) based on changing environmental enforcement trends.
- C_B: Cost of fines and environmental damage if violations occur—ranging from $1M to $6M, depending on lake eutrophication levels and legal penalties.

Option A: BNR Upgrade

- Capital and O&M cost: $12M
- If violation occurs (rare due to high efficiency): residual penalty of $0.2M

Option B: Constructed Wetlands

- Capital and O&M cost: $6M
- If violation occurs: full fine and remediation cost C_B

Objective: Select the most cost-effective long-term investment that ensures compliance and protects environmental quality.

Questions to Address

1. How do the expected lifecycle costs vary as fine probability, p, and cost C_B change?
2. For which combinations of p and C_B is Option A (BNR) or Option B (Wetlands) optimal?
3. What is the breakeven condition where both options have equal expected cost?
4. How sensitive is the decision to policy uncertainty and ecological response?
5. What implications does this analysis have for resilient and sustainable wastewater planning?
6. Should risk-averse environmental objectives (e.g., ecosystem protection) influence the final decision beyond cost alone?

Step 1: Define Expected Cost Equations

Option A: BNR System

- No violation (1−p): Cost = \$12M
- Violation (p): Cost = \$12M + \$0.2M = \$12.2M
- Expected cost:

$$EU_A = (1-p) \cdot 12 + p \cdot 12.2 = 12 + 0.2p$$

Option B: Constructed Wetlands

- No violation (1−p): Cost = \$6M
- Violation (p): Cost = \$6M + C_B
- Expected cost:

$$EU_B = (1-p) \cdot 6 + p \cdot (6 + CB) = 6 + p \cdot C_B$$

Step 2: Define the Parameter Grid

We select:

- Probability p: [0.01,0.03, 0.05,0.07, 0.10]
- Penalty cost C_B: [\$40M, \$50M, \$55M, \$60M, \$65M]

→ In this case, for illustrative purposes, we will consider a coarse 5 × 5 grid (25 scenarios)

Step 3: Calculate Expected Costs and Optimal Decision (Table 6.22)

So, Option B is preferred for all reasonable ranges of p and C_B, and Option A becomes optimal only if $C_B >$ \$60M and p = 0.10.

Step 4: Find the Decision Boundary

The boundary condition is determined by setting the expected costs of the two alternative treatment technologies equal to one another.

Table 6.22 Summary of expected costs and preferred treatment alternatives for various probabilities and costs of regulatory fines

p	C_B	$EU_A = 12 + 0.2p$	$EU_B = 6 + p \cdot C_B$	Preferred option
0.01	40	12.002	6.4	B (Wetland)
0.01	50	12.002	6.5	B (Wetland)
0.01	55	12.002	6.55	B (Wetland)
0.01	60	12.002	6.6	B (Wetland)
0.01	65	12.002	6.65	B (Wetland)
0.03	40	12.006	7.2	B (Wetland)
0.03	50	12.006	7.5	B (Wetland)
0.03	55	12.006	7.65	B (Wetland)
0.03	60	12.006	7.8	B (Wetland)
0.03	65	12.006	7.95	B (Wetland)
0.05	40	12.01	8	B (Wetland)
0.05	50	12.01	8.5	B (Wetland)
0.05	55	12.01	8.75	B (Wetland)
0.05	60	12.01	9	B (Wetland)
0.05	65	12.01	9.25	B (Wetland)
0.07	40	12.014	8.8	B (Wetland)
0.07	50	12.014	9.5	B (Wetland)
0.07	55	12.014	9.85	B (Wetland)
0.07	60	12.014	10.2	B (Wetland)
0.07	65	12.014	10.55	B (Wetland)
0.1	40	12.02	10	B (Wetland)
0.1	50	12.02	11	B (Wetland)
0.1	55	12.02	11.5	B (Wetland)
0.1	60	12.02	12	B (Wetland)
0.1	65	12.02	12.5	A (BNR)

$$
\begin{aligned}
EU_A &= EU_B \\
12 + 0.2p &= 6 + p \cdot C_B \\
p \cdot C_B &= 6 + 0.2p \\
C_B &= (6 + 0.2p)/p \\
C_B &= 6/p + 0.2
\end{aligned}
$$

Using this relationship, we are able to plot the regulatory fine amounts necessary for the two treatment technologies to have the same cost under a given probability of compliance failure (Table 6.23).

So, Option A is only justified when C_B>\$60.2 M at p = 0.10, which is well beyond normal fines or estimated environmental damages.

Table 6.23 Break-even values of damage cost (C_B) for BNR and wetland across different event probabilities

p	Threshold C_B
0.01	600.2
0.03	200.2
0.05	120.2
0.07	86.06
0.1	60.2

Step 5: Interpretation and Practical Recommendations

Answer to Question 1:

- Option A changes little with p (just +$0.2M over entire range).

  Option B is highly sensitive to both p and C_B, however, even at upper extremes ($p = 0.10$, $C_B =$ $6M), it is still the more economical treatment alternative in this scenario.

Answer to Question 2:

- Option B is optimal for all realistic combinations.
- Option A only becomes cost-effective at very high fines (> $60M) and strong regulatory enforcement ($p = 0.10+$), which is unlikely based on the provided information.

Answer to Question 3:

The breakeven formula ($C_B = 6/p + 0.2$) shows the decision is robust. Break-even C_B values are far above real-world ranges, confirming strong preference for Option B.

Answer to Question 4:

- Low sensitivity in decision boundary means:
  - The city does not need to invest in costly uncertainty reduction efforts.
  - But it should continue monitoring policy and lake health to anticipate any change in risk perception.

Answer to Question 5:

- Constructed wetlands are a cost-effective, nature-based solution that perform well under uncertainty.
- They offer resilience, co-benefits for biodiversity, and are publicly favored.
- The city should pair wetlands with adaptive monitoring to ensure compliance and sustainability.

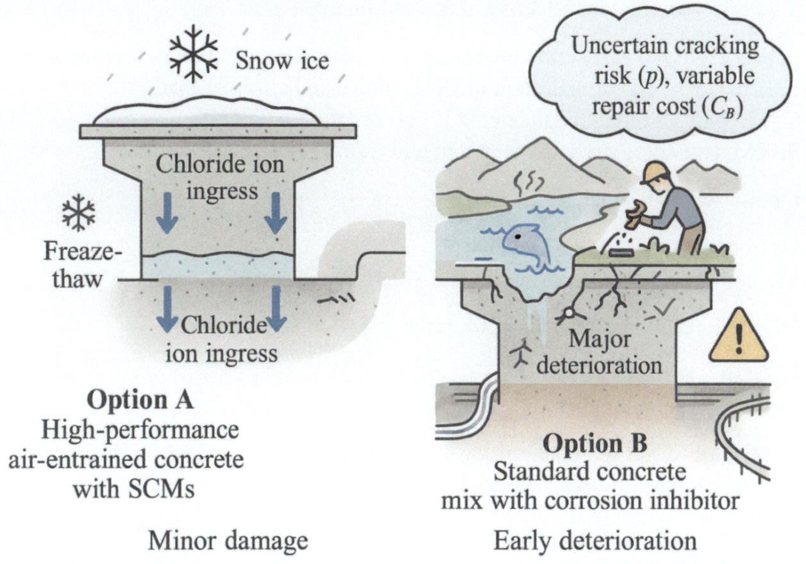

Fig. 6.23 Conceptual illustration of two concrete mix designs for a bridge deck in a cold climate

Answer to Question 6:

- If ecological conservation or reputational risk are valued more than just cost:
  - BNR may be worth considering as a zero-risk, high-performance option.
  - Decision-makers may use multi-criteria analysis that goes beyond expected monetary cost.

Two-Way Sensitivity Analysis for Concrete Mix Selection in Bridge Deck Design

A civil engineering firm must choose between two concrete mix designs for constructing a bridge deck exposed to freeze-thaw cycles, chloride ingress, and de-icing salts in a cold climate. The decision involves balancing upfront cost, service life performance, and the risk of premature deterioration. Figure 6.23 provides a visual comparison of how different concrete mixes perform under freeze-thaw and chloride exposure, highlighting the durability considerations and repair implications critical to bridge deck design in cold climates.

Two options are available:

- Option A: High-performance air-entrained concrete with supplementary cementitious materials (SCMs)
- Option B: Standard mix with chemical corrosion inhibitor

Two uncertain factors affect the expected lifecycle cost:

- p: Probability of early cracking/spalling in Option B within 20 years (ranges from 0.01 to 0.10), due to uncertain quality control and chloride loads.
- C_B: Repair and maintenance cost if deterioration occurs (ranges from \$0.5M to \$3.0M depending on damage extent and traffic disruption).

Option A: High-Performance Mix

- Upfront cost: \$1.8 M
- Maintenance cost if cracking occurs: \$0.1M (only surface treatment needed)

Option B: Standard Mix with Inhibitor

- Upfront cost: \$1.2 M
- Full-depth patching and deck closure if cracking occurs: C_B

Objective: Select the concrete mix with the lowest expected total lifecycle cost, considering long-term risk and repair implications.

Questions to Address

1. How do lifecycle costs vary as deterioration probability p and failure cost C_B change?
2. Which option is preferred across different combinations of these uncertainties?
3. What is the break-even line where both options yield equal expected cost?
4. How sensitive is the decision to variability in maintenance cost estimation and durability risk?
5. What should be prioritized in quality control, monitoring, or material testing to reduce uncertainty?
6. How would this decision change under more aggressive climate exposure or longer design life?

Step 1: Define Expected Cost Equations

Option A: High-Performance Concrete

- No deterioration: Cost = \$1.8 M
- Deterioration occurs: Cost = \$1.8M + \$0.1M = \$1.9 M
- Expected cost:

$$EU_A = (1 - p) \cdot 1.8 + p \cdot 1.9 = 1.8 + 0.1p$$

Option B: Standard Mix with Inhibitor

- No deterioration: Cost = \$1.2 M
- Deterioration occurs: Cost = \$1.2M + C_B
- Expected cost:

$$EU_B = (1 - p) \cdot 1.2 + p \cdot (1.2 + C_B) = 1.2 + p \cdot C_B$$

Table 6.24 Expected cost comparison between high-performance (Option A) and standard (Option B) material strategies under varying failure probabilities p and damage costs C_B, with preferred option based on lower expected utility cost

p	CB	$EU_A = 1.8 + 0.1p$	$EU_B = 1.2 + p \cdot C_B$	Preferred option
0.01	2	1.801	1.22	B (Standard)
0.01	4	1.801	1.24	B (Standard)
0.01	6	1.801	1.26	B (Standard)
0.01	8	1.801	1.28	B (Standard)
0.01	10	1.801	1.3	B (Standard)
0.03	2	1.803	1.26	B (Standard)
0.03	4	1.803	1.32	B (Standard)
0.03	6	1.803	1.38	B (Standard)
0.03	8	1.803	1.44	B (Standard)
0.03	10	1.803	1.5	B (Standard)
0.05	2	1.805	1.3	B (Standard)
0.05	4	1.805	1.4	B (Standard)
0.05	6	1.805	1.5	B (Standard)
0.05	8	1.805	1.6	B (Standard)
0.05	10	1.805	1.7	B (Standard)
0.07	2	1.807	1.34	B (Standard)
0.07	4	1.807	1.48	B (Standard)
0.07	6	1.807	1.62	B (Standard)
0.07	8	1.807	1.76	B (Standard)
0.07	10	1.807	1.9	B (Standard)
0.1	2	1.81	1.4	B (Standard)
0.1	4	1.81	1.6	B (Standard)
0.1	6	1.81	1.8	B (Standard)
0.1	8	1.81	2	A (High-Performance)
0.1	10	1.81	2.2	A (High-Performance)

Step 2: Define the Parameter Grid

Use a 5×5 grid:

- Deterioration probability p: [0.01,0.03, 0.05,0.07, 0.10]
- Failure cost C_B: [\$2M, \$4M, \$6M, \$8M, \$10M]

Step 3: Calculate Expected Costs and Preferred Option (Table 6.24)

Step 4: Find the Decision Boundary (Table 6.25)

$$EU_A = EU_B$$
$$1.8 + 0.1p = 1.2 + p \cdot C_B$$

Table 6.25 Threshold values of damage cost (C_B) at which the expected costs of high-performance and standard material strategies are equal, calculated across different probabilities pp of failure

p	Threshold C_B
0.01	60.1
0.03	20.1
0.05	12.1
0.07	8.67
0.1	6.1

$$p \cdot C_B = 0.6 + 0.1p$$
$$C_B = (0.6 + 0.1p)/p$$
$$C_B = 0.6/p + 0.1$$

Even at p = 0.10, the failure cost C_B must exceed $6.1M for Option A to be justified.

Step 5: Interpretation and Practical Recommendations

Answer to Question 1:

- Option A is very stable, increasing only slightly with p.
- Option B grows linearly with both p and C_B, but remains cheaper for all combinations tested.

Answer to Question 2: Option B is consistently preferred unless deterioration costs exceed ~$6M at the highest risk level (p = 0.10).

Answer to Question 3: Break-even formula: CB = 0.6/p + 0.1. For all practical ranges, the required CB is much higher than actual repair costs, confirming B is optimal.

Answer to Question 4: The decision is not sensitive to minor variations in repair cost. Focus should be placed on better durability risk forecasting if more extreme chloride exposure is expected (e.g., marine structures).

Answer to Question 5: Quality assurance testing (e.g., chloride diffusion tests, air void spacing) is more critical for Option B. If concrete durability can't be guaranteed through standard practice, Option A becomes more appealing.

Answer to Question 6: If design life is extended (e.g., from 20 to 75 years) or if future climate increases freeze-thaw cycles, the probability p would increase. Under more aggressive conditions, Option A may become justified or multi-criteria decision-making would be more appropriate.

Two-Way Sensitivity Analysis for Flood Risk Mitigation Strategy

A coastal municipality is evaluating three flood risk mitigation strategies to reduce the impacts of storm surges and sea level rise on residential and commercial zones.

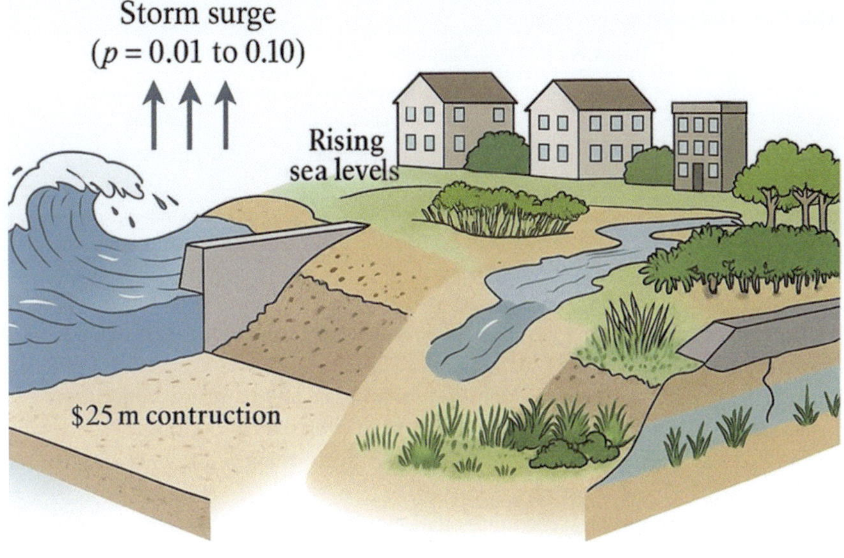

Fig. 6.24 Conceptual comparison of three coastal flood mitigation strategies: **A** a high-capital seawall offering strong defense, **B** low-cost nature based systems with higher performance uncertainty, and **C** a hybrid solution combining engineered and ecological elements

The options under consideration are:

- Option A: Build a permanent concrete seawall
- Option B: Implement nature-based solutions such as dunes, wetlands, and mangrove restoration
- Option C: Use a hybrid strategy combining smaller engineered barriers with nature-based buffers

The decision must account for uncertainties in future flood frequency and economic damage per event. Figure 6.24 illustrates the design context of the three flood risk mitigation options under consideration. These physical configurations help clarify how each strategy balances upfront investment with varying levels of protection under flood uncertainty.

Two key uncertain parameters:

- p: Annual probability of severe coastal flooding, projected between 0.01 (1%) and 0.10 (10%) (Table 6.26)
- C_D: Damage cost per flood event in millions of dollars, ranging from $5M to $30M depending on population exposure and infrastructure vulnerability

Select the option with the lowest expected lifecycle cost, incorporating both fixed costs and expected flood-related damages. In your answer, address the following questions:

Table 6.26 Mitigation options summary

Option	Base Cost (M$)	Damage if a flood occurs	Description
A—Seawall	$25M	$2M (residual overtopping/ maintenance)	High capital, strong defense
B—Nature-based	$12M	C_D (unpredictable performance)	Low cost, high variability
C—Hybrid	$18M	30% of C_D	Moderate cost and protection

1. How do expected total costs change across different flood probabilities and damage levels?
2. Which strategy is optimal under different climate and exposure conditions?
3. What are the break-even conditions between each pair of options?
4. How does the variability in nature-based performance influence the overall decision?
5. What policy or planning recommendations emerge from the sensitivity analysis?
6. Should non-monetary co-benefits (e.g., biodiversity, carbon storage) shift the decision toward greener solutions?

Step 1: Define Expected Cost Equations

Option A: Seawall

- Expected cost:

$$EU_A = 25 + p \cdot 2 = 25 + 2p$$

Option B: Nature-based

- Expected cost:

$$EU_B = 12 + p \cdot C_D$$

Option C: Hybrid

- Expected cost:

$$EU_C = 18 + p \cdot (0.3 \cdot C_D) = 18 + 0.3p \cdot C_D$$

Step 2: Define the Parameter Grid

- Flood probability p: [0.01,0.03, 0.05,0.07, 0.10]
- Damage per flood C_D: [$5M, $10M, $15M, $20M, $30M]
  - → For illustrative purposes a coarse 5 × 5 grid yielding 25 combinations is considered.

Step 3: Calculate Expected Costs and Preferred Option (Table 6.27)

Table 6.27 Expected cost comparison among three flood protection options under varying Flood Probabilities (p) and Damage Cost Values (C_D)

p	C_D	$EU_A = 25 + 2p$	$EU_B = 12 + p \cdot C_D$	$EU_C = 18 + 0.3p \cdot C_D$	Preferred option
0.01	50	25.02	12.5	18.15	B
0.01	100	25.02	13	18.3	B
0.01	150	25.02	13.5	18.45	B
0.01	200	25.02	14	18.6	B
0.01	250	25.02	14.5	18.75	B
0.03	50	25.06	13.5	18.45	B
0.03	100	25.06	15	18.9	B
0.03	150	25.06	16.5	19.35	B
0.03	200	25.06	18	19.8	B
0.03	250	25.06	19.5	20.25	B
0.05	50	25.1	14.5	18.75	B
0.05	100	25.1	17	19.5	B
0.05	150	25.1	19.5	20.25	B
0.05	200	25.1	22	21	C
0.05	250	25.1	24.5	21.75	C
0.07	50	25.14	15.5	19.05	B
0.07	100	25.14	19	20.1	B
0.07	150	25.14	22.5	21.15	C
0.07	200	25.14	26	22.2	C
0.07	250	25.14	29.5	23.25	C
0.1	50	25.2	17	19.5	B
0.1	100	25.2	22	21	C
0.1	150	25.2	27	22.5	C
0.1	200	25.2	32	24	C
0.1	250	25.2	37	25.5	A

Step 4: Identify Break-Even Points

Again, the break even points between the alternatives can be determiend by setting the expression for expected utility equal to one another.

Between A and B:

$$25 + 2p = 12 + p \cdot C_D \Rightarrow p(C_D - 2) = 13 \Rightarrow p = 13/(C_D - 2)$$

Between A and C:

$$25 + 2p = 18 + 0.3p \cdot C_D \Rightarrow p(0.3C_D - 2) = 7 \Rightarrow p = 7/(0.3C_D - 2)$$

Table 6.28 Break-even Damage Cost Values (C_D) at which the expected costs of mitigation options A & B, A & C, and B & C are equal

p	Threshold C_D		
	A & B	A & C	B & C
0.01	1302.00	2340.00	857.14
0.03	435.33	784.44	285.71
0.05	262.00	473.33	171.43
0.07	187.71	340.00	122.45
0.1	132.00	240.00	85.71

Between B and C:

$$12 + p \cdot C_D = 18 + 0.3p \cdot C_D \Rightarrow 0.7p \cdot C_D = 6 \Rightarrow p = 6/0.7C_D$$

These formulas can be used to determine the threshold across any value of C_D (Table 6.28).

Step 5: Interpretation and Practical Recommendations

Answer to Question 1:

- Option B has the lowest expected cost in most scenarios with low to moderate risk.
- Option C becomes preferable in high-risk, high-damage contexts.
- Option A is most robust, but cost-prohibitive unless flood probability and damage are very high.

Answer to Question 2:

- B is best when $p < 0.07$ and $C_D <$ \$20M
- C becomes optimal when $p > 0.07$ and $C_D >$ \$25M
- A is rarely optimal unless damages exceed \$50M and $p > 0.10$

Answer to Question 3: Use the break-even equations to set policy triggers (e.g., "If new projections raise annual flood probability to 0.08, switch to hybrid strategy.")

Answer to Question 4:

- Nature-based options have higher uncertainty in protective performance but excel under low enforcement or financial constraints.
- Including probabilistic performance ranges could further refine this model (e.g., 60–80% attenuation variability).

Answer to Question 5:

- Cities should pair early green infrastructure with adaptive retreat and zoning.
- Hybrid strategies offer a balance and support both resilience and biodiversity.

Answer to Question 6:

- Co-benefits of Option B (carbon sequestration, habitat creation, recreational spaces) may outweigh its expected cost for climate adaptation planning.
- Multi-objective frameworks should be used for final selection.

6.1.3.3 Threshold Analysis

Threshold analysis is a targeted form of sensitivity analysis used to determine the precise value of an uncertain parameter at which the preferred decision shifts from one alternative to another. Unlike one-way or two-way sensitivity analyses that explore how outcomes such as cost, performance, or risk vary over a range of input values, threshold analysis focuses on identifying tipping points—specific values of uncertain variables where the optimal decision reverses. This approach is particularly useful when dealing with binary or discrete decisions, such as whether to implement a high-cost, high-reliability infrastructure system or to opt for a more flexible but potentially risk-prone alternative. By isolating these critical values, decision-makers can test the robustness of their choices against uncertainty and better understand the minimum conditions under which a given alternative becomes superior.

The process of conducting threshold analysis involves several structured steps. First, the analyst defines the decision alternatives and establishes the performance metric of interest, typically cost, utility, environmental impact, or risk. Next, the uncertain parameter (or parameters) likely to influence the decision is identified—this may include climate variables, failure probabilities, interest rates, or construction material costs. Then, mathematical functions are formulated for each alternative, expressing the total cost or benefit as a function of the uncertain variable. The threshold is identified by setting these functions equal and solving for the uncertain parameter. This yields the exact point at which the two alternatives yield equal outcomes, beyond or below which one becomes preferable.

For example, consider a municipality choosing between building a permanent levee system or relying on temporary flood barriers to protect a neighborhood from river flooding. The cost of the permanent levee is fixed and known, while the cost of temporary barriers depends on the probability of flooding, which is uncertain due to climate variability. Threshold analysis allows the municipality to calculate the flood probability value at which the long-term costs of the temporary barrier approach match those of the levee system. If the actual flood probability exceeds this value, the permanent levee is the better option. This threshold becomes a policy-relevant boundary: if new climate projections indicate a higher likelihood of extreme floods, the decision can be confidently revisited.

One of the primary advantages of threshold analysis is its decision-oriented nature. Rather than analyzing broad outcome trends, it directly informs whether a decision should be reconsidered based on specific shifts in context or assumptions. This makes it highly valuable for decision-making under uncertainty, especially in infrastructure projects where stakes are high and misjudgments are costly. It is also computationally efficient. In many cases, the threshold can be found analytically, using simple algebraic equations, without requiring iterative simulations or extensive modeling. Furthermore, threshold analysis highlights which variables are most influential in altering the decision—helping engineers and planners focus monitoring efforts or further refine estimates for those critical parameters.

However, threshold analysis also has several limitations. Firstly, this approach typically handles one parameter at a time and does not explore how multiple uncertainties may interact. In complex systems with interdependent risks (e.g., flood probability and damage severity), a single-variable threshold may oversimplify reality. Additionally, threshold analysis does not provide information about how gradual changes in input values affect outcomes—it only identifies the break-even point. For broader understanding, one- or two-way sensitivity analysis may still be necessary to map the full behavior of the system across different scenarios. The advantages and limitations of threshold analysis are reflected in Fig. 6.25, which visualizes the break-even point at which one alternative becomes more favorable than another under varying assumptions. This figure exemplifies the decision-oriented nature of threshold analysis and computational simplicity, while also illustrating its inherent constraint of only being able to analyze a single parameter at a time without capturing interactions or gradual shifts in outcomes.

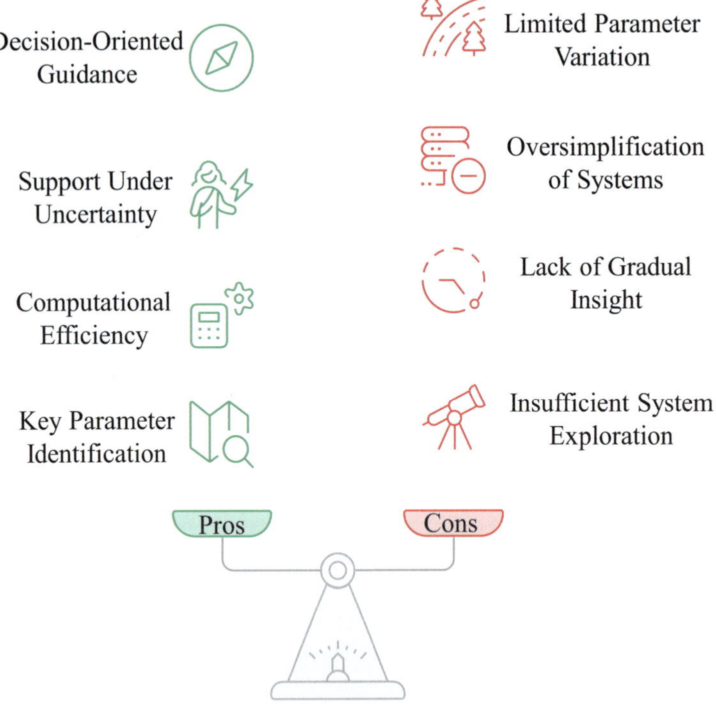

Fig. 6.25 The main advantages and limitations of the threshold analysis

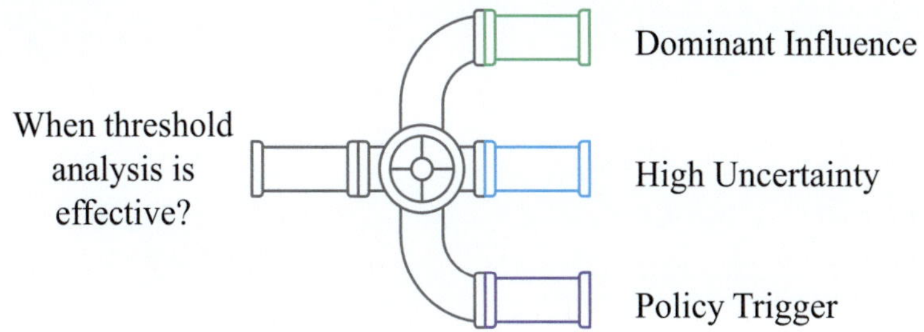

Fig. 6.26 Key conditions under which threshold analysis is most effective

As illustrated in Fig. 6.26, threshold analysis is particularly effective in situations where a single input parameter has a dominant influence on the decision outcome, is highly uncertain or expected to change over time—such as in response to climate variability or market trends—and represents a critical policy trigger, regulatory limit, or economic tipping point that could shift the preferred course of action.

Threshold analysis is widely applied in decision-making for infrastructure design, climate adaptation planning, and environmental compliance, where identifying the exact point at which an investment or action becomes justified is crucial. The procedure (Fig. 6.27) follows a clear logical path that begins with defining the competing decision options—such as permanent versus temporary flood protection—followed by developing cost, utility, or risk functions for each alternative in terms of the uncertain parameter. The threshold is then identified by equating these functions (e.g., Total Cost_A = Total Cost_B) and solving for the parameter value that balances the two. Once the threshold is established, the actual or projected value of the parameter is compared against it to determine the optimal course of action. This result is then translated into a decision rule (e.g., "If flood probability >2.5%, build levee; otherwise, deploy seasonal barriers") to guide practical implementation. Finally, scenario testing may be conducted to evaluate the robustness of the threshold under varying assumptions and uncertainties.

This technique helps civil engineers and planners establish clear decision boundaries and justify investments based on future conditions or risk exposure. It is especially useful when the outcome space is binary (e.g., invest or delay, upgrade or maintain) and when the cost of misjudging the uncertain parameter is high. Threshold analysis is also commonly used in regulatory frameworks (e.g., environmental concentration limits, resilience criteria) where action is required only when a variable exceeds a specified level.

Comparing Threshold Analysis with Traditional Sensitivity Methods

Contrasting threshold analysis with other forms of sensitivity analysis brings to light how each method addresses uncertainty and decision dynamics. One-way sensitivity analysis varies a single input over a plausible range and observes changes in the

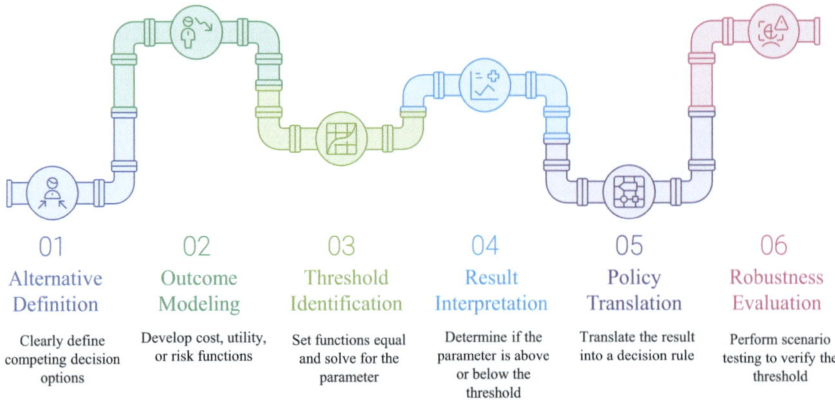

Fig. 6.27 Threshold analysis-based decision-making process

output, providing useful information about how sensitive the result is to individual parameters. It often produces tornado diagrams that show which variables have the greatest influence on outcome variability. Two-way sensitivity analysis takes this further by varying two parameters simultaneously, often visualized with heat maps or decision regions, revealing how combinations of uncertainties affect the preferred choice. In contrast, threshold analysis zooms in on the precise value at which decisions change. It does not show the shape of the outcome curve but instead answers the more focused question: "When does my decision stop being valid?"

By pinpointing when a decision should change, threshold analysis adds strategic value to traditional cost-benefit frameworks and complements broader sensitivity techniques like one-way and two-way analyses.

Application of Threshold Analysis in Civil and Environmental Engineering

Threshold analysis is particularly well-suited to strategic infrastructure planning, environmental management, and policy decision-making where a small change in one critical parameter can lead to a significant shift in the preferred solution. Its value lies in helping engineers and planners identify the exact conditions under which an action becomes economically justified, technically necessary, or environmentally imperative. The following examples illustrate how this method can be applied across various domains of civil and environmental engineering.

Flood Risk Management

In urban flood control projects, municipalities often choose between building permanent defenses (e.g., levees, floodwalls) and implementing temporary or adaptive solutions (e.g., deployable barriers, wetland buffers). The key uncertainty is typically the future probability of extreme rainfall or river discharge, influenced by climate change.

Threshold analysis helps determine the critical flood return period (or exceedance probability) beyond which the permanent infrastructure becomes more cost-effective over its lifespan. For instance, if the probability of a 100 year flood occurring in the next 30 years exceeds a calculated threshold, the city might justify the upfront investment in permanent structures.

Bridge Design and Load Forecasting

Bridge design often involves selecting materials and structural systems based on anticipated loads, maintenance costs, and life-cycle performance. A typical threshold analysis might examine the critical increase in traffic volume (due to population growth or economic development) at which a prestressed concrete design becomes more economical than a steel alternative. This threshold informs decisions about whether to overdesign a structure now or plan for retrofitting in the future, balancing capital investment with long-term flexibility.

Water Supply and Treatment Capacity

In water treatment or supply planning, engineers often face decisions about whether to expand treatment plant capacity now or defer expansion until demand increases. Here, the threshold value could be the population size, per capita water demand, or peak daily flow rate that justifies the investment. By identifying this threshold, municipalities can align infrastructure expansion with actual growth, avoiding both overbuilding and under-capacity risks.

Environmental Impact Mitigation

When evaluating environmental remediation strategies, such as soil decontamination or wetland restoration, threshold analysis can be used to assess pollutant concentration thresholds or land-use intensities at which the environmental cost of inaction surpasses the cost of intervention. For instance, a threshold level of nitrate in groundwater may trigger the need for buffer zone implementation or changes in agricultural practices to comply with health regulations or sustainability goals.

Renewable Energy and Building Retrofits

In sustainable building design, one common application of threshold analysis is evaluating the point at which investments in high-efficiency insulation or renewable energy systems (e.g., solar panels, geothermal heat pumps) become economically favorable. This threshold is often linked to energy price escalation rates, carbon tax levels, or equipment lifespan. Planners use this analysis to decide whether to invest in net-zero upgrades now or delay based on current economic and regulatory assumptions.

Pavement Design under Uncertain Climate Conditions

In pavement engineering, climate-related uncertainties such as increased freeze-thaw cycles or precipitation levels can degrade traditional asphalt performance. Threshold analysis can identify the critical number of freeze-thaw days per year beyond which using air-entrained concrete or polymer-modified asphalt becomes more economical than conventional designs. This is particularly important in regions experiencing rapid climate shifts or transitioning from dry to wet zones.

Construction Method Selection

Contractors and owners often choose between in-house construction and subcontracting based on cost, duration, and risk. A threshold analysis might calculate the labor cost inflation rate or project delay penalty at which subcontracting becomes preferable. This allows contractors to make forward-looking decisions based on economic scenarios rather than static cost comparisons.

In all these applications, threshold analysis provides a clear decision boundary, helping civil and environmental engineers translate uncertainty into actionable thresholds. It is especially useful when combined with other techniques, such as scenario analysis, cost-benefit analysis, or probabilistic risk assessment, to support comprehensive and resilient infrastructure decisions. By pinpointing when decisions should change, threshold analysis ensures that public and private sector investments remain both timely and justifiable under evolving conditions.

In civil engineering, threshold analysis is especially valuable for projects where a single uncertain factor has the potential to alter the optimal design or strategy. Unlike broader sensitivity methods that explore how output metrics vary, threshold analysis focuses on finding the exact value of a parameter at which one decision becomes more favorable than another. This makes it particularly suitable for scenarios involving climate uncertainty, cost escalation, or regulatory thresholds, where decision reversals are highly sensitive to one key variable.

Infrastructure systems, such as bridges, flood control structures, water treatment plants, and pavements, are often subject to discrete investment decisions (e.g., whether to upgrade, maintain, or defer). These decisions are heavily influenced by projected environmental conditions or market fluctuations. Threshold analysis helps engineers pinpoint critical conditions under which it becomes advantageous to shift strategies, thereby aligning investments with projected future states rather than current observations.

Detailed Case Studies in Threshold Decision-Making

These examples demonstrate how threshold analysis supports proactive and adaptive decision-making in civil engineering. By determining exactly when a decision flips from one alternative to another, engineers can prioritize monitoring, improve

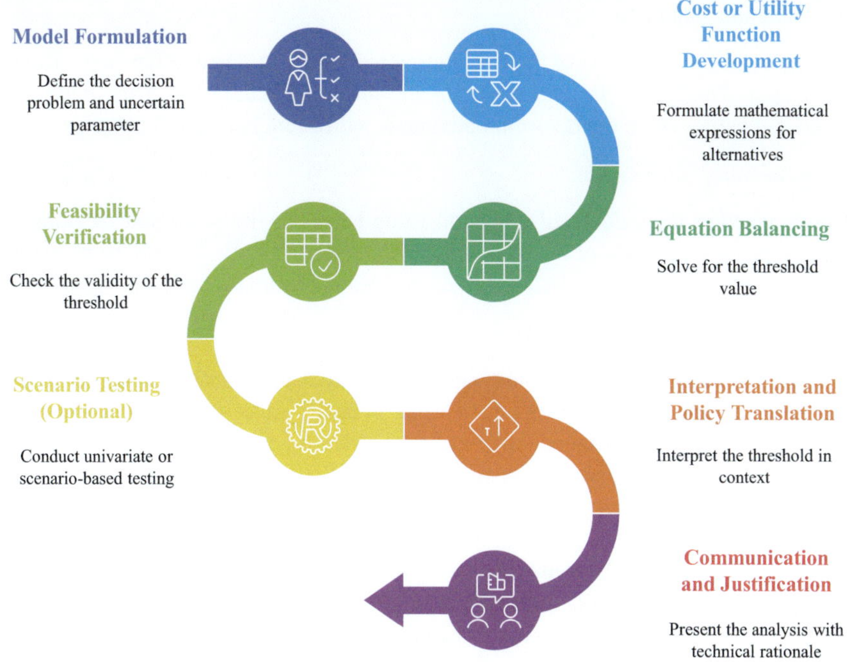

Fig. 6.28 Mastering threshold analysis: a step-by-step guide

forecasting accuracy, and communicate decisions in a clear, quantitative framework. Whether the concern is climate risk, urban expansion, material degradation, or environmental compliance, threshold analysis offers a precise and powerful way to convert uncertainty into action thresholds. The general workflow for applying threshold analysis involves the following steps (Fig. 6.28):

1. Model Formulation:

Define the decision problem clearly, specifying the competing alternatives (e.g., Option A vs. Option B) and identifying the key uncertain parameter that could influence the preferred choice. This parameter may relate to environmental conditions (e.g., flood probability, temperature extremes), economic factors (e.g., material costs, energy prices), or policy thresholds (e.g., regulatory limits, growth targets).

2. Cost or Utility Function Development:

Formulate a mathematical expression for the total cost, benefit, or utility of each alternative as a function of the uncertain parameter. These expressions should reflect life-cycle costs, maintenance, operational risks, or penalties associated with non-compliance or system failure.

3. Equation Balancing:

Set the outcome functions of the two alternatives equal to one another (e.g., Total_Cost_A = Total_Cost_B) and solve for the uncertain parameter. This yields the threshold value where the decision outcome is indifferent between the two options.

4. Feasibility Verification:

Check whether the calculated threshold value is practically valid and meaningful within the defined range of the parameter. If the threshold falls outside the plausible or acceptable range, it may indicate that one option dominates across all scenarios.

5. Scenario Testing (Optional):

If multiple variables influence the outcome, conduct univariate or scenario-based testing to examine how the threshold value behaves under different assumptions (e.g., changing discount rates, repair costs, or environmental regulations). This can add depth to the analysis and inform robustness checks.

6. Interpretation and Policy Translation:

Interpret the threshold in context: If the actual or projected value of the parameter exceeds (or falls below) the threshold, select the corresponding optimal alternative. Express the result in clear terms for decision-makers (e.g., "If annual flood probability exceeds 2.5%, build levee").

7. Communication and Justification:

Present the threshold analysis with appropriate technical rationale and visualization, if needed (e.g., a breakpoint chart). This ensures that stakeholders understand the logic behind the decision, enhancing transparency and support for future actions.

Example 1: Threshold analysis of coastal flood protection strategies

A coastal municipality is considering two flood protection strategies for a residential area vulnerable to storm surge events and sea-level rise:

- Option A: Build a permanent seawall. This option has a high up-front cost but offers full protection for the next 30 years.
- Option B: Continue with temporary flood protection using sandbags and mobile pumps. This option has lower initial costs but is subject to recurring flood risks and annual maintenance.

The uncertain parameter is the annual flood probability, denoted by p, due to climate change and evolving coastal dynamics.

The municipality wants to determine the threshold flood probability $p*$ at which Option A becomes more cost-effective than Option B over a 30 year analysis period (ignoring discounting for simplicity). Figure 6.29 introduces the threshold analysis framework, helping visualize how flood probability influences the preferred strategy between temporary flood defense and long-term seawall investment (Table 6.29).

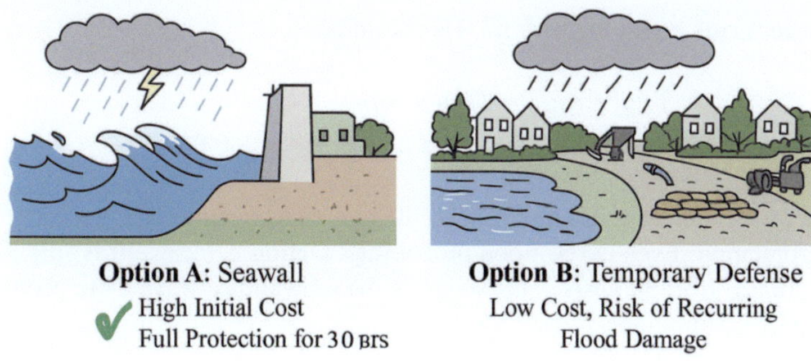

Fig. 6.29 Conceptual illustration of threshold analysis for selecting flood protection strategies

Table 6.29 Data for threshold analysis of coastal flood protection strategies

Parameter	Option A (Seawall)	Option B (Temporary Protection)
Initial cost	\$18 million	\$3 million
Annual maintenance	\$0.2 million/year	\$0.8 million/year
Expected flood damage per Event	\$0	\$12 million
Time horizon	30 years	30 years
Discounting	Ignored	Ignored

Questions to Address

1. What is the threshold value of annual flood probability p∗ where the total cost of both options is equal?
2. What happens if the actual flood probability is lower or higher than this threshold?
3. How does this analysis inform infrastructure investment and policy?
4. Is the decision sensitive to changes in flood damage or maintenance costs?

Step 1: Total Cost of Option A (Seawall)

$$\text{Total Cost}_A = \text{Initial Cost} + \text{Annual Maintenance} \times \text{Years}$$
$$Total\ Cost_A = 18 + (0.2 \times 30) = 18 + 6 = \$24M$$

Since the seawall offers full protection, no flood damage is expected.

Step 2: Total Cost of Option B (Temporary Protection)

- Initial Cost: \$3 million
- Annual Maintenance: \$0.8M/year × 30 years = \$24 million
- Expected Flood Damage per Year = p × 12 million

$$Total\ Cost_B = 3 + 24 + 30 \times 12p = 27 + 360pM\,\$$$

Step 3: Set Costs Equal to Solve for Threshold p∗

$$Total\ Cost_A = Total\ Cost_B$$
$$24 = 27 + 360p \Rightarrow p = -3/360 = -0.0083$$

This result is invalid, because probability cannot be negative.

Interpretation: Even at 0% flood probability, Option A (seawall) is still cheaper than Option B over 30 years. This suggests the seawall is always more economical under these assumptions.

Step 4: Test Alternative Assumptions

Let's test if higher seawall costs or lower damage costs would yield a realistic threshold.

Case 1: Seawall More Expensive

Suppose Option A's cost increases to $30 million:

$$Total\ Cost_A = 30 + 6 = 36 million$$

Set equal:

$$36 = 27 + 360p \Rightarrow p* = 0.025$$

Valid threshold, if flood probability is higher than 2.5% per year, the seawall becomes more cost-effective.

Case 2: Lower Flood Damage (e.g., $4M/event)

$$Expected\ flood\ damage\ over\ 30\ years = 30 \times 4p = 120p$$
$$Total\ Cost_B = 3 + 24 + 120p = 27 + 120p$$

Set equal:

$$24 = 27 + 120p \Rightarrow p = -3/120 = -0.025$$

Again, no realistic threshold—Option A dominates.

Let's keep the most reasonable case:

- Option A total cost = $36M (e.g., higher-cost seawall)
- Option B cost = 27+360

Threshold calculation:

$$36 = 27 + 360p \Rightarrow p* = 0.025\ or\ 2.5\%$$

1. What is the threshold value of annual flood probability p∗?

Under the adjusted cost scenario (more expensive seawall at \$30M initial cost), the threshold probability is:

$$p* = (36 - 27)/360 = 0.025(or 2.5\%)$$

This means that if flooding is expected more than once every 40 years, Option A (seawall) becomes more economical.

2. What happens if the actual flood probability is lower or higher than this threshold?

   - If $p < 2.5\%$ (e.g., floods occur once every 50+ years): Option B (temporary protection) is more economical.
   - If $p > 2.5\%$ (e.g., floods occur more frequently): Option A (seawall) is more economical.

   Hence, the optimal strategy depends on expected flood frequency.
3. How does this analysis inform infrastructure investment and policy?

   This analysis provides a quantitative decision rule:
   "If projected flood probability exceeds 2.5%, invest in a permanent seawall."
   It enables:

   - Forward-looking, evidence-based planning
   - Justification for proactive investments before damage occurs
   - Guidance for climate risk assessments and flood modeling efforts

4. Is the decision sensitive to changes in flood damage or maintenance costs?

   Yes, the threshold is very sensitive:

   - If expected flood damage is reduced (e.g., \$4M/event), the cost advantage of Option A increases, and no threshold exists—Option A always dominates.
   - If Option A's cost increases (e.g., from \$24M to \$36M), a valid threshold emerges.
   - If Option B's annual maintenance increases, the threshold probability shifts lower, again favoring the seawall.

Thus, the result depends critically on the balance of upfront cost vs. risk exposure.

This threshold analysis identifies 2.5% as the critical flood probability at which building a seawall becomes more cost-effective than relying on temporary protection. This helps guide investment under uncertainty and translates risk projections into actionable decisions. When combined with flood forecasting and climate modeling, threshold analysis provides a powerful tool for optimizing civil infrastructure planning in the face of environmental change.

Example 2: Threshold Analysis of Seismic Retrofit Strategies for Public Infrastructure

A city in a seismically active region is evaluating whether to retrofit a critical public building—an old municipal hospital—to meet current earthquake resistance standards. The building currently meets only minimal seismic safety criteria, making it vulnerable during moderate or large earthquakes.

The city is considering two alternatives:

- Option A: Perform a comprehensive seismic retrofit that brings the structure up to current building code standards.
- Option B: Defer retrofitting and continue routine maintenance, accepting the risk of potential damage during an earthquake.

The key uncertainty is the annual probability of a damaging earthquake, denoted as p, which is estimated using regional seismic hazard models but remains imprecise due to fault complexity.

The municipality wants to find the threshold earthquake probability above which investing in the retrofit becomes the more cost-effective and responsible option. Figure 6.30 visually outlines the decision framework for seismic retrofitting, showing how increasing earthquake probability affects the cost-effectiveness of investing in structural upgrades versus maintaining the current condition of a vulnerable public facility (Table 6.30).

Questions to Address

1. What is the threshold annual earthquake probability where both options have equal cost?
2. What happens if the actual probability is lower or higher than this threshold?
3. What implications does the result have for risk-informed investment in infrastructure?
4. How sensitive is the decision to variation in damage cost or retrofit cost?

Step 1: Total Cost of Option A (Seismic Retrofit)

$$\text{Total Cost}_\text{A} = \text{Initial Cost} + \text{Annual Maintenance} \times \text{Years}$$
$$\textit{Total Cost}_A = 12 + (0.1 \times 30) = 12 + 3 = \$15M$$

No earthquake damage expected due to full compliance with seismic design codes.

Step 2: Total Cost of Option B (Status Quo)

- Initial Cost: \$1 million
- Annual Maintenance: \$0.2M/year $\times$ 30 = \$6 M
- Expected Earthquake Damage per Year = p$\times$25 M

$$\textit{Total CostB} = 1 + 6 + 30 \times 25p = 7 + 750pM\,\$$$

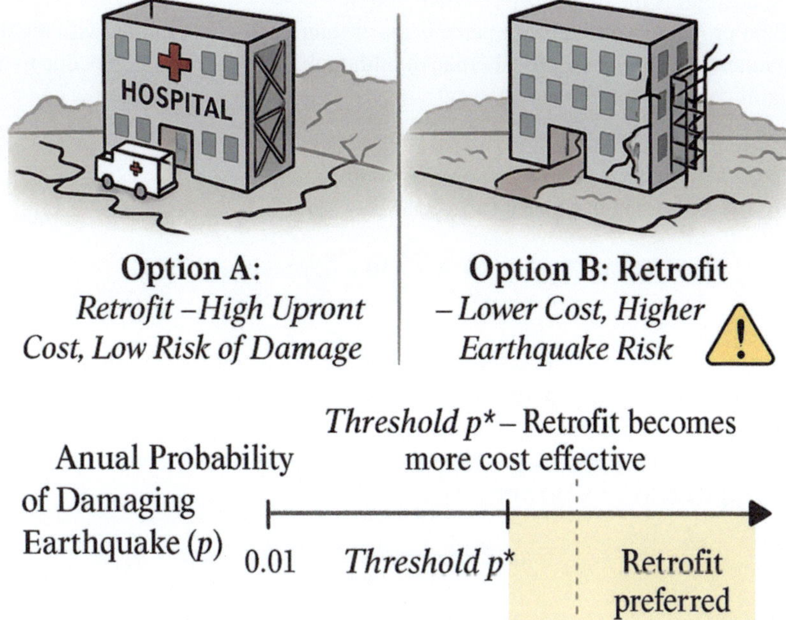

Fig. 6.30 Conceptual illustration of a threshold analysis comparing seismic retrofit versus deferred maintenance for a critical public hospital

Table 6.30 Data for threshold analysis of seismic retrofit strategies for public hospital

Parameter	Option A (Retrofit)	Option B (Status Quo)
Initial cost	\$12 million	\$1 million (routine maintenance)
Annual maintenance cost	\$0.1 million/year	\$0.2 million/year
Expected earthquake Damage	\$0 (fully retrofitted)	\$25 million per damaging event
Time horizon	30 years	30 years
Discounting	Ignored	Ignored

Step 3: Set Costs Equal to Solve for Threshold

$$Total\ Cost_A = Total\ Cost_B \Rightarrow 15 = 7 + 750p \Rightarrow p* = 0.0107 or 1.07\%$$

Step 4: Interpretation

- If the annual probability of a damaging earthquake exceeds 1.07%, then retrofitting the hospital is the more cost-effective option.
- If the probability of the earthquake is lower than 1.07%, continuing with regular maintenance is a less costly alternative (although, introduces more risk due to the significant cost per damaging event).

Step 5: Sensitivity Testing

Let's examine sensitivity to two key assumptions:

Case 1: Lower Damage Cost (e.g., \$15M)

$$Total\ Cost_B = 1 + 6 + \ (30 \times 15p) = 7 + 450p$$

Set equal:

$$15 = 7 + 450p \Rightarrow p* = 8/450 = 0.0178 or 1.78\%$$

A lower damage cost raises the threshold, making retrofitting less urgent.

Case 2: Retrofit Cost Increases (e.g., \$18M)

$$Total\ Cost_A = 18 + 3 = M21$$

Set equal:

$$21 = 7 + 750p \Rightarrow p* = 0.0187 or 1.87\%$$

A higher retrofit cost raises the threshold, reducing cost-effectiveness of Option A. Based on the results of this analysis we can discuss the questions provided in the problem statement:

1. What is the threshold value of annual earthquake probability?

   1. $p* = (15 - 7)/750 = 8/750 = 0.0107 = 1.07\%$

      If the probability of a damaging earthquake is higher than 1.07% per year (or roughly once every 93 years), Option A becomes more economical.

2. What happens if the actual earthquake probability is lower or higher than this threshold?

   - If p < 1.07%: Keep current strategy (Option B); lower cost, acceptable risk.
   - If p > 1.07%: Retrofit the hospital as Option A is cheaper and safer.

3. What implications does the result have for infrastructure investment and policy?

   This analysis gives a clear decision rule:

   "If the annual risk of earthquake damage exceeds 1.07%, invest in seismic retrofitting"

It supports:

- Proactive mitigation planning
- Justification for capital investment
- Compliance with public safety mandates
- Better allocation of limited resilience budgets

4. How sensitive is the decision to changes in cost assumptions?

- Lowering damage costs raises p∗, making retrofitting less urgent.
- Increasing retrofit costs also raises p∗.
- The decision is very sensitive to projected losses from earthquake events and cost of retrofitting.

Thus, accurate seismic risk modeling and realistic cost estimates are crucial.

Threshold analysis reveals that if the annual probability of a damaging earthquake is greater than 1.07%, then seismic retrofitting of the municipal hospital is the more cost-effective choice. This empowers city planners to make risk-informed decisions and align public safety investments with seismic hazard projections. The technique provides a transparent, quantitative foundation for infrastructure resilience planning in earthquake-prone regions.

Example 3: Threshold Analysis of Phosphorus Control Strategies in Wastewater Treatment

A regional wastewater authority is considering whether to upgrade its treatment plant with advanced nutrient removal (ANR) technology to reduce phosphorus discharge into a nearby lake. The lake is experiencing signs of eutrophication, threatening aquatic ecosystems and drinking water supplies.

The decision is between two alternatives:

- Option A: Upgrade the plant with ANR to consistently meet anticipated future discharge limits.
- Option B: Continue with standard treatment, risking non-compliance and potential fines or retrofit in the future.

The uncertain parameter is the future phosphorus discharge limit imposed by environmental regulators. The limit is expected to tighten, but its exact value is uncertain due to ongoing scientific reviews and policy negotiations.

The wastewater authority wants to determine the threshold phosphorus limit C* (in mg/L) below which it becomes more economical to upgrade now rather than risk future penalties and emergency retrofits. Figure 6.31 illustrates the wastewater authority's decision framework under regulatory uncertainty, showing how future phosphorus limits drive the threshold at which investing in advanced treatment becomes the economically preferable option (Table 6.31).

Questions to Address

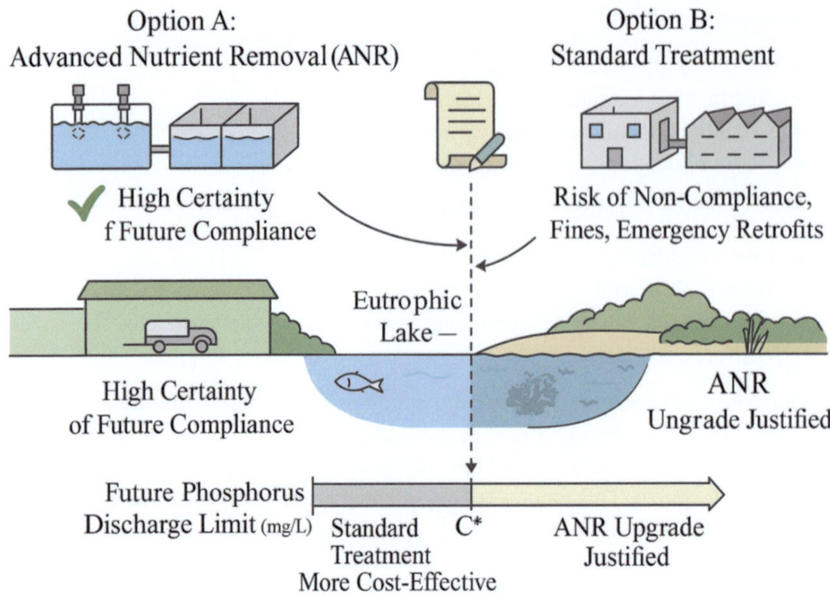

Fig. 6.31 Conceptual threshold analysis comparing advanced nutrient removal and standard wastewater treatment strategies under uncertain future phosphorus discharge limits

Table 6.31 Given data for threshold analysis of phosphorus control strategies in wastewater treatment

Parameter	Option A (Upgrade Now)	Option B (Delay and Retrofit)
Initial cost	\$10 million	\$1 million (minor upgrades)
Annual O&M cost	\$0.6 million	\$0.4 million
Retrofit cost (if forced later)	N/A	\$15 million
Non-compliance fine (per year if exceeded)	\$2 million	\$2 million
Phosphorus discharge (Option B, standard)	0.4 mg/L	0.4 mg/L
Phosphorus discharge (Option A, advanced)	0.1 mg/L	0.1 mg/L
Time horizon	20 years	20 years
Discounting	Ignored for simplicity	Ignored

1. What is the threshold phosphorus limit C* (mg/L) at which Option A becomes more cost-effective than Option B?
2. What happens if the future phosphorus limit is stricter or more lenient than the threshold?
3. How much do fines vs. retrofit costs contribute to the total cost of Option B?

4. How do changes in operation & maintenance (O&M) costs affect the decision?
5. How does this threshold analysis help support environmental compliance strategy and risk management?

Step 1: Define Total Cost of Option A (Upgrade Now)

$$Total\ Cost_A = Initial\ Cost + O\&M\ \times Years$$

$$Total\ Cost_A = 10 + (0.6 \times 20) = 22\ million\ dollars$$

Phosphorus discharge = 0.1 mg/L → Compliant under any expected regulatory scenario. No risk of fine or retrofit.

Step 2: Define Total Cost of Option B (Delay and Retrofit)
Let's define C as the future discharge limit in mg/L.

- If the regulatory limit is > 0.4 mg/L, Option B is compliant: → No fine, no retrofit:

$$Total\ Cost_B = 1 + (0.4 \times 20) = 1 + 8 = 9\ million\ dollars$$

- If the regulatory limit is ≤ 0.4 mg/L but > 0.1 mg/L, then Option B will incur annual fines for non-compliance:

$$Total\ Cost_B = 1 + (0.4 \times 20) + (2 \times 20) = 1 + 8 + 40 = 49\ million\ dollars$$

- If the regulatory limit is ≤ 0.1 mg/L, Option B must retrofit mid-way through (say, in year 10), and will incur fines until then:
  - O&M over 20 years: $0.4 \times 20 = 8$
  - Fine over 10 years: $2 \times 10 = 20$
  - Retrofit at year 10: $15M
  - Initial cost = $1M

$$Total\ Cost_B = 1 + 8 + 20 + 15 = 44\ million\ dollars$$

Step 3: Compare with Option A

- Option A total cost = $22M
- Option B total cost ranges from:
  - $9M if C > 0.4 mg/L
  - $49M if 0.1 < C ≤ 0.4 mg/L
  - $44M if C ≤ 0.1 mg/L

We want to find the threshold C* where:

$$Total\ Cost_A = Total\ Cost_B$$

Let's focus on the fines-only zone, where $0.1 < C \leq 0.4$. In this zone:

$$Cost_B = 9 + Fine\ component$$

Let the number of non-compliant years be n, with:

$$CostB = 9 + 2n(in\ millions) = 22 \Rightarrow 2n = 13 \Rightarrow n = 6.5years$$

Now convert n into a phosphorus discharge limit:

- If 6.5 years of fines occur before acceptable treatment is achieved, it implies regulation will tighten in year 13.5 (i.e., mid-life).
- Assume regulation tightens to 0.1 mg/L at year 13.5, requiring retrofit.
- So, threshold $C^*\approx 0.1$ mg/L: any stricter standard will force retrofit, making Option B more expensive.

Hence: $C^*=0.1$ mg/L
If $C \leq 0.1$ mg/L, Option A dominates.

1. What is the threshold phosphorus limit C^*?

   The threshold value is $C^*=0.1$ mg/L

   If future phosphorus standards are ≤ 0.1 mg/L, then upgrading now (Option A) becomes more cost-effective.
2. What happens if the future limit is stricter or more lenient?

   - If $C>0.4$ mg/L → No fine → Option B is cheaper ($9M)
   - If $0.1<C\leq 0.4$ mg/L → Annual fines of $2M → Option B cost rises to $49M
   - If $C\leq 0.1$ mg/L → Forced retrofit → Option B = $44M
     → In both these cases, Option A (Upgrade Now) is clearly better.
3. How much do fines vs. retrofit contribute to Option B's cost?

   - Fines alone: $40M over 20 years
   - Retrofit cost: $15M (one-time)
   - Together: $55M, showing that penalties dominate if discharge exceeds limits.

   This highlights the importance of planning before penalties accumulate.
4. How do O&M cost differences affect the threshold?

   If Option A's O&M increased from $0.6M to $0.9M/year:

   - New cost = $10 + (0.9 \times 20) =$ $28M

   → Closer to Option B with fines + retrofit ($44M), but still favorable for stricter standards.

   So, while O&M shifts the total, the threshold value is still governed by policy limit C.

5. How does this help with environmental compliance and risk management?

   This threshold analysis:

   - Ties capital investment directly to regulatory expectations
   - Helps justify early action versus waiting
   - Translates vague uncertainty into a clear trigger ($C \leq 0.1$ mg/L)
   - Supports resilience planning for long-term infrastructure strategy

Threshold analysis identifies a phosphorus discharge limit of 0.1 mg/L as the critical point where investing in advanced nutrient removal becomes more economical than delaying upgrades and risking fines or emergency retrofits. This empowers utility managers and environmental regulators to plan more strategically, minimize reactive spending, and align wastewater infrastructure investment with ecological protection goals.

Example 4: Threshold Analysis of Concrete Mix Selection in Cold Climate Construction

An engineering firm is selecting a concrete mix for the foundation of a mid-rise office building in a cold Canadian climate, where freeze-thaw durability is critical. The firm is deciding between:

- Option A: Use a standard 30 MPa concrete mix without air entrainment, which is cheaper but less resistant to freeze-thaw deterioration.
- Option B: Use a 35 MPa air-entrained concrete, with higher initial cost but significantly better long-term durability in cold regions.

The uncertain parameter is the number of annual freeze-thaw cycles at the site. This number varies annually and is expected to increase over the coming decades due to climate variability and increased temperature fluctuations.

The goal is to determine the threshold number of freeze-thaw cycles per year (N^*) beyond which air-entrained concrete (Option B) becomes more cost-effective over the structure's 50 year design life. Figure 6.32 illustrates how cold-region durability design depends on expected freeze-thaw exposure, with a threshold number of annual cycles (N^*) indicating when it becomes more economical to invest in air-entrained concrete for long-term structural performance (Table 6.32).

Questions to Address

1. What is the threshold number of freeze-thaw cycles per year N^* at which Option B becomes more economical than Option A?
2. What happens if the site experiences fewer or more freeze-thaw cycles than the threshold?
3. How much do freeze-thaw cycles amplify lifecycle maintenance costs?
4. How sensitive is the decision to initial material cost differences?
5. How can this threshold guide material selection in cold-climate concrete design?

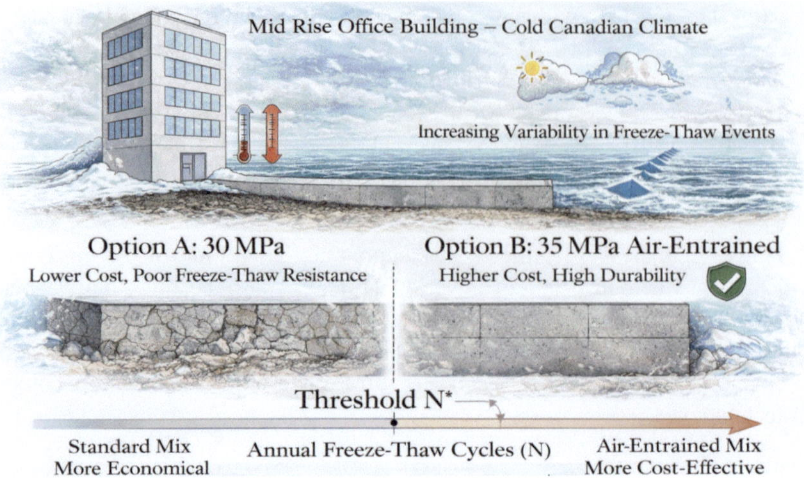

Fig. 6.32 Conceptual threshold analysis comparing standard and air-entrained concrete mixes for a cold-climate office building foundation

Table 6.32 Given data for threshold analysis of concrete mix selection in cold climate construction

Parameter	Option A (Standard concrete)	Option B (Air-entrained concrete)
Initial Cost per m^3	\$180	\$210
Annual maintenance/Repair (Base)	\$2/m^3	\$0.5/m^3
Freeze-thaw damage multiplier	$Maintenance_A$=2+0.2N	Maintenance B = 0.5 + 0.02
Volume of concrete	1,000 m^3	1,000 m^3
Design life	50 years	50 years
Discounting	Ignored	Ignored

Step 1: Calculate Total Cost of Each Option as a Function of N

Option A: Standard 30 MPa Concrete

- Initial Cost = \$180 × 1,000 = \$180,000
- Annual Maintenance = (2+0.2N) × 1,000
- Over 50 years:

$$Maintenance_A = 50 \times (2 + 0.2N) \times 1,000 = 100,000 + 10,000N$$

- Total Cost:

$$Total_A = 180,000 + 100,000 + 10,000N = 280,000 + 10,000N$$

Option B: Air-Entrained 35 MPa Concrete

- Initial Cost:$\textit{Initial Cost} = \$210 \times 1,000 = \$210,000$
- $\textit{Annual Maintenance} = (0.5 + 0.02N)(0.5 + 0.02N)(0.5 + 0.02N) \times 1,000$
- Over 50 years:

$$\textit{MaintenanceB} = 50 \times (0.5 + 0.02N) \times 1,000 = 25,000 + 1,000N$$

- Total Cost:

$$\textit{Total}\ \text{Cost}_B = 210,000 + 25,000 + 1,000N = 235,000 + 1,000N$$

Step 2: Set Total Costs Equal to Solve for Threshold N*

$$280,000 + 10,000N = 235,000 + 1,000N \Rightarrow 9,000N = -45,000 \Rightarrow N = -5$$

Invalid → Negative freeze-thaw cycles are not possible.

This implies that Option B (air-entrained) is always cheaper, regardless of the number of freeze-thaw cycles.

Step 3: Reverse Cost Structure for Realistic Threshold

Let's adjust initial cost of Option B to \$240/m^3:

- New Initial Cost = \$240,000
- Total Cost:

$$\textit{Total}\ \text{Cost}_B = 240,000 + 25,000 + 1,000N = 265,000 + 1,000N$$

Now set:

$$280,000 + 10,000N = 265,000 + 1,000N \Rightarrow 9,000N = -15,000 \Rightarrow N^* = -1.67$$

Still negative → still no realistic threshold

Try setting Option B cost at \$270/m^3:

- Initial Cost = \$270,000
- Total Cost:

$$270,000 + 25,000 + 1,000N = 295,000 + 1,000N$$

Now solve:

$$280,000 + 10,000N = 295,000 + 1,000N \Rightarrow N^* = 1.67$$

Valid threshold found.

1. What is the threshold number of freeze-thaw cycles per year N∗N^*N∗?

   When the air-entrained mix costs $270/m^3, the threshold is:

   $N^* = 1.67$ cycles per year

   So, if the site is expected to experience more than ~2 freeze-thaw cycles/year, air-entrained concrete becomes more cost-effective over 50 years.
2. What if the actual freeze-thaw frequency is lower or higher?

   - If $N<1.67$: Standard concrete is cheaper.
   - If $N>1.67$: Air-entrained concrete is better long-term.

     In cold regions with 10–50 cycles/year, air-entrained mixes are clearly justified.

3. How do freeze-thaw cycles affect total cost?

   Each additional cycle increases:

   - Option A cost by $10,000
   - Option B cost by $1,000

   Therefore, Option A is 10× more sensitive to freeze-thaw damage than Option B.
4. How sensitive is the decision to initial cost?

   - At $210/m^3 for Option B → it is always cheaper
   - At $270/m^3 → break-even at ~2 cycles/year
   - If price exceeds $280/m^3, standard mix might be viable only in warm climates

   Thus, the decision is very sensitive to both climate conditions and material cost variance.
5. How does this guide mix selection for cold climates?

   Threshold analysis shows that in any region with more than 2 freeze-thaw cycles per year, air-entrained concrete offers significant long-term savings.

   This justifies:

   - Upfront investment for increased durability
   - Preventive selection in Canadian, Nordic, or alpine regions
   - Performance-based mix design rather than cost-only decisions

Threshold analysis demonstrates that when freeze-thaw exposure exceeds approximately 2 cycles per year, air-entrained concrete is more cost-effective than standard mixes due to lower long-term maintenance and damage costs. This supports climate-adapted material selection and aligns infrastructure design with durability-based engineering practices.

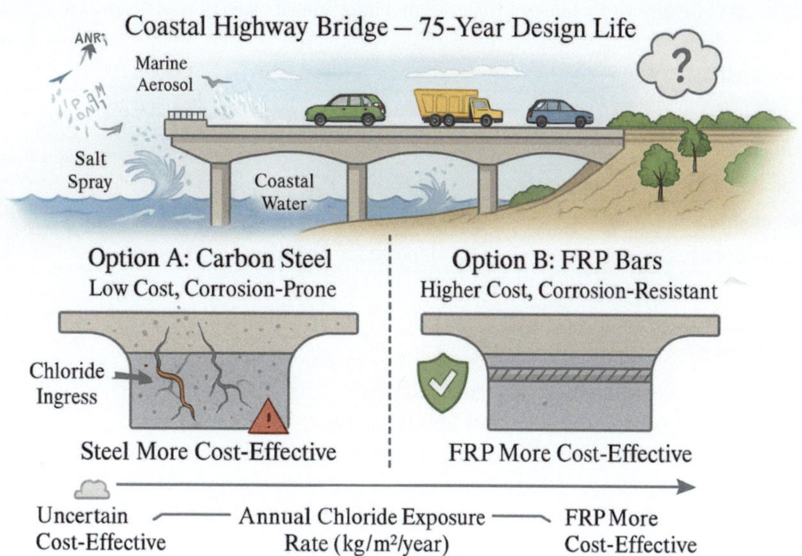

Fig. 6.33 Conceptual threshold analysis comparing conventional steel and FRP reinforcement options for a coastal bridge. As chloride exposure increases due to marine aerosols and de-icing salts, there exists a critical exposure level (E*) beyond which FRP becomes the more cost-effective long-term choice due to its superior corrosion resistance

Example 5: Threshold Analysis of Reinforcement Material Selection for Coastal Bridge Design

A coastal highway bridge project is at the material selection stage for its reinforced concrete components (e.g., girders and decks). The design team must choose between:

- Option A: Use conventional carbon steel reinforcement, which is cheaper but highly susceptible to chloride-induced corrosion.
- Option B: Use fiber-reinforced polymer (FRP) bars, which are more expensive up front but highly resistant to corrosion and chemical attack.

The uncertain parameter is the annual chloride exposure level (in kg/m^2/year), which depends on future marine aerosol intensity, de-icing salt use, and local climate evolution. Higher chloride exposure increases the deterioration rate of steel, leading to earlier maintenance or premature failure.

The goal is to determine the threshold chloride exposure rate E* beyond which using FRP becomes more cost-effective than conventional steel reinforcement over a 75 year service life. Figure 6.33 illustrates the material selection challenge in coastal bridge design, showing how increasing chloride exposure influences the lifecycle cost-effectiveness of FRP reinforcement versus traditional steel (Table 6.33).

Table 6.33 Given data for threshold analysis of reinforcement material selection for coastal bridge design

Parameter	Option A (Steel rebar)	Option B (FRP rebar)
Initial material cost (per span)	\$800,000	\$1,300,000
Annual maintenance base cost	\$10,000	\$2,000
Corrosion amplifier (per kg/m^2/year)	\$5,000	\$500
Design life	75 years	75 years
Discounting	Ignored for simplicity	Ignored

Questions to Address

1. What is the threshold chloride exposure rate E* (kg/m^2/year) at which the lifecycle cost of FRP becomes less than steel?
2. What are the relative effects of corrosion on lifecycle maintenance for steel vs. FRP?
3. What happens if chloride exposure is significantly below or above the threshold?
4. How sensitive is the decision to the initial material cost gap?
5. How can this result guide material selection policies in coastal infrastructure projects?

Step 1: Define Total Lifecycle Cost for Each Option
Let E be the annual chloride exposure rate in kg/m^2/year.
Option A: Steel Rebar

- Initial Cost: \$800,000
- Annual Maintenance:

$$Cost\ per\ year = 10,000 + (5,000 \times E)$$

- Over 75 years:

$$Total\ Maintenance = 75 \times (10,000 + 5,000E) = 750,000 + 375,000E$$

- Total Cost:

$$Total_A = 800,000 + 750,000 + 375,000E = 1,550,000 + 375,000E$$

Option B: FRP Rebar

- Initial Cost: \$1,300,000
- Annual Maintenance:

$$Cost\ per\ year = 2,000 + (500 \times E)$$

Over 75 years:

$$Total\ Maintenance = 75 \times (2,000 + 500E) = 150,000 + 37,500$$

- Total Cost:

$$Total_B = 1,300,000 + 150,000 + 37,500E = 1,450,000 + 37,500E$$

Step 2: Set Costs Equal to Find Threshold E*

$$1,550,000 + 375,000E = 1,450,000 + 37,500E \Rightarrow E = -0.296\,\text{kg/m}^2/\text{year}$$

Invalid result (negative exposure is not possible) means that FRP is already more cost-effective even at zero chloride exposure.
Step 3: Adjust Initial Costs for Realistic Threshold
Let's raise FRP cost to $1,700,000:

$$Total_B = 1,700,000 + 150,000 + 37,500E = 1,850,000 + 37,500E$$

Set equal:

$$1,550,000 + 375,000E = 1,850,000 + 37,500E \Rightarrow E = 0.889\ \text{kg/m}^2/\text{year}$$

Valid threshold found.

1. What is the threshold chloride exposure rate E∗E^*E∗?

   When FRP initial cost = $1.7M, threshold is E*=0.889 kg/m^2/year
   If the annual chloride deposition rate exceeds 0.889 kg/m^2, FRP becomes the better long-term choice.

2. How does corrosion affect lifecycle maintenance?

   Each 1 kg/m^2/year of chloride adds:

- $375,000 to steel lifecycle cost
- $37,500 to FRP lifecycle cost.

   Steel is 10× more sensitive to chloride-induced deterioration.

3. What happens if chloride exposure is below or above E*?

   - If E<0.889: Steel is cheaper.
   - If E>0.889: FRP is more cost-effective.

   In aggressive coastal environments (e.g., >1 kg/m^2/year), FRP clearly dominates.

4. How sensitive is the decision to material cost differences?

A $200,000 increase in FRP cost (from $1.5M to $1.7M) shifts the threshold from negative to nearly 1 kg/m²/year.

Thus, material pricing and market trends significantly influence lifecycle decisions.

5. How does this guide material policy in coastal projects?

This threshold provides an actionable rule:
"Use FRP reinforcement when chloride exposure exceeds 0.89 kg/m²/year."
It supports:

- Durable infrastructure planning
- Reduced long-term repair budgets
- Clear justification in design specifications for non-corroding alternatives.

This threshold analysis shows that in environments with moderate to high chloride exposure, FRP reinforcement becomes more economical than steel over a 75 year bridge lifespan. The method helps translate material performance and exposure data into practical, cost-optimized decisions in civil engineering materials selection.

Example 6: Threshold Analysis of Reservoir Expansion for Drought Resilience

A regional water utility is evaluating whether to expand an existing surface water reservoir to improve drought resilience and meet future demand during dry periods. Two alternatives are under consideration:

- Option A: Expand the reservoir now, increasing storage capacity by 30%. This option involves a high initial investment but provides buffer storage for future dry years and growing population demand.
- Option B: Delay expansion, relying instead on water restrictions and emergency groundwater pumping during drought years.

The uncertain parameter is the annual probability of a critical drought year, denoted as p. Due to climate change, this probability is expected to increase, but the exact future frequency remains uncertain.

The objective is to determine the threshold drought probability p* above which reservoir expansion becomes more cost-effective than continuing with emergency drought response. Figure 6.34 shows how the increasing likelihood of critical droughts affects the decision to expand reservoir capacity versus continuing with short-term emergency water strategies, highlighting the threshold probability p* where long-term investment becomes optimal (Table 6.34).

Questions to Address

- What is the threshold annual drought probability p* where the total cost of both options is equal?

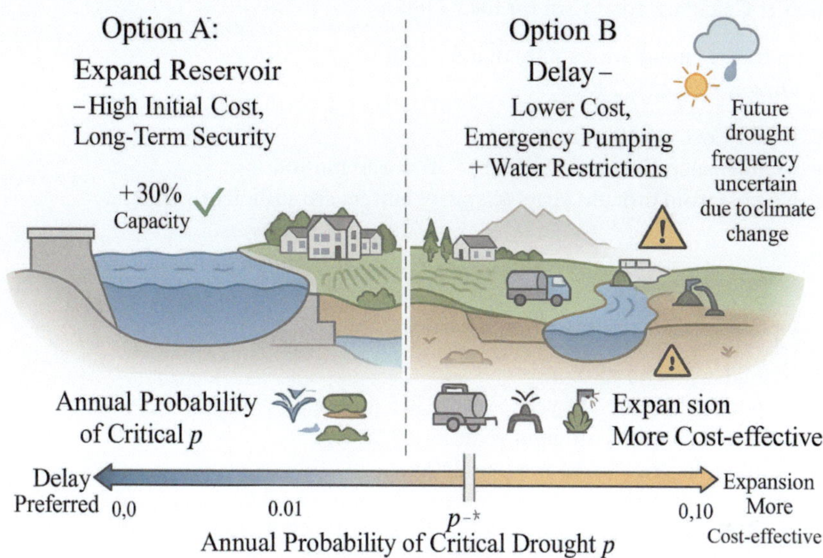

Fig. 6.34 Threshold analysis comparing proactive reservoir expansion and reactive drought response strategies. As the probability of severe drought years increases due to climate change, a critical threshold (p*) marks the point where investing in reservoir capacity becomes the more cost-effective and resilient choice

Table 6.34 Given data threshold analysis of reservoir expansion for drought resilience

Parameter	Option A (Expand reservoir)	Option B (Status Quo + Emergency response)
Initial cost	\$60 million	\$5 million (minor system improvements)
Annual maintenance cost	\$1 million	\$0.5 million
Cost per drought year	\$0 (fully protected)	\$8 million (water trucking + well pumping)
Time jorizon	30 years	30 years
Discounting	Ignored	Ignored

- What happens if droughts occur less or more frequently than the threshold?
- How does this support climate-resilient infrastructure investment?
- How sensitive is the decision to assumptions about drought cost per year or maintenance costs?
- What are the implications for water policy, demand management, and long-term planning?

Step 1: Calculate Total Cost for Each Option

Let p be the annual probability of a drought year.
Option A: Expand Reservoir

- Initial Cost = \$60 million
- Maintenance Cost = \$1M/year × 30 = \$30 million
- No cost from drought years (capacity buffers are sufficient)

$$Total\ Cost_A = 60 + 30 = 90\ million\ dollars$$

Option B: Delay + Emergency Measures

- Initial Cost = \$5 million
- Maintenance = \$0.5M/year × 30 = \$15 million
- Expected cost from drought years = p × 8 Mp
- → Over 30 years = 30 × 8 p = 240 p

$$Total\ Cost_B = 5 + 15 + 240p = 20 + 240p\ million\ dollars$$

Step 2: Solve for Threshold p*

Set $Total_A = Total_B$

$$90 = 20 + 240p \Rightarrow p* = 0.2917 or 29.17\%$$

1. What is the threshold annual drought probability p*?

$$p^* = 0.2917 \text{ or } 29.17\%$$

   If the probability of a drought year exceeds 29.17% (i.e., roughly 1 in 3 years), expanding the reservoir now is more cost-effective.
2. What if droughts occur less or more frequently than p∗p^*p∗?
   - If p<0.29: Rely on current system + emergency measures → lower total cost
   - If p>0.29: Expand the reservoir → saves money and ensures water reliability

   This supports data-driven investment based on projected climate trends.
3. How does this support climate-resilient infrastructure planning?

   Threshold analysis:

   - Converts climate uncertainty (future drought risk) into an investment rule
   - Justifies proactive investment when drought probability crosses a critical threshold
   - Aligns water supply design with long-term climate models

4. How sensitive is the decision to assumptions about drought cost or maintenance?
   - If drought cost increases (e.g., to \$12M/year):

$$30 \times 12p = 360p \Rightarrow p^* = 19.4\%$$

   → Reservoir becomes better even with fewer droughts
   - If Option A maintenance increases (e.g., to \$1.5M/year):

$$\begin{aligned} &\textit{Total Cost} = 60 + 45 = 105 \\ &105 = 20 + 240p \Rightarrow p^* = 35.4\% \end{aligned}$$

   Takeaway: Higher drought severity or lower Option A maintenance cost strengthens case for expansion.
5. What are the broader planning implications?
   - Threshold p* = 29.17% provides a decision boundary
   - Helps prioritize climate adaptation funding
   - Offers a way to communicate risk to stakeholders and justify ratepayer investments
   - Encourages integrated water management, combining infrastructure and demand-side solutions

This threshold analysis shows that when the annual probability of severe drought exceeds approximately 29%, it is more cost-effective to expand reservoir capacity now than to rely on emergency responses and water restrictions. This supports climate-informed water resource management, improves long-term system reliability, and ensures investments are aligned with evolving environmental risks.

Example 7: Threshold Analysis of Construction Method Selection Under Labor Cost Uncertainty

A construction firm is planning a 25-story residential high-rise in a densely populated urban area. Due to tight timelines, space constraints, and rising labor costs, they must choose the most cost-effective construction method.

Three mutually exclusive methods are available:

- Option A: Traditional cast-in-place concrete construction. Lower material costs but high labor demand and longer duration.
- Option B: Precast concrete panels, reducing labor on site and time but requiring factory production and crane logistics.
- Option C: Modular steel units, which are prefabricated and assembled rapidly but have high initial costs and require precise logistics.

The uncertain parameter is the average daily labor cost, L, due to inflation and labor shortages.

Fig. 6.35 Threshold analysis comparing three high-rise construction methods under labor cost uncertainty

Table 6.35 Data for Threshold Analysis of Construction Method Selection Under Labor Cost Uncertainty

Parameter	Option A (Cast-in-place)	Option B (Precast concrete)	Option C (Modular steel)
Base material cost	\$8 million	\$10 million	\$13 million
On-site labor days	800	500	250
Labor cost per day (variable)	L	L	L
Total cost formula	8 M + 800 L	10 M + 500 L	13 M + 250 L

The firm wants to determine the threshold labor cost values L_1^* and L_2^* that define which construction method becomes optimal under increasing labor cost scenarios. Figure 6.35 illustrates how construction method selection for an urban high-rise depends on labor cost scenarios, with threshold values L_1* and L_2* guiding when to shift from cast-in-place to precast or modular approaches (Table 6.35).

Questions to Address

1. What are the threshold labor cost values L_1^* and L_2^* that define when each option becomes cost-optimal?
2. For what range of L is each option the cheapest?
3. How do changes in labor market volatility impact the selection?
4. How does this help optimize cost-time trade-offs in urban construction?
5. What is the decision implication if labor costs increase by 15% per year?

Table 6.36 Optimal construction method as a function of daily labor cost

Labor cost, L	Optimal option
L<6,666.67	Option A (Cast-in-place)
6,666.67≤L<12,000	Option B (Precast)
L≥12,000	Option C (Modular)

Step 1: Define Total Cost Equations
Let L be the daily labor cost (in dollars):

- Option A:

$$C_A = 8,000,000 + 800L$$

- Option B:

$$C_B = 10,000,000 + 500L$$

- Option C:

$$C_C = 13,000,000 + 250L$$

Step 2: Find Thresholds Between Pairs of Options
Compare Option A vs. Option B

$$8,000,000 + 800L = 10,000,000 + 500L \Rightarrow \mathbf{L_1^*} = 6,666.67$$

So, if labor cost L<$6,666.67, then Option A is cheaper than B.
Compare Option B vs. Option C

$$10,000,000 + 500L = 13,000,000 + 250L \Rightarrow \mathbf{L_2^*} = 12,000$$

So, if labor cost L<$12,000, then Option B is cheaper than C.
Compare Option A versus Option C

$$8,000,000 + 800L = 13,000,000 + 250L \Rightarrow 550L = 5,000,000 \Rightarrow L = 9,090.91$$

This serves as a check: between A and C, Option A is better for labor costs below $9,090.91.

Step 3: Identify Optimal Strategy per Labor Cost Range (Table 6.36)

1. What are the threshold labor cost values?

   - Between Option A and B: L_1^*=$6,666.67

- Between Option B and C: L_2^*=$12,000.00

These define three distinct decision zones.

2. For what range is each option preferred?

- Option A: for labor costs below $6,666.67
- Option B: between $6,666.67 and $12,000
- Option C: for labor costs above $12,000

This allows planning based on labor market conditions.

3. How does labor market volatility affect choice?

- In regions with stable or low labor costs, cast-in-place remains optimal.
- In volatile or high-wage urban markets, precast or modular methods reduce labor exposure.
- Rapid wage inflation would push decisions toward Option C.

4. How does this help with cost-time trade-offs?

- Cast-in-place = cheapest, slowest
- Precast = moderate cost and faster
- Modular = fastest, highest upfront cost

Threshold analysis allows the cost of speed to be measured against labor inflation risk.

5. What if labor costs grow 15% annually?

Assuming current L=$6,000

- In 5 years:

$$L = 6,000 \times (1.15)^5 \approx 6,000 \times 2.01 = 12,060$$

Thus, Option C becomes optimal within 5 years—justifying long-term investment in modular capability.

This threshold analysis identifies that the optimal construction method shifts depending on the daily labor cost. Below $6,666/day, cast-in-place is cheapest; between $6,667 and $12,000, precast is ideal; and beyond $12,000/day, modular construction dominates. This helps decision-makers align construction methods with market conditions, project timelines, and strategic risk management in large-scale urban development.

Example 8: Threshold Analysis of Groundwater Remediation Strategies Under Contaminant Severity

An environmental consultant must select a remediation strategy to address chlorinated solvent contamination in groundwater at a former industrial site. The consultant is evaluating three different remediation methods:

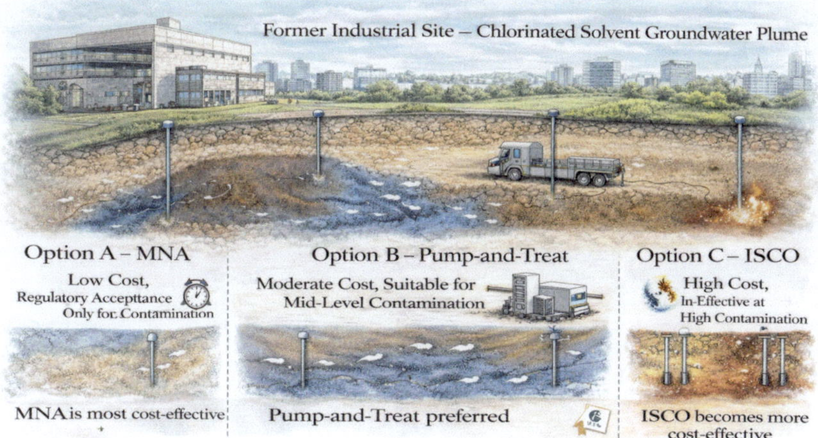

Fig. 6.36 Threshold analysis comparing groundwater remediation strategies under varying chlorinated solvent concentrations (C)

- Option A: Monitored Natural Attenuation (MNA)—minimal intervention, relying on natural degradation.
- Option B: Pump-and-Treat—active groundwater extraction with surface treatment.
- Option C: In-situ Chemical Oxidation (ISCO)—rapid but expensive destruction of contaminants in place.

The uncertain parameter is the initial contaminant concentration, C, in mg/L. The exact concentration varies spatially and temporally across the site and will influence the effectiveness and regulatory acceptance of each method.

The goal is to determine threshold concentration values C_1^* and C_2^* that define which remediation strategy is the most cost-effective depending on contaminant severity. Figure 6.36 illustrates how contaminant severity influences the selection of remediation technologies, showing the thresholds C_1* and C_2* that mark when to shift from passive to active and aggressive treatment methods for cost-effective groundwater cleanup (Table 6.37).

Questions to Address

1. What are the threshold contaminant concentration values C_1^* and C_2^* where the optimal remediation method changes?
2. Which method is most cost-effective for low, moderate, and high concentrations?
3. How does the risk of regulatory non-compliance affect the total cost?
4. How sensitive is the choice to assumptions about treatment time or penalties?
5. What guidance does this provide for adaptive site management?

Table 6.37 Data for Threshold Analysis of Groundwater Remediation Strategies Under Contaminant Severity

Parameter	Option A (MNA)	Option B (Pump and Treat)	Option C (ISCO)
Base cost (fixed)	\$300,000	\$500,000	\$1,000,000
Operating cst (per year)	\$30,000	\$50,000	\$20,000
Time horizon	10 years	10 years	10 years
Penalty or re-treatment cost if $C > C_{limit}$	\$500,000 if C>1.0	\$0	\$0
Effectiveness range	Up to 1 mg/L	Up to 5 mg/L	Any level

Step 1: Calculate Total Cost for Each Option

Option A: MNA

- Base cost: \$300,000
- Annual O&M: \$30,000/year × 10 = \$300,000
- If $C \leq 1.0$: No penalty

$$Total_A = 300,000 + 300,000 = 600,000$$

- If C>1.0: Regulatory rejection → \$500,000 re-treatment cost added

$$Total_A = 600,000 + 500,000 = 1,100,000$$

Option B: Pump-and-Treat

- Base cost: \$500,000
- Annual O&M: \$50,000/year × 10 = \$500,000
- Total:

$$Total_B = 500,000 + 500,000 = 1,000,000$$

Applies up to 5 mg/L contaminant level; no penalties expected.

Option C: ISCO

- Base cost: \$1,000,000
- O&M: \$20,000/year × 10 = \$200,000
- Total:

$$Total_C = 1,000,000 + 200,000 = 1,200,000$$

Effective at any concentration; no risk of failure.

Table 6.38 Decision rule summary for groundwater remediation strategies

Contaminant concentration C	Recommended remediation strategy
$C \leq 1.0$ mg/L	Option A: MNA
$1.0 < C \leq 5.0$ mg/L	Option B: Pump-and-Treat
$C > 5.0$ mg/L	Option C: ISCO

Step 2: Identify Threshold Concentration Values

Compare Option A versus Option B

- At C≤1.0: MNA = \$600K, Pump-and-Treat = \$1M → MNA preferred
- At C>1.0: MNA = \$1.1M, Pump-and-Treat = \$1M → Pump-and-Treat preferred

Threshold 1:

$$C_1^* = 1.0 \text{ mg/L}$$

Compare Option B versus Option C

- Both options are always valid (no penalty), but differ by cost:
  - Pump-and-Treat = \$1,000,000 (effective up to 5 mg/L)
  - ISCO = \$1,200,000 (effective at any level)

Thus, Option C is only chosen when C>5 mg/L (when Pump-and-Treat may underperform or fail).

Threshold 2 (Table 6.38):

$$C_2^* = 5.0 \text{ mg/L}$$

1. What are the threshold values?
   - C_1^* = 1.0 mg/L → above this, MNA incurs a penalty.
   - C_2^* = 5.0 mg/L → above this, Pump-and-Treat may not work effectively.
2. Most cost-effective options by concentration range:
   - Low contamination (<1.0 mg/L): MNA is cheapest and acceptable.
   - Moderate (1.0–5.0 mg/L): Pump-and-Treat avoids penalties and performs reliably.
   - High (>5.0 mg/L): Only ISCO is effective and regulatory-compliant.
3. How does compliance risk affect total cost?
   - MNA is the cheapest only if no penalties are incurred.
   - At C > 1.0, the \$500,000 penalty makes it more expensive than Pump-and-Treat.
   - Compliance rules directly influence strategy selection.

4. Sensitivity to assumptions:
   - If penalty increases to \$700,000 → MNA cost = 1.3 M, MNA becomes even less favorable.
   - If ISCO cost drops by 20% → total = \$960,000, making it competitive with Pump-and-Treat even for C ≈ 4–5 mg/L.
5. How does this support adaptive site management?
   - Provides concentration-based decision thresholds.
   - Helps stage interventions:
     - MNA in clean zones,
     - Pump-and-Treat in moderately contaminated areas,
     - ISCO in hot spots.
   - Enables cost-optimized remediation zoning and clear communication with regulators.

Threshold analysis shows that:

- MNA is ideal when contamination is low (<1 mg/L),
- Pump-and-Treat is optimal for moderate levels (1–5 mg/L),
- ISCO is necessary at high concentrations (>5 mg/L).

This analysis transforms environmental uncertainty into a clear, concentration-based remediation plan, enabling targeted investment and regulatory compliance across heterogeneous contaminated sites.

Example 9: Threshold AnalysisThreshold analysis of Pavement Maintenance Strategies Under Traffic Growth UncertaintyUncertainty

A regional transportation agency is planning a 25 year pavement management strategy for a newly developed arterial road. The decision centers on selecting the most cost-effective long-term approach for pavement maintenance and rehabilitation (M&R), given uncertainty in future traffic loading.

Three strategies are being considered:

- Option A: Routine maintenance only (e.g., patching and crack sealing). Lowest initial cost but requires frequent interventions under high traffic.
- Option B: Periodic resurfacing every 10 years. Moderate initial cost with balanced performance.
- Option C: Full-depth reconstruction with heavy-duty materials, high upfront cost but low degradation under traffic.

The uncertain parameter is the annual average daily traffic (AADT) growth rate rrr, expressed as a percentage. This affects the rate of pavement deterioration and the frequency of major repairs.

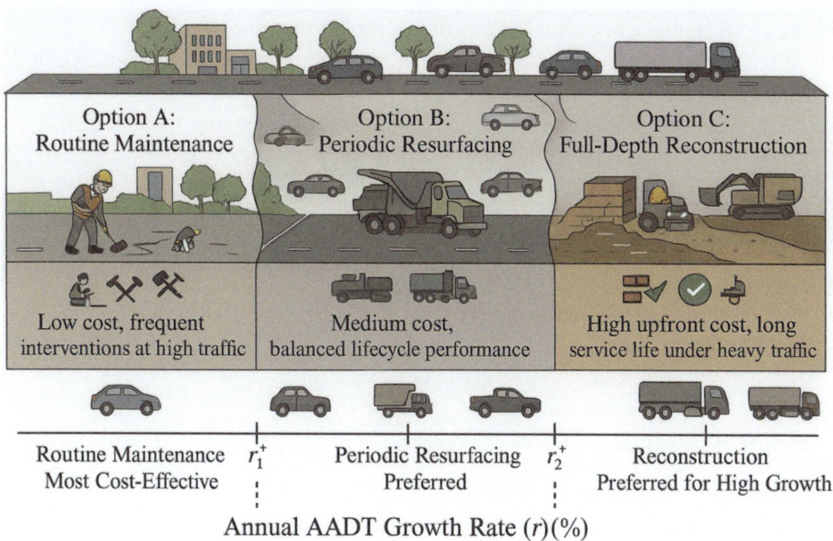

Fig. 6.37 Threshold analysis comparing pavement M&R strategies based on projected traffic growth rates. As AADT increases, the most cost-effective strategy shifts from routine maintenance to resurfacing, and eventually to full-depth reconstruction. Thresholds r_1* and r_2* mark the switching points for long-term cost efficiency

Table 6.39 Given data for threshold analysis of pavement maintenance strategies under traffic growth uncertainty

Parameter	Option A (Routine Maintenance)	Option B (Periodic Resurfacing)	Option C (Reconstruction)
Initial cost	\$0.5 million	\$1.5 million	\$4 million
Cost increase per 1% AADT growth	\$0.12M per %	\$0.08M per %	\$0.02M per %
Maintenance interval	Annual	Every 10 years	Every 25 years
Design life	25 years	25 years	25 years
Discounting	Ignored	Ignored	Ignored

The objective is to determine the threshold traffic growth rates r_1^* and r_2^* that separate the most cost-effective strategies. Figure 6.37 illustrates the traffic-dependent selection of pavement strategies, emphasizing the threshold growth rates (r_1* and r_2*) at which a shift in approach becomes optimal based on lifecycle performance and deterioration trends (Table 6.39).

Questions to Address

1. What are the threshold AADT growth rates r_1^* and r_2^* where the most cost-effective strategy changes?
2. Which strategy is optimal at low, medium, and high growth rates?

3. How do changes in traffic forecasting influence the result?
4. What is the impact of higher maintenance burden in routine maintenance under high growth?
5. How does this help with resilient and adaptive road network management?

Step 1: Express Total Cost as a Function of Growth Rate r

Let r = AADT growth rate (in % per year)

Option A (Routine Maintenance)

- Base Cost = \$0.5 M
- Deterioration cost increases by \$0.12 M per % of AADT growth

$$C_A = 0.5 + 0.12r$$

Option B (Resurfacing Every 10 Years)

- Base Cost = \$1.5M
- Traffic growth impact = \$0.08 M per % of AADT growth

$$C_B = 1.5 + 0.08r$$

Option C (Full-Depth Reconstruction)

- Base Cost = \$4 M
- Very low sensitivity to traffic growth = \$0.02M per %

$$C_C = 4 + 0.02r$$

Step 2: Solve for Threshold Points

Compare A versus B:

$$0.5 + 0.12r = 1.5 + 0.08r \Rightarrow r_1^* = 25\%$$

Compare B versus C:

$$1.5 + 0.08r = 4 + 0.02r \Rightarrow r_2^* = 41.67\%$$

Check A versus C directly:

$$0.5 + 0.12r = 4 + 0.02r \Rightarrow r = 35\%$$

Table 6.40 Decision rule summary pavement maintenance strategies

AADT growth rate r (%)	Most cost-effective strategy
$r < 25$	Option A: Routine maintenance
$25\% \leq r < 41.67$	Option B: Periodic resurfacing
$r \geq 41.67$	Option C: Full reconstruction

This confirms that Option C only becomes cheaper than A above 35% growth (Table 6.40).

1. What are the threshold growth rates?
   - $r_1^* = 25\%$: Routine → Resurfacing
   - $r_2^* = 41.67\%$: Resurfacing → Reconstruction
2. Strategy by AADT growth range:
   - Low growth (<25%): Routine maintenance is most economical.
   - Moderate growth (25–41.67%): Resurfacing balances cost and durability.
   - High growth (>41.67%): Reconstruction is justified due to high wear.
3. How does traffic forecasting influence the result?
   - Underestimating traffic leads to under-designed solutions → high life-cycle costs
   - Overestimating may lead to over-investment

   → Accurate projections are critical for cost-effective pavement management
4. How does rising traffic increase burden on Option A?
   - Routine maintenance has highest growth sensitivity ($0.12 M/%)
   - At 50% growth:

$$C_A = 0.5 + 0.12 \times 50 = \$6.5M$$
$$C_C = 4 + 0.02 \times 50 = \$5M$$

   → At high growth, routine maintenance becomes most expensive due to frequent patching and disruptions.
5. How does this support resilient infrastructure?
   - Thresholds help match strategy to growth scenarios
   - Enables adaptive planning:
     - Design flexible contracts for upgrade if traffic exceeds 25%
     - Prepare for phased upgrades as demand grows
   - Improves investment timing and avoids premature failure

This threshold analysis shows that:

- Below 25% growth, simple maintenance suffices.
- Between 25 and 41.67%, resurfacing strikes the best cost-durability balance.
- Beyond 41.67%, investing in full reconstruction is most efficient.

This framework helps road authorities align pavement strategies with actual usage trends, reducing long-term costs while maintaining performance under growing traffic demands.

6.2 Scenario-Based Decision Analysis

In the previous section, we explored a range of sensitivity analysis techniques—including one-way, two-way, and threshold analysis—which are powerful tools for understanding how variations in specific input parameters (such as cost, probability, or performance metrics) influence the outcomes of engineering decisions. These techniques are especially useful when the model structure is well-defined and when uncertainties can be reasonably quantified within known ranges.

However, in many real-world civil engineering problems, uncertainty goes beyond what can be captured through numerical variation alone. In situations involving long-term planning, climate change, demographic shifts, technological disruption, or evolving policy landscapes, we often face deep uncertainty—where not only the values of key variables are unknown, but also the relationships among variables or even the future context in which a decision will be made.

In this section, we introduce Scenario-Based Decision Analysis (SBDA), a strategic tool designed to address such complexity and ambiguity. Rather than varying parameters within a single model, SBDA evaluates decision alternatives across multiple, plausible, internally consistent future scenarios. Each scenario represents a distinct "world" that could emerge based on different assumptions about external drivers—such as environmental change, economic development, or regulatory shifts. Each scenario is treated as an internally consistent narrative or quantitative set of assumptions about how the world might evolve. These are not predictions or probabilistic forecasts, but exploratory constructs that help decision-makers test the performance of strategies under contrasting conditions. The aim is to identify options that are robust—performing reasonably well across all scenarios—or adaptive, allowing future adjustment as the real world unfolds.

SBDA is especially useful in civil engineering where long-term infrastructure investments must contend with uncertainties such as future rainfall intensity under climate change, sea level rise projections, demand growth or urban expansion, and regulatory or budgetary changes. By exploring how choices play out across multiple scenarios, engineers can better prepare for a wide range of possible futures, rather than optimizing for a single expected outcome that may never materialize.

6.2.1 *Why is Scenario-Based Decision Analysis Important in Civil Engineering?*

Civil engineering projects often have long time horizons, high capital investments, and significant consequences for public safety, environmental integrity, and societal wellbeing. Decisions about infrastructure, whether concerning flood defenses, wastewater systems, urban development, or transportation networks, must account for future conditions that cannot be precisely predicted. Traditional sensitivity analysis, while valuable, assumes a degree of predictability and control that is not always present in complex, uncertain systems.

SBDA offers a complementary approach by enabling engineers and planners to explore how decisions perform across a range of possible future environments, rather than focusing solely on variability in individual input parameters. The main advantages of SBDA are illustrated in Fig. 6.38. This method helps identify robust solutions that, while not necessarily optimal in any single scenario, perform reasonably well across multiple futures. It also enhances understanding of the vulnerabilities and trade-offs associated with each decision alternative, supporting the development of adaptive or flexible strategies that can evolve over time as future conditions become clearer.

Fig. 6.38 Key benefits of scenario-based decision analysis in engineering and planning

6.2.2 What is Scenario-Based Decision Analysis?

SBDA is a decision-support methodology that examines how different courses of action (e.g., design options, policy interventions, investment strategies) perform under a set of well-defined, contrasting scenarios. These scenarios may be constructed using expert judgment, stakeholder input, model projections, or historical analogs, and are typically qualitative or semi-quantitative in nature.

SBDA is built on several core features that distinguish it from traditional decision-making approaches. The foundational characteristics of SBDA are summarized in Fig. 6.39, highlighting its scenario evaluation, performance comparison, robustness criteria, and focus on long-term adaptability and resilience. The process of SBDA begins with the development of scenarios, each representing a possible future shaped by specific assumptions about key drivers such as climate variables, demographic trends, or regulatory shifts. Performance comparison is then carried out by estimating the outcomes of each decision alternative under every scenario, capturing a wide range of potential conditions. Rather than relying solely on expected values, SBDA uses decision criteria based on robustness metrics—such as minimax, minimax regret, or satisficing—to evaluate options. It should be noted that minimax selects the option that minimizes the worst-case outcome across all scenarios, minimax regret minimizes the maximum loss relative to the best scenario-specific outcome, and satisficing identifies alternatives that meet predefined acceptable performance thresholds rather than optimizing outcomes. Finally, SBDA emphasizes adaptability and resilience, promoting strategies that can either withstand adverse conditions or be adjusted over time as uncertainties evolve.

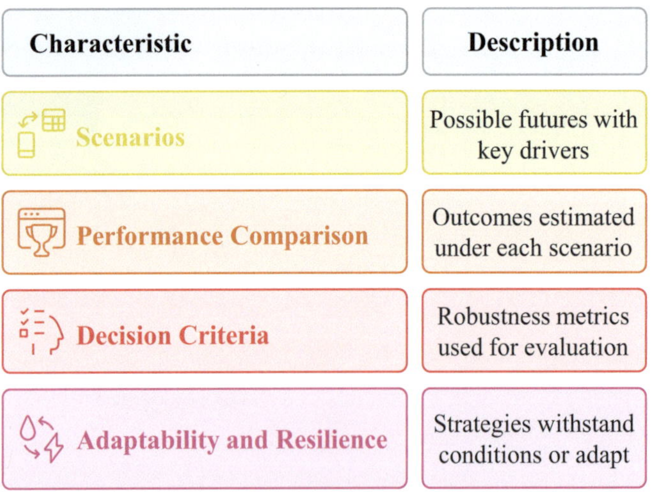

Fig. 6.39 Key characteristics of scenario-based decision analysis

SBDA is particularly useful in risk-prone or climate-sensitive projects, such as, coastal and riverine flood management, urban drainage and stormwater design, large-scale infrastructure development, and environmental and resource sustainability planning.

While sensitivity analysis explores how variations in input parameters affect the outcomes within a single model framework, SBDA goes a step further by testing the stability of decisions across a range of possible futures or model structures. Rather than focusing solely on parameter uncertainty, SBDA considers structural, environmental, or contextual uncertainties—such as climate shifts, policy changes, or socioeconomic developments—that may fundamentally alter the decision landscape. In this way,

- Sensitivity analysis answers:
  - *"What happens if this variable changes?"*
- Scenario-based analysis answers:
  - *"What happens if the world changes in this way?"*

In this way, SBDA complements sensitivity analysis by addressing deeper uncertainties and helping identify strategies that remain effective or adaptable across diverse long-term scenarios.

Both tools are valuable and often complementary in decision-making under uncertainty. Sensitivity analysis is best suited for situations where uncertainties are quantifiable, allowing decision-makers to explore how variations in specific parameters influence outcomes within a single model. In contrast, SBDA is more appropriate when uncertainties are structural or strategic, involving potential shifts in system behavior, policy environments, or long-term drivers. By combining both methods, analysts can capture a broader spectrum of uncertainty—from fine-tuned parameter changes to fundamentally different future conditions.

6.2.3 The Application of SBDA in Real-world Civil Engineering Problems

6.2.3.1 Flood Protection Strategy in a Coastal City

To ensure long-term flood resilience, a coastal city is evaluating three infrastructure investment options for protection against sea-level rise and increasing storm surge risks over a 30-year project horizon. The options, detailed in Table 6.41, differ in their design approach, upfront capital investment, and ongoing maintenance requirements.

To account for future uncertainty, three **climate scenarios** are defined (Table 6.42). These range from a conservative sea-level rise projection to a high-end scenario with more frequent and severe storm surges, based on mid-century (2055) forecasts. Figure 6.40 illustrates the strategic shift in flood protection planning based on

Table 6.41 The municipality's infrastructure options for flood protection

Option	Description	Initial cost ($M)	Maintenance ($M/year)
A	Reinforced levee system	100	2
B	Coastal wetland restoration	70	1.5
C	Hybrid system (levee + wetlands)	130	2.5

Table 6.42 Projected sea-level rise and storm surge frequencies by 2055 under different climate scenario

Scenario	Description	Sea-level rise by 2055	Storm surge frequency
S1	Conservative climate outlook	+0.3 m	1-in-100 year
S2	Moderate warming	+0.6 m	1-in-50 year
S3	High-end sea-level rise	+1.0 m	1-in-20 year

projected sea-level rise and storm surge severity. As climate risks intensify from S1 to S3, the hybrid infrastructure approach becomes increasingly justified in terms of long-term lifecycle costs.

Hydrodynamic modeling was used to estimate total lifecycle costs for each infrastructure option under each climate scenario. These costs include initial construction, long-term maintenance, and projected flood damages, as summarized in Table 6.43.

To identify the most suitable option under uncertainty, we apply two widely used decision-making criteria:

- Minimax Rule (Robustness Criterion):
  - This rule selects the option with the *lowest worst-case cost*. Looking at the total lifecycle costs across the three climate scenarios, the maximum cost for each option is:

    Option A: $230 M
    Option B: $310 M
    Option C: $220 M

  Since Option C has the *lowest maximum cost*, it is selected as the most robust choice under the Minimax rule.
- Minimax Regret Rule (Opportunity Loss Criterion):
  - A regret matrix is created by calculating the difference between each option's cost and the *lowest cost* in each scenario. The maximum regret for each option is then calculated as Table 6.44. According to the Minimax Regret rule, the best choice is the option with the *lowest maximum regret*, which is Option A.

By comparing both the robustness and regret-based decision criteria, planners can assess trade-offs between performance under worst-case conditions and potential for missed opportunities, supporting a more resilient and informed flood protection investment strategy.

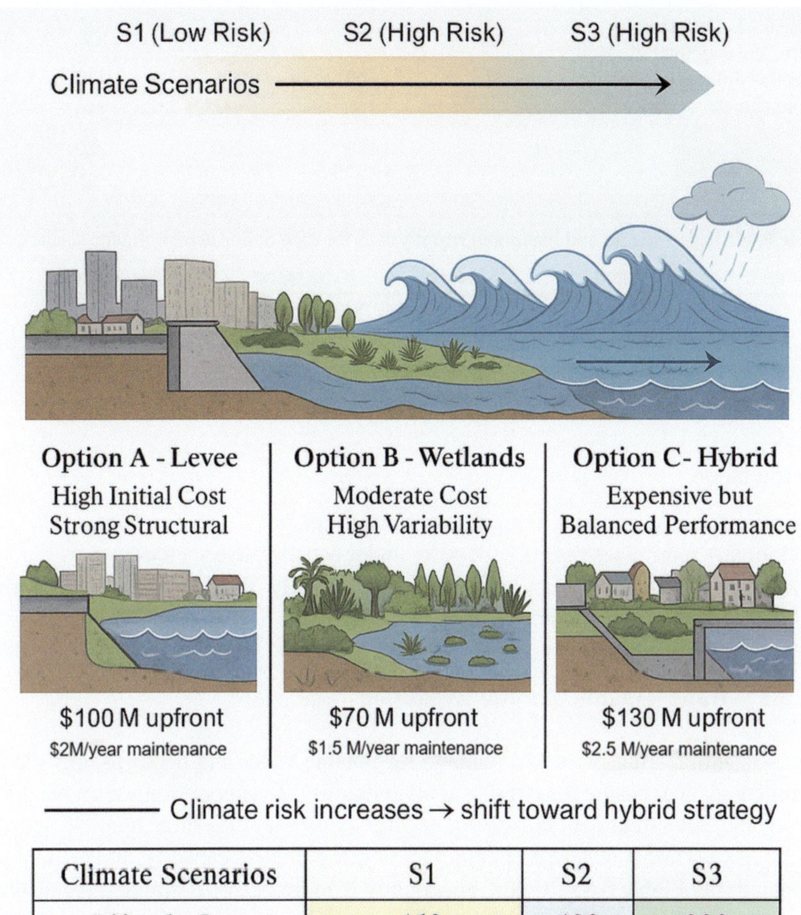

Climate Scenarios	S1	S2	S3
Lifeycle Cost	160	190	230
A	180	230	200
C	130	200	220

Climate risk increases → shift toward hybrid strategy

Fig. 6.40 Threshold analysis of flood protection strategies under three projected climate scenarios (S1 to S3)

Table 6.43 Performance matrix showing total lifecycle costs (in Millions of Dollars) across climate scenarios

Option	S1	S2	S3
A	160	190	230
B	180	230	310
C	190	200	220

Table 6.44 Regret matrix and maximum regret values for each option across climate scenarios

Option	S1 regret	S2 regret	S3 regret	Max regret
A	0	0	10	10
B	20	40	90	90
C	30	10	0	30

Interpretation:

- Option C is most robust to high-risk climate futures.
- Option A minimizes regret, especially under conservative scenarios.
- A staged or adaptive plan using Option B initially, then upgrading to C, may balance cost and resilience.

6.2.3.2 Water Treatment Plant Expansion

A city is anticipating increased demand for potable water and needs to upgrade its treatment plant to ensure long-term supply reliability. To support long-term planning for potable water treatment infrastructure, the city is evaluating three upgrade options over a 20 year horizon. These options vary in terms of cost, capacity, and technological flexibility (Table 6.45). Future uncertainty is addressed by defining three distinct planning scenarios that capture a range of possible conditions, including population growth, regulatory standards, and drought frequency. Figure 6.41 visualizes how uncertainties in population growth, regulatory pressure, and drought frequency influence the optimal investment in water treatment infrastructure.

To compare the performance of each upgrade option under different future scenarios, a performance matrix is constructed (Table 6.47). This matrix integrates both financial cost and compliance risk into a single composite score for each option across all three scenarios:

Table 6.45 Comparison of water teeatment plant upgrade options

Option	Description	Initial cost ($M)	Capacity (MLD)	Tech flexibility
A	Expand existing facility	40	25	Low
B	Build second plant	60	30	Medium
C	Modular membrane tech units	80	40	High

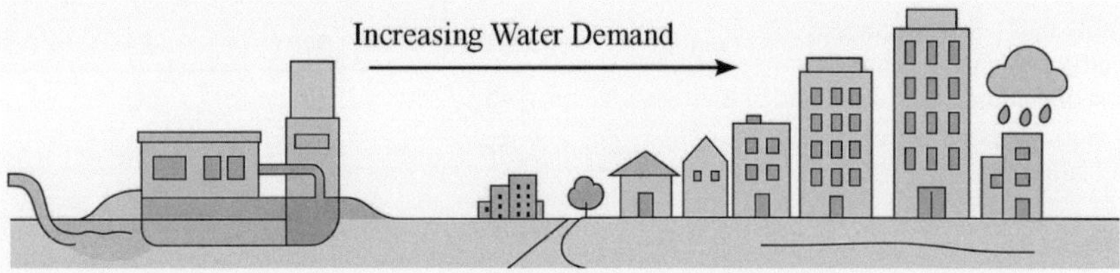

Fig. 6.41 Comparative threshold analysis of three water treatment upgrade strategies across future planning scenarios

- Scenario S1 represents a low-growth, low-risk future.
- Scenario S2 reflects moderate challenges, including tighter DBP standards.
- Scenario S3 assumes significant population growth and the strictest regulatory and environmental pressures (Table 6.46).

In general, lower performance scores indicate better overall outcomes, balancing cost and risk. This table enables the city to compare each option's strengths and vulnerabilities across potential futures. To select the most appropriate investment strategy, two decision-making criteria are typically applied:

1. Minimax Rule (Robustness Approach):

   This criterion selects the option with the lowest worst-case (maximum) score, prioritizing protection against the most adverse conditions.
2. Minimax Regret Rule (Opportunity Loss Approach):

   A regret matrix is derived by comparing each option's score in a scenario to the best (lowest) score for that scenario. The option with the lowest maximum regret across scenarios is then chosen, minimizing the potential for future regret in decision-making.

These structured approaches help ensure that the chosen upgrade is not only cost-effective but also resilient to future uncertainties in demand, regulation, and climate conditions. Performance of all options under different scenarios have been provided in Table 6.47.

To guide decision-making under uncertainty, we use the Minimax Regret Rule, which focuses on minimizing potential opportunity loss. The regret matrix in Table 6.48 quantifies the difference between each option's performance and the best possible performance within each scenario. For each option, the maximum regret

Table 6.46 Future planning scenarios for water teeatment needs

Scenario	Population growth	Contaminant regulations	Drought frequency
S1	Low	Current standards	Rare
S2	Medium	Stricter DBP limits	Occasional
S3	High	Advanced standards	Frequent

Table 6.47 Performance matrix showing combined cost and compliance risk scores across scenarios (Costs in Millions)

Option	S1	S2	S3
A	45	70	90
B	55	65	75
C	80	85	90

Table 6.48 Regret matrix and maximum regret scores for water treatment options

Option	S1	S2	S3	Max regret
A	0	5	15	15
B	10	0	0	10
C	35	20	15	35

across all scenarios is calculated in this table. According to the Minimax Regret criterion, the option with the lowest maximum regret is considered the most robust under uncertain future conditions. In this case, Option B has the smallest maximum regret of 10, making it the preferred choice using this decision rule.

Interpretation

- Option B balances risk and performance across all futures.
- While Option C is more future-proof, its high cost makes it less efficient unless S3 becomes reality.
- The city could start with Option B and monitor water quality and growth to decide on future modular upgrades.

6.2.3.3 Pavement design for highway expansion

A transportation agency is planning the construction of a 25 km highway and must select the most cost-effective pavement type over a 40 year lifecycle. Three pavement options are under consideration, varying in their initial costs and maintenance requirements (Table 6.49). Figure 6.42 illustrates how the preferred pavement strategy changes with scenario severity, highlighting a shift from asphalt to concrete as traffic, climate, and material price pressures increase.

Table 6.49 Comparison of pavement design options for highway construction

Option	Description	Initial cost ($M)	Annual rehab cost ($M/year)
A	Asphalt pavement	25	1.2
B	Concrete pavement	40	0.5
C	Composite (AC + PCC)	35	0.8

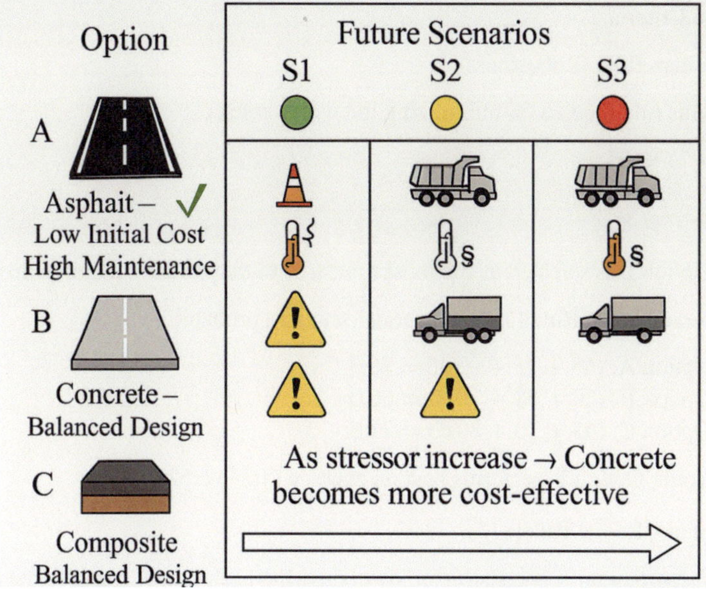

Fig. 6.42 cenario-based threshold analysis of three pavement design strategies over a 40 year lifecycle

Table 6.50 Future scenarios for highway design based on traffic, climate, and economic variables

Scenario	Traffic growth	Freeze-thaw gycles	Material price inflation
S1	Low	Mild	Stable
S2	Medium	Moderate	Moderate
S3	High	Severe	High

Table 6.51 Performance matrix showing total lifecycle costs (in millions of dollars) for each pavement option under varying future scenarios

Option	S1	S2	S3
A	65	75	95
B	60	68	70
C	63	70	80

To account for uncertainties in future conditions, three planning scenarios are defined in Table 6.50. These capture variability in traffic demand, climate stressors (freeze-thaw cycles), and material cost escalation over time:

Using engineering-economic modeling, the total lifecycle costs of each pavement option under all scenarios are presented in Table 6.51. The agency then applies multiple decision criteria to guide selection:

Decision Criteria:

1. Minimax Rule (Robustness):
   - This rule focuses on minimizing the worst-case cost:
     - A = 95
     - B = 70
     - C = 80

   So, Option B is selected as the most robust, with the lowest maximum cost.

2. Expected Value Rule (assuming equal scenario probabilities):
   - Option A: (65 + 75 + 95)/3 = 78.3
   - Option B: (60 + 68 + 70)/3 = 66.0
   - Option C: (63 + 70 + 80)/3 = 71.0

   So, again, Option B performs best on average (Table 6.52).

3. Minimax Regret Rule:
   - A regret matrix is constructed by comparing each option's cost to the lowest in each scenario (Table 6.52). The maximum regret is then computed for each option:
     - Option A: Max Regret = 25
     - Option B: Max Regret = 0
     - Option C: Max Regret = 10

   So, Option B is the best choice under this rule as well.

Final Recommendation: Across all three criteria—worst-case robustness, expected cost, and opportunity loss—Option B (Concrete pavement) consistently emerges as the most cost-effective and resilient investment for the highway project.

Interpretation

- Concrete pavement (Option B) is the most robust and lowest regret option under all scenarios.
- While more expensive upfront, it provides cost stability under uncertainty in traffic and climate.
- SBDA justifies higher initial investment in exchange for long-term resilience and predictability.

Table 6.52 Regret matrix and maximum regret values for pavement options

Option	S1	S2	S3	Max Regret
A	5	7	25	25
B	0	0	0	0
C	3	2	10	10

6.2.4 Metrics for Robustness and Regret Analysis

In SBDA, the goal is not always to optimize for a single most probable future, but rather to identify decisions that are robust across a wide range of plausible scenarios. Two prominent evaluation approaches used in this context are robustness analysis and regret analysis. Robustness analysis emphasizes resilience by focusing on the worst-case outcomes, ensuring that a chosen decision performs acceptably even under adverse conditions. On the other hand, regret analysis evaluates the potential cost of not having selected the best possible strategy in hindsight, aiming to minimize the gap between the chosen decision and the optimal one under each scenario. These tools allow civil engineers and planners to make decisions that are more resilient to uncertainty, especially in long-term infrastructure design, climate-sensitive projects, and resource allocation under ambiguity.

6.2.4.1 Robustness Metric (Minimax Rule)

The robustness of an alternative is measured by the worst outcome it produces across all scenarios. A more robust option is one that performs relatively well even in the most adverse conditions.

Let C_{ij} be the outcome (e.g., cost) of Option i under Scenario j. Then:

$$\text{Robustness Score}_i = \max_j (C_{ij}) \quad (6.2)$$

Select the option i^* such that:

$$i^* = \arg\min_i \max_j (C_{ij}) \quad (6.3)$$

In this context, lower scores indicate better performance, reflecting outcomes that are more favorable or less risky. The underlying assumption is that the decision-maker is risk-averse, meaning they prioritize strategies that minimize the likelihood of catastrophic failures. This approach is particularly effective in high-stakes scenarios where the consequences of failure are severe, such as dam collapses, uncontrolled flooding, or failures in critical infrastructure like water treatment plants. By focusing on avoiding worst-case outcomes, this interpretation aligns with a cautious and protective decision-making mindset.

6.2.4.2 Regret Metric (Minimax Regret Rule)

Regret is the difference between the outcome of a chosen alternative and the best possible outcome in each scenario. The maximum regret reflects the worst "what-if" loss had we known the true scenario in advance.

For each scenario j, identify the best outcome:

$$C_j^* = \min_i (C_{ij}) \tag{6.4}$$

Compute regret for each option:

$$R_{ij} = C_{ij} - C_j^* \tag{6.5}$$

Then, for each option:

$$\text{Max Regret}_i = \max_j (R_{ij}) \tag{6.6}$$

Choose the option i^* that minimizes regret:

$$i^* = \arg\min_i \max_j (R_{ij}) \tag{6.7}$$

This approach reflects the concept of opportunity loss incurred by not selecting the optimal strategy in hindsight. It is particularly suited for decision-makers with a moderate level of risk aversion who aim to reduce feelings of regret after the outcomes are realized. This method is especially valuable in decision-making contexts that involve political accountability, multiple stakeholders, or where public transparency is essential, as it promotes choices that are justifiable and balanced across various possible futures.

6.2.4.3 Additional Metrics (Optional in SBDA)

While robustness and regret are the most common metrics, SBDA may incorporate other analytical tools for specific contexts:

Satisficing

This approach ensures that a decision satisfies minimum acceptable criteria across the majority or all considered scenarios. It involves selecting options where the performance metric $C_{ij} \leq \text{threshold} \quad \forall j$, meaning that for each scenario j, the outcome remains within an acceptable limit. This method is particularly useful in multi-objective decision-making problems, such as those requiring a balance between economic cost and environmental performance, where no single optimal solution exists, but satisfactory performance across various dimensions is essential.

Trade-Off Plots

Trade-off plots are visual tools, typically presented as scatter plots (e.g., cost versus emissions), that illustrate how different decision options perform across multiple scenarios. They are particularly valuable in Multi-Criteria Decision Analysis (MCDA), as they help decision-makers identify and evaluate the compromises between conflicting objectives. By visualizing the distribution of outcomes, trade-off plots facilitate a clearer understanding of the relationships among competing criteria, supporting more informed and balanced decision-making.

Expected Value (with Probabilities)

When the probabilities of different scenarios (P_j) are known, the expected value for each decision option can be calculated using the formula $EV_i = \sum_j P_j \cdot C_{ij}$. This approach provides a weighted average of outcomes across all scenarios, reflecting both the performance and the likelihood of each scenario. The preferred decision is the one with the lowest expected value, indicating the most favorable average outcome over time. This method is particularly useful when reliable probabilistic information is available, allowing for a more quantitative and risk-informed decision-making process.

6.2.4.4 Metric Selection Guidance

Different decision metrics are best suited for different decision-making contexts (Fig. 6.43). Robustness is ideal when extreme or irreversible outcomes must be avoided, emphasizing safety and resilience. Regret is appropriate when stakeholders are particularly sensitive to missed opportunities and aim to minimize hindsight dissatisfaction. Satisficing is most useful when a minimum level of performance must be guaranteed across all future scenarios, ensuring essential thresholds are met. Lastly, Expected Value is the preferred metric when scenario probabilities are clearly defined and credible, allowing for probabilistic averaging to guide optimal choices.

6.2.5 Integration of SBDA with Adaptive Planning and Stakeholder Engagement

In complex civil engineering projects, the future is rarely static. Infrastructure systems must operate under dynamic conditions, including shifting climate patterns, urban expansion, evolving regulations, and changing community needs. While SBDA offers a structured method to evaluate decision alternatives under deep uncertainty, its

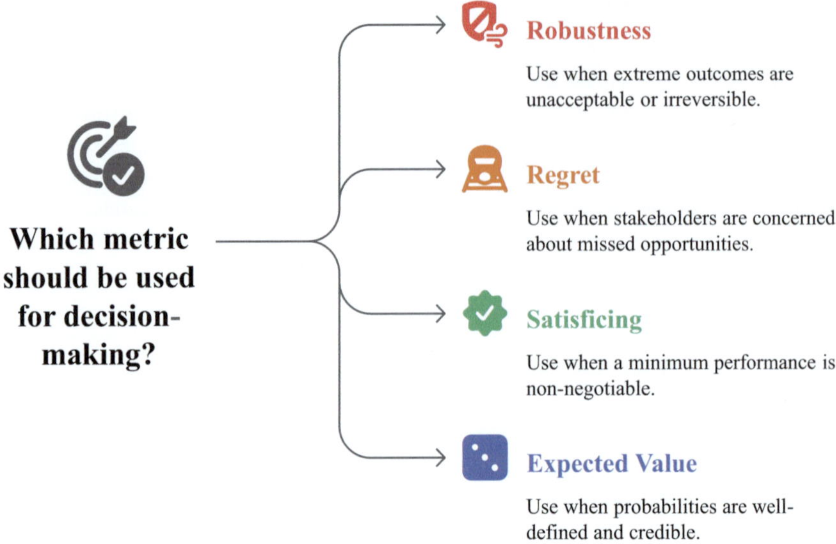

Fig. 6.43 Selection of decision-making metrics under uncertainty

full potential is realized when combined with adaptive planning frameworks and stakeholder engagement processes.

This integration supports iterative, flexible, and participatory decision-making, improving both the resilience of infrastructure systems and the legitimacy of decisions made in uncertain environments.

6.2.5.1 Adaptive Planning

Adaptive planning is a dynamic decision-making approach that emphasizes flexibility and the capacity to adjust actions over time in response to evolving information, unexpected events, or shifting conditions. Unlike static, one-time decisions, adaptive strategies are intentionally structured as sequences of steps that include predefined opportunities for revision, scaling, or redirection. In the context of civil engineering, adaptive planning can take several practical forms. For instance, infrastructure investments may be phased—such as constructing floodwalls in stages as risk levels change. Similarly, systems can be designed with modularity in mind, allowing components like water treatment plants to be expanded as demand grows. Additionally, contingency triggers may be embedded within the plan, such as setting specific thresholds that, once crossed, automatically prompt upgrades or further interventions. This approach enables more resilient, cost-effective, and responsive infrastructure development (Fig. 6.44).

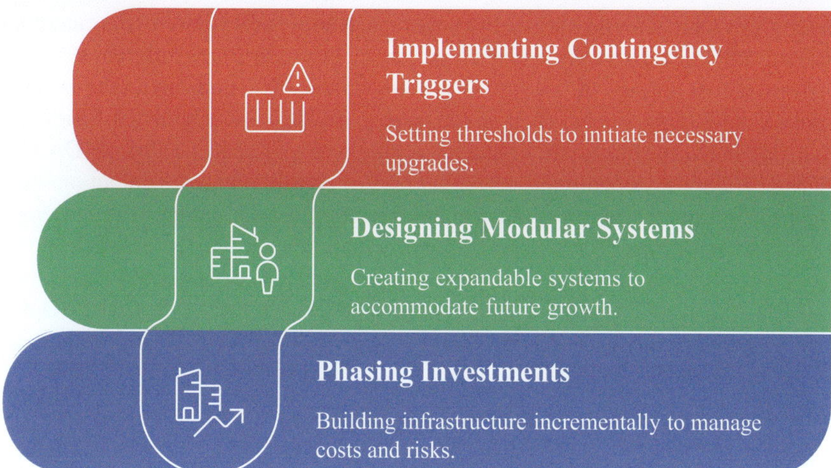

Fig. 6.44 Practical forms of the adaptive planning in civil engineering

6.2.5.2 How SBDA Supports Adaptive Planning

The SBDA naturally complements adaptive planning by providing critical insights that enhance flexibility and responsiveness. First, it reveals decision robustness across time horizons by allowing planners to evaluate how different strategies perform under both near-term and long-term scenarios. This helps identify which actions are reliably beneficial and safe to implement immediately, and which ones are better deferred or made conditional based on how the future unfolds. Second, SBDA identifies tipping points and triggers by highlighting scenario-specific thresholds—such as sea-level rise exceeding 0.6 m—where a current strategy might fail or require augmentation. These thresholds can then be used as adaptive triggers, guiding when to pivot to alternative strategies or increase system capacity, thereby strengthening long-term resilience and efficiency.

6.2.5.3 Engaging Stakeholders in Scenario-Based Decisions

Engineering decisions often affect a wide range of stakeholders, including local governments, utilities, residents, regulatory agencies, environmental groups, and funding bodies. SBDA facilitates inclusive, transparent, and credible decision-making in several key ways:(1) scenario co-development, (2) trade-off visualization, (3) clarifies uncertainty, and (4) supports consensus building.

SBDA enhances stakeholder engagement by integrating their diverse perspectives into the decision-making process. First, through scenario co-development, stakeholders actively participate in constructing plausible future scenarios based on their own knowledge, concerns, and priorities. This participatory approach not only makes the analysis more relevant but also increases acceptance and support for the outcomes. Second, SBDA enables trade-off visualization by illustrating how different strategies perform across a range of futures. This helps stakeholders understand the implications of conflicting values, such as balancing cost with environmental benefits or choosing between immediate savings and long-term resilience. Third, SBDA clarifies uncertainty by openly presenting a range of possible futures without claiming precision in forecasting. This transparency acknowledges the inherent limitations of modeling and fosters greater trust in the decision process. Finally, SBDA supports consensus building by offering a framework where, even in the absence of agreement regarding the most likely future scenario, stakeholders can support robust or adaptive strategies that are acceptable under multiple scenarios. This collaborative approach ultimately strengthens decision legitimacy and effectiveness.

6.2.5.4 Tools and Methods for Integration

To effectively integrate SBDA with adaptive planning and stakeholder engagement, a range of tools and frameworks can be employed to enhance decision-making under uncertainty (Table 6.53). Dynamic Adaptive Policy Pathways (DAPP) is used to map sequences of decision options over time, allowing planners to adapt as scenarios evolve. Robust Decision Making (RDM) focuses on evaluating how strategies perform across a wide array of possible futures by leveraging simulation ensembles, supporting decisions that remain effective under diverse conditions. Multi-Stakeholder Scenario Workshops facilitate collaborative scenario development and evaluation, incorporating both expert and community input to ensure relevance and legitimacy. Decision Scaling helps identify critical climate or system thresholds that can influence the success or failure of strategies, guiding more targeted and responsive planning. Finally, Monitoring and Feedback Loops provide mechanisms for real-time adjustments during implementation, ensuring that strategies remain aligned with emerging conditions and stakeholder priorities. Together, these tools support a flexible, inclusive, and resilient planning approach.

6.2.5.5 Step-By-Step Framework for Applying Scenario-Based Decision Analysis

Before generating specific examples, it is essential to understand the general workflow for applying SBDA, particularly when integrating adaptive planning and stakeholder engagement. This structured approach ensures that uncertainty is comprehensively addressed, decisions are flexible, and the process is inclusive and transparent. Below is a recommended nine-step procedure to apply SBDA effectively in civil engineering contexts (Fig. 6.45).

Table 6.53 Key tools and frameworks supporting the integration of SBDA with adaptive planning

Tool/Framework	Purpose
Dynamic Adaptive Policy Pathways (DAPP)	Maps sequences of decision options over time based on scenario evolution
Robust Decision Making (RDM)	Emphasizes performance across many futures, using simulation ensembles
Multi-stakeholder scenario workshops	Engages community and expert input in scenario development and evaluation
Decision scaling	Identifies climate or system thresholds that affect strategy success
Monitoring and feedback loops	Enables real-time adjustment of strategies during implementation

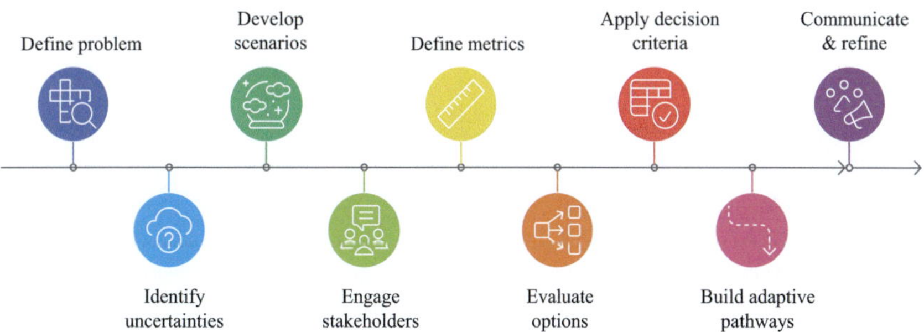

Fig. 6.45 Step-by-step framework for applying SBDA in civil engineering

Step 1: Define the Decision Problem

Begin by clearly articulating the decision problem, including its primary objectives—such as minimizing costs, reducing flood risks, or ensuring system reliability. This step involves identifying the set of viable decision alternatives, for example, traditional levees, green infrastructure, or hybrid solutions. It is essential to establish the relevant planning horizon, whether short-, medium-, or long-term, and to outline any operational or contextual constraints, including budgetary limits, environmental regulations, or technical feasibility. Clearly framing the problem sets the foundation for a structured and meaningful analysis.

Step 2: Identify Key Uncertainties

In this step, identify the external factors that may significantly affect decision outcomes. These uncertainties often fall into categories such as climate (e.g., rainfall variability, temperature changes, sea level rise), demographics (e.g., urbanization trends, population growth), economics (e.g., fluctuations in material costs, infrastructure demand), and policy or regulatory shifts (e.g., changes in zoning laws or emissions regulations). The focus should be on high-impact, high-uncertainty variables that introduce substantial unpredictability to the system's future state.

Step 3: Develop Plausible Scenarios

Once key uncertainties are identified, construct a set of three to five internally consistent scenarios that capture different plausible future conditions. These scenarios should combine various uncertainties to reflect complex realities—such as a future with both high rainfall and rapid urban development. To ensure validity and relevance, incorporate stakeholder input, expert judgment, and formal foresight tools like the Delphi method. These scenarios form the basis for evaluating how different strategies perform under varying future conditions.

Step 4: Engage Stakeholders Early

Stakeholder engagement should begin early in the process to ensure the analysis reflects diverse perspectives and concerns. Identify key stakeholders—such as municipal planners, utility managers, regulators, and community representatives—and actively involve them in scenario development, priority setting, and defining evaluation criteria. Early engagement fosters transparency, trust, and a sense of shared ownership in the decision-making process, ultimately leading to more widely supported outcomes.

Step 5: Define Performance Metrics

Establish a set of metrics to evaluate how each decision alternative performs across scenarios. These may include lifecycle costs, system reliability, environmental impact, social acceptability, and adaptability. The chosen metrics should be quantifiable or at least ordinally ranked and must be applicable to all alternatives being considered. Clear and consistent metrics allow for meaningful comparisons and support evidence-based decision-making.

Step 6: Evaluate Decision Alternatives Under Each Scenario

Using appropriate tools—such as engineering models, simulations, or expert estimations—evaluate each alternative's performance within each scenario. The results should be organized into a performance matrix, where each cell represents the outcome of a given alternative under a specific scenario. This structured representation enables easy comparison and forms the basis for robustness and regret analysis in subsequent steps.

Step 7: Apply Robustness and Regret Criteria

Assess the performance of alternatives using robust decision-making metrics. This may involve applying the minimax criterion to identify the option with the best worst-case performance, or the minimax regret approach to find the option with the lowest maximum regret across scenarios. The satisficing approach can also be used to identify options that meet minimum acceptable thresholds in all future scenarios. If probabilities are available, calculate expected values to support probabilistic decision-making. These evaluations help identify strategies that remain acceptable across a wide range of uncertain futures.

Step 8: Design Adaptive Pathways (if applicable)

Based on insights from the scenario analysis, develop adaptive strategies that unfold over time. This may include phased plans or conditional actions, such as "implement Option A now, and switch to Option C if demand exceeds a certain threshold by year Y." Identify specific triggers or tipping points within the scenarios that would prompt a reassessment or redirection of the strategy. When long-term flexibility is required, tools like the Dynamic Adaptive Policy Pathways (DAPP) method can be employed to structure and communicate these pathways.

Step 9: Communicate Results and Refine

Finally, present the analysis results in clear, accessible formats such as performance matrices, regret tables, scenario graphs, and adaptive pathway diagrams. These visual and tabular summaries help convey complex information effectively to stakeholders. Facilitate stakeholder review sessions to gather feedback and assess the need for refinement—this could include updating scenarios, introducing new alternatives, or adjusting the evaluation metrics. Iteration ensures the analysis remains relevant, credible, and aligned with stakeholder expectations.

6.2.6 *Applications of SBDA in Civil Engineering Practice*

6.2.6.1 Example 1—Flood Protection Strategy in a Coastal City

A coastal municipality is evaluating strategies to protect the downtown area from future coastal flooding. The city has experienced several storm surge events in the past two decades, and scientific reports suggest that sea-level rise (SLR) and storm surge intensity are expected to worsen under climate change. With a 30 year planning horizon and a limited infrastructure budget, the city must choose among three flood protection options provided in Table 6.54. Figure 6.46 provides a conceptual representation of the decision context. The city's objectives center on achieving a cost-effective, resilient, and adaptable infrastructure strategy. Specifically, the city aims to minimize the total lifecycle cost by reducing long-term maintenance expenses and potential damage, ensuring dependable protection against future extreme weather events while maintaining the flexibility necessary to adapt its infrastructure if future environmental conditions become more severe than currently anticipated. The key questions to be addressed are as follow:

1. What are the most relevant uncertainties affecting flood protection decisions?
2. What plausible future scenarios should be considered?
3. How does each flood protection option perform under each scenario?
4. Which option is most robust across all futures?
5. Which option has the least regret if future conditions turn out to be worse than expected?
6. Can an adaptive pathway be developed to allow flexible investment over time?

Step-by-Step Solution Using SBDA

Step 1: Define the Decision Problem

- Objective: Select a flood protection strategy that minimizes total lifecycle cost while providing reliable protection over a 30-year horizon.
- Alternatives:
  - A: Levee (high upfront, strong defense)
  - B: Wetlands (low-cost, eco-friendly, less reliable)

Table 6.54 Summary of coastal flood protection strategies evaluated by a municipality under climate change uncertainty

Option	Description	Initial cost ($M)	Annual maintenance ($M/year)
A	Reinforced concrete levee	100	2
B	Coastal wetland buffer restoration	70	1.5
C	Hybrid system (Levee + Wetlands)	130	2.5

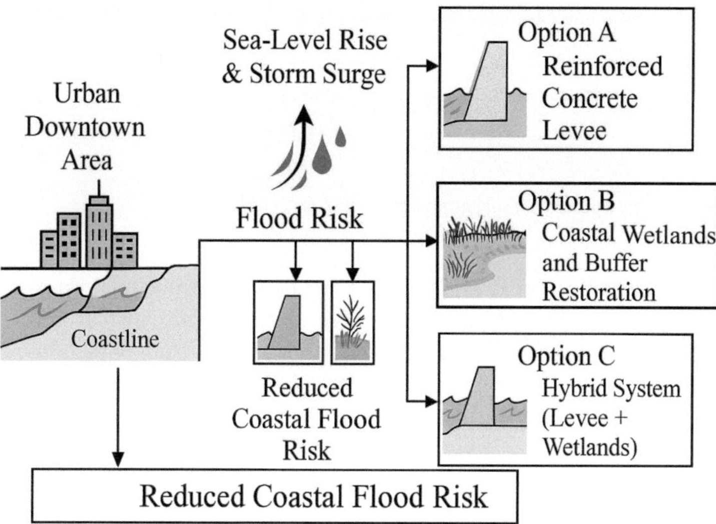

Fig. 6.46 Comparison of three coastal flood protection strategies over a 30-year planning horizon

  - C: Hybrid (moderate flexibility, higher cost)

- Constraints:
  - Budget is limited to $150M.
  - Must comply with federal environmental impact guidelines.
- Timeframe: 30 years

Step 2: Identify Key Uncertainties

- Sea-Level Rise (SLR): Unknown trajectory over the next 30 years.
- Storm Surge Frequency: Influenced by ocean warming and atmospheric variability.

These uncertainties cannot be assigned reliable probabilities due to model spread and climate variability.

Step 3: Develop Plausible Scenarios

Table 6.55 outlines three climate scenarios used to evaluate coastal flood protection strategies. Each scenario reflects varying assumptions about future sea-level rise and the frequency of extreme storm surge events, ranging from conservative to high-end projections aligned with climate change models.

Step 4: Engage Stakeholders

The city's objectives are focused on developing a resilient and forward-thinking infrastructure plan that balances financial prudence with robust performance. One of the primary goals is to minimize the total lifecycle cost, which includes not only

Table 6.55 Description of sea-level rise and storm surge scenarios for flood risk assessment

Scenario	Description	Sea-level rise	Storm surge recurrence
S1	Conservative climate Outlook	+0.3 m	1-in-100 year
S2	Moderate warming	+0.6 m	1-in-50 year
S3	High-end sea-level rise	+1.0 m	1-in-20 year

*Derived from regional IPCC projections and local coastal modeling

upfront investment but also long-term maintenance and potential damage costs. Another critical objective is to ensure reliable protection against increasingly frequent and intense extreme weather events. Recognizing the uncertainty of future climate conditions, the city also aims to preserve adaptability in its infrastructure choices, allowing for modifications if future risks prove more severe than anticipated.

To support these goals, a diverse group of stakeholders participated in a scenario workshop. The participants included municipal engineers, who prioritized cost-effectiveness and system reliability; environmental NGOs, who advocated for nature-based and ecologically sustainable solutions; residents and business owners, who expressed concerns about flood insurance rates and potential disruptions to their daily lives; and city council members, who emphasized the importance of public accountability and transparent decision-making. The workshop sought to validate three proposed planning scenarios, establish consensus on key decision metrics—namely cost and reliability—and stress the importance of adaptability as a guiding principle in long-term planning.

Step 5: Define Performance Metrics

- Total Lifecycle Cost = Initial + Maintenance + Flood Damage
- Maximum Acceptable Damage per Scenario = $100 M

All costs are in present values (no discounting).

Step 6: Evaluate Options Under Each Scenario

Table 6.56 summarizes the initial investment, maintenance costs, and projected flood damage costs for three flood protection options (A, B, and C) under three climate scenarios (S1–S3). The damage estimates reflect the varying severity of sea-level rise and storm surge events, supporting a comparative evaluation of total expected costs over the planning horizon. In addition, Table 6.57 presents the total lifecycle costs (in million dollars) for each flood protection option (A, B, and C) under three climate scenarios (S1–S3). These totals include initial construction, maintenance, and projected flood damage costs, enabling a holistic comparison of long-term financial implications

Table 6.56 Estimated lifecycle costs and flood damages for each flood protection option under different climate scenarios

Option	Initial ($M)	Maintenance ($M)	Damage S1		
			S1	S2	S3
A	100	60	0	20	70
B	70	45	20	70	120
C	130	75	0	10	50

Table 6.57 Total estimated lifecycle costs of flood protection options across climate scenarios

Option	S1	S2	S3
A	160	180	230
B	135	185	235
C	205	215	255

Step 7: Apply Robustness and Regret Criteria

A. Robustness (Minimax Rule)

Table 6.58 shows the worst-case total lifecycle cost for each flood protection option under the most adverse climate scenario. The minimax rule is applied to identify the most robust option by minimizing potential losses in the worst-case scenario. Option A is most robust (lowest worst-case cost).

B. Regret Matrix

Table 6.59 displays the regret values for each option under three climate scenarios (S1–S3), along with the maximum regret for each option. Regret is defined as the difference between an option's cost and the cost of the optimal option under the same scenario. The max regret column identifies the option with the least potential for hindsight-based decision regret. Option A also has lowest maximum regret under worst-case (S3) scenario

Step 8: Design Adaptive Pathways

The city has adopted an adaptive pathways approach to infrastructure design, structured as a two-stage plan to balance cost, environmental impact, and future uncertainty. The initial investment focuses on wetland restoration (Option B), with an estimated cost of $70 million. This nature-based solution serves as a foundational

Table 6.58 Robustness analysis using the minimax (Worst-Case) criterion

Option	Worst-case cost
A	230
B	235
C	255

Table 6.59 Regret analysis of flood protection options across climate scenarios

Option	Regret			
	S1	S2	S3	Max
A	25	0	0	25
B	0	5	5	5
C	70	30	20	70

measure, offering immediate environmental benefits while providing a moderate level of protection against flooding and sea level rise.

To support long-term adaptability, a robust monitoring plan will be implemented, tracking indicators such as the annual mean sea level and the frequency of storm surges exceeding 0.75 m. A critical trigger has been defined: if sea level rise surpasses 0.5 m or a 1-in-50 year storm event occurs within the next 15 years, the plan calls for an upgrade to a more robust hybrid infrastructure system (Option C), requiring an additional $60 million investment. This staged strategy spreads out financial commitments over time, helps avoid overdesign based on uncertain projections, and enhances the city's environmental credibility by prioritizing sustainable and responsive planning.

Step 9: Communicate Results and Refine

In the final step of the planning process, the focus shifts to synthesizing the findings and clearly conveying them to stakeholders for feedback and refinement. This stage involves the creation of key deliverables that illustrate both the technical and strategic aspects of the proposed adaptive plan. These include a performance matrix, presented in the form of a bar chart, which compares the outcomes of different options based on predefined decision metrics, and an adaptive pathway timeline that visually maps out the sequence and timing of potential infrastructure upgrades.

Additionally, a stakeholder summary is prepared to explain the rationale behind each option and support informed decision-making. This summary highlights why Option A is considered robust in terms of cost and reliability under moderate conditions, and why Option B offers greater flexibility in adapting to uncertain future changes. It also outlines the specific trigger conditions—such as thresholds for sea level rise or storm events—that would prompt an upgrade to a more advanced system. This communication step ensures transparency, promotes stakeholder engagement, and provides a foundation for refining the plan based on collaborative input.

Answers to the key questions

1. What are the most relevant uncertainties affecting flood protection decisions?

The two primary uncertainties that strongly influence the performance and reliability of the flood protection strategies are:

- Sea-Level Rise (SLR):
  - The magnitude of sea-level rise over the 30 year planning horizon is highly uncertain due to varying greenhouse gas emission trajectories, ocean thermal expansion, and polar ice melt rates. SLR directly affects whether a levee or natural buffer will be overtopped during future storm surge events.
- Storm Surge Frequency and Intensity:
  - Storm surge events are driven by a combination of atmospheric pressure, wind speed, and tidal conditions. With climate change, both the frequency and severity of extreme coastal flooding events are expected to increase, but exact projections remain uncertain.

These uncertainties are difficult to quantify probabilistically and are therefore represented using three plausible scenarios reflecting different climate pathways.

2. What plausible future scenarios should be considered?

Three internally consistent climate scenarios (Table 6.60) were developed through a combination of expert judgment, stakeholder engagement, and regional climate model outputs. These scenarios were chosen to span a wide range of possible futures without assigning exact probabilities.

3. How does each flood protection option perform under each scenario?

To assess the effectiveness of each flood protection option, a comprehensive evaluation was conducted across multiple future scenarios. The core performance metric used for comparison was the total lifecycle cost, a holistic measure that captures both short- and long-term financial implications. This includes the initial capital cost required to implement each option, projected maintenance expenses over a 30 year period, and the estimated costs associated with flood damages that could occur under each scenario.

By analyzing these three cost components together, decision-makers can better understand the trade-offs between upfront investment and long-term resilience. This approach allows for a realistic comparison of how each option performs not only in average conditions but also under extreme or uncertain future scenarios. It ensures that the selected strategy offers not just cost-effectiveness, but also durability and adaptability in the face of evolving climate risks.

Table 6.60 Climate change scenarios defined by SLR and storm surge frequency

Scenario	Sea-Level Rise (SLR)	Storm surge return period	Description
S1	+0.3 m	1-in-100 year	Conservative: minor climate change impacts
S2	+0.6 m	1-in-50 year	Moderate: middle-of-the-road projection
S3	+1.0 m	1-in-20 year	High-end: aggressive sea-level rise and more frequent flooding

Table 6.61 Total lifecycle costs of flood protection options under different climate scenarios

Option	Total cost (M $)		
	S1	S2	S3
A (Levee)	160	180	230
B (Wetlands)	135	185	235
C (Hybrid)	205	$215	255

Table 6.61 compares the total estimated lifecycle costs—including construction, maintenance, and flood damages—for three flood protection strategies (Levee, Wetlands, and Hybrid) under three climate scenarios (S1–S3). The values illustrate how each option performs financially as climate risks intensify.

The evaluation of the three flood protection options across different future scenarios revealed distinct strengths and trade-offs for each. Option A (Levee) delivers full protection under Scenario 1 (S1), incurs moderate losses in Scenario 2 (S2), and experiences significant but still manageable losses in the more extreme Scenario 3 (S3). Its performance highlights a balance between cost and protection in less severe futures, though it may require upgrades if conditions worsen.

Option B, which involves wetland restoration, stands out for having the lowest overall cost in Scenario 1. However, its performance deteriorates sharply in more severe conditions. In Scenario 2, it suffers large damages, and in Scenario 3, it incurs unacceptable flood losses totaling around $120 million—exceeding the city's defined risk tolerance. This makes Option B a flexible but potentially risky choice without planned upgrades.

On the other hand, Option C, a hybrid system combining structural and nature-based solutions, offers the highest level of protection across all scenarios. Yet, its substantial initial and maintenance costs render it less appealing under conservative or moderately adverse futures. This makes it best suited for high-risk scenarios, or as an upgrade path if trigger conditions are met.

4. Which option is most robust across all futures?

Robustness is assessed using the Minimax rule, which selects the option with the lowest worst-case cost (Table 6.58). Option A (Levee) is the most robust option—it minimizes the worst-case outcome across all scenarios. Even under the most extreme climate future (S3), it limits total cost to $230M, which is below the other options.

5. Which option has the least regret if future conditions turn out to be worse than expected?

Regret is calculated as the difference between each option's cost and the best-performing option for each scenario. Although Option A is most robust, Option B has the lowest maximum regret (Table 6.59). However, this is misleading because Option B fails to meet minimum performance thresholds in S3, incurring flood damage of $120M, which is beyond the city's tolerance of $100M.

While Option B presents the lowest theoretical regret based on cost-performance calculations, Option A emerges as the most practical low-regret choice when accounting for real-world constraints and risk tolerance. Crucially, Option A never exceeds the flood damage thresholds established by the city, ensuring consistent protection across all scenarios. Its regret in Scenario 1 is only moderate, where Option B is marginally less expensive, but this trade-off is outweighed by the reliability and robustness it offers in more severe futures. Moreover, Option A sustains acceptable performance even under extreme conditions, making it a resilient and dependable choice in the face of future uncertainty. Given its balance between cost, risk mitigation, and adaptability, Option A stands out as the least regretful viable solution, aligning both with practical implementation needs and long-term resilience goals.

6. Can an adaptive pathway be developed to allow flexible investment over time?

Yes. Based on the SBDA results and stakeholder priorities, the city can adopt a phased, adaptive investment pathway as follow:

- Initial Investment (Year 0–10):
  - Begin with Option B (Wetlands) for $70 M.
  - It provides cost-effective protection under Scenario 1 and builds environmental value.
- Monitoring System:
  - Track sea-level rise annually.
  - Use surge monitoring to detect increasing event frequency.
  - Install trigger points:
  - If SLR exceeds 0.5 m or 1-in-50 year events occur twice within 10 years, begin upgrading.
- Adaptive Upgrade (Year 10–15):
  - If triggers are met, transition to Option C (Hybrid).
  - Cost of upgrading from wetlands to hybrid is estimated at $60 M.
  - This preserves earlier investment and ensures readiness for severe future conditions.

An adaptive pathway combining Option B and C is feasible and strategically sound. It minimizes initial cost, satisfies environmental goals, and retains the flexibility to respond to worsening climate conditions. This pathway reflects robust, flexible, and stakeholder-supported planning.

6.2.6.2 Example 2—Wastewater Treatment Plant Upgrade under Environmental and Regulatory Uncertainty

A mid-sized city is currently operating a conventional activated sludge (CAS) wastewater treatment plant that is reaching the limits of both its hydraulic capacity and its ability to effectively remove nutrients. With population growth and increased development contributing to rising wastewater inflows, the existing infrastructure is under mounting pressure. Compounding the issue, new environmental regulations mandate more stringent effluent quality standards, particularly regarding the discharge of nitrogen and phosphorus—key nutrients that, in excess, contribute to eutrophication in receiving water bodies.

In response, the city's environmental engineering department must identify and implement an upgrade strategy that not only expands the plant's capacity to handle higher volumes of wastewater but also ensures compliance with enhanced treatment requirements. The planning horizon is 25 years. The chosen solution must strike a balance between technical effectiveness, operational feasibility, cost-efficiency, and long-term adaptability to future regulatory or environmental changes. The decision will be central to safeguarding public health, protecting local ecosystems, and supporting sustainable urban development. A conceptual overview of the strategic upgrade alternatives under uncertain regulatory and capacity conditions is illustrated in Fig. 6.47. This figure outlines the diverging pathways associated with each treatment option, highlighting their relative adaptability, resilience, and long-term suitability in response to increasing environmental pressure and wastewater inflow. Different options for wastewater treatment based on cost and upgrade flexibility have been presented in Table 6.62. The key questions that should be addressed are as follows:

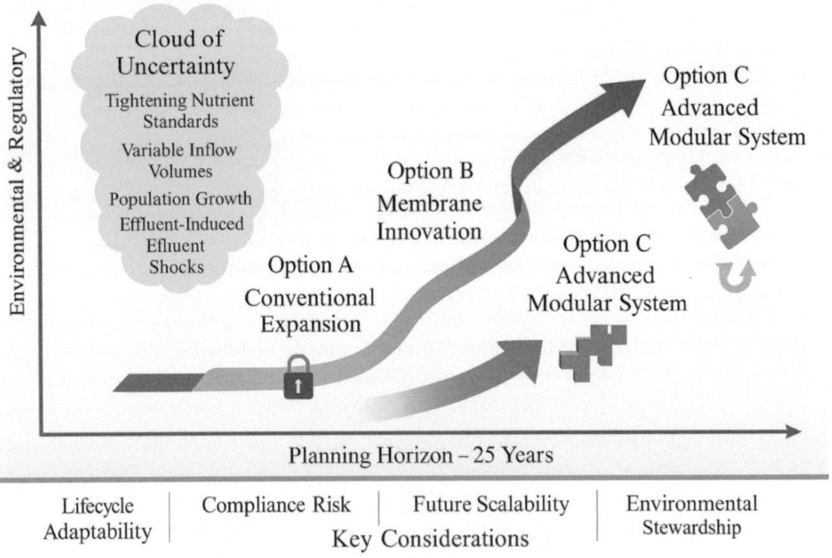

Fig. 6.47 Conceptual representation of strategic upgrade options for a municipal wastewater treatment facility under evolving environmental regulations and inflow demand

Table 6.62 Comparison of wastewater treatment options based on cost and upgrade flexibility

Option	Description	Initial cost ($M)	Annual O&M cost ($M)	Flexibility for future upgrades
A	Expand Existing Activated Sludge (AS)	45	5	Low
B	Membrane Bioreactor (MBR)	65	6.5	Medium
C	Modular Advanced Treatment Units (ATUs)	85	7	High (Scalable and upgradable)

1. What environmental and regulatory uncertainties affect wastewater treatment planning?
2. What future scenarios are plausible and should be considered?
3. How does each treatment option perform across those scenarios?
4. Which option is most robust to uncertainty?
5. Which option results in the least regret if regulations become more stringent?
6. Can an adaptive pathway be proposed to manage both environmental risk and financial constraints?

Step-by-Step Solution Using SBDA

Step 1: Define the Decision Problem

The city's environmental engineering department begins by clearly defining the decision problem to guide the selection of an optimal upgrade strategy for the wastewater treatment plant. The primary objective is to ensure long-term regulatory compliance with both current and anticipated effluent standards, particularly for nitrogen and phosphorus, while simultaneously minimizing total lifecycle costs under conditions of future uncertainty, including population growth, climate variability, and evolving regulations.

Three upgrade alternatives have been identified:

- Option A: Expand Activated Sludge (AS)—This is the lowest-cost option but offers limited flexibility to accommodate future changes or tightening standards.
- Option B: Adopt Membrane Bioreactors (MBR)—This represents a moderate-cost solution that offers moderate flexibility, improving both treatment performance and plant compactness.
- Option C: Deploy Advanced Treatment Units (ATUs)—Though the most expensive, this option provides high flexibility, making it better suited for long-term adaptability and more aggressive nutrient removal goals.

Key constraints shaping the decision include a budget cap of $100 million and the non-negotiable requirement to comply with both present and foreseeable effluent discharge regulations. The planning horizon for evaluating the performance and sustainability of each option is set at 25 years, ensuring that the selected strategy supports both immediate needs and future resilience.

Step 2: Identify Key Uncertainties

To make a robust and future-ready decision, the city must identify the critical uncertainties that could significantly influence the performance and suitability of each wastewater treatment upgrade option over the 25-year planning horizon.

- **Regulatory Stringency**

  One of the most pressing uncertainties is the future stringency of effluent regulations, particularly concerning nitrogen and phosphorus discharge limits. While current standards are known, environmental agencies may tighten these limits over time in response to evolving ecological concerns, watershed nutrient loading, or federal mandates. Stricter regulations would demand higher nutrient removal efficiency, potentially rendering low-flexibility options inadequate without costly retrofits.
- **Population Growth and Inflow Volume**

  Another key uncertainty involves future population growth and the resulting increase in wastewater inflow. Rising populations and urban development will increase both the hydraulic load and nutrient concentration entering the treatment plant. These changes can put pressure on the ability of the plant to meet effluent guidelines, especially if not anticipated in the design. The rate and scale of these changes are difficult to predict with precision, yet they have direct implications for treatment reliability and system resilience.

Addressing these uncertainties is essential in selecting a treatment option that not only meets today's needs but remains, viable and cost-effective under a range of possible future conditions.

Step 3: Develop Plausible Scenarios

Table 6.63 defines three potential future scenarios combining different levels of population growth and nutrient discharge standards for nitrogen (N) and phosphorus (P). Each scenario illustrates the resulting implications for treatment system performance, guiding infrastructure design and regulatory compliance strategies.

Step 4: Engage Stakeholders

To ensure that the decision-making process reflects diverse perspectives and practical constraints, the city organized a series of stakeholder engagement workshops. These sessions brought together a broad coalition of participants, including utility engineers and plant operators, who contributed technical insights on treatment feasibility and operational challenges; environmental regulators, who clarified current and anticipated compliance expectations; public health officers, who emphasized the importance of protecting community health and water quality; and representatives from the city council and finance department, who focused on fiscal responsibility and public accountability.

Table 6.63 Plausible future scenarios for wastewater treatment planning

Scenario	Description	Population growth	Nutrient limits (N, P mg/L)	Implication
S1	Conservative Growth + Current Standards	Low	10 (N), 1.0 (P)	Minimal change
S2	Moderate Growth + Stricter Standards	Medium	5 (N), 0.5 (P)	Requires better removal
S3	Rapid Growth + Ultra-Stringent Regulation	High	3 (N), 0.1 (P)	Requires advanced tech

Through these collaborative discussions, stakeholders collectively agreed on two primary performance metrics to guide the evaluation of upgrade options:

- Total Lifecycle Cost—encompassing capital investment, long-term operation and maintenance expenses, and any future retrofitting needs.
- Compliance Risk—representing the probability of failing to meet effluent standards over the 25 year planning horizon, particularly under scenarios of increasing regulatory stringency and rising inflows.

By establishing a shared framework for assessing alternatives, the stakeholder engagement process helped align technical, financial, regulatory, and public health priorities, laying the foundation for a transparent and balanced infrastructure decision.

Step 5: Define Performance Metrics

Based on the stakeholder engagement workshops, two two key performance metrics that integrate financial and regulatory considerations over the 25 year planning horizon were used to evaluate the feasibility of the different upgrade options.

1. **Total Lifecycle Cost**

This metric captures the comprehensive financial impact of each alternative by summing three cost components:

- Initial Capital Investment
- Operation and Maintenance (O&M) Costs over 25 years
- Penalties and Retrofit Costs incurred if the system fails to meet future regulatory standards

This approach ensures that decision-makers consider not only upfront expenditures but also the long-term economic consequences of each choice.

2. **Compliance Risk Score**

This qualitative metric translates regulatory reliability into a simple scale to reflect the risk of non-compliance:

- 0 = Full compliance with current and projected regulations (no action needed)
- 1 = Minor retrofit required to maintain compliance (manageable cost and effort)
- 2 = Major retrofit and/or fines due to significant non-compliance (high risk and cost)

These performance metrics enable a clear and structured comparison of options, facilitating the identification of strategies that balance cost-effectiveness with regulatory resilience.

Step 6: Evaluate Options Under Each Scenario

Table 6.64 presents lifecycle cost estimates (in millions of dollars) for three wastewater treatment options, including initial capital costs, 25 year operation and maintenance (O&M) expenses, and scenario-dependent retrofit costs. Total costs are calculated for each future scenario (S1–S3), reflecting the financial impact of regulatory tightening and population growth. Moreover, Table 6.65 summarizes the compliance risk scores for each treatment option under three future scenarios (S1–S3). Scores indicate the likelihood of failing to meet nutrient discharge standards, with higher values reflecting greater regulatory risk. Option C demonstrates the highest robustness with zero compliance risk across all scenarios.

Step 7: Apply Robustness and Regret Criteria

Table 6.66 presents the worst-case total lifecycle costs for each treatment option across all future scenarios. Applying the minimax rule, Option A emerges as the most robust choice in cost terms, offering the lowest maximum expenditure under uncertainty.

Table 6.67 shows the regret values for each treatment option under three future scenarios (S1–S3), with regret defined as the additional cost incurred compared to the lowest-cost option in each scenario. The maximum regret highlights the worst-case deviation from optimal cost. Option A demonstrates zero regret across all scenarios,

Table 6.64 Lifecycle cost estimates for treatment options under evolving scenario assumptions

Option	Initial	O&M (25 yrs)	Retrofit	Total		
				S1	S2	S3
A	45	125	+20 (S2), +40 (S3)	170	190	210
B	65	162.5	+10 (S3)	227.5	227.5	237.5
C	85	175	0	260	260	260

Table 6.65 Compliance risk scores for treatment options across future scenarios

Option	S1	S2	S3
A	0	1	2
B	0	0	1
C	0	0	0

Table 6.66 Robustness evaluation of treatment options using the minimax cost criterion

Option	Worst-case cost
A	210
B	237.5
C	260

Table 6.67 Regret analysis of treatment options based on lifecycle costs

Option	Regret			
	S1	S2	S3	Max.
A	0	0	0	0
B	57.5	37.5	27.5	57.5
C	90	70	50	90

Table 6.68 Risk-weighted composite cost scores incorporating compliance penalties

Option	S3 cost	Risk penalty	Composite
A	210	$2 \times 15 = 30$	240
B	237.5	$1 \times 15 = 15$	252.5
C	260	0	260

indicating superior cost performance—though this analysis excludes compliance risks.

Table 6.68 presents adjusted lifecycle costs for each treatment option under Scenario S3 by incorporating compliance risk penalties. A cost of $15 million is added per compliance risk point. The resulting composite scores balance financial cost with regulatory risk. While Option A remains the most cost-effective, Option C offers guaranteed compliance and may be favored by risk-averse stakeholders.

Step 8: Design adaptive pathway

To address uncertainty in future regulatory requirements and wastewater inflow, the city has developed an adaptive pathway, a flexible, phased approach that aligns investments with actual conditions over time rather than relying on a single, fixed infrastructure choice.

- Initial Phase (Years 0–10)

  The plan begins with constructing Option A (an expanded activated sludge system) at a cost of $45 million. This option meets immediate capacity and compliance needs without overcommitting resources. To prepare for potential changes, the city will install real-time nitrogen sensors and inflow meters to actively monitor treatment performance and plant loading. This data will serve as a foundation for informed, evidence-based upgrades.

- Trigger Condition for Upgrade

  A predefined trigger will initiate the next phase of investment if one of the following conditions is met:

  – Inflow exceeds 90% of hydraulic capacity, indicating rapid population growth or increased demand.
  – Stricter effluent regulations are enacted, requiring enhanced nitrogen or phosphorus removal.

- Upgrade Phase (Years 10–15)

  If triggered, the city will implement one of two upgrades depending on updated needs and financial flexibility:

  - Retrofit to Membrane Bioreactors (MBR) with an additional $25 million, providing a balance between cost and performance.
  - Add modular Advanced Treatment Units (ATUs) at a higher cost of $40 million, offering maximum flexibility and nutrient removal capability.

  This phased strategy provides multiple advantages:

- It avoids overdesign, reducing upfront financial burden while still enabling future scalability.
- It meets near-term regulatory and capacity requirements with a cost-effective initial investment.
- It retains the ability to adapt to evolving policies, technologies, and population growth, ensuring the system remains resilient and compliant well into the future.

Step 9: Communicate results and refine

In the final step of the decision-making process, the city's environmental engineering department focuses on clearly communicating the results of the analysis and gathering feedback for possible refinements. This step ensures transparency, builds stakeholder confidence, and supports informed decision-making across departments and the community.

Key Deliverables:

1. Performance Matrix

   A detailed matrix presents each upgrade option's total lifecycle cost alongside its compliance risk score, allowing stakeholders to compare trade-offs and identify the most resilient and cost-effective strategies.
2. Adaptive Pathway Timeline

   A visual timeline illustrates the phased approach over the 25-year horizon, showing when key investments (initial build, monitoring, upgrades) would occur based on observed conditions. This helps convey the flexibility and scalability of the proposed plan.

3. Stakeholder Briefing Document

   A concise summary is prepared for all involved parties—including city council, finance officials, and environmental regulators—explaining:

   - Why Option A is the most cost-effective choice in the short term, satisfying current needs within the budget.
   - Why Option B (MBR) or Option C (ATUs) may be required later to meet stricter effluent standards or rising inflows.
   - The trigger conditions that will prompt an upgrade decision, such as exceeding 90% plant capacity or the introduction of more stringent nutrient discharge regulations.

By presenting these deliverables, the city ensures that stakeholders are well-informed and that the upgrade strategy remains flexible, transparent, and responsive to both technical needs and public expectations.

Final Answers to the Questions

1. What environmental and regulatory uncertainties affect wastewater treatment planning?

   - Future nutrient standards (especially N, P)
   - Population and industrial growth affecting flow
   - Affects capacity and compliance

2. What future scenarios are plausible?

   - **S1**: Conservative growth, current limits (10 mg/L N, 1 mg/L P)
   - **S2**: Moderate growth, stricter (5 mg/L N, 0.5 mg/L P)
   - **S3**: Rapid growth, ultra-stringent (3 mg/L N, 0.1 mg/L P)

3. How do options perform?

   Table 6.69 provides a high-level comparison of the three treatment options based on total lifecycle cost range, compliance risk, and adaptability. Option A is the most cost-efficient but offers limited flexibility and higher regulatory risk. Option B represents a middle ground in both cost and performance. Option C, while the most expensive, stands out for its zero compliance risk and high adaptability, making it a strong candidate for future-proof infrastructure.
4. Most robust to uncertainty?

Table 6.69 Summary performance comparison of treatment options

Option	Cost range ($M)	Compliance risk	Adaptability
A	170–210	Moderate–High	Low
B	227.5–237.5	Low–Moderate	Medium
C	260	Zero	High

Option A (lowest worst-case cost), not robust to compliance risk under stricter futures.

5. Least regret if regulations become stricter?

   Option A demonstrates the lowest worst-case cost, making it the most attractive choice from a purely financial standpoint under uncertain future conditions. Its low initial investment and moderate operating costs ensure that even in adverse scenarios, the city avoids excessive spending—offering a degree of economic resilience.

   However, Option A is not fully robust when considering compliance risk in stricter regulatory futures. If nutrient discharge standards become significantly tighter or if inflow volumes increase beyond expectations, Option A may fail to meet performance requirements without substantial retrofitting. This exposes the city to the risk of non-compliance, potential fines, and emergency capital expenditures.

   Therefore, while Option A offers strong cost predictability under uncertainty, its limited flexibility and regulatory risk mean it should be pursued only within an adaptive framework—with monitoring in place and clear triggers for timely upgrades to more advanced options like MBR or ATUs.

6. Can an adaptive pathway be proposed?

   Yes, an adaptive pathway is not only possible but highly recommended given the uncertainty surrounding future regulatory requirements and wastewater inflow volumes. Indeed, an adaptive pathway starting with Option A and integrating conditional upgrades to Options B or C ensures the city remains both fiscally responsible and environmentally compliant, even in the face of future uncertainty.

6.2.6.3 Example 3—Foundation Strategy for a High-rise in a Seismic Zone

A construction firm is preparing to build a 25-storey commercial high-rise in a moderate-to-high seismic risk region. Due to site variability and uncertain seismic conditions over the building's 100 year design life, selecting an appropriate foundation system is critical to ensure both safety and cost-efficiency. Table 6.70 summarizes three foundation design options for a seismically sensitive structure, evaluating their initial construction cost, inherent seismic resistance, and potential for future upgrades. Given that seismic safety standards evolve with updates to national hazard maps, the selection must balance current performance with adaptability to future code requirements. The selection of an appropriate foundation system for high-rise construction in seismic zones is a multifaceted decision involving geotechnical, structural, and regulatory considerations. As illustrated in Fig. 6.48, the process integrates seismic risk assessment, soil variability, future design code evolution, and practical construction constraints to determine the most resilient and cost-effective foundation strategy. The key questions to be addressed are as follow:

Table 6.70 Foundation alternative options based on cost, seismic performance, and upgrade potential

Option	Description	Initial cost ($M)	Seismic resistance	Upgrade potential
A	Shallow mat foundation	20	Low	None
B	Deep Piles + Grade Beams	30	Medium	Moderate (retrofit possible)
C	Base-isolated raft foundation	45	High	High (scalable damping)

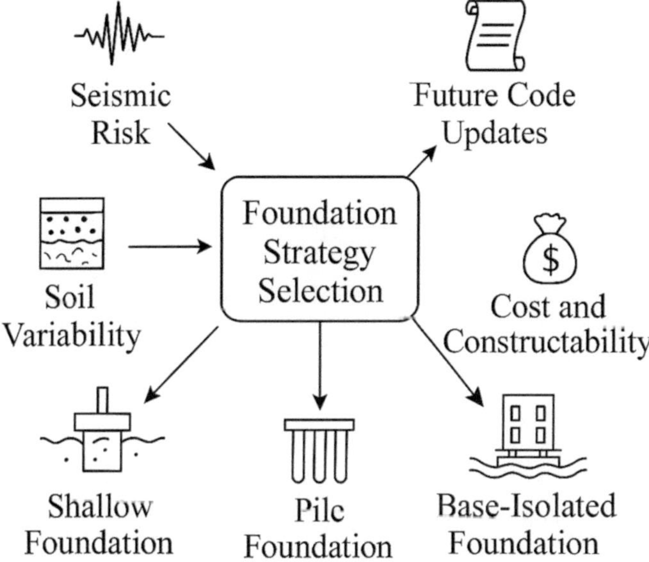

Fig. 6.48 Conceptual diagram of foundation system selection for a high-rise building under seismic uncertainty

1. What are the main uncertainties in choosing a foundation strategy in seismic regions?
2. What plausible future seismic or regulatory scenarios should be considered?
3. How do the foundation options perform under each scenario?
4. Which option is most robust to long-term seismic risk and regulation changes?
5. Which option minimizes regret if the worst-case scenario occurs?
6. Can a staged/adaptive foundation approach be implemented to manage risk and budget constraints?

Step-by-Step Solution Using SBDA

Step 1: Define the Decision Problem

The decision problem centers on selecting the most suitable foundation design for a critical infrastructure project, one that ensures long-term seismic safety, remains cost-effective over its lifecycle, and can adapt to future changes in regulatory standards or seismic hazard assessments. Given the structure's importance and its intended design life of 100 years, the foundation must be resilient not only to current seismic risks but also to potential shifts in hazard profiles due to climate change, tectonic re-evaluation, or evolving engineering standards.

The primary objective of this decision problem is to select a foundation design that ensures both immediate and long-term structural reliability in the face of seismic hazards. Specifically, the selected design must:

- Meet long-term seismic performance and safety requirements, ensuring structural stability and occupant protection throughout the structure's 100-year design life.
- Minimize total lifecycle cost, accounting not only for initial construction expenses but also for long-term maintenance and the potential need for retrofitting due to evolving seismic standards or unforeseen ground behavior.
- Accommodate future changes in regulations or hazard conditions, allowing for adaptability to shifts in the National Building Code of Canada (NBCC), updated seismic risk assessments, or site-specific hazard changes, thereby reducing the risk of obsolescence and unplanned interventions.

To address the project's long-term seismic and economic objectives, three foundation design alternatives have been identified, each with distinct characteristics in terms of performance, cost, and adaptability:

- Option A: Mat Foundation

  A low-cost, shallow-depth foundation ideal for low seismic zones or stiff soils. Offers limited performance in high seismic areas but is attractive from a budgetary standpoint.
- Option B: Deep Pile System

  A standard solution for moderate seismic risk zones, offering greater load-bearing and lateral resistance by anchoring into deeper, more stable strata. It balances performance and cost.
- Option C: Base-Isolated Raft

  A high-cost but high-performance solution using seismic base isolation to minimize earthquake-induced forces. Ideal for critical facilities requiring uninterrupted operation and maximum safety.

The foundation design decision must adhere to two critical constraints that shape the feasibility and acceptability of each alternative:

- Budget Constraint: The preferred foundation solution should cost under \$50 million to align with financial limitations.

- Regulatory Compliance: All options must meet or exceed the seismic provisions of the National Building Code of Canada (NBCC).

Given the 100 year design horizon and the evolving nature of seismic science and regulation, the selected option must provide both initial reliability and future adaptability, ensuring structural integrity and public safety over the long term.

Step 2: Identify Key Uncertainties

In the context of a 100 year design life, long-term planning must account for uncertainties that could significantly influence both the performance and regulatory acceptability of the foundation system over time. Two primary sources of uncertainty have been identified:

1. Seismic Hazard Level: Future ground motion intensities may increase due to tectonic activity or revised models.
2. Regulatory Standards: Future updates to NBCC may impose stricter foundation or damping system requirements for tall buildings.

Step 3: Develop Plausible Scenarios

Table 6.71 outlines three plausible future seismic design scenarios, incorporating projected peak ground acceleration levels and corresponding building code stringency. These scenarios reflect expected updates to seismic hazard classifications and regulatory standards aligned with upcoming National Building Code of Canada (NBCC) cycles, guiding resilient structural design planning.

Step 4: Engage Stakeholders

A comprehensive stakeholder engagement process was conducted to ensure that the foundation design decision reflects diverse priorities. The workshops and consultations involved key groups with vested interests in both the short-term construction and long-term performance of the structure:

- Structural engineers and geotechnical consultants, who provided insights into site-specific soil behavior, seismic performance, and technical feasibility of each foundation alternative.
- Regulatory authorities and insurers, who emphasized the importance of code compliance, liability mitigation, and risk-based assessment of design choices.
- Developers and investors, whose focus was on financial viability, lifecycle cost, and alignment with future market expectations.

Table 6.71 Seismic design scenarios based on ground acceleration and code evolution

Scenario	Description	Ground acceleration (g)	Code stringency
S1	Conservative Forecast	0.25g	Current Code
S2	Moderate Hazard Increase	0.35g	Revised Code (2025)
S3	Severe Earthquake Zone Reclassification	0.50g	Strict Code (2030+)

- Public safety advocates, who prioritized resilience, occupant safety, and system reliability during rare but high-impact seismic events.

Through these engagements, a clear consensus emerged: while initial construction cost remains an important consideration, it should not outweigh more critical priorities. Stakeholders agreed that safety, regulatory compliance, and the flexibility to retrofit in response to future seismic or code changes are essential criteria in the decision-making process. This shared understanding shaped the evaluation framework and highlighted the need for a robust yet adaptable foundation strategy.

Step 5: Define Performance Metrics

To objectively assess the foundation design alternatives, two primary performance metrics were established. These metrics provide a structured way to compare the trade-offs between cost, safety, and long-term resilience across varying future scenarios.

1. **Total Lifecycle Cost**

This metric captures the comprehensive financial burden of each foundation option over the 100-year design life. It includes:

- Initial construction cost
- Expected retrofit cost if future upgrades are needed to meet revised seismic codes or hazard levels
- Insurance premiums, which are adjusted based on the seismic performance and perceived risk of each foundation type

This approach ensures that the evaluation reflects not just upfront spending, but also the long-term economic implications of choosing a less adaptable or risk-prone design.

2. **Compliance Risk Score**

This qualitative metric assesses each option's ability to meet current and future NBCC seismic standards, under various uncertainty scenarios:

- 0 = Full compliance under all modeled hazard and regulatory scenarios (no retrofitting required)
- 1 = Partial compliance; may require moderate retrofitting to maintain compliance under stricter codes
- 2 = Non-compliant or at high risk of failure, necessitating major intervention or posing unacceptable life safety risks

Together, these metrics provide a robust decision framework that balances cost-effectiveness with seismic safety, adaptability, and regulatory foresight.

Step 6: Evaluate Options Under Each Scenario

A. Lifecycle Cost Estimates (in $M)

Table 6.72 presents total lifecycle cost estimates (in millions of dollars) for three foundation options across three seismic hazard scenarios (S1–S3). Costs include initial construction, projected retrofit expenses triggered by evolving code requirements, and long-term insurance premiums. Option C shows the highest upfront cost but avoids future retrofits and maintains the lowest insurance burden.

B. Compliance Risk Scores

Table 6.73 presents the compliance risk scores for each foundation option under three seismic design scenarios (S1–S3). Scores reflect the likelihood that each option will fail to meet future seismic code requirements, with higher values indicating increased regulatory risk. Option C demonstrates full compliance across all scenarios, offering the greatest code resilience.

Step 7: Apply Robustness and Regret Criteria

A. Robustness (Minimax Cost Rule)

Table 6.74 displays the worst-case total lifecycle cost for each foundation option under the most demanding seismic scenario. Applying the minimax rule, Option B emerges as the most robust choice in cost terms, offering the lowest maximum expenditure across all considered futures.

Table 6.72 Lifecycle cost estimates of foundation options under seismic scenarios

Option	Initial	Retrofit cost (S2/S3)	Insurance	Total		
				S1	S2	S3
A	20	+10 (S2), +25 (S3)	5	25	35	50
B	30	0 (S2), +5 (S3) only	3	33	33	38
C	45	0	1	46	46	46

Table 6.73 Compliance risk scores for foundation options across seismic scenarios

Option	S1	S2	S3
A	0	1	2
B	0	0	1
C	0	0	0

Table 6.74 Robustness assessment of foundation options using the minimax cost criterion

Option	Worst-Case Cost
A	50
B	38
C	46

B. Regret Matrix (Based on Cost)

Table 5.93 presents the regret values for each foundation option under three seismic scenarios (S1–S3), where regret represents the additional cost incurred compared to the least-cost option for each scenario. The maximum regret column highlights the worst-case deviation. Option B shows the lowest maximum regret, making it the most regret-averse choice under cost uncertainty (Table 6.75).

Step 8: Design Adaptive Pathway

To manage seismic uncertainty and future-proof the foundation system, a phased adaptive strategy has been proposed. This approach allows the city to minimize initial investment while preserving the ability to upgrade efficiently if seismic risks or regulations increase over time.

Phased Foundation Strategy

Phase 1 (Years 0–5): Immediate Implementation

- Construct Option B: Implement the deep pile system, a well-established and cost-effective solution suitable for current seismic risk levels and within the budget threshold.
- Install Monitoring Sensors: Integrate accelerometers and ground motion sensors to track real-time seismic activity and structural response.
- Design for Future Retrofit: Engineer pile caps and tie beams to be compatible with future base isolator installations, enabling smoother and more economical retrofitting if required.

Next phase will be initiated if either of the following conditions are met:

- The National Building Code of Canada (NBCC) releases updated seismic hazard maps indicating peak ground acceleration (PGA) $\geq 0.4g$ in the region.
- A regional seismic event occurs with a measured PGA $> 0.3g$, suggesting a shift in local hazard behavior or site amplification effects.

Phase 2 (Years 5–20): Conditional Retrofit

- Retrofit with Base Isolators: For an estimated cost of $10 million, base isolators are added to reduce lateral displacements during seismic events.
- This upgrade enhances structural resilience, reduces damage potential, and effectively extends the service life of the facility.

Table 6.75 Regret analysis of foundation options based on lifecycle costs

Option	Regret			
	S1	S2	S3	Max.
A	0	2	12	12
B	8	0	0	8
C	21	13	8	21

The adaptive foundation strategy offers several key advantages that make it a practical and future-ready solution for long-term infrastructure planning:

- Minimizes upfront capital expenditure by avoiding immediate investment in high-cost isolation systems.
- Maintains compliance and safety under current conditions while remaining ready for future updates.
- Preserves flexibility to adapt infrastructure based on real-world seismic data and evolving codes.

This adaptive pathway strikes a prudent balance between fiscal responsibility, seismic safety, and long-term resilience, ensuring that the foundation system remains robust, upgrade-ready, and regulation-compliant across its 100-year design horizon.

Step 9: Communicate Results and Refine

In the final step of the decision-making process, the project team focuses on effectively communicating the analysis results to key decision-makers and stakeholders, ensuring transparency, understanding, and the opportunity for refinement based on feedback. The goal is to facilitate an informed choice that aligns with technical, financial, and public safety priorities. To support this decision, the project team compiled and presented a set of targeted materials designed to clearly convey the trade-offs, adaptive strategies, and associated risks of each option as follow;

1. Cost + Risk Performance Matrix

   A comparative matrix outlining each foundation option's total lifecycle cost and compliance risk score, providing a clear visualization of the trade-offs between financial investment and regulatory/safety performance.
2. Adaptive Retrofit Flowchart

   A visual roadmap that illustrates the phased foundation strategy, including initial implementation, trigger conditions, and potential upgrade pathways. This helps stakeholders understand how flexibility is embedded into the design process.
3. Risk Communication Brief

   A concise summary explaining the strategic implications of each option:

   - Option A: Lowest initial cost but carries significant compliance and seismic performance risks in the long term.
   - Option B: Strikes a balanced approach, offering moderate cost, retrofit-ready design, and flexibility for future upgrades.
   - Option C: Provides the highest safety, but at a high upfront cost that may be unjustified unless extreme seismic conditions are confirmed.

Based on the presented evidence and discussions, stakeholders support Option B with built-in upgrade triggers, recognizing it as the solution that best balances long-term resilience, regulatory adaptability, and cost-effectiveness. This approach provides robust protection now while allowing strategic enhancement if seismic risks increase—ensuring the foundation system remains reliable and future-ready for its full 100-year lifespan.

Answers to the questions

1. What are the main uncertainties in choosing a foundation strategy?
   - Changes in PGA (Peak Ground Acceleration)
   - Updates to building code requirements
2. What plausible seismic scenarios should be considered?
   - S1: Conservative (0.25 g), current NBCC
   - S2: Moderate increase (0.35 g), 2025 code
   - **S3**: High risk reclassification (0.50 g), 2030+ code
3. How do options perform under each scenario?
   - A: Non-compliant in S3, high retrofit costs
   - B: Reliable and upgradable
   - C: Always compliant, most expensive
4. Which is most robust?

   Option B—lowest worst-case cost ($38M)
5. Which minimizes regret?

   Option B—lowest max regret ($8M), with compliance and retrofit flexibility
6. Can an adaptive approach be used?

   Yes—start with Option B, retrofit to C if risk increases.

6.2.6.4 Example 4—Urban Drainage System Upgrade Under Climate Change Uncertainty

A mid-sized city is increasingly experiencing more frequent and intense rainfall events, causing urban flooding and stressing existing stormwater infrastructure. The city needs to upgrade its drainage system, but there is deep uncertainty about how climate change will affect rainfall intensity, duration, and frequency over the next 30 years. Table 6.76 outlines three design options for upgrading a mid-sized city's stormwater drainage system in response to increasing flood risks. Each option is evaluated based on initial capital cost, annual maintenance cost, and climate adaptability. The alternatives range from conventional infrastructure improvements to advanced adaptive systems integrating green infrastructure and smart technologies, supporting resilience planning over a 30-year horizon. A conceptual overview of the alternative drainage strategies considered in this study is illustrated in Fig. 6.49. This figure emphasizes the differences in system design logic and climate adaptability—ranging from conventional upgrades to smart hybrid solutions—without committing to specific quantitative estimates. It serves to visually communicate the core characteristics of each approach within the broader context of deep climate uncertainty affecting urban flood resilience planning. The key questions that should be addressed are as follows:

1. What climate-related uncertainties impact the drainage design decision?
2. What future climate scenarios should be considered?
3. How does each design option perform across these scenarios?
4. Which option is most robust to climate variability?
5. Which option minimizes regret if the most extreme rainfall scenario occurs?
6. Is it possible to adopt a phased, adaptive drainage strategy?

Step-by-Step Solution Using SBDA

Table 6.76 Urban drainage upgrade alternatives under climate uncertainty

Option	Description	Initial cost ($M)	Annual maintenance ($M)	Climate adaptability
A	Conventional pipe network upgrade	40	2	Low
B	Distributed Green Infrastructure (GI)	55	2.5	Medium
C	Smart adaptive hybrid system	70	3	High (IoT + storage)

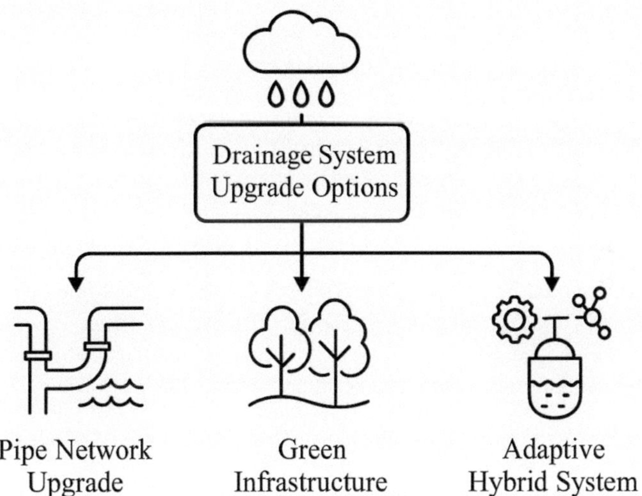

Fig. 6.49 Conceptual representation of urban drainage system upgrade alternatives under climate uncertainty

Step 1: Define the Decision Problem

The decision problem focuses on identifying the most suitable urban drainage solution to manage increasing flood risks associated with climate uncertainty, aging infrastructure, and stricter environmental standards. The solution must provide long-term flood resilience, maintain cost-effectiveness, and comply with water quality discharge regulations over a 30 year planning horizon. The main objective is to select a drainage solution that:

- Minimizes flood risk across a range of future climate scenarios (e.g., increased rainfall intensity, storm surges)
- Remains cost-effective, accounting for both capital expenditure (capEx) and ongoing operational and maintenance costs (opEx)
- Adapts to climate variability and evolving hydrological patterns over time

To reach the aforementioned objectives, three alternative drainage solutions have been identified:

- Option A: Conventional Pipe Resizing

  Involves upsizing stormwater pipes to handle greater flow volumes. This traditional approach is straightforward and relatively low-tech but can be costly and inflexible, especially under high-uncertainty futures.
- Option B: Green Infrastructure

  Uses nature-based solutions such as bioswales, rain gardens, and permeable pavements to slow runoff, promote infiltration, and improve water quality. This option is ecologically beneficial and cost-effective for small to mid-scale areas, though performance may vary with extreme events.
- Option C: Hybrid Smart System

  Combines physical infrastructure with real-time monitoring and control technologies (e.g., smart gates, sensors, adaptive storage) to dynamically manage flows. While initially more expensive and technologically complex, this option offers high adaptability and data-driven optimization under uncertain climate futures.

It should be noted that there are several key constraints that must be considered when evaluating and selecting the appropriate drainage solution. These constraints ensure that the chosen alternative not only meets the technical objectives but also aligns with financial, regulatory, and operational boundaries:

- **Budget**: The combined capital and operational cost must not exceed $100 million over the 30 year period.
- **Regulatory Compliance**: The selected system must meet water quality discharge guidelines, ensuring pollutants like sediment, nutrients, and heavy metals are adequately managed.
- **Timeframe**: The solution must provide effective service for a minimum of 30 years, supporting both short-term flood protection and long-term climate resilience.

Step 2: Identify Key Uncertainties

When planning a long-term urban drainage solution, it is critical to account for uncertainties that could significantly affect system performance and cost-effectiveness over the 30-year lifespan. Two major sources of uncertainty have been identified:

1. **Future Rainfall Extremes**

The most prominent uncertainty relates to changes in rainfall intensity, duration, and frequency, commonly represented by IDF (Intensity-Duration-Frequency) curves. Climate change is expected to alter precipitation patterns, potentially increasing the occurrence of short-duration, high-intensity storms or prolonged rainfall events. However, the magnitude and timing of these changes are difficult to predict with precision, making it challenging to size or design drainage infrastructure with confidence.

2. **Urban Development Density**

As the city continues to grow, population increases and urban densification will likely result in a higher percentage of impervious surfaces (e.g., roads, rooftops, parking lots). This change directly increases surface runoff and peak flow rates during storm events. The rate and distribution of urban development are uncertain, which adds complexity to forecasting future drainage loads and designing a system that remains effective under changing land use conditions.

These uncertainties reinforce the importance of selecting a drainage strategy that offers adaptability, allowing for performance optimization as more data becomes available and conditions evolve over time.

Step 3: Develop Plausible Climate Scenarios

Table 6.77 defines three future scenarios combining projected climate change impacts and urban development trends. Each scenario includes estimates of peak 24 h rainfall intensity and increases in impervious surface area, which influence runoff volumes and drainage system performance. These scenarios guide robust design and risk assessment for urban flood mitigation over the long term.

Table 6.77 Plausible climate and urbanization scenarios for stormwater planning

Scenario	Description	Peak 24 Hr Rainfall (mm)	Impervious area increase (%)
S1	Conservative warming	100	5
S2	Moderate warming and urban sprawl	125	15
S3	High-end warming and urbanization	150	30

Step 4: Engage Stakeholders

To ensure that the drainage solution reflects a broad range of perspectives and meets both technical and community needs, a comprehensive stakeholder engagement process was undertaken. This step was essential in identifying shared priorities, surfacing concerns, and building consensus around key performance criteria under future uncertainty.

The following stakeholder groups were actively engaged to ensure the proposed solution reflected a broad range of expertise, values, and community needs:

- City Planners and Engineers

  Provided input on infrastructure feasibility, integration with existing systems, and alignment with urban development goals. Their expertise helped frame realistic implementation pathways and maintenance considerations.
- Environmental NGOs

  Advocated for sustainable and nature-based solutions that support ecological health, improve water quality, and contribute to urban climate resilience. They emphasized minimizing environmental degradation and promoting green infrastructure.
- Local Residents and Businesses

  Raised concerns about flood risk, property protection, and service disruption during and after extreme weather events. Their input highlighted the importance of visible benefits, community co-benefits, and minimal long-term maintenance burdens.
- Climate Scientists and Modelers

  Contributed insights on future rainfall projections, urban climate trends, and hydrologic modeling. They helped quantify uncertainty in rainfall extremes and identify adaptive thresholds relevant for system design.

Through collaborative discussions, stakeholders supported a performance-driven approach that considers lifecycle cost, flood mitigation effectiveness, regulatory compliance, and adaptive capacity. Their engagement played a vital role in shaping a balanced and forward-looking drainage strategy.

During the engagement process, stakeholders collectively emphasized several key priorities that should guide the selection and design of the urban drainage solution. These priorities reflect a balance between long-term resilience, equity, and fiscal responsibility:

- Adaptability

  Stakeholders agreed that the chosen system must be flexible and responsive to future uncertainties, including climate change, regulatory updates, and urban expansion. Solutions that allow for modular upgrades, real-time monitoring, or scalable components were favored over rigid, one-time designs.

- Protection of Vulnerable Areas

  Ensuring the safety of flood-prone and socially vulnerable neighborhoods was identified as a top priority. These areas often face the greatest risk from stormwater flooding and have limited capacity to recover from disruptions. As such, drainage strategies must prioritize equity and risk reduction for these communities.
- Avoid Overspending on Overdesigned Systems

  Stakeholders cautioned against investing heavily in large-scale infrastructure that may prove unnecessary if extreme conditions do not materialize. Instead, they preferred cost-effective solutions that provide reliable performance now but retain the option for enhancement if and when future conditions demand it.

These priorities reinforce the need for an adaptive, balanced approach, one that delivers effective protection without locking the city into inflexible or financially burdensome infrastructure.

Step 5: Define Performance Metrics

To evaluate the drainage alternatives in a structured and transparent manner, two key performance metrics have been established. These metrics capture both economic and functional aspects of system performance over the 30-year planning horizon, allowing for informed comparisons under uncertainty.

1. Total Lifecycle Cost

This metric represents the full financial burden of each drainage option over its operational lifespan. It includes:

- Initial capital investment (construction and implementation costs)
- Ongoing maintenance and operational expenses
- Expected flood damage costs, which account for projected losses (e.g., property damage, service disruptions) from storm events that exceed the system's capacity

By integrating both upfront and long-term consequences, this metric ensures that decision-makers consider not only the cheapest option today, but the most economically sustainable solution over time.

2. Flood Failure Rate

This metric quantifies system reliability by calculating the percentage of modeled storm events (based on future climate scenarios) in which the drainage system is overwhelmed—resulting in surface flooding or system bypass. A lower failure rate indicates higher resilience and more consistent protection, particularly for vulnerable areas.

Together, these performance metrics provide a balanced lens through which to assess cost-effectiveness and flood protection reliability, both of which are essential for making a robust, future-oriented infrastructure decision.

Step 6: Evaluate Options Under Each Scenario

Table 6.78 presents the 30 year lifecycle cost estimates for three stormwater infrastructure upgrade options, including initial capital costs, maintenance, and scenario-dependent flood damage. Total costs are calculated for each climate scenario (S1–S3), reflecting how future rainfall intensification and urbanization impact financial performance. Option C, though highest in upfront cost, results in the lowest total cost under the most severe scenario (S3).

Table 6.79 shows the projected flood failure rates—defined as the probability of system underperformance resulting in flooding—for each stormwater management option across three climate scenarios (S1–S3). Option C demonstrates the highest reliability, maintaining low failure rates even under severe rainfall and urbanization conditions.

Step 7: Apply Robustness and Regret Criteria

Table 6.80 summarizes the worst-case (maximum) total lifecycle costs for each drainage upgrade option across all climate scenarios. According to the minimax rule, Option C is the most robust choice, offering the lowest maximum cost and thus the greatest protection against future uncertainty in rainfall intensity and urban growth.

Table 6.81 presents the regret values for each stormwater infrastructure option under three climate scenarios (S1–S3). Regret is calculated as the additional cost

Table 6.78 lifecycle cost estimates of urban drainage options across climate scenarios

Option	Initial	Maint. (30 yrs)	Flood damage (S1–S3)	Total		
				S1	S2	S3
A	40	60	20, 60, 100	120	160	200
B	55	75	10, 30, 60	140	160	190
C	70	90	5, 10, 15	165	170	175

Table 6.79 Estimated flood failure rates of drainage options under climate scenarios

Option	S1 (%)	S2 (%)	S3 (%)
A	10	25	40
B	5	15	25
C	2	4	7

Table 6.80 Robustness evaluation of drainage options using the minimax total cost criterion

Option	Worst-case cost
A	200
B	190
C	175

Table 6.81 Regret analysis of drainage options based on lifecycle costs

Option	Regret			
	S1	S2	S3	Max.
A	0	0	25	25
B	20	0	15	20
C	45	10	0	45

incurred relative to the lowest-cost option in each scenario. The maximum regret indicates the worst-case decision penalty. Option B demonstrates the lowest maximum regret, making it the most regret-averse choice under cost uncertainty.

Step 8: Design Adaptive Pathway

To manage uncertainty in future rainfall extremes and urban development, a two-phase adaptive drainage strategy has been developed. This approach allows the city to begin with a low-impact, cost-effective solution and scale up only if real-world conditions warrant additional investment—ensuring both resilience and fiscal efficiency.

Two-Phase Adaptive Strategy

Phase 1 (Years 0–10): Initial Implementation

- Begin with Option B: Green Infrastructure, focusing on distributed solutions such as bioswales, permeable pavements, and rain gardens.
- Employ a modular build approach, allowing for gradual expansion as needed.
- Install rainfall monitoring systems and initiate a tracking program for flood-related complaints to assess system performance over time.

Transition to Phase 2 if either of the following trigger conditions is met:

- Peak rainfall events exceeding 130 mm occur three or more times within a 10-year period
- Annual flood-related complaints exceed 10 incidents, indicating persistent performance gaps or growing public concern

These conditions signal a significant increase in system stress or community impact, justifying the need for a technological upgrade.

Phase 2 (Years 10–20): Conditional Retrofit

- Upgrade to Option C: Hybrid Smart System, incorporating:
  - Smart sensors for real-time monitoring of flow, depth, and rainfall
  - Detention tanks to temporarily store excess runoff
  - Control valves for dynamic flow regulation and system optimization
- Estimated cost for this upgrade phase is approximately $30 million, a fraction of the total budget cap, leaving room for further adaptation if necessary.

The adaptive drainage strategy offers several compelling benefits that align with the city's goals of resilience, fiscal responsibility, and regulatory compliance:

- Avoids overdesign by deferring high-cost upgrades until justified by observed trends
- Provides scalable protection that evolves with climate and urbanization
- Supports data-informed decision-making, enhancing both public confidence and regulatory alignment

This phased, flexible pathway ensures the drainage system remains cost-effective, environmentally sustainable, and resilient to future challenges over the 30-year design horizon.

Step 9: Communicate Results and Refine

The final step in the planning process involved presenting a comprehensive set of results to stakeholders and decision-makers to ensure transparency, support informed evaluation, and allow for refinements based on feedback. To facilitate this, the team prepared a series of deliverables as:

- Cost versus Risk Matrices

  Comparative visualizations that show how each drainage option performs in terms of total lifecycle cost and flood failure rate across different climate and urbanization scenarios. These matrices help stakeholders quickly understand trade-offs between investment and resilience.
- Floodplain Maps by Scenario

  Scenario-based geospatial maps illustrating projected flood extents under each option. These maps provide a visual reference for assessing the protection of vulnerable areas and the impact of different rainfall intensities or land-use changes.

During the stakeholder briefing, the results were summarized to clarify the implications of each alternative:

- Option A (Conventional Pipe Resizing) is the least resilient, offering limited adaptability and higher long-term risk under intensifying storm conditions.
- Option B (Green Infrastructure) is scalable and flexible, allowing for phased expansion and real-time monitoring, and performs well under moderate future scenarios.
- Option C (Hybrid Smart System) is the most robust, but its high initial cost may not be justified without clear evidence of worsening climate impacts.

Based on this analysis and stakeholder consensus, the city has decided to proceed with Option B, supported by an adaptive trigger plan that calls for upgrading to a smart hybrid system (Option C) if specific climate thresholds are exceeded in the future. This outcome strikes a strong balance between affordability, system flexibility, and long-term climate resilience.

Table 6.82 Overall performance summary of drainage design options

Option	Robustness	Cost efficiency	Flood risk Mitigation
A	Low	High (short-term)	Poor (fails in S2/S3)
B	Medium	Moderate	Moderate to good
C	High	Costly upfront	Excellent

Answers to the Key Questions

1. What climate-related uncertainties impact the drainage design decision?
   - Rainfall extremes: intensity, duration, and frequency
   - Urbanization pressures increasing impervious surface area
2. What future climate scenarios should be considered?
   - S1: Minor warming, light rainfall increase
   - S2: Moderate emissions + urban sprawl
   - S3: Severe climate with 150 mm rainfall and dense urbanization
3. How does each design option perform?
   Table 6.82 provides a qualitative assessment of each stormwater design option across three key criteria: robustness to future uncertainty, cost efficiency, and flood risk mitigation. Option A offers short-term cost advantages but lacks resilience under future climate scenarios. Option B provides a balanced middle-ground performance. Option C, while costlier upfront, excels in both robustness and long-term flood protection, making it the most resilient and future-ready option.
4. Which option is most robust to climate variability?
   Option C—robust even under worst-case (S3)
5. Which option minimizes regret?
   Option B—lowest maximum regret (20 M)
6. Is it possible to adopt a phased, adaptive drainage strategy?
   Yes—Start with Option B, upgrade to C only if climate extremes emerge.

6.2.6.5 Example 5—Municipal Solid Waste Management under Future Uncertainty

A regional municipality is tasked with selecting a long-term waste treatment strategy that addresses growing environmental, social, and regulatory pressures. The core objectives are to reduce dependence on landfilling, ensure compliance with environmental mandates, and advance broader sustainability goals such as resource recovery and emissions reduction.

This decision has become urgent due to several converging factors. First, the current landfill site is approaching its capacity limits, raising concerns about long-term waste disposal viability. Second, there is growing public opposition to continued landfill expansion, driven by concerns over environmental justice, odor, traffic, and land use. Lastly, the decision is complicated by key uncertainties, including future waste generation rates, which are influenced by population growth and consumption patterns; levels of participation in recycling programs, which fluctuate based on education, incentives, and convenience; and the enforcement and impact of future zero-waste policies, which remain difficult to predict but could drastically reshape waste streams.

Given these pressures and uncertainties, the municipality must identify a strategy that not only performs well under current conditions but also remains robust and adaptable over the coming decades.

Table 6.83 presents three municipal solid waste management options under consideration. Each alternative is assessed based on its initial investment, annual operation and maintenance (O&M) costs, potential for greenhouse gas emission reduction, and level of public acceptance. The comparison highlights trade-offs between economic feasibility, environmental performance, and societal support, aiding in the selection of a balanced and sustainable waste strategy.

The Key questions that should be addressed are as follows:

1. What uncertainties affect waste management decisions?
2. What scenarios should be considered?
3. How do options perform across scenarios?
4. Which is most robust to environmental/regulatory risk?
5. Which minimizes regret if zero-waste policies are enforced?
6. Can the system be made flexible and adaptive?

The selection of an appropriate municipal solid waste treatment strategy is complicated by uncertainties in future waste generation, citizen participation in recycling programs, and the degree of enforcement of zero-waste policies. To conceptually represent the decision context, Fig. 6.50 presents a schematic of the strategic alternatives under consideration. This visual framework emphasizes the influence of

Table 6.83 Evaluation of solid waste management alternatives based on cost, emissions, and social acceptance

Option	Description	Initial cost ($M)	O&M ($M)	Emission reduction	Public acceptance
A	Landfill Expansion + Basic Recycling	30	4	Low	Low
B	Mechanical-Biological Treatment (MBT)	60	5.5	Medium	Moderate
C	Waste-to-Energy (WTE, incineration)	90	6.5	High	Low–Moderate

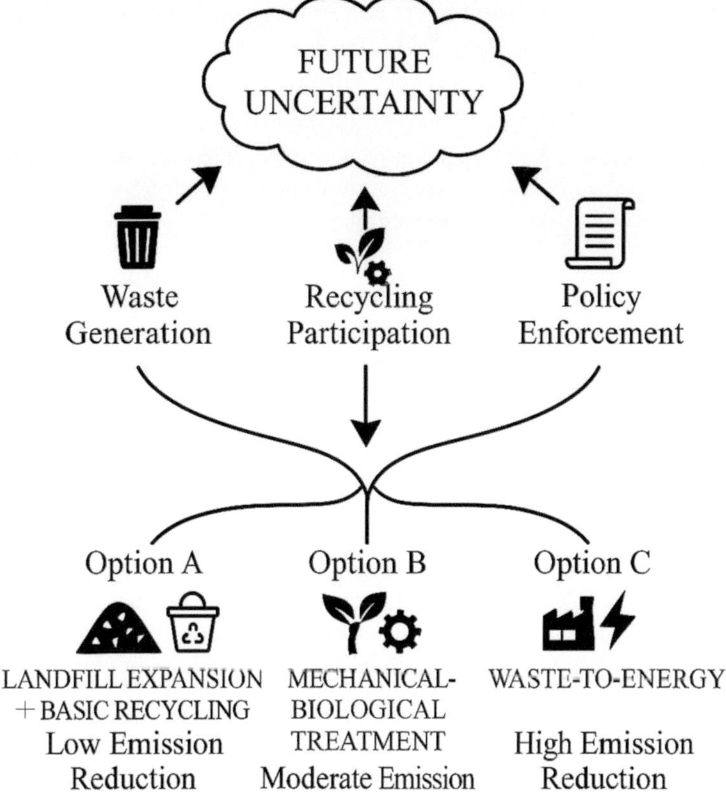

Fig. 6.50 Conceptual schematic of municipal solid waste management strategy selection under future uncertainty

systemic uncertainty on the relative performance and societal acceptability of landfill expansion, MBT, and WTE solutions.

Step-by-Step Solution Using SBDA

Step 1: Define the Decision Problem

The municipality faces a complex decision regarding the future of its regional waste management system. The primary objective is to select a strategy that minimizes environmental impact, manages long-term costs, and reduces the potential for social backlash, all while ensuring reliable waste treatment over a 25-year planning horizon. To achieve this, three alternative options are under consideration:

- Option A: Extend Landfill + Basic Recycling

  A continuation of the current model, involving modest landfill expansion and maintenance of existing recycling services. This is the most familiar and initially least expensive option, but it offers limited long-term resilience and is likely to face intense public opposition and increasing environmental scrutiny.

- Option B: Mechanical-Biological Treatment (MBT)

  Involves sorting mixed waste and processing organic fractions through biological stabilization (e.g., composting or anaerobic digestion). This option offers moderate diversion rates, is more environmentally friendly, and may reduce the volume of residual waste sent to landfills.

- Option C: Waste-to-Energy (WTE) Incineration with Power Generation

  Converts non-recyclable waste into usable energy through combustion, reducing landfill volume significantly. This option offers the highest diversion potential but may provoke public resistance due to emissions concerns and typically involves higher capital costs.

The decision is constrained by two key factors: the existing landfill has only 10 years of remaining capacity, and there is a possibility that a 75% waste diversion mandate will be enforced in the near future. These constraints necessitate the selection of a system that can either substantially reduce landfill dependence or enable rapid adaptation to meet stringent diversion targets.

Over the 25-year timeframe, the chosen system must perform reliably despite uncertainties in waste generation, public behavior, and future policy enforcement—making adaptability and long-term sustainability critical to the decision-making process.

Step 2: Identify Key Uncertainties

To make a resilient and forward-looking decision, the municipality must consider critical uncertainties that could significantly affect the performance, feasibility, and public acceptance of each waste treatment option over the 25-year horizon.

1. **Waste Generation Volume**

One major uncertainty lies in the future volume of municipal solid waste generated. This is closely tied to population growth, urban density, and consumer behavior, including packaging trends, product lifespans, and material choices. If waste generation increases faster than projected, the system may become overwhelmed; conversely, successful waste reduction initiatives could reduce the throughput required. Planning for a system that performs under both scenarios is essential.

2. **Regulatory Pressure**

Another key uncertainty involves the evolution of environmental regulations. The region may soon face stricter mandates, such as landfill bans on organics or recyclables, or the enforcement of a 75% diversion target. There is also the potential for carbon pricing, extended producer responsibility (EPR) programs, or zero-waste legislation, all of which could redefine operational and financial requirements for waste systems. The inability to meet such requirements could lead to fines, retrofits, or the loss of public and regulatory trust.

These uncertainties highlight the need for a solution that is not only cost-effective and low-impact today, but also flexible enough to adapt to shifting waste volumes and tightening regulations in the decades ahead.

Step 3: Develop Plausible Scenarios

Table 6.84 outlines three potential future scenarios combining changes in waste generation, diversion regulations, and policy frameworks. The scenarios range from stable growth with voluntary measures to a transformative zero-waste strategy involving strict regulatory enforcement and landfill bans. These projections support long-term planning and stress-testing of waste management alternatives under evolving environmental and policy conditions

Step 4: Engage Stakeholders

An inclusive stakeholder engagement process was conducted to ensure that the selection of a waste treatment strategy reflects both technical expertise and public priorities. Participants included waste engineers, who contributed operational and design perspectives; health and environmental regulators, who focused on compliance and long-term sustainability; local residents, who voiced concerns about quality of life and environmental justice; and recyclers, who emphasized material recovery and market viability.

Throughout the consultations, several core concerns emerged. Cost transparency was a consistent theme, with participants seeking clear, long-term financial implications of each option. Air quality concerns—especially related to emissions from Waste-to-Energy systems—were raised, particularly by residents and public health advocates. Legal compliance was flagged as a non-negotiable requirement, especially in light of potential diversion mandates and environmental enforcement. Finally, stakeholders stressed the importance of minimizing social backlash, noting that any solution perceived as unjust, polluting, or imposed without community input could face strong public resistance.

These discussions reinforced the need for a strategy that is technically sound, legally compliant, publicly acceptable, and financially clear, with room for future adaptation as regulatory and societal expectations evolve.

Table 6.84 Plausible future scenarios for solid waste management planning

Scenario	Description	Waste volume	Diversion regulation	Policy type
S1	Stable growth, Moderate policy	50,000 t	50% voluntary	Incentives-based
S2	Increased waste, Aggressive policy	70,000 t	75% mandatory	Regulatory
S3	Urban boom, Zero-waste transition	90,000 t	90% mandatory	Landfill Ban

Step 5: Define Performance Metrics

To evaluate the three waste treatment options in a structured and transparent manner, a set of performance metrics has been established. These metrics reflect the priorities identified by stakeholders—cost-efficiency, regulatory compliance, and social acceptance—and are designed to support robust comparison under future uncertainties.

1. Lifecycle Cost

This metric captures the total financial burden over the system's 25 year lifespan, calculated as:

Lifecycle Cost = Initial Capital Investment + Operations & Maintenance (O&M) Costs + Potential Penalties (e.g., for non-compliance with diversion mandates or emissions limits).

This comprehensive cost view ensures that both short-term and long-term financial implications are considered in the decision.

2. Compliance Risk Score

This qualitative score reflects each option's ability to meet current and potential future regulations. The scale is defined as:

- 0 = Fully Compliant under all projected regulatory scenarios
- 1 = Partially Compliant, requiring moderate upgrades or adjustments
- 2 = Non-Compliant under likely future mandates, posing legal and operational risksThis metric emphasizes regulatory resilience
- and helps flag options that may require costly or disruptive retrofits.

3. Public Acceptance

This metric assesses the anticipated level of community support or resistance, based on stakeholder feedback and public sentiment trends. Rated qualitatively as:

- Low: High potential for public opposition or legal challenge
- Medium: Some concerns or controversy, but generally manageable
- High: Broad support or strong alignment with public values

  This dimension is particularly important for identifying options likely to face social backlash, project delays, or reputational risks.

Together, these metrics provide a well-rounded framework for comparing the environmental, financial, regulatory, and social dimensions of each waste treatment strategy.

Step 6: Evaluate Options Under Each Scenario

Table 6.85 presents 25 year lifecycle cost estimates (in millions of dollars) for three solid waste management options, including initial capital expenditures, operation and maintenance (O&M) costs, and scenario-specific penalty costs associated with non-compliance to diversion or landfill regulations. Scenario S3 reflects the most stringent policy environment. Option C, while the most expensive upfront, incurs no penalties, offering full compliance across all scenarios.

Table 6.86 outlines the compliance risk associated with each waste management option under three future policy scenarios (S1–S3). Risk scores indicate the likelihood of failing to meet regulatory requirements, with higher values reflecting greater non-compliance risk. Option C maintains full compliance across all scenarios, demonstrating the highest regulatory resilience.

Step 7: Apply Robustness and Regret Criteria

A. Robustness (Minimax Cost Rule)

Table 6.87 presents the worst-case total lifecycle costs for each waste management alternative across all policy scenarios (S1–S3). Based on the minimax rule, Option A is the most robust in cost terms, offering the lowest maximum expenditure, though it carries higher compliance risks under stricter future regulations.

Table 6.88 shows the regret values for each waste management option under three policy scenarios (S1–S3), where regret represents the additional cost incurred relative to the lowest-cost option in each scenario. The maximum regret indicates the worst-case deviation from optimality. Option A has the lowest maximum regret—especially favorable if future regulations remain lenient—but may face compliance challenges under stricter scenarios.

Table 6.85 Lifecycle cost estimates of waste management alternatives under future policy scenarios

Option	Initial	O&M (25 yrs)	Penalty			Total		
			S1	S2	S3	S1	S2	S3
A	30	100	0	20	50	130	150	180
B	60	137.5	0	0	15	197.5	197.5	212.5
C	90	162.5	0	0	0	252.5	252.5	252.5

Table 6.86 Compliance risk scores for waste management options across policy scenarios

Option	S1	S2	S3
A	0	1	2
B	0	0	1
C	0	0	0

Table 6.87 Robustness evaluation of waste management options using the minimax cost criterion

Option	Worst-case cost
A	180
B	212.5
C	252.5

Table 6.88 Regret analysis of waste management options across policy scenarios

Option	S1	S2	S3	Max regret
A	0	0	30	30
B	67.5	0	15	67.5
C	122.5	55	0	122.5

Step 8: Design Adaptive Waste Management Pathway

To accommodate regulatory uncertainty and evolving waste generation trends, an adaptive two-phase waste management strategy has been proposed. This approach balances near-term feasibility with long-term flexibility, allowing the municipality to meet current needs while staying prepared for future mandates.

Phase 1 (Years 0–10): Initial Implementation

The pathway begins with Option B: Mechanical-Biological Treatment (MBT) at an estimated cost of $60 million. This system enables moderate diversion by mechanically sorting recyclables and biologically stabilizing organic waste. During this phase, the municipality will also expand recycling programs to increase diversion rates and reduce residual waste sent to landfill. MBT offers immediate environmental benefits and aligns with public preference for lower-impact solutions.

Trigger for Phase 2

A shift to the next phase is initiated if either of the following conditions occur:

- A diversion target greater than 75% is formally mandated
- A landfill ban on organics or recyclables is implemented

These conditions reflect tightening regulatory pressure that MBT alone may not be able to meet.

Phase 2 (Years 10–20): Conditional Upgrade

If the trigger is activated, the strategy transitions to a hybrid system by retrofitting the MBT facility with a Waste-to-Energy (WTE) combustion line, including advanced emissions control technology. This retrofit, estimated at $40 million, significantly enhances diversion by converting non-recyclable waste into energy, while reducing landfill dependency and satisfying stricter environmental regulations.

This adaptive pathway allows the city to avoid overinvesting upfront, while ensuring it can scale up strategically if regulatory demands intensify. It also balances cost, compliance, and community acceptance, maintaining a flexible framework for long-term waste management success.

Step 9: Communicate Results and Refine

The final stage focused on presenting the analysis to stakeholders and decision-makers through clear, accessible materials that support informed evaluation and iterative refinement. The team delivered a performance matrix comparing the three options across key metrics: lifecycle cost, compliance risk, and public acceptance.

This allowed for side-by-side evaluation of trade-offs under current and projected future conditions.

In addition, visual comparisons of compliance risks were provided, illustrating each option's vulnerability to regulatory shifts—particularly under stricter diversion mandates or landfill bans. To support flexible long-term planning, the team also shared an adaptive pathway diagram that outlined how the city could begin with Option B and transition to Option C if policy thresholds are crossed.

Following review and discussion, stakeholders reached consensus on proceeding with Option B (MBT) as the initial system, supported by a conditional upgrade plan to Option C (Waste-to-Energy) if future regulations or performance indicators warrant the shift. This approach was recognized for its balance of cost-efficiency, environmental responsibility, and adaptability to evolving policy and public expectations.

Answers to Key Questions

1. What uncertainties affect waste management decisions?

   The decision is shaped by two key uncertainties:

   - Waste volume growth, which depends on population, consumption trends, and behavioral changes in waste generation.
   - Regulatory tightening, including the potential introduction of stricter diversion mandates or landfill bans on recyclables and organics.

2. What scenarios should be considered?

   Three future scenarios were defined to assess system performance under varying conditions:

   - S1: Low growth with voluntary diversion programs (minimal policy pressure).
   - S2: Moderate growth paired with a 75% diversion mandate (likely near-future regulation).
   - S3: Urban boom accompanied by landfill bans and rapid regulatory tightening.

3. How do options perform across scenarios?

   A comparison of performance by flexibility, compliance risk, and cost is summarized in Table 6.89:

Table 6.89 Comparative summary of waste management options by flexibility, compliance risk, and cost

Option	Flexibility	Compliance risk	Cost
A	Low	High (Fails in S2/S3)	Low
B	Medium	Marginal in S3	Medium
C	High	Fully compliant	High

4. Which is most robust to environmental/regulatory risk?

   Option B is considered the most robust, offering a balanced combination of flexibility and cost-effectiveness while maintaining reasonable compliance potential across all scenarios.
5. Which minimizes regret if zero-waste policies are enforced?

   Option B again minimizes regret. It avoids non-compliance in Scenario 3 while preventing overspending in Scenario 1, where heavy infrastructure may not be necessary.
6. Can the system be made flexible and adaptive?

   Yes. The proposed strategy starts with Option B, integrates trend monitoring, and includes a conditional upgrade to Option C if triggers such as landfill bans or diversion mandates are enacted. This ensures both near-term feasibility and long-term adaptability under uncertain futures.

6.2.6.6 Example 6—Sustainable Material Selection for a Public Building Project

The municipal government's plan to construct a new multi-purpose public building, serving as a library, community center, and office space, presents an opportunity to align infrastructure development with long-term sustainability goals, specifically the target of carbon neutrality by 2050. The city must carefully select a material strategy that balances environmental impact, budgetary limits, and structural durability. The selected strategy should:

- Minimize embodied carbon and total lifecycle environmental impact,
- Remain within the allocated budget, ensuring cost-effectiveness for taxpayers,
- Maintain acceptable levels of durability, especially given the building's expected lifespan and functional demands.

The decision is influenced by several uncertainties:

- Future carbon pricing policies, which could significantly affect the cost and desirability of high-carbon materials.
- Technological advancements in low-carbon material production (e.g., carbon-sequestering concrete or more efficient steel recycling).
- Public pressure and policy incentives related to green building certification (e.g., LEED, Net-Zero).
- The long-term performance and weathering of newer materials like mass timber under the local climate.

The conceptual pathways explored in this project are illustrated in Fig. 6.51, highlighting material choices under uncertainty and their alignment with long-term sustainability goals.

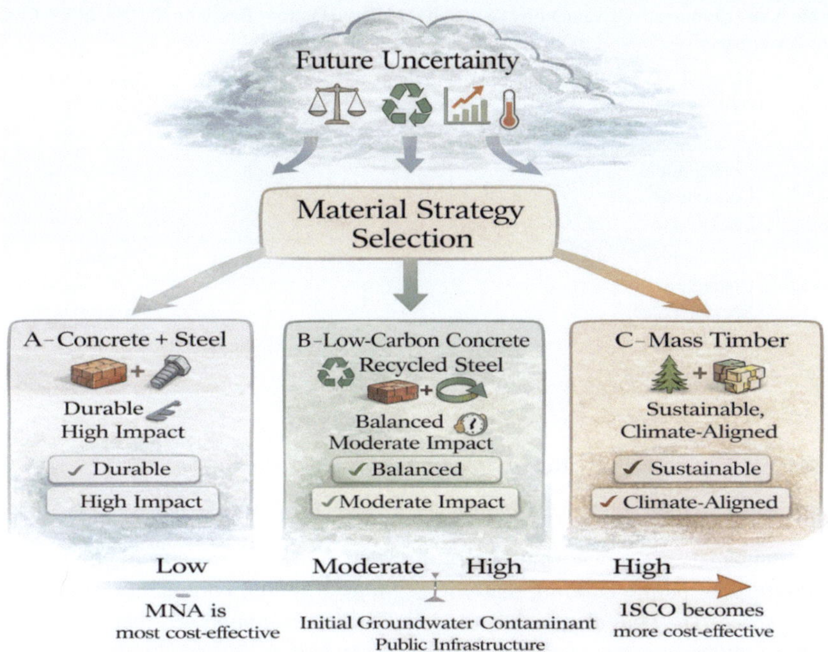

Fig. 6.51 Conceptual schematic of sustainable material strategy pathways for a public building

Table 6.90 compares three structural material strategies for a new construction project, assessing their initial costs, embodied carbon footprints, durability, and alignment with LEED and Net-Zero certification goals. While Option A is cost-effective and highly durable, it has the highest environmental impact. Option B offers a balanced compromise between emissions, cost, and certification potential. Option C, using mass timber, excels in sustainability performance and certification alignment, making it the most future-ready option despite higher initial cost and lower durability in certain conditions.

Key questions that should be addressed are as follows:

1. What sustainability-related uncertainties impact material selection?
2. What plausible future environmental and policy scenarios should be considered?
3. How do the material strategies perform across these scenarios?
4. Which option is most robust under climate policy risk?
5. Which option minimizes regret if carbon taxes or certifications become strict?
6. Can a phased or adaptive strategy be developed to allow flexibility in sustainable design decisions?

Table 6.90 Comparative Evaluation of Structural Material Options Based on Sustainability, Cost, and Performance

Option	Material strategy	Initial cost ($M)	Embodied carbon ($kgCO_2/m^2$)	Durability	LEED/Net-zero potential
A	Conventional Concrete + Steel	35	500	High	Low
B	Low-Carbon Concrete + Recycled Steel	40	320	Medium	Medium
C	Mass Timber (CLT, Glulam)	45	150	Medium–Low	High

Step-by-Step Solution Using SBDA

Step 1: Define the Decision Problem

The municipal government must select a material strategy for constructing a new multi-purpose public building that will serve as a library, community center, and office space. The core objective is to identify a material approach that aligns with both the city's environmental commitments, notably carbon neutrality by 2050, and financial constraints, while remaining resilient to evolving policies and expectations over a 50-year design life. The alternatives under considerations are as:

- Option A: Conventional Concrete + Steel

  The most traditional and affordable option, offering high structural durability but with the highest embodied carbon and limited alignment with sustainability targets.
- Option B: Low-Carbon Concrete + Recycled Steel

  A moderate step toward sustainability, this option reduces embodied carbon and improves LEED potential without major cost increases, while maintaining acceptable durability.
- Option C: Mass Timber (CLT, Glulam)

  The most environmentally progressive choice, with significantly reduced embodied carbon and high compatibility with LEED/Net-Zero certification, though at a higher upfront cost and moderate durability under local climate stressors.

  The constraints re also as follows:

- The total capital budget must remain below $50 million, covering materials, labor, and supporting infrastructure.

- The selected material must be capable of supporting future compliance with LEED or Net-Zero certification standards, which may become mandatory or highly incentivized within the planning horizon.

The challenge lies in selecting a strategy that meets current project needs while remaining adaptable and justifiable under future uncertainties in policy, technology, and material performance.

Step 2: Identify Key Uncertainties

Selecting a long-term material strategy for a public building requires careful consideration of uncertainties that could significantly influence both cost and sustainability performance over the 50-year design life. Three key uncertainties have been identified:

1. Carbon Tax Policy Evolution

   The trajectory of carbon pricing is uncertain. If carbon taxes or emissions penalties increase over time, materials with high embodied carbon—like conventional concrete and steel—may become substantially more expensive. Conversely, materials with low-carbon profiles may gain cost advantages or receive incentives. The financial viability of each option will be affected by how aggressively these policies evolve.
2. Green Certification Requirements

   It is unclear whether LEED, Net-Zero, or similar green certification standards will remain voluntary or become mandatory, either through regulation or as a prerequisite for public funding or community approval. If certification becomes a requirement, options that align poorly with these standards (such as conventional material mixes) may require costly retrofits or jeopardize project approval.
3. Material Performance Uncertainty

   Emerging materials like mass timber offer strong sustainability advantages, but their long-term performance under local climatic conditions (e.g., humidity, freeze-thaw cycles, fire resistance) may still carry unknowns. There is also uncertainty about maintenance requirements, material availability, and public perception over time. These factors could affect lifecycle cost, structural integrity, and user confidence.

These uncertainties underscore the need for a strategy that is flexible, future-proof, and adaptable, balancing today's constraints with tomorrow's expectations.

Step 3: Develop Plausible Sustainability Scenarios

Table 6.91 outlines three future scenarios reflecting varying levels of policy ambition related to carbon pricing, green building certification requirements, and adoption of sustainable construction innovations. These scenarios range from a continuation of current policies to a transformative pathway aligned with 2050 net-zero goals,

Table 6.91 Plausible future scenarios for sustainability-driven construction policy

Scenario	Description	Carbon tax (\$/ton CO_2)	Certification	Innovation adoption
S1	Current policies hold	\$50	LEED optional	Slow
S2	Moderate policy shift	\$100	LEED required	Medium
S3	Aggressive climate policy (2050)	\$200	Net-zero mandatory	Fast

supporting resilient material and design decision-making under deep sustainability uncertainty.

Step 4: Engage Stakeholders

To ensure the material selection process reflects a balance of technical expertise, fiscal accountability, and environmental responsibility, a diverse group of stakeholders was engaged in the decision-making process. Participants included architects, who emphasized design feasibility and material integration; contractors, who provided insights on sourcing, constructability, and schedule implications; finance department representatives, who focused on capital planning and cost oversight; and environmental advocacy groups, who championed sustainability and policy alignment.

Through workshops and consultations, stakeholders identified a set of shared priorities that should guide material strategy decisions:

- Embodied Carbon: Reducing the building's upfront environmental impact was seen as essential for aligning with long-term climate goals and minimizing future carbon-related liabilities.
- Transparency: Stakeholders called for clear, evidence-based comparisons of each material option's carbon footprint, cost, and long-term performance.
- Cost Responsibility: Ensuring the chosen material remains within the project's capital budget while accounting for long-term maintenance and potential regulatory costs was considered critical.
- Policy Risk Mitigation: With increasing uncertainty around carbon pricing and building performance standards, stakeholders supported a strategy that would minimize regulatory exposure and avoid costly retrofits in the future.

This input reinforced the importance of selecting a material strategy that balances environmental leadership with financial prudence, while remaining adaptable to evolving construction practices and policy landscapes.

Step 5: Define Performance Metrics

To evaluate the material strategy options in a structured and forward-looking manner, a set of performance metrics was defined. These metrics capture not only upfront costs and carbon impacts but also risks related to certification viability and material maturity, providing a balanced framework for comparison under future uncertainty.

1. Environmental Cost

This metric estimates the financial impact of embodied carbon, accounting for both direct material expenses and potential exposure to future carbon pricing. It is calculated as:

$$\text{Environmental Cost} = (\text{Embodied Carbon} \times \text{Projected Carbon Tax Rate}) + \text{Initial Material Cost}$$

This approach reflects how choices made today could result in higher future liabilities if carbon taxes increase over the building's life.

2. Certification Score

This qualitative metric assesses the alignment of each material strategy with green building certification systems such as LEED or Net-Zero. Rated as:

- High: Strong alignment; likely to support certification with minimal additional investment
- Medium: Partial alignment; may require design enhancements or material substitutions
- Low: Poor alignment; significant changes needed to meet certification thresholds

This score helps evaluate each option's readiness for potential policy shifts or funding requirements tied to environmental certification.

3. Construction Risk

This metric reflects the uncertainty associated with construction quality, material performance, and contractor familiarity. Scored on a scale of:

- 0 = Proven: Well-established material with widespread construction experience and minimal risk
- 1 = Moderate: Some innovation or site-specific challenges expected
- 2 = Emerging: Limited precedent; higher risk of delays, learning curve, or technical setbacks

Together, these metrics allow the project team to assess trade-offs between cost, policy resilience, and construction reliability, and to identify a material strategy that is both sustainable and practically achievable.

Step 6: Evaluate Options Under Each Scenario

Table 6.92 presents the environmental lifecycle costs of three structural material options, accounting for initial costs and carbon-related expenses under varying carbon tax scenarios (S1–S3). Carbon costs are calculated based on embodied emissions ($kgCO_2/m^2$) and scenario-specific tax rates. Option A remains the least expensive under current and moderate policy environments, but under aggressive climate

Table 6.92 Environmental lifecycle cost evaluation under carbon pricing scenario

Option	Initial	Carbon	S1 ($50)	S2 ($100)	S3 ($200)
A	35	500	42.5	50	65
B	40	320	46	52	64
C	45	150	52.5	60	75

Table 6.93 Certification alignment and risk scores across sustainability scenarios

Option	S1 Cert	S2 Cert	S3 Cert	Risk score
A	Low	Fail	Fail	0
B	Medium	Pass	Partial	1
C	High	Pass	Pass	2

Table 6.94 Robustness assessment using the minimax environmental lifecycle cost criterion

Option	Cost
A	65
B	64
C	75

policy (S3), its cost increases significantly, making Option B more competitive and Option C the most aligned with long-term sustainability goals despite its higher upfront investment.

Table 6.93 assesses each material option's alignment with building certification requirements under three future policy scenarios (S1–S3), alongside a qualitative risk score indicating long-term regulatory and reputational exposure. Option A lacks compliance under stricter future policies, while Option B shows moderate adaptability. Option C consistently meets certification standards across all scenarios and carries the highest sustainability assurance, reflected in its top risk score.

Step 7: Apply Robustness and Regret Criteria

A. Robustness (Minimax Environmental Cost)

Table 6.94 evaluates the worst-case (maximum) environmental lifecycle cost for each structural material option across all carbon pricing scenarios (S1–S3). Based solely on cost, Option A is the most robust, offering the lowest maximum expenditure. However, this does not account for regulatory or certification compliance risks, which may offset its cost advantage under aggressive climate policies.

B. Regret Matrix (Cost-Based)

Table 6.95 presents the regret values for each option under three carbon pricing scenarios (S1–S3), where regret is defined as the additional cost relative to the lowest-cost alternative in each scenario. The maximum regret highlights the worst-case cost

Table 6.95 Regret analysis of structural material options based on environmental lifecycle costs

Option	S1	S2	S3	Max regret
A	0	0	0	0
B	3.5	2	1	3.5
C	10	10	10	10

Table 6.96 Sustainability-weighted lifecycle cost evaluation with certification penalties

Option	S3 Total + Penalty	Result
A	65 + 5 = 70	✗
B	64 + 2.5 = 66.5	☑ Best Compromise
C	75 + 0 = 75	☑ Compliant but costly

penalty. Option A shows zero regret across all scenarios, making it the most regret-averse option in purely cost-based terms—though it carries significant sustainability compliance risks under future policy shifts.

C. Sustainability-Weighted Decision

Table 6.96 presents adjusted total costs under Scenario S3, incorporating penalties for non-compliance with LEED or Net-Zero certification—assumed at $5M per project. Option A remains the lowest-cost option but fails to meet sustainability criteria. Option C achieves full compliance but at the highest cost. Option B strikes the best balance, achieving partial compliance with moderate penalties, making it the most cost-effective and sustainability-aligned compromise.

Step 8: Design Adaptive Strategy

To manage policy uncertainty and maintain alignment with long-term sustainability goals, an adaptive material and energy strategy has been developed. This phased approach enables the city to move forward confidently within budget, while preserving the flexibility to respond to regulatory or funding changes without compromising project goals.

Phase 1: Design and Procurement

The project begins with *Option B: Low-Carbon Concrete + Recycled Steel* as the baseline material strategy. This option strikes a balance between moderate embodied carbon, durability, and cost control. The structural design will be optimized to accommodate partial wood substitution, ensuring compatibility with future upgrades.

A trigger condition has been defined:

- If LEED or Net-Zero certification becomes mandatory before the construction phase begins, the project will upgrade to Option C (Mass Timber).

- This would involve an estimated additional cost of $5 million, but would enable the city to qualify for green building grants, expedited permitting, or other policy-linked incentives that can offset cost increases and accelerate delivery.

Phase 2: Post-Occupancy Enhancements

To further advance carbon neutrality targets beyond material selection, the strategy includes a planned post-occupancy phase that incorporates renewable energy systems, such as:

- Photovoltaic (PV) panels for onsite electricity generation
- Battery storage systems to manage energy demand and grid interaction

These additions enhance the building's operational sustainability and help move the project toward Net-Zero energy performance, in alignment with the city's 2050 climate objectives.

This adaptive strategy allows the municipality to act decisively today while future-proofing the design against policy shifts and community expectations, ensuring both compliance and cost-effectiveness over the building's lifecycle.

Step 9: Communicate Results and Refine

In the final stage of the decision process, key findings were presented to stakeholders through a series of clear, comparative tools to support understanding, transparency, and informed feedback. The project team provided a suite of deliverables, including lifecycle carbon versus cost charts, which visualized the trade-offs between environmental impact and financial investment for each material option over the building's lifespan. These charts made it easier to contextualize future carbon cost exposure under varying policy scenarios.

In parallel, certification readiness heatmaps were shared to highlight how well each strategy aligned with evolving green building standards, such as LEED and Net-Zero. These visual tools identified Option C (mass timber) as the strongest performer in certification scenarios, with Option B offering moderate but improvable alignment. Finally, a phased decision tree was introduced to explain the adaptive pathway, showing how the project could transition from Option B to Option C if regulatory triggers—like a mandatory green certification—came into effect before construction.

As a result of this evidence-based communication, stakeholders approved Option B as the base strategy, with a phased upgrade to mass timber incorporated as a conditional pathway. This decision reflects a balanced response to environmental, financial, and policy considerations, and ensures the building remains flexible, cost-responsible, and aligned with long-term sustainability objectives.

Final Answers to the Key Questions

1. What are the most relevant uncertainties affecting flood protection decisions?
    - Carbon pricing, which will influence lifecycle cost of carbon-intensive materials.
    - Green certification regulations that may become mandatory.

- Performance reliability of mass timber under climate and fire exposure.

2. What plausible future scenarios should be considered?

   - S1: Minimal change in carbon tax and voluntary certification.
   - S2: Mandatory LEED for public buildings and \$100/ton CO_2.
   - S3: Net-zero building code and aggressive carbon pricing (\$200/ton).

3. How does each flood protection option perform under each scenario?

   - Option A is cost-effective but fails under S2 and S3 (non-compliance).
   - Option B balances cost and readiness.
   - Option C performs best in environmental and certification metrics but has the highest cost and material risk.

4. Which option is most robust across all futures?

   Option B is most robust when considering both lifecycle cost and policy compliance under future scenarios.
5. Which option has the least regret if future conditions turn out to be worse than expected?

   Option B has moderate regret in cost but avoids penalties under new certification standards.
6. Can an adaptive pathway be developed to allow flexible investment over time?

   Yes. Begin with Option B, and add mass timber components and energy upgrades only if regulations change.

   This adaptive pathway balances immediate feasibility with future sustainability compliance.

The decision-making perspective discussed in this chapter is grounded in established frameworks for uncertainty analysis, risk governance, resilience planning, and lifecycle infrastructure management widely adopted in civil engineering practice (Amiri et al., 2025; Der Kiureghian & Ditlevsen, 2009; Ebtehaj & Bonakdari, 2022a, 2022b; FEMA, n.d.; FHWA, 2023; Gholami et al., 2020; Haasnoot et al., 2013; Helton et al., 2004; HM Treasury, 2026; IEC, 2019; IPCC, 2022; ISO, 2017; ISO, 2018; ISO, 2022; Lempert et al., 2003; Marchau et al., 2019; NIST, 2016; OMB, 2023; Qureshi et al., 2025; Saltelli et al., 2008; Soltani et al., 2021; UNDRR, 2015; UNDRR, n.d.; United Nations General Assembly, 2016).

6.3 Summary

This chapter has demonstrated that decision trees constitute a rigorous yet intuitive decision-support framework for managing uncertainty in civil engineering projects across their entire lifecycle. It has emphasized that civil engineering decisions—ranging from design and construction to operation, maintenance, and end-of-life planning—are inherently uncertain due to environmental variability, data limitations, evolving regulations, and economic constraints. Within this context, decision trees have been shown to provide a structured and transparent approach for explicitly modeling uncertainty and evaluating competing alternatives using quantitative and probabilistic reasoning.

The chapter established the conceptual foundations of decision trees by describing their graphical structure and core components, including decision nodes, chance nodes, and outcome nodes. This structure enables engineers to decompose complex, multi-stage problems into manageable sequences of decisions and probabilistic events. By assigning probabilities to uncertain outcomes and utilities or costs to final consequences, decision trees were shown to transform subjective judgment into evidence-based analysis, thereby supporting rational, consistent, and defensible decision-making.

A central contribution of this chapter has been its systematic examination of decision-making under uncertainty across the various phases of civil engineering projects. During the design phase, decision trees were shown to support material selection, infrastructure sizing, and adaptability to future demands, particularly in the presence of climate change and urban growth uncertainty. In the construction phase, they were demonstrated to facilitate risk-informed responses to unforeseen site conditions, supply chain disruptions, and execution strategies. During operation and maintenance, decision trees were shown to guide performance monitoring, maintenance timing, emergency response planning, and adaptation to evolving usage patterns. The chapter further highlighted their relevance in end-of-life and adaptive reuse decisions, where long-term environmental, social, and economic consequences must be carefully balanced.

The chapter placed strong emphasis on expected utility theory as the analytical foundation of decision tree analysis. Utility was defined broadly to encompass both tangible factors—such as capital cost, maintenance expenditure, and service life—and intangible considerations, including safety, environmental impact, resilience, and social acceptance. By evaluating alternatives using expected utility as a probability-weighted measure of outcomes, the chapter demonstrated how engineers can compare competing options holistically, rather than relying on deterministic or worst-case assumptions. This approach was shown to be particularly valuable for high-stakes infrastructure decisions where long-term performance and risk tolerance are critical.

Through an extensive set of real-world civil engineering examples—including foundation design under soil uncertainty, flood protection system sizing, pavement material selection, stormwater basin design, seismic retrofitting, tunnel support systems, water and wastewater treatment technology selection, and energy-efficient building envelope design—the chapter demonstrated the practical application of decision trees across diverse engineering contexts. These examples illustrated how decision trees integrate technical analysis, probabilistic reasoning, and economic evaluation to support resilient, cost-effective, and defensible engineering solutions.

Finally, the chapter outlined a step-by-step methodology for implementing decision trees in engineering practice, encompassing problem formulation, alternative identification, uncertainty modeling, utility assessment, probability assignment, and comparison of expected utilities. By doing so, it equipped students and practitioners with a repeatable and transparent framework for navigating complex engineering decisions. Overall, this chapter has established decision trees as a foundational tool for modern civil engineering practice, enabling risk-aware, lifecycle-based, and sustainability-oriented decision-making under uncertainty.

References

Amiri, A., Ebtehaj, I., Soltani, K., Gumiere, S. J., & Bonakdari, H. (2025). Future soil erosion trends in Canadian agricultural lands from runoff and sustainability impacts. *Scientific Reports, 15*(1), 23184. https://www.nature.com/articles/s41598-025-05947-5

Der Kiureghian, A., & Ditlevsen, O. (2009). Aleatory or epistemic? Does it matter? *Structural Safety, 31*(2), 105–112. https://doi.org/10.1016/j.strusafe.2008.06.020

Ebtehaj, I., & Bonakdari, H. (2022a). A comprehensive comparison of the fifth and sixth phases of the coupled model intercomparison project based on the Canadian earth system models in spatio-temporal variability of long-term flood susceptibility using remote sensing and flood frequency analysis. *Journal of Hydrology, 617*, 128851. https://www.sciencedirect.com/science/article/pii/S0022169422014214

Ebtehaj, I., & Bonakdari, H. (2022b). A reliable hybrid outlier robust non-tuned rapid machine learning model for multi-step ahead flood forecasting in Quebec, Canada. *Journal of Hydrology, 614*, 128592. https://www.sciencedirect.com/science/article/pii/S0022169422011623

Federal Emergency Management Agency. (n.d.). *Benefit-Cost Analysis (BCA)*. Retrieved February 18, 2026, from https://www.fema.gov/grants/tools/benefit-cost-analysis

Federal Highway Administration. (2023, March). *Lifecycle Cost Analysis RealCost User Manual* (FHWA-HRT-23-021). U.S. Department of Transportation. https://highways.dot.gov/sites/fhwa.dot.gov/files/FHWA-HRT-23-021.pdf

Gholami, A., Bonakdari, H., Ebtehaj, I., & Khodashenas, S. R. (2020). Reliability and sensitivity analysis of robust learning machine in prediction of bank profile morphology of threshold sand rivers. *Measurement, 153*, 107411. https://www.sciencedirect.com/science/article/pii/S0263224119312783

Haasnoot, M., Kwakkel, J. H., Walker, W. E., & ter Maat, J. (2013). Dynamic adaptive policy pathways: A method for crafting robust decisions for a deeply uncertain world. *Global Environmental Change, 23*(2), 485–498. https://doi.org/10.1016/j.gloenvcha.2012.12.006

Helton, J. C., Johnson, J. D., & Oberkampf, W. L. (2004). An exploration of alternative approaches to the representation of uncertainty in model predictions. *Reliability Engineering and System Safety, 85*(1–3), 39–71. https://doi.org/10.1016/j.ress.2004.03.025

HM Treasury. (2026, February 5). *The green book: Appraisal and evaluation in central government*. GOV.UK. https://www.gov.uk/government/publications/the-green-book-appraisal-and-evaluation-in-central-government

Intergovernmental Panel on Climate Change. (2022). Summary for Policymakers. In H.-O. Pörtner, D. C. Roberts, E. S. Poloczanska, K. Mintenbeck, M. Tignor, A. Alegría, M. Craig, S. Langsdorf, S. Löschke, V. Möller, & A. Okem (Eds.), *Climate change 2022: Impacts, adaptation and vulnerability* (AR6 WGII). Cambridge University Press. https://doi.org/10.1017/9781009325844.001

International Electrotechnical Commission. (2019). *Risk management—Risk assessment techniques* (IEC Standard No. 31010:2019). https://www.iso.org/standard/72140.html

International Organization for Standardization. (2017). *Buildings and constructed assets—Service life planning—Part 5: Life-cycle costing* (ISO Standard No. 15686-5:2017). https://www.iso.org/standard/61148.html

International Organization for Standardization. (2018). *Risk management—Guidelines* (ISO Standard No. 31000:2018). https://www.iso.org/standard/65694.html

International Organization for Standardization. (2022). *Risk management—Vocabulary* (ISO Standard No. 31073:2022). https://www.iso.org/standard/79637.html

Lempert, R. J., Popper, S. W., & Bankes, S. C. (2003). *Shaping the next one hundred years: New methods for quantitative, long-term policy analysis* (MR-1626). RAND Corporation. https://www.rand.org/pubs/monograph_reports/MR1626.html

Marchau, V. A. W. J., Walker, W. E., Bloemen, P. J. T. M., & Popper, S. W. (Eds.). (2019). *Decision making under deep uncertainty: From theory to practice*. Springer. https://doi.org/10.1007/978-3-030-05252-2

National Institute of Standards and Technology. (2016). *Community Resilience Planning Guide for Buildings and Infrastructure Systems* (Vol. I) (NIST SP 1190). https://doi.org/10.6028/NIST.SP.1190v1

Office of Management and Budget. (2023). *Circular No. A-94: Guidelines and discount rates for benefit-cost analysis of federal programs*. Executive Office of the President. https://whitehouse.gov/wp-content/uploads/2023/11/CircularA-94.pdf

Qureshi, M. U. A., Amiri, A., Ebtehaj, I., Guimere, S. J., Cunderlik, J., & Bonakdari, H. (2025). Coupling HEC-RAS and AI for river morphodynamics assessment under changing flow regimes: Enhancing disaster preparedness for the Ottawa River. *Hydrology, 12*(2), 25. https://www.mdpi.com/2306-5338/12/2/25

Saltelli, A., Ratto, M., Andres, T., Campolongo, F., Cariboni, J., Gatelli, D., Saisana, M., & Tarantola, S. (2008). *Global sensitivity analysis: The primer*. Wiley. https://www.wiley.com/en-us/Global%2BSensitivity%2BAnalysis%3A%2BThe%2BPrimer-p-9780470059975

Soltani, K., Ebtehaj, I., Amiri, A., Azari, A., Gharabaghi, B., & Bonakdari, H. (2021). Mapping the spatial and temporal variability of flood susceptibility using remotely sensed normalized difference vegetation index and the forecasted changes in the future. *Science of The Total Environment, 770*, 145288. https://www.sciencedirect.com/science/article/pii/S0048969721003545

United Nations General Assembly. (2016). *Report of the open-ended intergovernmental expert working group on indicators and terminology relating to disaster risk reduction: Note by the secretary-general* (A/71/644). United Nations. https://digitallibrary.un.org/record/852089/files/A_71_644-EN.pdf

United Nations Office for Disaster Risk Reduction. (2015). *Sendai framework for Disaster Risk Reduction 2015–2030*. United Nations. https://www.undrr.org/publication/sendai-framework-disaster-risk-reduction-2015-2030

United Nations Office for Disaster Risk Reduction. (n.d.). *Sendai framework terminology on disaster risk reduction (DRR glossary)*. Retrieved February 18, 2026, from https://www.undrr.org/drr-glossary/terminology

Index

H. Bonakdari et al., *Hazard and Risk Assessment in Civil Engineering Design*, https://doi.org/10.1007/978-3-032-23600-5

S

MIX
Papier aus verantwortungsvollen Quellen
Paper from responsible sources
FSC® C105338

If you have any concerns about our products,
you can contact us on
ProductSafety@springernature.com

In case Publisher is established outside the EU,
the EU authorized representative is:
Springer Nature Customer Service Center GmbH
Europaplatz 3, 69115 Heidelberg, Germany

Printed by Libri Plureos GmbH
in Hamburg, Germany